Forest Development in Cold Climates

NATO ASI Series

Advanced Science Institutes Series

A series presenting the results of activities sponsored by the NATO Science Committee, which aims at the dissemination of advanced scientific and technological knowledge, with a view to strengthening links between scientific communities.

The series is published by an international board of publishers in conjunction with the NATO Scientific Affairs Division

A	**Life Sciences**	Plenum Publishing Corporation
B	**Physics**	New York and London
C	**Mathematical and Physical Sciences**	Kluwer Academic Publishers
D	**Behavioral and Social Sciences**	Dordrecht, Boston, and London
E	**Applied Sciences**	
F	**Computer and Systems Sciences**	Springer-Verlag
G	**Ecological Sciences**	Berlin, Heidelberg, New York, London,
H	**Cell Biology**	Paris, Tokyo, Hong Kong, and Barcelona
I	**Global Environmental Change**	

Recent Volumes in this Series

Volume 239—The Mammalian Cochlear Nuclei: Organization and Function
edited by Miguel A. Merchán, José M. Juiz, Donald A. Godfrey, and Enrico Mugnaini

Volume 240—Regulation of Gene Expression in Animal Viruses
edited by Luis Carrasco, Nahum Sonenberg, and Eckard Wimmer

Volume 241—Iodine Deficiency in Europe: A Continuing Concern
edited by F. Delange, J.T. Dunn, and D. Glinoer

Volume 242—Desmopressin in Bleeding Disorders
edited by G. Mariani, P. M. Mannucci, and M. Cattaneo

Volume 243A—Biological Effects and Physics of Solar and Galactic Cosmic Radiation, Part A
edited by Charles E. Swenberg, Gerda Horneck, and E. G. Stassinopoulos

Volume 243B—Biological Effects and Physics of Solar and Galactic Cosmic Radiation, Part B
edited by Charles E. Swenberg, Gerda Horneck, and E. G. Stassinopoulos

Volume 244—Forest Development in Cold Climates
edited by John Alden, J. Louise Mastrantonio, and Søren Ødum

Series A: Life Sciences

Forest Development in Cold Climates

Edited by

John Alden

Institute of Northern Forestry
United States Department of Agriculture, Forest Service
Fairbanks, Alaska

J. Louise Mastrantonio

Portland, Oregon

and

Søren Ødum

Royal Veterinary and Agricultural University
Hørsholm Arboretum, Denmark

Plenum Press
New York and London
Published in cooperation with NATO Scientific Affairs Division

Proceedings of a NATO Advanced Research Workshop
on Forest Development in Cold Climates,
held June 18–23, 1991,
in Laugarvatn, Iceland

NATO-PCO-DATA BASE

The electronic index to the NATO ASI Series provides full bibliographical references (with keywords and/or abstracts) to more than 30,000 contributions from international scientists published in all sections of the NATO ASI Series. Access to the NATO-PCO-DATA BASE is possible in two ways:

—via online FILE 128 (NATO-PCO-DATA BASE) hosted by ESRIN, Via Galileo Galilei, I-00044 Frascati, Italy

—via CD-ROM "NATO-PCO-DATA BASE" with user-friendly retrieval software in English, French, and German (©WTV GmbH and DATAWARE Technologies, Inc. 1989)

The CD-ROM can be ordered through any member of the Board of Publishers or through NATO-PCO, Overijse, Belgium.

Library of Congress Cataloging-in-Publication Data

Forest development in cold climates / edited by John Alden, J. Louise Mastrantonio, and Søren Ødum.
p. cm. -- (NATO ASI series. Series A, Life sciences ; v. 244)
"Proceedings of a NATO Advanced Research Workshop on Forest Development in Cold Climates, held June 18-23, 1991, in Laugarvatn, Iceland"--T.p. verso.
"Published in cooperation with NATO Scientific Affairs Division."
Includes bibliographical references and index.
ISBN 0-306-44480-1
1. Cold regions forestry--Congresses. I. Alden, John. II. Mastrantonio, J. Louise. III. Ødum, Søren. IV. NATO Advanced Research Workshop on Forest Development in Cold Climates (1991 : Laugarvatn, Iceland) V. North Atlantic Treaty Organization. Scientific Affairs Division. VI. Series.
SD410.38.F67 1993
634.9'0911--dc20 93-898
CIP

ISBN 0-306-44480-1

Printed in the United States of America

PREFACE

As forests decline in temperate and tropical climates, highly-developed countries and those striving for greater economic and social benefits are beginning to utilize marginal forests of high-latitude and mountainous regions for resources to satisfy human needs. The benefits of marginal forests range from purely aesthetic to providing resources for producing many goods and services demanded by a growing world population.

Increased demands for forest resources and amenities and recent warming of high-latitude climates have generated interest in reforestation and afforestation of marginal habitats in cold regions. Afforestation of treeless landscapes improves the environment for human habitation and provides for land use and economic prosperity. Trees are frequently planted in cold climates to rehabilitate denuded sites, for the amenity of homes and villages, and for wind shelter, recreation, agroforestry, and industrial uses. In addition, forests in cold climates reduce the albedo of the earth's surface in winter, and in summer they are small but significant long-lived sinks for atmospheric carbon dioxide. Finally, growth and reproductive success of forests at their geographic limits are sensitive indices of climatic change.

As efforts to adapt forests to cold climates increase, however, new afforestation problems arise and old ones intensify. Austral, northern, and altitudinal tree limits are determined by many different factors. Current hypotheses for high-latitude tree limits are based on low growing-season temperatures that inhibit plant development and reproduction. Forests, therefore, can be established beyond their natural limits only by either improving sub-marginal environments or by utilizing the inherent variation in tree-limit species to increase their fitness to harsh conditions. Both methods are necessary for forest development beyond natural tree limits, because environmental change requires corresponding change in survival and growth traits.

In June of 1987, ReTree International, Wilsonville, Oregon, U.S.A., aware of problems of survival and growth of trees beyond their natural limits, proposed a workshop to address practical aspects of afforestation and tree growth in cold regions of the world. The North Atlantic Treaty Organization Advanced Research Workshop, organized by ReTree, was held in Laugarvatn, Iceland, on June 18–23, 1991, to focus international research on strategies for establishing and growing trees in cold climates. International research is appropriate because fast-growing species of tree-limit genera such as Betula L., Populus L., and Larix Mill., and the evergreen conifers, have potential for circumpolar distribution. Afforestation of regions beyond the natural tree limits requires knowledge from disciplines as diverse as meteorology, forest ecology, forest genetics, and tree physiology. Methods are needed to assess and alleviate environmental constraints on forest establishment, to identify and conserve the gene pool of subarctic and alpine species and provenances, and to improve the fitness of inherently variable species to cold climates.

These tasks are difficult because they require communication across the boundaries of diverse disciplines and standardization of tree-limit terminology. This workshop brought together leading scientists and forest managers from many high-latitude countries to consolidate knowledge of forest development. The disciplines chosen were socioeconomics, to review the benefits of forestry in cold regions; climatology, biometeorology, and atmospheric sciences, to measure and predict environmental change in space and time; forest ecology, to identify the causes of tree limits and environmental constraints on forest establishment; biochemistry and tree physiology, to determine species' requirements and responses to environmental stimuli and to condition species to environments beyond their natural limits; paleobotany (palynology) and biogeography, to map temporal and spatial distributions, migration rates, and to identify species and provenances for adaptation to cold climates; forest genetics, to create new gene combinations and to accelerate adaptation of species to new environments; and forestry, to apply the methodology of the above disciplines to afforestation and reforestation of cold regions.

Many people contributed to the success of the conference and of this proceedings. The Scientific Director, Frank Lockyear, and the organizing committee members of the NATO-sponsored Advanced Research Workshop owe thanks to many organizations and individuals. Financial assistance for the workshop and editing the proceedings was provided by the Science Affairs Division of the North Atlantic Treaty Organization, Brussels, Belgium; the USDA Forest Service, Pacific Northwest Research Station, Portland, Oregon; the United States Department of the Army's Research Office, Research Triangle Park, North Carolina; and the Nordic Forest Research Council, Helsinki, Finland; and workshop manager Robert Wheelon, Director of ReTree International, Vancouver, Washington. Dr. Arni Bragason, Sigurður Blöndal, and the Icelandic Forest Service Organizing Committee handled workshop accommodations, field trips, and other arrangements that contributed to the success of the meeting. Finally, thanks are extended to each participant for preparing a chapter for this proceedings and to June Weinstock, Fairbanks, Alaska, for editorial contributions.

John N. Alden, Editor
USDA Forest Service

Frank H. Lockyear
President, ReTree International

FOREWORD

We Icelanders have a reputation for being retrospective. For centuries, we have spent considerable time measuring the present against the past, remembering and discussing what our forefathers did and how they lived.

One reason we have been able to do this is because Icelanders have always been eager to write down and record even the smallest events in our history. This national characteristic allows us to remember our origins as a nation that colonized the last unsettled country in Europe in the ninth century A.D. "First come, first served," the proverb says, and as the last nation "in queue" we cannot complain that we do not enjoy the most fruitful land, but rather have had to depend instead on the bounties of the sea surrounding our northerly island.

Early Icelanders recorded not only their own history, but also, with remarkable tolerance for that time, preserved the details of the beliefs, mythology, and cosmology of their pre-Christian ancestors—the only accounts of the Nordic pagan faith that survive today.

As a passionate advocate of both cultivating the past and cultivating the land for the future, I find it fascinating to recall that one of the main elements of the old Nordic cosmology was the vision of the world itself as a single, huge tree. Our great medieval poet and historian Snorri Sturluson describes this in wonderfully vivid language in his Prose Edda:

> The ash is the best and greatest of all trees; its branches spread out over the whole world and reach up over heaven. The tree is held in position by three roots that spread far out; one is among the Æsir [as the Great Gods were called], the second among the frost ogres where once was Ginnungagap [the Vacuum of the Cosmos], and the third extends over Niflheim [The World of Darkness], and under that root is the well Hvergelmir [The Bubbling Cauldron]; but Níðhögg [The Serpent; The Striker that destroys] gnaws at the root from below. Under the root that turns in the direction of the frost ogres lies the spring Mímir, in which is hidden wisdom and understanding; Mímir is the name of the owner of the spring. He is full of wisdom because he drinks from the spring out of the horn Gjöll. All-father [that is, Odin] came there and asked for a single drink from the spring, but he did not get it until he had given one of his eyes as a pledge. . . . A further root of the ash tree is in the sky, and under that root is the very sacred spring called the Spring of Urð [Destiny].

It is no surprise that when the first settlers came to Iceland from the forests of Norway, Scotland, and Ireland, they named places by what they saw. In many places in Iceland where there is today no sign of trees, old place names suggest that trees and vegetation once grew there. We have names like Skógar and Mörk (meaning "forest"); Holt (orginally the same as the German word "Holz," meaning "wood" or "forest," but now, interestingly, meaning a "rugged, rocky, and bare hill"). We have preserved from the old language the saying, "Oft er í holti heyrandi nær." ("In the woods, a listener is never far.") This is nonsense to the modern Icelander unless he knows that "holt" originally meant "wood" or "forest." We have Lundur (copse) and Viður (wood) in given place names, which among others are now preserved in the name of the now-treeless island of Viðey in the bay north of Reykjavík. As everyone who travels around Iceland knows, the land is an open book of place names telling us about the history of the country, the people who lived there, and the way they lived.

In the Book of Icelanders written in the early 12th Century, the historian Ari the Learned describes the quality of the land when the first settlers arrived. He says it was covered with trees from "mountains to shore." He was probably referring to low scrub or some kind of birchwood as we see today, for example in the East Fjords. (That is a daring declaration for me, a scientific layman, to make in the presence of such a learned audience!) I have often wondered at his choice of the word "viður" for "trees," because, in addition to his needing the "v" for alliteration in the phrase "viði vaxið milli fjalls og fjöru," its meaning in the modern Scandanavian languages is "wood that can be utilized; wood as a resource."

Mention of the beauty of the land or its vegetation is rare in our medieval literature; it is more often stated how good the land was to live from. People then as now were preoccupied with how they could use the land to make life more comfortable. From old records it is certain that Icelandic farmers, until very recently, considered woods and trees as extra bounty of the land they lived on, a sort of godsend or windfall. They used these trees largely for making charcoal and heating their homes rather than for building. Building timber was either imported from Norway (and used in constructing churches, for example) or taken from the plentiful driftwood the Gulf Stream has always brought to Iceland's shores.

Although much of the once healthy and flourishing woodland is now eroded, we have long realized that we cannot criticize our forebears for exploiting the natural vegetation of the country to the extent they did. They had to survive, not just taking their meager harvest from an almost barren soil, but also struggling against the elements themselves, even suffering volcanic eruptions that smothered the land with poisonous ash, and the floods from glaciers melted by these volcanic fires.

What we feel on reading ancient accounts of the virgin woodlands is not regret but hope, because they offer evidence that this country actually can sustain trees and become much greener with vegetation. This has created a national awakening to the value of our natural vegetation. It is precisely for leading this awakening that the 60-year-old Skógræktarfélag Íslands (the Icelandic Forest Association) was given the distinguishged Nordic Environmental award last week at a large Nordic environmental conference in Reykjavík. Even though the land was used so intensively in the past that vegetation is still on the retreat, we are determined to do everything in our power to reverse this trend and give back to the land what has been taken from it, and more. We eagerly await the time when we will see signs that we, the inhabitants and citizens of this land, have gained the upper hand by applying our knowledge and energies to this task. Then we will take pride in passing the land on to coming generations, and they can continue to improve the vegetation that covers it.

All over Iceland, young people are being brought up with a sharper awareness of the environment and the need to care for the land. One aspect of this is the large and very active afforestation campaigns in which hundreds of thousands of trees are planted each year.

Like our forefathers in the past, Icelanders have a future vision of the tree as the world. We believe that the symbol of all our lives could be trees and vegetation to grant shelter, trees that search for nutrition and sustenance through the roots they put down to drink from the spring of life. I will end with another quotation from Snorri Sturlson's Prose Edda.

> There is a beautiful hall near the spring under the ash tree, and from it come three maidens whose names are Urð, Verðandi, and Skuld—Past, Present, and Future. These maidens shape the lives of men.

Ladies and gentlemen, I propose a toast to this conference on "Forest Development in Cold Climates," a toast to the spring of knowledge that will feed the roots of the tree of the world.

Vigdís Finnbogadóttir
President of Iceland

CONTENTS

Socioeconomic Importance of Forests in Iceland 1
Sigurður Blöndal

BIOCHEMICAL AND PHYSIOLOGICAL REQUIREMENTS OF TREES FOR FOREST DEVELOPMENT BEYOND TREE LIMITS

Molecular Bases for Adaptation of Coniferous Trees to Cold Climates 15
Don J. Durzan

Growth and Development of Northern Forest Trees as Affected by Temperature and Light 43
Olavi Junttila and Jarle Nilsen

Testing Winter Desiccation Resistance for Species and Provenance Selection at Timberlines 59
J. Bo Larsen

Growth of Mountain Birch (Betula Pubescens Ehrh.) in Response to Changing Temperature 65
Oddvar Skre

Performance of Mountain Birch in Different Environments in Sweden and Iceland: Implications for Afforestation 79
Bjartmar Sveinbjörnsson, Mats Sonesson, Olle K. Nordell, and Staffan P. Karlsson

ENVIRONMENTAL LIMITATIONS OF FOREST DEVELOPMENT AT AND BEYOND TREE LIMITS

Causes of Alpine Timberline: A Review of the Hypotheses 89
P. Wardle

Physiology of Trees at Treeline 105
John Grace and Jennifer James

Treeline in Relation to Climate, with Special Reference to Oceanic Areas 115
Sakari Tuhkanen

Pliocene Fossil Nothofagus (Southern Beech) from Antarctica: Phytogeography, Dispersal Strategies, and Survival in High Latitude Glacial-Deglacial Environments 135
P.-N. Webb and D. M. Harwood

Predicting Afforestation Success during Climatic Warming at the Northern Limit of Forests 167
Allen M. Solomon and Darrell C. West

Climate Change as Seen by Trees and by Climate Modelers 189
Sue Ann Bowling

Effects of Wind on Boreal Forests 203
Alexander Robertson

Long-Term Biometeorological Monitoring at Two Forest Sites in Iceland and Newfoundland: Initial Results 227
H. Thorgeirsson, J. H. McCaughey, A. Robertson, and C. T. French

POTENTIAL TREE SPECIES AND PROVENANCES FOR FOREST DEVELOPMENT IN COLD CLIMATES

Genetic Diversity of Tree Populations at Their Arctic Limits 241
Peter M.A. Tigerstedt

Potential Species and Provenances for Forest Development in Cold Climates 251
Maximilian Hagman

Species and Provenance Choice at Northern Tree Limits in Maritime Climates 265
Roger Lines

Seed Collections from North American Trees for Marginal Sites in the Nordic Countries 277
Gisle Skaret and Ola Rosvall

The Uses of Lignoses in Horticulture in Iceland 293
Jóhann Pálsson

PROVENANCE SELECTION, TESTING, AND BREEDING TO ACCELERATE ADAPTATION AND GROWTH OF TREES AND SHRUBS

Accelerating Adaptation of Trees at Tree Limits by Selective Breeding 299
Dag Lindgren

Selection and Breeding of Scots Pine for Northern Sweden 321
Jan-Erik Nilsson

Provenance and Individual Variation in Climatic Hardiness of Scots Pine in Northern Finland 333
Jouni Mikola

Hybridization among Provenances of Lodgepole Pine 343
Anders Fries

Genetics, Cytogenetics, and Molecular Genetics of Icelandic Birch: Implications for Breeding and Reforestation 357
Kesara Anamthawat-Jónsson, J. S. Heslop-Harrison, Thorsteinn Tómasson

Chloroplast DNA Diversity, Phylogenetics, and Hybridization in *Picea* 369
Aðalsteinn Sigurgeirsson and Alfred E. Szmidt

Genetic Variation in Early Frost Tolerance of Spruce from Northwestern North America 383
Jon Dietrichson

CASE HISTORIES OF FOREST DEVELOPMENT BEYOND TREE LIMITS

History of Tree Planting on the Aleutian Islands 393
David Bruce

Afforestation of Lutz Spruce, Sitka Spruce, and Norway Spruce in the Vesteraalen Islands, Norway 427
Nils O. Kaasen, Gisle Skaret, and Tore Kaasen

Development of an Environmental Forest and Arboretum on the Outer Coast of Finnmark County, Norway 437
Skjalg Slettjord

Forest Development in Iceland 453
Jón Loftsson

Potential Treeline in the Faroe Islands 463
Tróndur G. Leivsson

Forest Trials at High Elevations in Britain 475
C. J. Nixon and A. L. Tyler

Wind and Wave Forests: A Case Study and Implications for Silviculture 491
Alexander Robertson

The Alpine Timberline of Tibet 511
Li Bosheng

Forestry in New Zealand's Southern High Country 529
N. J. Ledgard

Summaries of Working Group Sessions 549

Species Index 555

Subject Index 559

SOCIOECONOMIC IMPORTANCE OF FORESTS IN ICELAND

Sigurður Blöndal

Iceland Forestry Service
Hallormsstaður, Iceland

Although there is evidence that pre-settlement Iceland was abundantly forested, Iceland today is a barren land. Only one percent (1,250 square kilometers) of its area is woodland. In a forest survey between 1972 and 1975, the remaining woodlands were classified into four quality classes according to average tree height. Most of the native woodlands consisted of shrubby trees less than two meters in height, with downy birch (Betula pubescens Ehrh.) as the dominant species (Table 1).

Although accurate figures for plantations established during the last 40 to 45 years are not available because of the lack of statistics from private and corporate plantings, the area of planted woodland is believed to be about 15,000 hectares.

Table 1. Quality of birch woodlands in Iceland

Mean height (meters)	Area (hectares)	Percent of Total Area
0-2	101,328	80.8
2-4	18,970	15.1
4-8	3,020	2.4
8-12	2,151	1.7
TOTAL	125,469	100

(Source: Bjarnason and Sigurðsson, 1977)

Forest Development in Cold Climates, Edited by J. Alden *et al.*
Plenum Press, New York, 1993

The majority of the plantations are of exotic species, most of which have been established in the birch woodland.

Evidence is strong that Iceland was heavily forested when it was first settled in A.D. 874. In the Book of Icelanders written by Ari the Learned (1068-1148), wherein he describes the first settlers of Iceland, we find the following description: "At that time Iceland was covered by wood, from the shoreline to the hillsides. . ." (Benediktsson, 1968).

The extent of the birch forest at the time of settlement has long been a matter of dispute. According to one of the more reliable estimates, some 25 to 30 percent of the island was forested before it was settled.[1] This estimate was based on:

1. Written reports such as chronicles, church-deeds, and sagas from medieval times.
2. Place names, many of which give witness to woodlands where none exists today.
3. Charcoal pits, which are found on deforested as well as forested sites all over the country, even as high as 400 to 500 meters in elevation and on northern peninsulas.
4. Pollen analysis (Einarsson, 1963).
5. Isothermal curves for the minimum summer tetraterm (6.5 °C), which indicate that birch woodlands could have occupied at least 28,000 square kilometers (Sigurðsson, 1977).

Birch (Betula pubescens Ehrh.) woodlands were a very valuable resource to the early settlers of Iceland, who, however, through necessity, severely depleted the resource during the first few centuries of settlement. Charcoal-burning undoubtedly played a major role in the rapid destruction of the woodlands (Þórarinsson 1974). A survey of the agricultural resources between 1702 and 1712 showed that even then, only a small amount of the original woodlands remained (Magnússon and Vídalín, 1913-1946). Þórðarson (1955) gives an interesting example of the value of "raftskógur" (woodland in which trees provide rafters of at least three elns, equivalent to 188.3 centimeters) in Fljótsdalur district in eastern Iceland during the latter half of the 19th Century. Each "raftur" cost half a ríkisdollar, and a hundred "rafturs" yielded sufficient money to pay a laborer's wage for a year and to buy ten sheep. In his memoirs, Guðmundur Jónsson from Húsey (1983) describes the fate of a small woodland in east Iceland that suffered heavy grazing during an extremely harsh winter and gives a heart-rending epitaph: "The sheep lived but the forest died." As one can see, the forests were one of the most essential resources on which the peasant society depended.

Destruction of the woodland deprived the highly erodible loess soils of their protective forest cover, and large-scale erosion followed. Today only 25 percent of the country is covered with vegetation compared to the 60 percent that was covered at settlement (Bjarnason, 1942; Thoroddsen, 1911; Þorsteinsson and Blöndal, 1986). Furthermore, Þorsteinsson and Blöndal (1986) point out that "if the destruction of vegetation and soil over 11 centuries of settlement are considered together with the much poorer quality of the surviving vegetation, one can estimate that the remaining carrying capacity of the soil and vegetation is less than 20 percent of what it was at the time of settlement."

---

[1]Steindórsson, S., 1985, Skógur á Íslandi til forna (Forests in Iceland in Ancient Times), unpublished manuscript.

In addition to the colossal problem of soil erosion, removal of the woodlands also increased the area of wetlands, which in turn undoubtedly saved the country from large-scale desertification and further soil erosion.

In conclusion, the lack of forests in Iceland is due to the wholesale destruction of the original woodlands. A further contention, more relevant to a later part of this thesis, is that very few species of higher plants grow in Iceland because it is geographically isolated from the rich flora of Europe and North America.

In the only complete description of Icelandic plant societies, Steindórsson (1964) contends, on the basis of climatic data, that the Icelandic lowlands below 400 meters should be within the Circumpolar Boreal Coniferous Forest Zone even though conifers do not occur there naturally. Conifer seed is dispersed only short distances and is unable to reach Iceland from its glacial refugia on the European continent.

REBIRTH OF NATURAL BIRCH WOODLANDS

One of the major questions is whether it is possible or even desirable to regenerate the natural birch woodlands that originally occurred below 400 meters elevation, including areas that are still covered with vegetation and areas that are sparsely vegetated or entirely denuded.

Vegetated areas comprise some 25 percent of the land area, yet only ten years ago it was virtually impossible to obtain any vegetated land for forestry, whether to conserve natural woodland or for plantations. All uncultivated land was used, or at least farmers thought it should be used, as range for the 900,000 winter-foddered sheep and 60,000 horses. Only a few small, scattered areas were obtained for forestry. During the 1980s, considerable changes occurred in agricultural practice. These changes were gradual at first but increased rapidly in the latter half of the decade. Sheep numbers declined to 530,000 winter-foddered beasts, although horses have increased to 80,000. Horses, however, are usually kept in enclosures, and often on marshy land, whereas sheep are free grazing. Land that is to be converted to birch woodlands therefore must be fenced at considerable expense. Several uncultivated areas have become available recently and have been enclosed, both to protect the existing birch stands and to regenerate treeless areas. The largest of these new enclosures are:

1. The capital area (28,900 hectares, of which 9,700 are urban).
2. The Thórsmörk enclosure between the Mýrdals and Eyjafjalla glaciers (5,000 hectares).
3. The Ystutunga enclosure in Borgarfjörður (3,000 hectares).
4. The enclosure around the town of Akureyri (1,000 hectares).

These enclosures have considerable value as recreation areas for the urban population. Several other areas await enclosure.

The sparsely vegetated and barren areas of the Icelandic lowlands were previously birch woodlands. Such sites are very suitable for colonization by birch seedlings, provided a seed source is nearby or is created on the site (Magnússon and Magnússon, 1990). In the absence of a natural seed source, birch seedlings can be planted on the barren lowlands.

In 1990, the Icelandic Forestry Association initiated a pilot project for reforesting eroded land with Icelandic birch as the major species. Small quantities of Russian larch (Larix sukaczewii Dyl.), mountain pine (Pinus mugo Turra), and several Salix species were included in the trials. A project called "Operation Land-Reclamation Forest" succeeded in planting a total of 1.3 million trees on 76 different sites; altogether, 8,000 volunteers took part in the program. Apart from the recreation area surrounding Reykjavík, where the local forestry society practiced a similar type of forestry, and the state plantation of Hallormsstaður in east Iceland, a "Land-Reclamation Forest" was a new concept. The Iceland Forestry Service Research Branch examined the plantings for survival in the spring of 1991 and found that mortality after the first winter was only 10 percent, which is low considering that the planting stock was one-year-old containerized seedlings grown in 50 cubic centimeter plugs (Aradóttir, 1991). The high survival is very promising and indicates that we have the technical knowledge necessary to reforest large areas of devastated land. Even so, a long time will be required for the small population of Iceland to reforest the enormous areas devastated by centuries of abuse. Once a few birch nuclei are established, however, natural regeneration will complete the task if grazing animals are excluded from the plantations.

The method is simple. Small scattered islands of birch are planted over the area designated for afforestation, and the denuded land between them is allowed to regenerate by seeding from the mother trees. Recent research in establishing birch by direct seeding and planting indicates that the greatest success is achieved by planting many small islands of comparatively few trees (Aradóttir, 1991; Magnússon and Magnússon, 1990). In this way, "strategic bridgeheads" are established for growth of birch throughout the range it occupied following the retreat of Pleistocene glaciation in Iceland 9,000 to 10,000 years ago. Research on natural regeneration and other establishment techniques for birch is now one of the priorities for the Forest Research Branch.[2]

Restoration of the birch woodlands provides several benefits:

- Soil protection
- Enhancement of understory vegetation
- Forest products for local use
- Improved attractiveness of recreation areas
- Scenic value
- Camping sites

If managed for agroforestry, vegetation under tall forest is more variable and robust, especially if the forest is thinned to encourage understory vegetation. The understory also has higher grazing value than vegetation grown on exposed sites.

Recreation is probably the most important reason for the interest in birch woodland. This is especially so for the weekend cottage sites that have sprouted within a 150 kilometer radius of urban centers. For example, the owner of a 0.5-hectare summerhouse plot in birch woodland, 100 kilometers from Reykjavík, demands an initial fee of 700,000 Icelandic kronur (U.S. $11,200) for a 25-year lease and an annual rent of 30,000–40,000 ISK (U.S. $480-640).

---

[2]Anonymous, 1990, Verkefnaskrá 1991 (Research Projects for 1991), unpublished manuscript from the Forest Research Station, Iceland Forestry Service, Mógilsá, 49 p.

Table 2. Number of vascular plants in Iceland

	Category	Species (number)	Specification
IS	Native species, probably occurring before settlement	384	405 native species including hawkeye spp. (Hieracium L.)
IS?	Native species, probably older than settlement but disputed	21	
SI	Introduced, naturalized	64	64 species naturalized
SO	Unstable casuals, naturalized but not permanent, shifting and recurring between sites	29	167 casual aliens
SS	Rare casual aliens, some naturalized, others unstable	111	
SR	Cultivated plants that escape from cultivation and occasionally become naturalized for a short period	27	
	Total	636	

PLANT INTRODUCTIONS

According to a recent list made by the Akureyri Museum of Natural History, the following number of species of vascular plants occur in Iceland[3] (Table 2).

Of the native species that predate settlement, 214 probably survived Pleistocene glaciation (Steindórsson, 1962). The large number of permanent and unstable aliens in categories SO, SS, and SR shows that the native flora is poor in species considering the Icelandic climate. Geographic isolation is the major explanation of this poverty of species. Furthermore, a large portion of the native flora is arctic in affinity because it evolved in an arctic climate.

Success of species introductions molds the philosophy that we should deliberately import plant species that have potential for adaption. How far to go with plant introductions becomes a question of ethics, desire for diverse flora, and economics.

---

[3]Anonymous, 1990, Listi yfir allar villtar plöntur og slæðinga (háplöntur) (List of all native and casual vascular plants), unpublished manuscript from Náttúrufræðistofnun Norðurlands (The Akureyri Museum of Natural History), Akureyri, Iceland.

TREE INTRODUCTIONS

About 130 tree species of some 1,300 to 1,400 different provenances have been tested in Iceland. At least 14 exotic tree species have produced natural offspring and should be categorized as type SR casuals. None except European bird cherry (Prunus padus L.) is included in Table 2. Undoubtedly, species such as Siberian larch (Larix sibirica Ledeb.), Russian larch, and lodgepole pine (Pinus contorta Dougl. ex Loud.) will graduate to the SI category (naturalized exotics) as they invade their surroundings.

Several other tree species will become naturalized when they form an ecological niche among the Icelandic flora. Several spruce, pine, and broadleaf species have recently borne fertile seed and have potential for adaption and naturalization to Iceland. The result of tree introduction is presented in more detail by Loftsson (in press).

Fig. 1. Early desert reclaimed by European lymegrass (Leymus arenarius (L.) Hochst.), Salix lanata L., and Betula pubescens in northeast Iceland. The reclaimed area was opened for grazing five years before it was photographed.

If the goal is to enrich Iceland's flora, then a problem that has been under considerable discussion in neighboring countries arises: To what degree should plant introduction be reconciled with sensible nature conservation, to which aim all true foresters strive? By this, I do not mean the environmental "terrorism" against exotics practiced by small groups of extremists under the flag of nature conservation. In discussions of nature conservancy, it is often forgotten that ecosystems are dynamic. Plant species are bound not by political boundaries but by the suitability of habitats and by their powers of migration and regeneration. If oceans or other barriers prevent the natural migration of beneficial organisms, then we should lend a hand. This was the philosophy of Iceland's greatest plant introducer,

Hákon Bjarnason (1943), which he formulated early in his career, instilled with great force into his followers and colleagues, and pursued ardently during his long career. He was to Iceland what David Douglas was to Great Britain and set the example we follow today (Davies, 1979).

FORESTRY GOALS

Forestry objectives for Iceland have been set forth both as part of the "Land Reclamation Project for 1974 to 1978" (Anonymous, 1974) and as an operational guide for the Iceland Forestry Service (Blöndal, 1987). These objectives are to establish protection forests, production forests, amenity or recreation forests, and shelterbelts.

Protection Forests

The main objective of protection forests is to prevent erosion of exposed soils. Aeolian loess, which originated from volcanic tephra and glacial silt, is the dominant soil

Fig. 2. A 49-year-old plantation of Russian larch (Larix sukaczewii Dyl.) in East Iceland.

type of Iceland. Tree cover is the best way to protect loess soils from wind and water erosion. Tree foliage is also needed to protect some plant societies found only in the birch woodlands or forests of other light-demanding tree species such as the larches. Tree foliage, especially of deciduous species, nurses shade-tolerant understory vegetation and shelters other understory tree species. The third objective of protection forests is land reclamation; that is, to restore vegetation on sites where it has been destroyed. This is a rather recent concept in the afforestation program.

Downy birch (Betula pubescens Ehrh.) is the main species for protection forests. Downy birch can grow at elevations up to 300 to 400 meters elevation in the coastal areas and 400 to 500 meters in the valleys of north and east Iceland. At high elevation it is only a low shrub, but it nevertheless fulfills the objectives of protection forests.

Russian larch and Siberian larch have a special ability to grow on sparsely vegetated sites and will, therefore, play an important pioneer role in land reclamation. The same role applies to lodgepole pine (Pinus contorta var. latifolia Engelm.), shore pine (Pinus contorta Doug. ex Loud. var. contorta), and mountain pine, even if these species do not establish as quickly as the larches.

Two species of native willow, tea-leaf willow (Salix phylicifolia L.) and wooly willow (Salix lanata L.), will also play a minor role in site protection and rehabilitation.

It is expected that several alder species (Alnus Mill. spp.) will be useful in forest protection, especially on denuded sites, because of their ability to fix nitrogen. But use of alder has not been extensive so far.

In the protection forests, tree height and stem form are not critical characteristics, but the root system must be vigorous and forest canopies must be closed.

Fig. 3. A recreation forest of exotics and indigenous species surrounding a weekend cottage in a stony desert in the area of Reykjavík, southwest Iceland.

Production Forests

The objective of production forests is to produce wood for timber, firewood, and Christmas trees. The area suitable for timber production in Iceland is very limited because wood production requires higher standards for tree growth and quality (and thus better habitats) than do other forest uses. The main species planted for wood production are Russian and Siberian larch, Sitka spruce (Picea sitchensis (Bong.) Carr.), Lutz spruce (Picea xlutzii Little), lodgepole pine, and black cottonwood (Populus trichocarpa Torr. and Gray).

Four species are planted for Christmas tree production: Engelmann spruce (Picea engelmannii (Parry) ex Engelm.), Norway spruce (Picea abies (L.) Karst.), subalpine fir (Abies lasiocarpa (Hook.) Nutt.), and lodgepole pine.

Recreation Forests

The objective of recreation forests is to offer areas for short-term visits or day use close to urban areas and to provide for activities such as camping and hiking. No demands are placed on the production capacity of the forest and only very modest demands are placed on height growth and stem form. Shelter from wind is the main requirement. A forest atmosphere with rich understory vegetation is secondary. A diverse mixture of a large number of species is also desirable for recreation forests. All tree species that thrive and reach maturity are desirable, especially in urban areas.

Native birch and other native trees such as willows, rowan (Sorbus aucuparia L.), and quaking aspen (Populus tremula L.) are the principal species for recreation forests.

Fig. 4. A shelterbelt of beak willow (Salix borealis B. Flod.) at a dairy farm in the Akūreyri area, north Iceland.

The use of exotics varies greatly among recreational forests, depending on their adaptability to Iceland's variable climate. All of the exotics already mentioned are planted extensively, but the following can be added: common silver birch (Betula pendula Roth), balsam poplar (Populus balsamifera L.), American green alder (Alnus crispa (Ait) Pursh.), grey alder (Alnus incana (L.) Moench), Sitka alder (Alnus sinuata (Regel) Rydb.), sallow (Salix caprea L.), whych elm (Ulmus glabra Huds.), European larch (Larix decidua Miller), tamarack (Larix laricina (Du Roi) K. Koch), bristlecone pine (Pinus aristata Engelm.), stone pine (Pinus cembra L.), limber pine (Pinus flexilis James), dwarf Siberian pine (Pinus pumila (Pall.) Reg.), Scots pine (Pinus sylvestris L.), Douglas fir (Pseudotsuga menziesii (Mirb.) Franco), Korean arborvitae (Thuja koraiensis Nakai), western hemlock (Tsuga heterophylla (Raf.) Sarg.), and mountain hemlock (Tsuga mertensiana (Bong.) Carr.).

Shelterbelt and Agroforestry

The main objective of shelterbelts is to provide wind shelter for crop cultivation, domestic animals, and farmsteads, and to dampen traffic noise and provide wind shelter in urban areas. Initially, fast growing and hardy species are planted. Later, when shelterbelts are well developed, long-lived species can be added.

The main fast-growing and short-lived species used today are feltleaf willow (Salix alaxensis (Anderss.) Cov.), beak willow (Salix borealis B. Flod.), and black cottonwood. The long-lived and slower-growing species are downy birch, rowan, Russian and Siberian larches, and whych elm.

BENEFITS OF FORESTRY

Forests provide various intangible benefits that are difficult to put in monetary terms but are real nevertheless. Intangible benefits worth mentioning are new landscapes, diversified environments, vegetation (including the herbs, mosses, lichens, and fungi), and especially the new bird- and wildlife that accompany forests.

Shelter from the wind (which is all too common in Iceland's stormy climate) is probably one of the most important tangible benefits of shelterbelt forestry. The pride of living in a forested country which was previously barren and the opportunity to spend more leisure time outdoors when an amenity forest is nearby are important intangible benefits of afforestation.

THE ICELANDIC OBSESSION WITH TREE PLANTING

Tree planting in Iceland only became popular with the urbanization of Iceland in the mid-twentieth century. Afforestation was advocated by the poet and first Icelandic minister, Hannes Hafstein (1916), in his epic poem written in 1900 on the occasion of the new century:

> The time will come when tattered land regrows, empty villages refill, fields clothe the heaths, bread for her sons [grows] from mother earth's fertile soil, culture blossoms in the shade of woodland glades.

In accordance with his philosophy, the first forest laws were pushed through by Parliament in 1907. The Youths' Association (Ungmannafélag Íslands), a powerful nationalistic movement which had afforestation as one of its foremost aims, also was established in 1907. Despite the enthusiasm for forest renewal in Iceland generated by Minister Hafstein, however, little afforestation was achieved because broad public support and the knowledge and experience about planting techniques were lacking.

Today, however, most Icelanders are aware that forestry and tree planting are not just empty words and wild dreams but result from tasks they willingly embrace. The trees themselves have grown better than expected, and far more species have succeeded than imagined. Today, the work of a few pioneers and prophets who were regarded as eccentrics, even dissidents, and who preached to deaf ears, has matured into forests. One man stands out from this group of pioneers, Hákon Bjarnason, who introduced forestry to Iceland. The Icelandic Forestry Association, with a present membership of 7,000, was used by Hákon Bjarnason to bolster the small state forestry service, which he directed for 42 years.

The Icelanders' obsession for tree planting has been strengthened by the present president, Mme. Vigdís Finnbogadóttir, who, since the first day of her inauguration in 1980, has made tree planting a symbol of her office. Wherever she goes on her official visits in Iceland she plants three birch trees: "One for the daughter, one for the son, and one for the future of Iceland."

This tree-planting obsession is reflected in the diligence of the Icelandic Forestry Society and its 45 local branches, and by the 6,000 summerhouse owners who feel obliged to plant trees around their properties.

FUTURE FORESTRY

Forest plantations in Iceland are small and few compared to the potential forest area of the country. But they are extremely important as strategic bridgeheads for a massive advance in forest development.

The Iceland Forestry Service estimates that 10 percent of the land area (1.03 million hectares) is climatically suitable for forestry. Protection and recreation forests can be grown in most urban areas. How large these forests should be is a matter of political choice and reconciliation with other land uses.

About three percent of Iceland's land area is capable of growing production forests. In a report for the government paper on future resources, Gunnarsson et al. (1987) estimated that 405 square kilometers of plantations or only about one eighth of the area suitable for production forests would be sufficient to meet the nation's requirements for sawtimber. Hence, with the political climate of today, the government is willing to advance funds for establishing production forests. Production forests are a means of supporting the population in rural districts where sheep rearing has been dramatically reduced. Production forestry in a country such as Iceland, however, can only be achieved by financial aid from the central government.

Both technical knowledge and experience are available for establishing a network of shelterbelts in most rural districts. The nation's agricultural laws have a clause for subsidizing shelterbelt planting. Interest is growing among farmers for planting shelterbelts, and only encouragement and financial support from the government are needed to initiate a national shelterbelt afforestation program.

Icelandic forestry is now on a firm base. Afforestation is gradually changing the countryside surrounding many urban centers, and similar changes are occurring in several of the rural districts. If the government of Iceland stands firm on its promise to significantly increase the forestry budget, the countryside will be vastly changed in 50 years.

During the last three years, planting-stock production has risen from 1.5 million to 6.0 million seedlings annually. The increased interest in tree planting give hope that the words of the poet-politician Hannes Hafstein (1916)—"The time will come when tattered land regrows"—will come true.

REFERENCES

Anonymous, 1974, Landgræðsluáætlun 1974-1978 (Land Reclamation Project for 1974-1978), Landbúnaðarráðuneytið (Ministry of Agriculture), Reykjavík, 66-69.

Aradóttir, Á. L., 1991, Population Biology and Stand Development of Birch (Betula pubescens Ehrh.) on Disturbed Sites in Iceland, Texas A & M University, Austin, 104 p., Ph.D. dissertation.

Benediktsson, J., ed., 1968, Íslendingabók, Íslendingabók-Landnámabók (The Book of Icelanders), Hið íslenska fornritafélag (Icelandic Antiquarian Society), Reykjavík, 28 p.

Bjarnason, H., 1942, Ábúð og örtröð (Use and Abuse), Ársrit Skógræktarfélags Íslands (Yearbook of the Icelandic Forestry Association), Skógræktarfélag Íslands, Reykjavík, 8-40.

Bjarnason, H., 1943, Um ræktun erlendra trjátegunda (On the cultivation of exotic tree species), Ársrit Skógræktarfélags Íslands (Yearbook of the Icelandic Forestry Association), Skógræktarfélag Íslands, Reykjavík, 11-62.

Bjarnason, H. and Sigurðsson, S., 1977, Skóglendi á Íslandi (Woodlands in Iceland), Skógrækt ríkisins-Skógræktarfélag Íslands, Reykjavík (Publication of Iceland Forestry Service and the Icelandic Forestry Society), 38 p.

Blöndal, S., 1987, Möguleikar og markmið skógræktar á Íslandi (The potential and objectives of forestry in Iceland), Ársrit Skógræktarfélags Íslands (Yearbook of the Icelandic Forestry Association), Skógræktarfélag Íslands, Reykjavík, 47-55.

Davies, J., 1979, Douglas of the Forests, Paul Harris Publishing, Edinburgh, 192 p.

Einarsson, Þ., 1963, Pollen-analytical studies on the vegetation and climate history of Iceland in Late and Post-glacial times. In: Löve and Löve, eds., "North Atlantic Biota and their History," Pergamon Press, London, 355-365.

Gunnarsson, E., Guðmundsson, E., and Árnason, R., 1987, Hagkvæmni nytjaskógræktar (Feasibilty of Production Forestry), Skógræktarrit 10, (Sérprent úr ritinu Auðlindir um aldamót, Sérrit 3 úr Framtíðarkönnun ríkisstjórnar um Skógrækt), Forestry Report 10, Iceland Forestry Service, Reprint from the Government Paper No. 3, Resources for the Future, Reykjavík, 50 p.

Hafstein, H., 1916, Lóðabók (poems), Þorsteinn Gislason, Reykjavík, 13–15.

Jónsson, G., 1983, Viðarkol (Charcoal). In: Minningarþættir Guðmundar frá Húsey (Memoirs of Guðmundur from Húsey), Að vestan IV, Bókaútgáfan Skjaldborg, Akureyri, 172-176.

Loftsson, J., [In press], Forest Development in Iceland. In: Alden, J. N., Ødum, S., and Mastrantonio, J. L., eds., Forest development in cold climates, proceedings of an international symposium held in Laugarvatn, Iceland, 1991 June 18-24, NATO, Plenum, New York.

Magnússon, Á., and Vídalín, P., 1913-1943, Jarðabók I-XI (Farm Register I-XI), Hið íslenska fræðafélag (Icelandic Literary Society in Copenhagen), Copenhagen.

Magnússon, S. H., and Magnússon, B., 1990, Birkisáningar til landgræðslu og skógræktar (Seeding of birch for land reclamation and woodland establishment), Ársrit Skógræktarfélags Íslands (Yearbook of the Icelandic Forestry Association), Skógræktarfélag Íslands, Reykjavík, 9-18.

Sigurðsson, S., 1977, Birki á Íslandi (Útbreiðsla og ástand) (Birch woodlands in Iceland (Distribution and condition). In: "Skógarmál," [Private publisher,] Reykjavík, 146-172.

Steindórsson, S., 1962, On the Age and Immigration of the Icelandic Flora, Vísindafélag Íslendinga XXXV (Societas Scientarium Islandica XXXV), Reykjavík, 157 p.

Steindórsson, S., 1964, Gróður á Íslandi (Vegetation of Iceland), Almenna bókafélagið, Reykjavík, 186 p.

Thoroddsen, Th., 1911, Lýsing Íslands (Description of Iceland's Nature), Hið íslenska bókmenntafélag (Icelandic Litterary Society), Copenhagen, 422-444.

Þórarinsson, Þ., 1974, Þjóðin lifði en skógurinn dó (The people lived but the forests died), Ársrit Skógræktarfélags Íslands (Yearbook of the Icelandic Forestry Association), Skógræktarfélag Íslands, Reykjavík, 16-29.

Þórðarson, S., 1955, Úr sögu skóga á Austurlandi (History of the East Iceland Forests), Ársrit Skógræktarfélags Íslands (Yearbook of the Icelandic Forestry Association), Skógræktarfélag Íslands, Reykjavík, 19-30.

Þorsteinsson, I., and Blöndal, S., 1986, Gróðureyðing og endurheimt landgæða (Destruction and Rehabilitation of Vegetation), Námsgagnastofnun, Reykjavík, 24 p.

MOLECULAR BASES FOR ADAPTATION OF CONIFEROUS TREES TO COLD CLIMATES

Don J. Durzan

Department of Environmental Horticulture
University of California
Davis, California, United States

INTRODUCTION

With selected woody perennials, an analysis of molecular aspects of adaptive phenotypic expression is provided. Cold acclimation is examined in white spruce (Picea glauca (Moench) Voss) genotypes in the Canadian boreal forest. The adaptive behavior of spruce is contrasted with a genetic disorder in an almond cultivar (Prunus amygdalus (Batsch.) cv. Nonpareil) under Mediterranean climatic conditions in California. The disorder involves aberrations in growth during heat acclimation. All data are based on local field studies, with concepts being extended to boreal timberlines. Several new diagnostic approaches are developed in an effort to understand the behavioral aspects of tree fitness in relation to more flexible breeding and testing systems for clones and progeny.

Adaptation denotes any change in the genome of a tree species that allows it to better cope with the current and, possibly, future environment. Acclimation involves adaptation to new temperature, climate, or altitude (Tranquillini, 1979). The result is an harmonious adjustment to the change agent. For example, genome expressions may change, but not the genes themselves. A genetic constraint to adaptation is any aspect of gene inheritance, variation, and expression that prevents natural selection from resulting in the harmonious adjustment to the change agent (Agur and Slobodkin, 1986; Loeschcke, 1987). If there is little or no genetic variation, selection cannot produce evolutionary changes.

Genome reorganization and specialization often leads to new phenotypic expressions and enhanced fitness. If harmony with the environment is disrupted, new opportunistic modifications in structure, function, and life cycles may appear. For example, in the reproductive process, the adaptive processes inherent in polyembryony may increase the survival of the next generation (Klekowski, 1988). The adapted phenotype need not be inherited and is thought to represent a new organized, stable state. It is the "ability to adapt" that is inherited.

Adaptations at the molecular level should be interpreted from field experience (Tranquillini, 1979) and population genetics (Hagman, in press; Tigerstedt, in press; Wright, 1968). Cycles of rapid vegetative growth are a function of the built-in totipotency of

Forest Development in Cold Climates, Edited by J. Alden *et al.*
Plenum Press, New York, 1993

meristematic cells, usually in response to warm, moist summers (Axelrod, 1966; Steward, 1968). During the initial juvenile phase, competition with ground vegetation is severe, and the range of expression of heritable traits involved in adaptation becomes more specifically targeted at the environment. At maturity, and with a closed co-dominant canopy, the expression of genotypic characters again makes trees much more competitive (Eriksson and Lundkvist, 1986; Ford, 1984; Raymond and Lindgren, 1986).

In this review, adaptive phenotypic expressions are examined at the molecular level. Concepts derived from the responsiveness of species of known genetic backgrounds in stressful environments are related to methods of accelerating tree breeding and testing (Anamthawat-Jonsson et al., in press; Durzan, 1991) and of reverse genetics (Berg, 1991).

MOLECULAR ASPECTS OF TEMPERATURE ACCLIMATION

Graded responses by conifers to thermoperiodic stimuli are evident from many studies in controlled environments (Lang, 1963). Along clines in the natural environment, graded responses are most likely determined by wind-, frost-, and drought-sensitivity at low temperatures (Denny, 1988; Endler, 1977; Levitt, 1972; Mikola, in press; Woodward, 1987). Near the timberline at high latitudes, photoperiod also becomes important (Junttila and Nilsen, in press; Tranquillini, 1979).

To better understand the effects of temperature extremes on a genotype, temperature must be distinguished from heat in terms of molecular changes. Temperature measures the intensity of heat present in the system, that is, at the site, in the microenvironment, and in the tree (Thorgeirsson et al., in press). Temperature is a measure of the kinetic energy of a system and its subsystems—gaseous, molecular, and structural components of cells. Heat, expressed in calories or joules, is the total energy in each system. For responses at the molecular level, the temperature of the air and tree can be related to the heat in each system. The heat in each system will affect rates, equilibria and overall behavior of metabolic reactions.

Biochemical responses to low temperature involve the roles of ice, water, dissolved gases, and osmolytes (Yancey et al., 1982). Kinetic responses include protein denaturation, reduced lipid mobility, and variation in chemical composition (Levitt, 1972). The activation energy of molecules underlying these responses relates mass input, reaction velocities, temperature change, and the caloric changes per mole of new substance formed in the above- and below-ground parts of the tree.

At the physiological level, acclimation and resistance to stress involve osmotic adjustment and efficiency of water, carbon, nitrogen, and heat-use (Grace and James, in press; Larsen, in press). For example, water-use efficiency has been interpreted as the amount of carbon gained per unit of water lost (Osmund et al., 1980). At the metabolic level, diurnal response of plants to environmental factors is often mediated where carbon and nitrogen metabolism interact (Durzan and Steward, 1983). The nitrogen status of trees is reflected in activity of the transaminases and synthesis of proteins and storage N compounds (Durzan and Steward, 1983). The transaminases integrate carbon and nitrogen metabolism. Because available N limits the growth of boreal conifers, the free amino acid pool is a good indicator of how well individual trees are adapted to nitrogen deficiency (Durzan, 1971; 1975). Enzymes that assimilate and use available N efficiently are useful in selecting pioneering genotypes. However, in the alpine zone, primary production is not usually limited by insufficient nitrogen, but by climatic factors (Tranquillini, 1979). The energy for acclimation processes is derived from mitochondrial metabolism comprising a cytochrome pathway or an alternative pathway (Watt, 1986). When the cytochrome system is saturated or somehow restricted, the alternative pathway provides most of the energy for metabolic adjustments during environmental fluctuations (Lambers, 1985). A scheme for the hierarchical integration of the environmental-signal response system based on gene expression and the genotype's ability to adjust to external change is shown in Figures 1a and 1b.

Adaptation at the molecular level (Holliday, 1987) can be studied in terms of: (1) the mechanism of transmission of epigenetic factors from generation to generation (Nilsson, in press); (2) the mode of action of genes during the development of the tree from zygote to adult (Brink, 1962); and (3) the nature of gene expression and epigenetic changes as influenced by external factors (Dixon and Lamb, 1990; Sigurgeirsson, in press).

PHENOTYPIC PLASTICITY AND THE REACTION NORM

Changes in environmentally produced phenotypic expressions may or may not be adaptive (Via, 1987). Adaptive phenotypic plasticity is thought to evolve under natural selection in a variable environment. Phenotypic plasticity usually enables greater selective advantage through the production of appropriate stabilizing or destabilizing responses. Via (1987) illustrates the morphological, physiological, biochemical, and behavioral aspects of plasticity.

Phenotypic plasticity is arbitrarily defined as the difference between mean phenotypic values of adaptive traits. Selected traits in genetically identical and genetically different individuals are studied in several environments to obtain reaction norms. The reaction norm represents all phenotypes displayed by a single genotype under different environmental conditions (Van Noordwijk and Gebhardt, 1987). The reaction norm may also be defined as the entire range of the variant paths of development that occur in the carriers of a given genotype or idiotype,[1] and in all environments, favorable and unfavorable, natural and artificial (Reiger et al., 1976). In some metabolic studies, the reaction norm has been redefined as the average phenotype that is expressed in carriers of a genotype in a given environment or provenance (Figs. 2, 3). This phenotype is model-referenced to reaction norms of similar or different genotypes. Reaction-norm behavior can also be studied in the same environment or in different environments (Lindgren, in press).

The acclimating phenotype is described as a numerical data-set representing the continual departure from reaction norm or average metabolic behavior. The reaction norm may be arbitrarily comprised of a network of hierarchical pathways in the free amino acid pool, as determined from radioactive tracer studies (Durzan, 1987 and 1989). Metabolite behavior is usually represented as a change in concentration or flux of substrates and products. When taken together, departures from the reaction norm reveal the resilience of the system for discrete adaptive or developmental outcomes. Graphic displays of outcomes really represent a horizon of events (Figs. 2, 3). Events may be seen in relation to expressions of genetic variation and the effects of environmental change (Scharloo, 1987).

Reversible and irreversible molecular changes in reaction-norm behavior are diagnosed by matching the patterns of their graphic displays. These event horizons are designed to reveal changes in the functional expressions for a specific trait, changes in spatial-temporal

[1] Idiotype is the total of chromosomal and extrachromosomal genetic information, and is different from an ideotype, or "ideal" genotype for tree improvement purposes.

organization of these expressions, and kinetics that arise from instabilities in the metabolic system. Some instabilities in metabolic behavior may lead to new molecular and unexpected interactions and to changes in cellular organization. An important characteristic of these interactions and changes is their repeated behavioral patterns (Babloyantz, 1986).

For tree improvement programs, the reaction norms of new genotypes may be tested for during the critical phases of establishment in relation to diurnal and seasonal change. For adaptive traits having complex reaction systems, a single rate-limiting step may not be readily evident (Hartl, 1989). Control of an adaptive trait may be exerted by many enzymes, each of which may make relatively little contribution to the metabolic pathway (Kacser and Burns, 1973, 1981; Sauro et al., 1987). If material with different genetic

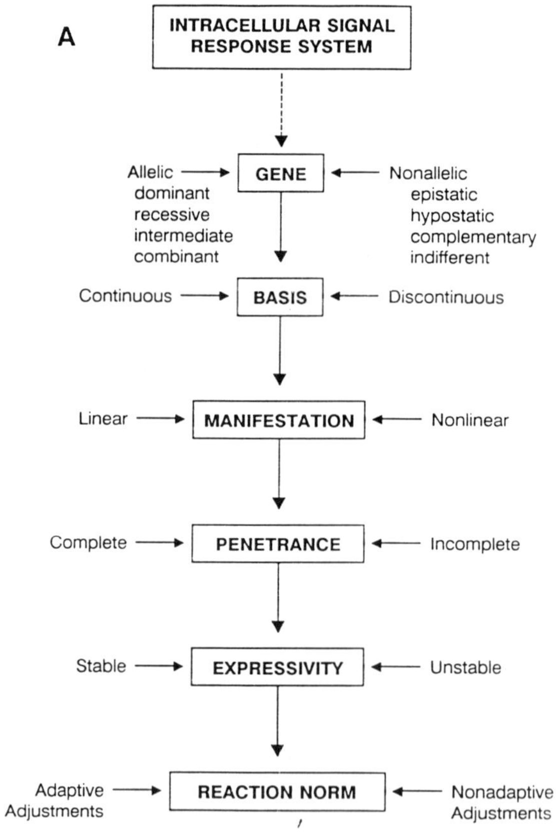

Fig. 1. Attributes of a hierarchical metabolic model for gene expression. Adaptive adjustment to external signals is modeled. (A) Genome-directed expressions lead to a reaction norm and to stable or unstable phenotypes under adaptive and non-adaptive conditions. Adaptation is also an inherent property of complex dynamic metabolic systems. Here, a property of self-organization adjusts interactions in metabolic networks to varying environments (Pahlich, 1991).

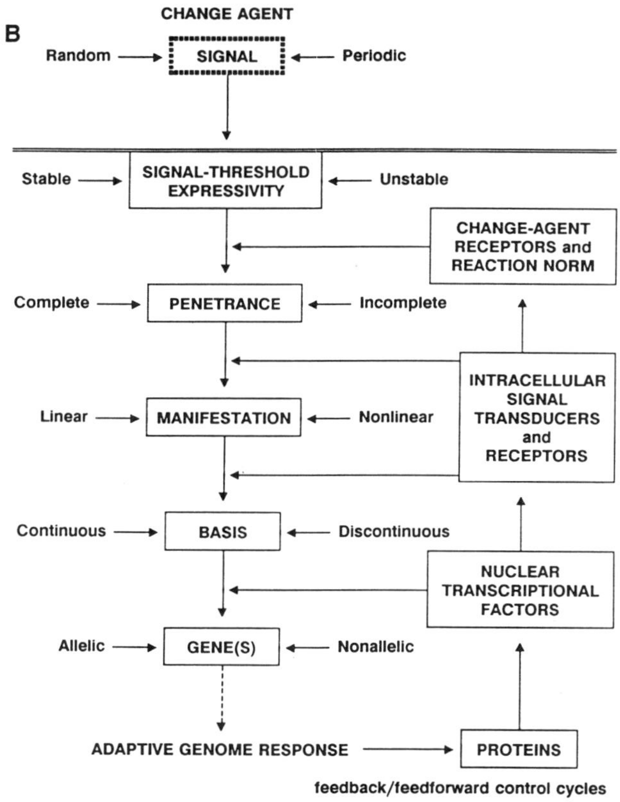

Fig. 1 (cont.). (B) Signal-response sequence in an adaptive metabolic system leads to adjustments in a reaction norm through genome-mediated controls. The adjustment involves a feedback and feedforward process. Many genes involved in adaptive adjustments undergo developmental as well as environmental regulation. Promoters associated with these genes are believed to determine the tissue- and cell-type-specific expressions seen in adaptive responses and in plant defense responses (Dixon and Lamb, 1990).

backgrounds were available for study and adaptive metabolic plasticity were recognizable (Fig. 4), global adaptive enzymatic behavior could be verified by examining the spreading of patterned responses in hierarchical, branched, and intercalated metabolic networks (Durzan, 1987, 1989; Pahlich, 1991; Stephanopoulos and Vallino, 1991).

METABOLIC TRACES IN ADAPTIVE ADJUSTMENTS

The ability of an organism to adapt to climate involves not only the constitutive genes (those with fixed or predetermined expressions), but also genes whose expression must be induced (Figs. 1, 4A). Here, induction must be distinguished from permissiveness. The net appearance or loss of structure and function may be caused by removal of a constraint (permissiveness) rather than a response to some external signal (induction).

Inductive enzymatic behavior and the spread of an adaptive response can be illustrated in white spruce seedlings by exposure to ^{14}C-urea (Durzan, 1973a,b,c) (Fig. 5). Contact with urea induces the de novo formation of urease in epidermal cells. Urease degrades urea to a number of new compounds that can be traced through several networking pathways. This technique provides evidence for a "metabolic trace" where none existed before. The

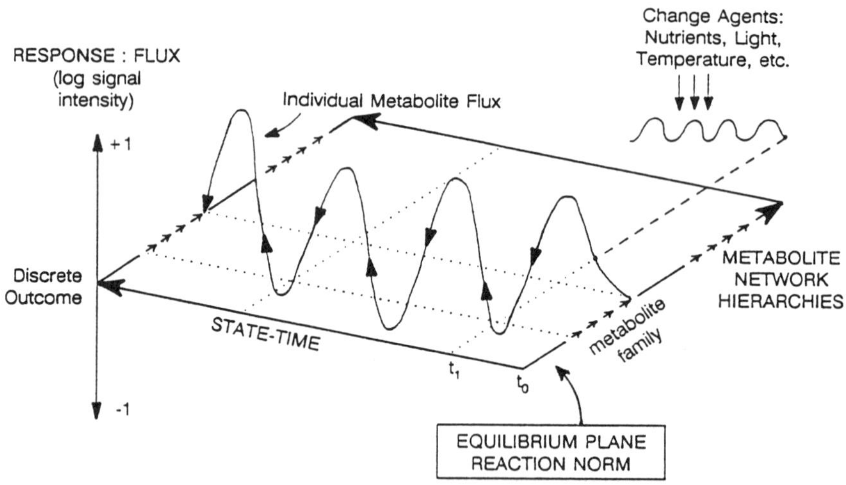

Fig. 2. Reaction norm for a growth rhythm is portrayed as an equilibrium plane or event horizon showing stability and homeostasis. The norm is comprised of an arbitrarily selected group of genetic loci that characterize metabolic families arranged in a hierarchical network. Each family has a port of entry for nitrogen from which metabolites are sequentially derived. A periodic flux for only one metabolite is illustrated. The sequential derivation of metabolites in a pathway is usually tracked as a set of fluxes involving known precursor-product relationships (See Fig. 8). Departures from mean behavior over time or physiological state for flux are represented by the vertical axis. The flux is perturbed and responsive to change agents in the environment that may or may not lead to acclimation. This response surface reflects the genetically programmed expression of "traits" having discrete outcomes. The response may be calibrated to the log of the signal intensity (external) and to the metabolic flux (internal state). The adaptive metabolic family behavior occurs over time (t_0, t_1 to a discrete outcome). Each point in time usually represents a different physiological state (state-time). The metabolite behavior, as indicated by the arrows on the curve, shows stability, that is, the ability to return to equilibrium after a temporary disturbance. This represents one type of homeostatic behavior.

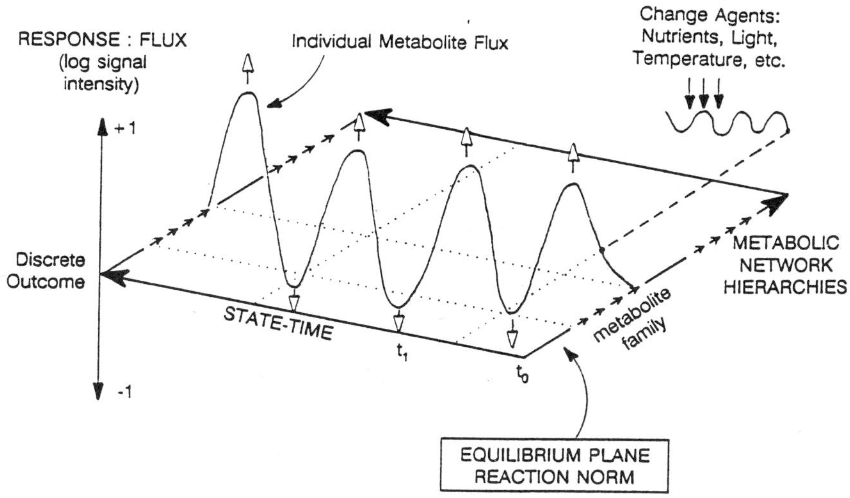

Fig. 3. Event horizon showing resilience around a reaction norm. Resilience (indicated by arrows departing from the stabilizing curve) is seen as the persistence of relationships within the overall system. It measures the ability of the metabolic system to absorb changes of physiological state over time and due to perturbing factors (change agents) and still persist, service, repair, restore, that is, remain adaptive. Near and far-from-equilibrium behavior is often evident in experimental data. This supports the notion that trees are not equilibrium machines. Metabolism, as seen by changes in nitrogen flux, shows regulated and adaptive behavior around the reaction norm. While metabolic plasticity may be optimal, resilience over the long run is needed to maintain fitness. Resilience can be used as a selection criterion for fitness by the tree breeder (Reiger et al., 1976).

trace represents a change in carbon and nitrogen flux that leaves a characteristic imprint on metabolism. Imprinted behavior, measured in this way, has been called an "engram" (Durzan, 1987). The metabolic trace provides the substrate upon which the efficiency of nitrogen (N) use may be calculated.

An insecticide, such as fenitrothion, when sprayed on spruce seedlings, can cause changes in the composition of chromosomal proteins (Pitel and Durzan, 1978). This is an example, at the structural level, of how a synthetic substrate (change agent) can leave metabolic traces.

BEHAVIORAL ASPECTS OF HOMEOSTASIS

Most genotypes adjust their life processes to physiological states that are either optimal (Parker and Maynard Smith, 1990), maximal (Ringpfeil, 1986), near-equilibrium, or far-from-equilibrium (Babloyantz, 1986; Morowitz, 1979; Watt, 1986). These adjustments can be referenced to reaction-norm behavior (Figs. 2-4). Homeostasis represents the tendency to restore a stable state relative to the reaction norm. Restoration of the stable state may involve interdependent and nondependent components of the reaction norm (White et al., 1984).

For rapid tree growth, homeostatic tendencies have four components: topological (topophysis), developmental (cyclophysis), microclimatic (periphysis), and genetic (variabili-

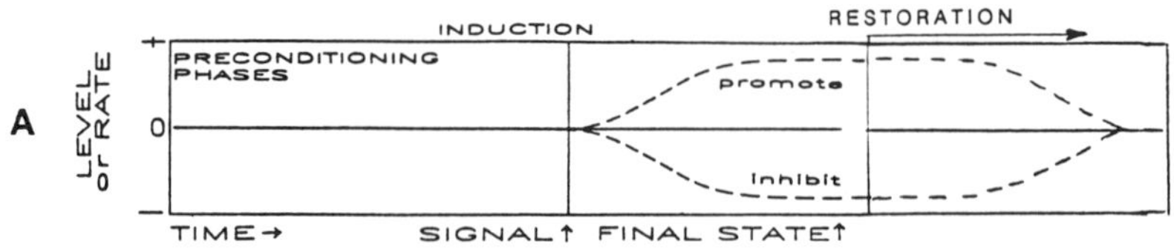

An induction event leads to a positive or negative adjustment.

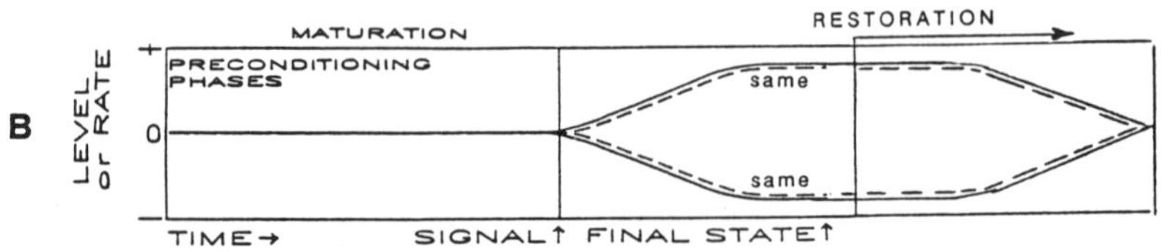

Adjustments related to maturation are genetically programmed and tend to be independent of preconditioning phases.

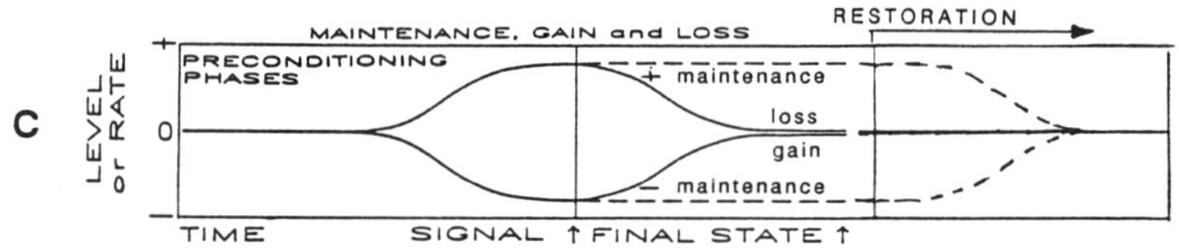

The reaction norm for a metabolite may drift during the preconditioning phase. At a signal, the response may be maintained or followed by a loss or gain.

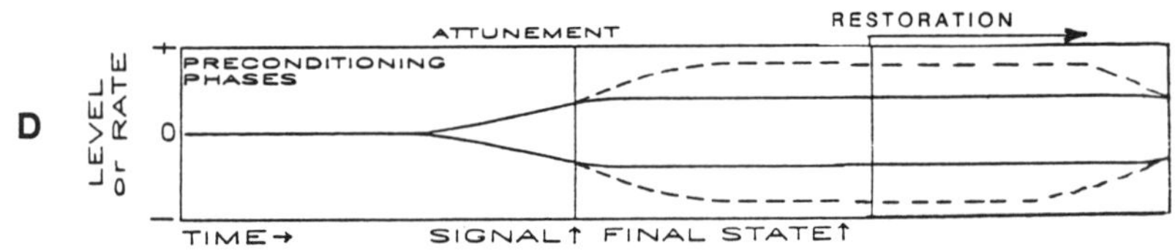

Attunement based on the reaction norm may occur during preconditioning, followed by a signalled enhancement to reach the fully adaptive response.

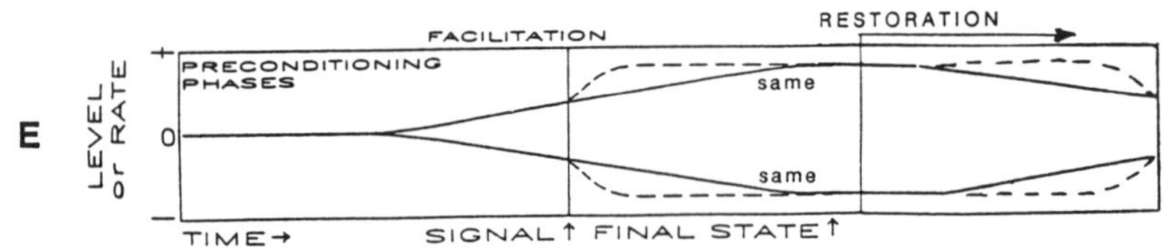

Facilitation affects the rate of expression of a metabolite flux after an environmental signal. The facilitated response is seen earlier (precociously) than in the absence of preconditions. Restoration of the reaction norm to the presignal state is shown.

Fig. 4. Adaptive behavior of a metabolite as a function of the presence or absence of preconditions. Broken lines show responses to preconditions. Responses based on the lack of preconditions are shown by the solid line. Preconditions represent a variable potential that is established, but not always immediately or fully expressed unless an environmental or climatic signal elicits the illustrated adjustment. The "final state" reached after the signal represents an end point of the response to signalled changes. Thereafter the predicted metabolite behavior that is needed to restore the original homeostatic reaction norm is indicated (restoration). The equilibrium plane or reaction norm is set at 0 level or rate. This also represents the mean metabolite behavior. Departures from the norm over time are identified as induction, maturation, maintenance, gain and loss, attunement, and facilitation (Modified from Durzan, 1990a,b).

ty throughout the life history). Over the long run, resilience in adaptive plasticity involves near-equilibrium and far-from-equilibrium behavior (Fig. 3). For example, in white spruce the annual and diurnal behavior of major free amino acids in the terminal buds of leader shoots indicates how intermediary nitrogen metabolism becomes attuned to the environment (Figs. 6, 7).

Two major components, arginine N and proline N, account for a significant portion of the free amino acid nitrogen pool. When periodic changes in these compounds are examined over a 12-month period by harmonic analysis (Brooks and Carruthers, 1953), 90 percent of the variation fits a fundamental harmonic oscillation. Using harmonic analysis, the ratio in amplitude of mean monthly air temperature and monthly hours of sunshine to

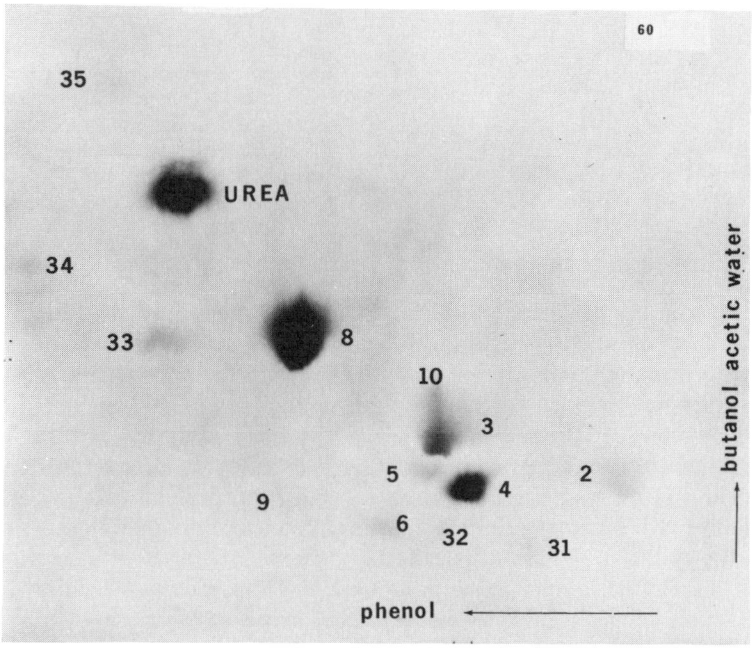

Fig. 5. Autoradiograph of radioactive metabolites formed after 60 minutes in white spruce seedling fed ^{14}C-urea. The distribution of carbon-14 into newly synthesized metabolites characterizes the "metabolic trace" that emerges in response to urea as a nutritional stimulus. The exposure of seedlings to ^{14}C-urea induces the enzyme urease in epidermal cells (See Durzan, 1973a for histochemical evidence). The metabolic trace represents assimilation products of a chemical signal and leads to the diffusion of new metabolites that may in turn activate or inhibit other reactions and enzymes. Compounds found in the metabolic trace were extracted in ethanol, separated by two-dimensional paper chromatography, and autoradiographed on X-ray film. Arrows show solvent directions from the origin of the chromatogram. Numbers identify radioactive products derived from urea, e.g. (2) aspartic acid, (3) glutamic acid, (4) serine, (5) glycine, (6) asparagine, (8) α-alanine, (9) glutamine, (10) carbamylaspartate; the others are unidentified. The material traces created by the assimilation of urea involve some of the metabolite adjustments illustrated in Figure 4. Several compounds (2, 3, 8, 9) are the starting points for the elaboration of metabolic families (networks) in nitrogen metabolism. The behavior of these families may be plotted as response surfaces (See Figs. 2, 3).

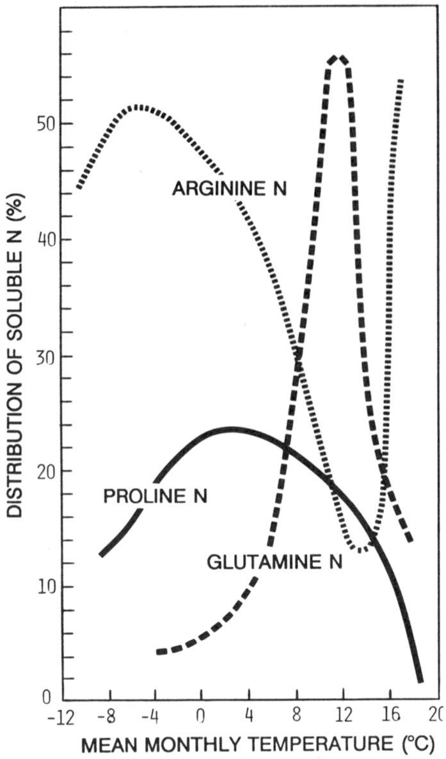

Fig. 6. Sequential ordering and dominance of free amino acid nitrogen in the alcohol-soluble N fraction of white spruce buds in response to mean monthly temperature. The distribution of nitrogen associated with each compound is shown as a percentage of the total soluble N in buds. Trees (ca 20-year-old) were situated at Chalk River, Ontario, Canada (45°47′ N. and 77°23′ W.). Samples were taken from three individual trees at temperatures of -30 °C and covered in ice. In some genotypes, levels of proline increased to 40% of the total soluble N, mainly at the expense of arginine N. Glutamine N levels increased mainly in late spring as buds swell and break. High levels of arginine N returned in July as shoot elongation ceased and new buds were set (Durzan 1968a,b,c, 1973c). The ordering of metabolites implies that temperature controls enzymatic activities that determine when and where the metabolic adjustment occurs. If saplings of spruce are continuously shaded to 13% of natural light, the composition of soluble N can be dramatically altered in above- and below-ground tissues (Durzan, 1971). In October, leaves increase their % arginine N content from ca. 4 to 16% of the total soluble N as shading increases from full to 13% of natural light.

arginine and proline N metabolism was estimated for each annual oscillation around the reaction norm. The contribution of arginine and proline N to the soluble nitrogen pool of buds and shoot tips changes 0.46- and 0.43-fold for any incremental change (±) in mean monthly air temperature, and 0.14- and 0.13-fold for any incremental change in mean monthly hours of sunshine (Durzan, unpublished). So far, it is not known how these values will be extended by trees situated beyond existing tree limits. In boreal forests, the response of conifers to short or prolonged growing seasons, however, depends on the available nitrogen and the ability of trees to store nitrogen as arginine and arginine-rich proteins. At the cellular level, the ability to remain fit is reflected by how efficiently proline N accumulates to offset metabolic stresses associated with drought, cold, and increased salinity (Pahlich, 1991). Homeostasis occurs at other levels. Developmental homeostasis or "canalization" (Waddington, 1957) denotes the stabilization of a tree's development; patterns of growth, differentiation, and morphogenesis are recognizable and discrete in the face of changes that tend to distort the genotypic reaction norm. Physiological homeostasis is a reflection of compensating adjustments in life-sustaining processes. Microenvironmental fluctuations around terminal shoots, for example, affect the ability of buds and leaves to accumulate arginine N in the fall (Fig. 7).

When all aspects of homeostasis and resilience are integrated, most adaptive systems are nonlinear, complex, and very difficult to predict. Phenotypic variability in whole-tree and organ-based responses may be large even when identical genotypes are compared in a common environment. Identical genotypes may produce different phenotypes if the adaptive behavior becomes asynchronous or correlations among physiological functions within the tree are somehow disrupted.

In some plants, the tendency for strong homeostasis is linked to the level of heterozygosity in the genome (Langridge, 1963). In this context, developmental homeostasis

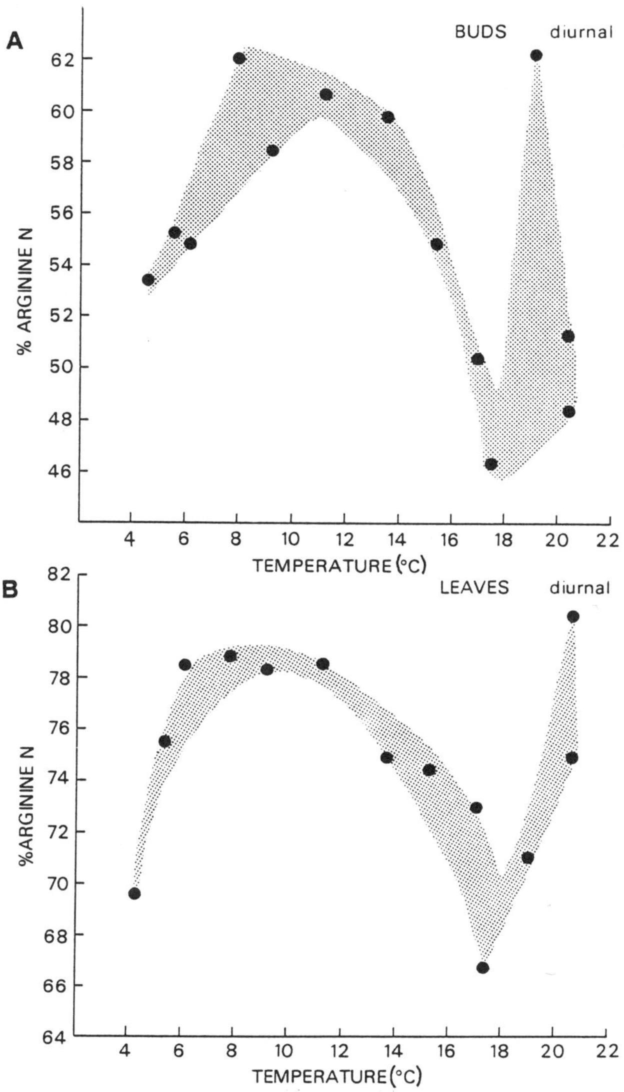

Fig. 7. Effect of mean hourly diurnal temperature (°C) in August on the distribution of soluble arginine N as a percent of the total soluble N in (A) buds and (B) leaves on terminal shoots of a 20-year-old tree at Chalk River, Ontario, Canada. A similar response profile is seen over the season for both organs, with leaves having a higher average reaction norm for arginine than buds. The amplitudes of these variations at critical times of the year are useful in selecting genotypes from genetic crosses aimed at adjusting the fitness of the next generation. Arginine in buds and leaves is converted to monosubstituted guanidines that contribute to winter dormancy. In the following spring, these compounds are broken down to urea and an amine as shoot growth begins (Durzan and Steward, 1983).

becomes a function of the inherent totipotency of the genome established by the segregation of alleles during meiosis and their recombination at fertilization (Li and Graur, 1991; Steward, 1968). The plasticity associated with the expression of alternative developmental fates must be somehow overridden by the homeostatic and stabilizing tendencies of each genetically predetermined system. This is especially evident in many conifers through their characteristic size, shape, recalcitrance to vegetative propagation, and long life cycle.

Opportunistic and flexible development may arise during the tree's reproductive phase as part of a survival strategy (Klekowski, 1988). One example is the expression of polyembryony of various types in seeds (Simak, 1973; Skaret, in press). Temperature-sensitive alleles are thought to be responsible for polyembryony. At the severe and erratic temperatures of high elevations and latitudes, adaptive defects in polyembryony are usually evident (Simak, 1973).

TEMPORAL CONSIDERATIONS

The adjustment time for adaptation at the molecular level is a critical factor in any response system. Time can be thought of in two ways: as a "quantity" measured by clocks, and as a vector that gives direction and spread to homeostatic adjustments in physiological states. The concepts of metabolic precedence, determination, and ordering must be introduced to better understand adaptive signalling networks and subsequent effects that become referenced to genes (Fig. 1). It is assumed that all critical, climatic and physiological change agents can be tracked and referenced to a standardized, universal, internal behavioral system. It is commonly thought that metabolic rhythms having an endogenous circadian periodicity are related to natural environmental cycles (Lang, 1963; Hamner, 1963). However, not all cells and tissues in a tree may be fully synchronized or "locked into" the expression of networks that signal the need for adaptive plasticity (Table 1).

Another major problem is whether repeatable patterns of adaptive metabolic behavior exhibit predetermined, regular, stochastic, or chaotic behavior. Methods for assessing the spread and loss of information from a chaotic time series (Wales, 1991) can be used to distinguish chaotic behavior from "background noise." The decay-time for adaptation is a parameter in models of living systems (Babloyantz, 1986; Bennett and Bourzutschky, 1991).

GENETIC ASSIMILATION

Some metabolic adjustments depend on genes for adaptive behavior that are amplified during the DNA replication process (Dawkins, 1982). In conifers, the geometrically replicated levels of DNA are thought to represent past adaptive behavior (Price et al., 1973). Miksche (1971) found 1.5 times more DNA in root tip cells of spruce from high latitude populations than in such cells from spruce at low latitudes. Dhillon (1987) suggested that redundant DNA could reflect a homeostatic response to environmental stimuli.

Flexible and adaptive growth patterns are often seen in individual juvenile trees. At timberlines, these adaptive shifts become fixed, as in mature Krummholz forms. Adaptations of this type have been termed "genetic assimilation" (Waddington, 1957), and are different from changes that result from genetically programmed maturation (Brink, 1962).

Table 1. Some events, configurations, and stimuli that may determine or lock-in physiological states reversibly or irreversibly and contribute to large-scale integration or misadjustments in growth and development. Examples may be interconnected.

Factor	Example
Stable intermediates in a reaction sequence	RNA polymerase-DNA-protein complexes in nuclei. Stable protein configuration in epigenetic situations. Daughter cell adhesion. Reinforcement centers in cell walls.
External environment	Water, light, temperature, pH, and colligative effects on surface enzyme activity.
Ionic milieu	Distribution of nutrients, physical forces (osmotic, surface tension). Trace metal chelates with amino acids.
Hierarchical and sequential changes in metabolic building blocks	Post-translational modification of proteins. Methylation of DNA to inactive genes. Modification of DNA base and signal sequences in proteins to schedule and allocate signals. Informational and organizational macromolecules in cellular networks and compartments.
Residence time of transient intermediates	Rates of hydrolysis, isomeration, ligation, transfer, oxidation, reduction, and assimilation. Auto-self-regulatory activities to keep genes "on." Cystolic metabolic arrays. Physiological preconditioning.
Accumulation of metabolic intermediates	Blocks in metabolism. Nutrient limitation. Stress responses in plants. Convolution of metabolic arrays and their connectivity. Water and nutrient-use efficiencies, source-sink relations.
Thresholds and metabolic control	Diffusion. Simple and complex bifurcating metabolic systems. Relationships among shape, volume, and metabolism. Cytochrome and alternate path respiration. Failure of error-prone repair systems, self-organizing and adaptive behavior.
Positional information and cell differentiation	Specialization of cells for metabolism, storage, structure, etc. Microtubular organization of cells. Physiological gradients. Deposition of materials extracellular to the cell wall and in the cell wall. Cell cycle activity.
Genetic control	Replicator activity, genome organization, mutations, somatic recombination. Promoter and enhancer DNA sequences. Boolean processes in enzyme-forming systems. Epigenetic factors.
Growth regulators and regulatory molecules	Messengers and signal molecules. Evolving transducing and catalytic activities of metabolic networks. Oligosaccharin activity. Low molecular weight transformations.

PHENOTYPIC PLASTICITY AT EXTREME TEMPERATURES

A comparative and developmental approach to reaction norms involving intermediary nitrogen metabolism is illustrated by two examples.

Extreme Heat: Bud-failure Syndrome

A single environmental variable such as a high temperature may limit allelic expression and reduce viability of a species. Conifers in cold climates rarely experience high summer temperatures, and metabolic data are lacking for conifers in hot, dry climates. Consequently, a genetic disorder called "noninfectious bud failure" of mature almond trees in the Mediterranean climate of Winters, California, is used as an example (Kester et al., 1986).

Almond species are from the lower slopes of mountains in central Asia and have unique physiological adaptations to subtropical, continental, Mediterranean climates. Limited rain in winter and spring, and hot conditions in summer, force adaptive adjustments in the almond. Bud-failure genotypes that are misadjusted may be inadvertently maintained by clonal propagation. This has provided an unexpected opportunity to compare how adjusted and misadjusted genotypes behave under field conditions of low moisture and high temperatures.

Kester et al. (1990) found that in the misadjusted genotype, high temperature causes necrosis, bud failure, and aberrant growth pattern in subsequent years. Bud failure is expressed through metabolic lesions and misadjustments in the free amino acid pool. The differences between normal trees and those prone to bud-failure are shown in maps of the flux behavior of amino acid families in relation to a metabolic reaction norm (Fig. 8). The glutamate pathway responsible for the formation of arginine via ornithine and citrulline, and from arginine to proline, was one of the more environmentally responsive metabolic families. In trees affected by bud failure, metabolites in this pathway were highly misadapted when compared to trees without bud failure. The other misadjustments (Fig. 9) are shown in terms of changes in concentration (rate or level) rather than fluxes as in Figures 4 and 8.

The vegetative propagation of trees bearing the bud-failure potential is a problem for the almond industry. Diagnostic methods are needed to rogue the bud-failure potential of trees before they leave the nursery.

Extreme Cold: Aberrations in Polyembryony

At timberline, phases of the reproductive cycle are often checked, completely interrupted, or stretched over several years (Tranquillini, 1979). This leads to a shortage of seed supply. Low temperatures and high-elevation tree limits tend to increase polyembryony, polygametophyty, and aberrant seed development (Simak, 1973). Proembryos and early embryos, derived by monozygotic or polyzygotic polyembryony, can now be rescued from immature seeds, mass-cloned in bioreactors, and tested in multi-plug trays (Durzan, 1989; 1991; in press). Local climatic and environmental variables can be imposed to precondition each batch of embryos targeted at specific sites for afforestation and reforestation (Durzan and Durzan, 1991). This clonal technology enables accelerated field-testing and fitness trials.

Climate has an impact on the composition of free amino acids, proteins, and other metabolites that nourish the developing embryos in seeds (Chalupa and Durzan, 1972;

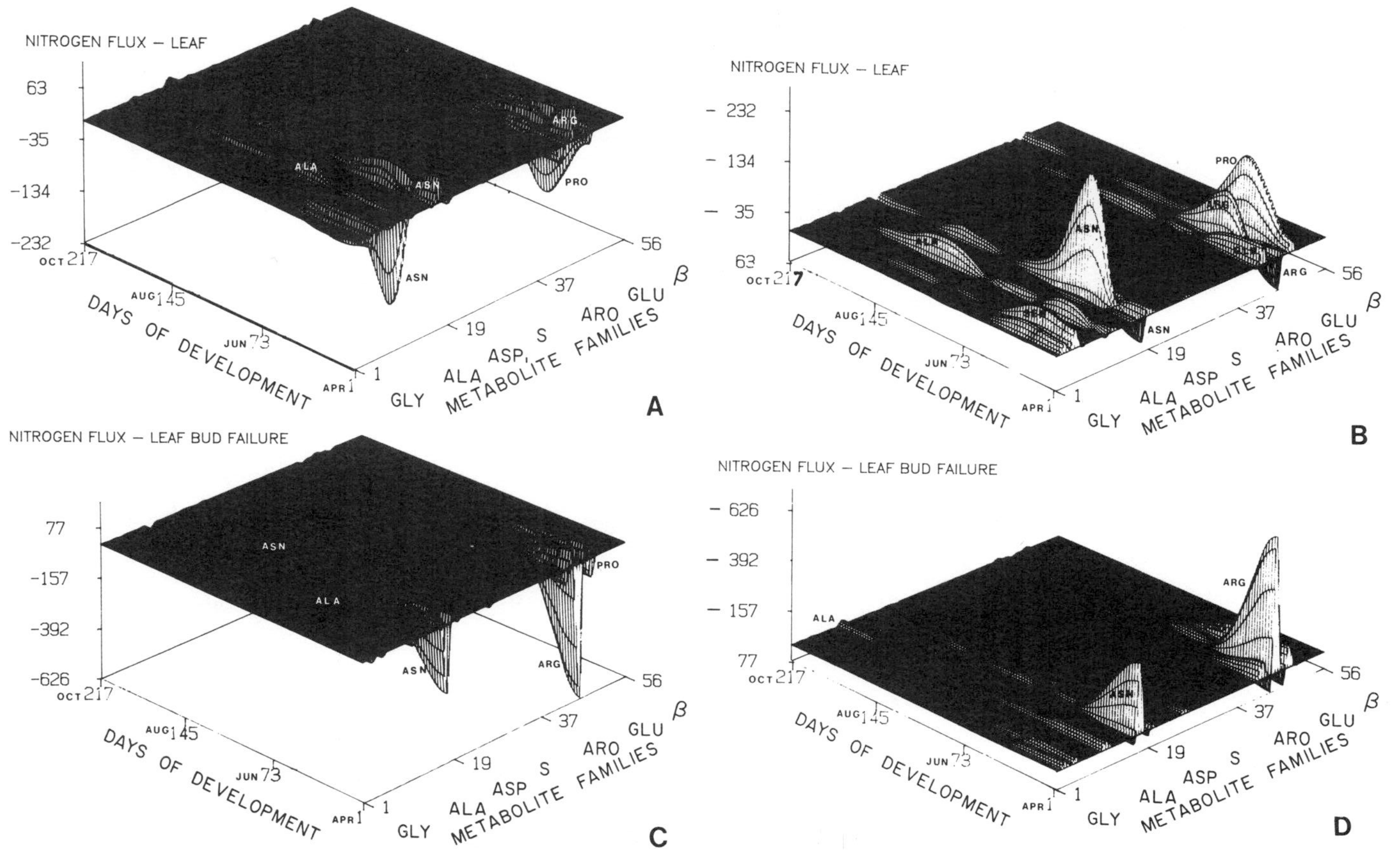

Fig. 8. Event horizons for the adaptive metabolite behavior of two almond variants exposed to the same hot, dry environment: (A) normal almond leaves (positive response surface), (B) normal leaves (negative response surface), (C) bud-failure leaves (positive response surface), (D) bud-failure leaves (negative response surface) (adapted from Durzan, D.J., and Kester, D.E., unpublished). Metabolite families are: gly, glycine; ala, α-alanine; S, sulfur amino acids; ARO, aromatic amino acids; glu, glutamate; β, beta amino acids. Individual metabolites are shown as nanomoles of amino acid per gram fresh weight (G F WT). The sequential positive and negative departures from the reaction norms can be used to estimate the misadjustments of metabolite families by matching normal and bud-failure behavior (for example, Figs 10A, 10B). Results with the almond variants (Kester *et al.*, 1986) indicate that we are dealing with continuous variants of the same allele correlated to the loss of heat resistance.

A

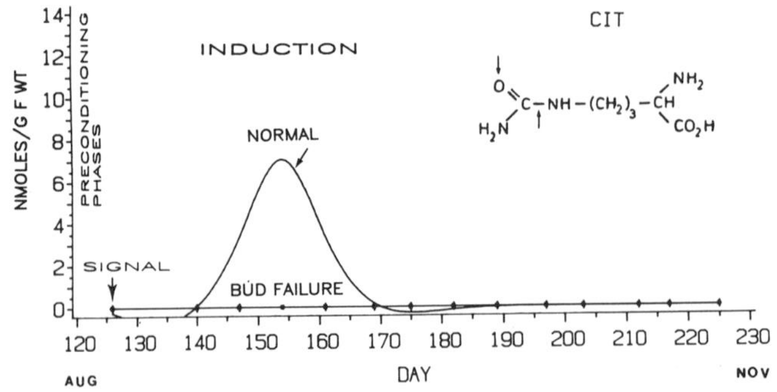

Induction of citrulline after a hot, dry summer (preconditioning phase) in buds of normal and bud-failure genotypes. Arrows on structure of citrulline indicate which bonds are used in subsequent metabolic steps leading to arginine biosynthesis and metabolism.

B

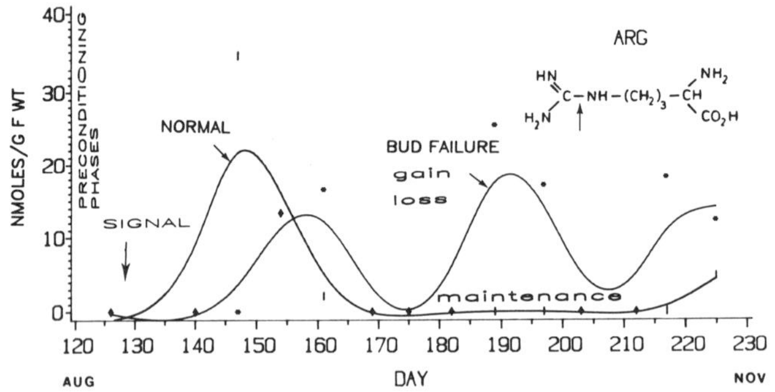

Maintenance of free arginine in the normal genotype and loss and gain in failed buds. Arginine, at this time of year, is derived mainly from protein turnover as leaves fall off trees. Arrow on structure of arginine shows the bond that is involved in one of several next possible major steps in arginine metabolism, that is, the cleavage of arginine by arginase to urea and ornithine.

C

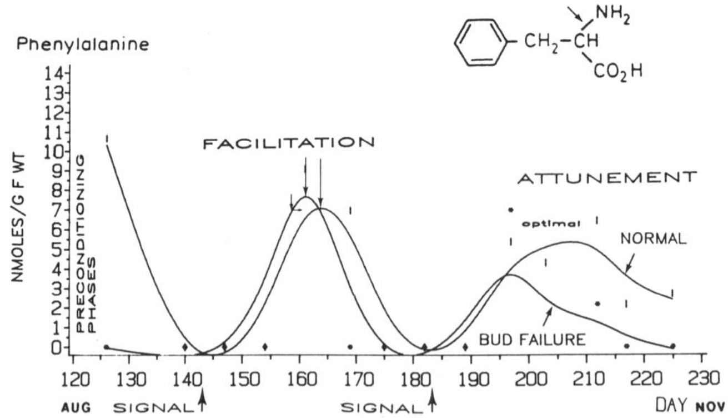

Adaptation by free phenylalanine levels in normal and bud-failure genotypes. Facilitation occurs between days 140 and 160. The bud failure response reaches a maximum at an earlier date. After 195 days, the levels of phenylalanine in normal buds are attuned to a higher rate than in the bud-failure genotype. The signal (arrow) at 180 days coincides with the longest day of the year.

Fig. 9. Adaptive metabolite behavior, using concentrations rather than departures from a reaction norm, in two genotypes (normal and bud failure) of Prunus amygdalus (Batsch.) cv. Nonpareil (almond) (Kester et al., 1990). Metabolites show a mid-summer to fall adaptation in a hot, dry Mediterranean climate (37° N. latitude, 122° W. longitude).

Durzan and Chalupa, 1968). In bioreactors, nutrients formulated on the composition of the female gametophyte may be used as a culture medium to support polyembryonic multiplication and embryo growth. The polyembryonic multiplication of embryos can be controlled by introducing stress-related growth regulators such as abscisic acid into the culture medium. In addition to existing seed supplies, the mass cloning of preconditioned embryos from specific crosses should now provide enough propagules for testing new genotypes at different sites.

At timberline, some conifers also regenerate vegetatively (Tranquillini, 1979). Lateral branches of older spruce trees develop adventitious roots that result in circular groups of trees. The cloning potential for these trees remains largely unknown.

CATEGORIES OF ADAPTIVE METABOLITE BEHAVIOR

Environmental effects may by passed from the mother tree to the next generation through the seed (Klekowski, 1988). The effects of preconditioning on response of the new generation to environmental signals is shown in Figure 4. Heritable, carry-over effects, reported for some plants (Lang, 1963), are not well documented in conifers because of their long, complex life cycle.

The immediate response of a tree to new conditions has been termed "direct" if it does not take advantage of preconditioning. Lang (1963) has listed direct effects of temperature and light cycles on growth. He suggests that direct effects may be distinguished by responses to cyclic and noncyclic conditions. The direct approach is based on a principle of *nonheredity*, that is, the future of a system depends only on its actual state at a given moment. The supposition of nonheredity is very restrictive and is contraindicated by many facts. Examples are numerous in forest development where the future productivity of a system depends on former conditions. For this reason, a "principle of *heredity*" has been formulated in a causal mechanistic, rather than in a genetic, sense.

I have extended Lang's (1963) classification of environmental effects beyond direct, inductive, conditioning, and carry-over effects with regard to adaptive plasticity. The new categories are: inductive and permissive, maturation, maintenance, gain and loss, facilitation, attunement (Fig. 4), and misadjustments (Figs. 10A, 10B). These categories recognize the importance of preconditioning behavior and are useful in identifying epidemiological effects, selecting causal agents, and demonstrating the relationship between a change-agent and the effect. Pattern-matching of metabolic fluxes around reaction norms in individuals bearing genetic differences often yields characteristic syndromes (Figs. 8, 10).

Inductive and Permissive Effects

With inductive and permissive effects, the response does not occur until after the environmental signal has acted (Figs. 4A, 9A). For example, seasonal exposure of spruce trees to brief periods of red light at night has been shown to inhibit the number of cones the following spring (Durzan et al., 1978). Lang (1963) subdivides inductive effects into delayed and short duration, transient and longer (but finite), and permanent.

Maturation Effects

Maturation refers to development of new capabilities (functions) through each phase of the life cycle. Maturation is genetically programmed and independent of prior physiological

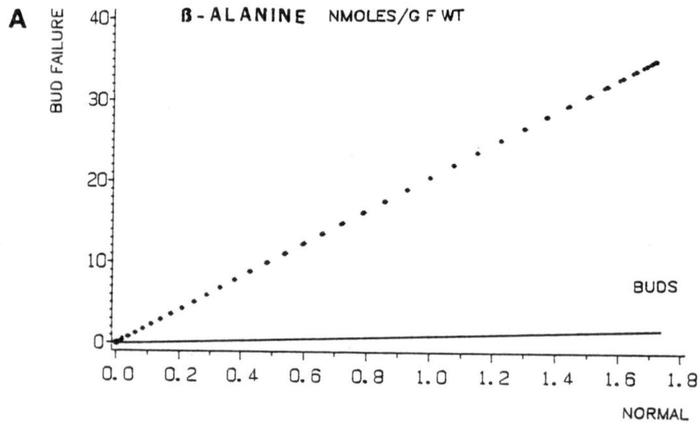

One-to-one rate correspondence in concentrations of β-alanine over time for normal buds (solid line) and heat-induced almond bud failure (dotted line). All axes (x and y) are in the same units (nmoles/gram fresh weight). A one-to-one correspondence indicates that behavior of the genotypes is not different over time. However, note that the dotted line, determined by experimentation indicates greater production of β-alanine in bud-failure than in normal trees. β-Alanine is a product of pyrimidine breakdown usually arising from the turnover of nucleic acids.

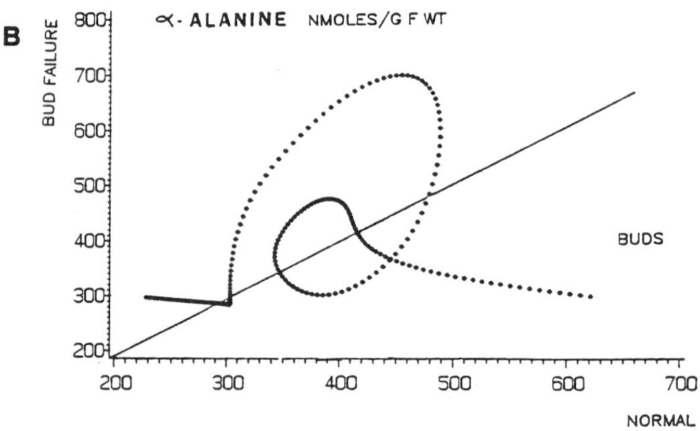

A lack of correspondence (misadjustment) in the relation between metabolite concentration (α-alanine) in normal (solid line) and bud-failure (dotted lines) buds. The dotted line shows the degree of misadjustment of α-alanine in the bud-failure syndrome. This amino acid is one of the first nitrogenous products of photosynthesis and is derived from pyruvic acid.

Fig. 10. Misadjustments of metabolite concentrations in buds of normal and bud-failure almond genotypes (data of Durzan, D. J., and Kester, D. E.).

conditions (Fig. 4B). If certain environmental stimuli and physiological states alter genome expression, however, new variations in maturation may appear.

Maintenance, Loss, and Gain

This type of adaptive behavior requires full development of the adaptive trait during the preconditioning phase (Fig. 4B). One example is the loss of cold hardiness (Vegis, 1963). Preconditioning maintains or alters the level of a metabolite after an environmental signal. In Figure 9B, the level of free arginine is maintained in normal almond buds from 170 to 212 days, while levels in failed buds fluctuate (gain or loss). Because maintenance, loss, and gain reflect tendencies toward stable and unstable behavior (Figs. 2, 3), these parameters are useful in breeding strategies that employ "disruptive selection" to enhance fitness of the new generation (Tigerstedt, in press).

Facilitation

Adaptive plasticity is considered "facilitated" when the rate of change follows an environmental signal earlier than expected. Change is prescribed by the preconditioning factors, and the response is realized early. For example, prolepsis is a morphological indication of early growth cessation and hardening that is common in crosses of lodgepole pine (*Pinus contorta* var. *latifolia* Engelm.) from northern provenances (Fries, in press). Also, compared to Sitka spruce (*Picea sitchensis* (Bong.) Carr.), Norway spruce *Picea abies* (L.) Karst.) tends to be more susceptible to damage by warm winter periods in maritime climates (Lines, in press). Although warmth facilitates early growth, too precocious an adaptation may cause the browning of foliage in spring if temperatures fluctuate widely.

In almond buds, a maximum in free phenylalanine (days 155 to 165) appears earlier in bud-failed trees (Fig. 9C). The phenylalanine pathway leads to formation of wood, phenol and lignin as tree form is altered by local environmental conditions.

Attunement

Attunement refers to situations in which preconditioning factors adjust the full expression (promotion or inhibition) of a response after an environmental signal. Figure 9C illustrates how free phenylalanine is attuned between 195 to 225 days in almond buds both in trees that are affected and in those that are not affected by bud failure. With bud failure, buds show lower levels of phenylalanine in response to summer heat after the longest day. Pine buds show adjustments in metabolism of ribonucleic acid as they become attuned to conditions in a climate chamber (Lindfors and Kupila-Ahvenniemi, 1988).

Misadjustments

To what extent does adaptive behavior actually pursue environmental change? Pursuit represents a physiological adjustment or misadjustment between two points: the internal metabolite behavior and the environmental variable causing the changes. In the simplest case, the path of pursuit is a straight line between the corresponding points on response curves referenced to the reaction norm and to the environmental parameters (Figs. 2, 3). In this situation, linear pursuit leads to possible "conditions of capture," that is, a return to a more stable situation in the relationship between the selected variables. This notion assumes that the tree's metabolic system always adjusts and spreads toward changes in environmental parameters. In some models the relationship between variables may remain as a fixed ratio. If not fixed, then the relationship must be corrected to account for the adjustment in behavior. For example, a hysteresis in the relationship between climatic factors may occur and establish a limit cycle (Fig. 11A). A limit cycle is an isolated, closed trajectory (curve) for the relationship between two selected variables (phase plane). Limit cycles are fundamental in the theory of oscillations of nonlinear energy-absorbing and dissipative systems.

The relative behavior of metabolites (for example, precursors vs. products) can be described graphically with reference to a limit cycle (Figs. 11, 12). Every possible trajectory near a limit cycle may wind itself upon the limit cycle or unwind from it. If all nearby trajectories approach the limit cycle, then the limit cycle is stable. If the trajectories unwind from it, then the cycle is unstable. Combinations of trajectory behavior can also lead to semi-stable conditions.

Adaptive behavior can by represented by a "forcing function" around the limit cycle near or far from the reaction norm, that is, the metabolic system is forced periodically and asymptotically to the limit by an external source, such as temperature or photoperiod (Figs. 11B, 12A). The selected "indicator" metabolites in Figure 12B are closely linked in the glutamate pathway. The seasonal changes between August and December force an internal change in the relationship (trajectory) between soluble arginine and proline N. Evidence for phenotypic plasticity is also found in the soluble and insoluble phases (Figs. 13, 14).

Chaotic and/or self-organized behavior may occur at far-from-the equilibrium state or reaction norm. The tendency for chaotic behavior in the response-system usually takes the form of an increased amplitude or phase change (Gleick, 1987).

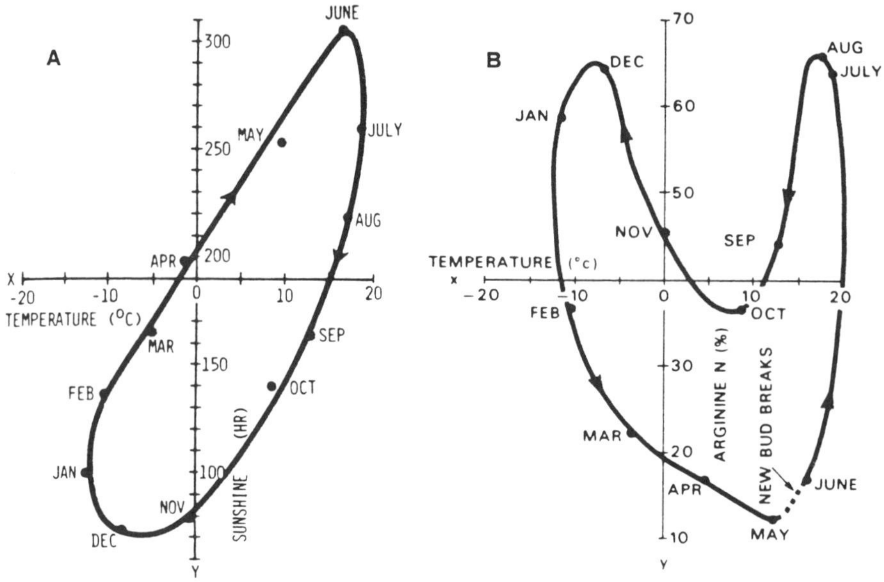

Fig. 11. Seasonal fluctuation in climatic factors and in amino acid N content of spruce shoot tips of an elite phenotype or "plus" tree establish limit cycle behavior (After Durzan and Steward, 1983).

(A) Climatic factors. Air, temperature and total monthly hours of sunshine attain minimum values in winter and maximum values in summer. The extremes of temperature lag about one month behind those of sunshine and cause the elliptical trajectory of the curve (limit cycle).

(B) Soluble arginine N content as a percentage of the total soluble N and mean monthly temperature establish an annual limit cycle. There are two maxima in arginine N content: during winter, when arginine N accumulates as a storage product at low temperatures, and during late summer, when arginine N accumulation marks the onset of dormancy. The minimum values observed relate to the use of stored arginine by the shoot tips (i) for the formation of glutamine during the onset of growth in May and (ii) in its conversion in October to proline, which accumulates during the onset of dormancy in the bud. Data of D.J. Durzan (unpublished). Limit cycle parameters may be reformulated to compare two parental types selected for breeding purposes. One parent may have better rates of growth at lower temperatures, whereas the other may be selected to use and store its N as arginine. The relation between these two traits may be tracked as a limit cycle to evaluate potential field performance.

Another estimate of the misadjustment of metabolite performance may be based on the excess coefficient of variation (CV) in two genetic variants in the same environment. In the case of the almond, the excess CV for each metabolite is estimated by the difference between normal and bud-failure tree performance over several seasons. To calculate the percent misadjustment for this metabolite, the excess CV in deteriorating and destabilizing

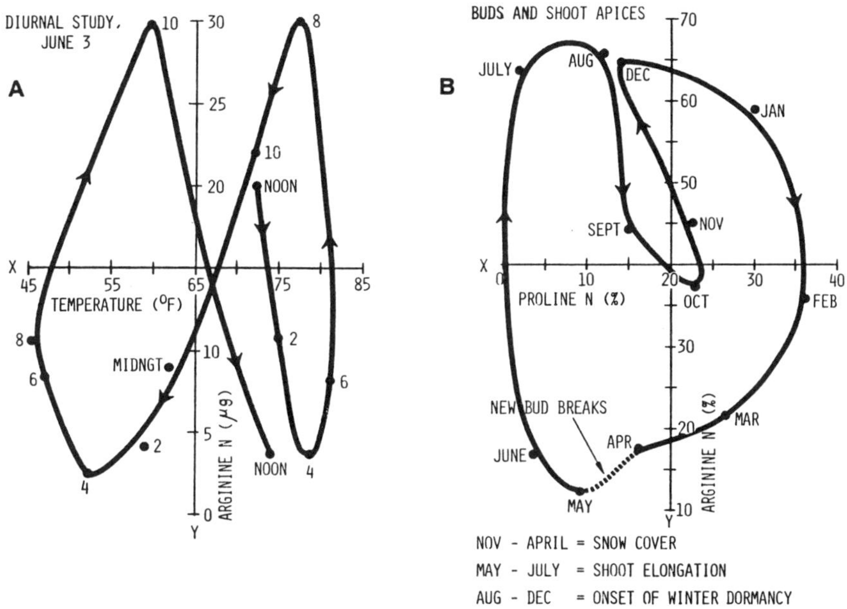

Fig. 12. Complex trajectories in metabolite behavior in spruce buds and shoot apices.

A. Diurnal temperature and arginine N (µg N per g fresh wt) fluctuations in young leaves of breaking buds on white spruce trees under field conditions. Numbers on curve indicate two-hour intervals when arginine N levels (µg N per gram fresh weight) were estimated.

B. Relation between percent soluble proline N and arginine N in buds and shoot apices in an individual white spruce tree under field conditions. Note the dominance of enzymatic events from August to December when the initial dormant phase is stabilized by synthesis of proline from arginine. This limit cycle relates the behavior of individual amino acids derived from the same metabolic family (*viz.* glutamate family).

tissues (bud failure) is divided by the minimum CV for the same metabolite in normal almond trees. The potential for bud failure is usually hidden in the nursery, but in our case was traced to cells and tissues from which cultivars were vegetatively propagated (Kester et al., 1986). Severe temperatures under field conditions shifted the metabolic patterns and revealed the unsuspected susceptibility of the bud-failure variant.

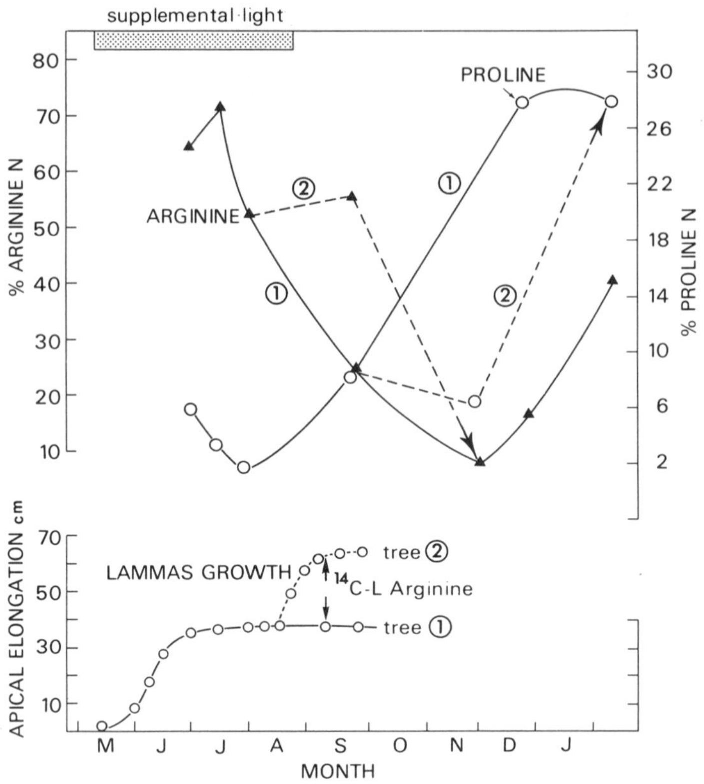

Fig. 13. Buds on white spruce trees exposed to continuous, supplemental light under field conditions added a flush of growth (Lammas growth) in August, rather than enter rest. The percentage of proline N increased as arginine in the soluble N pool declined (solid line) as "untreated control" buds (tree 1) entered winter rest. Buds exposed to continuous light (tree 2) show precocious bud break with the temporary restoration of higher arginine N levels and a temporary drop in proline N. Lights were placed 3 to 4 feet above trees of about 5 meters high. Results indicate how shoot growth (apical elongation in cm of leader bud) adjusted to the photoperiodic signal in terms of facilitated and attuned metabolite behavior.

CONCLUSIONS

At the molecular level, adaptive phenotypic plasticity may be arbitrarily characterized as an hierarchical, spreading series of metabolic networks that comprise a reaction norm. The graphic or tabular display of the reaction norm represents an event horizon that can be used for diagnostic purposes involving genotype × environment interactions. Diagnostic reasoning may be extended to recognize specific heritable traits and extremes in topological, developmental, and microclimatic aspects of phasic development. New variables, based on physiological preconditioning, can be introduced to characterize near and far-from equilibrium adaptive behavior in trees under field conditions. Categories of metabolite behavior are identified and interpreted in relation to limit cycles found in the environment.

In forestry, intensive cultural systems are often based on ideotypes that must be adapted to sites entirely different from the habitat of origin. Potential misadjustments to environmental change may be carried through populations of meristematic cells. These

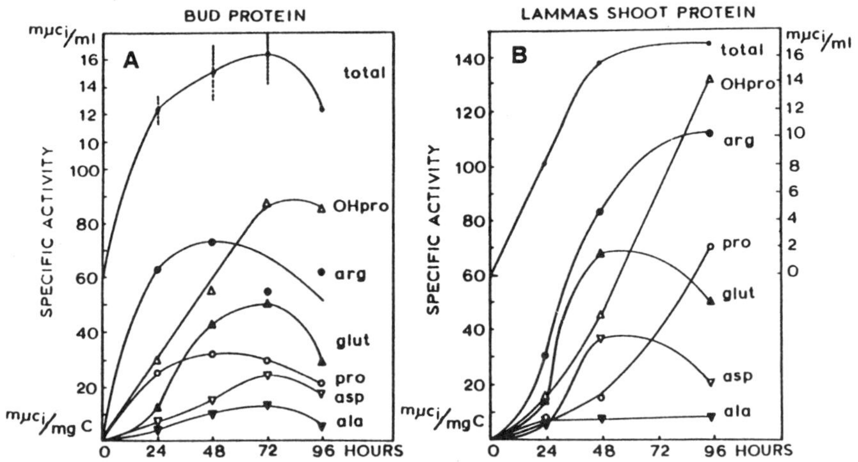

Fig. 14. The impact of protein metabolism on the alcohol-insoluble, structural products of white spruce buds. The metabolic relations between proline and arginine, when followed by ^{14}C-L-arginine (uniformly labelled), reveal the formation of ^{14}C-proline and the subsequent incorporation of arginine-derived metabolites into the buds. Each chart shows the specific activities (mμCi/mg carbon) of the ethanol-insoluble compounds derived from uniformly labelled ^{14}C-arginine. ^{14}C-Arginine was applied to (A) the terminal bud of dormant shoots and to (B) the apex of Lammas shoots of white spruce. (OHpro, hydroxyproline; arg, arginine; glut, glutamic acid; pro, proline; asp, aspartic acid; ala, alanine.) The insoluble fraction contains numerous proteins that can be detected by cytochemical dyes. Structural and functional proteins serve as "markers" for adaptions to external and extraordinary conditions. The decline in specific activity of carbon for some amino acid residues in protein at 96 h indicate metabolic turnover, that is, proteins are being synthesized and then broken down at different rates. These proteins may comprise the basis and control cycles in Figures 1A, 1B. Total radioactivity is shown on a relative basis as mμCi per ml of acid hydrolysate relative to 1 gram of alcohol-insoluble residue.

potential misadjustments are easily and inadvertently perpetuated through clonal propagation -- rooting, grafting, and micropropagation.

Diagnostic tools are now available for selecting useful genetic variation and roguing out potential misadjustments to extreme climatic conditions. To ameliorate the seed supply problem, individual genotypes, selected for forest development at or above timberline, can now be preconditioned and mass-cloned in a true-to-type and model-referenced way before their fitness is evaluated by widespread and long-term field tests.

ACKNOWLEDGEMENTS

The author thanks Drs. D. Kester and H. Lieth, and the editors, for useful comments and criticisms.

REFERENCES

Agur, Z., and Slobodkin, L. B., 1986, Environmental fluctuations: How do they affect the topography of the adaptive landscape?, Journal of Genetics, 65:45-54.

Anamthawat-Jonsson, K., Heslop-Harrison, J., and Tomasson, T., [In press], Genetic diversity of Icelandic birch, Betula pubescens Ehrh., and implications for agroforestry. In: Alden, J. N., Odum, S., and Mastrantonio, J. L., eds., Forest Development in Cold Climates, Proceedings of an International Symposium, 1991, June 18-24, [Laugarvatn, Iceland], NATO, Plenum, New York, New York.

Axelrod, D. I., 1966, Origin of deciduous and evergreen habits in temperate forests, Evolution, 20:1-15.

Babloyantz, A., 1986, Molecules, Dynamics and Life, John Wiley-Interscience, New York, 345 p.

Bennett, C., and Bourzutschky, M. S., 1991, "Life" not critical?, Nature, 350:468.

Berg, P., 1991, Reverse genetics: Its origins and prospects, Bio/Technology, 9:342-343.

Brink, R. A., 1962, Phase change in higher plants and somatic cell heredity, Quarterly Review of Biology, 37:1-22.

Brooks, C. E. P., and Carruthers, N., 1953, Handbook of Statistical Methods in Meteorology, Air Ministry Meteorological Office, London, England, Her Majesty's Stationery Office, 412 p.

Chalupa, V., and Durzan, D. J., 1972, Differences in the seed size and in the biochemical composition of seed of various provenances of jack pine (Pinus banksiana Lamb.), Lesnictvi (Czech. Forestry), 18:1089-1106.

Dawkins, R., 1982, The Extended Phenotype. The Gene as a Unit of Selection, W. H. Freeman, San Francisco, 307 p.

Denny, M. W., 1988, Biology and the Mechanics of the Wave-Swept Environment, Princeton Univ. Press, Princeton, New Jersey, 329 p.

Dhillon, S. S., 1987, DNA in tree species. In: Bonga, J. M., and Durzan, D. J., eds., Cell and Tissue Culture in Forestry, Martinus Nijhoff, Dordrecht, The Netherlands, Vol. 1:298-313.

Dixon, R. A., and Lamb, C. J., 1990, Molecular communication in interactions between plants and microbial pathogens, Annual Review of Plant Physiology, 41:339-367.

Durzan, D. J., 1968a, Nitrogen metabolism of Picea glauca, I. Seasonal changes of free amino acids in buds, shoot apices, and leaves, and the metabolism of uniformly labelled ^{14}C-L-arginine by buds during the onset of dormancy, Canadian Journal of Botany, 46:909-919.

Durzan, D. J., 1968b, Nitrogen metabolism of Picea glauca, II. Diurnal changes of free amino acids, amides, and guanidino compounds in roots, buds, and leaves during the onset of dormancy of white spruce saplings, Canadian Journal of Botany, 46:921-928.

Durzan, D. J., 1968c, Nitrogen metabolism of Picea glauca, III. Diurnal changes of free amino acids, amides, protein, and chlorophyll in leaves of expanding buds, Canadian Journal of Botany, 46:929-937.

Durzan, D. J., 1971, Free amino acids as affected by light intensity and the relation of responses to the shade-tolerance of white spruce and shade intolerance of jack pine, Canadian Journal of Forest Research, 1:131-140.

Durzan, D. J., 1973a, The metabolism of ^{14}C-urea by white spruce seedlings in light and darkness, Canadian Journal of Botany, 51:1197-1211.

Durzan, D. J., 1973b, The incorporation of tritiated water into amino acids in the presence of urea by white spruce seedlings in light and darkness, Canadian Journal of Botany, 51:351-358.

Durzan, D. J., 1973c, Nitrogen metabolism of Picea glauca, V. Metabolism of uniformly labeled ^{14}C-L-proline and ^{14}C-L-glutamine by dormant buds in late fall, Canadian Journal of Botany, 51:359-369.

Durzan, D. J., 1975, Nutrition and water relations of forest trees: A biochemical approach. In: Reid, C. P. P., and Fechner, C. H., eds., Third North American Conf. Tree Biology, Colorado State Univ., Fort Collins, Colorado, p. 15-63.

Durzan, D. J., 1987, Physiological states and metabolic phenotypes. In: Bonga J. M., and Durzan, D. J., eds., Cell and Tissue Culture in Forestry, 2nd Edition, Volume 2, Principles and Methods, Martinus Nijhoff/Dr. W. Junk, p. 405-439.

Durzan, D. J., 1989, Performance criteria in response surfaces for metabolic phenotypes of clonally propagated woody perennials. In: Dhawan, V., ed., Application of Plant Biotechnology in Forestry and Agriculture, Plenum Press, New York, p. 181-203.

Durzan, D. J., 1990a, Genetic manipulation of forest trees, Implications for physiological processes. In: Rodriguez, R. R., Sanchez-Torres, R., and Durzan, D. J., eds., Molecular Basis of Plant Aging, Plenum Press, New York, p. 303-309.

Durzan, D. J., 1990b, Molecular phenogenetics as an aid to fruit breeding, Proc. Intl. Symp. In Vitro Culture and Horticultural Breeding, May 30-June 3, 1989, Cesena, Italy, Acta Horticulturae, 280:547-556.

Durzan, D. J., 1991, Improving tree growth and quality through biotechnology, Kemia-Kemi (Finnish Chemistry), 18:192-199.

Durzan, D. J., [In press], Reproductive adaptation by polyembryony of coniferous forest trees as revealed by the metabolism of tritiated water. In: Methodologies to Assess the Impact of Climatic Changes on Vegetation: Analysis of Water Transport in Plants and Cavitation of Xylem Conduits.

Durzan, D. J., and Chalupa, V., 1968, Free sugars, amino acids, and soluble proteins in the embryo and female gametophyte of jack pine as related to climate at the seed source, Canadian Journal of Botany, 46:417-428.

Durzan, D. J., and Durzan, P. E., 1991, Future technologies: model-reference control systems for the scale-up of embryogenesis and polyembryogenesis in cell suspension cultures. In: Debergh P. C., and Zimmerman, R. H., eds., Micropropagation, Kluwer Academic, Netherlands, p. 289-423.

Durzan, D. J., and Steward, F. C., 1983, Nitrogen metabolism. In: Bidwell R. G. S., and Steward, F. C., eds., Plant Physiology: An Advanced Treatise, Academic Press, New York, p. 255-265.

Durzan, D. J., Campbell, R. A., and Wilson, A., 1978, Inhibition of female cone production in white spruce by red light treatment during night under field conditions, Environmental and Experimental Botany, 19:133-144.

Endler, J. A., 1977, Geographic Variation, Speciation, and Clines, Princeton Univ. Press, Princeton, New Jersey, 246 p.

Eriksson, G., and Lundkvist, K., 1986, Adaptation and breeding of forest trees in boreal areas. In: Lindgren, D., ed., Provenances and Forest Tree Breeding for High Latitudes, Proc. Frans Kempe Symp. 6, Umeå, Sweden, p. 67-80.

Ford, E. D., 1984, Increasing forest productivity and value by exploiting climatic variability. In: Forest Potentials. Productivity and Value, Proc. Weyerhaeuser Science Symp., 1984 August 20-24, p. 75-99.

Fries, A., [In press], Hybridization among provenances of lodgepole pine, *Pinus contorta* var. *latifolia* Engelm. In: Alden, J. N., Odum, S., and Mastrantonio, J. L., eds., Forest Development in Cold Climates, Proceedings of an International Symposium, 1991, June 18-24, [Laugarvatn, Iceland], NATO, Plenum, New York, New York.

Gleick, J., 1987, Chaos. Making a New Science, Penguin Books, New York, 354 p.

Grace, J., and James, J., [In press], The physiology of trees at treeline. In: Alden, J. N., Odum, S., and Mastrantonio, J. L., eds., Forest Development in Cold Climates, Proceedings of an International Symposium, 1991, June 18-24, [Laugarvatn, Iceland], NATO, Plenum, New York, New York.

Hagman, M., [In press], Potential species and provenances for forest development in cold climates. In: Alden, J. N., Odum, S., and Mastrantonio, J. L., eds., Forest Development in Cold Climates, Proceedings of an International Symposium, 1991, June 18-24, [Laugarvatn, Iceland], NATO, Plenum, New York, New York.

Hamner, K., 1963, Endogenous rhythms in controlled environments. In: Evans, L. T., ed., Environmental Control of Plant Growth, Academic Press, New York, p. 215-230.

Hartl, D. L., 1989, The physiology of weak selection, Genome, 31:183-189.

Holliday, R., 1987, The inheritance of epigenetic defects, Science, 238:163-170.

Junttila, O., and Nilsen, J., [In press], Physiological response of forest trees to low temperature and extreme light conditions. In: Alden, J. N., Odum, S., and Mastrantonio, J. L., eds., Forest Development in Cold Climates, Proceedings of an International Symposium, 1991, June 18-24, [Laugarvatn, Iceland], NATO, Plenum, New York, New York.

Kacser, H., and Burns, J. H., 1973, The control of flux, Symposia Society Experimental Biology, 27:65-104.

Kacser, H., and Burns, J. H., 1981, The molecular basis of dominance, Genetics, 97:639-666.

Kester, D. E., Liu, L., Fenton, C.A.L., and Durzan, D. J., 1986, Almonds (Prunus dulcis [Miller] D.A. Webb). In: Bajaj, Y. P. S., ed., Biotechnology in Agriculture and Forestry, Springer-Verlag, Berlin, Heidelberg, 1:414-430.

Kester, D. E., Durzan, D. J., and Shakel, K., 1990, Relation between phenotypic expression of noninfectious bud-failure (NBF) in almond and the seasonal growth-dormancy-rest cycle in almond shoots, Proc. XXIII International Horticultural Congress, Florence, Italy, Abstract 2226.

Klekowski, E. J., Jr., 1988, Mutation, Developmental Selection and Plant Evolution, Columbia Univ. Press, New York, 373 p.

Lambers, H., 1985, Respiration in intact plants and tissues: Its regulation and dependence on environmental factors. Metabolism and invaded organisms, Encyclopedia of Plant Physiology, New Series, 18:418-473.

Lang, A., 1963, Achievements, challenges, and limitations of phytotrons. In: Evans, L., ed., Environmental Control of Plant Growth, Academic Press, New York, p. 405-435.

Langridge, J., 1963, The genetic basis of climate response. In: Evans, L., ed., Environmental Control of Plant Growth, Academic Press, New York, p. 367-378.

Larsen, B., [In press], Testing winter desiccation resistance for species and provenance selection at timberlines. In: Alden, J. N., Odum, S., and Mastrantonio, J. L., eds., Forest Development in Cold Climates, Proceedings of an International Symposium, 1991, June 18-24, [Laugarvatn, Iceland], NATO, Plenum, New York, New York.

Levitt, J., 1972, Responses of Plants to Environmental Stresses, Academic Press, New York, 697 p.

Li, W.-H, and Graur, D., 1991, Fundamentals of Molecular Evolution, Sinauer Inc, Sunderland, Mass., 284 p.

Lindfors, A., and Kupila-Ahvenniemi, S., 1988, Metabolic and structural responses of pine buds under climate chamber conditions, Physiologia Plantarum, 72:407-413.

Lindgren, D., [In press], Provenance selection, testing, and breeding to accelerate adaptation and growth of trees and shrubs at tree limits. In: Alden, J. N., Odum, S., and Mastrantonio, J. L., eds., Forest Development in Cold Climates, Proceedings of an International Symposium, 1991, June 18-24, [Laugarvatn, Iceland], NATO, Plenum, New York, New York.

Lines, R., [In press], Aspects of species and provenance choice at northern tree limits in maritime climates. In: Alden, J. N., Odum, S., and Mastrantonio, J. L., eds., Forest Development in Cold Climates, Proceedings of an International Symposium, 1991, June 18-24, [Laugarvatn, Iceland], NATO, Plenum, New York, New York.

Loeschcke, V., 1987, Introduction. In: Loeschcke, V., ed., Genetic Constraints on Adaptive Evolution, Springer-Verlag, New York, p. 1-2.

Mikola, J., [In press], Provenance and individual variation in climatic hardiness of Scots pine, Pinus sylvestris (L.), in northern Finland. In: Alden, J. N., Odum, S., and Mastrantonio, J. L., eds., Forest Development in Cold Climates, Proceedings of an International Symposium, 1991, June 18-24, [Laugarvatn, Iceland], NATO, Plenum, New York, New York.

Miksche, J. P., 1971, Intraspecific variation of DNA per cell between Picea sitchensis (Bong.) Carr provenances, Chromosoma, 32:343-352.

Morowitz, H. J., 1979, Foundations of Bioenergetics, Academic Press, New York, 235 p.

Nilsson, J.-E., [In press]. Selection and breeding of Scots, Pinus sylvestris L., for northern Sweden. In: Alden, J. N., Odum, S., and Mastrantonio, J. L., eds., Forest Development in Cold Climates, Proceedings of an International Symposium, 1991, June 18-24, [Laugarvatn, Iceland], NATO, Plenum, New York, New York.

Osmund, C. B., Björkman, O., and Anderson, D. J., 1980, Physiological Processes in Plant Ecology, Springer-Verlag, Berlin, 320 p.

Pahlich, E., 1991, Metabolic network homeostasis and plant stress sensitivity, Proc. VIII Reunion Nacional de Botanica, 1991 October 16–19, Santiago, Chile, p. 54.

Parker, G. A., and Maynard Smith, J., 1990, Optimality theory in evolutionary biology, Nature, 348:27-33.

Pitel, J. A., and Durzan, D. J., 1978, Changes in the composition of the soluble and chromosomal proteins of jack pine (Pinus banksiana Lamb.) seedlings induced by fenitrothion, Environmental and Experimental Botany, 18:153-162.

Price, H. J., Sparrow, A. H., and Nauman, A. F., 1973, Evolutionary and developmental considerations of the variability of nuclear parameters in higher plants. Genome volume, interphase chromosome volume, and estimated DNA content of 236 gymnosperms. In: Basic Mechanisms in Plant Morphogenesis, Brookhaven Natl. Lab. Symp. No. 25, June 4–6, p. 390-421.

Raymond, C. A., and Lindgren, D., 1986, A model for genetic flexibility. In: Lindgren, D., ed., Provenances and Forest Tree Breeding for High Latitudes, Proc. Frans Kempe Symp., Umeå, Sweden, 159-177.

Reiger, R., Michaelis, A., and Green, M. M., 1976, Glossary of Genetics and Cytogenetics, 4th ed., Springer-Verlag, Berlin, 647 p.

Ringpfeil, M., 1986, The maximum concept, Trends in Biotechnology, Aug.:199-200.

Sauro, R. M., Small, J. R., and Fell, D. A., 1987, Metabolic control and its analysis. Extensions to the theory and matrix method, European Journal of Biochemistry, 165:215-221.

Scharloo, W., 1987, Constraints in selection response. In: Loeschcke, V., ed., Genetic Constraints on Adaptive Evolution, Springer-Verlag, Berlin, 125-149.

Sigurgeirsson, A., [In press], Chloroplast DNA diversity, phylogenetics, and hybridization of spruce, Picea A. Dietr. In: Alden, J. N., Odum, S., and Mastrantonio, J. L., eds., Forest Development in Cold Climates, Proceedings of an International Symposium, 1991, June 18-24, [Laugarvatn, Iceland], NATO, Plenum, New York, New York.

Simak, M., 1973, Polyembryonal seeds of Pinus silvestris in arctic regions, Institutionen för Skogsföryngring Research Notes, 45:2-14.

Skaret, G., [In press], Seed collections from North American trees for tests on marginal sites in Nordic countries. In: Alden, J. N., Odum, S., and Mastrantonio, J. L., eds., Forest Development in Cold Climates, Proceedings of an International Symposium, 1991, June 18-24, [Laugarvatn, Iceland], NATO, Plenum, New York, New York.

Stephanopoulos, G., Vallino, J. J., 1991, Network rigidity and metabolic engineering in metabolite overproduction, Science, 252:1675-1681.

Steward, F. C., 1968, Growth and Organization in Plants, Addison-Wesley, Reading, Massachusetts, 564 p.

Thorgeirsson, H., McCaughey, J. H., and Alexander, R., [In press], Energy fluxes of plantations and forests in the boreal region. In: Alden, J. N., Odum, S., and Mastrantonio, J. L., eds., Forest Development in Cold Climates, Proceedings of an International Symposium, 1991, June 18-24, [Laugarvatn, Iceland], NATO, Plenum, New York, New York.

Tigerstedt, P. M., [In press], Genetic diversity of tree populations at their arctic limits. In: Alden, J. N., Odum, S., and Mastrantonio, J. L., eds., Forest Development in Cold Climates, Proceedings of an International Symposium, 1991, June 18-24, [Laugarvatn, Iceland], NATO, Plenum, New York, New York.

Tranquillini, W., 1979, Physiological Ecology of the Alpine Timberline, Springer-Verlag, Berlin, 136 p.

Van Noordwijk, A. J., and Gebhardt, M., 1987, Reflection on the genetics of quantitative traits with continuous environmental variation. In: Loeschcke, V., ed., Genetic Constraints on Adaptive Evolution, Springer-Verlag, Berlin, p. 73-90.

Vegis, A., 1963, Climatic control of germination, bud bread and dormancy. In: Evans, L. T., ed., Environmental Control of Plant Growth, Academic Press, New York, p. 265-284.

Via, S., 1987, Genetic constraints on the evolution of phenotypic plasticity. In: Loeschcke, V., ed., Genetic Constraints on Adaptive Evolution, Springer-Verlag, Berlin, 48-71.

Waddington, C. H., 1957, The Strategy of Genes, Allen and Unwin, London, 195 p.

Wales, D. J., 1991, Calculating the rate of loss of information from chaotic time series by forecasting, Nature, 350:485-488.

Watt, W. B., 1986, Power and efficiency as indexes of fitness in metabolic organization, American Naturalist, 127:629-653.

White, I. O., Mottershead, D. N., and Harrison, S. J., 1984, Environmental Systems. An Introductory Text, Allen and Unwin, London, 495 p.

Woodward, F. I., 1987, Climate and Plant Distribution, Cambridge Univ. Press, Cambridge, 174 p.

Wright, S., 1968, Evolution and the Genetics of Populations, Vols. 1 to 4, Univ. Chicago Press, Chicago.

Yancey, P. H., Clark, M. E., Hand, S. C., Bowlus, R. D., and Somero, G. N., 1982, Living with water stress: Evolution of osmolyte systems, Science, 217:1214-1222.

GROWTH AND DEVELOPMENT OF NORTHERN FOREST TREES AS AFFECTED BY TEMPERATURE AND LIGHT

Olavi Junttila and Jarle Nilsen

Department of Plant Physiology and Microbiology
University of Tromsø
Tromsø, Norway

INTRODUCTION

Temperature and light are the two main environmental factors influencing plant growth and development. Most basic metabolic reactions in plants are strictly dependent on temperature. Light is the ultimate source of energy for plants and, in addition, the morphogenic development of plants is strongly affected by light conditions. For trees growing at the northern, marginal areas of their range, temperature is the main limiting factor. Low temperature during the growing season affects metabolism, photosynthesis, and respiration. Studies with herbaceous plant species strongly suggest that resource acquisition is generally not a limiting factor for plants growing under arctic and alpine conditions. In their review on plant growth under cold conditions, Körner and Larcher (1988) concluded that developmental processes, especially mitosis, play a significant role in adaptation to low temperature.

In this review, we will discuss developmental aspects of tree growth under extreme northern conditions, with emphasis on shoot growth. Temperature also has a significant influence on diameter growth, but the effects of environmental factors on wood production are partly indirect and related to the regulation of shoot and leaf growth (Larson, 1964). Questions of cold hardiness have been treated in several recent reviews and books (Guy, 1990; Junttila, 1989; Palta, 1990; Sakai and Larcher, 1987), and are also covered in other chapters of this volume. Only certain aspects of low temperature survival of trees are discussed here. Photoinhibition is another important problem, especially for evergreen conifers growing under northern conditions. For this topic, we also refer to recent comprehensive presentations (Gillies and Vidaver, 1990; Krause, 1988; Öquist, 1987; Strand, 1987). Several aspects of the effects of temperature on growth and carbon dioxide (CO_2) exchange in birch are dealt with by Skre (in press). In arctic environments the level of available mineral nutrients is low and can limit growth. This aspect, however, is not dealt with in this review.

Forest Development in Cold Climates, Edited by J. Alden *et al.*
Plenum Press, New York, 1993

SHOOT ELONGATION

Species with Free Growth

In tree species with a typical free-growth pattern, primordia (node/stem unit) initiation and internode elongation occur simultaneously during shoot elongation. The first process is located in the apical meristem proper, and the second in the subapical meristem (Sachs, 1965). Free growth is strongly modified by several environmental factors, especially temperature. Figures 1 and 2 illustrate the effects of temperature on the rate of shoot elongation in seedlings of Norway spruce (Picea abies (L.) Karst.), willow (Salix pentandra L.) and mountain birch (Betula pubescens Ehrh.). Results for Picea abies have been calculated from data of Heide (1974), those for Salix from Junttila (1980). Shoot elongation was examined under continuous light; Betula and Salix were grown under natural light conditions at latitude 69°39′ N. In these examples, the Q_{10} (the factor by which the rate of growth increases with a 10 °C increase in temperature) between 12 °C and 18 °C for shoot elongation was 3.8 for seedlings of Picea abies. For Betula, the Q_{10} between 9 °C and 15 °C averaged 5.2, and for Salix, it was as high as 6.3. Within these ranges of temperature, shoot elongation for all species showed a statistically significant linear correlation with temperature. Extrapolation of the regression lines for the growth response to temperatures below 10 °C shows that zero growth would occur at 7 to 7.5 °C for seedlings of Picea abies, Salix pentandra, and Betula pubescens. Results for northern populations of Betula also indicate that plants from low elevation sites have a higher growth rate than those from higher elevations, even at 9 °C and 12 °C (Fig. 2). Based on studies of Norway spruce at timberline in Scandinavia, Skre (1972) estimated that 8.8 °C is an approximate minimum temperature required for growth. Temperature required for seed maturation is generally higher than for shoot growth. Minimum temperatures for maturation of Picea abies seed is estimated to be 9.7 to 10.0 °C (Skre, 1972).

In nature, temperature has a diurnal fluctuation. Certain plants seem to respond to accumulated heat sum and to daily mean temperature. In some cases, fluctuating temperatures have specific effects on plant growth and development that are significantly different

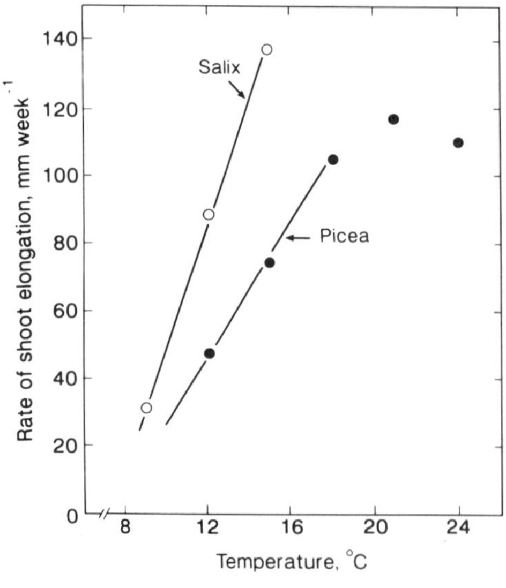

Fig. 1. Effect of temperature on the rate of shoot elongation in first-year seedlings of Salix pentandra L. and Picea abies (L.) Karst. grown under a 24-hour photoperiod. Results for Salix are means of two provenances (Junttila, 1980); those for Picea represent means of four provenances (from Heide, 1974). Growth rate was calculated for the period of linear shoot elongation.

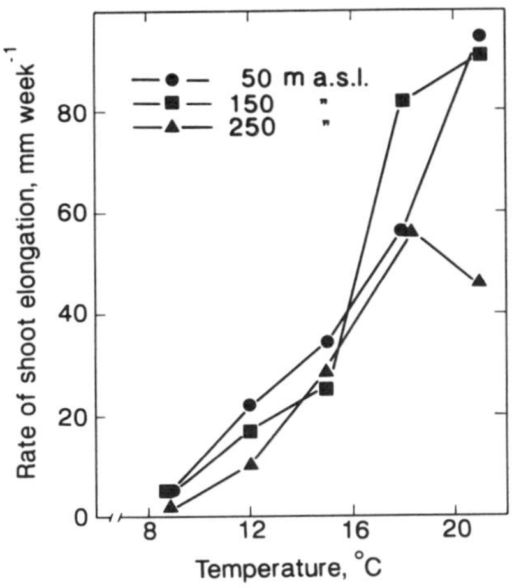

Fig. 2. Effect of temperature on the rate of shoot elongation in first-year seedlings of three altitudinal (50, 150, and 250 meters above sea level) populations of Betula pubescens Ehrh. from latitude 69° N. Seedlings were grown in a phytotron under a natural 24-hour photoperiod at latitude 69°39′ N. Growth rate was calculated for the period of linear shoot elongation.

from those observed at corresponding constant temperatures. Results presented by Junttila and Skaret (1990) indicate that shoot elongation in first-year seedlings of Picea abies, Lutz spruce (P. xlutzii Little), and white spruce (P. glauca (Moench) Voss) is stimulated by alternating day-night temperature. Exact information about specific temperature effects are important in developing growth models related to temperature regulation.

For shoot elongation to occur, photoperiod must be above the critical limit for cessation of growth. Rate of shoot elongation increases rapidly with increasing photoperiod above the critical daylength for shoot elongation. The northern provenances of Picea abies under continuous natural daylight at latitude 69°39′ N., Betula pubescens, and Salix pentandra have a maximum elongation rate under continuous light (Håbjørg, 1972a; Heide, 1974; Junttila, 1980).

Species with Fixed Growth

In tree species with fixed growth, the influence of environmental factors on shoot elongation is not as strong as for species with free growth. As illustrated for Scots pine (Pinus sylvestris L.) (Fig. 3a), temperature of the current season influences the rate of shoot elongation in plants grown under northern summer light conditions with 24-hour photoperiod and gives a Q_{10} of 3.6 (calculated between 10.5 and 15 °C). Current season temperature, however, has only a minor effect on the total shoot length. These results, like those published earlier (Junttila, 1986), indicate that alternating day and night temperatures do not effect shoot elongation in Pinus sylvestris. The elongation rate, and to a lesser extent the period of elongation, are also influenced by the number of stem units in the bud.

In contrast, initiation of stem units is strongly influenced by temperature. In Pinus sylvestris, 9 °C is close to the threshold for initiation of primordia (Fig. 3b). Two different experiments gave 9.9 and 11.8 as Q_{10}-values, calculated on the basis of shoot length between 9 and 15 °C. These are much higher than the Q_{10} for shoot elongation. This difference reflects the high sensitivity of cell division and organogenesis to temperature. The rate of cell division is generally closely dependent on temperature (Francis and Barlow, 1988). Primordial initiation in Pinus sylvestris shows a strong positive response to fluctuating day-night temperature, and of the combinations tested in our experiments, 21 °C for 12 hours and 12 °C for 12 hours was the most favorable combination under continuous natural daylight at latitude 69°39′ N.

The optimum photoperiod for shoot growth in Pinus sylvestris is about 18 hours (Ekeberg et al., 1979; Wareing, 1950a, 1950b). Also, seedlings of northern ecotypes of Pinus sylvestris exhibited greater shoot growth when exposed to 20-hour photoperiods than did seedlings exposed to 24-hour photoperiod, although those ecotypes experience continuous light during their natural shoot-elongation period. Experiments with southern pine species have shown that initiation of primordia and internode elongation may have different photoperiod requirements. Under certain threshold photoperiods, growth is not interrupted and resembles the free-growth pattern (Nagata and Yurugi, 1986; Yurugi and Nagata, 1987). Such results have not been reported for pines in the Temperate Zone.

LEAF GROWTH

Figure 4a illustrates some effects of temperature on leaf growth, expressed as maximum leaf area, in seedlings of Betula pubescens from three elevations at latitude 69°39′ N. Leaf size tended to decrease with increasing altitude of origin. For temperatures below 15 °C, leaf area was linearly related to temperature.

Long photoperiod is needed for needle elongation in Pinus sylvestris (Ekberg et al., 1979; Wareing, 1950a; 1950b). Under the light climate at high latitudes, this requirement is satisfied, and needle elongation is strongly dependent on temperature (Junttila, 1986; Junttila and Heide, 1981). The optimum temperature for needle elongation is slightly above 20 °C, with no apparent specific response to fluctuating temperatures (Fig. 4b). To our surprise, we have observed a negative aftereffect of the previous year's temperature on needle length the following year. In one experiment, needle length of seedlings grown at

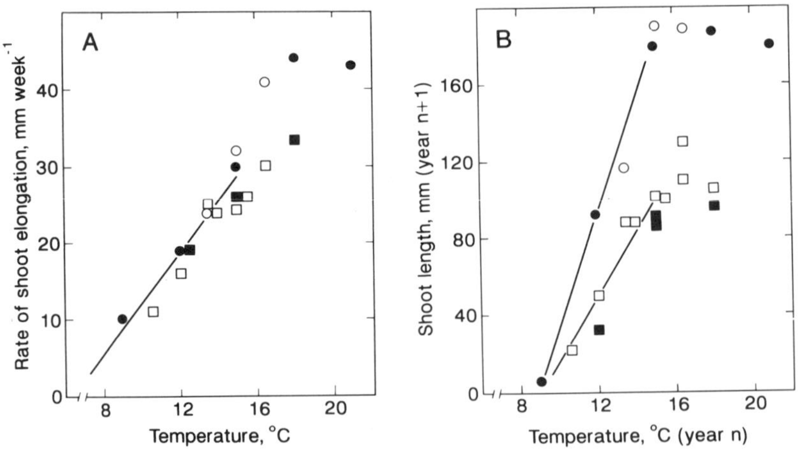

Fig. 3. (a) Effect of temperature on the rate of shoot elongation in second- and third-year seedlings of Pinus sylvestris L. from latitude 69° N. The regression line, calculated for the range from 9 to 15 °C, was: Y=3.22X-20.32, r^2=0.88, P < 0.0001; X=temperature (°C), Y=growth rate (mm/week); (b) effect of previous-year temperature on shoot length in third- and fourth-year seedlings of Pinus sylvestris L. from latitude 69° N. Seedlings were grown in a phytotron under natural 24-hour photo-period at latitude 69°39′ N. The regression lines, calculated for the range from 9 to 15 °C, were as follows: third-year seedlings, Y=29.3X-261, r^2=0.98, P < 0.002; fourth-year seedlings, Y=16.6X-151, r^2=0.91, P < 0.0001. X=temperature (°C), Y=shoot length (mm).

12 °C and 21 °C during the previous season averaged 72 and 41 millimeters, respectively. The current season temperature was 12 °C. Reasons for this response are unclear, but they could be related to the distribution of available resources among the developing needle fascicles. In that experiment, the number of needle pairs of the annual shoot was 65 and 202, respectively, for seedlings grown at 12 °C and 21 °C.

CESSATION OF SHOOT GROWTH

For trees growing in northern climate, cessation of shoot growth is of great significance for survival. Cessation of shoot elongation is a prerequisite for proper winter hardening in trees (Weiser, 1970). Several studies have clearly shown a close relationship between onset of growth cessation and development of cold hardiness (Junttila, 1989; Junttila and Kaurin, 1990; Junttila and Skaret, 1990). In species with a free-growth pattern, selection pressure for the regulation of growth cessation is especially strong. For those species, photoperiod, or more correctly, night length, is the primary signal for cessation of growth. The presence of photoperiodic ecotypes in most temperate-zone tree species is well documented (Dormling, 1971; Dormling et al., 1968; Heide, 1974; Håbjørg, 1978; Junttila, 1980; Vaartaja, 1954). The critical photoperiod for cessation of shoot growth in ecotypes at latitude 69° N. is approximately 22 hours. Under local light regimes, seedlings of northern ecotypes stop growing about middle of August. Although the critical photoperiod for growth cessation depends on latitude of the seed source, comparable provenances of various species may have slightly different critical photoperiods (Junttila and Skaret, 1990). In addition, the rate of response to photoperiod can vary among species. The rate of response is important for cold acclimation, and should be studied more thoroughly.

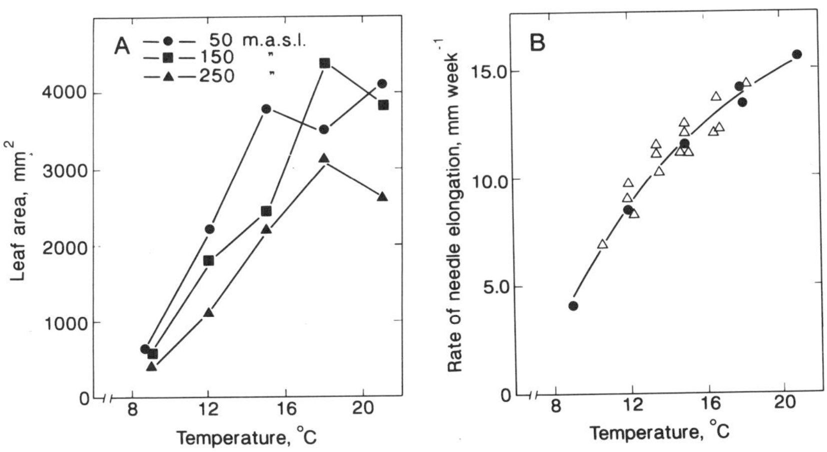

Fig. 4. (a) Effect of temperature on leaf area in first-year seedlings of three altitudinal (50, 150, and 250 meters above sea level) populations of Betula pubescens Ehrh. from latitude 69° N. Results are means of ten largest leaves from ten seedlings. (b) Effect of temperature on the rate of needle elongation in two- and three-year old seedlings of Pinus sylvestris L. from latitude 69° N. Seedlings were grown in a phytotron under natural 24-hour photoperiod at latitude 69°39′ N. Growth rate was calculated for the period of linear needle elongation. Regression line for needle growth rate was $Y=-12.77+2.41X-0.05X^2$; $r^2=0.94$, $P < 0.0001$; X=temperature (°C), Y=needle elongation (mm/week).

Photoperiodic responses in trees are quantitative characters showing intermediate distribution in the F_1 generation (Eriksson et al., 1978; Hummel et al., 1982; Junttila, 1982). The basic genetic aspects of photoperiodic control of growth cessation are not known, however. Photoperiodic sensitivity is generally reduced with increasing age of the tree, and control of growth becomes more dependent on endogenous than on exogenous factors. These aspects are also unexplored.

Within a temperature range of 12 to 24 °C, the critical photoperiod is independent of temperature (Heide, 1974). Also, results with three altitudinal ecotypes of Betula pubescens (Fig. 5) indicate just a minor effect of temperature on the time for cessation of apical growth. There is, however, an interaction of photoperiod and temperature on growth cessation. In Populus xeuramericana, bud set was delayed by sub- or super-optimal night temperatures. The optimal night temperature for bud set was 25 °C, and the critical day length decreased as night temperature decreased or increased from optimal (Nagata et al., 1990). Night temperature of 10 °C delayed bud set in seedlings of Picea abies, and the critical photoperiod was decreased more than one hour (Dormling, 1979). In addition, low night temperature (about 5 °C) can compensate for short photoperiods and may induce growth cessation even under continuous light (Heide, 1974; Junttila, 1980).

Temperature can affect the rate of bud development and other processes connected with cessation of growth. In general, high temperature accelerates the short day response (Håbjørg, 1972a; Heide, 1974; Junttila and Skaret, 1990). For example, a Picea abies provenance from latitude 64° N. grown in a daylight phytotron at latitude 59°40′ N. reached 95 percent of the final shoot length approximately 10 days earlier at 18 °C than at 12 °C (Heide, 1974). Koski and his coworkers have developed a model for regulation of growth cessation based on interaction of photoperiod and temperature (Koski and Sievänen, 1985).

In species with a fixed-growth pattern, cessation of growth is regulated internally and is less affected by environmental factors. Because the rate of shoot elongation is to some degree affected by temperature, however, the date for bud set is delayed when the growing-season temperature is low. For example, in second-year seedlings of Pinus sylvestris grown in a daylight phytotron at latitude 69°39′ N., cessation of elongation growth (defined as a date for 95 percent of final shoot length) was observed after 17, 21, 26, 36, and 58 days at 21, 18, 15, 12, and 9 °C, respectively.

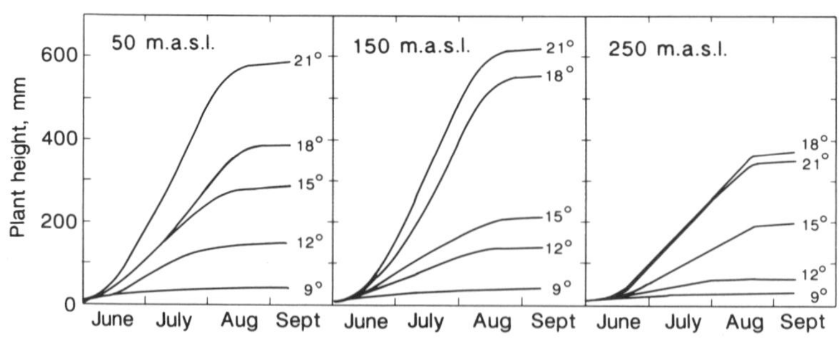

Fig. 5. Effect of temperature on cumulative height growth in first-year seedlings of three altitudinal populations (50, 150, and 250 meters above sea level) of Betula pubescens Ehrh. from latitude 69° N. Seedlings were grown in phytotron under natural 24-hour photoperiod at latitude 69°39′ N.

Adaptation to photoperiod is probably the most prominent feature in latitudinal ecotypes of temperate tree species. But there are also significant differences between latitudinal ecotypes in regard to their responses to spectral energy distribution and sensitivity to irradiance. For example, use of light sources with a high red:far red ratio during the night extension did not produce a long-day effect on a northern (latitude 69°39′ N.) but did so on a southern ecotype of Salix pentandra (Junttila and Kaurin, 1985). Northern ecotypes at high latitude are rather insensitive to a night-break treatment, and require much higher irradiances than the southern ecotypes for a long-day response. Junttila and Kaurin (1985) found that irradiance of 1.5 mW m^{-2} nm^{-1} at 660 nanometers from incandescent lamps was enough to prevent growth cessation in a southern ecotype of Salix, while irradiances up to 30 mW m^{-2} nm^{-1} at 660 nanometers were required for a northern ecotype. Håbjørg (1972b) has shown similar differences among latitudinal provenances of Betula. Further studies are needed on these aspects of light adaptation in northern tree species.

BUD DORMANCY

Bud dormancy is part of the adaptive mechanism for winter survival in climates with pronounced annual cold periods. Dormancy in Temperate Zone tree species is induced following cessation of apical growth. The depth of dormancy in temperate species is affected by various factors, including temperature conditions prevailing during bud development. The relationship between initiation of primordia in the bud and development of bud dormancy, however, is not well known. Development of cold acclimation occurs more or less parallel to induction of dormancy (Cannell, 1990), although there is not necessarily a direct dependence between these two processes at physiological or biochemical levels.

The cold period required to break dormancy is dependent on genotype, environmental conditions during induction of dormancy, temperature during chilling and, perhaps, also on temperature after chilling (Hänninen, 1990a). According to Hänninen and Pelkonen (1988), seedlings of Picea abies and Pinus sylvestris have a chilling requirement of 6-8 weeks and a rest period with a distinct end point. Flushing is primarily regulated by temperature, but if the chilling requirement is not fulfilled, daylength can have a significant influence. With changing global climate, there is now an increasing interest in studies on temperature regulation of tree phenology, including acclimation and de-acclimation. Theoretical models for these processes have been developed (Cannell, 1990; Hänninen, 1990a, 1990b; Repo et al., 1990) but, for most tree species, extensive experimental data on regulation of dormancy/growth and acclimation/de-acclimation cycles are needed for further model development. De-acclimation is, in many cases, a much faster process than acclimation (Burr et al., 1989), and stability of frost resistance under fluctuating winter temperature conditions is an important area for further study.

CONSEQUENCES OF CLIMATE CHANGE

Global climatic conditions show various types of fluctuations, both over short and long time intervals. We are presently concerned about global climate change caused by human activities. Although such changes, including a global rise in temperature, are expected to occur, both the extent and the type of changes are debated. Based on some recent studies, Briffa et al. (1990) estimated that summer temperatures in northern Fennoscandia could increase between 0.9 and 1.5 °C within the next 40 years. Much higher values have been mentioned, however.

If major changes in precipitation patterns do not occur, a temperature increase in northern Fennoscandia would enhance tree growth and development and cause trees to migrate northward and to higher elevations. The most critical processes that would be affected if treeline is extended to higher latitudes would be photoperiodically-regulated cessation of growth and initiation of cold acclimation in the fall. Treeline ecotypes in northern areas are already adapted to such extreme light conditions that the proper timing of growth cessation would probably not be altered significantly. If an ecotype with a critical photoperiod of 20 hours is moved to a latitude of 70°39′ N., it will receive the signal for growth cessation about one week later than the local ecotype with 22 hours as a critical daylength (Fig. 6). This delay of growth cessation would not have dramatic effects, especially if the drop in fall temperature is delayed as well. Furthermore, even a rather large increase in the annual temperature of these areas would not significantly alter the annual cycle of dormancy and flushing, or of acclimation and de-acclimation.

The situation will be different, however, if climatic change leads to large, irregular temperature variations with alternating periods of warmth and cold or frost. Such changes could create problems for northern tree species, especially if temperature variations occur late in the winter after the chilling requirement is fulfilled. The warm periods, however, must be long enough to induce bud break; until that stage, northern tree species have considerable cold hardiness.

HORMONAL REGULATION OF GROWTH

Many physiological aspects of cell division and cell elongation are known, but the mechanisms behind regulation of growth by temperature or photoperiod are still unresolved. There is reason to believe, however, that plant hormones are involved in the regulation of various aspects of growth and development in trees, and that environmental control can be mediated through plant hormones. The literature on plant hormones in woody plants is

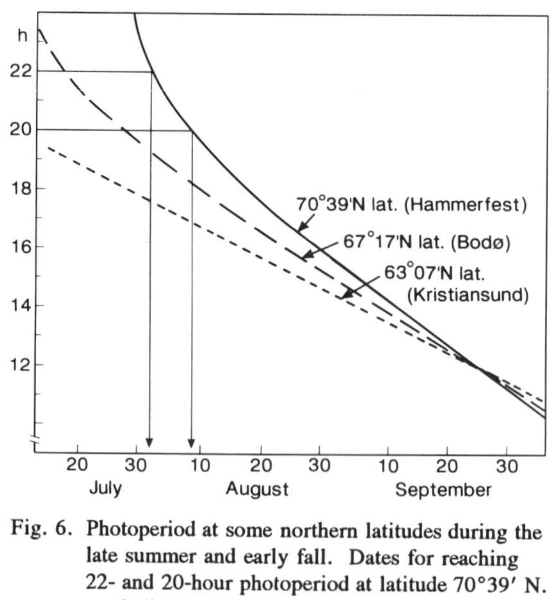

Fig. 6. Photoperiod at some northern latitudes during the late summer and early fall. Dates for reaching 22- and 20-hour photoperiod at latitude 70°39′ N. are indicated.

extensive, and various aspects have been presented in comprehensive reviews (Kossuth and Ross, 1987; Powell, 1987).

Various plant hormones are involved in regulating dormancy, but present knowledge is based on rather vague experimental evidence. This is aptly summarized by Lavender and Silim (1987): "Although we do not know which plant growth regulators (PGR) are controlling induction, depth or termination of dormancy, the weight of evidence is that PGRs are the most likely mechanism for such control. They are capable of modifying certain phases of dormancy unlike any other mechanism studied, including mineral nutrition, substrates, enzymes, or specific metabolic pathways. Therefore, it is quite probable that the coupling between environmental factors and mechanisms of dormancy involve PGRs."

Several recent studies reveal interesting aspects in regard to regulation of shoot elongation. Studies with single gene dwarf mutants of maize, rice, and pea strongly indicate that gibberellin A_1 (GA_1) is the effector gibberellin (GA) for elongation growth (Graebe, 1987; Phinney, 1984). In vegetative tissues of higher plants, GA_1 is primarily synthesized via the early 13-hydroxylation pathway (Fig. 7) composed of GA_{53}, GA_{44}, GA_{19}, GA_{20}, and GA_1 (Graebe, 1987). All of these GAs may show biological activity, but such activity is a result of their metabolic conversion to GA_1. Gibberellin A_1 can be inactivated by 2ß-hydroxylation to GA_8. Also, GA_{19} and GA_{20} can be inactivated by an irreversible conversion to GA_{17} and GA_{29}, respectively.

Experiments with a number of woody species have shown that the application of GA can substitute for long day and prevent cessation of growth in seedlings kept under short day conditions (Junttila, 1991). Analytic and metabolic studies with Salix strongly indicate that early 13-hydroxylation can be the main pathway for biosynthesis of GAs in Salix, and perhaps in other hardwoods as well (Junttila, 1991).

Recent results suggest that GA_1 is the active GA for stem elongation in Salix (Junttila et al., 1991, 1992). It has been shown that growth-retarding effect of a new type of plant growth regulator is antagonized by GA_1 but not by GA_{19} or GA_{20}. These compounds inhibit especially the conversion of GA_{20} to GA_1 (Nakayama et al., 1990). Complete lack of activity by GA_{20}, when applied to seedlings of Salix treated with a compound BX 112, implies that GA_{20} is not active per se. All three gibberellins, GA_{19}, GA_{20}, and GA_1, are equally active when applied to seedlings treated with growth retardant, which inhibits growth by blocking early steps of GA biosynthesis (before GA_{53}). Further, GA_9 is highly active in stimulating shoot elongation in Salix, but metabolic studies strongly indicate that this activity is caused by conversion of GA_9 to GA_1, via GA_{20}, and that GA_9 is not active per se (Junttila et al., 1992).

Studies on environmentally-controlled stem growth in herbaceous plants indicate that photoperiodic control of shoot elongation can be related to control of GA_1 biosynthesis via the early 13-hydroxylation pathway. Steps converting GA_{53} to GA_{44} and GA_{19} to GA_{20} can be controlled by photoperiod (Gilmour et al., 1986; Graebe, 1987) (Fig. 7). This suggests that cessation of elongation growth in seedlings of Salix induced by short days could be related to photoperiodic regulation of GA_1 biosynthesis, possibly because of a short-day-induced block in the conversion of GA_{19} to GA_{20} (Junttila and Jensen, 1988). Short-day induced growth inhibition is effectively antagonized by GA_{20} and GA_1, but not by GA_{19} or other members of the early 13-hydroxylation pathway.

Another interesting aspect of GA is that applied GA stimulates elongation growth, especially at low temperature, and may decrease the temperature required for elongation growth. Such effects have been demonstrated in several perennial grass species. GA-deficient mutants of maize and rice also have a higher growth-constraining temperature than corresponding normal types (Stoddart and Lloyd, 1986). After an application of GA to such mutants, the temperature required for growth was close to that observed in normal genotypes. On the other hand, GA-insensitive mutants of wheat did not show such a response (Stoddart and Lloyd, 1986). Thus, differences in the ability of plants to grow at low temperature could be partly related to their GA metabolism. This aspect needs to be investigated with seedlings of trees as well. These results support the idea that genetically regulated growth potential could be related to hormone metabolism (Rood et al., 1988). Indications of such dependencies have been presented for tree species (Rood and Pharis, 1987), but much research is required before conclusions can be drawn. Genetic regulation of GA biosynthesis and sensitivity is now well established for several plant species, opening the possibility that such differences could be related to the growth potential in general, and to the ability for growth under limiting temperature conditions in particular.

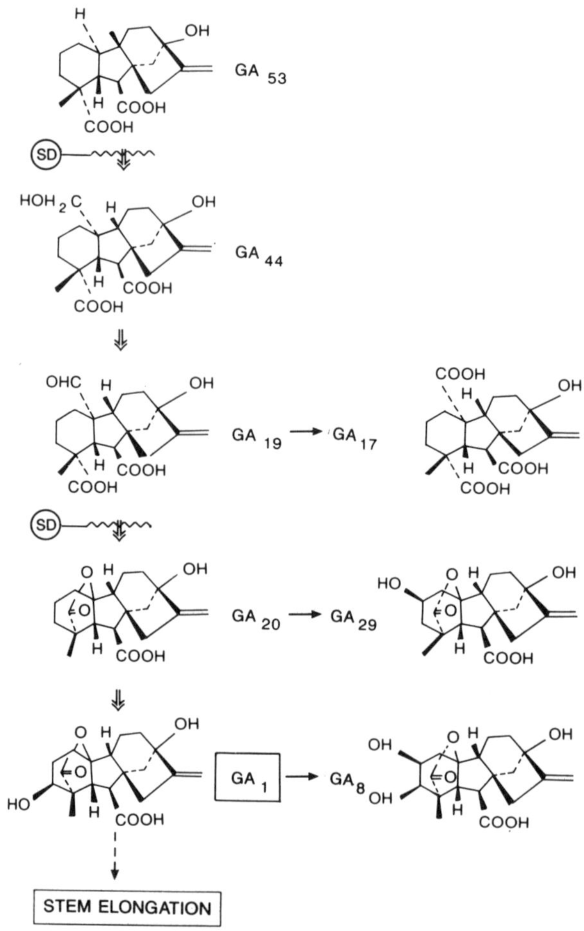

Fig. 7. The early 13-hydroxylation pathway for biosynthesis of gibberellin A_1. SD: Steps suggested to be controlled by photoperiod (Gilmour et al., 1986; Graebe, 1987; Junttila and Jensen, 1988).

The subapical meristem is probably the site of action for GA_1 (Sachs, 1965). This is supported by studies on dwarf watermelon seedlings which show that GA increases the proportion of rapidly dividing cells in the subapical meristem by increasing the probability that slowly cycling or nonproliferative cells will enter the proliferative pool (Edelman and Loy, 1987). Very little is known, however, about the site of synthesis, translocation, or about the mode of action of GA_1.

The situation for tree species with determined growth pattern is much less clear. GAs identified from several conifer species are somewhat different than GAs from hardwoods, although GA_1 is also present (Junttila, 1991). The active GA for stem elongation for conifers is not known.

A predetermined growth pattern with primordia initiation and stem-unit elongation as separate processes obviously implies complex changes in hormone metabolism. Hormonal control of primordia differentiation is poorly understood and would be an important and interesting area for further study. Cytokinins in developing buds would be particularly worth studying. Recent studies on cytokinins in buds of *Pinus sylvestris* (Häggman, 1991) suggest considerable metabolic changes in endogenous cytokinins during the winter. These changes may well be related to other metabolic changes observed in the buds during winter (Kupila-Ahvenniemi et al., 1987; Nuotio et al., 1990).

FUTURE RESEARCH

Further research is needed on the response of trees to the environment. We believe several problems related to environmental regulation of tree growth could now be investigated very effectively by organizing an international and collaborative research program. Such a program could include phytotron studies, field experiments, and theoretical modeling. Both the facilities and the competence to initiate such a program are available.

Biological research is currently in a stage of revolutionary development, and our understanding of basic biological processes is rapidly moving to a molecular level. Environmental regulation of growth in trees is a complex process involving interactions between different factors at various levels of organization. Further progress, however, is strongly dependent on the basic understanding of molecular regulation of cell division and cell elongation. Plant hormones could be part of the regulatory mechanism and are an important area for further study. Such studies should be concentrated on the regulation of hormone metabolism and the mode of action, including receptors and transduction pathways. Use of mutants, and especially the use of genetically transformed plants, provide quite extraordinary experimental systems for such studies. Experiments with genes for biosynthesis or sensitivity for plant hormones, using promoters that are tissue specific and environmentally controlled, create many possibilities for research. Genetic transformation of several hardwoods is already possible, and technology for transforming conifers is under rapid development.

Phytochrome is supposed to be the central pigment system in photoperiodic responses, and the mode of phytochrome action, including gene regulation, is an important area for research. Photoperiodic responses in trees are inherited as polygenic characters, but the nature of physiological or genetic differences observed between various latitudinal ecotypes is entirely unknown.

Another complex area for further study is the mechanism for climatic adaptation in trees. Progress in this field is closely dependent on the knowledge of physiological and biochemical mechanisms for climatic responses in trees. The possibility for natural selection of gametes (Mulcahy et al., 1986) requires careful study. If such selection occurs, for example, in respect to climatic conditions, it would facilitate rather rapid changes in tree populations.

ACKNOWLEDGEMENT

Thanks are due to Dr. Ronald Robberecht, Department of Range Resources, University of Idaho, Moscow, Idaho, U.S.A., for valuable comments on the manuscript, and to Mrs. Hilkka Falkseth for drawing the figures.

REFERENCES

Briffa, K. R., Bartholin, T. S., Jones, P. D., Karlén, W., Schweingruber, F. H., and Zetterberg, P., 1990, A 1,400-year tree-ring record of summer temperatures in Fennoscandia, Nature, 346:434-439.

Burr, K. E., Tinus, R. W., Wallner, S. J., and King, R. M., 1989, Relationships among cold hardiness, root growth potential and bud dormancy in three conifers, Tree Physiol., 5:291-306.

Cannell, M. G. R., 1990, Modelling the phenology of trees. In: Jozefek, H., ed., Modelling to Understand Forest Functions, Silva Carelica, Vol. 15, Joensuu, Finland, University of Joensuu, 11-27.

Dormling, I., 1971, Photo- und thermoperiodische Reaktionen bei Fichte, Kiefer und Gerste, Industr. Pflanzenbau, Wien, Band IV:205-218.

Dormling, I., 1979, Influence of light intensity and temperature on photoperiodic response of Norway spruce provenances. In: IUFRO Norway Spruce Meeting, 24 September--1 October, 1979, Bucharest, Romania, Lower Saxony Forest Research Institute, Department of Forest Tree Breeding, Escherode, Germany, 398-408.

Dormling, I., Gustafsson, Å., and von Wettstein, D., 1968, The experimental control of the life cycle in Picea abies (L.) Karst. I. Some basic experiments on the vegetative cycle, Silvae Genet., 17:44-64.

Edelman, L., and Loy, J. B., 1987, Regulation of cell division in the subapical shoot meristem of dwarf watermelon seedlings by gibberellic acid and polyethylene glycol 4000, J. Plant Growth Regul., 5:149-161.

Ekberg, I., Eriksson, G., and Dormling, I., 1979, Some photoperiodic reactions in conifer species, Holarctic Ecol., 2:255-263.

Eriksson, G., Ekberg, I., Dormling, I., and Matern, B., 1978, Inheritance of bud-set and bud-flushing in Picea abies (L.) Karst., Theor. Appl. Genet., 52:3-19.

Francis, D., and Barlow, P. W., 1988, Temperature and the cell cycle. In: Long, S. P., and Woodward, F. I., eds., Plants and Temperature, Symp. Soc. Exp. Biol., Vol. 42, The Company of Biologists Limited, Cambridge, United Kingdom, 205-218.

Gillies, S. L., and Vidaver, W., 1990, Resistance to photodamage in evergreen conifers, Physiol. Plant., 80:148-153.

Gilmour, S. J., Zeevaart, J. A. D., Schwenen, L., and Graebe, J. E., 1986, Gibberellin metabolism in cell-free extracts from spinach leaves in relation to photoperiod, Plant Physiol., 82:190-195.

Graebe, J. E., 1987, Gibberellin biosynthesis and control, Annual Rev. Plant Physiol., 38:419-465.

Guy, C. L., 1990, Cold acclimation and freezing stress tolerance: Role of protein metabolism, Annual Rev. Plant Physiol. Plant Mol. Biol., 41:187-223.

Håbjørg, A., 1972a, Effects of photoperiod and temperature on growth and development of three latitudinal populations of Betula pubescens Ehrh., Meld. Norges Landbr. Høgsk., 51(2):1-27.

Håbjørg, A., 1972b, Effects of light quality, light intensity and night temperature on growth and development of three latitudinal populations of Betula pubescens Ehrh., Meld. Norges Landbr. Høgsk., 51(26):1-17.

Håbjørg, A., 1978, Photoperiodic ecotypes in Scandinavian trees and shrubs, Meld. Norges Landbr. Høgsk., 57(33):1-20.

Häggman, J., 1991, Cytokinins in developing buds and tissue cultures of Scots pine and their role in xylem production of Loblolly pine, Acta Univ. Ouluensis, Serie A, 220:1-68.

Hänninen, H., 1990a, Modelling bud dormancy release in trees from cool and temperate regions, Acta Forest Fennica, 213:1-47.

Hänninen, H., 1990b, Modelling the growth rhythm of trees: Conceptual, experimental and applied aspects. In: Jozefek, H., ed., Modelling to Understand Forest Functions, Silva Carelica, Vol. 15, University of Joensuu, Joensuu, Finland, 35-45.

Hänninen, H., and Pelkonen, P., 1988, Effects of temperature on dormancy release in Norway spruce and Scots pine seedlings, Silva Fennica, 22:241-248.

Heide, O. M., 1974, Growth and dormancy in Norway spruce ecotypes (Picea abies), I. Interaction of photoperiod and temperature, Physiol. Plant., 230:1-12.

Hummel, R. L., Ascher, P. D., and Pellett, H. M., 1982, Inheritance of photoperiodically induced cold acclimation response in Cornus serica L., Red-osier dogwood, Theor. Appl. Genet., 62:385-394.

Junttila, O., 1980, Effect of photoperiod and temperature on apical growth cessation in two ecotypes of Salix and Betula, Physiol. Plant., 48:347-352.

Junttila, O., 1982, The cessation of apical growth in latitudinal ecotypes and ecotype crosses of Salix pentandra L., J. Exp. Bot., 33:1021-1029.

Junttila, O., 1986, Effects of temperature on shoot growth in northern provenances of Pinus sylvestris L., Tree Physiol., 1:185-192.

Junttila, O., 1989, Physiological responses to low temperature, Ann. Sci. For. (Suppl.), 46:S604-S613.

Junttila, O., 1991, Gibberellins and regulation of shoot elongation in woody plants. In: Takahashi, N., MacMillan, J., and Phinney, B. O., eds., Gibberellins, Springer-Verlag, Berlin, Heidelberg, New York, 199-210.

Junttila, O., and Heide, O. M., 1981, Shoot and needle growth in Pinus sylvestris as related to temperature in northern Fennoscandia, For. Sci., 27:423-430.

Junttila, O., and Jensen, E., 1988, Gibberellins and photoperiodic control of shoot elongation in Salix, Physiol. Plant., 74:371-376.

Junttila, O., and Kaurin, Å., 1985, Climatic control of apical growth cessation in latitudinal ecotypes of Salix pentandra. In: Kaurin, Å., Junttila, O., and Nilsen, J., eds., Plant Production in the North, Oslo, Norway, Norwegian University Press, 83-91.

Junttila, O., and Kaurin, Å., 1990, Environmental control of cold acclimation in Salix pentandra, Scand. J. For. Sci., 5:195-204.

Junttila, O., and Skaret, G., 1990, Growth and survival of seedlings of various Picea species under northern climatic conditions. Results from phytotron and field experiments, Scand. J. For. Res., 5:69-81.

Junttila, O., Jensen, E., and Ernstsen, A., 1991, Interaction of growth retardants and gibberellins on shoot elongation in seedlings of Salix pentandra, Physiol. Plant., 83:17-21.

Junttila, O., Jensen, E., Pearce, D., and Pharis, R. P., 1992, Stimulation of shoot elongation in Salix by gibberellin A_9; activity appears to be dependent upon hydroxylation to GA_1 via GA_{20}, Physiol. Plant., 84:113-120.

Körner, Ch., and Larcher, W., 1988, Plant life in cold climate. In: Long, S. P., and Woodward, F. I., eds., Plants and Temperature, Symp. Soc. Exp. Biol., Vol 42, The Company of Biologists Limited, Cambridge, United Kingdom, 25-57.

Koski, V., and Sievänen, R., 1985, Timing of growth cessation in relation to the variations in the growing season. In: Tigerstedt, P. M.A., Puttonen, P., and Koski, V., eds., Crop Physiology of the Trees, Helsinki University Press, Helsinki, Finland, 167-193.

Kossuth, S. V., and Ross, S. D., eds., 1987, Hormonal control of tree growth, Plant Growth Regulation, 6:1-216.

Krause, G. H., 1988, Photoinhibition of photosynthesis. An evaluation of damaging and protective mechanisms, Physiol. Plant., 74:566-574.

Kupila-Ahvenniemi, S., Lindfors, A., and Häggman, H., 1987, Seasonal variations in the properties of ribosome assemblies from vegetative buds of Scots pine in different season, Physiol. Plant., 71:55-60.

Larson, P. R., 1964, Some indirect effects of environment on wood formation. In: Zimmermann, M. H., ed., The Formation of Wood in Forest Trees, Academic Press, New York, United States, 345-364.

Lavender, D. P., and Silim, S. N., 1987, The role of plant growth regulators in dormancy in forest trees, Plant Growth Regul., 6:171-191.

Mulcahy, D. L., Bergamini Mulcahy, G., and Ottaviano, E., eds., 1986, Biotechnology and Ecology of Pollen, Springer-Verlag, New York, Berlin, Heidelberg, Tokyo, 528 p.

Nagata, H., and Yurugi, Y., 1986, Induced continuous shoot growth without branching in Pinus densiflora, J. Jap. For. Soc., 68:397-405.

Nagata, H., Yurugi, Y., and Horiuchi, Y., 1990, Studies on dormancy in woody plants (V). Effects of temperature on terminal bud formation of Populus x euramericana cv. I-214, The Bulletin of the Faculty of Bioresources, Mie University, Tsu, Japan, 4:157-164.

Nakayama, I., Kamiya, Y., Kobayashi, M., Abe, H., and Sakurai A., 1990, Effects of plant-growth regulator, Prohexadione, on the biosynthesis of gibberellins in cell-free systems derived from immature seeds, Plant Cell Physiol., 31:1183-1190.

Nuotio, S., Häggman, H., and Kupila-Ahvenniemi, S., 1990, Changes in gene expression of Scots pine buds during the winter and under experimentally altered light and temperature conditions, Physiol. Plant., 78:511-518.

Öquist, G., 1987, Light stress at low temperature. In: Kyle, D. J., Osmond, C. D., and Arntzen, C. J., eds., Photoinhibition, Elsevier Science Publishers, Amsterdam, The Netherlands, 67-87.

Palta, J. P., 1990, Stress interactions at the cellular and membrane levels, Hort. Science, 25:1377-1381.

Phinney, B. O., 1984, Gibberellin A_1, dwarfism and the control of shoot elongation in higher plants. In: Crozier, A., and Hillman, J. R., eds., The Biosynthesis and Metabolism of Plant Hormones, Cambridge University Press, Cambridge, United Kingdom, 17-42.

Powell, L. E., 1987, The hormonal control of bud and seed dormancy in woody plants. In: Davies, P. J., ed., Plant Hormones and their Role in Plant Growth and

Development, Dordrecht, Martinus Nijhoff/Dr Junk Publishers, The Netherlands, 539-552.

Repo, T., Mäkelä, A., and Hänninen, H., 1990, Modelling frost resistance of trees. In: Jozefek, H., ed., Modelling to Understand Forest Functions, Silva Carelica, Vol. 15, University of Joensuu, Joensuu, Finland, 61-74.

Rood, S. B., and Pharis, R. P., 1987, Hormones and heterosis in plants. In: Davies, P. J., ed., Plant Hormones and their Role in Plant Growth and Development, Dordrecht, Martinus Nijhoff/Dr Junk Publisher, The Netherlands, 463-473.

Rood, S. B., Buzzell, R. I., Mander, L. N., Pearce, D., and Pharis, R. P., 1988, Gibberellins: a phytohormonal basis for heterosis in maize, Science, 241:1216-1218.

Sachs, R. M., 1965, Stem elongation, Annu. Rev. Plant Physiol., 16:73-96.

Sakai, A., and Larcher, W., 1987, Frost survival of plants. In: Billings, W. D., Golley, F., Lang, O. L., Olson, J. S., and Remmert, H., eds., Ecol. Studies 62, Springer Verlag, Berlin, Germany, 340 p.

Skre, O., 1972, High temperature demands for growth and development in Norway spruce (Picea abies (L.) Karst.) in Scandinavia, Meld. Nor. landbr. Høgsk., 51(7):1-29.

Skre, O., [In press], Growth of mountain birch, Betula pubescens Ehrh., in response to variable temperatures. In: Alden, J. N., Ødum, S., and Mastrantonio, J. L., eds., Forest Development in Cold Climates, Proceedings of an International Symposium, 1991 June 18-24, [Laugarvatn, Iceland], NATO, Plenum, New York, New York.

Stoddart, J. L., and Lloyd, E. J., 1986, Modification by gibberellin of the growth-temperature relationship in mutant and normal genotypes of several cereals, Planta, 167:364-368.

Strand, M., 1987, Photosynthetic responses of seedlings of Scots pine (Pinus sylvestris L.) to low temperature and excessive light, Department of Plant Physiology, University of Umeå, Umeå, Sweden, ISBN 91-7174 308 1, 105 p. Ph.D. thesis.

Vaartaja, O., 1954, Photoperiodic ecotypes of trees, Can. J. Bot., 32:329-399.

Wareing, P. F., 1950a, Growth studies in woody species, I. Photoperiodism in first-year seedlings of Pinus sylvestris, Physiol. Plant., 3:258-279.

Wareing, P. F., 1950b, Growth studies in woody species, II. Effects of daylength on shoot growth in Pinus sylvestris after the first year, Physiol. Plant., 3:300-314.

Weiser, C. J., 1970, Cold resistance and injury in woody plants, Science, 169:1269-1278.

Yurugi, Y., and Nagata, H., 1987, Induced continuous shoot growth without branching on 12-hr photoperiod in Pinus luchuensis, J. Jap. For. Soc., 9:236-239.

TESTING WINTER DESICCATION RESISTANCE FOR SPECIES AND PROVENANCE SELECTION AT TIMBERLINES

J. Bo Larsen

The Royal Veterinary and Agricultural University
Frederiksberg, Denmark

INTRODUCTION

High cuticular water loss leading to tree desiccation and leaf mortality is a predominant ecophysiological factor in determining plant growth and survival at timberlines (Hadley and Smith, 1990; Larcher, 1957; Sakai, 1970). Resistance to winter desiccation might, therefore, be important for the survival and development of trees at the alpine and boreal timberline, and an important trait for species and provenance selection in tree breeding at timberlines.

Resistance to drought damage can be divided into two components (Levitt, 1972):

1. Drought avoidance—the ability to avoid desiccation by lowering water loss or increasing uptake.

2. Drought tolerance—the ability to tolerate desiccation and survive extended water losses.

In order to test a large number of genetic units (provenances, families, clones), adequate methods for testing (screening) desiccation resistance are needed. Earlier studies were able to test only drought avoidance, mainly by measuring the cuticular drying rate of detached twigs under standard conditions. The cuticular transpiration rate is computed as the slope of the linear regression equation between duration of drying and the water content of the twig during the phase of cuticular transpiration (Hygen, 1951). Large differences in resistance (avoidance) between species and provenances were demonstrated (Larsen et al., 1981; Michael, 1967; Rossa and Larsen, 1980; Saetersdal, 1963).

Studies of winter desiccation injury are based on the assumption that differences in drought resistance among various genetic entities are caused mainly by avoidance. Hence, a method for testing both components of desiccation resistance is needed.

Forest Development in Cold Climates, Edited by J. Alden *et al.*
Plenum Press, New York, 1993

The following method simulates winter desiccation and tests both drought avoidance and drought tolerance. Results indicate the importance of testing both aspects of desiccation resistance to effectively screen the selection potential on the species, provenance, and single-tree level.

METHODS

The winter desiccation test is carried out using the equipment developed and described by Larsen (1981). For the test, plants are placed in a winter-acclimated growth chamber under standard -5 °C soil, 12 °C air, 40 percent relative humidity, and a 10-hour photoperiod at 1,200 microEinsteins (μE). By keeping the root medium frozen, the uptake of water through the roots is interrupted, and the desiccation process develops under defined and controlled conditions.

Drought stress is followed by daily measurement of the water potential (xylem pressure potential) with the pressure chamber technique (Waring, 1970) immediately prior to the predawn light phase. The vitality or extent of damage to each plant is tested as soon as the water potential is determined, and the desiccation process is stopped by sprinkling the shoot with water. Subsequently, the roots are thawed and the plants are allowed to recover in a greenhouse under optimal conditions. Drought damage is evaluated after bud burst the following spring in 10 percent intervals of bud injury (0 percent = no damage; 100 percent = no living buds).

From the development of water potential and mortality rate over time, drought resistance and its two components, avoidance and tolerance, are calculated.

The relation between drought stress measured as water potential in bars and the corresponding vitality in percent mortality has the shape of the Gauss cumulative probability function. By means of probit analysis (Weber, 1972), the drought stress or water potential that statistically produces 50 percent mortality is calculated. The water potential at 50 percent mortality (WP_{50}) is an expression of drought tolerance (Fig. 1). The relation between the duration of drought stress and the water potential is approximately linear. The

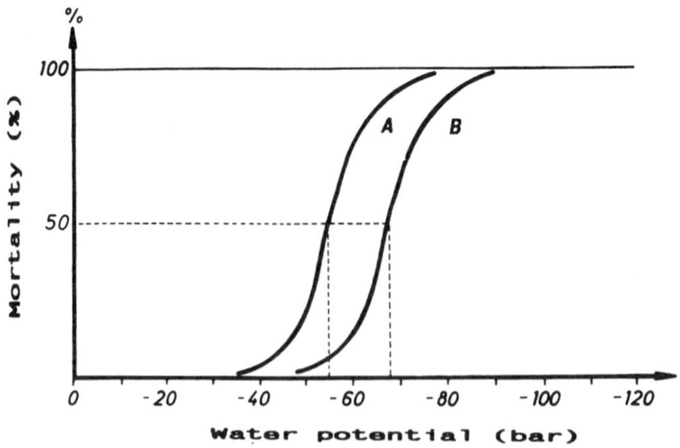

Fig. 1. Bud mortality for twigs of genetic entities A and B as a function of water potential.

slope (b) of the regression line y = a + bx therefore expresses that part of the drought resistance that is avoidance against desiccation (drought avoidance) (Fig. 2).

By means of the probit WP_{50} and the regression analysis, duration in days of the drought treatment to 50 percent mortality is calculated for the standard test conditions. This value (DAY_{50}) is an estimate of the total drought resistance.

Figures 1 and 2 illustrate the desiccation resistance relations of two different genetic units (A and B). Genetic entity B displays higher drought avoidance (Fig. 2)—that is, lower desiccation rate (-11.1 bars/day)—than genetic entity A (-21.8 bars/day), as well as a higher drought tolerance (-67 bars against -55 bars) (Fig. 1). The total winter desiccation resistance, or DAY_{50} of B (6.5 days), is more than twice as high as A (2.8 days) (Fig. 1).

DOUGLAS-FIR PROVENANCES

In a study of 40 Douglas-fir (Pseudotsuga menziesii (Mirb.) Franco) provenances from the 1967 to 1969 IUFRO-collection, differences in drought resistance among seed sources were large (Larsen, 1981). The provenances were divided into three groups (Fig. 3):

1. Provenances from the southern interior parts of the natural range (Colorado, eastern Arizona, and New Mexico) showed the highest desiccation avoidance and tolerance. Desiccation avoidance ranged between -3.5 and -8.2 bars/day, and tolerance ranged between -101 and -108 bars. Total desiccation resistance ranged from 15.6 to 23.3 days and averaged 19.9 days to 50 percent mortality.

2. Provenances from the interior parts of British Columbia, Idaho, and western Arizona were intermediate in drought avoidance and tolerance. Drought avoidance ranged between -8.2 and -11.5 bars/day and tolerance ranged between -85 and -102 bars. Total desiccation resistance ranged from 6.5 to 9.7 days and averaged 7.9 days to 50 percent mortality.

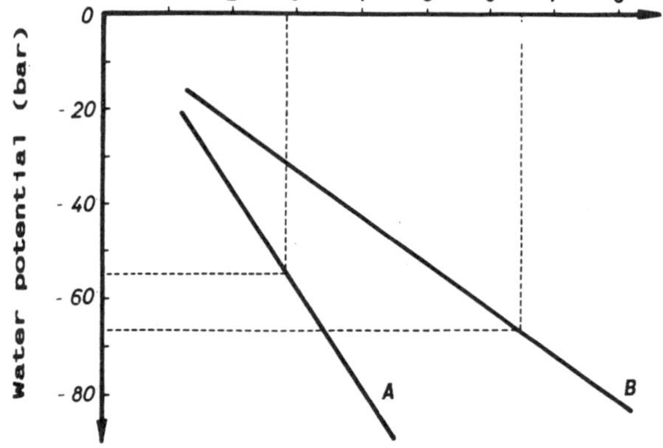

Fig. 2. The water potential for twigs of genetic entities A and B as a function of drought stress.

3. Provenances from the coastal areas of British Columbia, western Washington, western Oregon, and northwest California were least drought resistant. Desiccation avoidance ranged from -15.0 to -22.9 bars/day, and tolerance ranged from -66 and -82 bars. Desiccation resistance varied from 2.5 to 4.0 days to 50 percent mortality. Among the Cascade provenances of central British Columbia, Washington, and Oregon, total desiccation resistance increased significantly with altitude (Larsen, 1981).

The correlation between desiccation avoidance and tolerance for all 40 provenances was positive and highly significant (r = 0.829). It is interesting to note, however, that within each of the three groups, desiccation avoidance and tolerance were not related (Fig. 3).

NORWAY SPRUCE CLONES

In order to study the variation in winter desiccation resistance of single trees, 10 Norway spruce (<u>Picea</u> <u>abies</u> (L.) Karst.) clones were tested for desiccation avoidance and tolerance.

Within this limited material, variation in desiccation resistance was large (Fig. 4). Desiccation avoidance averaged -10.2 and varied between -6.7 and -14.0 bars/day. Desiccation tolerance varied between -65 and -90 bars and averaged -78.1 bars to 50 percent bud mortality. Total desiccation resistance varied between 4.0 and 10.6 days, and averaged 6.8 days to 50 percent mortality. Desiccation avoidance and tolerance were not correlated.

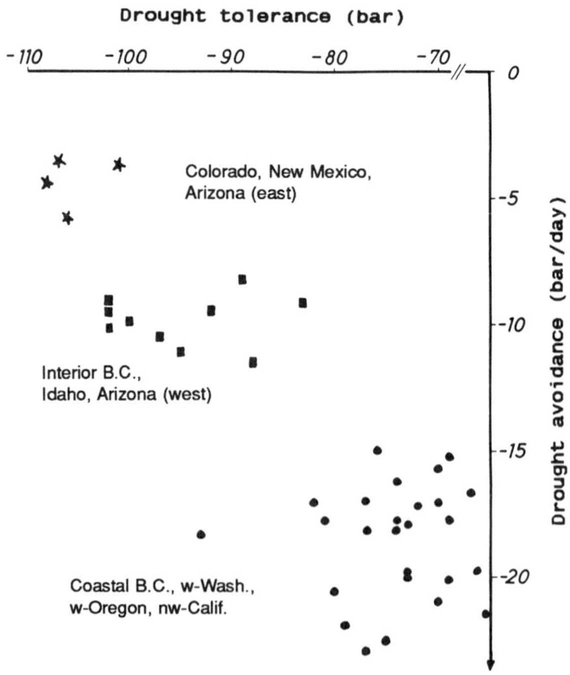

Fig. 3. Winter desiccation resistance (Douglas-fir provenances).

DISCUSSION

Although no single ecological factor seems to be responsible for forming treelines (Grace, 1989), the most common damage to trees at boreal and alpine timberlines is winter desiccation of exposed shoots. High cuticular water loss of shoots above the snow cover desiccates trees in the timberline-to-tundra transition zone during late winter, and leads to mortality of twigs and leaves (Tranquillini, 1979, 1982). Extreme desiccation conditions are exaggerated by incomplete deposition of wax and development of cuticular layers on leaf and twig surfaces from insufficient maturation of new shoots (Delucia and Berlyn, 1984; Hansen-Bristow, 1986; Sowell et al., 1982). Further, high wind speed in combination with ice crystals can damage shoot and leaf surface layers and accelerate moisture loss (Hadley and Smith, 1986, 1989).

Earlier results (Larsen et al., 1981; Rossa and Larsen, 1980) demonstrate that variation between different populations of Douglas fir and grand fir (Abies grandis (Dougl.) Lindl.) in cuticular transpiration rate is large. Further, close correlations between drought avoidance and cuticle thickness pinpointed the ecological significance of the development of layers on the leaf surface for cuticular transpiration.

Results obtained by Larsen (1981), as well as the present findings, prove that drought avoidance, as measured by cuticular transpiration rates, and drought tolerance are responsible for the resistance of conifers to desiccation injury during winter. The absence of correlations between desiccation avoidance and tolerance within broad provenance regions of Douglas fir and between clones of Norway spruce underlines the importance of testing both resistance components in screening species, provenances, families, clones, and other

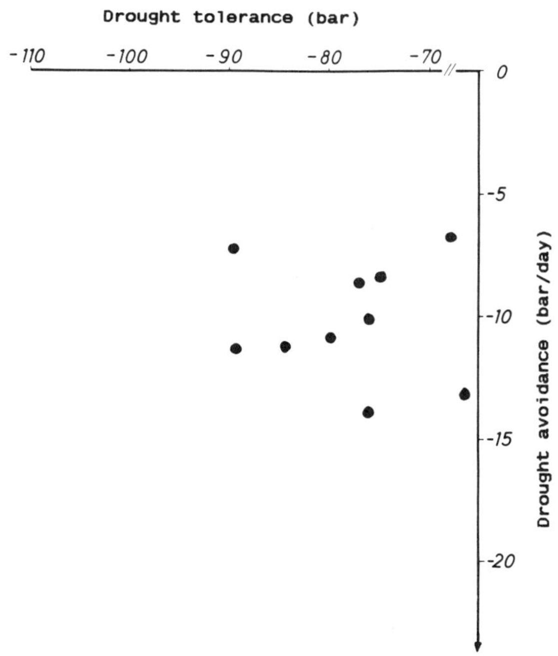

Fig. 4. Winter desiccation resistance (Norway spruce clones).

genetic entities for winter desiccation resistance. Furthermore, the large variation in winter drought resistance of Douglas fir provenances (Fig. 3) and Norway spruce clones (Fig. 4) reveals the potential for selecting and breeding this trait.

REFERENCES

Delucia E. H., and Berlyn G. P., 1984, The effect of increasing elevation on leaf cuticle thickness and cuticular transpiration of balsam fir, Can. J. Bot., 62:2423-2431.

Grace, J., 1989, Tree lines, Phil. Trans. R. Soc. Lond. B, 324:233-245.

Hadley, J. L., and Smith W. K., 1986, Wind effects on needles of timberline conifers seasonal influence on mortality, Ecology, 67:12-19.

Hadley, J. L., and Smith W. K., 1989, Wind erosion of leaf surface wax in alpine timberline conifers, Arct. Alp. Res., 21:392-398.

Hadley, J. L., and Smith W. K., 1990, Influence of leaf surface wax and leaf area-to-water content ratio on cuticular transpiration in western conifers, USA, Can. J. For. Res., 20:1306-1311.

Hansen-Bristow, K., 1986, Influence of increasing elevation on growth characteristics at timberline, Can. J. Bot., 64:2517-2523.

Hygen, G., 1951, Studies on plant transpiration, I., Physiol. Plant., 4:57-183.

Larcher, W., 1957, Frosttrocknis an der Waldgrenze und der alpinen Zwergstrauch-Heide auf dem Patscherkofel bei Innsbruck, Veroff. Ferdinandeum Innsbruck, 37:49-81.

Larsen, J. B., 1981, Geographic variation in winter drought resistance of Douglas-fir (Pseudotsuga menziesii (Mirb.) Franco), Silvae Genetica, 30:109-114.

Larsen, J. B., Magnussen, S., and Rossa, M. L., 1981, Untersuchungen über die Trockenresistenz und den Wasserhaushalt verschiedener Herkünfte von Abies grandis (Dougl.) Lindley, Schriftenreihe der Forstlichen Fakultät Göttingen, 71:122-155.

Levitt, J., 1972, Responses of plants to environmental stresses, Academic Press, New York and London.

Michael, G., 1967, Untersuchungen über die winterliche Dürreresistenz einiger immergrünen Gehölze im Hinblick auf eine Frosttrocknisgefahr, Flora, B., 156:350-370.

Rossa, M.-L., and Larsen, J. B., 1980, Die winterliche Austrocknungsrate verschiedener Herkünfte der Douglasie (Pseudotsuga menziesii) und deren Abhängigkeit von der Ausbildung der Cuticula und der Spaltöffnungstiefe, Allg. Forst- und Jagdzeitung, 151:137-147.

Saetersdal, L. S., 1963, The rate of drying of young excised plants of various provenances of Norway spruce and Douglas-fir, Medd. f. Vest. Forst. Forsoeks., 38:1-86.

Sakai, A., 1970, Mechanisms of desiccation damage of conifers wintering in soil-frozen areas, Ecology, 51:657-664.

Sowell, J. B., Koutnik, D. L., and Lansing, A. J., 1982, Cuticular transpiration of whitebark pine (Pinus albicaulis) within a Sierra Nevadan timberline ecotone, Arc. Alp. Res., 14:97-103.

Tranquillini, W., 1979, Physiological Ecology of the Alpine Timberline, Ecological Studies, 31, Springer Verlag, Berlin.

Tranquillini, W., 1982, Frost-drought and its ecological significance. In: Encyclopedia of plant physiology, Vol. 12b, Springer Verlag, Berlin.

Waring, R. H., 1970, Die Messung des Wasserpotentials mit der Scholander-Methode und ihre Bedeutung für die Forstwissenschaft, Forstwiss. Cbl., 89:195-200.

Weber, E., 1972, Grundriss der biologischen Statistik, Gustav Fischer Verlag, Jena.

GROWTH OF MOUNTAIN BIRCH (BETULA PUBESCENS EHRH.) IN RESPONSE TO CHANGING TEMPERATURE

Oddvar Skre

Norwegian Forest Research Institute
Fana, Norway

TREELINE FORMATION

In alpine and arctic areas, where climate regularly changes between cold winters and warm growing seasons, plants have evolved different survival strategies. Selection pressures in alpine and arctic climates are determined by abiotic factors rather than competition (Kallio, 1984). Plants that tolerate freezing and drying stress have an advantage over plants that have not evolved resistance to extreme cold and drought.

Over the course of a year, plants are subjected to many kinds of damage. To repair the injuries and protect living tissue against new damage, energy is needed. Energy is also required for active uptake of nutrients and water by the roots, and for growth and reproduction. The source of that energy is hydrolysis of adenosine triphosphate (ATP), which is produced through photosynthesis and respiration (Fig. 1). As plants approach their northern and altitudinal limits, low temperatures limit their ability to complete annual growth and dormancy processes and to produce viable seeds (Heikinheimo, 1932; Langlet, 1960).

The distribution of forests in Fennoscandia is limited by the arctic and alpine treelines. The most important treeline species are mountain birch (Betula pubescens Ehrh.), Norway spruce (Picea abies L. Karst.), and Scots pine (Pinus sylvestris L.). The development of flower and vegetative buds (Hustich, 1944; Junttila and Heide, 1981; Romell, 1925) in spruce and pine is determined by the previous summer climate. After pollination and fertilization, embryo and storage tissue develop in the immature seed. Hagem (1917) and Heikinheimo (1921) found that temperatures during seed maturation determine seed viability. Both seeds and embryos from treeline areas were often weak and incompletely developed. According to these authors, the alpine spruce and pine treelines in Scandinavia are mainly seed maturation limits. Depending on wind and weather conditions, however, viable seed may be dispersed more than 100 meters above the seed parents in good seed-bearing years. The same ability is also shown by mountain birch. Seedlings growing outside the climatic tree limits of a species are often sterile and unable to produce viable seeds (Skre, 1972). When the seedlings are established, they often develop polycormic, scrub-like life forms that reproduce vegetatively (Kallio et al., 1983). According to Elkington (1968) and Vaarama

Forest Development in Cold Climates, Edited by J. Alden *et al.*
Plenum Press, New York, 1993

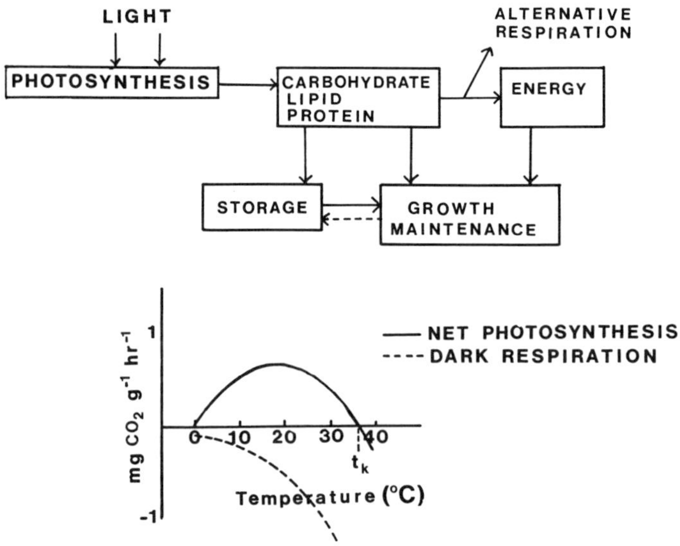

Fig. 1. The flow of assimilates and energy in plants (top) and net photosynthesis and dark respiration rates as milligrams carbon dioxide per gram dry weight and hours (mg CO_2 g^{-1} hr^{-1}) in a subarctic moss, Sphagnum subsecundum (Skre and Oechel, 1979), as a function of temperature (bottom). t_k = compensation temperature. Reproduced from Skre (1991a).

and Valanne (1973), the polycormic form in birch results from inbreeding, hybridization, and introgression of dwarf birch (Betula nana L.) in mountain birch. Norway spruce also reproduces vegetatively at low temperatures by layering of branches, which produces the "krummholz" growth form (Tranquillini, 1979). Scots pine does not layer, but compensates for its inability to reproduce vegetatively with high seed production. Hagem (1917) found that stands of Scots pine trees were often even-aged near treeline altitudes in southeast Norway, indicating that the trees originated from one or a few good seed-bearing years. If the time lapse between seed crops is too long, the old trees eventually die without establishing new seedlings.

To maintain the position of the forest-tundra ecotone, seed from parents below treeline must eventually replace the clones and genets of spruce and polycormic birch above treeline. Because of abundant seed production, Scots pine, like birch, will easily reproduce from seed in the favorable microclimate of burned areas (Kujala, 1927).

FACTORS RESTRICTING GROWTH

During different stages of plant development, growth is concentrated in some organs while others remain inactive. As a rule, annual growth is initiated with stem elongation, followed by leaf growth and secondary radial growth in the stem. Root growth usually starts in the spring, is interrupted during the period of shoot growth, and then continues into fall as long as soil temperature is favorable (Hagem, 1947; Rutter, 1957).

The growth of an organ may be divided into three stages or phases. As photosynthetic and respiratory tissues develop in Phase I growth of a leaf, for example, dry weight increases exponentially according to the law shown in Equation 1 (Blackman, 1919):

$$W_1(t) = W_0\, e^{rt}, \tag{1}$$

where W_1 is the dry weight at time (t) and W_0 is the initial dry weight of tissues growing at a rate r. This law applies to the first phase of organ growth and lasts until the metabolic tissue of the plant is fully developed (West et al., 1920). Then a linear growth phase (Equation 2) occurs, followed by a third phase during which leaf growth decreases exponentially according to Equation 1 and finally stops (Fig. 2).

$$W_2(t) = W_0 + rt \tag{2}$$

This scheme may also apply to the growth of other organs such as stems and roots, and to total leaf and shoot growth. During the second phase, growth is a linear function of time, and its relation to external factors can be studied. In spruce shoots, the period from 20 percent to 80 percent of the final shoot growth is characterized by Phase II growth (Mork, 1941; Romell, 1925).

Based on data from Norway spruce, one mechanism suggested for the effects of low-temperature on plant distribution is inhibition of dark respiration (Dahl and Mork, 1959; Skre, 1972). Through photosynthesis, solar energy is transformed to chemical energy and stored as different organic compounds. Differentiation and growth involve the synthesis of new plant tissue. This is an energy-consuming process and depends on the supply of organic matter from photosynthesis as well as chemical energy released during respiration (Fig. 1). During dark respiration, energy is released in small increments through the hydrolysis of ATP and other energy-rich compounds, which are translocated to sites where growth takes place. This part of the respiration process, called growth respiration (Penning de Vries, 1972), occurs in the mitochondria. The light-dependent photorespiration (Zelitch, 1966), and the alternative or cyanide-resistant respiration (Lambers, 1980), do not involve ATP formation.

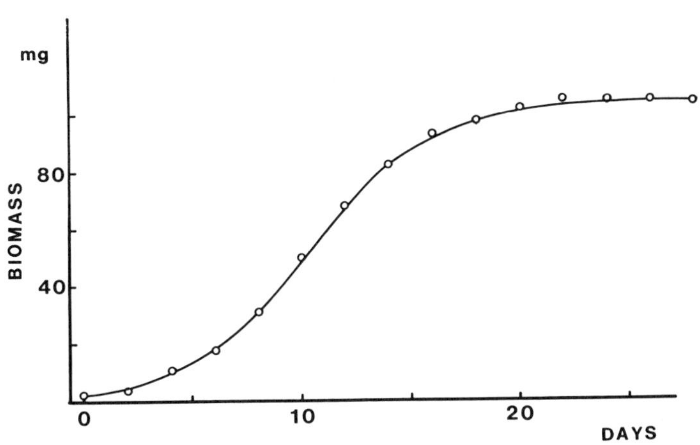

Fig. 2. Growth in leaf biomass of Scotch elm (Ulmus glabra Huds.) in milligrams (mg) per leaf. Reproduced from Skre (1991b).

Several authors have examined net photosynthesis (the difference between gross photosynthesis and photorespiration and dark respiration) as a function of temperature by measuring gas exchange and dry matter increase in a number of alpine plants, namely, Norway spruce (Aalvik, 1939; Hagem, 1947), black spruce (Picea mariana (Mill.) B.S.P.) (Vowinckel et al., 1975), and mountain birch (Sveinbjørnsson, 1983). They found that net photosynthesis increased with increasing temperature up to an optimum value (Fig. 1). If the plants were grown close to their compensation temperature for a long time, growth could be limited by the supply of organic matter.

In the field, the supply of organic matter in conifers may be limited during exceptionally mild winters when net photosynthesis and compensation temperatures are lowered due to light limitation (Hagem, 1947; Printz, 1933). Because of respiration loss and winter damage on evergreen needles, buds, and stems, plants experience a net loss of organic matter until

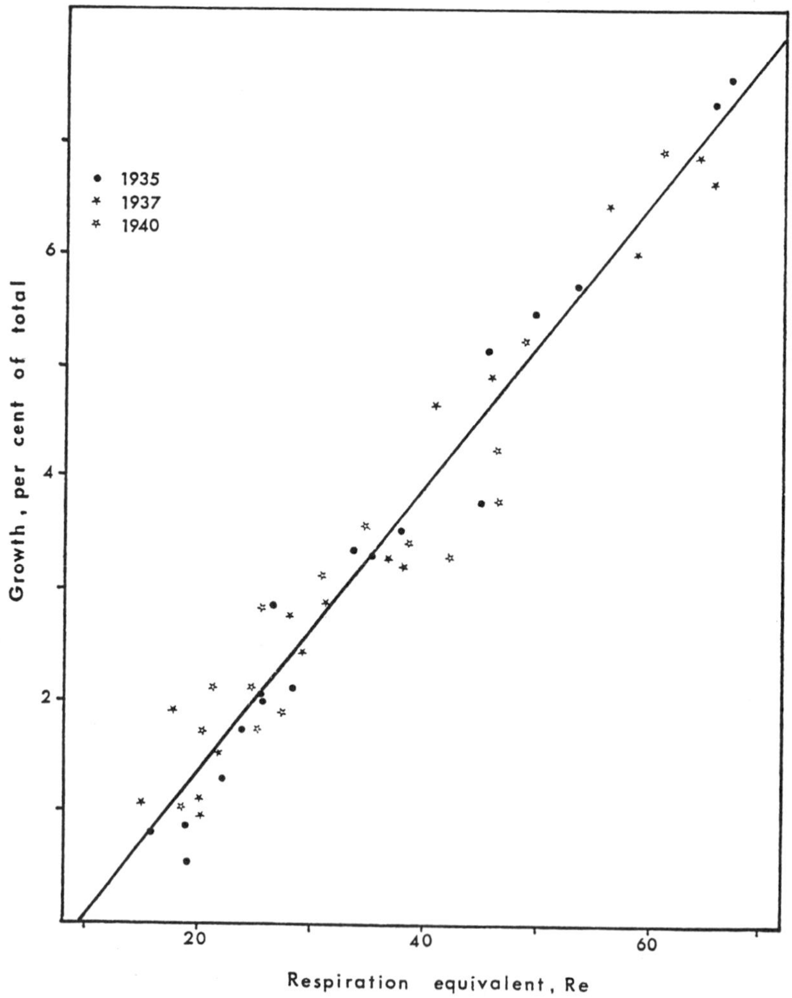

Fig. 3. The relationship between daily apical growth in Norway spruce and the daily accumulated respiration in a subalpine area at Hirkjölen, southeastern Norway (Dahl and Mork, 1959).

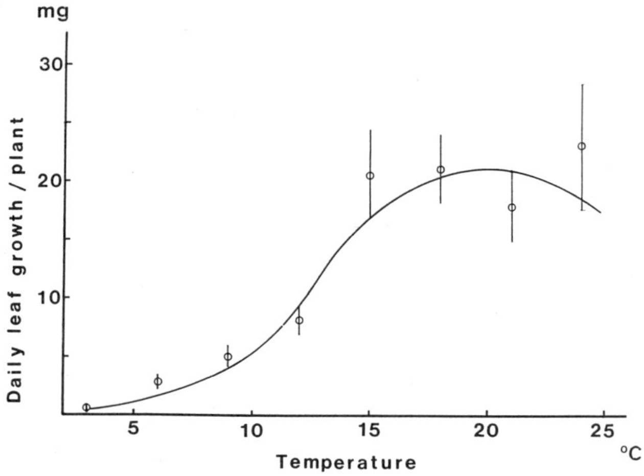

Fig. 4. Growth in leaf biomass per plant in milligrams (mg) dry weight per day during the linear growth phase of mountain birch (Betula pubescens Ehrh.) seedlings from Ås, southern Norway (95 meter altitude), as a function of temperature at 24-hour photoperiod and 300 microEinsteins per square meter per second (μE m^{-2} sec^{-1}) light intensity. Means with ± 2 standard errors (Skre, 1991b).

new leaf tissue is formed (Kozlowski and Gentile, 1958). The evergreens are able to maintain a certain rate of photosynthesis even during the winter, whereas deciduous trees are completely dependent on the new leaves as a carbon source. The period when net photosynthesis is negative in conifers usually lasts only for one or two months of the growing season (Rutter, 1957). With improving light conditions, carbon balance is shifted and the leaves will produce photosynthates for new growth that soon compensate for the loss of organic matter. As the temperature drops, dark respiration decreases exponentially while net photosynthesis is still high (Fig. 1). At low and medium temperatures above the freezing point, dark respiration may therefore be a limiting factor for growth. In this case, the relationship between accumulated respiration and growth is linear (Dahl and Mork, 1959). By interpreting measurements on shoot elongation in spruce from a subalpine area in southeastern Norway, Mork (1941) and Dahl and Mork (1959) found that the daily apical growth during the constant growth phase was closely correlated (r=0.978) with the daily accumulated respiration, or respiration equivalent (Fig. 3). Dark respiration as a function of temperature was measured in separate experiments using the Warburg manometric technique (Dahl and Mork, 1959).

In accordance with the respiration curve of Dahl and Mork (1959), the daily growth in leaf biomass during the linear growth phase of mountain birch seedlings grown at 24-hour day length increased at an approximately exponential rate with temperature up to about 15 °C (Fig. 4). At higher temperatures, growth followed a typical temperature-optimum function similar to the photosynthesis-temperature relationship in Figure 1. The implication is that dark respiration may limit growth at low temperatures, and photosynthesis may limit growth at high temperatures.

Nitrogen availability, however, is probably the most important limiting factor for plant growth in cold climates, because low temperatures slow the decomposition of organic matters in soil (Ågren, 1985). Arctic plants adapt to low soil nitrogen concentrations by increasing their relative root biomass and by adopting a slow-growth strategy to maintain high metabolic activity (Chapin, 1979; Crawford, 1989).

GROWTH STRATEGY AT DIFFERENT TEMPERATURES

The expected two or three degree Celsius increase in summer temperature caused by the greenhouse effect (Skre, 1990) is expected to raise the altitudinal tree limits of subalpine species by about 400 meters. Because respiration rates increase exponentially with temperature while ambient temperatures decrease linearly with altitude, the difference in altitude corresponding to one monthly accumulated respiration unit (the accumulated respiration corresponding to one month at 10 °C) increases with altitude. As an example, the altitude limit of European alder (Alnus glutinosa (L.) Gaertn.), a lowland species with high temperature requirements, is expected to increase only by 300 meters (Skre, 1990).

The greenhouse effect is also expected to increase winter temperatures and the winter respiration loss. The increase in the length of the frost-free period is expected to be much larger at oceanic treelines than at continental treelines (Skre, 1990). Because winter temperatures are expected to increase more than summer temperatures, the major effect of climatic change at tree limits would be a reduction in the dormancy period.

To investigate the influence of the expected climatic change on respiration losses during the winter and the physiological status of the trees, Skre (1990) grew mountain birch and lowland birch (Betula pendula Roth) seedlings in fertilized peat at an oceanic lowland site at Fana, Norway (at 50 meters elevation) and at a nearby treeline site at Kvamskogen (450 meters) (Fig. 5). Both localities are located at 59° N. latitude and 5° E. longitude. A comparison between sites showed that growth rates were approximately two times faster at Fana than at Kvamskogen (Fig. 5a). The difference between the lowland and treeline sites increased as the plants grew older, because birch, like many other deciduous species (Prudhomme, 1983), are rather opportunistic, and invest a relatively large part of the energy from respiration into growth of structural and photosynthetic tissues. By increasing their photosynthetic capacity, plants are able to replace some of the carbohydrates that are consumed in the growth process (Thornley, 1972). The increase in photosynthetic capacity is usually strongest in the fast-growing southern ecotypes. Growth rates at Fana had increased to about four times the growth rates at Kvamskogen (1987) in two-year-old seedlings of lowland birch (ecotype "BAL," Fig. 5a) and southern mountain birch (BAM) ecotypes. In the northern ecotype from Finnish Lapland (BJ), on the other hand, altitude had almost no effect on growth, which was much slower, and photosynthetic products were invested as storage products and stem and root growth. Arctic populations may avoid growth in order to maintain high nutrient concentrations in their tissue (Chapin, 1979). Higher respiration loss during the winter at the lowland site paralleled the higher growth rate during the summer. The amount of non-structural carbohydrates in roots dropped from 40 to 15 percent (dry weight basis) during the winter at Fana and decreased from near 40 percent to only 25 percent during the winter at Kvamskogen (Fig. 5b), indicating that the source strength (carbon storage capacity) is lower at the low-altitude site. The expected temperature rise due to the greenhouse effect would not only result in much higher growth rates but would reduce carbohydrate and energy resources as well, particularly in slow-growing species and ecotypes.

If temperatures increased too fast, plants would not have adequate time to adapt to the new climate. Photosynthesis will be the growth-limiting process at ambient temperatures, and many arctic plants will be restricted not only by their upper altitude limits but by a lower limit as well. Printz (1933) suggested that high winter temperature might restrict

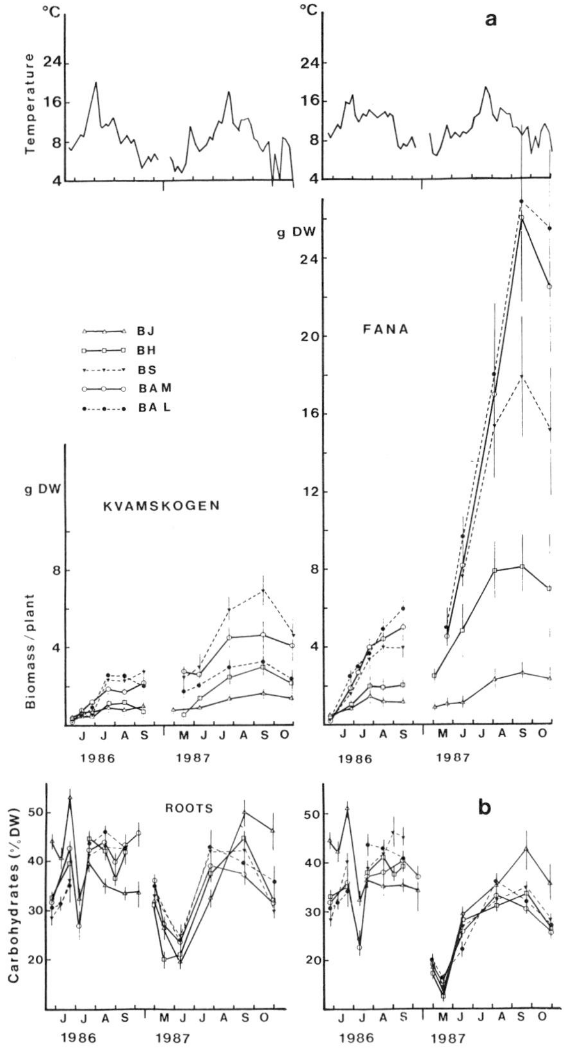

Fig. 5. (a) Biomass as grams dry weight (gDW) per plant of mountain (Betula pubescens Ehrh.) and lowland birch (Betula pendula Roth) seedlings grown in fertilized peat from the 4-leaf stage during the 1986 and the 1987 seasons at Fana (50 meters) and Kvamskogen (450 meters) at about 59° N. latitude in western Norway. Means with ± 2 standard errors. Five-day means of temperature (°C) are shown on top (Skre, 1990); (b) Total non-structural carbohydrates as percentage dry weight in roots of birch seedlings at Fana and Kvamskogen during the 1986/87 season (Skre, 1990). Means with ±2 standard errors. (See Figure 5a for explanation).

The ecotypes are:

Betula pubescens

BJ	Kevo, northern Finland (200 meters)	68° N.	26° E.
BH	Blefjell, eastern Norway (750 meters)	59° N.	9° E.
BS	Fana, western Norway (50 meters)	59° N.	5° E.
BAM	Løten, eastern Norway (200 meters)	61° N.	11° E.

Betula pendula

BAL	Løten, eastern Norway (200 meters)	61° N.	11° E.

spruce growth in western Norway, but his theory was later rejected (Hagem, 1947). Rapid climatic warming, however, might again actualize a situation like the one predicted by Printz (1933).

COMPENSATION MECHANISMS

To demonstrate the effect of temperature on the growth of roots and leaves of mountain birch, seedlings from the same populations shown in Figure 5 were grown at varying temperatures in continuous light, and their growth was followed by non-destructive and destructive methods (Skre 1991b, 1991c). The shoot-to-root ratio increased with temperature from 10 °C to at least 25 °C (Fig. 6). A similar increase was found in the leaf area ratios and specific leaf area (Fig. 7). The definitions of these growth parameters are:

Specific leaf area (SLA) = LA/LW
Leaf area ratio (LAR) = LA/W,

where LA = leaf area in square centimeters (cm^2),
LW = leaf weight in milligram (mg), and
W = total plant weight in milligrams (mg).

Mountain birch seedlings seemed to compensate for reduced photosynthesis at high temperatures by increasing stem elongation rates and leaf area in order to compete more efficiently for available light (Warren-Wilson, 1972). At medium temperatures (10 to 15 °C), however, photosynthetic rates were high and the plants seemed to invest more energy in root growth and structural leaf tissue than when temperatures were less favorable for photosynthesis.

Mountain birch seedlings were also grown at varying temperatures and photoperiods to determine photosynthesis and dark respiration rates (Skre, 1989) (Fig. 8). Temperature had a strong acclimation effect on photosynthesis. Photosynthesis rates of plants grown at high temperatures increased and their optimum and compensation temperatures for photosynthesis were higher than in plants grown at low temperatures.

Increased respiration rates are partly responsible for the lower rates of photosynthesis in low-temperature treated plants. Results agree with similar studies on Oxyria digyna L.

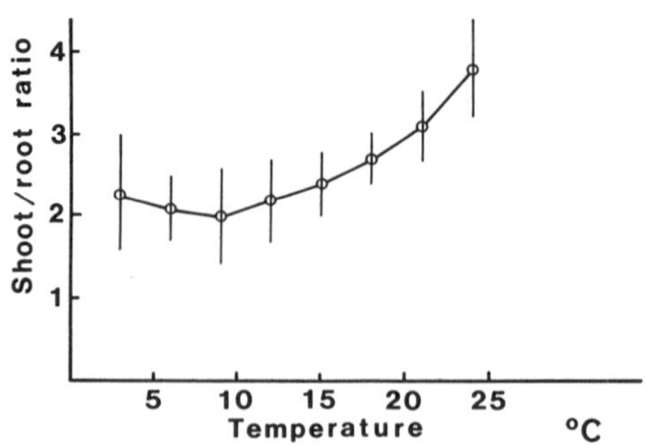

Fig. 6. Shoot-to-root ratios with ±2 standard errors as a function of temperature for 6-week old subalpine mountain birch seedlings from southern Norway (Skre, 1991b).

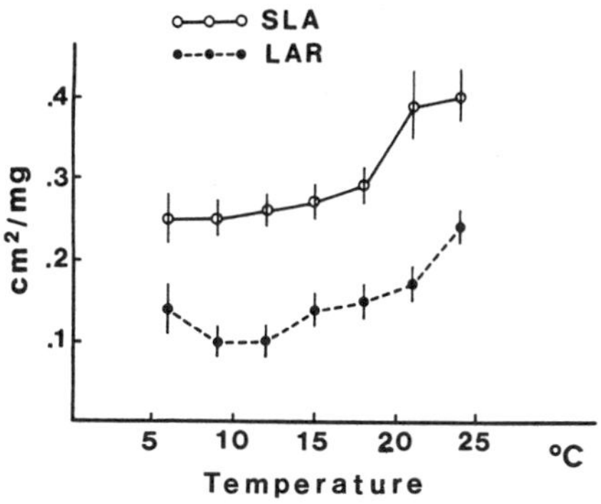

Fig. 7. Leaf area ratio (LAR) and specific leaf area (SLA) as square centimeters per milligram (cm^2 mg^{-1}) with ± 2 standard errors in 6-week old subalpine mountain birch seedlings from southern Norway (Skre, 1991b).

(Mooney and Billings, 1961). At high temperatures, Mooney and Billings' experiments demonstrated that arctic populations of Oxyria digyna had slightly higher photosynthesis and dark respiration rates than southern populations. The same effect of temperature on photosynthesis and dark respiration of low- and high-altitude mountain birch seedlings is shown in Figure 8.

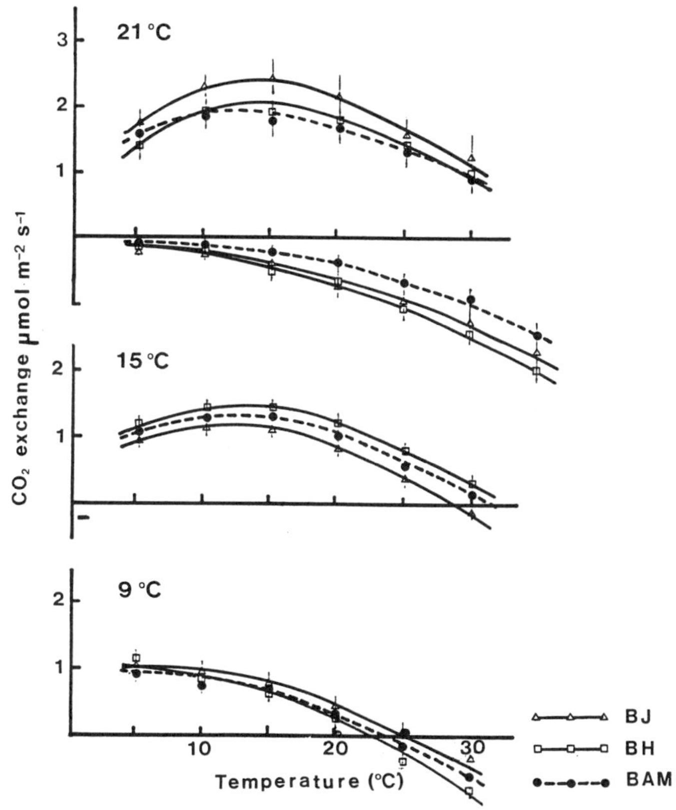

Fig. 8. Photosynthesis rates (measured in micromoles carbon dioxide per square meter per second) (μM CO_2 m^{-2} sec^{-1}) as a function of temperature in discs of mature leaves from three mountain birch populations. Leaves were held three weeks at 18-hour photoperiods and temperatures of 21, 15, and 9 °C, respectively. Dark respiration of the 21 °C sample is shown as negative CO_2 (Skre, 1989). Irradiance is 100 microEinsteins per square meter per second (μE m^{-2} sec^{-1}). For explanation of symbols, see Figure 5

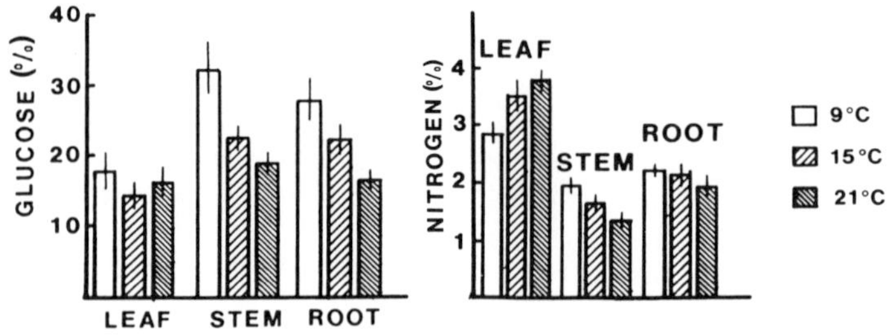

Fig. 9. Percentages (%) of total non-structural carbohydrates as glucose equivalents (left) and total nitrogen (right) with ± 2 standard errors in leaf, stem, and root tissue of 6-week-old subalpine mountain birch seedlings from southern Norway grown at 9, 15, and 21 °C, when averaged over 12-, 18-, and 24-hours photoperiods. The effects of photoperiod were insignificant.

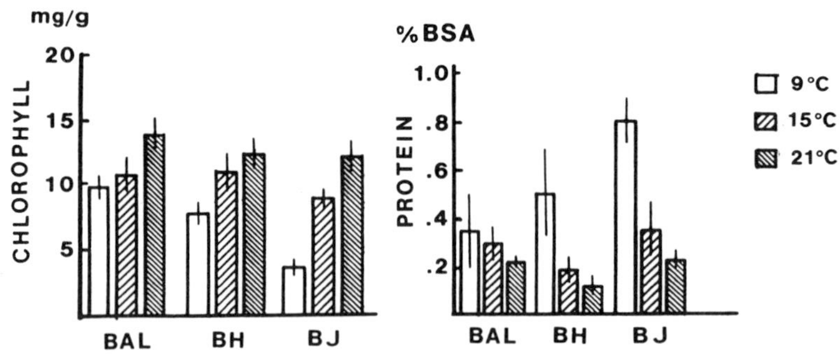

Fig. 10. Chlorophyll concentrations as milligram dry weight (mg/g) (left) and percentage dry weight (%) of water-soluble protein content as bovine serum albumin (BSA) equivalents (right) with ± 2 standard errors in leaves of three birch populations (See Figure 5 for explanation of symbols) grown at 9, 15, and 21 °C, when averaged over 12-, 18-, and 24-hour photoperiods (Skre, 1991d). The effects of photoperiod were small or insignificant.

Finally, chemical analysis of birch tissue grown at different temperatures (Skre, 1991d) showed that mountain birch seedlings react to increased temperatures by increasing the total nitrogen and chlorophyll concentration in their leaves (Figs. 9, 10). On the other hand, the total concentration of nitrogen and carbohydrate in stem and root tissue (Fig. 9), and the soluble carbohydrates (Skre, 1991d) and protein nitrogen (Fig. 10) in leaf tissue decreased, indicating that frost resistance was reduced (Parker, 1963; Volger and Heber, 1975). The accumulation of soluble proteins in leaves of seedlings grown at low temperature was highest in mountain birch populations from northern Finland and lowest in populations of lowland birch from southern Norway (Fig. 10).

Results (Figs. 6-10) indicate that mountain birch is able to compensate for decreased photosynthesis and energy losses resulting from increased temperatures by increasing photosynthesis rates, optimum and compensation temperatures for photosynthesis, shoot elongation, leaf area ratio, and the leaf nitrogen and chlorophyll concentrations. On the other hand, birch tissue is more likely to be damaged by frost when grown at high than at low temperatures (Fig. 9).

According to Ågren (1985), nitrogen availability limits plant growth and affects survival in cold climates more than any other factor. Due to slow decomposition of organic matter, the available nitrogen and phosphorus in arctic soils is very low. Northern plants adapt to low soil-nutrient content by reducing their growth rates at low temperatures rather than producing nutrient-deficient tissue (Chapin, 1979). The results of Skre (1991d) support this hypothesis (Fig. 9). In addition, plants respond to low temperatures when grown at more than 12 hours photoperiod by decreasing their shoot-to-root ratio. The shoot-to-root ratio is lower in northern than southern populations (Fig. 11). High rates of shoot growth are not an advantage in arctic climates (Crawford, 1989). Instead, plants direct their resources to growth of root tissues to increase the absorption of nitrogen and phosphorus, thereby partly compensating for nutrient limitations (Chapin and Tryon, 1982).

The results of Dahl and Mork (1959) on Norway spruce, and Skre (1990, 1991b) on mountain birch, strongly suggest that as long as plants are well supplied with water and

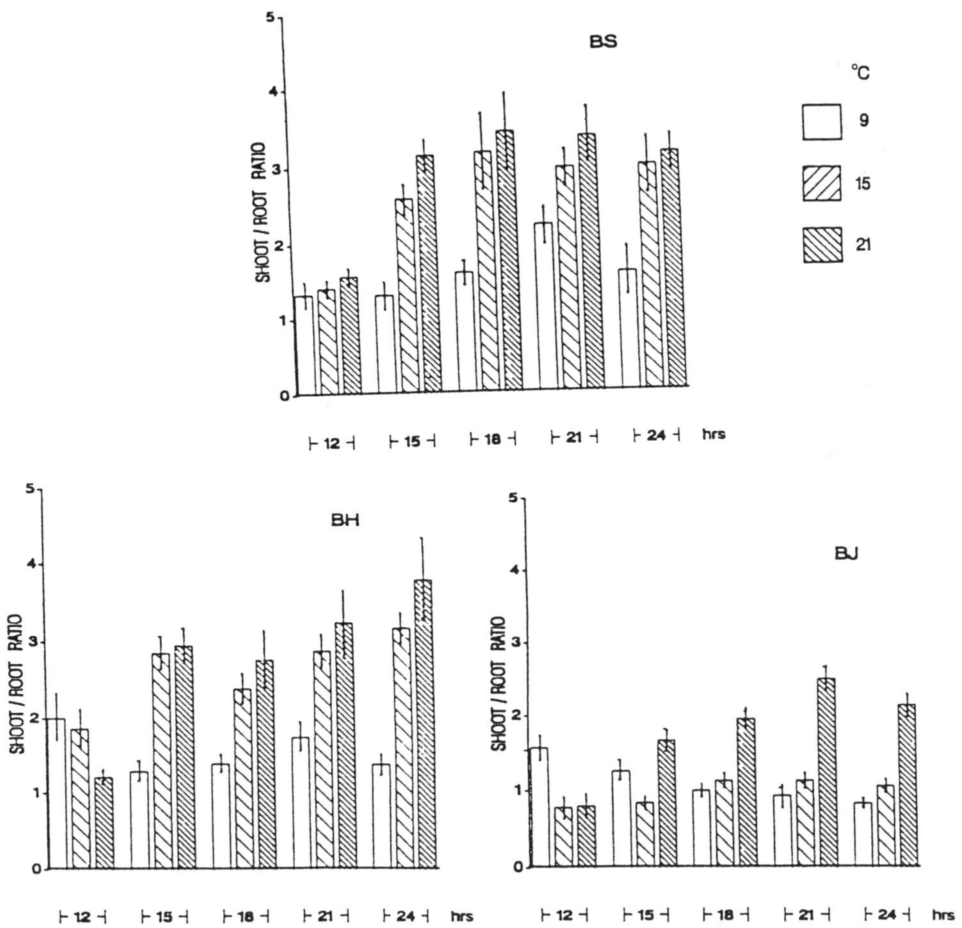

Fig. 11. Shoot-to-root ratio in 6-week old seedlings from three populations of mountain birch (see Figure 5 for explanation of symbols) grown in different temperatures and photoperiods at 500 microEinsteins per square meter per second (μE m^{-2} sec^{-1}) irradiance. Means with ± 2 standard errors, n = 8 (Skre, 1991c).

nutrients, dark respiration is the main limiting factor for growth at low temperatures, and photosynthesis at high temperatures (Figs. 3 and 4). Penning de Vries (1972), who measured the growth of maize embryos on glucose in darkness at low temperatures, found good correlation between growth and respiration measured as uptake of glucose units above a certain maintenance respiration. The existence of an alternative respiration that is not linked to ATP production and growth (Lambers, 1980) further demonstrates that dark respiration limits growth at low temperatures. A low-growth strategy that maintains metabolic activity for completing growth and reproduction cycles is the best way for plants to survive in an arctic climate with low nutrient supply, (Chapin, 1979). One function of alternative respiration may be removal of carbon and reducing power that is not necessary for plant development (Lambers, 1980). Preliminary studies (Skre, 1992) have shown that alternative respiration is more important as a growth-regulating mechanism in arctic ecotypes of mountain birch than in southern ecotypes, and that the proportion of alternative respiration in plant tissue is highest in plants growing at low temperatures and nutrient levels.

Figure 5b indicates that low temperatures increase the carbohydrate level in roots and consequently the carbon storage capacity of plants (Skre, 1990); this tendency is strongest in arctic ecotypes. With a 3 °C increase in mean annual temperatures, equivalent to a 400 meter altitude increase in isotherm (Fig. 5b), carbohydrates will be mobilized in the roots about one month earlier at present tree limits. Global warming is expected to shorten the dormancy period and increase respiration throughout the dark season; it will therefore favor plants that are adapted to a warm climate, grow rapidly (Fig. 5a), and are able to compensate for the expected carbon loss by increasing their photosynthetic capacity.

REFERENCES

Aalvik, G., 1939, Über Assimilation und Atmung einiger Holzgewächse im westnorwegischen Winter (Assimilation and respiration in selected tree species in western Norway in winter), Meddelelser fra Vestlandets forstlige Forsøksstasjon, 6(4):1-266.

Ågren, G. I., 1985, Limits to plant production, Journals of Theoretical Biology, 113:89-92.

Blackman, V. H., 1919, The compound interest law and plant growth, Annals of Botany, 33:353-360.

Chapin, F. S. III, 1979, Nutrient uptake and utilization by tundra plants. In: Underwood, L. S., Tieszen, L. L., Callahan, A. B., and Folk, G. E., eds., Comparative mechanisms of cold adaptation, Academic Press, New York, London, Toronto, 215-234.

Chapin, F. S. III, and Tryon, P. R., 1982, Phosphate absorption and root respiration of different plant growth forms from northern Alaska, Holarctic Ecology, 5:164-171.

Crawford, R. M. M., 1989, Studies in plant survival. In: Andersson, D. J., Greig-Smith, P., and Pitelka, F. A., eds., Studies in Ecology, Vol. 11, Blackwell Scientific Publications, Oxford, 296 p.

Dahl, E., and Mork, E., 1959, Om sambandet mellom temperatur, ånding og vekst hos gran (Picea abies (L.) Karst.) (On the relationship between temperature, respiration, and growth in Norway spruce), Meddelelser fra det norskeskogforsøksvesen, 16:81-93. English summary.

Elkington, T. T., 1968, Introgressive hybridization between Betula nana L. and B. pubescens Ehrh. in northwest Iceland, New Phytologist, 67:109-118.

Hagem, O., 1917, Furuens og granens frøsetning i Norge (Seed production in Scots pine and Norway spruce in Norway), Meddelelser fra Vestlandets forstlige forsøksstasjon, 1(2):1-188.

Hagem, O., 1947, The dry matter increase in coniferous seedlings in winter, Meddelelser fra Vestlandets forstlige forsøksstasjon, 8(1):1-317.

Heikinheimo, O., 1921, Die Waldgrenzwälder Finlands und ihre künftige Nutzung (The timberline forests of Finland and their future use), Metsätietaro Tutkimuslaitos Julkessa, 4(3):1-71.

Heikinheimo, O., 1932, Über die Besamungsfähigkeit der waldbäume (On the fertility of forest trees), Metsätietaro Tutkimuslaitos Julkessa, 17(3):1-61.

Hustich, I., 1944, Några synspunkter på skogsgränserna i nordligste Skandinavien (Some views regarding timberlines in northern Scandinavia), Svenska Skogsvårdsföreningens Tidsskrift, 42:132-141.

Junttila, O., and Heide, O. M., 1981, Shoot and needle growth in Pinus silvestris as related to temperature in Northern Fennoscandia, Forest Science, 27:423-430.

Kallio, P., 1984, The essence of biology in the North, Nordia, 18(2):53-65.

Kallio, P., Niemi, S., Sulkioja, M., and Valanne, T., 1983, The Fennoscandian birch and its evolution in the marginal forest zone. In: Morisset, P., and Payettes, S., eds., Tree-line ecology, Proceedings of the Northern Quèbec Tree-Line Conference in Poste-de-la-Baleine, Northern Quèbec, 1981 June 22–July 1, Centre d'etudes nordique, Laval University, Québec City, Québec, 101-110.

Kozlowski, T. T., and Gentile, A. C., 1958, Respiration of white pine buds in relation to oxygen availability and moisture content, Forest Science, 4:147-152.

Kujala, V., 1927, Untersuchungen über den Bau und die Keimfähigkeit von Kiefern- und Fichtensamen (Investigations on the morphology and germination capacity in seeds of Scots pine and Norway spruce), Metsätietaro Tutkimuslaitos Julkessa, 12(6):1-106.

Lambers, H., 1980, The physiological significance of cyanide-resistant respiration in higher plants, Plant, Cell and Environment, 3:293-302.

Langlet, O., 1960, Mellaneuropeiska gransorter i svensk skogbruk (Central European provenances of Norway spruce in Swedish forestry), Kungliga Skogs-och Lantbruks-akademiens Tidskrift, 99:259-329. English summary.

Mooney, H. A., and Billings, W. D., 1961, Comparative physiological ecology of arctic and alpine populations of Oxyria digyna, Ecological Monographs, 31:1-29.

Mork, E., 1941, Om sambandet mellom temperatur og vekst, Undersøkelser av de daglige variasjoner i granens høydetilvekst (On the relationship between temperature and growth; investigations on the daily variations in shoot elongation in Norway spruce), Meddelelesr fra det norske skogforsøksvesen, 8:1-89.

Parker, J., 1963, Cold resistance in woody plants, Botanical Review, 29:123-205.

Penning de Vries, F. W. T., 1972, Respiration and growth. In: Rees, A. R., Cockshull, K. E., Hand, D. W., and Hurd, J. R., eds., Crop processes in controlled environments, Academic Press, New York, London, 327-347.

Printz, H., 1933, Granenes og furuens fysiologi og geografiske utbredelse (On the physiology and geographical distribution of spruce and Scots pine), Nyt Magasin for Naturvidenskapene, 73:169-219.

Prudhomme, T. I., 1983, Carbon allocation to antiherbivore compounds in a deciduous and an evergreen subarctic shrub species, Oikos, 40:344-356.

Romell, L. G., 1925, Växttidsundersøkningar på tall och gran (Phenological investigations in Scots pine and Norway spruce), Meddelelser fra Statens Skogförsöksanstalt, 22:45-124.

Rutter, A. J., 1957, Studies in the growth of young plant of Pinus sylvestris L., I. The annual cycle of assimilation and growth, Annals of Botany, 21:399-425.

Skre, O., 1972, High temperature demands for growth and development in Norway spruce (Picea abies (L.) Karst.) in Scandinavia, Meldinger fra Norges Landbrukshøgskole, 51(7):129.

Skre, O., 1989, Adaptation to cold conditions and restriction on growth in mountain plants: A literature survey, Blyttia, 47:135-142.

Skre, O., 1990, Consequences of possible climatic temperature changes for plant production and growth in alpine and subalpine areas in Fennoscandia. In: Holten, J., ed., Effects of climate changes on terrestrial ecosystems. Report from a seminar in Trondheim (January 16, 1990), NINA, 18-37.

Skre, O., 1991a, Physiology of plant survival under cold conditions, with particular reference to dark respiration as a factor limiting growth at timberline; A literature review, Meddelelser fra Skogforsk, 44(1):1-34.

Skre, O., 1991b, Temperature responses on the growth and development of mountain birch (Betula pubescens Ehrh.), elm (Ulmus glabra Huds.), and maple (Acer platanoides L.) seedlings in continuous light, Meddelelser fra Skogforsk, 44(5):1-44.

Skre, O., 1991c, Growth experiments with ecotypes of mountain birch (Betula pubescens Ehrh.) and lowland birch (Betula pendula Roth) at varying temperatures, light and daylengths, Meddelelser fra Skogforsk, 44(6):1-41.

Skre, O., 1991d, Chemical composition of birch seedlings grown at varying temperature, light and photoperiod, Meddelelser fra Skogforsk, 44(8):1-28.

Skre, O., 1992, [In press], Nutrient limitation in birch ecotypes and the possible function of alternative respiration as a growth-regulating mechanism.

Skre, O., and Oechel, W. C., 1979, Moss production in a black spruce (Picea mariana) forest with permafrost near Fairbanks, Alaska, as compared with two permafrost-free stands, Holarctic Ecology, 2:249-254.

Sveinbjørnsson, B., 1983, Bioclimate and its effect on the carbon dioxide flux of mountain birch (Betula pubescens Ehrh.) at its altitudinal tree-line in the Torneträsk area, northern Sweden. In: Morisset, P., and Payette, S., eds., Tree-line ecology, Proceedings of the Northern Quèbec Tree-Line Conference in oste-de-la-Balein, Northern Quèbec, June 1981, Centre d'etudes nordiques, Laval University, Quèbec City, Quèbec, 111-122.

Thornley, J. M. H., 1972, A balanced quantitative model for root:shoot ratios in vegetative plants, Annales of Botany, 26:431-441.

Tranquillini, W., 1979, Physiological ecology of the alpine timberline, Tree existence at high latitude with special reference to the European Alps, Ecological Studies 31, Springer Verlag, Berlin, Heidelberg, New-York, 137 p.

Vaarama, A., and Valanne, T., 1973, On the taxonomy, biology and origin of Betula tortuosa Ledeb., Reports from Kevo Subarctic Research Station (Finland), 10:70-84.

Volger, H. B., and Heber, U., 1975, Cryoprotective leaf proteins, Biochemica et Biophysica Acta, 412:335-349.

Vowinchel, T., Oechel, W. C., and Boll, W. G., 1975, The effect of climate on the photosynthesis of Picea mariana at the subarctic line, 1. Field measurements, Canadian Journal of Botany, 53(7):604-620.

Warren-Wilson, J., 1972, Ecological data on dry matter production by plants and plant communities. In: Bradley, E. F., and Denmead, O. T., eds., The Collection and Processing of Field Data, CSIRO Symposium, Interscience Publishers, New York, London, Sydney, 77-123.

West, C., Briggs, G. E., and Kidd, F., 1920, Quantitative analysis of plant growth, New Phytologist, 19:200-207.

Zelitch, J., 1966, Increased rate of net photosynthetic carbon dioxide uptake caused by the inhibition of glycolate oxidase, Plant Physiology, 41:1623-1631.

PERFORMANCE OF MOUNTAIN BIRCH IN DIFFERENT ENVIRONMENTS IN SWEDEN AND ICELAND: IMPLICATIONS FOR AFFORESTATION

Bjartmar Sveinbjörnsson

Department of Biological Sciences
University of Alaska–Anchorage
Anchorage, Alaska

Mats Sonesson, Olle K. Nordell, and Staffan P. Karlsson

Abisko Scientific Research Station
Abisko, Sweden

INTRODUCTION

This paper gives a general overview of factors that limit the northern distribution of trees, and a specific account of those factors that affect the distribution of mountain birch (Betula pubescens Ehrh.) in northern Scandinavia and Iceland.

Treeline is a distinct vegetation boundary. In this paper it refers to the zone of scattered trees beyond the forest border. This zone is often diffuse, with the forest gradually thinning into a forest–tundra ecotone or Kampfzone of shrub-like trees. In the Alps, treeline is often related to human impacts (Tranquillini, 1979)

The transect from forest to non-arboreal vegetation is related to climatic, edaphic, and biological variation. Treeline has been correlated with mean air temperatures for the warmest month or months (Köppen, 1936), number of days with air temperatures above a certain mean (Hopkins, 1959), and the number of degree-days above certain temperature thresholds (Hare, 1968). The forest creates its own microclimate (Benjey, 1974) as a result of surface roughness and canopy characteristics that govern air movement and influence snow distribution patterns, evapotranspiration, and temperature. The forest microclimate affects growth of plants on the forest floor, and often results in significantly different ground vegetation inside and outside the forest. Trees and the ground vegetation produce different litters, both in quantity and quality, which result in different soil characteristics that further affect tree growth.

Tree performance close to treeline is affected by nutrient levels and by the direct effect of physical factors on metabolism. Atmospheric carbon dioxide decreases with elevation,

Forest Development in Cold Climates, Edited by J. Alden *et al.*
Plenum Press, New York, 1993

leading to a 10 to 20 percent reduction in net photosynthesis between 800 to 2,000 meters above sea level in the Alps (Tranquillini, 1979). Because elevational treelines are generally below 1,000 meters in subarctic-boreal areas, the decrease in carbon dioxide concentration is usually less than five percent. There are, however, other possible complications. Soil carbon mineralization decreases with elevation because of decreasing soil temperature, and wind near the ground is greater at high elevations than low elevations, thus decreasing the amount of carbon dioxide available to tree seedlings and saplings at treeline (Bazzaz and Williams, 1991).

Soil minerals are also required for tree growth and reproduction. These might increase or decrease with elevation, depending on the balance between litter input and decomposition. Nutrient content in soils increases with elevation in the Bavarian Alps (Ehrhardt, 1961), but a decrease was reported across the treeline boundary on Mount Fuji in Japan (Masuzawa, 1985). Mycorrhizal associations, important in securing nutrients for trees, decrease with elevation in heaths, perhaps leading to reduced nutrient uptake. Tissue nutrient concentration, however, is generally considered a good indicator of nutrient availability (Rencz, 1976) and is consistently higher in high alpine plants (Körner, 1989) than in plants at lower elevations. This applies to trees as well (Ehrhardt, 1961). Tissue nutrient concentrations of arctic treeline species appear to increase with latitude (Marchenko and Karlov, 1962).

Wind lowers leaf temperature and net photosynthesis, and together with a shorter photosynthetic season, can directly limit growth by causing plants to produce insufficient production of carbohydrates (Vowinckel, 1975). Alternatively, reduced meristematic temperature (Grace et al., 1989) is believed to limit growth respiration and utilization of assimilates for tissue expansion (Dahl and Mork, 1959).

Low soil temperature reduces seedling and sapling root growth and hence the volume of soil explored by roots for nutrient acquisition (Chapin, 1986). Low soil temperature increases the viscosity of water and reduces the flow of transpirational water with dissolved nutrients to root surfaces. Nutrient uptake by roots is temperature dependent, and hence temperature limited, although less so in saplings in the taiga than in more southern sites (Chapin, 1986).

Like plant growth, reproduction also is reduced at distribution boundaries. It is well known that the time interval between years of great seed fall increases with altitude (Allen and Platt, 1990). The time between seed crops in northern forests also increases with latitude. Seed banks could provide for non-mast years, but seeds of most species are viable for only a limited number of years. Seed germination is limited to a few microsites at tree limits and to short periods when temperatures are sufficiently warm and moisture is adequate. Seedling survival is generally low close to treeline; drought and heat kill many seedlings in the summer, and frost heaving dislodges them in winter (Black and Bliss, 1980).

Established trees in marginal forests are adversely affected by both summer and winter conditions. In windy areas such as mountains, blowdown is common, sometimes creating definite patterns such as waves. Blowdown may also hasten the successional replacement of the canopy of one tree species by saplings of another, as, for example, the replacement of paper birch (Betula papyrifera var. neoalaskana (Sarg.) Raup.) by white spruce (Picea glauca (Moench) Voss) in Alaska. Heavy snow can damage both limbs and trunks (Eurola and Huttonen, 1984). Tissue loss is important in evergreens because they store most of their organic and inorganic reserves in their leaves. Tissue loss, although directly caused by winter conditions, is also associated with summer conditions. For example, cuticular layers are thinner at higher elevations than at lower elevations because of shorter summers, predisposing the trees to water loss and needle death (Tranquillini, 1979).

MOUNTAIN BIRCH

Mountain birch is the dominant tree at high latitudes and elevations in Scandinavia, Iceland, and southern Greenland. It varies greatly in growth habit, partly from frequent hybridization and introgression with the shrubby dwarf birch (Betula nana L.) and partly from natural selection in diverse habitats.

Tree growth generally decreases with elevation. Annual height growth (75 to 200 millimeters) at Hallormsstadur and Fnjóskadalur, Iceland is significantly greater than the annual height growth (25 to 70 millimeters) at Nuolja and Stenbacken in Sweden. Allowing for elevation differences, however, tree growth is similar in both Swedish Lapland and Iceland. The maximum potential annual height growth of about 200 millimeters near sea level decreases about 34 millimeters per 100 meters in elevation (Fig. 1).

Under comparable conditions, elevational differences in instantaneous rates of photosynthesis and dark respiration were not significant for either abscised or in situ tree branches at Nuolja and Stenbacken, northern Sweden, and at Mùgilsá and Fnjóskadalur, Iceland (Sveinbjörnsson, 1983; author's unpublished data). Whole-plant carbon balance must be corrected for the surface area and biomass distribution of the plants, however, and this may vary with elevation. Leaf area decreased only slightly with elevation in saplings planted at different elevations in Fnjóskadalur, Iceland. The increase in the number of leaves (Fig. 2) partially compensated for the slight decrease in leaf size. The absolute root surface area or weight is unknown, but root-to-shoot ratios generally increase with environmental severity as is the case, for example, with Alaska white spruce (Abadie, 1991). In addition, extant

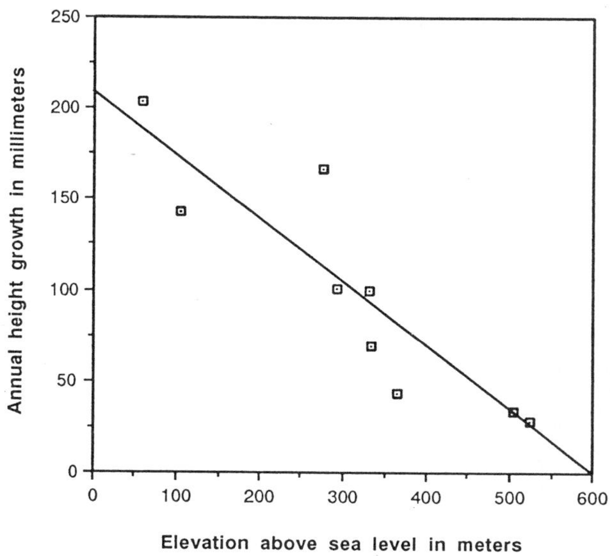

Fig. 1. Annual height growth of mountain birch in Iceland and Swedish Lapland in 1988 in relation to elevation.

temperature conditions may limit carbon capture or use, whereas water relations and irradiance do not vary significantly with elevation. The annual growth of mountain birches near Kevo, Finnish Lapland, was not related to the length of the leafy season (Haukioja et al., 1985).

Application of nitrogen and phosphorus fertilizers on mountain birch in Swedish Lapland (Sveinbjörnsson et al., 1991) demonstrated that while nitrogen greatly stimulated growth, phosphorus had no effect. The growth response to nitrogen remained significant, although diminished, five years after application of fertilizer. The effect increased strongly with elevation and was greatest at treeline despite an increase in foliar nitrogen of the control trees (Sveinbjörnsson et al., 1991). Application of nitrogen fertilizer in Iceland also significantly increased growth of birch; again, the increase was greatest at the treeline and forest limit where the control trees also had the highest foliar nitrogen concentration (Fig. 3).

We conclude that a strong nitrogen deficiency exists in the mountain birch ecosystems of Swedish Lapland and Iceland, and that the deficiency cannot be predicted from the foliar nitrogen concentration. On one hand, nitrogen concentrations were higher in the leaves of Icelandic birch than in the leaves of Swedish birch (Betula pendula Roth), and the Icelandic

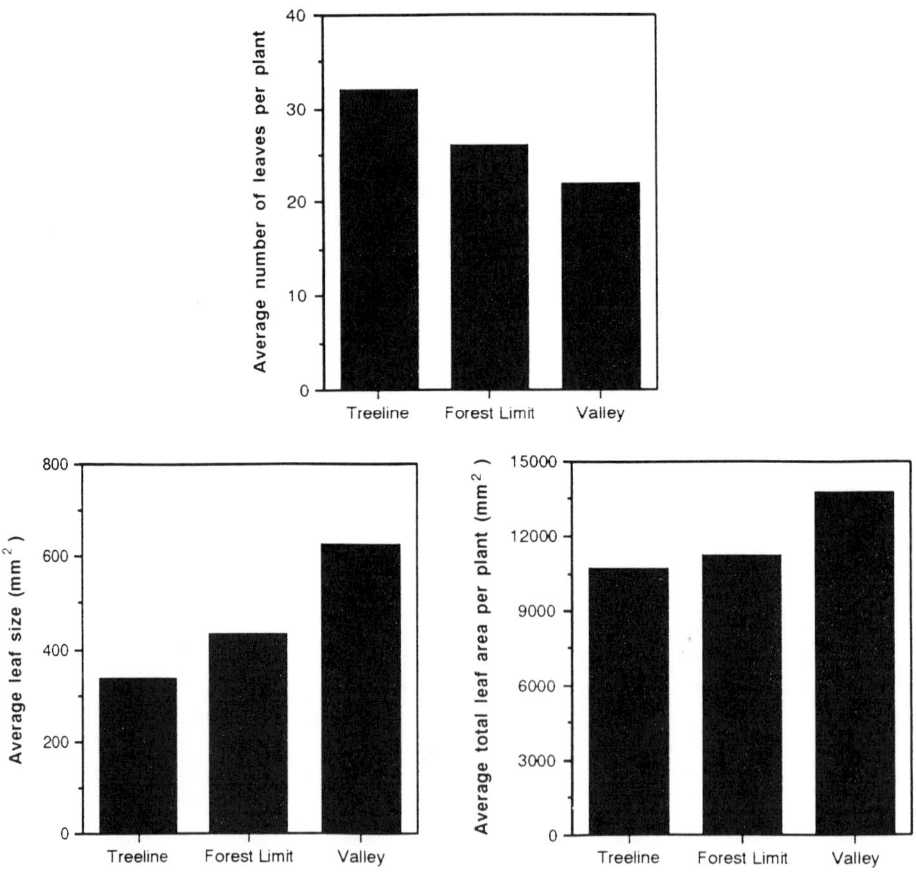

Fig. 2. Leaf area, size, and number per seedling in mountain birch grown for one year in different elevational zones in Fnjóskadalur, northern Iceland.

birches grew faster. Along local elevational gradients, however, tree growth decreased with increasing foliar nitrogen concentrations.

Photosynthetic rate and foliar nitrogen concentrations in Swedish Lapland were positively correlated (Karlsson and Nordell, 1988). The relationship between foliar nitrogen and net photosynthetic rate, however, did not result in an elevational increase in photosynthesis as might be expected, because leaf conductance decreased as foliar nitrogen concentration increased with elevation.

Reproduction of mountain birch is episodic in Swedish Lapland. Recruitment of seedlings at high and low elevations at the end of the 1930s and the beginning of the 1940s was high. Two or three good periods of reproduction have occurred in the Torneträsk area since the early 1940s (Sonesson and Hoogesteger, 1981). Seed production, measured as seed rain, diminished from the valley to the altitudinal forest limit by about two orders of magnitude, whereas reproductive effort, measured as number of trees with female inflorescences, remained relatively constant. Application of a complete fertilizer stimulated male inflorescence, but female inflorescence was little affected by either nitrogen, phosphorus, or complete fertilizers.

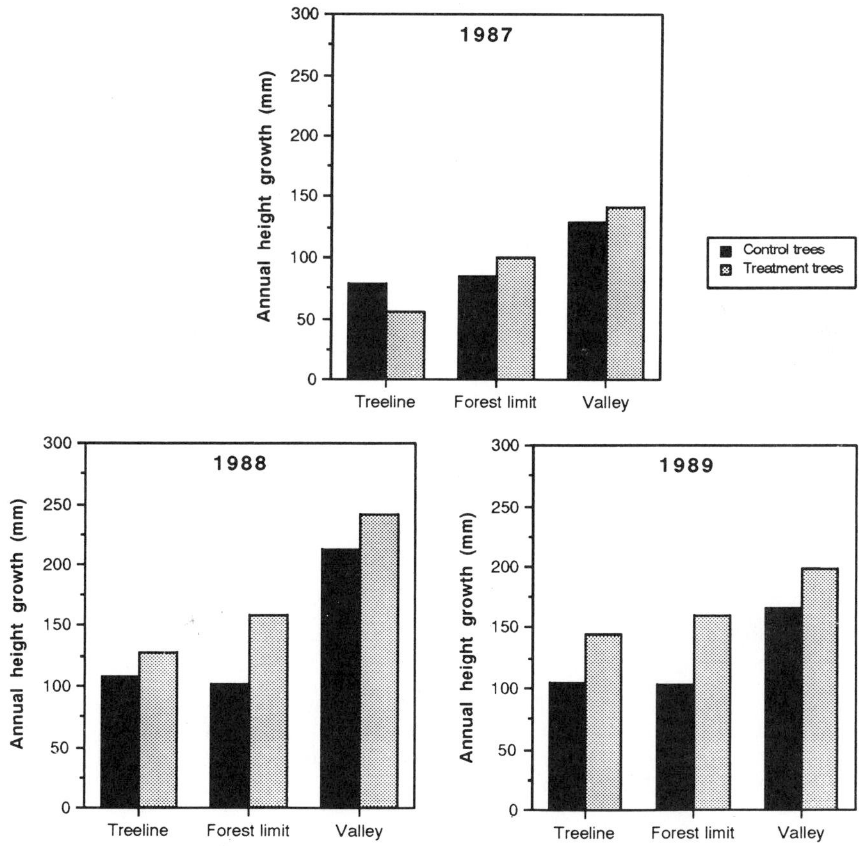

Fig. 3. Annual height growth of mountain birch trees in different elevation zones in Fnjóskadalur, northern Iceland, before fertilization in 1987 and after fertilization in 1988 and in 1989.

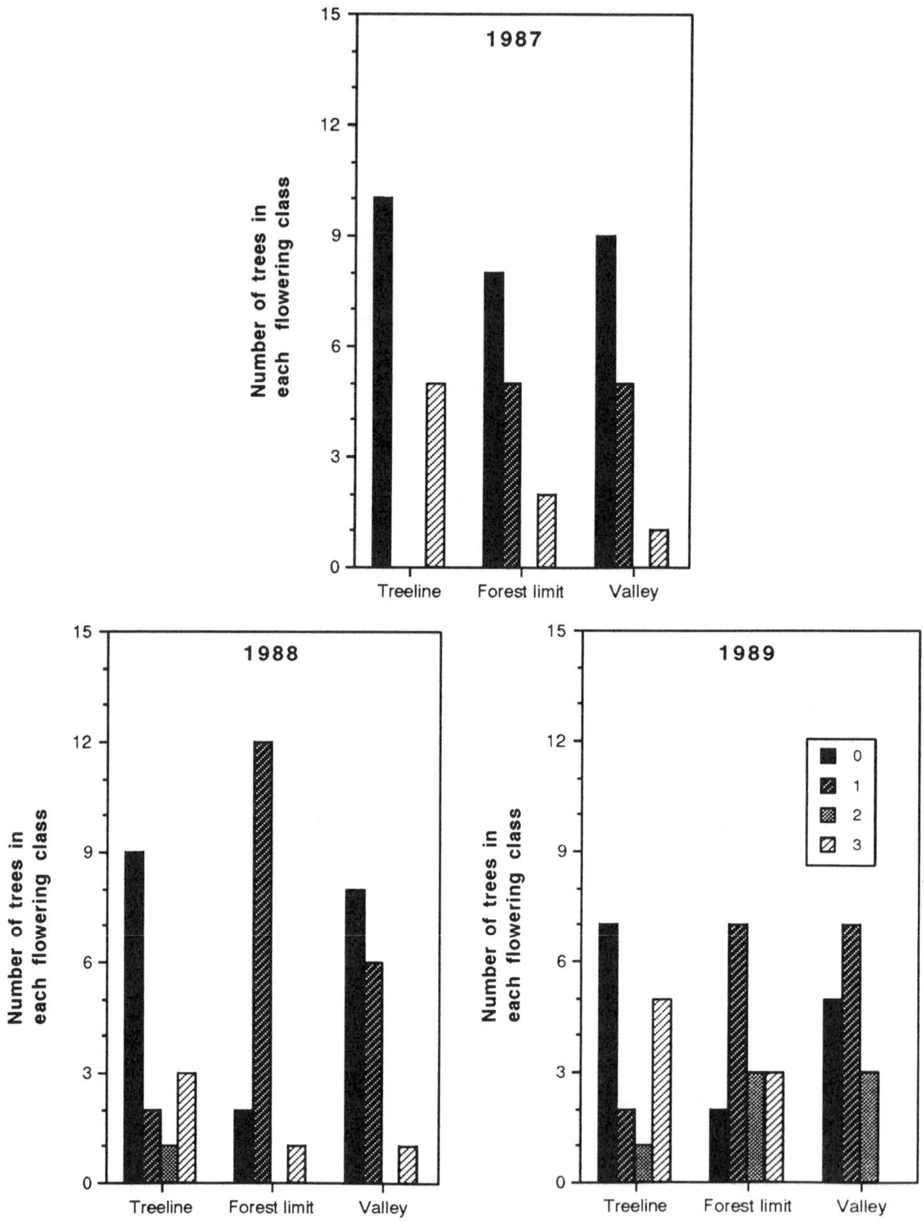

Fig. 4. Flowering frequency (number of trees out of 15 with female flowers) and flowering intensity (flowering classes: 0= no flowers, 1= <10, 2= 10 to 100, and 3= >100 flowers per tree) in different elevation zones in Fnjóskadalur, northern Iceland, before fertilization in 1987 and after fertilization in 1988 and in 1989.

In Fnjóskadalur, Iceland, the number of flower-producing trees decreased with elevation (Fig. 4). Thus, the treeline had more non-flowering trees and fewer high flower-producing trees than did the forest limit and lower elevations. Nitrogen fertilization reduced the number of non-flowering trees at all elevations, but most noticeably at the forest limit. Flowering is often related to growth (Rohmeder, 1967; Gross, 1972), but no relationship between these processes was evident in the trees studied in Iceland.

Damage estimates are only cursory. In Swedish Lapland, where birch forests are defoliated to different degrees at approximately 11-year intervals by the autumn moth (Eppirritera autumnata Bkh., formerly known as Oporinia autumnata Bkh.) (Tenow, 1975), middle elevation forests are most severely affected, because overwintering insect eggs are not exposed to the lethal low temperatures of high and low elevations. Drought, insects (other than E. autumnata), and fungal damage were noted in the birch forest along elevational transects in Sweden and Iceland. At treeline in Iceland where injury was least, about one-third of the saplings sustained top damage, and at the forest limit where injury was most common, two-thirds of the plants sustained top damage.

IMPLICATIONS FOR FORESTRY

Several of these findings have implications for afforestation efforts in Iceland and other northern areas:

1. Because the birch treeline in Swedish Lapland is close to equilibrium with elevation at 550 meters, it follows that the birch treeline in northern and northeastern Iceland also should be about 550 meters above sea level rather than 300 meters as it is at present. A natural upward migration of the birch forest has been observed after fencing to exclude livestock.

2. Tree growth is limited by lack of nitrogen, especially at higher elevations, so application of nitrogen fertilizer would increase growth. Addition of nitrogen and possibly phosphorus will increase flowering, and presumably seed production and forest expansion as well. Research is needed to determine if nutrients other than nitrogen also limit birch growth.

3. Beyond the forest limit, natural regeneration is almost entirely limited to sites where ground vegetation has been disturbed. Because many naturally disturbed sites are found in Iceland, seed bed preparation might not limit afforestation and natural forest regeneration. Where the vegetation cover is intact, surface scarring and sowing of seeds are recommended.

4. Sheep grazing is the greatest threat to birch forests and birch afforestation in Iceland. Preferential grazing of birch at Hallormsstadur demonstrates the destructive force of sheep (Thorsteinsson, 1986). Reduction in sheep numbers or their exclusion from afforestation areas is essential. Introduction and intermixing of tree and shrub species not used by sheep, such as alder (Alnus Mill.), represent a viable strategy for soil protection and nitrogen improvement. Legumes and other plants not used by sheep also could be introduced.

5. An increase in tree and shrub diversity may increase the resistance of forests to pests. Alternative species and provenances adapted to cold climates could replace less shade-, cold-, and pest-resistant species, as they do through natural succession in boreal forests elsewhere. Introduction of mycorrhizal fungi associated with foreign tree species and provenances in wild populations may also stimulate productivity and hasten succession in artificial forests at tree limits.

GLOBAL CHANGE

The concentration of carbon dioxide in the atmosphere is increasing steadily at a rate of two parts per million per year (Gribbin, 1981). The concentration is expected to double by the middle of the next century. Carbon dioxide, a greenhouse gas, is transparent to short wavelength radiant energy but absorbs long wavelength energy re-emitted from the earth. Increasing atmospheric concentration of carbon dioxide is expected to increase global temperature, especially at northern latitudes, where annual temperatures may increase more than 5 °C by 2050 (Manabe and Weatherald, 1980; Seidel and Keyes, 1983). The concentration of carbon dioxide in the atmosphere is currently suboptimal for photosynthesis in C-3 plants, which therefore should respond favorably to short-term increases in atmospheric carbon dioxide. However, the long-term effects of higher carbon dioxide concentrations on photosynthesis and other physiological functions of C-3 trees are unknown. Tree growth at high elevations could be enhanced by the increase in atmospheric carbon dioxide (LaMarche et al., 1984).

Global warming might not affect all areas at high latitudes equally. The effects of cloud cover on the exchange of energy between earth and space are complicated and poorly understood (Wang et al., 1985). Above-normal temperatures after leaf drop by deciduous trees in the autumn would lead to the loss of carbohydrate reserves (Crawford, 1989). If warmer springs were accompanied by greater snowfall, the onset of bud burst would be delayed further, forcing the trees to draw on carbohydrate reserves. Earlier and warmer summers would greatly stimulate tree growth and increase afforestation success at and beyond tree limits.

ACKNOWLEDGEMENTS

This research was funded by the United States National Science Foundation, the Swedish Council on Natural Sciences, the Icelandic Science Foundation, and the Royal Swedish Academy of Sciences.

REFERENCES

Abadie, W. D., 1991, Growth and carbon flux of white spruce at different elevations in the Chugach Mountains, Alaska. M.Sc. Thesis, University of Alaska-Anchorage, 97 p.

Allen, R. B., and Platt, K. H., 1990, Annual seedfall variation in Nothofagus solandri (Fagaceae), Canterbury, New Zealand, Oikos, 57:199-206.

Bazzaz, F. A., and Williams, W. E., 1991, Atmospheric CO_2 concentrations within a mixed forest: Implications for seedling growth, Ecology, 72:12-16.

Benjey, W. G., 1974, Seasonal energy balance trends in the subarctic: the microclimates of broken spruce forest and clearing, Kluane Lake, Yukon, 1970-1971. Ph.D. Thesis, Michigan.

Black, A. R., and Bliss, L. C., 1980, Reproductive ecology of Picea mariana (Mill.) B.S.P. at tree line near Inuvik, Northwest Territories, Canada, Ecol. Monogr., 50:331-354.

Chapin, F. S., III., 1986, Controls over growth and nutrient use by taiga forest trees. In: Van Cleve, K., Chapin F. S., III, Flanagan, P. W., Viereck, L. A., and Dyrness, C. T., eds., Forest ecosystems in the Alaskan taiga, A synthesis of structure and function, Springer-Verlag, New York, 96-111.

Crawford, R. M. M., 1989, Studies in plant survival, Blackwell Scientific Publications, Oxford, 296 p.

Dahl, E., and Mork, E., 1959, Om sambandet mellom temperatur, andning, og vekst hos gran (Picea abies (L.) Karst.), Medd. fra det norske Skogforsoksves., 16:81-93.

Eurola, S., and Huttonen, A., 1984, Riisintunturi, Oulanka Univ. Rep., 5:65-67.

Ehrhardt, F., 1961, Untersuchungen über den Einfluss des Klimas auf die Stickstoffnachlieferung von Waldhumus in verschiedenen Hohenlagen der Tiroler Alpen, Forstwiss. Zentralbl., 80:193-215.

Grace, J., Allen, S. J., and Wilson, C., 1989, Climate and the meristem temperatures of plant communities near the tree line, Oecologia, 79:198-204.

Gribbin, J., 1981, The politics of carbon dioxide, New Sci., 90:82-84.

Gross, H. L., 1972, Crown deterioration and reduced growth associated with excessive seed production by birch, Can. J. Bot., 50:2431-2437.

Hare, F. K., 1968, The Arctic, Q. J. R. Meteorol. Soc., 94:439-459.

Haukioja, E., Niemelä, P., and Sirén, S., 1985, Foliage phenols and nitrogen in relation to growth, insect damage, and ability to recover after defoliation, in the mountain birch Betula pubescens ssp. tortuosa, Oecologia, 65:214-222.

Hopkins, D. M., 1959, Some characteristics of the climate in forest and tundra regions in Alaska, Arctic, 12:214-220.

Karlsson, P. S., and Nordell, K. O., 1988, Intraspecific variation in nitrogen status and photosynthetic capacity within mountain birch populations, Holarct. Ecol. 11:293-297.

Köppen, W., 1936, Das geografische System der Klimate. In: Köppen, W., and Geiger, R., eds., Handbuch der Klimatologie, Bornträge, Berlin, 44 p.

Körner, C., 1989, The nutritional status of plants from high altitudes: A worldwide comparison, Oecologia, 81:379-391.

LaMarche, V. C., Graybill, D. A., Fritts, H. C., and Rose, M. R., 1984, Increasing atmospheric carbon dioxide: Tree-ring evidence for growth enhancement in natural vegetation, Science, 225:1019-1021.

Manabe, S., and Weatherald, R. T., 1975, The effects of doubling the CO_2 concentration on the climate of a general circulation model, J. Atmos. Sci., 32:3-15.

Marchenko, A., and Karlov, Y. M., 1962, Mineral exchange in spruce forests of the northern taiga and forest tundra in the Archangelsk Oblast, Soviet Soil Science, 7:722-734.

Masuzawa, T., 1985, Ecological studies on the timberline of Mt. Fuji, I. Structure of plant community and soil development on the timberline, Bot. Mag. Tokyo, 98:15-28.

Rencz, A., 1976, The level and distribution of biomass and nutrients in a subarctic lichen woodland. M.Sc. thesis, McGill University.

Rohmeder, E., 1967, Beziehung zwischen Frucht--bzw. Samenerzeugung und Holzerzeugung der Waldbäume, Allg. Forstzeitschr., 22:33-39.

Seidel, S., and Keyes, D., 1983, Can we delay greenhouse warming? The effectiveness and feasibility of options to slow build-up of CO_2 in the atmosphere, Strategic Studies Staff, Office of Policy Studies, U. S. Environmental Protection Agency, Washington, D. C. #20460.

Sonesson, M., and Hoogesteger, J., 1981, Recent tree-line dynamics (Betula pubescens Ehr. ssp. tortuosa (Ledeb.) Nyman) in northern Sweden, Nordicana, 47:47-54.

Sveinbjörnsson, B., 1983, Bioclimate and its effect on the carbon dioxide flux of mountain birch (Betula pubescens Ehrh.) at its altitudinal tree-line in the Torneträsk area, northern Sweden, Nordicana, 47:111-122.

Sveinbjörnsson, B., 1991 [See p. 4, "Sveinbjörnsson et al."]

Tenow, O., 1975, Topographical dependence of an outbreak of Oporinia autumnata Bkh. (Lep., Geometridae) in a mountain birch forest in northern Sweden, Zoon, 3:85-110.

Thorsteinsson, I., 1986, The effect of grazing on the stability and development of northern rangelands: a case study of Iceland. In: Gudmundsson, Ó., ed., Grazing research at northern latitudes, Plenum, New York, 37-43.

Tranquillini, W., 1979, Physiological ecology of the alpine timberline. Tree existence at high altitudes with special references to the European Alps, Springer-Verlag, New York, 137.

Vowinckel, T., 1975, The effect of climate on the photosynthesis of Picea mariana at the subarctic tree line. Ph.D. thesis, McGill University.

Wang, W. C., Wuebbles, D. J., and Washington, W. M., 1985, 6. Potential climatic effects of perturbations other than CO_2. In: MacCracken, M. C., and Luther, F. M., eds., Projecting the Climatic Effects of Increasing CO_2, (DOE/ER-0237), Office of Energy Research, Carbon Dioxide Research Division, Washington, D. C., 136-191.

CAUSES OF ALPINE TIMBERLINE: A REVIEW OF THE HYPOTHESES

P. Wardle

Department of Scientific and Industrial Research
Christchurch, New Zealand

INTRODUCTION

Timberline, or the low-temperature limits of forest and tree growth, is one of the most fundamental ecological boundaries. Below timberline, foliage and reproductive structures are successfully produced at the height of tree canopies; above timberline, plants grow only in favoured environments close to the ground. Timberlines are also relatively well-defined, even where altitudinal environmental gradients are gradual. Most other sharp boundaries in nature reflect environmental discontinuities such as shore lines, or disturbances such as forest clearance.

In temperate and arctic regions, the most tangible aspect of the timberline environment is the harshness of winter. Some earlier attempts to determine the causes of timberline emphasized the effects of winter wind in drying out foliage, and of snow in burying plants or breaking stems. It was also noted, however, that forests and trees reach higher elevations in inland regions which have warm summers and cold winters than in coastal regions that experience more equable climates. Much of the effort in timberline research has attempted to reconcile this apparently contradictory evidence for summer versus winter control.

As information on the ecology and physiology of timberline has accrued, it has become evident that explanations are complex, and differ in detail from region to region. Current knowledge is summarized by Tranquillini (1979). This review is not as comprehensive. Rather, it attempts to integrate our present information, with emphasis on insights that timberlines in the Southern Hemisphere may provide.

Phytogeography of Alpine Timberline

At north temperate timberlines, needle-leaved Pinaceae are overwhelmingly dominant. These are drawn from only five genera: the evergreen pines (Pinus L.), firs (Abies Mill.), hemlocks (Tsuga (Endl.) Carr.), and spruces (Picea A. Dietr.), and the deciduous larches (Larix Mill.). Some species and clusters of related species extend over large areas although

Forest Development in Cold Climates, Edited by J. Alden *et al.*
Plenum Press, New York, 1993

such wide-ranging entities are typically variable at the ecotype and provenance level. Junipers (Juniperus L.) also are common at timberline, but mainly as shrubby species. Among deciduous broad-leaved trees, birches (Betula L.) are the most important, especially in arctic Scandinavia. In the European Alps, European beech (Fagus sylvatica L.) reaches timberline in places. North of the tropics, evergreen broad-leaved trees are represented at timberline only in the Himalaya, by an oak (Quercus semecarpifolia Sm.) and Rhododendron L. species, but the latter are shrubs rather than trees.

Timberlines of the tropics and Southern Hemisphere vary greatly in physiognomy and floristic composition, in contrast to the similarities that prevail in northern regions. It is appropriate to begin with New Zealand because more research has been done there than at other southern and tropical timberlines. Through most of the New Zealand archipelago, that is, from latitude 37°30' S. to 46°30' S., the subalpine forests are formed by one of two species of evergreen southern beech (Nothofagus Blume) which ascend above 1,500 meters in warm inland districts, but only to 900 meters in southern coastal districts. In most places, closed-canopy forests of erect trees abut sharply on shorter vegetation, which usually consists of dense scrub or grassland of tall, evergreen tussocks, or a mixture of the two. Only some 200 to 300 meters above the forest limit do these grade into more typically alpine vegetation of short sward and turf communities and barren, rocky terrain.

Nothofagus is absent from some sites, such as valley-head cirques, and here too, scrub and tussock grassland prevail. Much larger districts, notably the western flank of the Southern Alps, also lack Nothofagus; in these, the mid-mountain slopes support mixed, multi-tiered, evergreen forest of southern conifers and broad-leaved trees. This grades upwards into low forest and tall, dense scrub containing smaller conifers and broad-leaved species, including southern heaths (Epacridaceae) and woody Asteraceae. Groves of either of two Hoheria A. Cunn. species (Malvaceae), which are among the few native deciduous trees, occupy debris slopes. At an altitude equivalent to the upper limit of Nothofagus, the communities of tall shrubs or small trees grade, in turn, to shorter scrub and tussock grassland. In the Southern Alps, the mixed mid-slope forests ascend to around 800 meters, and the transition to low scrub and tall grasses occurs around 1,200 meters.

The mountains of Tasmania extend only a little higher than the climatic tree limit, but they support a unique woody vegetation, including more tree and shrub conifers, six of which are endemic. A deciduous southern beech (Nothofagus gunnii (Hook. f.) Oerst.) forms thickets of interlocking stems. There are southern heaths and woody Asteraceae as in New Zealand, as well as Proteaceae. Much of this vegetation has been burnt, and replaced by fire-tolerant plants, notably species of Eucalyptus L'Herit. On mainland Australia, only the mountains of the south east rise to timberline, which lies at 1,850 to 2,000 meters. Here, rather open stands consisting solely of snow gum (Eucalyptus pauciflora Sieber) shelter an understory of tussock grasses and wiry shrubs.

Nothofagus also predominates in the forests of temperate South America, forming timberlines that decline in altitude from 1,650 meters at 40° S. to 450 meters at 56° S. The timberline species lenga (Nothofagus pumilio (Poepp. & Endl.) Krasser) and ñirre (Nothofagus antarctica Oerst.) are deciduous, and in contrast to New Zealand, usually grow as low scrub at their upper limits. In some localities, however, tall, erect araucaria (Araucaria araucana (Mol.) K. Koch) trees tower above low beech canopies. Nothofagus stands at 56° S., with limits set ultimately by land's end against the ocean, are the southernmost forests in the world. In these, the two deciduous species are accompanied by the evergreen Magellanic coihue (Nothofagus betuloides Blume).

In the tropical Andes, dense mixed forests extend from the lowlands into the cloud belt at about 3,000 meters. Above the cloud belt, remnant groves of coloradito (Polylepis sericea Wedd.), a rosaceous tree, survive on mountain slopes, to reported altitudes of 4,900 meters, perhaps the highest reached anywhere by trees of "conventional" growth form. On frosty high plains, there are scattered plants of Espeletia Matis ex Bonpl. species, members of the Asteraceae, in which a stout, unbranched, or sparingly branched stem is crowned by a rosette of large, spreading leaves.

On the high volcanoes of Mexico, rather open forests of Hartweg pine (Pinus hartwegii Lindl.) form a timberline that can exceed 4,000 meters. Andean alder (Alnus acuminatus H.B. & K.), representing another northern hemisphere genus, occurs at high altitudes from Mexico through the Andes to northern Argentina, where it can form on upper forest limit at less than 3,000 meters.

Evergreen Nothofagus species are important in New Guinea but only as mid-altitude stands. The upper forests, which are a mixture of southern conifers and broad-leaved trees, peter out into woodland and shrubland, in which Australasian elements are accompanied by northern genera, including Rhododendron, Vaccinium L. and Prunus L. The better drained parts of extensive depressions and shallow gullies about the altitude of the tree limit support tussock grassland dotted with thick-stemmed tree ferns (Cyathea Smith species). These tree ferns are frost tolerant, and are also favored by the fires that have reduced the upland forests of New Guinea.

In Africa, temperate zone mountains in the north and south are high enough to support a climatic tree limit, but in these summer-dry regions any semblance of natural timberline has long since been removed by fire. In east Africa, on the other hand, the high volcanoes still carry areas of dense forest on their humid flanks. On Mt. Kenya, the upper limits are formed at 3,300 meters by Hagenia abyssinica (Bruce) Gmel. (in the Rosaceae) and tree heath (Erica arborea L.). Above the closed forest, and especially in high, formerly glaciated valleys, tussock grassland is dotted with Lobelia Plum. ex L. and Dendrosenecio (Hedb.) R. Nordenstam species that form giant rosettes. Dendrosenecio keniodendron R.E. & Th. C.E. Fries is a sparingly branched "tree groundsel" that extends above 4,600 meters.

The essential unity among northern timberlines is related to the disposition of the great mountain belts, that link with each other directly or via the arctic. The contrasting diversity of southern and tropical timberlines can be attributed mainly to their isolation on fragments of the ancient Gondwana continent that separated during the late Mesozoic or early during the Cenzoic era. Some southern genera, notably Nothofagus, were certainly present at Gondwanan timberlines. Other genera probably spread from the north, for example, Pinus and Alnus in the Americas, Rhododendron in Asia, and Erica (Tourn.) L. in Africa. Still other high-altitude trees of the southern regions and tropics must have evolved from ancestors that lived at lower altitudes.

Timberlines also exist on oceanic islands such as Hawaii, where māmane (Sophora chrysophylla L.) forms short woodlands up to 3,000 meters on the volcanoes. A plantation of tall North American conifers thriving at the upper limit of S. chrysophylla demonstrates that the altitude and physiognomy of oceanic timberlines is limited by genetic resources of the native flora.

Floristic Diversity of Timberlines

As a rule, warm, moist regions support richer floras than do cold, dry regions, perhaps because harsh environments lead to slower turnover of generations and therefore slower evolution, and because they present an evolutionary test that relatively few organisms can pass. Further, because most high-altitude environments are geologically young, there has not been as much time for floristic diversity to develop.

Timberline regions provide some examples. In the Front Range of Colorado where dry, cold winds prevail and the snow pack is generally thin, the only tree-sized conifers at timberline are Engelmann spruce (*Picea engelmannii* Parry ex Engelm.), subalpine fir (*Abies lasiocarpa* (Hook.) Nutt.), limber pine (*Pinus flexilis* James), and in the south, bristlecone pine (*Pinus aristata* Engelm.). In the Coast Ranges of the Pacific Northwest where the climate is moist and comparatively windless and a deep snow pack provides winter protection, corresponding species are Sitka spruce (*Picea sitchensis* (Bong.) Carr.), *Abies lasiocarpa* and whitebark pine (*Pinus albicaulis* Engelm.), but these are accompanied by mountain hemlock (*Tsuga mertensiana* (Bong.) Carr.) and—only just reaching timberline—Alaska-cedar (*Chamaecyparis nootkatensis* (D. Don) Spach) and Pacific silver fir (*Abies amabilis* Dougl. ex Forbes).

The mountain beech (*Nothofagus solandri* (Hook. f.) Oerst.) forests that reach timberline to the east of the main divide in the South Island of New Zealand are often almost devoid of vascular species other than the canopy dominant. Silver beech (*N. menziesii* (Hook. f.) Oerst.) forests on more western ranges experience much higher precipitation, humidity, and cloud cover, and usually have well developed understories with many species. In the districts where *Nothofagus* is absent, the mixed forest and scrub that occurs instead is also richer in species in milder, wetter areas. For example, the small trees of *Dracophyllum traversii* Hook. f. (in the Epacridaceae) and *Olearia lacunosa* Hook. f. (Asteraceae) are abundant on the western flanks of the Southern Alps, but rare on the eastern flanks.

In the northern Andes and on the East African volcanoes, high altitudes are characterised by low precipitation and almost nightly severe frosts; few tree species exist there, although these few include rosette trees that are a unique evolutionary response to the exacting climate. The mountains of New Guinea, in contrast, are wet and cloudy to their summits, and the uppermost forests, woodlands, and shrublands are comparatively rich in species.

The patterns of floristic diversity have implications for afforestation in cold climates. The main one is that in continental regions with low precipitation and harsh winters, few tree species, whether native or exotic, are available for afforestation; where environments are moister and milder, the choice expands considerably.

Timberline in Relation to the Growing Season

A broad correlation between the altitude of north temperate timberlines and the isotherm of 10 °C for the mean temperature of the warmest month of the year, usually July, is well established. Zotov (1938) demonstrated the same correlation in New Zealand, except that in the Southern Hemisphere the warmest month is usually January. The figure of 10 °C must be only an approximation, however, because the climatic data points are mostly extrapolated from distant climate stations, and the components of timberline (forest limit, tree limit, or tree-species limit, and whether these limits are extremes or averages) tend not to be clearly or uniformly stated. Moreover, a correlation of this kind cannot allow for microclimatic variation, the enhanced temperatures often attained by plant tissues, or differing ecological potential among species.

Many attempts have been made to devise more sensitive correlations between timberline and climate. For instance, in the east Asian tropics, Ohsawa (1990) showed that the upper limit of broad-leaved evergreen forests coincides with both annual and monthly mean temperatures of 6 to 7 °C, because of the non-seasonal temperature climate. In Japan, the 10 °C isotherm for the warmest month corresponds, not with the upper limit of forest, but with that of dwarf Siberian pine (Pinus pumila Regel) scrub (Yanagimachi and Ohmori, 1991). In Tierra del Fuego, the Nothofagus timberline coincides with a January isotherm of only 6 °C (Tukhanen, in press), which probably reflects frequent Antarctic squalls. Despite greater precision in estimating ambient temperatures and altitudes of timberlines, it is unlikely that there is any precise formula that universally relates altitude of timberline to temperature.

The rise of both summer isotherms and timberlines from coastal districts towards continental interiors is clearly related to increasing continentality, in which summer and winter temperatures increasingly diverge. Added to this is the trend for summer, winter, and mean annual isotherms to lie at higher altitudes, as the average elevation of land masses increases. These effects can be demonstrated by comparing the maximum altitude of timberline in the European Alps, about 2,000 meters, with the 3,500 meters attained at a similar latitude in the Rocky Mountains. On oceanic islands, at the other end of the gradient, altitudinal vegetation belts are extremely depressed.

Along the gradient from the arctic zone to the tropics, the temperature differences between summer and winter narrow. At equatorial latitudes, such seasonal temperature differences are small and relate more to contrast between wet and dry seasons than to solar elevation; dry seasons have less cloud cover, and because of that, greater diurnal temperature range.

At an experimental area in New Zealand, the upper limit of Nothofagus solandri coincides with a mean January temperature of 10 °C, whereas several exotic species are growing vigorously at an altitude where the January mean is only 8.5 °C. Evidently Nothofagus and, by implication, other native trees do not match the climatic capacity of species from high altitudes in the North Temperate Zone (McCracken et al., 1985). The ability of the latter to exceed the 10 °C January isotherm in New Zealand reflects a growing season that is cooler but longer than at comparable altitudes on northern continents (Benecke et al., 1981).

Carbon Assimilation and Growth

In both seedlings and maturing trees, height and diameter growth decrease with increasing altitude. The onset of bud burst and shoot extension is progressively delayed, and the onset of winter inactivity is hastened. The ability to photosynthesize and accumulate carbon also decreases, largely because of the decreasing length of the season when air and leaf temperatures are high enough to permit a rate of photosynthesis that compensates the annual loss from respiration. Moreover, with increasing altitude, this period is increasingly subject to interruption by cold snaps that tend to depress photosynthesis for some time. In seedlings, deep, late-lying snow can delay the beginning of photosynthesis and growth in the spring.

Timberline could represent the altitude at which the annual carbon balance no longer permits development of the arborescent habit (Boysen-Jensen, 1949). The argument for this hypothesis is that development of a tall, stout trunk represents a demand on carbon gain that cannot be met above the climatic timberline. Low-growing plants do not have this demand on the carbon balance, and can therefore grow at greater altitudes. Although some studies have indicated a precarious carbon balance in timberline trees (for example, Schulze et al., 1967, for Pinus aristata), others have demonstrated a strongly positive carbon balance right up to timberline (for example, Tranquillini, 1959, for Swiss stone pine (P. cembra L.).

Comparison of growth rates among tree species also suggests that timberlines do not universally indicate the highest altitude at which trees can achieve a positive carbon balance. Although some species increase extremely slowly in height and diameter at timberline, others grow vigorously, at least during the sapling and young tree-stages. Such contrasting species can grow in close proximity, both at timberline and at lower altitudes. Nothofagus solandri with two to 17 growth rings per centimeter of radius, and the podocarp Halocarpus biformis (Hook.) Quinn with 25 to 50 rings are New Zealand examples (Wardle, 1963). Although average diameter growth rates in Nothofagus decrease with increasing altitude, trees at the upper forest edge can have wide growth rings because of less competition from neighbouring trees. Grace and Norton (1990) ascribe wide rings in Scots pine (Pinus sylvestris L.) at the upper forest edge in Scotland to better insolation and warming of tissues.

Winter Desiccation of Shoots

One of the most striking aspects of northern timberlines is the winter discoloration and drying of needles and shoots; the most widely accepted explanation of timberline is that growing seasons are too short and too cool for shoots to fully develop and harden against winter conditions. For some regions and some species, such as Pinus cembra in the European Alps, this explanation seems well substantiated. Severe, annually recurring die-back results in a distinct "kampfgürtel," or belt of krummholz, at the level where the balance between growth and die-back are critical.

Erect, well-shaped conifers below the forest limit enter winter with well-developed needles that maintain a healthy water balance throughout winter, even while the soil and trunks are frozen. Their water balance is maintained because closed stomata and well-developed leaf cuticles and wax layers provide a nearly perfect seal against transpiration. Drying can affect trees below the forest limit, but only during occasional winters. In contrast, needles of stunted plants above the forest limit, especially those on the windward sides of plants or exposed shoots, are likely to be small and yellowish, with thin cuticles. They enter winter with unseasonably high water content, but rapidly dry out and die once soil and stems freeze. Feedback between inadequate growing season and repeated winter damage heightens the contrast between krummholz plants and the erect trees below.

Winter desiccation is not confined to windy situations; indeed, desiccation injury is evident at some sheltered timberlines, where comparative calm is reflected in symmetrical "islands" of krummholz. Because theory and experiment indicate that high wind speeds should not directly cause high rates of evapotranspiration, windward orientation of desiccation injury and deformation must be caused either by other effects of strong winter winds, or by wind-cooling during the growing season that predisposes exposed foliage to winter injury.

The Role of Frost

Rapidly-growing young shoots are vulnerable to summer frosts, which even in hardy timberline species of Picea, Abies, and Larix need to fall only to -1 to -2 °C to inflict damage. In New Zealand, unseasonably cold weather has occasionally caused frost damage to young shoots of Nothofagus throughout the canopy. Episodic frosting of canopies probably keeps Nothofagus well below timberline in New Guinea. Until recently, however, frost injury has not been proposed as a factor involved in setting timberlines, because it occurs intermittently over a wide range of altitudes below timberline, where recovery is usually complete.

Freezing injury is most likely to affect small seedlings in the open, because frosts are most severe within a shallow layer of temperature inversion. In New Zealand, both the upper limit of Nothofagus forest and the "inverted" timberline against valley floors coincide with levels of frost intensity that exceed the tolerance of Nothofagus seedlings (Wardle, 1985). Frost damage to newly flushed shoots also contributes to the poor performance of Larix and Picea seedlings planted above timberline in New Zealand; this contrasts with infrequent damage to seedlings of lodgepole pine (Pinus contorta Doug. ex Loud.), a species that grows vigorously at high altitudes in New Zealand even though it seldom reaches timberline in its native North America.

As shoots mature, their cold tolerance increases, and falling temperatures during autumn induce further increases in hardiness, to levels that are normally more than adequate to withstand winter cold. Nothofagus solandri saplings, however, like those of many temperate broad-leaved trees, are apt to produce a late summer flush of growth that is vulnerable to autumn frost. The effect of the frost may show up later as apparent winter desiccation (McCracken et al., 1985). Mature trees do not produce late summer growth, and usually survive winter unscathed. In a few winters, however, the foliage of marginal trees becomes brown and dry; the reason seems to be episodes of strong, very cold wind (Baylis, 1959). In trees and young plants bordering open, frosty depressions, the foliage within a meter or so of the ground is also frosted during some winters.

The snow pack at New Zealand timberlines is generally less than 20 centimeters, but where snow drifts into the timberline to depths of a meter or more, marginal Nothofagus trees develop several stunted stems instead of a single, erect trunk, and shoots that project above the snow pack are subject to repeated winter die-back. This vulnerability is reinforced from year to year, as winter-damaged shoots tend to commence growth late and lack hardiness when autumn frosts begin. Similar die-back can contribute to stunting of Nothofagus on windy ridges; even in the absence of die-back, however, wind-exposed trees can become low and wind-trained, probably because assimilation and growth of shoots on the windward side are reduced through cooling.

Where the deciduous Fagus sylvatica reaches timberline in the European Alps, the uppermost plants are stunted and subject to winter die-back of immature shoots. Tranquillini and Plank (1989) suggest that the primary cause is freezing damage.

In Picea engelmannii, needles of krummholz shoots undergoing winter desiccation are often sharply differentiated into proximal green, turgid portions, and distal dead, dehydrated portions; green segments can even occur distally to dry segments. Similar patterns develop in overwintering needles of other timberline conifers. Seemingly identical patterns were obtained by experimental freezing of P. engelmannii needles, which suggests that, at least in some circumstances, visible winter desiccation may have been preceded by lethal freezing

(Wardle, 1981). Perkins et al. (1991) suggest that winter desiccation in red spruce (Picea rubens Sarg.) also follows low temperature damage, perhaps after shoots have been dehardened by unseasonal warmth.

Effects of Snow and Ice

At north temperate and Andean timberlines, deep drifts may persist in hollows well beyond the summer solstice, and reduce the growing period until it is too short for tree seedlings to survive. Conditions beneath isothermal snow packs in spring encourage fungi that attack tree foliage. In very snowy regions such as the Coast Ranges of British Columbia, variations in the time that the snow pack melts are directly related to the distribution of forest and more tolerant herbaceous vegetation. In New Zealand, in contrast, the snow pack retreats above timberline as early as September or October, long before Nothofagus begins to flush. At tropical timberlines, snow falls are light, and seldom persist through a day, let alone form an enduring snow pack.

In a benign role, snow protects foliage from winter desiccation. This is particularly evident in krummholz of needle-leaved conifers just after the snow pack has melted. Within the protected zone, foliage that has not been buried too long emerges green and healthy. Stems that overwintered above the snow pack may bear twigs only on their leeward side, and the twigs may bear live needles only on their lee. On very exposed sites, all of the foliage above the snow pack may be killed.

Above the snow pack, bark and dead stems are abraded by wind-driven ice particles, although sand grains may be responsible for much of the abrasion attributed to ice. Abrasion of leaf cuticles by ice particles, thereby weakening the barriers to winter desiccation, has been reported by some workers (Hadley and Smith, 1983), but questioned by others (Turner, 1968). Shoots rubbing against each other may also damage the cuticle and wax layer of leaves (Grace and James, in press). Apparent impact lesions on needles of Picea engelmannii during winter were initiated during the preceding summer (Wardle, 1968), possibly by sap-sucking insects.

Sparse foliage borne on stunted stems projecting above the snow pack becomes warmer than the surrounding air through daytime reflection of radiation from the snow surface, which leads to cuticular water loss and desiccation (Baig and Tranquillini, 1980). At a Nothofagus solandri timberline in New Zealand, daytime warming caused daily changes in the water stress of shoots of krummholz plants above the snow pack, but values approached lethal levels only in shoots that had been damaged by autumn frost (McCracken et al., 1985).

In a related study, I could not find significant abrasion among krummholz plants of Nothofagus solandri. Moreover, desiccation of plants shaded through winter by a high bluff was as severe as desiccation of plants in full winter sunshine (Wardle, 1985). On calm evenings, temperatures above the snow pack fell very rapidly because of radiation to a clear sky. Such rapid heat loss seems likely to lead to supercooling of water in leaf tissues. Should a strong wind rise, the impact of ice particles might shock supercooled cell contents into sudden freezing, thereby irreversibly damaging cell membranes and leading to death and desiccation. The idea has not been tested, but if substantiated, could explain some puzzling features about winter desiccation at timberline.

Heavy snowfall, especially if wet, can cause havoc below the forest limit by breaking crowns and limbs; this can give entry to pathogens and provide fuel for fire. On the western side of the Japan Alps, intolerance of heavy snow is regarded as the primary factor in setting the upper limits of erect coniferous trees against extensive scrub of _Pinus pumila_ or broad-leaved hardwoods.

The Appalachian Mountains are notorious for glaze storms that strip leaves and shoots from trees, as rain and fog freeze on contact with their foliage. According to Marchand (1980), such loss of leaves and shoots at timberline creates a critical carbon imbalance.

Aspect and Timberline

Timberlines tend to be higher on southern aspects than on northern aspects in the Northern Hemisphere, with the reverse applying in the Southern Hemisphere. If higher timberlines on sunny aspects result directly from greater input of solar energy, the difference between northern and southern aspects should increase with increasing latitude. On oceanic and tropical mountains east-facing slopes are likely to be sunnier than west-facing slopes, because of orographic cloud formation during the afternoon.

Direct effects of insolation on the altitude of timberline, however, tend to be obscured by other aspect differences, especially the duration of snow cover and exposure to wind. In New Zealand, _Nothofagus_ timberlines reach maximum altitudes on spurs and ridges irrespective of aspect. Because the only microclimatic parameter likely to match this pattern is the distribution of minimum temperatures on clear, calm nights, this is in keeping with the proposition that the length of the frost-free season ultimately determines the altitude of _Nothofagus_ timberlines.

Roots and Mycorrhizae

Root extension rates and root-to-shoot ratios both decrease with decreasing soil temperature. From around 5 °C, low soil temperatures also begin to restrict photosynthesis, probably by reducing water uptake and thereby causing stomatal closure. Soil temperatures at most timberlines are likely to be well above limiting levels through the growing season, and generally higher immediately above the forest limit than within closed forest. On the other hand, tree limit is higher than usual near hot springs in Alaska (B. Sveinbjörnsson, Department of Biological Science, University of Alaska, personal communication). Low temperature combined with poor aeration must severely inhibit root functioning and be an important factor in the treelessness of wet areas near timberline.

Many mountains in the humid tropics and oceanic temperate localities have a very high incidence of cloud cover, for example, around 90 percent of afternoons on the western flanks of the Southern Alps in New Zealand. The wetness and reduced sunshine should, on the one hand, decrease mineralization of nutrients and increase the rate of leaching, and on the other, reduce transpiration and the rate of nutrient uptake. These effects slow growth and contribute to the low stature of trees toward some upper forest limits.

The most important timberline genera and families possess ectotrophic mycorrhizae. Moser (1966) suggested that ectotrophic mycorrhizae are more effective than endotrophic mycorrhizae at extracting nutrients from the slowly decaying, acid humus ("mor") that is characteristic of forests growing in cold climates. In New Zealand, the native _Nothofagus_ and introduced Pinaceae, both of which bear ectotrophic mycorrhizae, achieve far greater growth rates and stature at timberline than the endomycorrhizal genera that form low, diffuse timberlines in the absence of _Nothofagus_.

The altitude of timberline is usually thought to be unaffected by soil nutrient status (Tranquillini, 1979), but Sveinbjörnsson et al. (in press) found that raising the level of available nitrogen in the soil increased growth and seed production of Betula at tree limit.

Critical Stages of Tree Growth

As the upper limit of trees is approached, the quantity and percent viability of seed and the frequency of seed years decrease. Nevertheless, even at the uppermost limit of tall trees, enough seeds and seedlings are produced to maintain stands, at least in the absence of catastrophic destruction or adverse influences such as severe grazing pressure. In krummholz belts above the erect-tree limit, stems rarely bear reproductive structures, let alone viable seed. In the timberline environment, however, where strong winds can carry seed long distances, a lack of seed is unlikely to be an absolute barrier to seedling establishment. Moreover, some species benefit from transport by mammals and birds.

Nothofagus seed will germinate freely above timberline, but the interval between germination and onset of autumn frosts decreases with increasing altitude. The seedlings are relatively small and delicate, and rarely occur within dense herbaceous and shrubby vegetation. They appear more frequently on patches of bare ground in the open, but nearly all succumb by the end of their first winter from frost, heating of the soil surface, water stress, and herbivorous insects. Seedlings surviving longer than the first winter usually remain unthrifty, partly because they seldom develop effective mycorrhizal infections when situated beyond the rhizospheres of existing Nothofagus trees. If a seedling becomes established above the natural timberline, its shoots will be subjected to severe winter desiccation once it grows above the height of surrounding shelter.

Within the forest, seedlings are protected from these hazards. Seeds germinate and seedlings persist many years in quite heavy shade, although they scarcely grow. Once the canopy opens, shaded seedlings are released. In New Zealand, significant areas of Nothofagus krummholz occur only where the forest limit is locally depressed because the environment at potential canopy height is more adverse for Nothofagus than the environment close to the ground; in contrast to the usual situation at the Nothofagus timberline, seedlings become established but do not attain tree size.

Pinus contorta seeds are much heavier than those of Nothofagus solandri, and produce robust seedlings. These seedlings grow more rapidly in full sunlight than in shade, are scarcely affected by the frosts of the New Zealand winter, and strong taproots enable them to avoid drought and frost heaving by needle ice. P. contorta plants develop into erect trees for 150 meters above the natural limits of Nothofagus solandri, and stunted plants subject to winter dieback extend at least 400 meters above. Unless controlled, P. contorta wildings are likely to form a new timberline at much higher altitudes than those formed by native species. In North America, P. contorta scarcely reaches timberline, in contrast to pines such as P. aristata that become massive trees among timberline krummholz of Picea and Abies.

Seedlings of Picea engelmannii are chlorotic in full sunlight, and are susceptible to early summer frosts, both in New Zealand and in their native Rocky Mountains. In the Rockies, however, P. engelmannii, together with Abies lasiocarpa, forms most of the timberline, and both species ascend as islands of low krummholz to the extreme limits of woody vegetation. Seedlings that give rise to krummholz islands presumably become established in micro-habitats that provide a precise combination of shade, shelter, and winter snow cover. Once established, the plants spread through layering of lower branches, and are potentially immortal. In wind-swept locations, each island traps its own protective mound of winter snow. On the windward side, the snow is apt to blow away, exposing the plant to desiccation and death, whereas snow accumulating in the lee of the plant allows shoots to extend in that direction. In this way, islands of low krummholz "migrate" across tundra landscapes.

Regeneration and the Role of Disturbance

Tree stands near timberline are usually wind firm, but many are vulnerable to avalanches. Snow and wind can break off large branches. Effects of pathogens range from local damage to widespread death of stands. Fire is another hazard, especially at forest margins that abut flammable scrub or grassland.

Partial canopy loss is usually recovered by canopy spread, vegetative reproduction, or seedlings, but catastrophes such as fire, landslides, and volcanic eruptions often destroy the forest environment and provide open sites that can be as severe as environments above timberline. Moreover, because seed sources for colonization of denuded sites are more distant, the comparatively poor seed production near timberline becomes critical for forest regeneration, and species with efficient dispersal are favored. Under these circumstances, forest is likely to spread slowly from surviving forest margins and isolated trees, and until mutual protection develops, the colonists are likely to form sparse, stunted stands. The time lag may allow denuded surfaces to be clothed by shorter vegetation that resists entry of seedlings of the original forest dominants. For instance, in New Zealand large areas of subalpine *Nothofagus* forest, burnt since the arrival of humans about 1,000 years ago, have not been reoccupied by *Nothofagus*. Instead, vegetation similar to that in other *Nothofagus*-free areas, that is, low mixed forest grading into mixed scrub which in turn merges into tussock grassland, has developed.

Many areas supporting low forest and scrub without *Nothofagus* can be accounted for in terms not of fire but of local climate. The best examples are patterns of cold air drainage and temperature inversion that exclude *Nothofagus* from formerly glaciated hanging valleys. For the major *Nothofagus*-free districts in New Zealand, however, there are no obvious explanations in terms of either fire or present environment. Usually they are explained in terms of slow spread of *Nothofagus* on to landscapes that lost their forest during major glaciations, and were then preempted by other forest and scrub communities during subsequent warming. The distribution of beech forest in the central North Island is partly explicable in terms of vegetation response to cataclysmic volcanic eruptions.

Natural Timberlines—Abrupt or Gradual?

An ongoing debate among timberline ecologists is whether abrupt timberlines can be "natural" or merely represent the loss of a "kampfgürtel" in which trees become shorter and sparser with increasing altitude. In New Zealand, the upper limit of undisturbed *Nothofagus* forest is normally abrupt. Although there can be scattered young or stunted plants for a few meters above, occurrence of krummholz signifies depression of the forest limit. The position of the timberline often coincides with a change from steep to gentle slopes, which results in sluggish drainage of water and cold night air, but sharp timberlines also cross apparently uniform slopes.

It is probably more usual for trees to become gradually sparser and/or more stunted where undisturbed timberlines lie across gradual environmental gradients. The prevalence of abrupt *Nothofagus* timberlines in New Zealand seems exceptional, although *N. pumilio* locally forms abrupt timberlines in the southern Andes. In most of Eurasia and in northern

Andean regions, millennia of intensive human occupation have greatly affected the configuration of timberline. Even in the Australian Alps, the rather sharp upper limit of Eucalyptus woodland is possibly anthropogenic because the flammability of Australian vegetation has made up for the low density of people.

Nevertheless, special environmental circumstances often give rise to sharp forest limits with little or no krummholz above. On the Front Range in Colorado, where westerly winds prevail, closed Picea-Abies forest ascends to an abrupt limit on the western flanks in places where trees downwind shelter those at the forest margin; only a few stunted krummholz plants grow in the windswept tundra beyond. On the eastern flanks, the wind blows down into the timberline, producing a wide krummholz belt through complex interactions between wind, snow cover, and vegetation.

A General Theory of Timberline

It is well established that timberlines on the main land masses in north temperate regions tend to correlate with a certain level of summer warmth, although the correlation may not be exact. In New Zealand, the upper limits of native tall woody species coincide quite closely, indicating that they are governed by the same environmental factors which, again, appear to correlate with summer warmth. Yet exotic species, especially needle-leaved conifers from the North Temperate Zone, can be successfully sown or planted well above the natural New Zealand timberline. The most likely explanation lies in the differences in low-temperature tolerance (Sakai et al., 1981).

Frost-tolerance has two broad components: the ability of cells to undergo supercooling without lethal ice formation, and the facility for water to move from cells into intercellular or inter-organ spaces where it can freeze harmlessly (Sakai and Larcher, 1987). It is likely that all timberline species have some ability to supercool without damage. In continental, temperate, and Arctic regions, tissues of timberline species are frozen most of the winter, and even in the oceanic climate of New Zealand the twigs and the outer layers of trunks freeze at times. The almost unlimited cold tolerance shown by many conifers and deciduous hardwoods of high northern latitudes depends on avoiding intracellular freezing through intense dehydration of cells. Timberline trees probably have two orders of cold-hardiness, one to withstand the relatively mild winters of south-temperate timberlines and unseasonal frosts during growing seasons, and the other, presumably evolved over a far longer time span, to withstand extreme winter cold at northern continental and Arctic timberlines.

Despite the complexities, a useful hypothesis can be formulated for nearly all Temperate Zone timberlines (Fig. 1). First and foremost, growing seasons must be long and warm enough for trees to develop shoots to maturity and harden them against winter adversity. This emphasizes the time needed to complete morphological and physiological development, rather than the quantity of carbon fixed or biomass added. Nevertheless, to maintain the forest limit, trees must accumulate sufficient carbon to offset the winter period when negative carbon balance and loss of tissues through abscission and injury are likely, and to provide a surplus of assimilates for reproduction.

For species to reach tree size, the appropriate growing conditions must prevail in the "free" air at the height of tree canopies. Above the limit of erect trees, plants growing close to the ground can benefit from shelter by topography and snow, and from solar warming of the soil surface and plant tissues. That environment is exploited by both shrubs and stunted trees, although the species that succeed there must be able to tolerate adverse factors such as frost and strong insolation.

Fig. 1. Factors that determine or influence the upper limits of trees of forests.

Plants at tropical timberlines must withstand daily freeze-thaw cycles that are of great amplitude in some localities. Rosette trees are the most striking evolutionary response to such conditions; Dendrosenecio keniodendron is the most studied of these species. During daytime, its leaf rosettes act as heat-absorbing parabolas. At night, the mature leaves close over the meristematic leaves and protect them from freezing, while water in the mature leaves freezes extra-cellularly (Beck et al., 1984). Stems are clad in an insulating coat of dead leaves.

Trees and shrubs of normal architecture prevail on tropical mountains where slopes are steeper and less frost-prone, and at lower altitudes. Towards the upper limit of tree and shrub growth in New Guinea, stems die back at apparently irregular intervals, possibly because they fail to maintain a positive carbon balance, and are replaced by epicormic shoots (Wardle, 1971).

Recent decades have been characterized by significant warming, and occurrence of healthy seedlings above the forest limit may indicate that timberlines have begun to rise accordingly. This implies that tree establishment trials may now succeed at altitudes where tree growth was formerly impossible.

REFERENCES

Baig, M. N., and Tranquillini, W., 1980, The effects of wind and temperature on cuticular transpiration of Picea abies and Pinus cembra and their significance in desiccation damage at the alpine treeline, Oecologia (Berlin), 47:252-456.

Baylis, G. T. S., 1959, An example of winter injury to silver beech at moderate altitude, Proceedings of the New Zealand Ecological Society, 6:21-22.

Beck, E., Schulze, E. D., Sensor, J., and Scheibe, R., 1984, Equilibrium freezing of leaf water and extracellular ice formation in afroalpine "giant rosette" plants, Planta, 162:276-282.

Benecke, U., Schulze, E. D., Matyssek, R., and Havranek, W. M., 1981, Environmental control of CO_2-assimilation and leaf conductance in Larix decidua Mill., Oecologia (Berlin), 50:54-61.

Boysen-Jensen, P., 1949, Causal plant geography, Dansk Vidernskabernes Selskab, Biologiske Meddelelser, 21:1-19.

Grace, J., and James, J. [In press], The physiology of trees at the tree line. In: Alden, J. N., Ødum, S., and Mastrantonio, J. L., eds., Proceedings of an international symposium; 1991 June 18-24, [Laugavatn, Iceland], NATO, Plenum, New York.

Grace, J.. and Norton, D. A., 1990, Climate and growth of Pinus sylvestris at its upper altitudinal limit in Scotland: Evidence from Tree Growth Rings, Journal of Ecology, 78:601-610.

Hadley, J. L., and Smith, W. K., 1983, Influence of wind exposure on needle desiccation and mortality for timberline conifers in Wyoming, USA, Arctic and Alpine Research, 15:127-135.

Marchand, P., 1980, Causes of coniferous timberlines in the northern Appalachian Mountains. In: Mountain environments and tree growth, Proceedings of IUFRO Workshop, 1979 November, Christchurch, New Zealand, New Zealand Forest Service, Wellington, 231-246.

McCracken, I. J., Wardle, P., Benecke, U., and Buxton, R. P., 1985, Winter water relations of tree foliage in New Zealand and Switzerland. In: Establishment of tending subalpine forest: research and management, Proceedings of IUFRO Workshop, Riederalp, Switzerland, 1984 September 3–5, Swiss Federal Institute of Forestry Research, Birmensdorf, 85-93.

Moser, M., 1966, Die ektotrophe Ernahrungsweise an der Waldgrenze, Allgemeine Forstzeitung, 77:120-127.

Ohsawa, M., 1990, An interpretation of latitudinal patterns of forest limits in south and east Asian mountains, Journal of Ecology, 78:326-339.

Perkins, T. D., Adams, G. T., and Klein, R. M., 1991, Desiccation or freezing? Mechanisms of winter injury to red spruce foliage, American Journal of Botany, 78:1207-1217.

Sakai, A., and Larcher, W., 1987, Frost survival of plants, Responses and adaptation to freezing stress, Springer Verlag, Berlin, Heidelberg, 321 p.

Sakai, A., Paton, D. M., and Wardle, P., 1981, Freezing resistance of trees of the south temperate zone, especially subalpine species of Australasia, Ecology 62:563-270.

Schulze, E. D., Mooney, H. A., and Dunn, E. L., 1967, Wintertime photosynthesis of bristlecone pine (Pinus aristata) in the White Mountains of California, Ecology, 48:1044-1047.

Sveinbjörnsson, B., Sonesson, O. N., and Karlsson, S., [In Press], Mountain birch performance in Sweden and Iceland, Implications for afforestation. In: Alden, J. N., Ødum, S., and Mastrantonio, J. L., eds., Proceedings of an international symposium, 1991 June 18-24, [Laugavatn, Iceland], NATO, Plenum, New York.

Tranquillini, W., 1959, Die Stoffproduktion der Zirbe (Pinus cembra L.) an der Waldgrenze während eines Jahres, Planta, 54:107-151.

Tranquillini, W., 1979, Physiological ecology of the alpine timberline, Springer Verlag, Berlin, 137 p.

Tranquillini W., and Plank, A., 1989, Ökophysiologische Untersuchungen an Rotbuchen (Fagus sylvatica L.) in verschiedenen Höhenlagen Nord- und Südtirols, Centralblatt für das gesamte Forstwesen, 106:225-246.

Tukhanen, S., [In press], Treeline in relation to climate, with special reference to oceanic areas. In: Alden, J. N., Ødum, S., and Mastrantonio, J. L., eds., Proceedings of an international symposium; 1991 June 18-24, [Laugarvatn, Iceland], NATO, Plenum, New York.

Turner, H., 1968, Über Schneeschliff in den Alpen, Wetter und Leben, 20:192-200.

Wardle, P., 1963, Growth habits of New Zealand subalpine shrubs and trees, New Zealand Journal of Botany, 1:18-47.

Wardle, P., 1968, Engelmann spruce (Picea engelmannii Engel.) at its upper limits on the Front Range, Colorado, Ecology, 49:483-495.

Wardle, P., 1971, An explanation for alpine timberline, New Zealand Journal of Botany, 9:549-554.

Wardle, P., 1981, Winter desiccation of conifer needles simulated by artificial freezing, Arctic and Alpine research, 13:419-423.

Wardle, P., 1985, New Zealand timberlines, 2. A study of forest limits in the Crow Valley, near Arthur's Pass, Canterbury, and 3. A synthesis, New Zealand Journal of Botany, 23:235-261; 263-271.

Yanagimachi, O., and Ohmori, H., 1991, Ecological status of Pinus pumila scrub and the lower boundary of the Japanese alpine zone, Arctic and Alpine Research, 23:424-435.

Zotov, V. D., 1938, Some correlations between vegetation and climate in New Zealand, New Zealand Journal of Science and Technology, 19:474-487.

PHYSIOLOGY OF TREES AT TREELINE

John Grace and Jennifer James

Institute of Ecology and Resource Management
University of Edinburgh
Edinburgh, United Kingdom

INTRODUCTION

Evidence strongly supports the hypothesis that the elevational and northern treeline in the boreal region is determined more by summer temperature than by other climatological factors. The traditional view is that treeline coincides with the 10 °C summer isotherm, and therefore that trees may simply be unable to grow at lower summer temperatures (Köppen, 1931). Historical and prehistorical data show clearly the advance of treeline in warm periods (Kullman, 1981, 1988; Pears, 1988). In addition, growth rings of trees near the treeline usually show a correlation between ring width and summer temperature (Mikola, 1962). Finally, several studies relate height growth to summer temperatures, either in forests or in artificial, controlled environments (Junttila, 1986; White, 1974).

Many studies suggest, however, that treeline is determined by winter desiccation. At the alpine treeline, winter desiccation is inferred from the presence of dead, brown leaves on conifers in the spring, a phenomenon that is frequently observed and demonstrated by the presence of severe water stress in the leaves during winter (Michaelis, 1934; Tranquillini, 1979; Wardle, 1971). This water stress exists because transpiration occurs when soil water is frozen in late winter or early spring.

The observation that winter conditions are important in determining the tree limit seems to contradict the traditional view that summer temperature is the key factor. Both Wardle and Tranquillini, however, proposed a hypothesis to reconcile these views. They said that summer weather determined the ripening or thickening of the cuticle, so that after relatively cold summers the leaves begin the winter with a poorly-developed cuticle and so are less able to conserve water (Tranquillini, 1979; Wardle, 1971).

Recent work at other sites casts doubt on that hypothesis. In this paper, we present data from Scots pine (Pinus sylvestris L.) treeline in the Cairngorm Mountains of Scotland, where the climate is much more oceanic than in the Alps of continental Europe. The elevation of the treeline is 600 to 620 meters above sea level, and the latitude is 57° N. The

Forest Development in Cold Climates, Edited by J. Alden *et al.*
Plenum Press, New York, 1993

site is described in several earlier publications (Pears, 1967, 1968; Watt and Jones, 1948; Wilson et al., 1987), and the general characteristics of climate in the British uplands are outlined in Grace and Unsworth (1988).

TREELINE IN THE CAIRNGORM MOUNTAINS

General Characteristics

The Scots pine population in the Cairngorm Mountains is a relic of the ancient Caledonian pinewood that once covered the Scottish highlands. Because of its remoteness and elevation, it has never been cut and may be regarded as truly wild. Above the treeline,

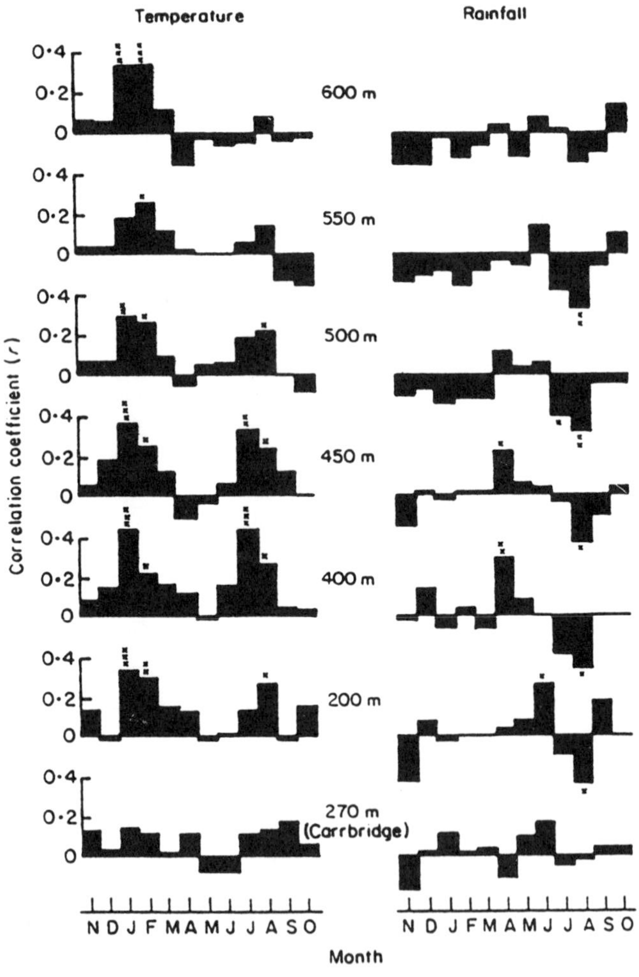

Fig. 1. Growth of Scots pine along an altitudinal gradient from the Spey Valley to treeline in the Cairngorm Mountains. The histograms show the correlation between width of growth rings of Pinus sylvestris and the temperature and rainfall in each month of the year. Asterisks denote statistical significance: *= P<0.05; **= P>0.01; ***= P<0.001. Source: Grace and Norton, 1990.

there are numerous young trees that were established during recent warm summers (Pears, 1988). At the treeline, many of the pines have a krummholz growth form, being dwarf and multistemmed, although a few are spire-like. Below treeline (250 to 500 meters), the trees are generally tall and straight. Trees range in age from 100 to 330 years, with one 430-year-old that may be the oldest recorded Scots pine in Scotland (Grace and Norton, 1990).

Growth and temperature

To assess the impact of climatological variables on growth of this population, core samples were taken from tree trunks at different altitudes and the ring widths were correlated with weather variables using conventional dendrochronological methods (Grace and Norton, 1990). Temperature proved much more significant than rainfall in determining the thickness of the tree rings, and closer to the treeline, the importance of the summer temperature diminished while temperatures in late winter became progressively more important (Fig. 1). The temperatures during the previous year had no statistically significant effect on ring width.

Vegetative Characteristics

The weight, length, and specific leaf area of pine needles declined with altitude, particularly beyond 600 meters where the plants are saplings and may lose more than half their needles as well as their leading shoot in the winter (Fig. 2).

Needle death is conspicuous in early summer, even though the site has an oceanic climate, and soil does not freeze in mild periods during the winter. The number of needles produced each year decreases with elevation from 60 fascicles per shoot at 270 meters to only 35 fascicles at 450 meters. At the treeline, however, 60 fascicles are produced (Grace and Norton, 1990). In saplings above the treeline, many of the needles are lost during the

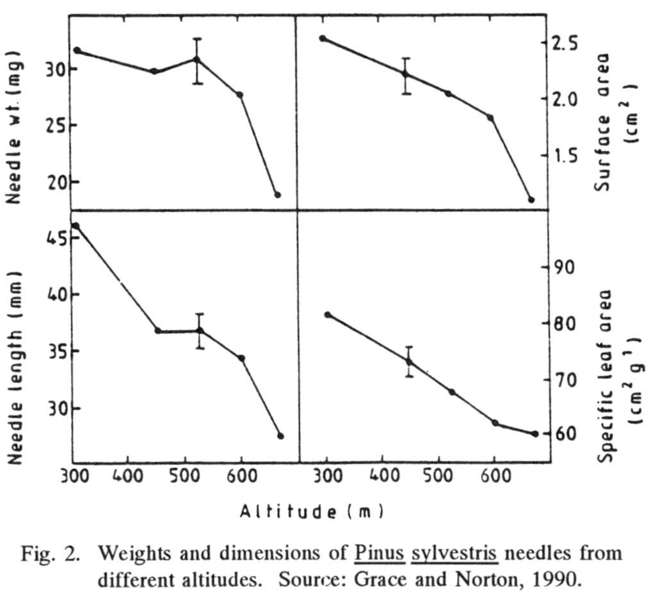

Fig. 2. Weights and dimensions of Pinus sylvestris needles from different altitudes. Source: Grace and Norton, 1990.

winter. Otherwise, leaves are retained for up to four years, with considerable loss after two years.

Cuticular Development and Transpiration

Cuticular transpiration was assessed by weighing detached leaves on a microbalance in an atmosphere of known humidity (Grace, 1990). Cuticular conductance was calculated and plotted against relative water content (Fig. 3). The needles from the two highest elevations showed the greatest conductances.

Thickness of the leaf cuticle has previously been assessed by staining sections of needle with Sudan Black or Sudan Red to visualise the lipid-rich, and therefore water-impermeable, regions of the cuticle (Baig and Tranquillini, 1976, 1980; Lange and Schulze, 1966; Sowell et al., 1982). This method has severe drawbacks, however. First, the cuticle is often very thin, sometimes too thin to accurately measure with a light microscope. Second, the cuticle as revealed by staining merges into the cellulose cell walls. Even the anticlinal walls may be cuticularised so that measurement of thickness in transverse section is practically impossible. Third, the cuticle is unevenly developed over the leaf surface, thus contributing to sampling errors for thin sections. These problems are overcome by weighing a piece of cuticle of known size. The cuticle is isolated by incubating a 10 millimetre length of needle in a mixture of chromic and nitric acids. The resulting preparation appears as an undamaged, isolated piece of cuticle when viewed with the scanning electron microscope. We have also tried isolating cuticle with enzyme preparations, but these are not effective for Pinus, perhaps because the phenolic materials interfere with enzyme action.

Results indicate that mass per unit area of cuticle does not vary much with altitude, and certainly not as much as indicated by Baig & Tranquillini (Grace, 1990) (Fig. 4). Our recent data confirm this conclusion and show that most of the cuticular development occurs in the early part of the growing season. Thus, for Pinus sylvestris at this site, cuticular development is not very important in determining sensitivity to winter conditions.

If cuticles are well developed at the treeline, why do needles show elevated cuticular water loss? The answer may lie in the interpretation of the weight-loss curves of detached leaves, a technique pioneered by Hygen (1951) and still extensively used. Hygen recognised two phases of water loss: an initial phase in which the stomata are open and water loss is rapid, and a second phase when the stomata are shut and water loss is very slow. The flaw in this technique is that it is not possible to determine whether the stomata are fully shut in the "cuticular phase." Elsewhere we have argued that the transpiration and conductance during this phase should be called minimum conductance, not cuticular conductance (van Gardingen et al., 1991). In an experiment on Pinus and Picea, it was shown that the

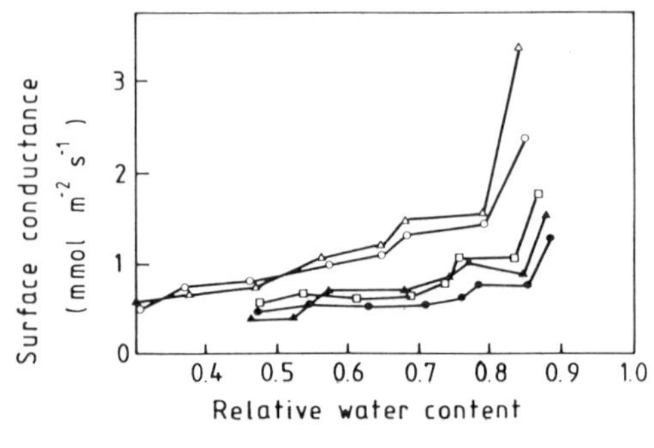

Fig. 3. Surface conductance of pine needles taken from various altitudes, plotted against relative water content: ▼—300 meters; ●—450 meters; □—525 meters; △—600 meters; O—670 meters. Source: Grace, 1990.

minimum conductance increased substantially following exposure in a wind tunnel, and damage to the leaf surface resulted from collisions between adjacent leaves (van Gardingen et al., 1991). Such abrasive damage is likely to disrupt the turgor relations within the epidermis and cause stomatal dysfunction. Abrasive damage occurs in grasses, herbaceous plants, and broadleaved trees and is a major cause of elevated water loss in windy environments (Ambrust et al., 1974; Grace, 1974, 1977; MacKerron, 1976; Wilson, 1984). Abrasive damage may be a particular feature of treeline, because young plants are exposed to wind-blown ice and soil particles much more than they would be as seedlings in the forest. North American studies have also stressed the importance of abrasive damage (Hadley and Smith, 1983, 1986; Holtmeier, 1980; Marchand and Chabot, 1978). Such damage is frequently observed on stems as well as leaves.

The fact that the cuticles of conifers are so well developed, and that a hypodermis occurs as well as an epidermis, may be regarded as circumstantial evidence that natural selection has favoured individuals able to withstand mechanical abrasion.

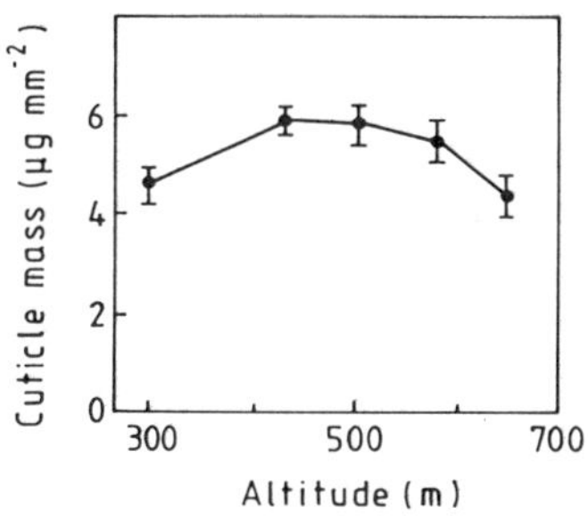

Fig. 4. Cuticular mass of pine needles at different altitudes in the Cairngorm Mountains, Scotland. Source: Grace 1990.

The correlations with temperature as shown in Figure 1 may now be reinterpreted. The mean January temperature of the site is 1.5 °C. Cold winters in the Cairngorm Mountains are associated with hail and ice crystals, both of which cause abrasive damage to the leaf and stem surface, and thus high transpiration rates. Needles are lost and considerable damage occurs to the bark. Branches may also break due to the weight of the snow. These effects will cause direct and indirect loss of carbon assimilation, and thus reduce tree growth the following summer. The hypothesis of Wardle and Tranquillini is not supported; if it were, there would be a positive correlation between the temperature in the previous year, which supposedly controls thickening or ripening of the cuticle, and width of growth ring. Although such correlations have been observed at some treelines (Norton, 1984), they were not found in the Cairngorms (Grace and Norton, 1990).

An unexpected result from the tree-ring correlations was that the correlation between ring width and July temperature weakened as the treeline was approached. This may be partly explained by the pattern of temperatures along the altitudinal gradient. Tissue temperatures were assessed by inserting thermojunctions into the apical meristems at four stations across the treeline. Above the treeline, the temperatures were measured on the dwarf shrubs of <u>Arctostaphylos</u>. At the treeline, the thermojunctions were inserted in the tissues of krummholz pines; at low altitudes, the trees were more than 10 meters tall. The difference between tissue and air temperature increased as the height of the vegetation declined (Fig. 5).

These data are analysed elsewhere (Grace et al., 1989). The point is that trees at the treeline do not necessarily have the lowest tissue temperatures; they apparently benefit from short stature. It may be for similar reasons that the trees at treeline produce larger buds and greater diameter growth than the trees just below treeline (Grace and Norton, 1990).

Carbon Balance

The photosynthetic production per plant is the product of the leaf area and the rate of photosynthesis per unit of area. The leaf area present at the start of the growing season depends on the extent to which leaves have survived the winter, and on their size and number. Needle size may be influenced by temperature during needle expansion, and the number of needles may depend on the temperature in the previous year (Junttila, 1986). The net rate of photosynthesis per area of leaf may depend on several other variables, including radiation, water and nutrient supply, and temperature. These factors have not yet been thoroughly investigated, but it seems that photoinhibition is a widespread phenomenon in conifers in cold climates (Öquist and Huner, 1991) and may limit the rate of photosynthesis, particularly at the start and end of the growing season. Recent data (author's unpublished data) on Scots pine in the Cairngorms show differences in the rate of photosynthesis between valley and treeline that can be related to phenology and temperature.

The partitioning of assimilates has never been investigated at the treeline, but it is known that few fertile seeds are produced (Miller and Cummings, 1982). The advance of the treeline in warmer years may require two consecutive years with warm summers, one to provide viable seeds, the next to permit germination and establishment.

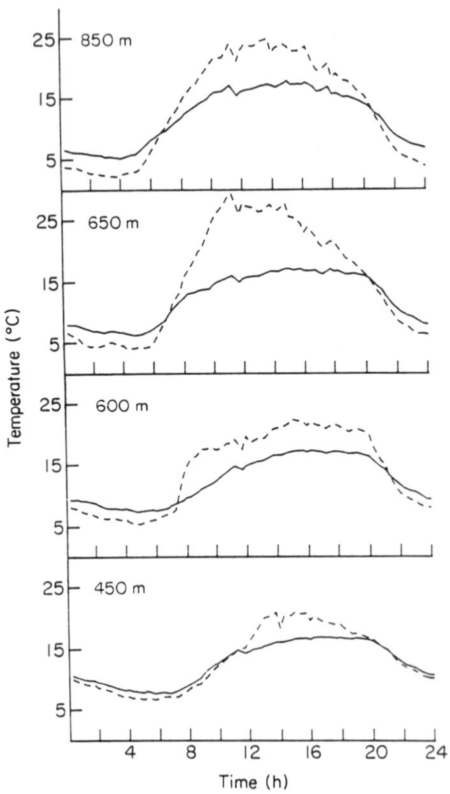

Fig. 5. Air and tissue temperatures at different altitudes extending from tall forest at 450 meters, through krummholz at 600 meters, to dwarf shrub communities above the treeline at 650 and 850 meters. Solid lines are air temperatures; broken lines are temperatures obtained from thermocouples inserted into terminal meristems. Values plotted are the mean diurnal curves for the month of June. (Source: Wilson et al., 1987.)

The whole-plant carbon balance of treeline plants at the treeline has not been determined. It is possible that the ratio of "dark respiration" to photosynthesis is high, because there is a lot of climatically-damaged tissue that does not participate in photosynthesis, and also a lot of wound respiration (Crawford, 1989; Skre, 1991).

DISCUSSION AND CONCLUSIONS

Temperature

Low temperature reduces plant performance in many ways, but perhaps most notably by its influence on cell division and expansion (Hari and Leikola, 1974). Temperature directly influences the expansion of leaf area and photosynthetic capacity above ground, and the expansion of root area below ground. Most species show a near-linear relationship between the rate of leaf expansion and temperature over their range during the growing season near treeline. Consequently, the old concept of "heat sum" (Grace, 1988), measured as degree days above a certain threshold, is likely to be very useful when assessing sites for tree growth. Indeed, it seems to work adequately in the case of needle extension of *Pinus sylvestris* in Scotland.

Low temperature subsequently affects the biological processes that determine the uptake of resources (carbon, water, and nutrients) by leaves and roots. Most notably, photosynthetic rate is a strong function of temperature in the range 5 to 15 °C, and photosystems may be prone to photoinhibition at low temperatures. Thus, quite small differences between altitudes or aspects may exert a disproportionate influence on photosynthesis (Benecke and Nordmeyer, 1983; Hasler, 1984).

Uptake rates of water and nutrients by roots have not been investigated, and in common with other ecosystems, studies of the below-ground environment have lagged behind those above ground. A few conclusions can be drawn, however, from other situations. Water becomes viscous at temperatures well above freezing, so water potential is likely to be generally reduced when plants are cold. This may influence the allocation of resources by trees at treeline. It is likely, for example, that the xylem is a major source of water at low temperatures when frost-drought occurs, water being withdrawn by the process of cavitation (Pena and Grace, 1986). Trees may respond by allocating more photosynthate to stem tissues, as seems to happen in the krummholz habit. It is premature, however, to claim that the krummholz habit is adaptive; the dwarf form may simply be the result of damage in harsh winters, rather than a genetically-determined suite of traits.

Uptake of nutrients most likely depends on mycorrhizas and other associations, but very little is known about their performance at low temperatures or their integrity after cycles of wetting, drying, and freezing.

Wind

Wind speed influences several processes at the treeline. Its influence on leaf and meristem temperatures is of great biological significance. For small trees, there may be an increase in meristematic temperatures of two or three degrees in a sheltered versus exposed microsite, given bright sunshine (Grace et al., 1989). Also, mechanical stimulation has a direct effect on growth—the so-called thigmomorphogenesis (Grace, 1977). In pine, mechanical stimulation has been found to exert a 10 to 20 percent reduction in growth (Rees and Grace, 1980), although it seems possible that all plants in the natural situation would receive adequate mechanical stimulation to elicit thigmomorphogenesis. In pine, the response involves a reduction in height growth and lateral extension growth, and a small increase in girth. Finally, the most important way wind influences trees at treeline is in its direct mechanical action on the plant structure. In this instance, a series of thresholds of wind speed may be envisaged. The lowest threshold causes damage at the microscale (damage to plant epidermis); the highest causes stems to break.

Contrasts Between Oceanic and Continental Climates

Almost all natural treelines coincide approximately with the 10 °C isotherm of mean temperature in the warmest month, but it seems unlikely that all treelines are determined by frost-drought following the failure of the cuticle to thicken and ripen, as suggested by Tranquillini (1979) from his study in the continental climate of the Austrian Alps. In oceanic climates, winters are relatively warm. Thus, the soil is not always frozen and frost-drought is less prevalent. Plant surfaces, however, may be damaged any time of year by wind and wind-borne particles, although such damage is more common in winter when hail is frequent. Loss of leaves in the winter adversely influences carbon gain the following summer.

Wind speed is a much more significant factor in oceanic climates, so departure of the treeline from the 10 °C isotherm may be considerable, depending on local topography.

Damage through abrasion is significant. It may cause water stress, defoliation, death of needles, and a consequent reduction in the photosynthetic area of the plant. Whether the thickness of cuticle is a major determinant of susceptibility to damage is open to question. The classical views about cuticle ripening are not well supported by recent studies of cuticular development. It seems more likely that abrasion occurs at specific points of contact between needles, these points being large enough to cause a major increase in cuticular water loss, and perhaps also to prevent stomatal closure due to disruption of turgor relations of the epidermis. Such abrasion may also serve as a route for entry of pollutants (Fowler et al., 1989).

REFERENCES

Ambrust, D. V., Paulson, G. M., and Ellis, R., 1974, Physiological responses to wind by sandblast-damaged winter wheat plants, Agronomy Journal, 66:421-423.

Baig, M. N., and Tranquillini, W., 1976, Studies on upper timberline: morphology and anatomy of Norway spruce (Picea abies) and stone pine (Pinus cembra) needles from various habitat conditions, Canadian Journal of Botany, 54:1622-1632.

Baig, M. N., and Tranquillini, W., 1980, The effects of wind and temperature on cuticular transpiration of Picea abies and Pinus cembra and their significance in desiccation at the alpine treeline, Oecologia, 47:252-256.

Benecke, U., and Nordmeyer, A. H., 1983, Carbon uptake and allocation by Nothofagus solandri var. cliffortioides and Pinus contorta ssp. contorta at montane and subalpine altitudes. In: Carbon uptake and allocation in subalpine ecosystems as a key to management, Waring, R. H., ed., Forest Research Laboratory, Oregon State University, Corvallis, Oregon.

Crawford, R. M. M., 1989, Studies in Plant Survival, Blackwell Scientific Publications, Oxford, 296 p.

Fowler, D., Cape, J. N., and Unsworth, M. H., 1989, Deposition of atmospheric pollutants on forests, Philosophical Transactions of the Royal Society, B324:247-265.

Grace, J., 1974, The effect of wind on grasses. 1. Cuticular and stomatal transpiration, Journal of Experimental Botany, 25:542-551.

Grace, J., 1977, Plant response to wind, Academic Press, London, 204 p.
Grace, J., 1988, Temperature as a determinant of plant productivity. In: Long, S. P., and Woodward, F. I., eds., Plants and Temperature, Cambridge University Press, Cambridge, 91-107.
Grace, J., 1989, Tree Lines, Philosophical Transactions of the Royal Society, B324: 247-265.
Grace, J., 1990, Cuticular water loss unlikely to explain treeline in Scotland, Oecologia, 84:64-68.
Grace, J., Allen, S. J., and Wilson, C., 1989, Climate and the meristem temperatures of plant communities near the tree-line, Oecologia, 79:198-204.
Grace, J., and Norton, D. A., 1990, Climate and growth of Pinus sylvestris at its upper altitudinal limit in Scotland: evidence from growth rings, Journal of Ecology, 78:601-610.
Grace, J., and Unsworth, M. H., 1988, Climate and microclimate of the uplands. In: Ecological Change in the Uplands, Usher, M. B. A., and Thompson, D. B. A., eds., Blackwell Scientific Publications, Oxford, 137-150.
Hadley, J. L., and Smith, W. K., 1983, Influence of wind exposure on needle desiccation and mortality for timberline conifers in Wyoming, USA, Arctic Alpine Research, 15:127-135.
Hadley, J. L., and Smith, W. K., 1986, Wind effects on needles of timberline conifers: seasonal influence on mortality, Ecology, 67:12-19.
Hari, P., and Leikola, M., 1974, Further development of the dynamic growth model of plant height growth, Flora, 163:357-370.
Häsler, R., 1984, CO_2 uptake in European larch (Larix decidua Mill.) near treeline in Switzerland (Stillberg/Davos), Eidgenoessische anstalt fuer das forstliche versuchwesen, 270:113-122.
Holtmeier, F-K., 1980, Influence of wind on tree physiognomy at the upper timberline in the Colorado Front Range. In: Beneke, U., and Davis, M. R., eds., Mountain Environments and Subalpine Tree Growth, Forest Research Institute, New Zealand Forest Service, Wellington, New Zealand, 247-261.
Hygen, G., 1951, Studies in plant transpiration. 1. Physiologia Plantarum, 4:57-183.
Junttila, O., 1986, Effects of temperature on shoot growth in northern provenances of Pinus sylvestris L., Tree Physiology, 1:185-192.
Köppen, W., 1931, Grundriss der Klimakunde, De Gruyter, Berlin.
Kullman, L., 1981, Recent tree-limit dynamics of Scots Pine (Pinus sylvestris L.) in the southern Swedish Scandes, Wahlenbergia, 8:3-67.
Kullman, L., 1988, Holocene history of the forest-alpine tundra ecotone in the Scandes Mountains (central Sweden), New Phytologist, 108:101-110.
Lange, O. L., and Schulze, E.-D., 1966, Untersuchungen über die Dickenentwicklung der kutikularen Zellwandschichten bei der Fichtennadel, Fortwiss. Zentr., 85:27-38.
MacKerron, D. K. L., 1976, Wind damage to the surface of strawberry leaves, Annals of Botany, 40:351-354.
Marchand, P. J., and Chabot, B. F., 1978, Winter water relations of treeline plant species on Mount Washington, New Hampshire, Arctic Alpine Research, 10:105-116.
Michaelis, P., 1934, Okologische Studien an der alpinen Baumgrenze. 5. Osmotischer Wert und Wassergehalt wahrend des Winters in den verschiedenen Hohenlagen, Jahrbucher fur Wissenschaftliche Botanik, 34:336-362.
Mikola, P., 1962, Temperature and tree growth near the northern timberline. In: Kozlowski, T. T., ed., Tree Growth, Ronald Press, New York, 265-274.

Miller, G. R., and Cummins, R. T., 1982, Regeneration of Scots pine Pinus sylvestris at the natural treeline in the Cairngorm Mountains, Scotland, Holarctic Ecology, 5:27-34.

Norton, D. A., 1984, Tree-growth-climate relationships in subalpine Nothofagus forests, South Island, New Zealand, New Zealand Journal of Botany, 22:471-81.

Öquist, G., and Huner, N. P. A., 1991, Effects of cold acclimation on the susceptibility of photosynthesis to photoinhibition in Scots pine and in winter and spring cereals: a fluorescence analysis, Functional Ecology, 5:91-100.

Pears, N. V., 1988, Scots pine (Pinus sylvestris) seedling survival above the tree-line in the Cairngorm Mountains, Scotland, Forestry, 61:61-71.

Pears, N. V., 1967, Present treelines of the Cairngorm Mountains, Scotland, Journal of Ecology, 55:815-880.

Pears, N. V., 1968, Post glacial tree-lines of the Cairngorm Mountains, Scotland, Transactions of the Botanical Society of Edinburgh, 40:361-394.

Pena, J., and Grace, J., 1986, Water relations and ultrasound emissions of Pinus sylvestris L. before, during and after a period of water stress, New Phytologist, 103:515-524.

Rees, D. J., and Grace, J., 1980, The effect of wind on the extension growth of Pinus contorta (Douglas), Forestry, 53:146-156.

Skre, O., 1991, Physiology of plant survival under cold conditions, with particular reference to dark respiration as a factor limiting growth at timberline, A literature review, Meddelelser fra Skogforsk, Vol. 44, Institutt for Skogfag, Agricultural University of Norway, Uppsala, 34 p.

Sowell, J. B., Koutnik, D. L., and Lansing, A. J., 1982, Cuticular transpiration of whitebark pine (Pinus albicaulis) within a Sierra Nevada timberline ecotone, USA, Arctic Alpine Research, 14:97-103.

Tranquillini, W., 1979, Physiological ecology of the alpine timberline, Springer-Verlag, Berlin, 137 p.

van Gardingen, P., Grace, J., and Jeffree, C. E., 1991, Abrasive damage by wind to the needle surfaces of Pinus sylvestris L. and Picea sitchensis (Bong.) Carr., Plant Cell & Environment, 14:185-193.

Wardle, P., 1971, An explanation for alpine timberline, New Zealand Journal of Botany, 9:371-402.

Watt, A. S., and Jones, E. W., 1948, The ecology of the Cairngorms. 1. The environment and altitudinal zonation of the vegetation, Journal of Ecology, 36:283-304.

White, E. J., 1974, Multivariate analysis of tree height increment on meteorological variables, near the altitudinal tree limit in northern England, International Journal for Biometeorology, 8:199-210.

Wilson, J., 1984, Microscopic features of wind damage to leaves in Acer pseudoplatanus L. and its relationship with season, leaf age and wind speed, Annals of Botany, 46:303-311.

Wilson, C., Grace, J., Allen, S., and Slack, F., 1987, Temperature and stature: a study of temperatures in montane vegetation, Functional Ecology, 1:405-414.

TREELINE IN RELATION TO CLIMATE, WITH SPECIAL REFERENCE TO OCEANIC AREAS

Sakari Tuhkanen

Department of Geography
University of Turku
Turku, Finland

INTRODUCTION

Treeline is a very traditional, still fascinating, phytogeographical complex of problems. One of the problems is the climatic characterization of treeline. Proposals have appeared in the scientific literature for more than a hundred years concerning the climatic indicators that best denote the location of treeline. Often, however, insufficient emphasis is placed on the difficulties of delineating treeline. Difficulties include the indefiniteness of the concept of treeline, its highly variable nature both physiognomically and taxonomically, and the sparseness of representative climatological records from treeline environments. Some authors stress the diversity of treeline; others consider it to be almost ecologically equivalent world wide.

Oceanic regions at upper-middle and higher latitudes have often posed problems for the biogeographer because of their anomalous nature. Opinions differ widely on how to assign the highly oceanic corners of the circumpolar zone to biogeographical regions. Treeline (or forest line) is often given a decisive role in determining the boundary between the Arctic and boreal zones, but that may lead to questionable solutions, especially in oceanic areas.

TREELINE TERMINOLOGY

The terminology referring to the treeline belt, that is, the transition between continuous forest and treeless area, is ambiguous. A number of concepts are needed. The treeline terminology suggested by Hustich (1966) is well-known and appreciated (Fig. 1).

Economic forest line constitutes the boundary for secure timber exploitation. Until this line, forest regeneration is active and continuous on an annual basis. This line is difficult to locate precisely, because the annual regeneration line shifts in relation to fluctuating weather conditions. The second limit is the physiognomic (also called phytosociological, empirical, biological, or vegetative) forest line, that is, the limit of continuous or nearly continuous forest cover. At this line, forest regeneration is slow and unsure, and seed years are rare (Mikola, 1978; Pohtila, 1980).

Forest Development in Cold Climates, Edited by J. Alden *et al.*
Plenum Press, New York, 1993

Often an operational definition of forest line is needed, for example, in terms of spacing of tree stems (Braathe, 1977) or the coverage by tree canopies of the soil surface (Larsen, 1965). Definitions applicable to all forest lines in different regions are not possible because they depend on species-specific characteristics, among other things.

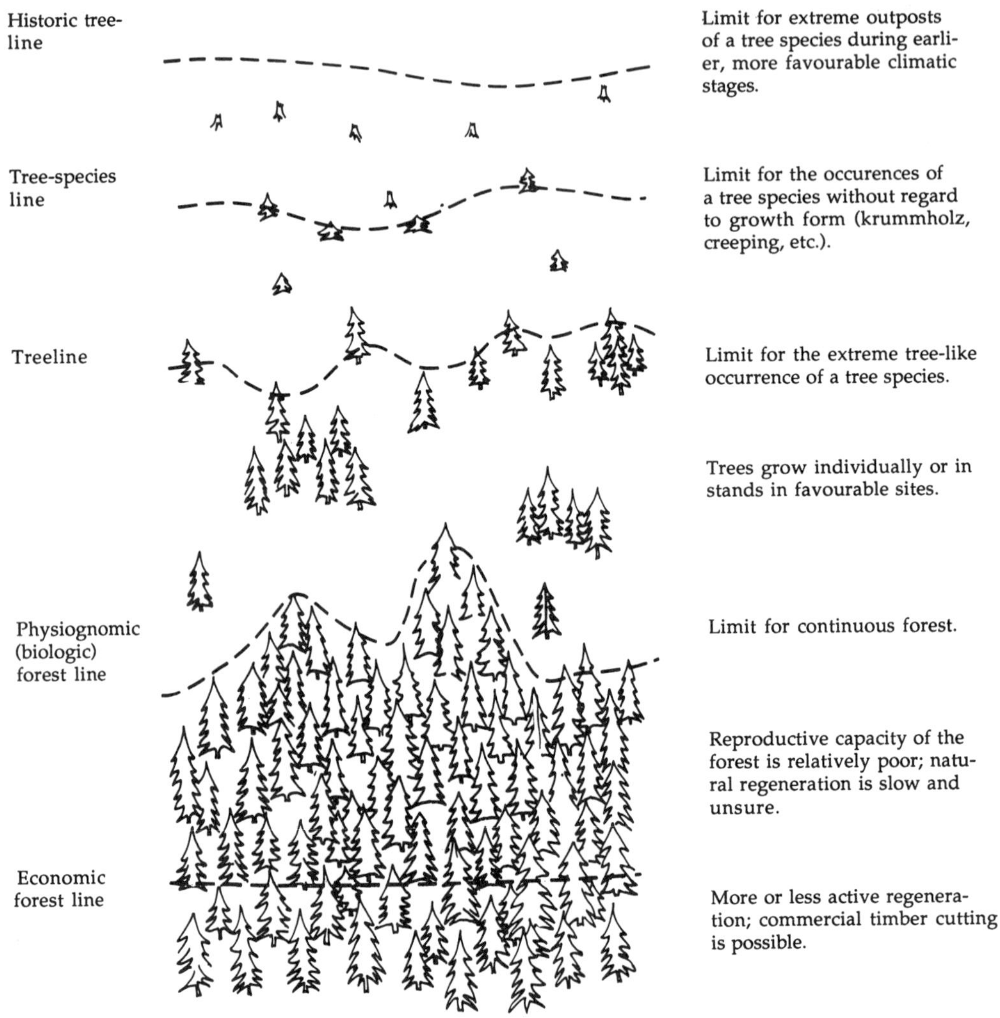

Fig. 1. Different tree and forest lines, according to Hustich (1966), modified after Tuhkanen (1982) and Heikkinen (1984).

Isolated trees and small stands of trees, even forest patches, characterize the area between physiognomic forest line and treeline proper which is the extreme boundary where trees still achieve arboreal form and size. Eventually the limit of tree species is reached. This represents the furthermost outpost of one or more tree species without regard to growth form. This line is probably the easiest limit to locate on a map and the one that allows the least margin for interpretation.

The term "timberline" has been variously used. It may refer to the economic forest line or even to treeline proper. Frnklin and Dyrness (1973), for instance, used the term "timberline region" to cover the entire altitudinal belt from the forest line to the scrub line. According to Hustich (1979), timberline is synonymous with physiognomic forest line, but he later defined it as "the economical or generative forest line" (Hustich, 1983).

I will use the term treeline in a general sense to refer to the transition from forests to treeless vegetation. The treeline belt is an ecotone between two different habitats.

It has been suggested that under natural conditions, the treeline would be sharper than a transitional belt from forest line to treeless vegetation. In some cases, that is true. In the southern beech (Nothofagus Blume) forests of New Zealand or southern South America, for example, the forest line and treeline often coincide (Wardle, 1974). In many mountain areas as well as at northern latitudes, however, the natural treeline belt is a more or less wide ecotone, often called "Kampfzone" (zone of struggle), that extends from closed forest to isolated dwarfed scrub-like trees or krummholz (Holtmeier, 1985a).

Different opinions have also been put forward on what constitutes a tree (Eidem, 1955; Hustich, 1979). The word is generally understood to mean a plant of a certain size with a woody stem that is "tree-like" in shape, and obviously not a shrub. According to Braathe (1977), the minimum height for a tree is 3 meters at the forest line and 2.5 meters at treeline proper.

When working on a large scale, it makes little difference whether a tree is considered to be the height of a man, 2, 2.5, 3, or perhaps 5 meters. Problems may arise, however, from the requirement of "treelikeness." In some papers, for instance, the mountain birch (Betula pubescens Ehrh. ssp. tortuosa (Ledeb.) Nyman) in northern Fennoscandia is not included as a resident of the forest and treelines. Mountain birch can be relatively tall and arboreal, but it is often low (although exceeding two or three meters) and bush-like with multiple stems. The mountain birch forests of northern Fennoscandia, occurring beyond and above coniferous forests, should be included in the boreal zone proper (Ahti et al., 1968), and not in forest tundra, as is sometimes done.

I will not go into detail on zonational terminology (forest tundra, scrub tundra, subarctic, etc.), because there are almost as many definitions and landscape subdivisions as there have been ecologists working with the treeline ecotone. Löve (1970) and Hustich (1979) proposed a standardization of nomenclature; this process is far from complete (Ahti, 1980; Alexandrova, 1970; Atkinson, 1981; Payette, 1983; Pignatti, 1980).

Certain attributes are commonly attached to tree and forest lines. In many cases, there is evidence that ancient trees grew north of the treeline apparent today. Hence, a definition of present and historic may be needed. Historic treeline is frequently considered to be the highest postglacial treeline, 5,000 to 9,000 years B.P., or the period when the climate in northwestern Europe was about 2 °C warmer than it is today (Grace, 1989).

Treelines are often delineated as potential or actual. Potential implies the limits that are controlled mainly by climatic conditions and possibly other natural environmental factors such as topographic and edaphic conditions. In reality, however, human activity has forced treeline to retreat extensively. The lines presently encountered are considered actual treelines. Potential is often natural, but not always, and natural is not always potential. It is possible, for instance, that the trees and forest have simply not had time to return to the Alaskan Peninsula and the Aleutians after the last glaciation. The Faeroes and the Falklands are naturally treeless, but there is no doubt that they were not within the potential treeline. The terms natural and anthropogenic may also be useful.

Latitudinal and altitudinal treelines correspond to horizontal and vertical lines. The vertical and horizontal treelines coincide at high latitudes. Polar treelines are encountered in the Northern and Southern Hemispheres, hence the attributes Arctic and Antarctic are used. This may be misleading, for the treeline towards treeless regions in the north or south can hardly be taken as a definite criterion for the limits of the Arctic and Antarctic phytogeographical zone. There are some regions, oceanic regions in particular, that should be included in the boreal (or in the Southern Hemisphere, antiboreal, or austral or cool temperate) phytogeographical zone, but they are treeless. Simply, northern or southern treeline would be more neutral terms.

As far as the altitudinal limit of forests is concerned, the concept of upper or lower treeline is relevant. The upper tree-line is often called the alpine treeline, but the term upper treeline is more useful in a worldwide sense (Troll, 1973). There may also be lower treeline in the mountains, resulting from decreasing rainfall, cold-air drainage, or anthropogenic factors.

There is a difference in the character of treeline between pronounced oceanic and continental regions. If the climate is highly oceanic, as in western Norway, for example, the polar treeline, especially that of coniferous forests, turns far south and is characterized by the term oceanic or maritime treeline. Parallel phenomena to the oceanic and continental trends of the latitudinal treeline are the increase in altitude of the upper treeline in coastal ranges from the coast inland. The increase in altitude of treeline from the edges of a mountain range to its inner parts is known as the "Massenerhebung effect," that is, the influence of a mountain massif on the elevation of the upper treeline (Arno and Hammerly, 1984; Brockmann-Jerosch, 1919; Fægri, 1971).

Tree and forest lines are also caused by aridity and exist where trees give way to grassland and steppe. This kind of drought limit is not reviewed. Only polar and upper treelines caused by low thermal energy are discussed in the following sections.

NORTHERN TREELINE

In the Northern Hemisphere, the northernmost point of the northern treeline (in a strict sense) (Fig. 2) is located on Taimyr Peninsula at latitude 72°40′ N. (Tikhomirov, 1970). The southernmost points of the northern treeline are a matter of interpretation; areas on the coast of Newfoundland are treeless at 48° N. and the treeless Aleutian Islands reach 52° N.

The northern treeline is formed by evergreen and deciduous conifers, especially in continental areas, and by broad-leaved species usually in the oceanic areas. Evergreen broad-leaved species do not form the northern treeline, and they are extremely rare as treeline species even elsewhere in the extra-tropical Northern Hemisphere (Ohsawa, 1990; Wardle, 1974).

If we distinguish between treelines formed by broad-leaved species, larch (Larix Miller) species and evergreen conifers, the tree limits almost coincide in the central part of the continents and diverge most obviously in the extreme east and west in Eurasia and North America (Fig. 3).

In western Norway, the treeline turns south. The treeless heaths of western Norway are probably climatically conditioned, although they have been extended considerably by human activity (Lindemann, 1970).

The Faeroe Islands have been treeless throughout post-glacial time, but planting experiments have shown that they lie well within the potential treeline (Ødum, 1979; Tuhkanen, 1987). All three Fuegian southern beech (Nothofagus Blume) species and other Fuegian tree species such as the Winter's bark (Drimys winteri J. R. et G. Forster) and the firebush (Embothrium coccineum J. R. et G. Forster) are among those southern South American tree and scrub species that have been grown successfully on the Faeroes (Ødum, 1991; Tuhkanen, 1987; Ødum et al., 1989). In contrast, Iceland once had extensive birch forests that were later mostly destroyed by human activity (Grøntved, 1942). Even under natural conditions, however, birch probably avoids the most oceanic parts in southwestern Iceland (Grøntved, 1942). The field-layer and ground-layer flora points clearly to the

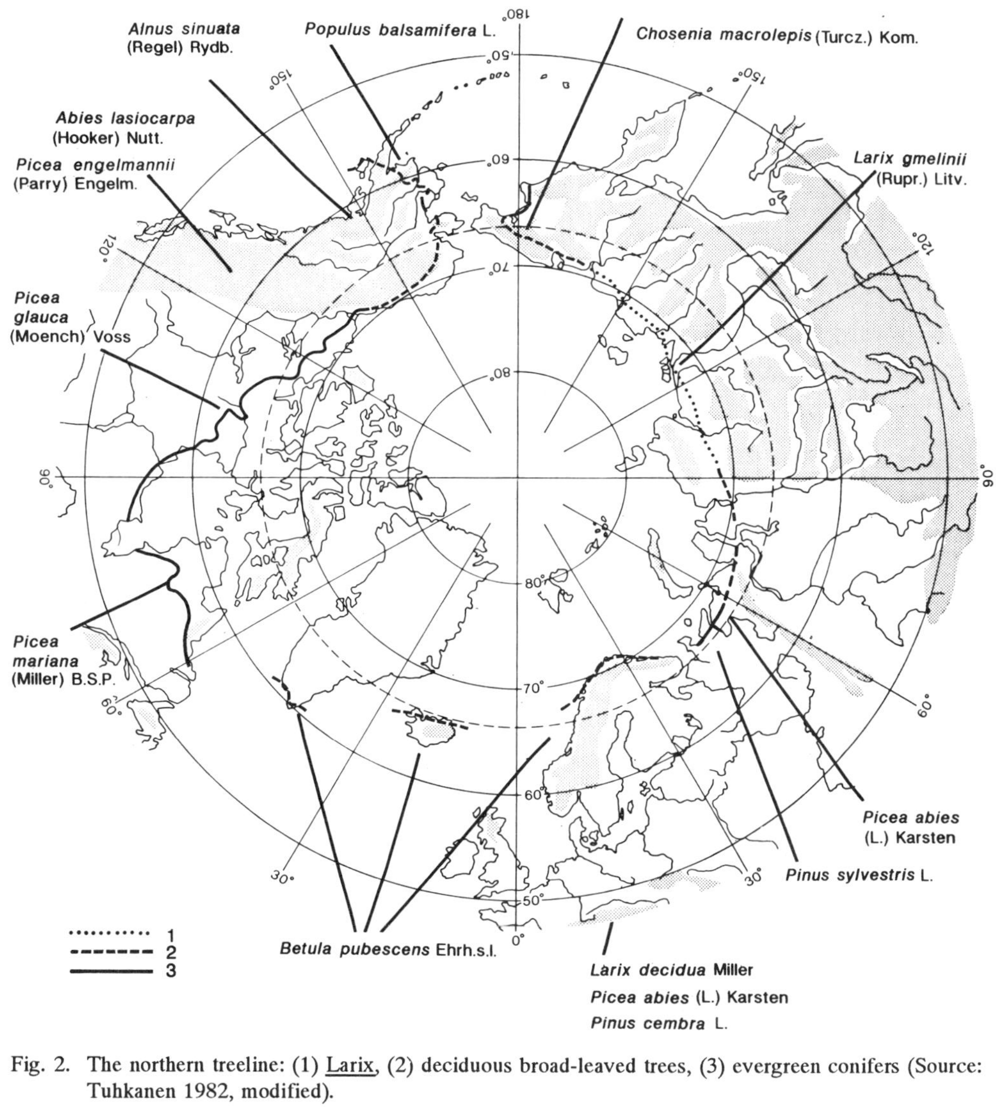

Fig. 2. The northern treeline: (1) Larix, (2) deciduous broad-leaved trees, (3) evergreen conifers (Source: Tuhkanen 1982, modified).

northern half of the boreal zone, bearing a resemblance to that of northwestern and northern Norway. There is, however, a marked Arctic element which is largely attributed to the highly oceanic climate and the dispersion history of the plants (Jónsson, 1905; McVean, 1955; Steindórsson, 1962). In spite of the superficially tundra-like physiognomy of the Icelandic vegetation, inclusion of Iceland in the Arctic zone, except at higher elevations, cannot be justified. Considerable tree planting has been done and many different species tried. Among the most successful species are Sitka spruce (Picea sitchensis (Bong.) Carr.) in the southwest and Siberian larch (Larix sibirica Ledeb.) in sheltered valleys of the south-eastern part of the country (Bjarnason, 1967; Blöndal, 1987).

Southwest Greenland has a small area in the northern boreal zone. Local climatic differences are great in this region with its highly oceanic, windswept coasts and its sheltered inland valleys in which some relatively continental spots can be found. Planting

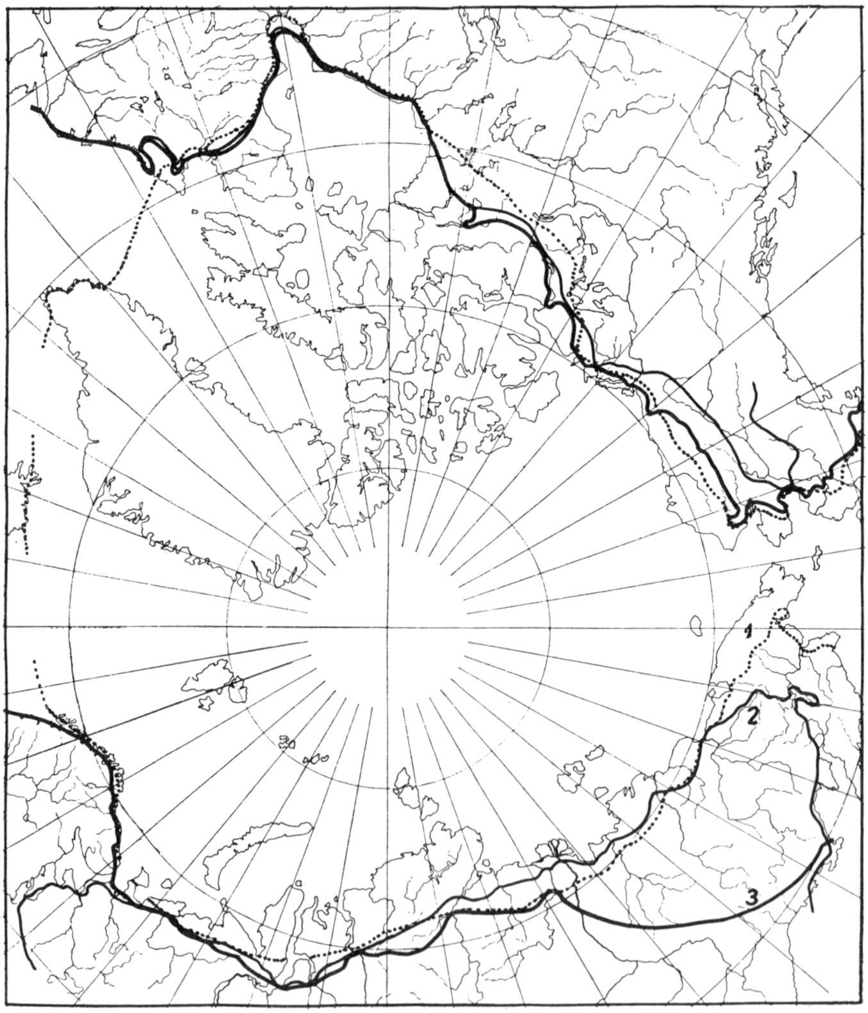

Fig. 3. A comparison of (1) the northern treeline of broadleaved trees (formed by tree-sized Betula L., Populus L., and Chosenia Nakai), (2) the northern treeline of Larix Miller, and (3) evergreen conifers (except Pinus pumila (Pall.) Regel) and Juniperus L. spp. (Hustich, 1966).

experiments have been carried out also on Greenland using a variety of tree species such as Siberian larch (Larix sibirica), subalpine fir (Abies lasicorpa (Hooker) Nutt.), lodgepole pine (Pinus contorta Douglas ex Loud.), Sitka spruce (Picea sitchensis), and white spruce (P. glauca (Moench) Voss), some of which have been reasonably successful (Böcher, 1979; Ødum, 1979, 1990, 1991). Areas that lie within the potential treeline for conifers are limited in southwestern Greenland. Apart from the natural birch forest, some of the treeless heaths also belong to the northern boreal zone (Böcher, 1979).

On the eastern coast of Labrador and Newfoundland, the treeline, especially that formed by coniferous species, turns sharply southward. The effect of cold ocean currents is clearly reflected in the climate. The growing season is cool, damp, and windy, and the winters are much colder than on western coasts. Coastal heaths resemble those in northwest Europe (Meades, 1983). At lower altitudes, especially in Newfoundland, these heaths are mostly anthropogenic in origin (Ahti, 1959).

On both sides of the Bering Sea, the treeline turns southward when approaching coastal areas. There it is possible to delimit an amphi-Bering heath formation, corresponding to the amphi-Atlantic one, and comprising the greater part of the area of the Kurile Islands, part of eastern Kamchatka, the Aleutians, and southwest Alaska (Lavrenko, 1950). The treeless nature of some of these areas may be attributed to historical factors or anthropogenic causes. The treeless Aleutian Islands and southwest Alaska are often classified as Arctic or "oceanic tundra" (Hustich, 1966; Young, 1971). The dominant plant communities close to sea level, however, are not characterized by Arctic species (Hämet-Ahti, 1979a; Tatewaki, 1958), and the area can be included in the northern boreal zone (Hämet-Ahti, 1979a, 1981; Tuhkanen, 1984).

In the opinion of Griggs (1934), the forest has simply not had time to return to the Aleutian Islands after the last glaciation; treeline has certainly been shown to have advanced westward on Kodiak Island during this century (Beals, 1966; Bruce and Court, 1945). Planting experiments have been successfully carried out with Sitka spruce (Picea sitchensis) at Unalaska Bay, for instance, and the trees have regenerated naturally (Alden and Bruce, 1989).

On the southern Alaskan coast and the western coast of the Alaska Panhandle, forests are composed of Sitka spruce (Picea sitchensis) in exposed, windy sites, western red cedar (Thuja plicata Donn ex D. Don) and western hemlock (Tsuga heterophylla (Raf.) Sarg.) in more sheltered places. Exposed patches that are naturally treeless, however, can probably be found even here (Fægri, 1969).

SOUTHERN TREELINE

In the Southern Hemisphere, the southernmost point of treeline (Fig. 4) is in the Cape Horn Archipelago at latitude 56° S. and the northernmost point is between Ile Amsterdam and Ile St. Paul at latitude 38°30′ S. The general thermal level of the extra-tropical Southern Hemisphere is lower than that of the Northern Hemisphere because of Antarctic ice (Weischet, 1978). Southern treeline is oceanic throughout, the degree of oceanity being generally higher than anywhere on the northern treeline. In both hemispheres, the difference between the northernmost and southernmost point of treeline is about 20 degrees of latitude.

The climate of the south Chilean archipelago is very humid and windy, and soil drainage is poor. Moorland formation dominates, but in protected sites, patches of forest occur even on the outermost islands (Dollenz, 1980; Pisano and Venegas, 1984; Young, 1972). The physiognomy of the moorland formation may superficially resemble that of tundra or forest tundra ("tundra magallánica" or "Magellanic tundra"), but such a name is hardly appropriate from an ecoclimatic and phytogeographical point of view.

There is no real latitudinal treeline in Tierra del Fuego. Altitudinal treeline is formed by two deciduous *Nothofagus* species, the high deciduous beech or lenga (*N. pumilio* (Poeppig et Endl.) Krasser) and the low deciduous beech or ñirre (*N. antarctica* (G. Forster) Oersted). The evergreen beech or guindo (*N. betuloides* (Mirbel) Oersted) may form the treeline in the moorland region of southwest Chile.

The Falkland Islands are treeless. There is probably no climatic reason, however, why *Nothofagus* could not grow on the Falklands as indicated by tree plantings (Low, 1986). Efforts to establish trees on the Falklands are limited, but several different species have been tried, among them guindo (*Nothofagus betuloides*), lenga (*N. pumilio*) and ñirre (*N. antarctica*), and they can obviously grow on suitable sites. In gardens, a few individuals of such oceanic, temperate species as Sycamore maple (*Acer pseudoplatanus* L.), holly (*Ilex aquifolium* L.), laburnum (*Laburnum anagyroides* Med.), and Swedish whitebeam (*Sorbus intermedia* (Ehrh.) Pers.) can be seen. All of these grow naturally in northwestern Europe or are planted in the Faeroes (Tuhkanen, 1987).

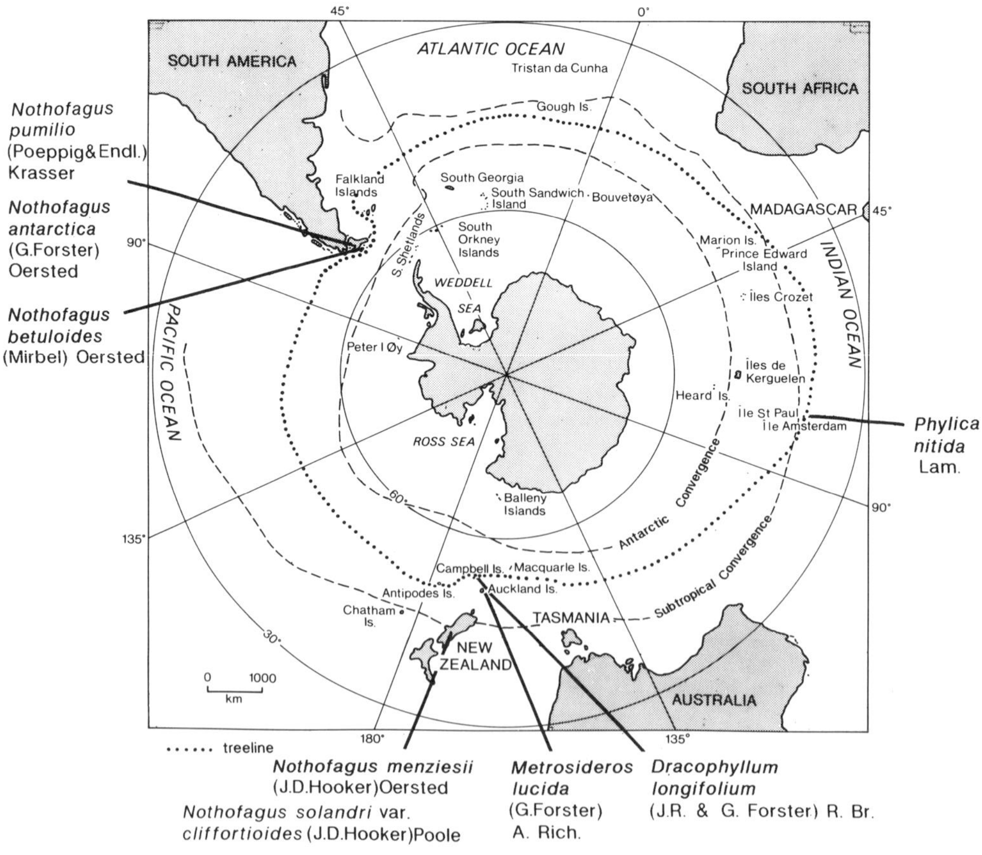

Fig. 4. The southern treeline (Brockmann-Jerosch, 1928; Troll, 1978, modified).

A large number of trees and shrubs from the Cape of Good Hope were planted near the meteorological station on Marion Island in the late 1940s. The choice of trees (cluster pine, Pinus pinaster Ait., Lawson false cypress or white cedar, Chamaecyparis lawsoniana (A. Murray) Parl., and karo, (Pittosporum crassifolium Banks et Sol. ex A. Cunn.) was such that they survived only a few years (La Grange, 1954). Other species might have produced better results.

Ile St. Paul is treeless, but on Ile Amsterdam, Phylica nitida Lam. (Rhamnaceae) grows and even forms small patches of forest (Brockmann-Jerosch, 1928). Macquarie Island is treeless. On Campbell Island, there are two tall shrubs, Dracophyllum scoparium J. D. Hooker and the inaka or grass tree, D. longifolium (J. R. et G. Forster) R. Br. The latter can probably be regarded as a tree. On the Lord Auckland Islands, there are irontree or rata forests of Metrosideros lucida (G. Forster) A. Rich (Burrows et al., 1979). The Antipode Islands are treeless (Falla, 1948).

In most mountain ranges in New Zealand, evergreen Nothofagus species (silver beech, N. menziesii (J. D. Hooker) Oersted, or mountain beech, N. solandri var. cliffortioides (J. D. Hooker) Poole.) form the upper treeline. This line is usually abrupt and the treeline proper and the forest line coincide (Wardle, 1965). In certain circumstances there is, however, a krummholz zone. South American Nothofagus treelines are physiognomically more or less similar to those in New Zealand, but the forests grade into dense krummholz more often.

This brief review shows how great is the taxonomic diversity of trees at treeline. Many families of angiosperms and gymnosperms, hardwoods and conifers, deciduous and evergreen species are represented, including two genera of tree ferns (Hermes, 1955; Ohsawa, 1990; Troll, 1973; Wardle, 1974).

FACTORS REGULATING TREELINE

The location of treeline is controlled primarily by climate. Locally, numerous other factors may be important: human activities, catastrophic events, topographic, edaphic, and biotic factors.

Many theories and hypotheses exist regarding treeline (Daubenmire, 1954; Larcher, 1984; Pears, 1985; Tuhkanen, 1982). Such theories may stress the significance of excessive light in the mountains, light deficiency, carbon dioxide deficiency, amount of nutrients, high relative humidity, deep snow cover, wind, desiccation during winter cold, heat deficiency, and other factors. None of these seems to adequately explain the worldwide phenomenon of the treeline.

The processes and mechanisms of plant physiology that could explain treeline are being studied in a number of places. Many of the classic investigations have been made in the Alps and New Zealand. These studies suggest two primary causes: negative carbon dioxide budget and developmental phases (completion of annual growth and reproduction cycles) closely related to poor climatic resistance to winter conditions (Tranquillini, 1979; Wardle, 1971).

Everything considered, it is very difficult to determine which climatic factors are critical for tree growth. In extra-tropical climates, treeline seems to be a result of many unfavourable factors, relating primarily to temperature during the growing season. In the transitional belt between closed forest and the upper limit of a tree species, exposure to wind and radiation is the most decisive site-controlling factor and influences temperature conditions, depth and duration of snow cover, soil moisture, evapotranspiration, and other factors (Holtmeier, 1985b). Trees suffer from frost-drought, frost, ice abrasion, and creeping snow.

CLIMATIC CHARACTERISTICS AND TREELINE

Many attempts have been made to delimit and characterize the location of treeline with the aid of various climatic parameters and indices. The oldest and best-known climatic indicator for treeline is the 10 °C (50 °F) isotherm for the warmest month of the year (Andersson, 1902; Köppen, 1919; Supan, 1884). Temperature sums and lengths of various warm seasons have also frequently been used to characterize treeline.

An approach to the problem in terms of air-mass climatology has been offered by Bryson (1966) and Krebs and Barry (1970). The arctic front in summer is located along a line that correlates well with the boundary between the boreal forest and tundra in central Canada (Mitchell, 1973); a comparable situation is seen in Eurasia.

Hare and Ritchie (1972) noted that the values for many radiation parameters (annual global net radiation, net radiation during the growing season) rise sharply at treeline, and are associated with the zonation of boreal and arctic vegetation. Radiation shows, however, a close correlation with summer temperature conditions, especially the temperature sum for growing season (Grigor'yev and Budyko, 1960) and potential evapotranspiration (Hare, 1964). In addition, radiation data are very sparse, especially from treeline environments, which decreases their usefulness.

It seems somewhat surprising that the occurrence of permafrost does not correlate with either the northern treeline or the boreal vegetation zonation. In northwestern Canada, the forest line coincides fairly closely with the southern limit of continuous permafrost (Brown, 1960, 1970). In Eurasia, however, permafrost extends deep into the boreal forest zone in the most continental regions (Tikhomirov, 1970).

A general principle is that the northern treeline is primarily a thermal limit, although it may be significantly affected by wind, especially in highly oceanic climates. Moisture is of secondary importance (Hare, 1950, 1954). Similarly, it seems that the severity of winter is not a critical factor for the northern treeline (Fig. 5). Winter conditions may, however, be significant in highly oceanic areas where high wind speeds and thin snow cover are the rule.

For the period 1931-1960, the 11 °C (51.8 °F) isotherm for the mean temperature of the warmest month of the year (either July or August) is closely related to northern treeline in the relatively maritime area of eastern Canada, and the 12 °C (53.6 °F) isotherm follows closely to northern treeline in the continental interior of North America (Fig. 6). The curve for the 15 °C (59 °F) mean daily maximum temperature in the warmest month of the year does not coincide with the treeline any better than do the isotherms for mean temperature. The location of the isotherms described here varies greatly from year to year, and the average values based on a normal period of 30 years are only an abstraction. The 10 °C isotherm for the warmest month of the year, for instance, varies by a distance of some 1,000 kilometers in Canada over a period of no more than 10 or 20 years (Tuhkanen, 1980).

Within the entire northern circumpolar region, correspondence between the climatic indices and treeline (Fig. 7) is similar in Eurasia and North America with the exception that

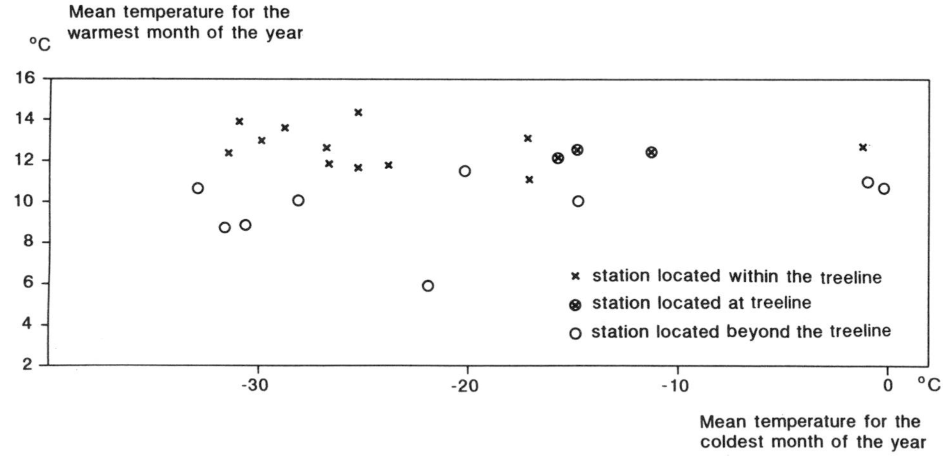

Fig. 5. **Summer and winter temperatures at climatic stations located in the vicinity of treeline in Alaska and Canada. Mean temperature for the warmest month of the year at treeline is relatively constant at 11-12 °C (51.8-53.6 °F), but the mean temperature for the coldest month of the year varies by tens of degrees (Tuhkanen, 1982).**

the treeline passes north of climatic isopleths at many points in extremely continental areas of the Soviet Union. This can be seen clearly around the Bay of Khatanga, where trees survive and grow in 75-day growing season and monthly growing degree temperature sum (threshold +5 °C) of only 11.2 °C. The mean July temperature is 12.4 °C (54.3 °F) (Norin, 1978), however, which is relatively warm for latitude 72° N.

On the basis of existing data, it is difficult to say that any of the parameters or indices is better than the others. All are satisfactory indicators of northern treeline. The Holdridge biotemperature, roughly the sum of the monthly mean temperatures above 0 °C divided by 12 (Holdridge 1966, 1967), however, may have some advantages. For example, in Tierra del Fuego, the mean temperature for the warmest month of the year at upper treeline (upper limit of Nothofagus krummholz) is only 6 to 6.5 °C (42.8 to 43.7 °F), but the biotemperature is 2.5 to 3 °C. Growing season in a highly oceanic climate is relatively long and may compensate low mean temperatures. Nevertheless, I agree with Professor Wardle (in press) that no formula has universal validity for all extra-tropical regions.

In general, the isopleths described so far pass to the north of interesting "anomaly" regions such as Iceland, the Faeroe Islands, the Aleutian Islands, and southwest Greenland. Natural forests survive in Iceland and in the southwest corner of Greenland. The Faeroe Islands are naturally treeless, but plantings indicate they are well inside the potential treeline (Ødum, 1979, 1990). The Aleutian Islands may also be treeless for reasons associated with the distribution history of trees following the most recent glaciation (Griggs, 1934). Plantings of Sitka spruce (Picea sitchensis) have been made with modest success (Alden and Bruce, 1989). In any case, the Aleutians should not be included in the phytogeographical Arctic zone, because the composition of plant communities at lower altitudes is predominantly boreal (Hämet-Ahti, 1981).

Widely different opinions exist on the assignment of the highly oceanic corners of the Northern Hemisphere circumpolar zone to biogeographical regions and zones. The same is

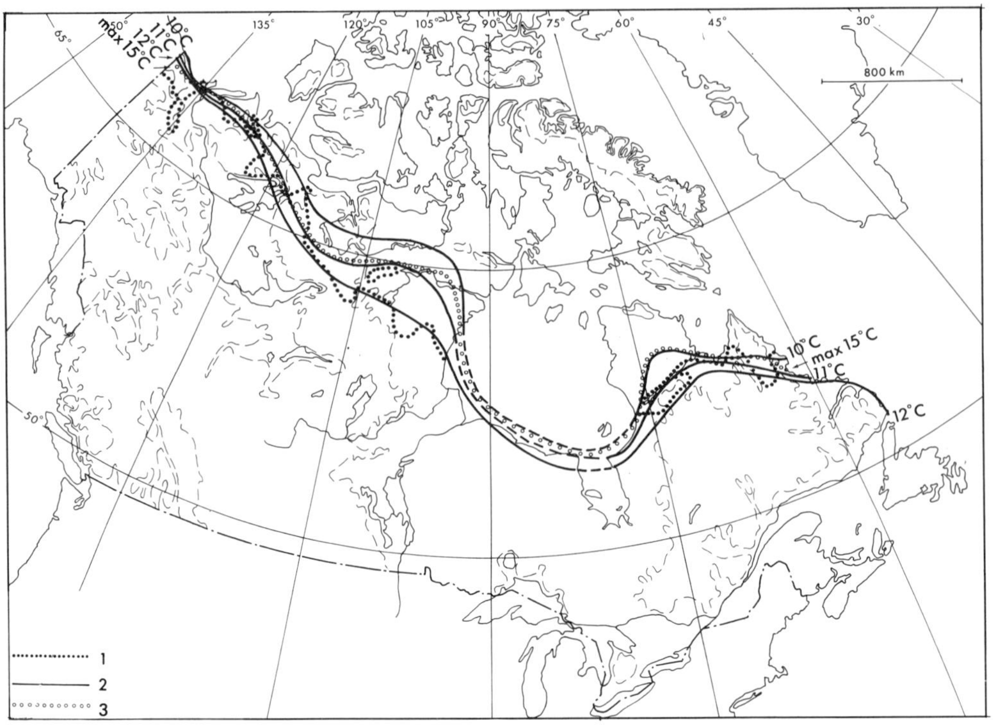

Fig. 6. A comparison between: (1) treeline (Rowe, 1972), (2) 10, 11, and 12 °C (50, 51.8, and 53.6 °F) isotherms for the warmest month of the year (July or August), and (3) 15 °C (59 °F) isotherm for mean daily maximum temperature in the warmest month of the year (Tuhkanen, 1980). The climatic data used are from the Canadian climatic summaries from 1931-60.

true for Tierra del Fuego or the Auckland and Campbell Islands, for example, in the Southern Hemisphere.

Vegetation of the ocean coasts and islands, and that of the hypercontinental areas, differs significantly in many respects from the "average" plant cover of the majority of the boreal zone. The anomalies are primarily sectorial rather than zonational, however, and can be interpreted as normal, representing the two extremes of the climatic continentality oceanity continuum within the boreal zone.

Accordingly, treelessness cannot be automatically considered as a criterion for the Arctic or Antarctic phytogeographical zone. Each of the boreal subzones has its own treeless, or partly treeless, oceanic sector. The corresponding zone in the Southern Hemisphere, the "antiboreal" zone, is also treeless in part. Similarly, the birch and alder forests found in humid, oceanic areas in all the corners of the northern circumpolar zone can be regarded as sectorial, not zonational, features (Hämet-Ahti, 1981). There is no reason to over-emphasize the role of trees in phytogeographical zonations.

DEVIATION OF CLIMATE ISOPLETHS FROM TREELINES

There are several reasons to suggest that treeline would not coincide perfectly with the isopleth of any climatic indicator, even if it could be located unambiguously and exactly on a map, and even if it were possible to construct climatic isopleths more accurately.

First, we are dealing only with climatic indicators and characteristics, and not with the actual determining factors for the positioning of treeline.

Treeline is not a stable phenomenon. Treelines react even to relatively short-term fluctuations in climate. In the 1930s, treelines were reported to be advancing in many parts of the northern circumpolar zone. Later, in the 1950s and 1960s, treelines were retreating. Recent reports from Scandinavia indicate that treelines continue to recede (Gorchakovsky and Shiyatov, 1978; Kullman, 1979, 1990; Scott et al., 1987). The climate of the North Atlantic region in particular has been cooling since the early 1950s (Alexandersson and Eriksson, 1989; Jones et al., 1987; Koutaniemi, 1990). On the other hand, altitudinal expansion of forests is reported locally (Sonesson and Hoogesteger, 1983).

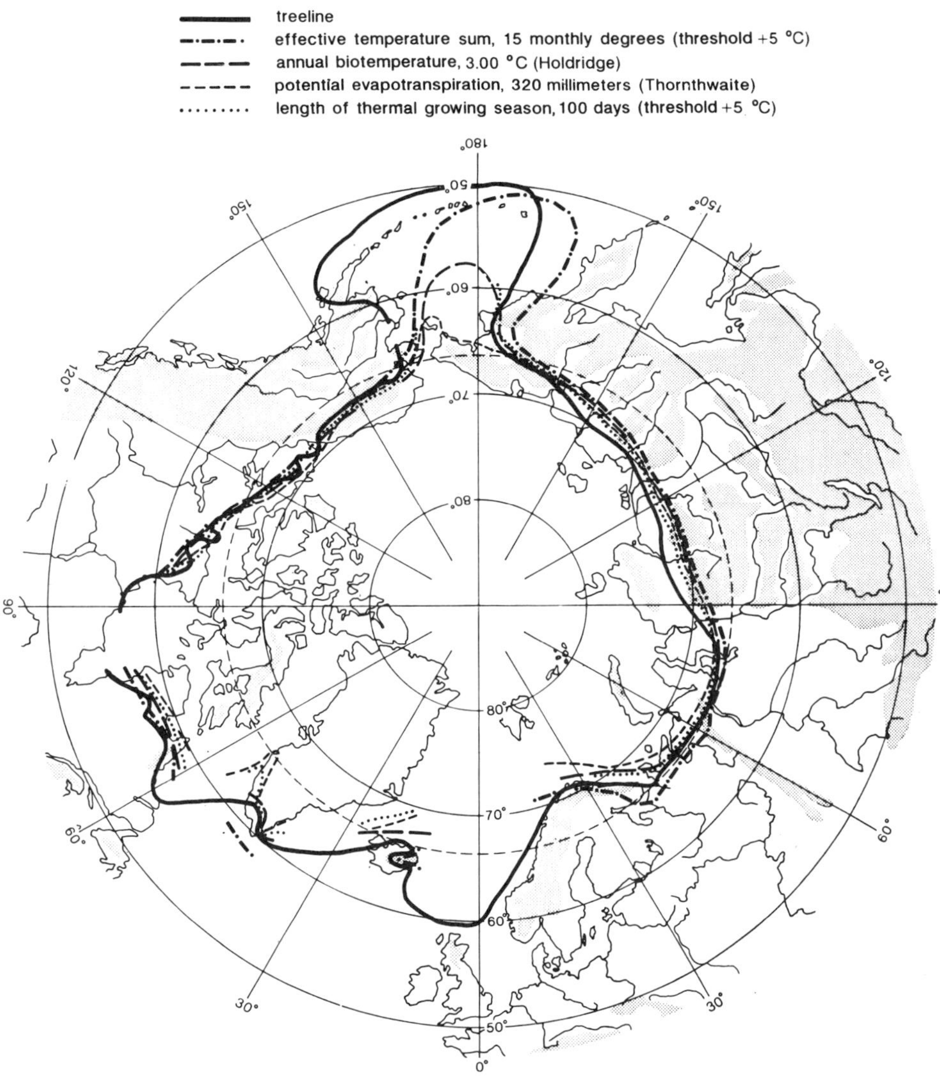

Fig. 7. A comparison between treeline and the isopleths of some climatic indices in the northern circumpolar area. Biotemperature according to Holdridge (1966, 1967), potential evapotranspiration according to Thornthwaite (1948). Climatic data are from Alaskan, Canadian, Danish, Icelandic, Finnish, and Soviet climatic summaries, mainly for 1931-60.

Research from central Canada suggests that the treeline should be regarded as a relict (Elliot, 1979; Larsen, 1972; Nichols, 1976). In southwestern Alaska and western Norway, however, spruce has not had time to reach its climatic limit (Griggs, 1934; Tallantire, 1972).

The treeline is probably capable of advancing rapidly in response to an improvement in climate provided there is no topographic impediment. It is slower to retreat in response to deterioration in climate, because trees can live for hundreds of years without reproducing (LaMarche and Mooney, 1967; Payette et al., 1985). Kullman (1990) does not, however, support this view. According to him, responses to climatic cooling may be as rapid as to warming. Obviously, species characteristics, such as longevity, have considerable effect on treeline stability (Scuderi, 1987).

Treeline, however, is not always determined by climate. The northern treeline has retreated in many places as a consequence of human activities, such as reindeer grazing and timber felling (Brockmann-Jerosch, 1919; Hustich, 1948, 1966; Ritchie, 1960; Tikhomirov, 1961). Frequently, human activities trigger rapid retreat of treelines because trees are vulnerable to small changes in the environment at their limits. Similarly, catastrophic events such as forest fires, avalanches, and volcanic eruptions may also initiate retreat of the treeline (Hämet-Ahti, 1963; Hustich, 1941, 1957, 1966; Nichols, 1976; Payette and Gagnon, 1979; Veblen et al., 1977), which may be able to recover naturally after these events (Shankman, 1984).

Certain biotic factors such as insect pests, for example, a geometrid (Oporinia autumnalis (Bkh.)), have caused the treeline of mountain birch (Betula pubescens ssp. tortuosa) to retreat in places in northern Fennoscandia (Kallio and Lehtonen, 1973; Nuorteva, 1963; Seppälä and Rastas, 1980).

Edaphic and topographic factors also are important in determining treelines. In rough and rocky terrain and in wet mires, trees may not be able to thrive and treeline is often forced to withdraw. The oldest known tree, the bristlecone pine (Pinus longaeva Bailey) grows about 100 meters higher in nutrient-rich dolomite and limestone regions than in regions where sandstone and shale form the nutrient-poor substrates (LaMarche, 1973).

The effects of all non-climatic environmental factors are likely to cause deviations from the potentially climatic treeline in the order of tens of kilometers, possibly 100 kilometers in some extreme cases, and 100 or 200 meters in the Alps.

An interesting point is that different tree species form treeline in different parts of the circumpolar region. From a global point of view, taxonomic diversity of treeline is much higher. Of tens of thousands of tree species in the world, only a small fraction, possibly a few hundred, reach the thermally-controlled treeline. Nevertheless, they belong to different genera and different families and have different kinds of ecomorphological adaptations. In light of this, it is surprising that some climatic isopleths do indicate the position of northern treeline relatively well. The same is true for both latitudinal and altitudinal treelines in most extra-tropical regions. Daubenmire (1954, 1978) suggested that some fundamental limiting factor for arborescent growth can override differences in the ecological requirements of the various species that form timber line around the world. He wrote (Daubenmire, 1954):

> Because a great many genetically distinct trees contribute different segments of a timberline pattern that has a remarkable geographic conformity, the hypothesis is suggested that a major autecologic principle is involved that may be analogous to the wilting coefficient of the soil, in which some environmental complex abruptly exceeds the tolerance of all trees regardless of variation among them.

A certain dichotomy exist in the literature on this point. Some researchers support the idea of a general principle behind the treeline phenomenon (Dahl, 1986; Daubenmire, 1954, 1978; Tranquillini, 1979), but others (Hämet-Ahti, 1979b; Hermes, 1955; Troll, 1973) disagree, emphasizing instead the taxonomical and ecological diversity of treeline.

I do not want to exaggerate the predictability of treeline in terms of any particular climatic indicator. There are many deviations. For instance, natural treeline formed by the Siebold's beech (Fagus crenata Blume) in certain inner parts of Honshu Island, Japan, is interpreted as being in the orotemperate zone and not the oroboreal zone (Hämet-Ahti et al., 1974). In the tropics, there are still more striking examples. In the Serra do Mar Mountains of Brazil, for example, the upper treeline occurs at 2,000 meters elevation. This treeless summit area, however, experiences only slight frost even in winter, and 10 months of the year have mean temperatures above 10 °C (Arno and Hammerly, 1984). Of course, it is possible to infer that the climatic indicators could show the potential extreme treeline if the local species assembly were different.

Treeline is an important and highly interesting biogeographical phenomenon but, as Professor Leena Hämet-Ahti, a Finnish botanist, has emphasized, it cannot be used as a "zero line" for intercontinental and transcontinental comparisons of vegetation (Hämet-Ahti, 1979b). Therefore, the term "subalpine" should similarly be avoided because it may give an erroneous impression of parallelism. For instance, the subalpine forests of the subalpine fir (Abies lasiocarpa (Hooker) Nutt.) and Engelmann spruce (Picea engelmannii (Parry) Engelm.) of eastern British Columbia, and the subalpine forests of the Siebold's beech (Fagus crenata) of Honshu are ecologically very different and cannot be compared. Moreover, do the subalpine forests include the transition belt between treeline proper and the forest line, or is this parkland the "subalpine" forest (Löve, 1970)? How is the lower altitudinal limit of these forests determined? The use of dominant tree species as the only criterion is hardly enough in comparative studies. Accordingly, the contents of the terms "subalpine" and "alpine," and possibly "subarctic" and "arctic," may not be any more than a reference to the location of a vegetation formation in relation to the local forest or treeline.

REFERENCES

Ahti, T., 1959, Studies on the caribou lichen stands of Newfoundland, Annales Botanici Societatis Zoologicae Botanicae Fennicae 'Vanamo', 30(4):1-44.

Ahti, T., 1980, Definition and subdivision of the subarctic: A circumpolar view, Canadian Botanical Association Bulletin, 13(2)(Suppl.):3-10.

Ahti, T., Hämet-Ahti, L., and Jalas, J., 1968, Vegetation zones and their sections in northwestern Europe, Annales Botanici Fennici, 5:169-211.

Alden, J. and Bruce, D., 1989, Growth of historical Sitka spruce plantations at Unalaska Bay, Alaska, General Technical Report PNW-236, Portland, OR., U.S. Department of Agriculture, Forest Service, Pacific Northwest Research Station, 18 p.

Alexandersson, H., and Eriksson, B., 1989, Climate fluctuations in Sweden 1860-1987, SMHI Reports, Meteorology and Climatology, 58:1-54.

Alexandrova, V. D., 1970, The vegetation of the tundra zones in the USSR and data about its productivity. In: Fuller, W. A., and Kevan, P. G., eds., Productivity and conservation in northern circumpolar lands, IUCN Publications, 16:93-114.

Andersson, G., 1902, Zur Pflanzengeographie der Arktis, Geographische Zeitschrift, 8:1-23.

Arno, S. F. and Hammerly, R. P., 1984, Timberline. Mountain and Arctic forest frontiers, Seattle, Wash., The Mountaineers, 304 p.

Atkinson, K., 1981, Vegetation zonation in the Canadian Subarctic, Area, 13:13-17.

Beals, E. W., 1966, Vegetation of cottonwood forests of Kodiak Island, Canadian Field-Naturalist, 80:166-171.

Bjarnason, H., 1967, Forestry in Iceland, Journal of the Royal Scottish Society, 22:55-60.

Blöndal, S., 1987, Afforestation and reforestation in Iceland, Arctic and Alpine Research, 19(4):526-529.

Böcher, T. W., 1979, Birch woodlands and tree growth in southern Greenland, Holarctic Ecology, 2:218-221.

Braathe, P., 1977, Biological and technological aspects of forest management in Norway. In: North American forest lands at latitudes north of 60 degrees. Proceedings of a Symposium, 1977 September 19-22, at the University of Alaska, School of Agriculture and Land Resources Management and Agriculture Experiment Station, Fairbanks, Alaska, p. 35-54.

Brockmann-Jerosch, H., 1919, Baumgrenze und Klimacharacter, Beiträge zur Geobotanischen Landesaufnahme der Schweiz, 6:1-255.

Brockmann-Jerosch, H., 1928, Die südpolare Baumgrenze. Festschrift Hans Schinz, Beiblatt zur Vierteljahrschrift der Naturforschenden Gesellschaft in Zürich, 73:705-718.

Brown, R. J. E., 1960, The distribution of permafrost and its relations to air temperature in Canada and the U.S.S.R., Arctic, 13:163-177.

Brown, R. J. E., 1970, Permafrost as an ecological factor in the Subarctic. In: Ecology of the subarctic regions, Écologie des régions subarctiques, UNESCO, Proceedings of the Helsinki Symposium, Actes du colloque d'Helsinki, Ecological Conservation, 1:129-140.

Bruce, D., and Court, A., 1945, Trees for the Aleutians, Geographical Review, 35:418-423.

Bryson, R. A., 1966, Air masses, streamlines and the boreal forest, Geographical Bulletin, Ottawa, Canada, 8:228-269.

Burrows, C. J., McQueen, D. R., Esler, A. E., and Wardle, P., 1979, New Zealand heathlands. In: Specht, R. L., ed., Heathlands and related scrublands, Ecosystems of the World, Amsterdam, The Netherlands, Elsevier, 9A:333-364.

Dahl, E., 1986, Zonation in arctic and alpine tundra and fell-field ecobiomes. In: Polunin, N., ed., Ecosystem theory and application, New York, NY, John Wiley & Sons, 35-62.

Daubenmire, R., 1954, Alpine timberlines in the Americas and their interpretation, Butler University Botanical Studies, 11:119-136.

Daubenmire, R. F., 1978, Plant geography, with special reference to North America, New York, Academic Press, 338 p.

Dollenz, O., 1980, Estudios fitosociológicos en el archipiélago Cabo de Hornos. I. Relevamientos en caleta Lientur, isla Wollaston y surgidero Romanche, isla Bayly, Anales del Instituto de la Patagonia, 11:225-238.

Eidem, P., 1955, En oversikt over skoggrenseundersøkelser i Norge, Norsk Geografisk Tidsskrift, 15:159-172.

Elliot, D. L., 1979, The current regenerative capacity on the northern Canadian trees, Keewatin, N.W. T., Canada: some preliminary observations, Arctic and Alpine Research, 11:243-251.

Fægri, K., 1969, A note on the maritime forest limit in southeast Alaska, Årbok for Universitetet i Bergen, Matematisk-naturvitenskaplig Serie, 1968(5):1-20.

Fægri, K., 1971, Geo-ökologische Probleme der Gebirge Skandinaviens. In: Troll, C., ed., Geoecology of the highmountain regions of Eurasia, Wiesbaden, Germany, Franz Steiner Verlag, Erdwissenschaftliche Forschung, 4:98-106.

Falla, R. A., 1948, The outlying islands of New Zealand, New Zealand Geographer, 4:127-154.

Franklin, J., and Dyrness, C. T., 1973, Natural vegetation of Oregon and Washington. General Technical Report PNW-8, Portland, OR., United States, 417 p.

Gorchakovsky, P. L., and Shiyatov, S. G., 1978, The upper forest limit in the mountains of the boreal zone of the USSR, Arctic and Alpine Research, 10:349-363.

Grace, J., 1989, Tree lines, Philosophical Transactions of the Royal Society of London, B, 324:233-245.

Griggs, R. F., 1934, The edge of the forest in Alaska and the reasons for its position, Ecology, 15:80-96.

Grigor'yev, A. A., and Budyko, M. I., 1960, Classification of the climates of the U.S.S.R., Soviet Geography, 1:3-24.

Grøntved, J., 1942, The Pteridophyta and Spermatophyta of Iceland. In: The botany of Iceland IV, Part I. Copenhagen, Einar Munksgaard & London, Humphrey Milford, 427 p.

Hämet-Ahti, L., 1963, Zonation of the mountain birch forests in northernmost Fennoscandia, Annales Botanici Societatis Zoologicae Botanicae Fennicae 'Vanamo', 34(4):1-127.

Hämet-Ahti, L., 1979a, The vegetational phytogeographic zonation across Bering Strait-Bering Sea, Paper (unpublished), The XIV Pacific Science Congress, Khabarovsk, 6 p.

Hämet-Ahti, L., 1979b, The dangers of using timberline as the "zero line" in comparative studies on altitudinal vegetation zones, Phytocoenologia, 6:49-54.

Hämet-Ahti, L., 1981, The boreal zone and its biotic subdivision, Fennia, 159:69-75.

Hämet-Ahti, L., Ahti, T., and Koponen, T., 1974, A scheme of vegetation zones for Japan and adjacent regions, Annales Botanici Fennici, 11:59-88.

Hare, F. K., 1950, Climate and zonal divisions of the boreal forest formation in eastern Canada, Geographical Review, 40:615-635.

Hare, F. K., 1954, The boreal conifer zone, Geographical Studies, 1:4-18.

Hare, F. K., 1964, New light from Labrador-Ungava, Annals of the Association of American Geographers, 54:459-476.

Hare, F. K., and Ritchie, J. C., 1972, The boreal bioclimates, Geographical Review, 62:333-365.

Heikkinen, O., 1984, The timber-line problem, Nordia, 18:105-114.

Hermes, K., 1955, Die Lage der oberen Waldgrenze in den Gebirgen der Erde und ihr Abstand zur Schneegrenze, Kölner Geographische Arbeiten, 5:1-277.

Holdridge, L., 1966, The life zone system, Adansonia, 6:199-203.

Holdridge, L., 1967, "Life zone ecology," San José, Costa Rica, Tropical Science Centre, 206 p.

Holtmeier, F.-K., 1985a, Die klimatische Waldgrenze—Linie oder Übergangsraum (Ökoton)?, Erdkunde, 39:271-285.

Holtmeier, F.-K., 1985b, Climate stress influencing the physiognomy of trees at the polar and mountain timberline, Eidgenössische Anstalt für das Forstliche Versuchswesen, Berichte, 270:31-40.

Hustich, I., 1941, Tallens tjocklekstillväxt som klimatisk indikator, Terra, Helsinki, Finland, 53:42-54.

Hustich, I., 1948, The Scotch pine in the northernmost Finland, Acta Botanica Fennica, 42:1-75.

Hustich, I., 1957, On the phytogeography of the subarctic Hudson Bay lowland, Acta Geographica, Helsinki, Finland, 16:1-48.

Hustich, I., 1966, On the forest-tundra and the northern treelines, A preliminary synthesis, Annales Universitatis Turkuensis A II 36, Reports from the Kevo Subarctic Research Station, Finland, 3:7-47.

Hustich, I., 1979, Ecological concepts and biogeographical zonation in the North: The need for a generally accepted terminology, Holarctic Ecology, 2:208-217.

Hustich, I., 1983, Tree-line and tree growth studies during 50 years: some subjective observations, Collection Nordicana, 47:181-188.

Jones, P. D., Wigley, T. M., Folland, C. I., and Parker, D. E., 1987, Spatial patterns in recent worldwide temperature trends, Climate Monitor, 16:175-180.
Jónsson, H., 1905, Vegetationen i Syd-Island, Botanisk Tidsskrift, 27:1-82.
Kallio, P., and Lehtonen, J., 1973, Birch damage caused by Oporinia autumnata (Bkh.) in 1965-66 in Utsjoki, N. Finland, Annales Universitas Turkuensis A II 53, Reports from the Kevo Subarctic Research Station, Finland, 10:55-69.
Köppen, W., 1919, Baumgrenze und Lufttemperatur, Petermanns Geographische Mitteilungen, 65:39-42.
Koutaniemi, L., 1990, Suomen lämpö- ja sadeolojen viimeaikainen kehitys, Recent trends in temperature and precipitation changes in Finland, Terra, Helsinki, Finland, 102:275-283.
Krebs, J. S., and Barry, R. G., 1970, The arctic front and the tundra—taiga boundary in Eurasia, Geographical Review, 60:548-554.
Kullman, L., 1979, Change and stability in the altitude of the birch tree-limit in the southern Swedish Scandes 1915-1975, Acta Phytogeographica Suecica, 65:1-121.
Kullman, L., 1990, Dynamics of altitudinal tree-limits in Sweden: a review, Norsk Geografisk Tidsskrift, 44:103-116.
La Grange, J. J., 1954, The South African station on Marion Island, Polar Record, 7:155-158.
LaMarche Jr., V. C., 1973, Holocen climatic variations inferred from treeline fluctuations in the White Mountains, California, Quaternary Research, 3:632-660.
LaMarche Jr., V. C., and Mooney, H. A., 1967, Altithermal timberline advance in western United States, Nature, 213:980-982.
Larcher, W., 1984, Klima und Pflanzenleben im Gebirge, Universitas, Universität Innsbruck, Austria, 39:371-402.
Larsen, J. A., 1965, The vegetation of the Ennadai Lake area, N.W. T.: Studies in Subarctic and Arctic bioclimatology, Ecological Monographs, 35:37-59.
Larsen, J. A., 1972, Growth of spruce at Dubawnt Lake, N.W. T., Arctic, 25:59.
Lavrenko, E. M., 1950, Osnovnye cherty botaniko-geograficheskogo razdeleniya SSSR i sopredel'nykh stran, Problemy Botaniki, 1:530-548.
Lindemann, R., 1970, Das Problem der Waldlosigkeit der norwegischen Atlantküste, Deutscher Geographentag Kiel 1969, Tagungsberichte und Wissenschaftliche Abhandlungen, Wiesbaden, Germany, p. 61-72.
Löve, D., 1970, Subarctic and subalpine: where and what?, Arctic and Alpine Research, 2:63-73.
Low, A. J., 1986, Tree planting in the Falkland Islands, Forestry, 59:59-64.
McVean, D. N., 1955, Notes on the vegetation of Iceland. In: Transactions and Proceedings of the Botanical Society of Edinburgh, 36:320-338.
Meades, W. J., 1983, Heathlands. In: South, G. R., ed., Biogeography and ecology of the island of Newfoundland, Monographiae Biologicae, 48:267-318.
Mikola, P., 1978, Consequences of climatic fluctuation in forestry, Fennia, 150:39-43.
Mitchell, V. L., 1973, A theoretical tree line in central Canada, Annals of the Association of American Geographers, 63:296-301.
Nichols, H., 1976, Historical aspects of the northern Canadian treeline, Arctic, 29:38-47.
Norin, B. M., ed., 1978, Ary-Mas, Prirodnye usloviya, flora i rastitel'nost' samogo severnogo v mire lesnogo massiva, Leningrad, Nauka, 192 p.
Nuorteva, P., 1963, The influence of Oporinia autumnata (Bkh.) (Lep., Geometridae) on the timberline in subarctic conditions, Annales Entomoloci Fennici, 29:270-277.
Ødum, S., 1979, Actual and potential tree-line in the North Atlantic region, especially in Greenland and the Faroes, Holarctic Ecology, 2:222-227.
Ødum, S., 1990, Afforestation experiments reflecting the tree-line conditions in Southwest Greenland, Meddelelser om Grønland, Bioscience, 33:43-61.

Ødum, S., 1991, Choice of species and origins for arboriculture in Greenland and the Faroe Islands, Dansk Dendrologisk Årsskrift, 9:1-78.

Ødum, S., Hansen, L., and Rasmussen, S., 1989, Commented inventory of trees and schrubs in gardens and plantations. In: Højgaard, A., Johansen, J., and Ødum, S., eds., A century of treeplanting in the Faroe Islands, Annales Societatis Scientiarum Faeroensis, Supplementum, 14:85-124.

Ohsawa, M., 1990, An interpretation of latitudinal patterns of forest limits in South and East Asian mountains, Journal of Ecology, 78:326-339.

Payette, S., 1983, The forest tundra and present tree-lines of the northern Québec-Labrador Peninsula, Collection Nordicana, 47:3-23.

Payette, S., Filion, L., Gauthier, L., and Boutin, Y., 1985, Secular climate change in old-growth tree-line vegetation in northern Québec, Nature, 315:135-138.

Payette, S., and Gagnon, R., 1979, Tree-line dynamics in Ungava Peninsula, northern Quebec, Holarctic Ecology, 2:239-248.

Pears, N., 1985, Basic biogeography, 2nd edition, Longman, London, 358 p.

Pignatti, S., 1980, Zum Problem der Höhenstufen und Vegetationszonen, Phytocoenologia, 7:52-64.

Pisano, E., and Venegas, C., 1984, Primera colección botánica y esquema fitogeográfico de la Isla Noir, Tierra del Fuego, Anales del Instituto de la Patagonia, Serie Ciencias Naturales, 15:25-35.

Pohtila, E., 1980, Climatic fluctuations and forestry in Lapland, Holarctic Ecology, 3:91-98.

Ritchie, J. G., 1960, The vegetations of Northern Manitoba. V: Establishing the major zonation, Arctic, 13:210-229.

Rowe, J. S., 1972, Forest regions of Canada, Publications of the Department of the Environment, Canadian Forest Service, 1300:1-172.

Scott, P. A., Hansell, R. J., and Fayle, D. C., 1987, Establishment of white spruce populations and responses to climatic change at the treeline, Churchill, Manitoba, Canada, Arctic and Alpine Research, 19:175-186.

Scuderi, L. A., 1987, Late-Holocene upper timberline variation in the southern Sierra Nevada, Nature, 325:242-244.

Seppälä, M., and Rastas, J., 1980, Vegetation map of northernmost Finland with special reference to subarctic forest limits and natural hazards, Fennia, 158:41-61.

Shankman, D., 1984, Tree regeneration following fire as evidence of timberline stability in the Colorado Front Range, U.S.A., Arctic and Alpine Research, 16:413-417.

Sonesson, M., and Hoogesteger, J., 1983, Recent tree-line dynamics (Betula pubescens Ehrh. ssp. tortuosa (Ledeb.) Nyman) in northern Sweden, Collection Nordicana, 47:47-54.

Steindórsson, S., 1962, On the age and immigration of the Icelandic flora, Vísindafélag Islendinga, Societas Scientiarum Islandica, 3:1-157.

Supan, A., 1884, Grundzüge der physischen Erdkunde, Leipzig, Veit et Comp., 492 p.

Tallantire, P. A., 1972, The regional spread of spruce (Picea abies (L.) Karst) within Fennoscandia, A reassessment, Norwegian Journal of Botany, 19:1-16.

Thornthwaite, C. W., 1948, An approach toward a rational classification of climate, Geographical Review, 33:233-255.

Tikhomirov, B. A., 1961, The changes in biogeographical boundaries in the north of U.S.S.R. as related with climatic fluctuations and activity of man, Botanisk Tidsskrift, 56:285-292.

Tikhomirov, B. A., 1970, Forest limits as the most important biogeographical boundary in the North. In: Ecology of the subarctic regions, Écologie des régions subarctiques, UNESCO, Proceedings of the Helsinki Symposium, Actes du colloque d'Helsinki, Ecological Conservation, 1:35-40.

Tranquillini, W., 1979, Physiological ecology of the alpine timberline. Ecological Studies 31, Springer-Verlag, Berlin, Heidelberg, and New York, 137 p.

Troll, C., 1973, The upper timberlines in different climatic zones, Arctic and Alpine Research, 5:A3-A18.

Troll, C., 1978, Der asymmetrische Vegetations- und Landschaftsbau der Nord- und Süd-halbkugel. In: Troll, C., and Lauer, W., eds., Geoecological relations between the southern temperate zone and the tropical mountains, Proceedings of the symposium of the International Geographical Union Commission on High-Altitude Geoecology, 1974 November 21-23 at Franz Steiner Verlag, Erdwissenschaftliche Forschung, Mainz, Wiesbaden, Germany, 11:10-28.

Tuhkanen, S., 1980, Climatic parametres and indices in plant geography, Acta Phytogeographica Suecica, 67:1-110.

Tuhkanen, S., 1982, Trädgränsen och klimatet, Naturen, 1982:82-94.

Tuhkanen, S., 1984, A circumboreal system of climatic phytogeographical regions, Acta Botanica Fennica, 127:1-50.

Tuhkanen, S., 1987, The phytogeographical position of the Faeroe Islands and their eco-climatic correspondences on the other continents: Problems associated with highly oceanic regions, Annales Botanici Fennici, 24:111-135.

Veblen, T. T., Ashton, D. H., Schlegel, F. M., and Veblen, A. T., 1977, Plant succession in a timberline depressed by vulcanism in south-central Chile, Journal of Biogeography, 4:275-294.

Wardle, P., 1965, A comparison of alpine timber lines in New Zealand and North America, New Zealand Journal of Botany, 3:113-135.

Wardle, P., 1971, An explanation for alpine timberline, New Zealand Journal of Botany, 9:371-402.

Wardle, P., 1974, Alpine timberlines. In: Ives, J. D., and Barry, R. G., eds., Arctic and Alpine environments, Methuen, London, 371-402.

Wardle, P., [In press], Causes of Alpine timberline; a review of the hypotheses. In: Alden, J. N., Ødum, S., and Mastrantonio, J. L., eds., Forest Development in Cold Climates, Proceedings of the NATO Advanced Research Workshop, 1991 June 18-23, [Laugarvatn, Iceland], Plenum, New York.

Weischet, W., 1978, Die ökologisch wichtigen Charakteristika der kühl-gemässigten Zone Südamerikas mit vergleichenden Anmerkungen zu den tropischen Hochgebirgen. In: Troll, C., and Lauer, W., eds., Geographical relations between the southern temperate zone and the tropical mountains, Proceedings of the symposium of the International Geographical Union Commission of High-Altitude Geoecology, 1974 November 21-23 at Mainz, Wiesbaden, Germany, Franz Steiner Verlag, Erdwissenschaftliche Forschung, 11:255-280.

Young, S. B., 1971, The vascular flora of St. Lawrence Island with special reference to floristic zonation in the Arctic regions, Contributions from the Gray Herbarium of Harvard University, Cambridge, Massachusetts, 201:11-115.

Young, S. B., 1972, Subantarctic rain forest of Magellanic Chile: distribution, composition, and age and growth rate studies of common forest trees, Antarctic Research Series (U.S.), 20:307-322.

PLIOCENE FOSSIL NOTHOFAGUS (SOUTHERN BEECH) FROM ANTARCTICA: PHYTOGEOGRAPHY, DISPERSAL STRATEGIES, AND SURVIVAL IN HIGH LATITUDE GLACIAL-DEGLACIAL ENVIRONMENTS

P.-N. Webb[1] and D. M. Harwood[2]

[1]Department of Geological Sciences and Byrd Polar Research Center
The Ohio State University
Columbus, Ohio, United States

[2]Department of Geology
University of Nebraska
Lincoln, Nebraska, United States

INTRODUCTION

A recently discovered vegetational biome in the Pliocene Sirius Group of the Dominion Range, Transantarctic Mountains, Antarctica (Figs. 1 and 2) is discussed. The flora is dominated by the southern beech tree genus Nothofagus. The geological occurrence, phytogeographic implications, dispersal, survival, climatic significance, and eventual demise of this flora are discussed, as are recent experiments in which species of Nothofagus have been transplanted from South American and Australasian forests to a variety of Northern Hemisphere alpine and Arctic post-glacial environments. The ability of Nothofagus to survive hostile glacial-deglacial conditions at very high southern latitudes for almost the entire duration of the Cenozoic Era (66 million years) is documented.

Is a discussion of the Late Cenozoic fossil vegetation of Antarctica of interest to a meeting convened in Iceland and devoted to forest development in cold climates? In answering this question we must consider several points. First, paleoclimate data and modelling experiments point to the likelihood of significant bipolar warming at high latitudes during the next century. Parts of the time interval discussed here, the Pliocene Epoch (5.2 to 1.6 million years before present), were periods of relative warmth in the marine and terrestrial environments of both the Arctic and Antarctic regions. Budyko and Izrael (1987) suggest that summer air temperatures in the Arctic Circle were 6 °C to 10 °C warmer than now during the period 4.3 to 3.3 million years before present. Antarctic data indicate similar or even more pronounced temperature increases in the early and middle Pliocene (Webb and Harwood, 1991). This period of geological time has been termed the "Pliocene Climatic Optimum," and in some scenarios of global warming, the interval is

Forest Development in Cold Climates, Edited by J. Alden *et al.*
Plenum Press, New York, 1993

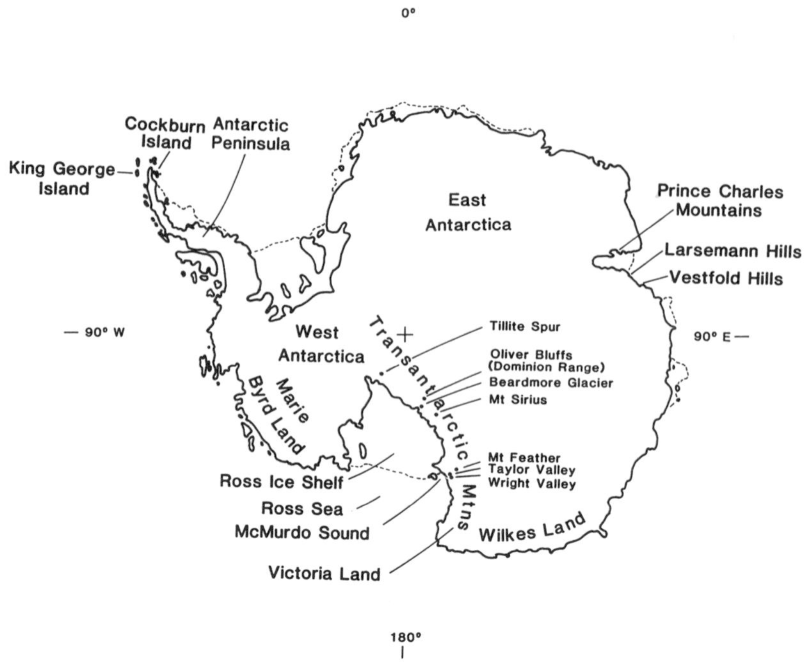

Fig. 1. Important Pliocene terrestrial and marine localities in Antarctica.

suggested as a possible analog for climate conditions in the mid twenty-first century (Houghton et al., 1990). If this degree and rapidity of climate change were to occur in future, polar biomes would be restructured, with the potential for re-establishment of tree communities in the present taiga and polar desert terrains of both polar regions.

The Antarctic paleobotanical data discussed here are not derived from the "Pliocene Climatic Optimum" record, but from what we term the "post-Pliocene Climatic Optimum," an interval of time approximately equivalent to the late Pliocene (3 to 1.6 million years before present). This is a transitional cooler interval in Antarctica between Late Cenozoic "temperate to subpolar" glaciation with relatively small and dynamic ice sheets, and Pleistocene "polar" glaciation with massive and relatively stable continent or sub-continent-sized ice sheets (Webb and Harwood, 1991) (Fig. 3). We discuss three principal issues: (1) the survival of the southern beech (Nothofagus) in Antarctica over millions of years of Cenozoic time, (2) the response of Nothofagus to climate deterioration in Antarctica and adjacent landmasses near the end of the Pliocene, and (3) the potential for the geographic range of Nothofagus to expand and re-establish communities, with human assistance, in the southern high latitudes.

THE CONCEPT OF POLAR LATITUDE FORESTS

To many researchers, the possibility of tree biomes in Antarctica is unlikely, even in the geologically recent past. For example, Wace (1965, p. 205) writes, "The very existence of the remains of large trees near the present position of the South Pole probably implies some latitudinal movement of the rocks containing them, since it is difficult to envisage the survival of such plants through long periods of darkness in winter months."

AGE	GROUP	FORMATION	DESCRIPTION	PALEOENVIRONMENTS AND EVENTS
LATE PLEISTOCENE		Beardmore drift III, II, I	Gravel drift/moraines	Drift sheets and lateral moraines derived from Mill and Beardmore glaciers Local erosion and deflation of Sirius Group Faulting and uplift of Sirius Group and older rocks Glacial erosion of Sirius Group
		Beardmore Erosion Surface		
? EARLY PLEISTOCENE	SIRIUS GROUP	Mt Mills Formation 50m	Non-fossiliferous, lithified/poorly lithified diamictite, breccia, conglomerate, sandstone	Glaciogene sedimentary basin formed within Beardmore Valley
LATE PLIOCENE [<2.5-3.1 Ma]		Meyer Desert Formation 150m	Alternating units of diamictite (dominant), fluvioglacial sandstone and siltstone, and finely-bedded glaciolacustrine sandstone and mudstone. In situ roots and stems of Nothofagus and rudimentary soils on upper surfaces of diamictites. Allochthonous twigs and leaves of Nothofagus abundant in fluvial and rare in lacustrine sediments. Fragmented moss cushions in diamictites and fluvial sediments. Fine fractions of all facies contain in situ vascular palynomorphs, and allochthonous Cretaceous-early Pliocene palynomorphs, and siliceous (diatoms, silicoflagellates, radiolarians) and calcareous (foraminifera), microfossils.	Subaerial terrain of glacial streams and lakes, periodically overrun by glacier ice. Lakes both ice-covered and ice-free. Erosion surfaces & vegetated paleosols developed between aqueous and glacier cycles. Duration of cycles unknown.
		Dominion Erosion Surface		High relief glacial erosion within ancestral Beardmore Valley system
JURASSIC	FERRAR DOLERITE & BEACON SUPERGROUPS		Quartz diorite dikes and sills	
TRIASSIC			Quartzose sandstone, mudstone, carbonaceous shale, coal	

Fig. 2. Mesozoic-Cenozoic terrestrial stratigraphy for the southern Transantarctic Mountains. Only the youngest part of the Cenozoic record, the latest Pliocene-Pleistocene, is known in any detail.

The paleontological record suggests otherwise (Figs. 4 and 5). For example, during the Late Cretaceous, New Zealand was situated close to Antarctica (between 68° S. and 78° S. latitude) yet supported a podocarp and beech (Nothofagus) flora under conditions that included low temperatures, long periods of darkness, high rainfall, and cloud cover (Mildenhall, 1980).

A number of Southern Hemisphere plant biogeographers have written with extraordinary confidence and a degree of prophetism on the subject of polar paleoclimate and propagation of polar forests. Van Steenis (1971, p. 77) writes:

> It is difficult to visualize, but nevertheless beyond any reasonable doubt, that the paleoclimate of the Cretaceous and the large part of the Tertiary up to the Pliocene was very much milder. This must have been caused by much diminished (or even absent) polar ice caps. At least the borderlands of Antarctica could easily have carried a temperate to cold-temperate forest, suitable for the growth of Nothofagus. A climate shift of 15 latitudinal degrees from Fuegia where the deciduous N. antarctica is found today to about 70° S. would allow Nothofagus to grow in Antarctica.

Van Steenis (1971, p. 87) also provides a reasonably accurate forecast of the Pliocene vegetation to be recovered from Antarctica fourteen years later in 1985 (Webb and Harwood, 1986, 1987) when he wrote that "the antarctic tree limit conditions at 70° S. would have been similar to those northern hemisphere regions, where dense and tall forests occur along the coastal regions but gradually become more stunted and widely spaced as the treeline is approached."

Creber (1990, p. 40) comments:

> ...the only necessary factor for trees to grow at very high latitudes at the present day is an increase in the average ambient temperature. All other factors required for the evolution of forest polar trees are within the scope of adaptations that are known to be available to modern trees, such as photoperiodic ecotypes, higher rates of cambial activity during the growing season, and an appropriate crown architecture. Additionally, the trees should have availed themselves of the facility to adopt the deciduous habit as an adaptation to the long winter darkness, although there is evidence that the evergreen habit might have been sustainable in a long dark winter.

PHYTOGEOGRAPHIC NOMENCLATURE

The Arctic taiga-tundra ecotone transition zone is a critical boundary in considerations

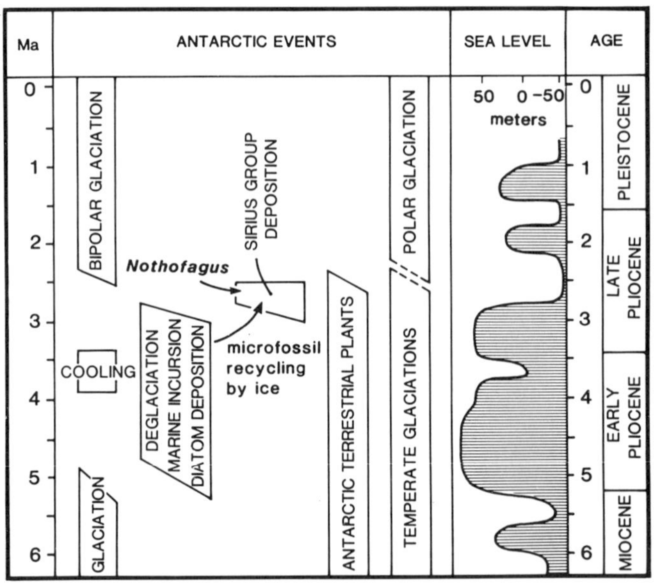

Fig. 3. Age relationships between Sirius Group Nothofagus flora and glaciological and marine history.

of poleward advances by tree communities, but we question whether Arctic biome nomenclature is applicable in describing the vegetation of pre-Pleistocene Antarctica. Nordenskjöld (1928) applied the terms "grass tundra" and "desert tundra" to communities in the present day subantarctic islands, and the term "polar desert" to the whole of Antarctica. These Arctic vegetational concepts have also been used by authors in discussing Antarctic Cenozoic paleobotany (Truswell, 1990), but characterization and comparison of past bipolar vegetation patterns remain at a rudimentary level.

ARCTIC AND ANTARCTIC COMPARISONS

Much can be learned from recognizing and understanding Late Cenozoic vegetational events in Antarctica, and applying these data to forecasting future trends in the northern and southern high latitudes. It is important though, to emphasize major points of difference between the Earth's polar extremities. In Antarctica, a major continent occupies a totally polar position, with land areas spanning latitudes 90° S. to 63° S. In the Arctic, land areas do not extend northward of 80° N. to 83° N. and many physiographic regions are separated by the Arctic Ocean. Parts of Antarctica exhibit extreme topographic contrast with high alpine and lowland landscapes developed close to coastlines. Mountain-building episodes, an important feature of the circum-Pacific margin as a whole in the Late Cenozoic, are also significant in the Antarctic. In contrast, huge expanses of the Arctic terrestrial margin involve relatively subdued low-elevation topography. The glacial-deglacial record of ice sheets in Antarctica has been traced back to the latest Eocene and Oligocene (Hambrey and Barrett, in press; Wise et al., 1991). Ehrmann and Mackensen (1992) provide ice-rafting evidence from two regions off East Antarctica which suggests the earliest glaciation reaching the coastline of Antarctica occurred at about 45.5 million years B.P., or middle Eocene. Widespread glaciation in the Arctic is generally traced from the Pliocene, although there is evidence for late Miocene marine ice cover in the Arctic Ocean (Thiede et al., 1990). Because of the major ocean expanses between Antarctica, the sub-Antarctic islands, and temperate latitude landmasses and islands, Cenozoic vegetation excluded from Antarctica for climatic reasons has found natural re-establishment difficult or impossible on the continent, even with the return to more favorable climatic conditions during the Holocene. By contrast, in many areas of the Arctic, the polar treeline has been able to respond to climate oscillations, shifting naturally both vertically and especially latitudinally. While the reasons for deforestation in Antarctica, and Arctic regions such as Iceland and southern Greenland have different origins, both regions necessitate anthropogenic intervention,

Antarctic Floral History (Cretaceous-Cenozoic)					
Geologic Time	Floral Phases	Cryptogams, Gymnosperms, Angiosperms, Nothofagus	Major Plant Groups	Plant Events	Geologic Events
Holocene	V		Cryptogam dominated vegetation (algae, lichens, fungi, mosses); Minor alien angiosperms	Recolonization of lowland and montane terrain. Appearance of Neoaustral flora	Deglaciation and increase in exposed rock area
mid-late Pleistocene	IV		No known cryptogam, gymnosperm or angiosperm record	?	Major glaciations, East and West Antarctic Ice Sheets; regional cooling
early Pleist. late Pliocene	III		? Nothofagus wood, roots, leaves and pollen; sedge, conifers, mosses, pond weeds	Local extinction of angiosperms and gymnosperms in Antarctica; Last surviving trees in Antarctica (Sirius Group)	Glacial-deglacial conditions, diamicton, fluvial and lacustrine sediments
mid Mio-e. Plio., early Miocene, Oligocene, Eocene, Paleocene, Late Cretaceous	II	? ? ? ?	Nothofagus, angiosperm and gymnosperm pollen; Gymnosperms, angiosperms (incl. Nothofagus), ferns	No in situ terrestrial macroflora known; Widespread distribution in West and East Antarctica of Paleoaustral flora; First appearance of Nothofagus in Antarctica	Separation of Sth Am. & West Antarctica; First widespread glaciations in Antarctica; Separation of Antarctica and Australia; Separation of Antarctica and New Zealand
Early Cretaceous	I			First appearance of angiosperms in West and East Antarctica	Separation of East Antarctica and India

Fig. 4. The major floral phases, floral groups, plant events and geologic events for the Cretaceous-Cenozoic of Antarctica. Floral Phase III, of principal interest here, is shaded.

including perhaps man-induced climate warming, to assist in forest re-establishment. Finally, major taxonomic contrasts exist between the Cenozoic and present day vegetation in the two polar areas, and similar phytogeographic responses may not accompany parallel climate trends at Earth's polar extremities.

The data base on high latitude Antarctic fossil vegetation is important in considerations of future forest development north of the present Arctic treeline, whether or not afforestation is assisted by the projected climate warming. Of principal interest, then, is the recent discovery of a cool to cold rainforest vegetation that survived near the South Pole in Antarctica until about two million years ago, and the maintenance of this isolated tree vegetation through many cycles of glaciation and deglaciation following the disruption of the land linkages between southern South America and Antarctica in the late Oligocene-early Miocene, 30 to 22 million years before present (Barker and Burrell, 1976), and between Tasmania and Antarctica in the late Eocene, about 40 million years before present (Kennett et al., 1975).

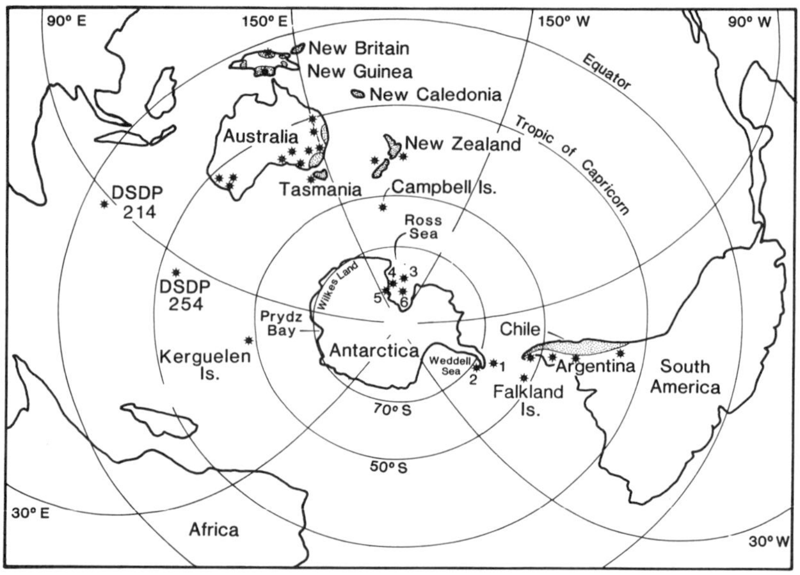

Fig. 5. Present day (stippled) and fossil Cretaceous-Cenozoic (asterisks) distribution of the southern beech (<u>Nothofagus</u>), modified and updated from J. Wardle (1984). Fossil occurrences include <u>in situ</u> and allochthonous roots, wood, leaves, and palynomorphs. The numerous <u>Nothofagus</u> fossil occurrences (mostly palynomorphs) from New Zealand, the continental shelf surrounding New Zealand, Tasmania, and the South Tasman Rise south of Tasmania, are represented schematically. DSDP (Deep Sea Drilling Project) Sites 214 and 254, on Ninetyeast Ridge, contain fossil pollen with affinities to <u>Nothofagus</u>. Fossil localities in Antarctica refer to (1) King George Island, (2) Seymour Island, Snow Hill Island, Vega Island, (3) DSDP Site 270, (4) CIROS-1 drillhole, (5) Oliver Bluffs, Dominion Range, and (6) RISP Site J-9. Localities in the Weddell Sea, Ross Sea, and Wilkes Land regions, where recycled <u>Nothofagus</u> palynomorphs have been recovered from Quaternary sediments, are not plotted. Data derived from Askin (1989), Birkenmajer and Zastawniak (1989), Birnie and Roberts (1986), Carlquist (1987), Costin (1981a, 1981b), Dettmann (1989), Dettmann and Thomson (1987), Howard (1981), Kemp and Harris (1975), Mildenhall (1980, 1989), Suggate (1978), Truswell (1990), Van Steenis (1953, 1972), J. Wardle (1984), Webb and Harwood (1987), and Wrenn and Beckmann (1982).

ANTARCTIC VEGETATION DURING THE CRETACEOUS AND CENOZOIC

Figure 4 summarizes the major phases of floral development, content of floral groups, botanical events, and geological events for the Cretaceous Period and Cenozoic Era in Antarctica. Floral Phases I and II mark the evolution and dispersal of a Paleoaustral (Fleming, 1962, 1963a) vegetation across much of West and East Antarctica. Vascular vegetation is thought to have declined during the early Miocene but no in situ plant data are yet available for the middle-late Miocene to early Pliocene interval (younger part of Floral Phase II). The transition from Floral Phase III to IV represents a floral crisis and marks the extinction of the tree biome in Antarctica, including the demise of the last surviving southern beech (Nothofagus) forests in the Transantarctic Mountains. The vegetational and geological history for the Pleistocene (younger part of Phase III, and Phase IV) is almost totally unknown but may yet provide a cryptogam and depauperate phanerogam record.

Phase V (post glacial/Holocene) is characterized by a redevelopment of the cryptogam flora along with a few alien vascular taxa in coastal regions. Phase V vegetation may represent a southernmost region in Fleming's (1962, 1963a) Neoaustral biogeographic element.

The composition and distribution of the present day flora (Fig. 4) is reviewed by Greene et al. (1967) and Smith (1984). Rudolph (1966, p. 109) summarises this flora thus: "...Antarctica is truly a land of cryptogamic plants, one where the vegetation is composed of species of lichens, algae, mosses and fungi."

TAXONOMY, PHYLOGENY, ECOLOGY AND BIOGEOGRAPHY OF NOTHOFAGUS (SOUTHERN BEECH)

An extensive literature on fossil and living southern hemisphere Nothofagus (Fagaceae) has evolved during the past century. Useful articles and reviews, all of which provide reference to earlier significant literature on this genus, include Hill and Read (1989), Humphries (1981), and Van Steenis (1953) for taxonomy; Humphries (1981) for phylogeny; Costin (1981a, 1981b), Howard (1981), J. Wardle (1984), and P. Wardle (1971; in press) for ecology; and Cranwell (1963), Darlington (1965), Dettmann (1989), and Fleming (1962, 1963a, 1963b, 1975) for biogeography. The distribution of fossil and extant Nothofagus is shown in Figure 5.

CENOZOIC PALEOBOTANICAL RECORD IN THE TRANSANTARCTIC MOUNTAINS

The Transantarctic Mountains extend for some 2,500 kilometers across the continent of Antarctica, form one margin of the East Antarctic craton, are up to 250 kilometers wide, and in places reach elevations of more than 4,000 meters (Fig. 1). What little is known of the terrestrial record is deduced largely from analysis of drill cores in the offshore marine basins of the Ross Sea and the paleofjords which penetrate the Transantarctic Mountains in a few places. These data suggest that the Transantarctic Mountains existed in the late Cretaceous-early Cenozoic (Paleocene-Eocene), and that alpine and possibly ice-sheet-scale glaciation reached the coastline at a few localities in the early Oligocene (Barrett et al., 1989). At various times these mountains served as a source of sediments for adjacent marine and terrestrial basins, as a climatic barrier, as a barrier to marine invasions, as a nucleation center for alpine and continental ice, and as a containment dam for a succession of ice sheets that originated in East Antarctica. Before major Pleistocene glaciation and during deglacial periods of the Cenozoic, the Transantarctic Mountains provided habitats, north-south distribution pathways, and possibly refugia for vegetation and perhaps animal life (Fig. 6).

During the Paleocene and Eocene, the Transantarctic Mountains, other parts of East Antarctica, and West Antarctica supported a rainforest vegetation dominated by the southern beech Nothofagus, other angiosperms, and gymnosperms (Askin, 1989; Birkenmajer and

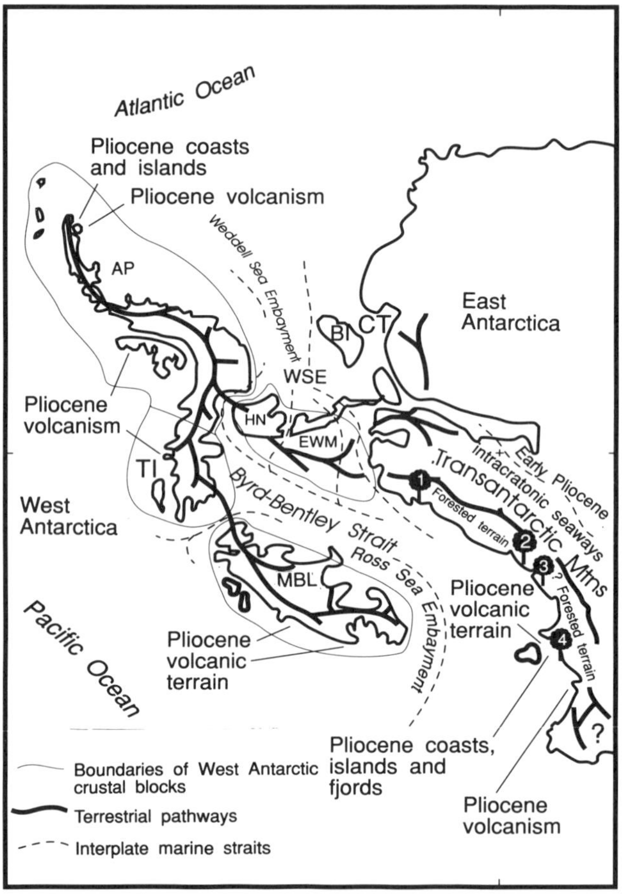

Fig. 6. Paleogeographic map of West Antarctica and part of East Antarctica. The crustal microplates of West Antarctica provide for an almost continuous 5000 kilometer "Transantarctic Land Bridge" between the northern tip of the Antarctic Peninsula and northernmost Victoria Land. Numbered sites in the Transantarctic Mountains refer to localities at which confirmed or suspected Pliocene macroplant and palynological material has been documented. (1) Tillite Spur, (2) Oliver Bluffs, (3) Mount Sirius, (4) Mount Feather and Lake Vanda (Wright Valley). Delineation of crustal microplates is taken from Storey (1991). Abbreviations are: AP = Antarctic Peninsula, TI = Thurston Island, MBL = Marie Byrd Land, HN = Haag Nunatak, WSE = Weddell Sea Embayment, and EWM = Ellsworth Mountains.

Zastawniak, 1989; Kemp, 1975; Kemp and Barrett, 1975; McIntyre and Wilson, 1966; Mildenhall, 1989; Truswell, 1990; Wrenn and Beckmann, 1982) (Fig. 4). From a consideration of the palynological content of Oligocene glaciomarine sediments of the south western Ross Sea, Mildenhall (1989) suggested "that the coastal Antarctic vegetation was essentially a beech forest with podocarps, proteas and other shrubby angiosperms. Conditions must have been temperate, if not at times warm temperate, and humid to allow such plants to survive." At the CIROS-1 drillhole in McMurdo Sound, a well-preserved leaf of the beech (Nothofagus aff. gunnii (Hook. f.) Oerst.) (Hill, 1989) was recovered from Oligocene interglacial sediments which also contain pollen of the Nothofagus fusca, Nothofagus brassii, and Nothofagus menziesii groups (Mildenhall, 1989) (Fig. 4).

Truswell (1990, p. 82), noting a gradual diminution of taxa throughout the Oligocene and early Miocene, suggests that increasing stress on the vegetation was related to intensification of glaciation, and that there is little evidence that the early Tertiary forests of greater Antarctica declined through a tundra or scrubland phase with increased glaciation.

A long-held view, based on deep sea proxy-isotope data, is that a major ice sheet developed in Antarctica during the middle Miocene (approximately 14 million years before present), and that it was long-lived in East Antarctica and overrode the Transantarctic Mountains several times during the Miocene, Pliocene, and Pleistocene (Denton et al., 1984; Shackleton and Kennett, 1975). Acceptance of this scenario makes it difficult to visualize the survival of terrestrial vascular vegetation in most parts of the continent during the Miocene and Pliocene.

No records exist of Miocene terrestrial environments in the Transantarctic Mountains except for erosion surfaces and subaerial and subglacial volcanics (LeMasurier and Thomson, 1990; Stump et al., 1980). Where palynomorphs have been recovered from Miocene marine sediments, they have generally been ascribed to recycling from much older sources rather than having an origin in a Miocene terrestrial landscape (Truswell, 1990). Marine micropaleontological studies point to the presence of open marine waters in the Ross embayment sedimentary basins during the Miocene (Harwood et al., 1989). Open waters also occupied coastal areas adjacent to the Transantarctic Mountains as well as fjords which penetrated these mountains (Ishman and Webb, 1988). While much of the Nothofagus and podocarp palynoflora reported by Brady and Martin (1979) from Miocene sediment matrix in the south central Ross Sea (Site J9) may be recycled from older sediments (Truswell, 1990), a recent palynological study based on diatomaceous clasts (Harwood et al., 1989) from within these sediments points to the existence of an early Miocene terrestrial vegetation (Jiang and Harwood, in press). This flora includes a significant proportion of Nothofagus pollen. The vegetation from which this palynoflora was derived could have been located in the Transantarctic Mountains, Ellsworth Mountains, Marie Byrd Land, and other regions of West Antarctica (Figs. 1, 5, and 6). The eventual recovery of in situ Antarctic Miocene vascular plant macrofossils remains a critical scientific objective. Such a discovery would help address the problem posed by the present >20 million year gap in the vascular plant land record, between upper Oligocene—lower Miocene macrofloras of the Ross Sea and Antarctic Peninsula regions, and the Pliocene macroflora at Oliver Bluffs in the Dominion Range of the Transantarctic Mountains (Birkenmajer and Zastawniak, 1989; Carlquist, 1987; Webb and Harwood, 1987) (Figs. 1, 4, 5, and 6).

THE PLIOCENE SIRIUS GROUP

In situ roots and stems, and allochthonous stems, twigs, and leaves of the southern beech genus Nothofagus occur in outcrops of the Sirius Group at Oliver Bluffs in the

Dominion Range, Transantarctic Mountains (Figs. 1 to 7). This locality is located at latitude 85° S., 500 kilometers from the South Pole and approximately 200 kilometers inland up the Beardmore Glacier from the point where it joins the Ross Ice Shelf. The present elevation of the fossil sites at Oliver Bluffs is approximately 1760 meters above sea level and has probably been uplifted tectonically during the past two million years (Behrendt and Cooper, 1991; McKelvey et al., 1991; Webb et al., 1986). The latitude and longitude of the site have remained unchanged during this time.

Discussions on the occurrence, stratigraphic relationships, paleontology, and age of the Sirius Group in general and the Oliver Bluff's plant material in particular are to be found in Askin and Markgraf (1986), Carlquist (1987), Harwood (1986a), Harwood and Webb (1990), Hill and Truswell (in press), Hill et al. (1991), McKelvey et al. (1991), Mabin (1986), Mercer (1986), Webb (1990, 1991), Webb and Harwood (1986, 1987, 1991), and Webb et al. (1984, 1986, 1987). Burckle and Pokras (1991) offer stimulating counterviews on the subject of the Sirius Group, especially its age, fossil content, and climatic significance. We find little or no support for their conclusions.

Paleobotanical material discussed here was collected in the Meyer Desert Formation, one of two formations that constitute the Sirius Group in the Dominion Range (McKelvey et al., 1991). The formation consists of a succession of pebbly sandstone and mudstone (diamictite) and sandstone-mudstone members (Fig. 2). Contacts between these lithofacies units (members) are sharp, and subaerial exposure, aqueous and glacial erosion, and non-deposition events coincide with many of the boundaries between members. Primitive soils occur at some of these contacts. The outcrops at Oliver Bluffs have a total thickness of approximately 100 meters and are part of a widespread glacigene succession deposited along the southern wall of a pre-Pleistocene Beardmore valley system, a major drainage feature that bisects the Transantarctic Mountains and links the interior of East Antarctica with the Ross embayment (Fig. 1). Meyer Desert Formation sediments were deposited under

Fig. 7. Artist's impression of a low-standing Pliocene krummholz forest of *Nothofagus* aff. *gunni*, developed on a glacial rock debris terrain during Sirius Group times, and exposed on the southern slopes of the Beardmore valley system at a latitude of 85° S. in the Transantarctic Mountains. Drawing by John Nagy, The Ohio State University.

temperate to subpolar conditions during several cycles of glacier expansion and recession. Ice contact, fluvial, and lacustrine environments are represented (McKelvey et al., 1991; Webb and Harwood, 1991; Webb et al., 1987).

The span of time represented by the successions at Oliver Bluffs is unknown. Collectively, the sediments of the several glacial and sedimentary cycles might represent thousands to tens of thousands of years. The intervening erosional and non-depositional disconformities possibly represent about the same amount of time. A detailed examination of bed-by-bed stratigraphy, especially in the aqueous facies, reveals numerous relatively minor sedimentary hiatuses. More late Pliocene time is probably unrepresented than is represented in the Oliver Bluffs successions.

The age of the sedimentary successions that crop out at Oliver Bluffs is constrained by the stratigraphic range of recycled Pliocene marine diatoms present in these sediments. The diatom floras were derived from the Wilkes or Pensacola basins. These marine basins are part of an intracratonic seaway which developed inland of the Transantarctic Mountains in East Antarctica during the Cenozoic and is presently obscured by the East Antarctic Ice Sheet (Figs. 1 and 6). Diatom biostratigraphic data indicate that the Sirius Group and its in situ and near in situ macroplants are younger than 3.1–2.5 million years (late Pliocene) (Harwood, 1983, 1986a, 1986b; Webb and Harwood, 1991) (Figs. 2 and 3). This diatom-based age determination has been recently confirmed by Potassium/Argon and Argon/Argon dating of volcanic ash from correlative marine sediments of the CIROS-2 drillhole in western McMurdo Sound (Barrett et al., in press).

SIRIUS GROUP PLANT MATERIAL

Plant material recovered from Oliver Bluffs consists of in situ roots and stems and allochthonous stems and leaves and pollen of Nothofagus, along with moss cushions. Root structures are associated with rudimentary paleosols at the upper surface of diamictites (boulder-pebble sandstones and mudstones). These penetrate to no more than five centimeters vertically into the substrate and may extend laterally up to 30 centimeters. Woody material consisting of krummholz-like stems and branching twigs up to 10 centimeters long and two centimeters in diameter is most common in fluvial sandstones and siltstones, just above paleosol horizons. Woody material is also present in the higher levels of the fluvial facies. Wood specimens vary from brittle and slightly mineralized to supple, occur with and without the bark preserved, and may be burned. In cross section, stems range from circular to strongly compressed. Carlquist (1987) detected growth rings and early and late wood in thin sections. He assigned the wood to the southern beech genus Nothofagus and considered the anatomy of the Oliver Bluffs wood to closely match that of the present day Chilean-Argentinean-Fuegian Nothofagus betuloides (Mirb.) Oerst., and the Tasmanian alpine Nothofagus gunnii (Hook. f.) Oerst. Jane Francis, of The University of Leeds (personal communication), reports more than 50 growth rings in selected stems, suggesting mature rather than young plants.

The most striking plant material consists of dense leaf litter mats of Nothofagus in the fluvial sandstones. The best-preserved material is found where leaf litter mats are compressed against the underside of boulders and cobbles (usually Jurassic intrusive dolerite), although single leaves also survive on the bedding planes of fluvial sandstones and siltstones. Rare but well-preserved Nothofagus leaves are also present in the laminated lacustrine sandstones and mudstones.

With regard to leaf material, Hill and Truswell (in press) and Hill et al. (1991) note that preservation is good, with organic preservation of high quality. The venation pattern has clear to high resolution and cell structure can be observed under scanning electron microscopy. Leaf length ranges up to six centimeters with most being between three to four centimeters. Leaves have prominent craspedodromous venation, with ridges between the secondary veins and running parallel with them. This suggests plicate vernation. Among the living Nothofagus, only the deciduous species have plicate vernation. The leaf assemblage is monospecific and Dr. R. Hill (University of Tasmania, personal communication) compares the leaf assemblage to the extant Tasmanian subalpine to alpine Nothofagus gunnii (Hook. f.) Oerst. Hill and Truswell (in press) note that although there are similarities between the Oliver Bluffs leaf material and present day N. gunnii and N. pumilio (Poepp. & Endl.) Krasser, they are not closely related, belong to different subgenera, and produce different pollen types. Leaves show no sign of predation, fungal infestation, or physical damage. The large leaf size suggests relatively mild conditions and possibly also relatively low light intensity, which could be associated with shade or continuous cloud cover. Hill et al. (1991) suggest that the dense leaf mat recovered from Oliver Bluffs represents a single season's leaf fall and that the leaves have not been transported far from the point of origin.

Seasonal temperature estimates based on living Nothofagus growth rates and freezing limits indicate that the minimum mean summer and winter temperatures were above 5 °C and -20 °C, respectively, for the Pliocene Nothofagus at Oliver Bluffs. An artist's impression of the largely monospecific vegetation is provided in Figure 7.

Organic material recovered in palynological preparations consists almost entirely of parenchymatous and woody tissue (Askin and Markgraf, 1986). The woody tissue is presumably derived from degraded or fragmented Nothofagus, while the parenchymatous tissue might be derived from herbaceous vegetation. Pollen grains of Nothofagus (fusca group—common), angiosperms (probable Polygonaceae or Labiatae—common), and podocarpaceous conifer (probable Dacrydium—rare) have been identified. Noting the presence of rare recycled bisaccate probable Mesozoic pollen in Sirius Group preparations and the propensity for Nothofagus pollen to be recycled, Askin and Markgraf (1986) were not emphatic in their belief that the Nothofagus (fusca group) pollen at Oliver Bluffs was derived from a late Pliocene Nothofagus vegetation. They were positive, however, about the angiosperm pollen origins and suggested that it was derived from vegetation that was extant at the time of deposition of the Pliocene Sirius Group sediments. In commenting on the rarity of conifer (?Dacrydium) pollen in the Sirius Group, Askin and Markgraf noted that the "Pilgerodendron uvifera Dwarf-Heath" and "Creeping Dacrydium fonckii Heath" communities of the Magellanic Tundra (Pisano, 1983) also produced little pollen.

In a recent examination of pollen from the Oliver Bluffs samples, Hill and Truswell (in press) compare the single Nothofagus pollen morphotype present to Nothofagidides lachlanae (Couper), which belongs to the same pollen group as N. gunnii. They argue that there is, therefore, a consistent affinity between the Oliver Bluffs leaf and pollen material, and the pollen is in situ and not recycled as suggested by Askin and Markgraf (1986).

It is highly improbable that the Oliver Bluffs (Dominion Range) flora (latitude 85° S.) is the lone occurrence of Pliocene vegetation in the entire Transantarctic Mountains or elsewhere in Antarctica. We propose that a low-diversity vascular and cryptogam vegetation extended across other parts of East Antarctica, including the lower latitude regions of central and northern Victoria Land (central Transantarctic Mountains) and perhaps the coastal regions of Wilkes Land (Figs. 1 and 6). If this were the case, the range of Pliocene plants would have extended northward to latitudes as low as 65° to 70° S., where annual temperature ranges and photoperiods would have been more favorable.

In support of the argument for widespread vascular vegetation, we note that Harwood (1986a) documented woody tissue, similar to that from Oliver Bluffs, in Sirius Group diatom preparations from latitudes as far north in the Transantarctic Mountains at latitude 77° S. (Mt. Feather), and also at Mt. Sirius, other parts of the Dominion Range, and at Tillite Spur in the southern Transantarctic Mountains. Fragmentary wood, probably Nothofagus, was also reported from Mt. Sirius (latitude 84° S.) by Hagen et al. (1989). Hall (1975) reported Nothofagus pollen (fusca and menziesii groups), monolete and trilete fern spores, and algal spheres in Late Cenozoic sediments of Dry Valley Drilling Project drillhole 4-4A, Lake Vanda, Wright Valley (latitude 77°30′ S.), also in the central part of the Transantarctic Mountains. Hall noted that the Nothofagus pollen grains varied between well preserved to broken and degraded. Because Cenozoic geology of this area was so poorly understood at the time, he attributed this well-preserved material to long distance wind-borne pollen from Australia, New Zealand, or South America. This now seems unlikely. Based on what we now understand of Miocene-Pliocene geology in the Wright and Taylor valleys area, the pollen probably originated in local forests of late Miocene or Pliocene age.

Further evidence for vascular vegetation in the dry valley region (Victoria, Wright, and Taylor valleys) is documented in the recent work of Matsumoto et al. (1991). These authors extracted chemical fossils from Quaternary sediments in Don Juan Pond (Wright Valley) and Lake Bonney (Taylor Valley), and concluded from organic component analysis that the predominance of recycled long-chain compounds pointed to the presence of vascular plant development in the region. Matsumoto et al. (1991) suggested several Eocene-Holocene sources for these compounds, including the Pliocene Sirius Group.

In summary, direct and indirect evidence exists for low-diversity Pliocene tree and cryptogamic vegetation over wide areas of the Transantarctic Mountains. We should not overlook the possibility that this vegetation persisted in some areas of Antarctica during the poorly studied early Pleistocene.

FOREST EVENTS IN THE PLIOCENE TRANSANTARCTIC MOUNTAINS

As stated earlier, several stratigraphically distinct vegetational (Nothofagus) and associated pedological events are represented in the Sirius Group succession at Oliver Bluffs, and collectively, these might span thousands or possibly tens of thousands of years. Several types of vegetational responses might be represented in the Oliver Bluff record. It is unclear whether these events signify the continued existence of Nothofagus in the immediate area, involve a succession of advances and retreats from the Beardmore valley system, or whether they represent repeated immigrations into the Transantarctic Mountains from some remote refugium. There may also have been local or even regional variations in the elevation of the snowline, treeline, and krummholz zones, with a healthier, but as yet undiscovered, beech vegetation developed at lower topographic levels in the Beardmore valley. Alternatively, the several forest events evident at Oliver Bluffs may not represent kampfgürtel tree limits, but might have been part of a dwarf polar flora growing at or just above sea level. Lastly, the Oliver Bluffs Nothofagus forests may also represent refugia or terminal stands of trees for Antarctica as a whole. The most critical requirement is that a

Miocene-Pliocene dispersal center or centers had to be maintained in favorable areas, somewhere in West or East Antarctica, for this late Pliocene forest biome to exist at all.

Although the demise of this Antarctic tree record appears to be a relatively abrupt event, one should not assume there was an immediate catastrophic deterioration in terrestrial climate. Only a degree or two drop in temperature during the summer months would have been sufficient to halt reproductive capability and lead to extinction of Nothofagus. Further, climate deterioration and eventual ice sheet development may not have occurred for tens of thousands of years after extinction of tree groups.

MICROCLIMATE, MICROENVIRONMENTS, AND POLAR VEGETATION

No part of present-day Antarctica experiences a summer (December-February) month mean temperature of 10 °C or more. Therefore, the area lies beyond the present day limit for upright tree growth.

Survival of tree species at temperatures lower than 10 °C is possible. Nordenskjöld (1928) noted that ground-hugging dwarf (krummholz) tree species may find warmer microenvironments at ground level on rock and moraine exposures and survive temperatures colder than 10 °C. Tuhkanen (in press) observed a mean monthly temperature of 6 to 6.5 °C for the warmest month of the year at krummholz limits for Nothofagus in Tierra del Fuego.

Nordenskjöld (1928) observed that in northeastern Greenland, soil temperatures at a depth of four centimeters are as much as 6 °C higher than the July (summer) mean temperature of 4.4 °C. He also noted that the low air temperature is greatly compensated by strong insolation, which in Greenland may last all day and night, thereby permitting the temperature of the ground and of the air about the plants to rise markedly.

Further, Nordenskjöld noted that Gunnar Andersson observed temperatures in plant clumps to be as much as 10 °C higher than the surrounding air. Similar temperature differences are noted for air and the surface of moss mats at Signy Island (61° S.), South Orkney Islands (Holdgate, 1964). At Cape Hallett in northern Victoria Land, Antarctica, Rudolph (1966) documented summer (December–January) soil temperatures at 0.5 centimeters as 6 to 7 °C warmer than air temperatures at two feet (0.6 meters), and 10 °C warmer than at five feet (1.5 meters).

A series of major Pleistocene glaciations in Antarctica destroyed Cenozoic interglacial soils that might have served as loci for present-day vascular recolonization. Moss carpets serve as a surrogate soil in many parts of the Antarctic Peninsula and support vascular plants (Fleming et al., 1938; Follman, 1964; Skottsberg, 1954). Moss carpets may be up to a meter in depth and in some localities extend over many hectares. The co-occurrence of Nothofagus aff. gunnii and moss in the Pliocene Sirius Group reinforces our belief that the beech-bryophyte association is essential in the development, propagation, and survival of the southern polar fossil forests.

The role of symbiotic ectotrophic mycorrhiza in modern Nothofagus is well known but yet to be considered in highly-stressed polar fossil populations. Van Steenis (1971, p. 73) states that "ectotrophic micorrhiza play an essential role in Nothofagus forests, and must have had that role also in past geological ages. It could well be that through them they are capable to maintain their abundance, virility, and dominance."

PHYTOGEOGRAPHIC IMPLICATIONS

The occurrence of a Nothofagus-dominated Pliocene flora in the southern Transantarctic Mountains poses a number of significant problems (Webb, 1990, 1991; Webb and Harwood, 1991). These include a major gap (early Miocene to early Pliocene, or >20 million years) in the vascular record of East Antarctica, the relationship between Miocene and Pliocene forest development and glacial histories in both East and West Antarctica, the precise time for extinction of Nothofagus and its associated flora in the latest Pliocene, the taxonomic impact of geographic isolation of the Antarctic flora following the separation of Antarctica (Antarctic Peninsula) from South America in the late Oligocene-early Miocene (30-22 million years before present), and the location of dispersal and survival centers for vascular plants within Antarctica during the last 25 to 30 million years. Only the phytogeographic and paleoecological implications of this flora will be explored here in detail.

As noted earlier, coastal glaciation in East Antarctica was in effect at about the time of the Eocene-Oligocene transition (35 to 40 million years ago), or even earlier (middle Eocene, 45.5 million years B.P.) (Ehrmann and Mackensen, 1992). It is also believed that glaciation in West Antarctica was a later development (occurring intermittently during the Miocene and Pliocene), and that major Antarctic late Paleogene and Neogene ice sheets were not permanent features of the landscape but passed through many phases of expansion and contraction (Harwood, 1986a; Webb and Harwood, 1991). It is unclear how geographically extensive Neogene ice sheet development was, how many ice centers evolved, and whether the regional glacial-deglacial histories of these ice centers were in or out of phase with each other in terms of growth and recession. Until these questions are resolved, we accept the probability that cold subpolar rainforests survived and perhaps even thrived in extensive ice-free areas, both in West and East Antarctica.

Because of the apparent later inception of major glaciation in archipelago-like West Antarctica, this area is favored as the locus (refugium) for propagation, dispersal, and survival of the Miocene-Pliocene tree flora. The geology and geography of this vast region is best comprehended by reference to studies of the structural geology (Storey, 1991) and sub-ice sheet topography of West Antarctica (Drewry, 1983). West Antarctica consists of five major crustal blocks (Fig. 6), some of which have present-day elevations of 2,000 to 3,000 meters above sea level. The blocks are separated by faulted and eroded depressions, some over 2,500 meters below sea level. Large areas on the flanks of these blocks now lie at depths of 0 to 500 meters below sea level. If the present West Antarctic Ice Sheet was removed and a crustal adjustment of 500 meters made (Drewry, 1983), geographic distinctions between the boundaries of these crustal blocks would be less apparent. For instance, in the absence of significant ice during a major phase of deglaciation, continuous or near continuous terrestrial communication would be possible between the Antarctic Peninsula, Thurston Island, and Marie Byrd Land blocks; and between the Antarctic Peninsula, Haag Nunatak, Ellsworth Mountain blocks and the southern Transantarctic Mountains (Fig. 6). Marine and terrestrial data point to the presence of an archipelago terrain of Paleozoic and Mesozoic rocks and Miocene-Pliocene subaerial and subglacial volcanoes, separated by marine channels, straits, and interior seaways (Harwood et al., 1989; LeMasurier and Thomson, 1990; and Scherer, 1991). This physiographic setting was alluded to by Taylor (1914), Fairbridge (1952), Kennett (1978), and Webb (1979).

The implications of terrestrial dispersal routes within West and East Antarctica for propagation, distribution, and survival of Miocene-Pliocene Nothofagus are clear. Present-day Nothofagus is unable to cross even relatively narrow stretches of marine water expanses (Fleming, 1963b). Therefore, a key issue in our argument centers on the proximity of the several structural elements of our proposed Miocene-Pliocene "Transantarctic Land Bridge." The hypothesized land bridge extends over much the same terrain as the well-trodden highway utilized by biogeographers to foster vertebrate migrations between South America and Australia during the early Paleogene (Simpson, 1940; Skottsberg, 1953).

Of the crustal plate linkages identified here, the Antarctic Peninsula–Haag Nunatak–Ellsworth Mountain–Transantarctic Mountain dispersal route is more relevant for exchanges of terrestrial biota (plant, and potentially invertebrate and vertebrate taxa) between the Antarctic Peninsula and the northern Transantarctic Mountains and vice versa. This hypothesized land bridge provides a pathway of approximately 5,000 kilometers between the northernmost Antarctic Peninsula and northernmost Victoria Land, as well as dispersal routes into other regions of East Antarctica (Figs. 1 and 6).

Local and regional geological and glaciological events might have facilitated or obstructed biotic flow along the length of the "Transantarctic Land Bridge" during the Miocene and Pliocene, promoting periods of widescale communication and periods of isolation among the terrestrial floral and faunal communities. This biogeographic scenario provides for the development of a very similar or highly contrasted terrestrial biota along different parts of the land bridge at various times during the last thirty million years. During the same interval of time, glacial and deglacial events also impacted marine communities and communication between the Antarctic interior basins and the Southern Ocean (Harwood, 1991; Webb, 1991).

The Oliver Bluffs Nothofagus-dominated biome is located at a very high latitude (85° S.) on this pathway. Coeval floras would likely have been more diverse towards the lower latitudes, that is, towards the northern Antarctic Peninsula and northern Victoria Land.

COLONIZATION AND SURVIVAL OF NOTHOFAGUS

An extensive literature has evolved on the biology and ecology of modern Nothofagus, and on the phytogeography of both the modern genus and its Cretaceous–Cenozoic forebears. The lessons of these endeavors should be understood when considering Late Cenozoic biogeographic disjunction and the possibility of a geologically youthful land bridge across Antarctica.

While modern Nothofagus is capable of establishing and maintaining depauperate to dense forest stands under extreme climatic and environmental conditions which often involve very low temperatures, low light, high winds, snow cover, physical disturbance, and rudimentary soils, it exhibits a limited potential for widespread pollen and seed dispersal (J. Wardle, 1984). For example, Fleming (1963b, p. 379) writes, "New Zealand botanists and foresters who know living Nothofagus are almost unanimous in maintaining that Nothofagus requires land for its dispersal." Major land bridge connections were core elements in the Nothofagus-based biogeographic arguments of Van Steenis (1962, and other papers cited herein) and others.

Attempts to circumvent the problems associated with the Cretaceous-Paleogene land bridge hypotheses and the eventual acceptance of continental drift and sea floor spreading led many authors to promote other seed, pollen, and plant dispersal mechanisms, such as avian vectors (gut and adherence to feathers), atmospheric circulation, ocean currents, island stepping, and iceberg rafting (Barber et al., 1959; Birnie and Roberts, 1986; Costin, 1965; Darlington, 1965; Fleming, 1962, 1963b; Humphries, 1981; Preest, 1963; Smith, 1984; Van

Steenis, 1953, 1962, 1971, 1972; and J. Wardle, 1984). For example, it was noted that stranded trunks of South American Nothofagus transported by the West Wind Drift in the Southern Ocean have been found on Macquarie Island (Barber et al., 1959) and Marion Island (Van Zinderen Bakker, 1971), while Nothofagus trunks of South American origin have been found on South Georgia (Matthews, 1931; Smith, 1984), South Shetlands Islands (Smith, 1984), and the South Sandwich Islands (Longton, 1977). In no instance was the genus able to propagate from these remnants. These dispersal mechanisms probably played little or no role in promoting the development of the Pliocene Sirius Group Nothofagus flora. Even combinations of dispersal modes, such as dispersal by wind across narrow water gaps (Fleming, 1963b), do not provide a satisfactory explanation.

Although arguments for post-Cretaceous-Paleogene stepping stones and land bridges between South America–Antarctica–New Zealand and Australia lapsed with the advent of seafloor spreading theory and final acceptance of continental drift, the probability of a more restricted polar biotic land bridge through West Antarctica and East Antarctica during the Miocene and Pliocene seems to be the most plausible explanation for the presence Pliocene flora in the Transantarctic Mountains.

CENOZOIC VEGETATION EVENTS IN NEW ZEALAND, THE SOUTHWEST PACIFIC, AND ANTARCTICA

The response of Late Cenozoic Nothofagus to long-term climate cooling and short-term climate fluctuations is well studied in New Zealand. These data are helpful in interpreting the plant record of the southern Transantarctic Mountains. Harris (in Suggate, 1978, p. 724) notes that in New Zealand, pollen of Nothofagus was reduced from six distinctive types in the Miocene, to four at the Pliocene-Pleistocene boundary, and to two above this boundary and to the present day. Similar floral trends in New Zealand during this period are also discussed by Mildenhall (1980) and Mildenhall and Pocknall (1984). Nothofagus was present on Campbell Island (latitude 52°30′ S.) during the Pliocene, after which it became extinct (Fleming, in Suggate, 1978, p. 533), presumably because of replacement of cool wet conditions by colder subpolar environments. In the Wanganui area of the North Island of New Zealand, early Quaternary Nothofagus exhibits an oscillating record, appearing during periods of wetter and cooler climates equivalent to southernmost New Zealand today (Fleming, in Suggate, 1978, p. 561). The propagation abilities and responses of present-day Nothofagus to changing climate and geography are discussed by a number of authors. Fleming (1963b, p. 379) notes that in New Zealand, "Areas without persistent seed sources, such as Mount Egmont, Stewart Island, and parts of Westland, have not been repopulated by Nothofagus since the deforestation of the last glaciation." P. Wardle (1963) deduced that Nothofagus had survived through Quaternary glaciations in southernmost New Zealand in regions considered by Willett (1950) to have been a tundra zone. These New Zealand examples of climate response by Nothofagus involve alpine settings, the presence of alpine ice caps, extensive areas of periglacial terrain, and importantly, very active late Cenozoic mountain building associated with the Kaikoura Orogeny (Suggate, 1978).

Park (1909) believed that the greater part of the South Island of New Zealand was covered by a northerly extension of a south polar ice sheet originating in Antarctica, a notion dismissed as unlikely by Willett (1950). To what extent do the Pliocene-Pleistocene floral events in New Zealand reflect some combination of effects wrought by mountain building in the Southern Alps and a deterioration of climate in distant Antarctica? Fleming (1963a, Figs. 1–4) believes there was a definite relationship between Cenozoic terrestrial environments of Antarctica and New Zealand (Fig. 8).

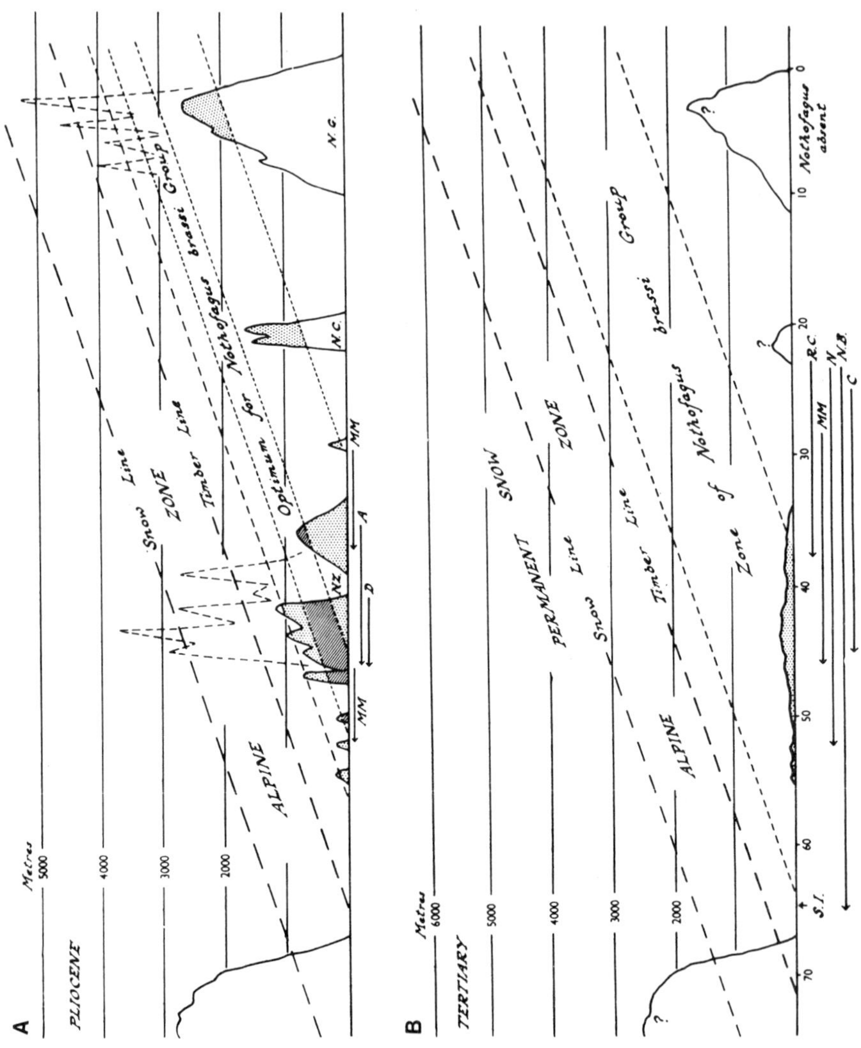

A
PLIOCENE
Metres
5000
4000
3000
2000
Snow Line
ZONE
Timber Line
Optimum for Nothofagus brassi Group
ALPINE
N.G.
N.C.
NZ
MM
A
D
B
TERTIARY
Metres
6000
5000
4000
3000
2000
PERMANENT SNOW ZONE
Snow Line
Timber Line
ALPINE
Zone of Nothofagus brassi Group
Nothofagus absent
S.I.
R.C.
MM
N
N.B.
C
0
10
20
30
40
50
60
70

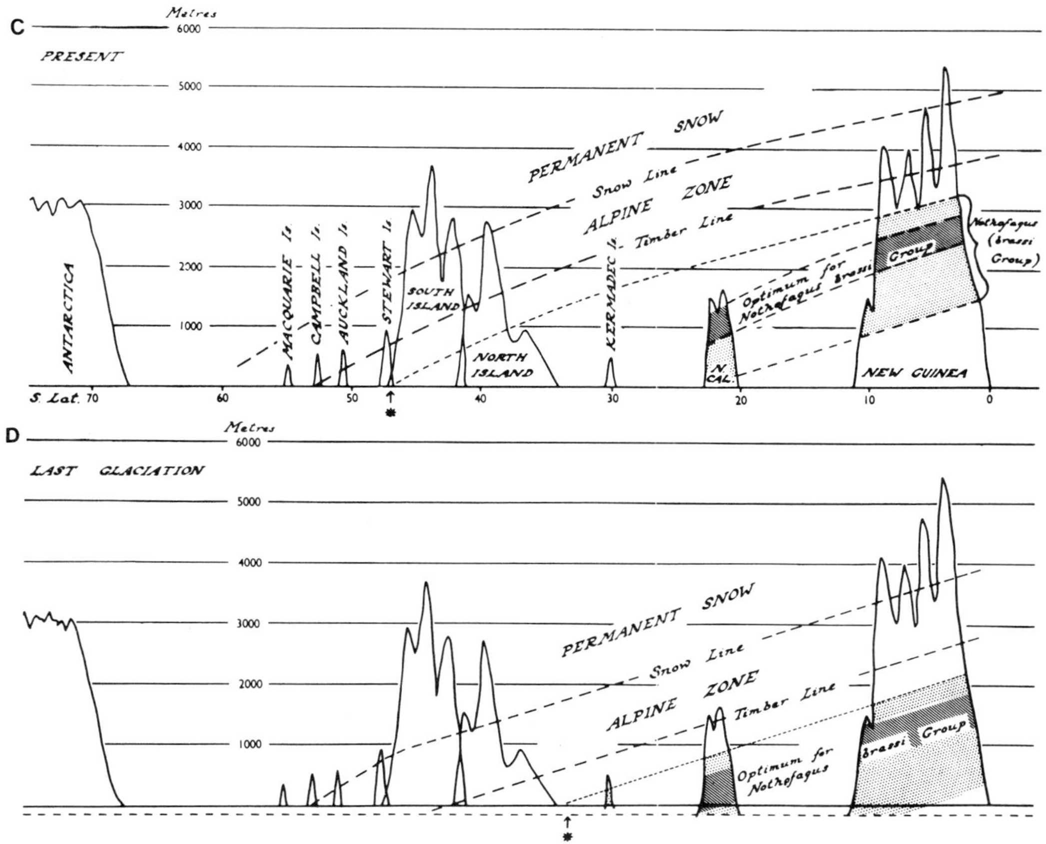

Fig. 8. Latitudinal and elevational modification of vegetation boundaries between the Miocene and present day, along a transect Antarctica-Subantarctic islands-New Zealand-New Caledonia-New Guinea (from Fleming, 1963a, 1975). (a) Miocene; (b) Pliocene; (c) last glaciation; (d) present day. Explanation of abbreviations and other details for each figure are provided in the text.

The elevation–longitude vegetation and geomorphologic models developed by Fleming (1963a, Figs. 1–4; 1975), Smith (1986), Taylor (1926), Willett (1950), Van Steenis (1962), and Zotov (1938) for southern hemisphere landmasses, including New Guinea, New Caledonia, Australia, New Zealand, and Antarctica, are especially pertinent to the present discussion. These schemes portray variations of the Nothofagus timberlines, alpine vegetation zones, and snowlines from the mid-Tertiary (Miocene) to the present day, a time of progressive southern hemisphere cooling. The two principal variables in these models entail an interplay of fluctuating vegetational zone elevations and changing elevations (usually uplift) of landmasses upon which the vegetation developed.

Fleming's (1963a) time-slice cross sections are reproduced in Figure 8a-d. Several important points emerge from these cross sections. First, his views on southern high latitude phytogeography are prophetic in that they were advanced more than twenty years before the discovery of late Pliocene vascular fossils, including tree species, in Antarctica. Second, Fleming recognized, as did Holdgate (1961), the cryospheric climatic and biotic crises in post-Pliocene Antarctica from evidence available outside Antarctica. Third, he recognized that biogeographic boundaries may also move southward (poleward), and that in island settings terrestrial environments and their floras may not be in phase with each other. Fleming's analysis predated considerations of seafloor spreading and the fact that in the Miocene and Pliocene, New Zealand occupied a higher latitude site, closer to the Ross Sea–Transantarctic section of Antarctica. The series of cross-sections also incorporate the concept of mountain building during the late Tertiary, a significant phenomenon common in all Southern Hemisphere circum-Pacific landmasses, including Antarctica. In Figure 8a (Tertiary) he determined a 15 degrees-of-latitude southward displacement of mid-Tertiary (Miocene) temperature zones at sea level in New Zealand, based on the distribution of coral reefs (RC), warm water marine mollusca (MM), Nautiloids (N), coconuts (C), and Nothofagus brassii group pollen (NB). In Figure 8a, he also extends his "Alpine Zone" and "Timber Line" south of latitude 70° S. into Antarctica, and the southern limit of Nothofagus brassii to 64° S. latitude, south of Campbell Island. While it is tempting to credit Fleming with having anticipated the presence of tree floras and relatively warm deglacial climates in and near the Ross Sea–Transantarctic Mountains sector during the Late Cenozoic, it is likely that he extrapolated Paleogene paleobotanical data into the region from the Seymour Island area of the Antarctic Peninsula. In Figure 8b (Fleming's Pliocene), he portrays a major northward shift (cooling) in marine and terrestrial biogeographic boundaries across all latitudes, with the southern limit of the treeline positioned between Antarctica and the subantarctic islands south of New Zealand. Given the presence of late Pliocene Nothofagus at latitude 85° S. in Antarctica, Fleming's view of southern biogeographic boundaries during the Pliocene is correct if his southern treeline limit is extended a relatively short distance into Antarctica. This change may be accommodated by invoking climate amelioration and an accompanying re-interpretation of treeline gradients at high latitudes. The dramatic northward shift of the treeline out of Antarctica following the Pliocene is illustrated in his cross section for the last glaciation (Fig. 8c). During the latest Pliocene–earliest Pleistocene, the Nothofagus treeline moved, apparently quite abruptly, approximately 4,000 kilometers northward (85° S. to 46° S.), from the region of the Beardmore Glacier to southernmost New Zealand. At this time, also, the southern beeches become extinct throughout the Transantarctic Mountains, probably in the rest of Antarctica, and on the intermediate subantarctic islands such as Campbell Island. A major restructuring in Antarctic Ross Sea inshore benthic foraminiferal faunas also occurs between the Pliocene and Pleistocene (Webb, 1988). Fleming's cross-section for the present day (Fig. 8d) highlights a post-glacial (Holocene) southward movement of vegetation zones and the inability of climatically displaced beech (Nothofagus) species to re-enter either New Zealand or Antarctica, naturally, from more northerly sites.

Similar conclusions occurred to other biogeographers in this and other regions of the Southern Hemisphere. Holdgate (1961, p. 638) comments:

> Fossil evidence has now established that from the Cretaceous until about the Pliocene, the southern beeches and many Podocarpaceae had a wider distribution in the Antarctic and Subantarctic than they do now. They both occurred in Antarctica itself and later grew in Kerguelen (Island). . . . Towards the end of the Pliocene a northward movement seems to have occurred: the Nothofagus brassii group, which may be assumed to be the most warmth-loving of the southern beeches, disappears abruptly from the rocks of New Zealand and Australia but begins to appear in New Guinea, where it now flourishes. Simultaneously, the Nothofagus fusca and N. menziessii groups, which have smaller leaves and appear more cold-adapted, become prominent in New Zealand where they are dominant today. It is hard to avoid the conclusion that this northward movement was connected with the cooling of Antarctica...

Research on Miocene, Pliocene, and Pleistocene Nothofagus from New Zealand and other sites points to climatically- and probably tectonically-driven expansion and contraction in geographic and elevational range on scales of thousands, hundreds of thousand, and millions of years. Unfortunately, in the case of Antarctic Nothofagus we are unable to match our records to those of New Zealand at a resolution better than about the million year level. During climate amelioration following late Pleistocene glaciation in New Zealand, Nothofagus spread from the Nelson Lakes south to northern Westland and south to Canterbury, in 5,000 to 6,000 years (Moar, 1973). We assume that the repeated forest events at Oliver Bluffs involve a similar recovery time. In a highly stressed setting such as obviously existed in the Transantarctic Mountains, a variety of factors of lesser importance in New Zealand would have been important in governing both survival and rates of geographic expansion and contraction. These factors would have included the absence of tree and other vascular competitors, the presence of rudimentary soils and a minimum level of nutrients, adequate moisture at root level, perhaps the presence of moss mats to serve as root-freezing inhibitors and as a surrogate soil, absence of diseases, favorable geographic orientation with relation to insolation, adequate annual microenvironmental and macroenvironmental temperatures, and an ability to produce pollen and distribute seed.

GLOBAL WARMING AND ANTARCTIC VEGETATION RESPONSES

In the preceding discussion we considered phytogeographic (both vertical and lateral) and ecological responses by Antarctic and New Zealand Nothofagus to changing physical and climatic conditions over time spans ranging from thousands to millions of years. Can we estimate the effects of the projected global warming on vegetation in Antarctica during the next century? How closely should southern high-latitude forestry and botanical forecasting be influenced by the dramatic vegetation modelling experiments conducted at high latitudes in the northern hemisphere (Holten, 1990), where some similar but also many environmental differences prevail?

For example, in the case of North America, Overpeck et al. (1991) suggest that if increases in atmospheric gas concentrations warm the global average temperature 1.5 °C to 4.5 °C in the next century, certain plant ranges might extend 500 to 1,000 kilometers poleward in the next 200 to 500 years. Holten (1990) describes a similarly dramatic analysis for Norway. He suggests that vegetational changes accomplished between 18,000 and 6,500 years before present will be achieved within the next 100 years under the scenario of carbon dioxide doubling. Treelines of some taxa are projected to migrate 350 kilometers northward and to gain between 200 to 400 meters in elevation.

ANTHROPOGENIC DISPERSAL OF NOTHOFAGUS

The potential for Nothofagus to become re-established in some future partially deglaciated Antarctica or sub-Antarctic islands might be better understood if we examine past forestry experiments. The record for recent attempts in Antarctica itself is not positive. Corte (1961) reports the introduction of two species of Nothofagus from Tierra de Fuego to Cierva Cove (64° S.), Antarctic Peninsula, which did not survive.

Nothofagus has been transplanted from the Southern Hemisphere to the Arctic with varying success. Ødum (1990) notes that Nothofagus pumilio (Poepp. & Endl.) Krasser transplanted to Greenland survived for one year. Ødum (1989) reports that Nothofagus antarctica (Forst.) Oerst. (deciduous), N. pumilio (deciduous), and N. betuloides (Mirb.) Oerst. (evergreen) were successfully transplanted to the Faero Islands (62° N.) from Ushaia, Tierra del Fuego (56° S., treeline 450 meters), and Lago Tromen-Volcano (Lanin region, Province Neuquen (40° S., treeline 1,600 to 1,800 meters). These provenances were collected between sea level and the treeline, with the highest success occurring where trees were taken close to treeline.

John Alden (Institute of Northern Forestry, Fairbanks, Alaska) has drawn our attention to attempts to establish Nothofagus under greenhouse conditions in Alaska. More than forty specimens collected by Soren Ødum in Tierra del Fuego were sent from the Horsholm Arboretum, Denmark, to Alaska in August 1981. The three Fuegian species represented were Nothofagus pumilio, N. antarctica, and N. betuloides. All material was collected slightly south of latitude 54° S., and at sites between 50 and 200 meters above sea level. Greenhouse temperature was depressed from 60 to 70 °F at the time of arrival in Alaska, to 32 to 36 °F some 70 days later. Greenhouse temperatures ranged occasionally from 28 to 44 °F due to outside climate influences. All trees were dead or dying by the spring and summer of 1982, six to nine months after transplantation from Denmark. John Alden offered the comment that the plants were entering the growing season at the time of their transport from Denmark to Alaska and failed to adapt to temperature depression.

The deciduous Tasmanian Nothofagus gunnii (Hook. f.) Oerst. survived three to four years in the Faero Islands without producing significant growth (Ødum, 1989, p. 134). Ødum also reported that evergreens from Tasmania (Nothofagus cumminghamii (Hook. f.) Oerst.) and New Zealand [mountain beech (Nothofagus solandri var. cliffortioides (Hook. f.) Poole) and silver beech (Nothofagus menziesii (Hook. f.) Oerst.)] had a poor post-transplantation record. Benecke (1972) notes that Nothofagus solandri var. cliffortioides from New Zealand was unable to survive at the 1,900 meter conifer treeline near Innsbruck, Austria. Mitchell (1974) and Tuley (1980) document the successful development of transplanted Nothofagus in England, Ireland, and Scotland.

What role, if any, might vegetative propagation have played in the maintenance of fossil communities such as that at Oliver Bluffs, Beardmore Glacier? Ødum's (1989) observation, that cuttings of Nothofagus antarctica were easily rooted but rooting of N. dombeyi is more difficult, is of interest. It seems possible that in the future, vegetative krummholz might survive on subantarctic islands and perhaps even on the Antarctic Peninsula.

Ødum (1990, p. 61) notes that the potential treeline in southwest Greenland and Faero Isles is situated at approximately 150 meters above sea level and that specimens of Nothofagus from near the southern hemisphere high latitude treeline adapt best to these northern hemisphere sites. Future Antarctic treelines will also be, initially, situated close to sea level and transplant provenances from near-treeline habitats, perhaps from Tierra del Fuego, southern New Zealand, and Tasmania, might be the most successful candidates for reforestation. John Alden (personal communication) notes that although Nothofagus may occupy colder sites at natural tree limits than Northern Hemisphere tree limit species, it may be less hardy than subarctic tree species when introduced anthropologically beyond natural tree limits in both Northern and Southern Hemispheres.

As climate ameliorates, Antarctic terrains are likely to become seriously out of phase with their potential to support tree species such as Nothofagus and other plants. This situation occurs in New Zealand, where the climatically expelled Nothofagus brassii group could live and propagate in New Zealand today, "if they had been able to return after their Pleistocene extinction here" (Fleming, 1963a). Ødum (1990) also notes that vegetation in southwest Greenland, Iceland, and the Faroe Isles is out of phase with the present climate.

The performance of species of South American and Australasian Nothofagus in the Faero Islands and other Northern Hemisphere experiments may be a good indicator of the growth, propagation, and survival potential for this genus during interglacial conditions on the Antarctic continent, the Antarctic offshore islands, and the sub-Antarctic islands.

ANTARCTIC AND ARCTIC PLIOCENE VASCULAR VEGETATION

Although beyond the scope of this discussion, we draw attention to the successful development of Northern Hemisphere Arctic forest tundra and heathland during the late Pliocene–early Pleistocene. Fossil material includes tree leaves, wood, pollen, other angiosperms and crytogamic plants, as well as insects and mammals. These floras and faunas occur in the Cape Deceit Formation of western Alaska, Gubik Formation of northern Alaska, Beaufort Formation of Northwest Territories, Kap København Formation of northeast Greenland, and localities in eastern Siberia and Kamchatka Peninsula. Comprehensive discussions and reviews of these Arctic terrestrial and associated marine successions are provided by Bennike (1991), Bennike and Böcher (1990), Gladenkov et al. (1991), and Matthews and Ovenden (1990).

The Antarctic late Pliocene–earliest Pleistocene plant and animal biota is depauperate, whereas that of the Arctic is remarkably rich. This might be attributed to contrasting land-sea relationships, the availability of terrestrial connections to lower latitude landmasses, topographic relief, and differing glacial histories. The markedly longer Antarctic late Paleogene–Neogene glacial history is thought to have progressively filtered out all but the hardiest floral taxa. The Oliver Bluffs Nothofagus is a remnant of this long selection process.

POLAR VEGETATION AND THE UNIFORMITARIAN DICTUM

The communities of the polar biologists, botanists, and foresters and the community of polar paleontologists have much to offer each other. But one must caution against too ready acceptance of the "Uniformitarian" dictum that the "present is the key to the past," or too eager projection of the past and present into the future. The first-mentioned group of scientists has amassed large biological data bases based on historical time scale of a few decades, a period of time that is insignificant in terms of global climatic cycles. The latter group works with data bases that often span hundreds of thousands and million years, periods of time that involve major and sometimes catastrophic extremes in global climate. The relationship between the biological and paleontological world is well expressed by Spicer (1990, p. 27):

> The present 'icehouse' world furnishes us with living vegetation that can be regarded only as a partial model for ancient plant ecosystems. Critically, modern plants growing at today's cold high latitudes provide a poor analog of polar vegetation at times of global warmth. In order to understand these ancient systems, we have to turn to the fossil record and take care in extrapolations from living plants. We have to utilize fully the more time-stable aspects of morphologies and anatomies related to plant physiology and the physical processes of sedimentation, and make minimal use of taxon-specific tolerances.

A similar sentiment was expressed much earlier in New Zealand by Cockayne (1926). Van Steenis (1962, p. 294) states, "The late Dr Cockayne used to feel very strongly that the present is often a poor guide to the past." To which we might add, the present may also be a poor guide to the future.

We hope the contents of this paper re-emphasize the fact that the polar latitudes are especially prone to major climatic and biological perturbations, and the unexpected is almost commonplace.

ACKNOWLEDGEMENTS

Peter Webb thanks the NATO Science Committee for financial assistance, which allowed him to attend the NATO Advanced Research Workshop On Forest Development in Cold Climates, at Laugarvatn, Iceland, in June 1991, and also acknowledges help and encouragement provided by Frank Lockyear and Robert Wheelon of ReTree International.

We are most grateful to Dr. John Alden, Institute of Northern Forestry, Fairbanks, Alaska, for his thorough review of drafts of our manuscript, and for information on modern Nothofagus forestry experiments. Dr. Robert Hill, University of Tasmania, provided comments on a draft of the manuscript.

John Nagy, Byrd Polar Research Center, and Beth Daye, Department of Geological Sciences, The Ohio State University, provided assistance in the preparation of illustrations. We completed field and laboratory aspects of this study with assistance from National Science Foundation grant DPP8919910. Peter Webb acknowledges assistance provided by an Humboldt Foundation Research Award during the final phase of manuscript preparation at the Alfred Wegener Institute for Polar and Marine Research, Bremerhaven, Federal Republic of Germany.

REFERENCES

Askin, R. A., 1989, Endemism and heterochroneity in the Late Cretaceous (Campanian) to Paleocene palynofloras of Seymour Island, Antarctica: Implications for origins, dispersal and paleoclimates of southern floras. In: Crame, J. A., ed., Origins and

Evolution of the Antarctic Biota, Geological Society, London, Special Publication No. 47:107-119.

Askin, R. A., and Markgraf, V., 1986, Palynomorphs from the Sirius Formation, Dominion Range, Antarctica, *Antarctic Journal of the United States*, 21(5):34-35.

Barber, H. N., Dadswell, H. E., and Ingle, H. D., 1959, Transport of driftwood from South America to Tasmania and Macquarie Islands, *Nature*, 184:203-204.

Barker, P. F., and Burrell, J., 1976, The opening of Drake Passage, *Marine Geology*, 25:15-34.

Barrett, P. J., Adams, C. J., McIntosh, W. C., Swisher, C. C., and Wilson, G. S., 1992, Radiometric ages support Antarctic deglaciation around three million years ago, *Nature*, 359:816-818.

Barrett, P. J., Hambrey, M. J., Harwood, D. M., Pyne, A. R., and Webb, P.-N., 1989, Synthesis. In: Barrett, P. J., ed., Antarctic Cenozoic History from the CIROS-1 Drillhole, McMurdo Sound, Department of Scientific and Industrial Research, Wellington, New Zealand, *Bulletin*, 245:245-251.

Behrendt, J. C., and Cooper, A. K., 1991, Evidence of rapid Cenozoic uplift of the shoulder escarpment of the Cenozoic West Antarctic rift system and a speculation on possible climate forcing, *Geology*, 19:315-319.

Benecke, U., 1972, Wachstum, CO_2-Gaswechsel und Pigmentgehalt einiger Baumarten nach Ausbringung in verschiedene Höhenlagen, *Angewandte Botanik*, 46:117-135.

Bennike, O., 1991, The Kap København Formation: Stratigraphy and paleobotany of a Plio-Pleistocene sequence in Peary Land, North Greenland, *Meddelelser om Grønland Geoscience*, 23:85.

Bennike, O., and Böcher, J., 1990, Forest-tundra neighboring the North Pole: Plant and insect remains from the Plio-Pleistocene Kap København Formation, North Greenland, *Arctic*, 43(4):331-338.

Birkenmajer, K., and Zastawniak, E., 1989, Late Cretaceous-early Tertiary floras of King George Island, West Antarctica: Their stratigraphic distribution and paleoclimatic significance. In: Crame, J. A., ed., Origins and Evolution of the Antarctic Biota, Geological Society, London, Special Publication No. 47:227-240.

Birnie, J. F., and Roberts, D. E., 1986, Evidence of Tertiary forests in the Falkland Islands (Islas Malvinas), *Palaeogeography, Palaeoclimatology, Palaeoecology*, 55:45-53.

Brady, H. T., and Martin, H., 1979, Ross Sea region in the Miocene: A glimpse of the past, *Science*, 203:437-438.

Budyko, M. I., and Israel, I., 1987, Anthropogenic climate change, Gidrometeorizdat, 406 pp.

Burckle, L. H., and Pokras, E. M., 1991, Implications of a Pliocene stand of *Nothofagus* (southern beech) within 500 kilometres of the South Pole, *Antarctic Science*, 3(4):389-403.

Carlquist, S., 1987, Pliocene *Nothofagus* wood from the Transantarctic Mountains, *Aliso*, 11:571-583.

Cockayne, L., 1926, The Vegetation of New Zealand, "Die Vegetation der Erde," Engelmann, Leipzig, 456 pp.

Corte, A., 1961, La primera fanerogama adventicia hallada en el continente antartico, *Instituto Antartico Argentino Contribucion*, No. 62:1-14.

Costin, A. B., 1965, Long distance seed dispersal to Macquarie Island, *Nature*, 206:317.

Costin, A. B., 1981a, Alpine and sub-alpine vegetation. In: Groves, R. H., ed., Australian Vegetation, Cambridge University Press, Cambridge, 361-376.

Costin, A. B., 1981b, Vegetation of high mountains in Australia. In: Keast, A., ed., Ecological Biogeography of Australia, W. Junk, The Hague, 719-731.

Cranwell, L. M., 1963, *Nothofagus*: living and fossil. In: Gressitt, J. L., ed., Pacific Basin Biogeography, Bishop Museum, Hawaii, 387-400.

Creber, G. T., 1990, The South Polar Forest Ecosystem. In: Taylor, T. N., and Taylor, E. L., eds., Antarctic Paleobiology, Springer Verlag, New York, 37-41.

Darlington, P. J., 1965, Biogeography of the Southern End of the World, Harvard University Press, Cambridge, 236 pp.

Denton, G. H., Prentice, M. L., Kellogg, D. E., and Kellogg, T. B., 1984, Late Tertiary history of the Antarctic ice sheet: Evidence from the Dry Valleys, Geology, 12:263-67.

Dettmann, M. E., 1989, Antarctica: Cretaceous cradle of austral temperate rainforests. In: Crame, J. A., ed., Origins and Evolution of the Antarctic Biota, Geological Society, London, Special Publication No. 47:89-105.

Dettmann, M. E., and Thomson, M. R. A., 1987, Cretaceous palynomorphs from the James Ross Island area, Antarctica—A pilot project, British Antarctic Survey, Bulletin, 77:13-59.

Drewry, D. J., 1983, Antarctica: Glaciological and Geophysical Folio, Cambridge University Press, Cambridge.

Ehrmann, W. U., and Mackensen, A., 1992, Sedimentological evidence for the formation of an East Antartic ice sheet in Eocene/Oligocene time, Palaeogeography, Palaeoclimatology, Palaeoecology, 93:85-112.

Fairbridge, R. W., 1952, The Geology of the Antarctic. In: Simpson, F. A., ed., The Antarctic Today, A. H. and A. W. Reed, Wellington, 56-101.

Fleming, C. A., 1962, New Zealand Biogeography: A Paleontologist's Approach, Tuatara, 10(2):53-108.

Fleming, C. A., 1963a, Age of the alpine biota, Proceedings of the New Zealand Ecological Society, 10:15-18.

Fleming, C. A., 1963b, Paleontology and southern biogeography. In: Gressitt, J. L., ed., Pacific Basin Biogeography, Bishop Museum Press, Hawaii, 369-385.

Fleming, C. A., 1975, The geological history of New Zealand and its biota. In: Kuschel, G., ed., Biogeography and Ecology In New Zealand, W. Junk, The Hague, 1-86.

Fleming, W. L. S., Steenson, A., Roberts, B. B., and Bertram, G. C. L., 1938, Notes on the scientific work of the British Graham Land Expedition, 1934-37, Geographic Journal, 91:508.

Follman, G., 1964, Das Pflanzenleben des Antarktis: Botanische Ergebnisse der 17. Chilenischen Antarktis-Expedition, Umschau, 64(4):100-103.

Gladenkov, Y. B., Barinov, K. B., Basilian, A. E., and Cronin, T. M., 1991, Stratigraphy and Paleoceanography of Pliocene Deposits of Karaginsky Island, Eastern Kamkatka, U.S.S.R., Quaternary Science Reviews, 10(2/3):239-245.

Greene, S. W., Gressitt, J. L., Koob, D., Llano, G. A., Rudolph, E. D., Singer, R., Steere, W. C., and Ugolini, F. C., 1967, Terrestrial Life of Antarctica. In: Bushnell, V. C., ed., Antarctic Map Folio Series, American Geographical Society, New York, Folio 5, 24 p.

Hagen, E. H., Faure, G., and Jones, L. M., 1989, Isotopic studies of calcite, pyrite and wood from glacial deposits in the Beardmore Glacier area, Transantarctic Mountains, Antarctic Journal of the United States, 24(5):67-68.

Hall, S. A., 1975, Palynologic investigations of Quaternary sediment from Lake Vanda, Antarctica, Antarctic Journal of the United States, 10(4):173-174.

Hambrey, M. J., and Barrett, P. J., [In press], The Cenozoic sedimentary and climate record from the Ross Sea region of Antarctica. In: Kennett, J. P., ed., American Geophysical Union, Antarctic Research Series, Washington, D. C.

Harwood, D. M., 1983, Diatoms from the Sirius Formation, Transantarctic Mountains, Antarctic Journal of the United States, 18(5):98-100.

Harwood, D. M., 1986a, Diatom biostratigraphy and paleoecology and Cenozoic history of Antarctic ice sheets, The Ohio State University, 592 pp. Ph. D. Dissertation.

Harwood, D. M., 1986b, Recycled siliceous microfossils from the Sirius Formation, Antarctic Journal of the United States, 21(5):101-103.
Harwood, D. M., 1991, Cenozoic diatom biogeography in the southern high latitudes: Inferred biogeographic barriers and progressive endemism. In: Thomson, M. R. A., Crame, J. A., and Thomson, J. W., eds., Geological Evolution in Antarctica, Cambridge University Press, Cambridge, 667-673.
Harwood, D. M., Scherer, R. P., and Webb, P.-N., 1989, Multiple marine productivity events in West Antarctica as recorded in upper Miocene sediments beneath the Ross Ice Shelf(J9), Marine Micropaleontology, 15:91-115.
Harwood, D. M., and Webb, P.-N., 1990, Early Pliocene glaciation of the Antarctic ice sheet and Late Pliocene onset of bipolar glaciation, EOS, Transactions, American Geophysical Union, 71(1):538-539.
Hill, R. S., 1989, Fossil Leaf. In: Barrett, P. J., ed., Antarctic Cenozoic History from the CIROS-1 Drillhole, McMurdo Sound, Department of Scientific and Industrial Research, Wellington, New Zealand, Bulletin, 245:143-144.
Hill R. S., and Read, J., 1989, A revised infrageneric classification of Nothofagus, Botanical Journal of the Linnean Society, 105:37-72.
Hill, R. S., and Truswell, E. M., [In press], Nothofagus fossils in the Sirius Group, Transantarctic Mountains: Leaves and pollens and their climatic implications. In: Kennett, J. P., ed., American Geophysical Union, Antarctic Research Series, Washington, D. C.
Hill, R. S., Harwood, D. M., and Webb, P.-N., 1991, Last remnant of Antarctica's Cenozoic flora: Pliocene Nothofagus of the Sirius Group, Transantarctic Mountains, 8th Gondwana Subcommission Symposium, Hobart, Australia (abstract).
Holdgate, M. W., 1961, Biological routes between the southern continents, New Scientist, No. 239:636-638.
Holdgate, M. W., 1964, An experimental introduction of plants to the Antarctic, British Antarctic Survey Bulletin, 3:13-16.
Holten, J. L., ed., 1990, Effects of Climate Change on Terrestrial Ecosystems, Norsk Institutt For Naturforskning, Notat 4, 82 p.
Howard, T. M., 1981, Southern closed-forests. In: Groves, R. H., ed., Australian Vegetation, Cambridge University Press, Cambridge, 102-120.
Houghton, J. T., Jenkins, G. J., and Ephraums, J. J., 1990, Climate Change: The IPCC Scientific Assessment, Cambridge University Press, Cambridge.
Humphries, C. J., 1981, Biogeographical methods and the southern beeches. In: Forey, P. L., ed., The Evolving Biosphere, Cambridge University Press, Cambridge, 283-297.
Ishman, S. E., and Webb, P.-N., 1988, Late Neogene benthic foraminifera from the Victoria Land Basin margin, Antarctica: Application to glacio-eustatic and tectonic events, Revue de Paleobiologie, Special Volume, No. 2, Part 2:523-551.
Jiang X., and Harwood, D. M., [In press], Palynomorphs from Site J9 diatomite: A clear view of mid Cenozoic Antarctic vegetation. In: Kennett, J. P., ed., American Geophysical Union, Antarctic Research Series, Washington, D. C.
Kemp, E. M., 1975, Palynology of Leg 28 drillsites, Deep Sea Drilling Project. In: Hayes, D. E., and Frakes, L. A., eds., Initial Reports of the Deep Sea Drilling Project, U. S. Government Printing Office, Washington, 28:599-623.
Kemp, E. M., and Barrett, P. J., 1975, Antarctic glaciation and early Tertiary vegetation, Nature, 258:507-508.
Kemp, E. M., and Harris, W. K., 1975, The vegetation of Tertiary Islands on the Ninetyeast Ridge, Nature, 258:303-307.
Kennett, J. P., 1978, The development of planktonic biogeography in the Southern Ocean during the Cenozoic, Marine Micropaleontology, 3:301-345.
Kennett, J. P., Houtz, R. E., et al., 1975, Cenozoic paleoceanography in the southwest Pacific Ocean, Antarctic glaciation and the development of the circum-Antarctic

current. In: Hayes, D. E., and Frakes, L. A., eds., Initial Reports of the Deep Sea Drilling Project, U. S. Government Printing Office, Washington, D. C., 29:1155-1169.

LeMasurier, W. E., and Thomson, J. W., eds, 1990, Volcanoes of the Antarctic Plate and the Southern Ocean, Antarctic Research Series, American Geophysical Union, Washington D.C., 48, 487 pp.

Longton, R. E., 1977, A Nothofagus log stranded on Candlemas Island, South Sandwich Islands, British Antarctic Survey Bulletin, 45:148-149.

Mabin, M. C. G., 1986, Sirius Formation basal contacts in the Beardmore Glacier region, Antarctic Journal of the United States, 21(5):32-33.

Matsumoto, G. I., Yoshida, Y., Watanuki, K., and Torii, T., 1991, Chemical fossils of vascular plant activity in the McMurdo Dry Valleys during the pre- and inter-glacial periods, Antarctica, Sixth International Symposium on Antarctic Earth Sciences, Saitama, Japan (abstract), 391-396.

Matthews, L. H., 1931, South Georgia, the British Empire's sub-antarctic outpost, Simpkim Marshall Ltd, 292 pp.

Matthews, J. V., and Ovenden, L. E., 1990, Late Tertiary macrofossils from localities in Arctic/Subantarctic North America:A review of the data, Arctic, 43(4):364-392.

McIntyre, D. J., and Wilson, G. J., 1966, Preliminary palynology of some Antarctic Tertiary erratics, New Zealand Journal of Botany, 4:314-321.

McKelvey, B. C., Webb, P.-N., Harwood, D. M., and Mabin, M. C. G., 1991, The Dominion Range Sirius Group—A record of Late Pliocene–Early Pleistocene Beardmore Glacier. In: Thomson, M. R. A., Crame, J. A., and Thomson, J. W., eds., Geological Evolution of Antarctica, Cambridge University Press, Cambridge, 675-682.

Mercer, J. H., 1986, Southernmost Chile; a modern analog of the southern shores of the Ross embayment during Pliocene warm intervals, Antarctic Journal of the United States, 21(5):103-105.

Mildenhall, D. C., 1980, New Zealand Late Cretaceous and Cenozoic plant biogeography: A contribution, Palaeogeography, Palaeoclimatology, Palaeoecology, 31:197-233.

Mildenhall, D. C., 1989, Terrestrial palynology. In: Barrett, P. J., ed., Antarctic Cenozoic History from the CIROS-1 Drillhole, McMurdo Sound, Department of Scientific and Industrial Research, Wellington, New Zealand, Bulletin, 245:119-127.

Mildenhall, D. C., and Pocknall, D. T., 1984, Paleobotanical evidence for changes in Miocene and Pliocene climates in New Zealand. In: Vogel, J. C., ed., Late Cainozoic Palaeoclimates of the Southern Hemisphere, A. A. Balkema, Rotterdam, 159-171.

Mitchell, A., 1974, A Field Guide to the Trees of Britain and Northern Europe, W. Collins and Sons.

Moar, N. T., 1973, Late Pleistocene vegetation and environment in southern New Zealand. In: van Zinderen Bakker, E. M., ed., Palaeoecology of Africa and of the surrounding islands and Antarctica, A. A. Balkema, Capetown, 8:179-198.

Nordenskjöld, O., 1928, A general characterization of polar nature. In: Joerg, W. L. G., ed., The Geography of Polar Regions, American Geographical Society, New York, 359 pp.

Ødum, S., 1989, A ten year trial with South American trees and shrubs with special regard to the Nothofagus. In: Hojgaard, A., Johansen, J., and Ødum, S., eds., A Century of Tree-planting in the Faroe Islands, Annales Societatis Scientiarum Faeroensis Supplementum, 14:125-156 pp.

Ødum, S. 1990, Choice of species and origins for arboriculture in Greenland and the Faroe Islands, Dansk Dendrologisk Arsskrift, 9, 77 pp.

Overpeck, J. T., Bartlein, P. J., and Webb, T., 1991, Potential magnitude of future vegetation change in eastern North America: Comparisons with the past, Science, 254:692-695.

Park, J., 1909, Geology of the Queenstown Subdivision, New Zealand Geological Survey Bulletin, No. 7.

Pisano, E., 1983, The Magellanic tundra complex. In: Gore, A. J. P., ed., Mire: Swamp, Bog, Fen, and Moor—Regional Studies, Ecosystems of the World, 4B, Elsevier, Amsterdam.

Preest, D. S., 1963, A note on the dispersal characteristics of the seeds of the New Zealand podocarps and beeches, and their biogeographical significance. In: Gressitt, J. L., ed., Pacific Basin Biogeography, Bishop Museum Press, Honolulu, 415-224.

Rudolph, E. D., 1966, Terrestrial vegetation of Antarctica: Past and present studies. In: Tedrow, J. C. F., ed., Antarctic Soils and Soil Forming Processes, Antarctic Research Series, No. 8, American Geophysical Union, Washington, D. C., 109-124.

Shackleton, N. J., and Kennett, J. P., 1975, Paleotemperature history of the Cenozoic and initiation of Antarctic glaciation: Oxygen and Carbon isotope analyses in DSDP Sites 277, 279, and 281. In: Kennett, J. P., and Houtz, R. E., eds., Initial Reports of the Deep Sea Drilling Project, U. S. Government Printing Office, Washington, D. C., 29:743-755.

Scherer, R. P., 1991, Quaternary and Tertiary microfossils from beneath Ice Stream, B: Evidence for a dynamic West Antarctic Ice Sheet history, Palaeogeography, Palaeoclimatology, Palaeoecology, 90:395-412.

Simpson, G. G., 1940, Antarctica as a faunal migration route, Proceedings of the 6th Pacific Science Congress, 2:755-768.

Skottsberg, C., 1953, Influence of the Antarctic continent on the vegetation of southern lands, Proceedings of the 7th Pacific Science Congress, 5:92-99.

Skottsberg, C., 1954, Antarctic flowering plants, Svensk Botanisk Tidskrift, 51:330-338.

Smith, J. M. B., 1986, Origins of Australasian tropicalpine and alpine floras. In: Barlow, B. A., ed., Flora and Fauna of Alpine Australasia, Chapter 7, Commonwealth Scientific and Industrial Organization, 109-128.

Smith, R. I. L., 1984, Terrestrial plant biology of the sub-antarctic and Antarctic. In: Laws, R. M., ed., Antarctic Ecology, Academic Press, London, Volume 1:61-162.

Spicer, R. A., 1990, Reconstructing high-latitude Cretaceous vegetation and climate: Arctic and Antarctic compared, In: Taylor, T. N., and Taylor, E. L., eds., Antarctic Paleobiology, Its Role in the Reconstruction of Gondwana, Springer Verlag, New York, 27-36.

Storey, B. C., 1991, The crustal blocks of West Antarctica within Gondwana: reconstruction and break-up model. In: Thomson, M. R. A., Crame, J. A., and Thomson, J. W., eds., Geological Evolution of Antarctica, Cambridge University Press, Cambridge, 587-592.

Stump, E., Sheridan, M. F., Borg, S. G., and Sutter, J. F., 1980, Early Miocene subglacial basalts, the East Antarctic Ice Sheet, and the Uplift of the Transantarctic Mountains, Science, 207:757-759.

Suggate, R. P., ed., 1978, The Geology of New Zealand, New Zealand Government Printer, Wellington, 820 pp.

Taylor, G., 1914, Physiography and glacial geology of East Antarctica, Geographic Journal, 44:365-382.

Taylor, G., 1926, Glaciation in the south west Pacific, Proceedings of the Third Pan-Pacific Science Congress, 1819-1825.

Thiede, J., Clark, D. L., and Herman, Y., 1990, Late Mesozoic and Cenozoic paleoceanography of the northern polar oceans. In: Grantz, A., Johnson, L., and Sweeney, J. F., eds., The Geology of North America, Volume L, The Arctic Ocean Region, The Geological Society of America, Boulder, 427-458.

Truswell, E. M., 1990, Cretaceous and Tertiary vegetation of Antarctica: A palynological perspective. In: Taylor, T. N., and Taylor, E. L., eds., Antarctic Paleobiology, Its Role in the Reconstruction of Gondwana, Springer Verlag, New York, 71-88.

Tuhkanen, S., [In press], Treeline in relation to climate, with special reference to oceanic areas. In: Alden, J. N., Ødum, S., and Mastrantonio, J. L., eds., Forest Development in Cold Climates, Proceedings of an International Symposium, 1991, June 18-24, [Laugarvatn, Iceland], NATO, Plenum, New York, New York.

Tuley, G., 1980, Nothofagus in Britain, Forestry Commission, Forest Record 122, London.

Van Steenis, C. C. G. J., 1953, Results of the Archibold Expeditions Papuan Nothofagus, Journal of the Arnold Arboretum, 34(4):300-374

Van Steenis, C. C. G. J., 1962, The land-bridge theory in botany, Blumea, 11(2):235-371.

Van Steenis, C. C. G. J., 1971, Nothofagus, key genus of plant geography, in time and space, living and fossil, ecology and phylogeny, Blumea, 19(1):65-98.

Van Steenis, C. C. G. J., 1972, Nothofagus, key genus to plant geography. In: Valentine, D. H., ed., Taxonomy, Phytogeography, and Evolution, Academic Press, New York, 275-288.

Van Zinderen Bakker, E. M., 1971. In: van Zinderen Bakker, E. M., Winterbottom, J. M., and Dyer, R. S., eds., Marion and Prince Edward Islands, A. A. Balkema, Capetown, 1-15.

Wace, N. M., 1965, Vascular Plants. In: van Mieghem, J., and Oye, P., eds., Biogeography and Ecology in Antarctica, W. Junk, The Hague, 201-266.

Wardle, J. A., 1984, The New Zealand Beeches, Ecology, Utilization and Management, The Caxton Press, Christchurch, 444 pp.

Wardle, P., 1963, Growth habits of New Zealand subalpine shrubs and trees, New Zealand Journal of Botany, 1:18-47.

Wardle, P., 1971, An explanation for alpine timberline, New Zealand Journal of Botany, 9(3):371-402.

Wardle, P., [In press], Causes of alpine timberline: A review of the hypothesis. In: Alden, J. N., Ødum, S., and Mastrantonio, J. L., eds., Forest Development in Cold Climates, Proceedings of an International Symposium, 1991, June 18-24, [Laugarvatn, Iceland], NATO, Plenum, New York, New York.

Webb, P.-N., 1979, Paleogeographic evolution of the Ross Sector during the Cenozoic. In: Nagata, T., ed., Proceedings of the Seminar III on Dry Valley Drilling Project, 1978, National Institute of Polar Research, Tokyo, Special Issue 13:206-212.

Webb, P.-N., 1988, Upper Oligocene-Holocene Foraminifera of the Ross Sea region, Revue de Paleobiologie, Special Volume, No. 3, Part 2:589-603.

Webb, P.-N., 1990, The Cenozoic history of Antarctica and its global impact, Antarctic Science, 2(1):3-21.

Webb, P.-N., 1991, A review of the Cenozoic stratigraphy and paleontology of Antarctica. In: Thomson, M. R. A., Crame, J. A., and Thomson, J. W., eds., Geological Evolution of Antarctica, Cambridge University Press, Cambridge, 599-607.

Webb, P.-N., and Harwood, D. M., 1986, Major Pliocene climate amelioration in the Transantarctic Mountains, Annual Meeting, Geological Society of America, San Antonio, November 1986, Abstracts with Programs, 18(6):783.

Webb, P.-N., and Harwood, D. M., 1987, The terrestrial flora of the Sirius Formation: Its significance in interpreting late Cenozoic glacial history, Antarctic Journal of the United States, 22(4):7-11.

Webb, P.-N., and Harwood, D. M., 1991, Late Cenozoic glacial history of the Ross embayment, Antarctica. In: Cronin T., and Dowsett, H., eds., Pliocene Climates, Quaternary Science Reviews, 10(2/3):215-223.

Webb, P.-N., Harwood, D. M., McKelvey, B. C., Mercer, J. H., and Stott, L. B., 1984, Cenozoic marine sedimentation and ice volume variation on the east antarctic craton, Geology, 12:287-291.

Webb, P.-N., Harwood, D. M., McKelvey, B. C., Mabin, M. G. C., and Mercer, J. H., 1986, Late Cenozoic tectonic and glacial history of the Transantarctic Mountains, Antarctic Journal of the United States, 21(5):99-100.

Webb, P.-N., Harwood, D. M., McKelvey, B. C., Mabin, M. G. C., and Mercer, J. H., 1987, Sirius Formation of the Beardmore Glacier region, Antarctic Journal of the United States, 22(1):8-13.

Willett, R. W., 1950, The New Zealand Pleistocene snow line, climatic conditions, and suggested biological effects, New Zealand Journal of Science and Technology, B32:18-47.

Wise, S. W., Breza, J., Harwood, D. M., and Wei W., 1991, Paleogene glacial history of Antarctica. In: Muller, D. W., Mackenzie, J. A., and Weissert, H., eds., Controversies in Modern Geology, Academic Press, London, 133-171.

Wrenn, J. H., and Beckmann, S. W., 1982, Maceral, total organic carbon, and palynological analyses of Ross Ice Shelf Project Site J-9 cores, Science, 216:187-189.

Zotov, V. D., 1938, Some correlations between vegetation and climate in New Zealand, New Zealand Journal of Science and Technology, 19:474-487.

PREDICTING AFFORESTATION SUCCESS DURING CLIMATIC WARMING AT THE NORTHERN LIMIT OF FORESTS

Allen M. Solomon

Environmental Research Laboratory
U. S. Environmental Protection Agency
Corvallis, Oregon

Darrell C. West

Environmental Sciences Division
Oak Ridge National Laboratory
Oak Ridge, Tennessee, U.S.A.

INTRODUCTION

Global temperature increases from greenhouse gases are expected during the 21st Century, possibly as early as the next decade. Warming is predicted to be greatest at highest latitudes. Initial attempts to document climate change will be hampered by the great inter-annual variability in weather at high latitudes and the scarcity of long-term weather records. Certain key properties of these environmental changes, however, can be defined, despite uncertainty concerning results from present climate models. A group of scientists meeting at Villach, Austria, in 1987 (Jaeger, 1988) agreed that 3 °C was a moderate estimate of average global warming during the next century, while at high latitudes (60 to 90° N.), 0.6 to 0.7 °C per decade in winter and 0.1 to 0.2 °C per decade in summer are more probable climate change expectations. Results from the most recent climate model results are not very different and range from 0.4 to 0.8 °C per decade (Mitchell et al., 1990). In addition, results from both the Villach group and the more recent compilations agree that winter precipitation and soil moisture could increase at high latitudes as more precipitation falls, and as more precipitation falls as rain instead of snow. These climate changes are quite intense and rapid. For the next few decades, they may result in a much wider range of temperature and precipitation extremes than has ever been recorded in cold regions.

The warming projections imply a significantly longer growing season, proportionately greater amounts of annual heat units (degree days), retreat of permafrost, and a decline in annual high-albedo snow cover, without significant change in mean July temperatures. The warming trend in cold regions is fundamentally different from previous warming periods. The warming of 9,000 to 11,000 years ago at northern latitudes produced about the same

Forest Development in Cold Climates, Edited by J. Alden *et al.*
Plenum Press, New York, 1993

annual and summer temperature increases, that is, about 3 °C (Ritchie, 1984). At that time, summer insolation increased about 7 percent from astronomical forcing alone, and winter insolation decreased 7 percent.

The increased intensity of the hydrological cycle should enhance cloudiness and precipitation, further reducing the weak CO_2-induced summer warming, and result in long, cool, maritime-like growing seasons in continental boreal woodlands. Today's continental vegetation under July mean temperatures of 10 to 12 °C is shrub-tundra, but a maritime version of these temperatures produces dense stands of conifers around Valdez, Alaska, or Bergen, Norway.

Predicted climate changes in many cold regions would enhance tree growth, decrease mortality among trees and seedlings, and increase production of viable seeds. Those considering the task of afforestation in cold regions adjacent to high-latitude treeline (shrub tundra or forest-tundra transition) should be pleased about the potential effects of global warming.

Still, it would be inadvisable to proscribe new tree planting regimes and selections of species and provenances in any specific area based on projections of global warming. Current projections, based on general circulation models of the atmosphere (GCMs), may be only coincidentally correlated with the conditions that actually prevail in any given location where forest growth is being managed (Bowling, in press). The spatial resolution of the GCMs (2 to 10° of longitude and latitude) is so coarse, and their inability to reproduce known patterns of precipitation and cloudiness is so poor, that local conformity with global predictions is unlikely. It is more likely that global climate changes will be redefined by local landforms, topography, soils, wind, and precipitation patterns. Processes not yet formalized in models, or those of small regional extent, could have great effects on the climate change expected. The only GCM to interact with an ocean model that contains realistic ocean current geography, for example, indicates that Iceland, northern Scandinavia, and northwestern Russia could become up to 6 °C cooler in winter while the rest of the world warms (Washington, 1990; Washington and Meehl, 1989).

The strong seasonality of expected warming, combined with the "distortion" of global-scale changes induced by local substrate and topographic conditions, requires that any scheme used to predict success of tree planting and afforestation cannot be based on linear extrapolation of current conditions. Instead, forecasts must be based on natural histories of species involved, including those properties of each phase of tree life cycles affected by seasonal changes. In addition, forecasts must involve local landscape features, such as soil moisture, slope and aspect, and topographic control of moisture and temperature.

Most empirically-based predictive schemes are inappropriate. The use of site index or yield tables, for example, assumes a dynamic equilibrium climate rather than one trending in one direction or another during a tree's lifetime. The linkage of process-related growth models to site index, as in STEMS (Belcher et al., 1982), or the use of more complex regression-based growth and yield models, such as FORCYTE (Kimmins and Scoullar, 1981, 1983), is unsuitable for the same reasons, even when adapted to cold regions (Yarie, 1986). The suite of mathematical models of tree growth and succession at the stand level, termed gap models (Botkin, et al., 1972; Shugart, 1984; Shugart and West, 1977), however, appear better suited to these considerations. Such models use known characteristics of growth, reproduction, and mortality by species to predict tree succession as constrained by competition for sunlight and effects of temperature, precipitation, and site factors.

MODEL DESCRIPTION

The stand simulators described here are modified from the initial concept presented by Botkin and his coworkers two decades ago (Botkin et al., 1970, 1972). The original and subsequent models are based on the idea that gaps in forest canopies caused by death of large individuals are the fundamental units of reproduction in closed forest communities (Watt, 1947) in the absence of catastrophic mortality induced by wildfire or windthrow. A simulated forest succession emerges as tree species compete for sunlight based on their differing natural history traits and on environmental conditions. Obviously, the closed-canopy forest is not the most common physiognomy in forested lands in cold regions, and mixed-age, mixed-species stands may be less common than single-species and single-age groupings. This model is quite suitable for cold region silviculture because simulated growth among individuals in open woodlands and in pure stands is determined by the same requirements for sunlight, warmth, water, and nutrients, as are closed-canopy, multiple-age, and mixed-species stands.

Model Development

Several versions of the initial gap model are in common use and will be discussed. The initial model, JABOWA (Botkin et al., 1972), was revised by a graduate class in ecological modeling at the University of Tennessee. Results were published as FORET (Shugart and West, 1977). FORET differed primarily in the absence of the least valid JABOWA routines (effect of soil moisture on seedlings; effect of seed source strength on stand composition), the addition of stochasticity to tree growth, and more efficient program organization. The "family tree" of the gap models can be useful in ascertaining many of the characteristic capabilities of model variants, particularly those derived directly from either JABOWA or FORET.

The original JABOWA was the direct progenitor of a fire-dominated Sierra Nevada forest variant, SILVA (Kercher and Axelrod, 1984). JABOWA was also supplemented with nutrient cycling routines to produce FORTNITE (Aber et al., 1978; Aber and Melillo, 1982). FORET was similarly modified to produce FORNUT (Weinstein et al., 1982). Pastor and Post (1985, 1986) combined routines from FORET and FORENA with nutrient approaches in FORTNITE, including major modifications of the latter, to produce LINKAGES, a gap model dominated by both nutrient cycling and climate effects. Prentice (1986) and Leemans and Prentice (1987, 1989) used both JABOWA and FORET as guides in producing FORSKA, a true second-generation gap model that differs in several fundamental ways but which does not consider climate effects. FORSKA is the basis for a mixed life-form version, FLAM (Fulton, 1991), which recently was supplemented with climate effects (Fulton, In press). FORSKA has been redeveloped as a landscape-level model including climate as a driver of stand dynamics in FORSKA-2 (Prentice et al., 1991 and in press).

Numerous commonly used model versions are direct modifications of FORET. BRIND (Noble et al., 1980) developed detailed wildfire forcing and response routines in Australian *Eucalyptus* forests. KIAMBRAM (Shugart et al., 1980) included some 125 tropical species and explored the role in tropical notophyll forests of vines and strangler trees. FORENA (Solomon and Shugart, 1984; Solomon, 1986) added tree-growth forcing by soil moisture and winter low temperatures, allowing the model to be used throughout eastern North America. CLIMACS (Dale and Hemstrom, 1984; Dale et al., 1986) unraveled the problem of simulating stand development in the very dense, vertically deep and dark forests of the Pacific Northwest. ZELIG (Urban, 1990; Urban et al., 1991) ran several contiguous FORET stands to simulate effects of seedfall in adjacent stands. Most recently, Bonan (1988) developed a boreal forest model (LOKI) that provides detailed treatment of physical environmental factors in high latitude forests (permafrost dynamics, low sun angle effects) and of ecological processes unique to high latitude forests (such as interactions among lichen mats, fire, trees, and permafrost).

Model Strategy

Gap models treat three aspects of annual forest dynamics in mixed-age and mixed-species stands: seedling establishment (usually as saplings), growth of established trees, and mortality of established trees. Establishment and mortality are treated as stochastic processes. They depend on random values drawn from specified statistical distributions. Tree growth is deterministic, being defined by quantitative, functional relationships between intrinsic and extrinsic forces and tree responses.

The deterministic growth relationships are stochastic in FORET and its variants because external control of growth by temperature and precipitation is also stochastic. Each monthly or annual temperature and precipitation value used in the models is drawn at random from a normal distribution. The distribution is usually defined by data from individual 30-year weather stations or multiple-station region-level weather data. Hence, weather is different every year. This model feature is particularly attractive for use in cold regions where irregular episodes of extreme weather, quantifiable from long weather records, may be critical determinants of tree growth and survival.

The gap is normally considered to be located within a matrix of closed-canopy forest; thus, the effects of forest edges are eliminated. This concept could be an advantage in cold maritime regions, where trees are planted densely for mutual shelter effects. On the other hand, frequent mortality among older and larger trees can produce open woodlands in cold regions, rather than the closed-canopy forests most appropriate for the model concept.

The spatial location of trees on the plot is not considered in the models, but rather, any effects of a tree on the plot are averaged over the area of the plot. The vertical structure of the plot is tabulated, however, so the model tracks the number and size of stems and estimates leaf area by vertical increment (usually 0.1 to 1.0 meter increments).

The gap size considered varies. In JABOWA, the gap is 1/100 hectare while subsequent versions of the models (FORET, LINKAGES, and others) use larger plot sizes (1/12 to 1/10 hectare). Smaller plots come to be dominated by single trees, then when the modeled dominant dies, plots receive 100 percent of full sunlight. This produces a highly oscillatory stand density and an unrealistically reduced species diversity in the gaps (Shugart and West, 1979). Considering that cold region forest plantings often form open woodlands, the small plots of JABOWA are probably less appropriate in cold than in temperate regions.

Modeled Tree Establishment

Efforts to relate the frequency of established seedlings to seed production and germinability by species have been largely unsuccessful both in the field and in gap models. Hence, seedling establishment has been treated as a largely stochastic process in the gap models. The most common approach has been to select several seedlings at random from among the species eligible to enter the plot (Fig. 1, lower left). Eligibility is defined by comparing the natural history of each species to relevant annually variable conditions on the plot. For example, species requiring mineral soil would be selected if mineral soil was present as defined by recent tree mortality (tip-up mound) or by less than a specified amount of stand biomass (too few trees to produce litter). Species requiring leaf litter, direct sunlight or shade, or specified minimum spring temperatures would be selected if those conditions prevailed, as would species that are immune to these conditions. Each species is tested for its reliance on the plot conditions of the year. An annual pool of eligible species is formed from which the random number of seedlings is drawn.

This approach is particularly appropriate for use in afforestation of cold regions. The models can easily be programmed with new birth switches representing processes not previously considered. The random selection of species or seedling numbers previously has been biased to constantly favor certain species or to favor them at irregular intervals corresponding with mast years (as in JABOWA and LINKAGES) or fire and insect attack (LOKI). Absence of any of the requisite conditions, or their presence only once every few decades, excludes or limits simulated tree reproduction in a very realistic fashion. Also, the seedling establishment routine can be replaced by specified sapling densities and species mixes to mimic planting regimes in the intensely managed forests of many cold regions.

One common way establishment has been biased toward certain species is through an independent birth routine designed to mimic the addition of root sprouts in FORET; this appears in most subsequent versions of the gap model (Table 1, subroutine BIRTH). If the tree that dies is a species that generates root sprouts, and if its diameter was within the range that can generate sprouts, one to three new saplings are added to the plot. A similar "layering" routine for trees that continue to live while sprouting is also implemented in Bonan's LOKI model. These two processes are probably particularly important in tree reproduction in cold regions, where they may be the only source of regeneration.

<u>Modeled Tree Growth</u>

The strategy for simulating growth of established trees is to define an optimum growth for each species at each age, then reduce growth each year because of non-optimal growth conditions. Optimum tree growth (diameter) on the simulated plot would follow a sigmoid growth curve: modest during early years, rapid in midlife, slowing to an asymptote as maximum possible diameter is approached, and ending at the maximum known age of each species (Fig. 1, center right; Table 1, subroutine GROW). Thus, the limits to growth are defined by the greatest age and diameter ever measured in forest-grown trees, usually found in common silvics manuals, for example, Burns and Honkala (1990a, 1990b).

Such a strategy appears germane to modeling tree growth in cold regions. Certainly, annual and cumulative growth of trees in such areas is much less than optimum. For example, compare the largest exotic trees growing in Iceland (Blöndal, 1987) with those growing in more optimal conditions in their home ranges. Russian larch (<u>Larix sukaczewi</u> N. Dyl.) in Iceland averaged about 3/8 meter per year of height growth over 48 years, and about one meter per year in the Soviet Union (Nikolov and Helmisaari, 1991). Sitka spruce (<u>Picea sitchensis</u> (Bong.) Carr.) in Iceland grew 14 meters in 50 years, and on average sites in southeast Alaska, 27 meters in 50 years (Harris, 1990). Black cottonwood (<u>Populus trichocarpa</u> Torr. Gray) grew 0.5 meters per year in 35 years in Iceland, and 2.1 meters per year in 10 years in the lower Fraser River Valley of British Columbia (DeBell, 1990).

The only model that departs from the constraints of age and diameter in defining optimum growth is FORSKA. Diameter is replaced with the maximum known height of each tree species, and then an asymptotic relationship between height and diameter is used to derive the diameter of each tree. Initial growth of young trees is constrained by the

Table 1. FORENA stand growth simulator equations

Process	Equations	Comments
Subroutine GROW		
Growth of each tree under optimal conditions	$d[D^2H] = R\ LA\ 1 - \frac{DH}{D_{max}H_{max}}$ R = growth rate parameter, LA = leaf area of tree, D = diameter at breast height, H = height of tree, D_{max} = maximum diameter for a particular species, D^2H = index of tree volume	See Botkin et al. (1972) for rationale.
Height/diameter relation	$H = 137 + b_2D - b_3D^2$ $b_2 = 2({}_{max} - 137)/D_{max}$, $b_3 = (H_{max} - 137)/D^2_{max}$	Equation from Ker and Smith (1955). b_2 and b_3 determined by setting $H = H_{max}$, $dH/dt = 0$ when $D = D_{max}$. Empirical equation from Sollins et al. (1973).
Leaf area of tree	$LA = aD^b$ $a = 0.1193$, $b = 2.939$	Empirical equation from Sollins et al. (1973).
Extinction of light as a function of leaf area in forest canopies	$Q(h) = Q_0 \exp[-k \int_h^{\infty} LA(h')dh']$ $k = 0.25$	Empirical equation from Kasanaga and Monsi (1954), Loomis et al. (1967), and Perry et al. (1969).
Reduction of photosynthesis due to shading	$r_s(AL) = 1 - \exp[-4.64(AL - 0.05)]$ $r_I(AL) = 2.24\ 1 - \exp[-1.136(AL - 0.08)]$ AL = available light by substituting sum of leaf areas for taller trees in above equation for light extinction, r_s = shade-tolerant, r_I = shade-intolerant	Empirical relations from Kramer and Kozlowski (1960). These equations multiply the right-hand side (R.H.S.) of the optimal growth equation.
Growth-reducing effect of temperature	$T(DEGD) = \frac{4(DEGD - DEGD_{min})(DEGD_{max} - DEGD)}{(DEGD_{max} - DEGD_{min})^2}$ DEGD = growing degree-days, $\left.\begin{matrix}DEGD_{min}\\DEGD_{max}\end{matrix}\right\}$ = extremes of growing degree-days corresponding to species range	From Botkin et al. (1972). This equation multiplies the R.H.S. of the optimal growth equation. Note: DEGD is chosen from the normal distribution with mean and variance for the location of interest. From Botkin et al. (1972). This equation multiplies the R.H.S. of the optimal growth equation.

Growth-reducing effect related to stand biomass (competition)	S(BAR) = 1 - BAR/SOILQ BAR = total biomass of stand, SOILQ = maximum biomass recorded	
Growth-reducing effect of moisture stress†	AW = (FC - dry) SD (AW = available water, FC = field capacity, Dry = wilting point, SD = soil depth)	Evapotranspiration computed monthly by Thornthwaite equation. Rainfall input simulated from monthly means and variances. Growth is reduced as a linear function of the days in growing season where AW = 0.
Growth reducing effect of winter low temperatures†	If $T_{jan} = T_{species}$ $\frac{d[D^2H]}{dt} = 0$ (T_{jan} = January temperature $T_{species}$ = minimum January temperature tolerated by species)	Simulates loss of growth due to frost damage to buds.
<u>Subroutine</u> BIRTH		
New saplings are entered on plot	Selects species from eligible set based on the uniform distribution. Number of each of these species entered on plot is also chosen from the uniform distribution	See Shugart and West (1977).
Stump and root sprouting	Similar to BIRTH	See Shugart and West (1977).
<u>Subroutine</u> KILL		
Intrinsic tree mortality	$p = 1 - (1 - \epsilon)^n$ p = probability of mortality at year n, ϵ is chosen such that p = 0.99 when n = AGE_{max} (the maximum age for the species)	From Botkin et al. (1972).
Mortality of trees with suppressed growth	If growth is less than a critical value for the species, p = 0.368.	From Botkin et al. (1972).

*Detailed parameter values are provided in Shugart and West (1977).

†Detailed description is provided in Solomon et al. (1984).

maximum annual height recorded in the literature on early growth of each tree species. These data also can be obtained from plantations, and seem to be equally valid when applied to provenances or to species. In fact, either formulation appears appropriate for use in afforestation in cold regions, although FORSKA's lack of reliance on a constant value for maximum diameter, a highly variable property, probably provides an advantage. On the other hand, FORSKA's requirement for specialized information on sapwood area and photosynthetic compensation point may make implementation difficult.

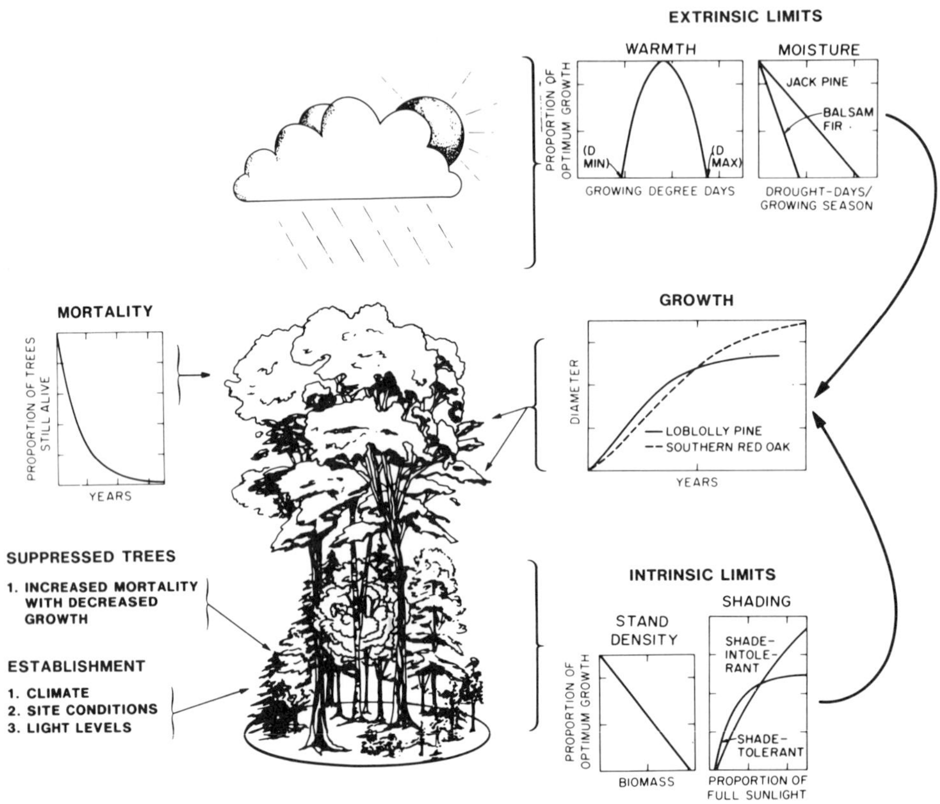

Fig. 1. Conceptual structure and function of the FORENA gap model (Source: Solomon and Webb, 1985).

The optimum tree growth described above rarely, if ever, occurs. Instead, optimum tree growth is reduced by intrinsic constraints on growth (competition for sunlight or soil nutrients, stand density, and so on) or by constraints extrinsic to stands (soil moisture and growing-season warmth). In the original JABOWA and FORET models, the growth multiplier values (usually < 1), representing individual constraints, are multiplied by each other, then by the amount of optimal growth produced by the fundamental growth equation (Table 1, subroutine GROW). FORENA and LINKAGES select only the single greatest stress value as a multiplier of optimum growth, based on Liebig's Law of the Minimum. In either case, the approach closely mimics the situation in cold regions where most seasons are less than optimum and environmental constraints are unlikely to ever allow optimum growth for more than a short time during the life of any individual tree.

Intrinsic Constraints. The most important growth multiplier results from competition for sunlight. The gap models accumulate the leaf area of all trees at and above each one-half or one meter elevation, and divide this quantity by the area of the plot, to produce an average leaf area at each level. Measured light extinction as a function of leaf area (Kasanaga and Monsi, 1954; Monsi et al., 1973) defines the resulting percentage of full sunlight at each level (Table 1, subroutine GROW). In all gap models except FORSKA, the top of each tree is assumed to hold its entire leaf area. Hence, leaf area approximates a disk. In FORSKA, leaf area forms a cylinder, with leaf area distributed between the top of the tree and the top of the bole (the light level required to reach or exceed the photosynthetic compensation point for each species). Although the "disk" may produce correct answers to species composition predictions, the "cylinder" approach of FORSKA is especially appropriate in predicting stand structure at high latitudes where some trees may photosynthesize on lowermost branches their entire lifetime.

The response to non-optimum light differs by the tolerance class of trees (Fig. 1, and Table 1, subroutine GROW). Shade-intolerant species have a growth advantage over shade-tolerant species when leaf area is small, and a growth disadvantage when leaf area is great. Although all gap models use very similar light extinction and shading response equations (Table 1, subroutine GROW), the number of shade classes has varied up to five, with tolerance in more than two classes being a weighted average between the two initial classes. The low diversity of species in northern regions suggests that just two tolerance classes are probably adequate to characterize tree growth in response to sunlight.

It is not obvious that the foregoing approach to defining competition for light will be accurate for predicting tree growth at high latitudes. Billings (1987) points out that light competition is of greatest importance in the low-latitude tropics, and is of decreasing significance to plant growth with increasing latitude. The low stem densities and low sun angles of cold regions argue for an approach that considers lateral competition for sunlight, that is, a spatially-explicit leaf-area model. The FORSKA model nullifies part of the problem because it defines the vertical distribution of leaf area as a function of available sunlight, simulating self-pruning. Like all other gap model versions, however, FORSKA assumes sunlight enters the plot at 90° from horizontal. Bonan and Shugart (1989) argue that, in reality, scattering of low angle sunlight at high latitudes produces a diffuse light above closed forest stands, making shading a vertical phenomenon. In open woodlands, the model merely utilizes the sunlight available. If this is always a high proportion of full sunlight, then the simulations will merely reflect the weak competition for light that characterizes these open stands and open-grown trees. Only forest edge simulation is absent from the model's capabilities, and this may be of minor importance.

Other intrinsic constraints on tree growth include competition for nutrients, and mineral-cycling phenomena that define nutrient availability. In most forms of the model (JABOWA, FORET, FORENA, FORSKA), a proxy growth multiplier for the competition for nutrients is stand density. Annual growth of all trees on the plot is reduced linearly as biomass increases, until no growth occurs at the highest stand biomass measured (Fig. 1, lower right, and Table 1, subroutine GROW). The linear relationship between stand biomass and annual growth is not documented in the literature. This relationship provides only a rough proxy for the competition for water and nutrients.

In several stand simulators, this routine has been replaced with procedures that define the demand by individual species for nutrients (nitrogen) from fertilizer trials (Mitchell and Chandler, 1939), and that partially provide for nutrient demand by modeling the long-term decay, oxidation, and release of nitrates from annual litter increments deposited by trees in the simulated gap. Models that include nitrogen cycling were initiated with FORTNITE and FORNUT (Weinstein et al., 1982) and were improved with LINKAGES (Pastor and Post, 1985). Growth of trees in all three models declines as an inverted J-function of decreasing available nitrogen, such that a wide range of plentiful nitrogen values produces little effect on growth. Growth of each species varies based on tolerance to nutrient deficiencies. Thus, nutrient-demanding species (usually late successional trees) have growth advantages when nitrogen is plentiful, and species that tolerate nutrient deficiencies (usually shade-intolerant, pioneer trees) have growth advantages when nitrogen is limited.

LINKAGES also includes the effects of climate variation on nutrient cycling. Precipitation and temperature determine soil moisture variation which, in turn, defines the annual amount of nutrients released. This feature makes LINKAGES particularly useful in future afforestation efforts in cold regions under continuously shifting climates, although the direct effect of temperature on nitrogen release, and of wildfire on nutrient availability (not included in any of the model versions), should also be developed.

Extrinsic Constraints. Extrinsic factors controlling growth include heat units (Growing Degree Days—GDD—measured from a 5 °C base) in JABOWA, FORET, FORTNITE, FORENA, and LINKAGES; soil moisture in the latter three; and winter low temperatures in the latter two. FORSKA uses neither heat units, winter cold, nor soil moisture, although a very recent version (FORSKA-2; Prentice et al., 1991 and in press) rectifies this deficiency.

Growth response to moisture in FORENA and LINKAGES is defined by "drought days," that is, the number of days in the growing season during which soil moisture is at or below the wilting point (-15.5 bars). The soil is considered as a bucket with a volume (available water capacity) defined by the difference between minimum (wilting point in centimeters) and maximum (field capacity in centimeters) moisture in a square centimeter column of specified (rooting) depth, usually measured in local soil profiles. In all model versions except FORSKA-2 and LOKI, soil moisture is calculated and updated monthly from precipitation and temperature values that are entered into a Thornthwaite equation (Thornthwaite and Mather, 1957) in order to define evapotranspiration. This formulation is particularly arbitrary at high latitudes. FORSKA-2 uses a Penman-Monteith expression (Cramer and Prentice, 1988) for a less empirical and more physically-based estimate of soil moisture. LOKI uses a Preistly-Taylor expression, taking the low sun angles of high latitudes into account in calculating solar radiation, and also tracking seasonal water locked up in ice, both as snow and as permafrost. Inexplicably, Bonan's LOKI model uses potential evapotranspiration rather than the more suitable actual evapotranspiration favored in the other versions of soil moisture dynamics.

If there is more precipitation than evaporation in a given month, excess water runs off and no drought days are registered by the model. If monthly precipitation does not exceed or equal evapotranspiration, soil moisture is still not limiting to growth until the available water capacity of the soil has also been evaporated; at that point, drought day tabulation begins. The total drought days accumulated during the growing season are compared with the species-specific drought day curve (Fig. 1 and Table 1, subroutine GROW) to calculate the soil-moisture growth multiplier for the year. The most extreme drought-day value in the geographic range of each species provides the x-intercept. This value was determined by mapping the variation of drought days in the geographic range of each species. The equation by Thornthwaite or, for FORSKA, by Penman and Monteith, is applied with standard soil moisture capacities of an upland clay-loam to all county-level monthly temperature and precipitation records in the eastern United States. These values were mapped as annual drought-day distributions for comparison with geographic ranges of each species. In the absence of evidence to the contrary, these calculations appear suitable for use in cold regions, although evapotranspiration at high latitudes with low sun angles is poorly replicated by either Thornthwaite or Penman-Monteith expressions. Hence, the boreal model of Bonan is probably most useful where effects of soil moisture variation are of primary concern.

Growth response to heat units is modeled as a parabola, with minimal growth at the northern (minimum heat units tolerated) and southern (maximum heat units tolerated) geographic range of each species, changing rapidly to optimal growth at the center of the range (Fig. 1 and Table 1, subroutine GROW). Evidently, phenological events in temperate and boreal trees are a direct function of heat units (Ebell and Schmidt, 1964) and photosynthetic responses by trees follow a parabolic function of temperature (for example, Larcher, 1983, p. 114). Although Jacoby and Ulan (1981) found slightly better correlation of tree-ring width with summer growing degree-days than with summer temperature, there is little more than intuitive logic to suggest that annual heat units control annual growth and do so as a function of a parabola. Because temperature responses are critical to afforestation in cold regions, these assumptions require closer inspection.

The physical conditions that limit forest distribution in cold regions are among the best known of bioclimatic phenomena. Ødum (1990), along with several authors of papers in this volume (Junttila and Nilsen; Skre; Wardle; all in press), emphasize the importance of growing-season warmth for tree growth in cold regions; we may conclude that summer warmth accounts for most variance in tree growth at the northern forest border. Wardle (in press) suggests that the commonly observed coincidence of the 10 °C July isotherm with the northernmost forests was as accurate a delimiter of forests as any; Kauppi and Posch (1985) found similarly good correspondence for 600 growing degree days above 5 °C during the growing season. The detailed reviews of treeline bioclimatology by Tuhkanen (1980, 1984) suggested that, in addition, moisture indices during the growing season can permit one to distinguish between the effects on tree growth of cold, wet maritime regions and cold, dry continental ones. Winter conditions, particularly those that cause "frost-drought," are also known to restrict tree growth (Grace and James, in press; Larsen, in press), while winds in maritime regions and in alpine situations are also indicated (Robertson, in press; Leivsson, in press; Lines, in press). Finally, Tigerstedt (in press) examines the role of "bottleneck years," that is, the occurrence of uncommon extreme weather events, also documented by Lindgren (in press), that are critical in defining success or failure of afforestation efforts.

How well do the stand simulators account for responses to these critical conditions of summer warmth and winter cold? The models described all use growing-degree days, and in FORENA and LINKAGES, winter-low temperature as well, with stochastic distributions that can mimic effects of both normal and uncommon extreme weather events. The modeled growth, however, responds only to annual growing-degree days and to threshold values of winter-low temperatures. Hence, the impact of irregular events such as frost during the growing season is not modeled. Also, modeled warmth early in the growing season is as important as warmth at the end of the season despite the documented importance of spring warmth (Fritts, 1976).

Furthermore, the characterization of physical conditions may not be as precise as necessary for predicting afforestation success. The edges of the geographic range of tree species provide locations where temperature values thought to control ranges can be measured. This "realized range," however, is frequently smaller than the "hypothetical" range in which species can actually become established, grow, and successfully reproduce, for example, Pinus sylvestris L. (Bennett, 1984). If a parabola can represent tree response to heat sums, the parabola covers a greater range of growing-degree day values than can be (and has been) deduced from the continuous area of tree distributions. The models could probably be improved for use in cold regions if current parabolic growing-degree day functions are replaced by response functions (Bartlein et al., 1986) that compare long-term measurements of the environmental variables with tree-ring increments at highest latitudes. For example, Jacoby and Ulan (1981) compared tree-ring widths to monthly temperature, growing-degree days, and precipitation at several forest-tundra sites in western North America. They concluded that April, June, and July temperatures and growing-degree days are of varying, but critical, importance in tree growth.

In addition, even where the realized and hypothetical ranges do coincide, growth of established trees is not normally limiting to completion of tree life cycles. Although Wardle (in press) expresses little enthusiasm for the idea, reproduction and seedling establishment normally limit tree success in cold regions where saplings and mature trees grow well. This is evident from afforestation efforts described by Mikola (in press) and exemplified in the Icelandic tree plantation experiences examined at Skorradalur and Haukadalur by participants of this workshop. Hence, examination of simulated seedling establishment may be of more importance. The gap models all are quite general in their representation of seedling establishment; only saplings of 0.5 to 2.5 centimeters diameter at breast height are considered for entry to simulated plots, and these are considered to be results of random processes. Still, the model's definitions of establishment conditions, and their application as either present or absent, are probably adequate to predict seedling success in specific locations of potential interest. Despite certain gross distortions of reality, the establishment routines in the stand simulators may be quite appropriate to assess future afforestation success.

Modeled Tree Mortality

Once trees enter the plot, they are subject to a probability of mortality with increasing age such that no more than one percent ever reach the oldest known age for the species. The probability of death follows a backwards J curve (Fig. 1 and Table 1, subroutine KILL), generating high mortality at the youngest ages and declining mortality with age. As a result, stand-age structures also approximate a backwards J as they normally do in mixed-age and mixed-species stands. This age-dependent mortality is repeated in all versions of the stand simulator, although in FORSKA, age structure is an emergent property of the stand, resulting from increased difficulty of obtaining water with increasing tree height and leaf area and decreasing cross-sectional area of sapwood.

All versions of the model also contain an age-independent mortality routine that is designed to mimic suppression mortality (Table 1, subroutine KILL). In JABOWA and most of its derivatives, an annual diameter increment below one millimeter in any tree of any species enhances the probability of mortality such that the tree has only a 1 in 10 chance of surviving the following 10 years. We found the routine unnaturally killed trees that were vigorous and dominant when subjected to a single year of difficult conditions, prematurely killed new saplings with small annual increments, and unrealistically killed trees adapted to slow growth such as drought-tolerant trees, preventing them from assuming dominance on droughty soils (Solomon et al., 1984). Therefore, in FORENA, we modified the age-independent mortality routine to require at least two successive years of weak growth before the probability of mortality increases (Solomon and Bartlein, in press; Solomon and Shugart, 1984). The new routine also defined weak growth as being 10 percent of the optimum growth possible for each tree of each species at its current age. This change allowed slow-growing trees to do well under high stress conditions that kill more rapidly-growing species. The modified age-independent mortality routine has since been incorporated in LINKAGES and FORSKA.

As far as they go, the mortality routines appear to be sound and suitable for application in the assessment of future success of afforestation efforts. Not all model versions, however, account for large-scale or episodic tree death events, such as may occur from insect infestation, disease, or wildfire. These events occasionally have been incorporated as separate subroutines. For example, Bonan (1988) has included a simple but effective spruce budworm infestation routine based on data in MacLean (1980). The probability of budworm outbreaks is a function only of time since the previous outbreak, while the effect of outbreaks is to enhance the probability of mortality by category of tree ages, tree diameters, and intensities of attack. The linkage of an analogous routine relating effects of green spruce aphid (Elatobium abietinum Walker) to the minimum winter temperatures already produced in FORENA or LINKAGES would allow accurate prediction of the high mortality that occurs in Iceland's Sitka spruce plantations in coastal areas where winter temperatures rarely fall below -10 °C (Blöndal, 1987). Mortality of larch induced by the larch canker (Potebniamyces coniferarum (Hahn) Smerlis) during cool summers in Greenland (Ødum, 1990) and comparable sources of weather-disease mortality can be similarly simulated.

Some routines to mimic episodic mortality processes have originated multiple times. Models to describe fire frequency and intensity versus response by tree growth and reproduction have been developed independently: FORAR (Mielke et al., 1979), BRIND (Noble et al., 1980), CLIMACS (Dale and Hemstrom, 1984), SILVA (Kercher and Axelrod, 1984), and LOKI (Bonan, 1988). Positive responses by growth, and indirectly, mortality, to CO_2 "fertilization" have emerged from work by Botkin et al. (1973), Shugart (1984), and Solomon and West (1986). All of these are empirical and simple compared with the process-oriented fertilization routine of Prentice et al. (in press) which represents a large step toward physiologically-driven stand model processes. When relevant field data become available, additional subroutines can be constructed to deal with such locally unique phenomena as the fungus-induced mortality in Siberian larch in Iceland following episodic defoliation by late-spring frosts.

The Collective Gap Model

So far, we have described the appropriateness of specific gap model subroutines in predicting afforestation success under chronic climate change. Now, we examine a characteristic of gap models as a whole that may limit or promote their use. The models are designed to avoid simulating inappropriate forest behaviors by the imposition of strict limits on species size, age, and geographic ranges of allowable climate variables. As a result, constraining the model with the climate values of a given region almost always produces a forest that is normal for the region, but one that is incorrect for any given location within the region. The model is simply incapable o reproducing specific forest stands because stand structure and composition depend on the history of weather and disturbance events, the fortuitous match of seed sources and seed beds when light becomes available on the forest floor, and so on. The model substitutes output from a random number generator for these events and, therefore, it cannot replicate the resulting stand characteristics, especially during the first 50 or 100 years when random events are most important to forest stand properties.

One approach that allows application of the model to specific locations is to utilize as much of the known history of weather, disturbance, and land use as possible, and to use natural history data on the tree ecotypes (provenances) rather than of the species as a whole. The only example available is from data and simulations of experimental watersheds on Piedmont soils at Coweeta Hydrological Laboratory, Otto, North Carolina (Solomon, 1988). The watersheds were selectively logged in the 1890s, and were covered by dense forest stands dominated by American chestnut (Castanea dentata L.) when the chestnut blight arrived in the early 1930s. This disturbance destroyed approximately 70 percent of the forest biomass, providing a convenient point from which to simulate annual forest succession. Stem numbers, diameters, heights, and species (including the standing dead chestnuts) were surveyed in 1936, 1972, and 1986.

That history of forest growth was simulated with FORENA. The values for maximum height and diameter, normally measured in the geographic range of each species, were replaced with the greatest values measured in the southeastern Piedmont region (Virginia, North Carolina, South Carolina), approximating ecotypes. The simulations were then constrained with monthly temperature and precipitation values measured at Coweeta, in place of the random values the model normally uses. The history of Coweeta land use and disturbance was also replicated. Fifty-seven model runs were begun with the numbers, diameters, and heights of each stem of each species present on 57 Coweeta plots in 1936. Simulation results were averaged, then compared with averages of data from surveys of the 57 Coweeta plots conducted in 1972 and 1986.

Results are illustrated in Figure 2. The similarity of measured and simulated biomass on the plots is striking; only chestnut is inaccurately reproduced by the model. The 1936 survey included about 70 tonnes of chestnut, all as standing dead trees. The plot survey of 1972 revealed a few chestnuts as root sprouts, and in 1986, fewer still. None of the chestnut biomass, however, was simulated because the model was programmed to kill all chestnut trees and seeds after 1936.

The model accurately reproduced the increase to dominance of chestnut oak (Quercus prinus L.) which replaced the dead American chestnuts; the 1936 chestnut oak biomass doubled by 1986 both in the simulations and on the plots. At the same time, biomass of red maple (Acer rubrum L.), red oak (Quercus rubra L.), scarlet oak (Q. coccinea Muenchh.), and pignut hickory (Carya glabra (Mill.) Sweet) assumed subdominant status both in measurements and simulations, even though biomass changes among these species were drastically different. Simulated biomass accounted for 94 percent of the variance in measured biomass values in 1986, although the 1936 measured biomass values were very different from 1986 values, having less than two percent of their variance in common. Total above-ground biomass in 1986 was simulated at 168 tonnes per hectare, and was measured at 165 tonnes per hectare. Predictions were probably more accurate than necessary to project future growth, considering that estimates of future weather are likely to be much less reliable.

CONCLUSIONS

Gap models appear to be the most appropriate method available to predict success of afforestation efforts in cold regions under chronically changing climate. They have no serious flaws for this objective. Although the light competition that drives them is not very important in high-latitude cold regions, the models work well under conditions of low shade and high sunlight. In addition, afforestation in cold regions during gradual warming should produce a steadily increasing competition for light as trees grow more densely, requiring a

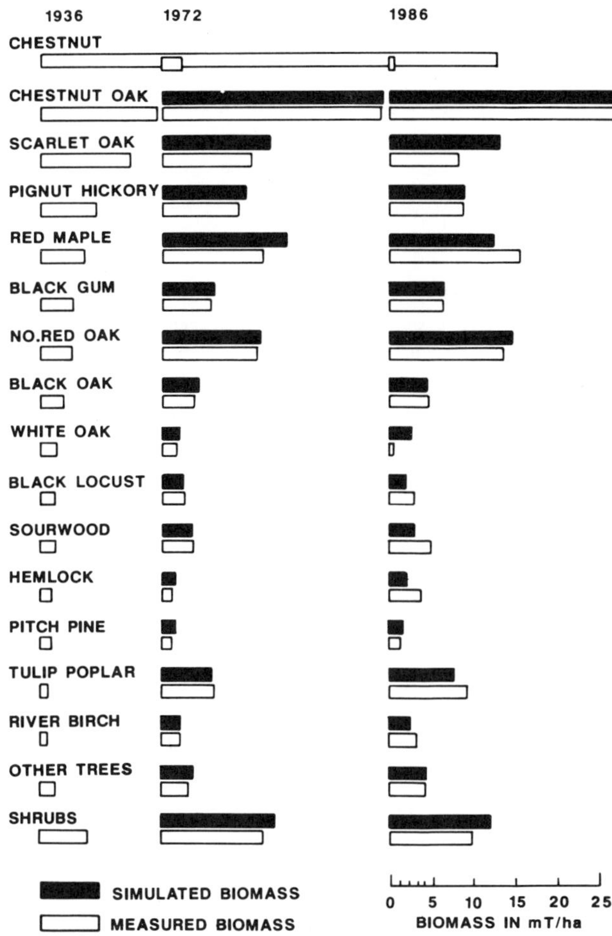

Fig. 2. Simulated and measured average biomass on 57 plots at Coweeta Hydrological Laboratory, Otto, North Carolina. Stands were measured in 1936, 1972, and 1986. The stand model FORENA was initiated using 1936 measurement data, and subsequent monthly weather data, to simulate forest growth through 1972 and 1986. (Source: Solomon, 1988, based on unpublished data of D. C. West and M. L. Tharp)

model that can accurately mimic forest growth across the entire range of available sunlight values.

Gap models have several fundamental strengths which are based on known physiological and ecological principles. Such models treat seedling establishment as a set of processes independent of tree growth and mortality. Dynamics in each life stage are constrained by the life history of each species. As a result, the individual peculiarities of natural history that determine successful growth and reproduction in cold regions are easily included in the model.

Gap models are a strong tool for afforestation assessments in a changing world because weather information can easily be integrated with growth, death, and reproduction in different tree species (by year in JABOWA and FORET, by month in FORENA, LINKAGES, and FORSKA-2). This feature alone would be enough to recommend the models for this purpose. That is because the response of species to chronic and episodic weather events is a pivotal determinant of success in afforestation. There is also considerable validity in the simulated growth strategy in which optimum growth is defined, then reduced by variables intrinsic and extrinsic to the stands. The more recent addition of effects by soil moisture deficits and nutrient availability have considerably enhanced the validity of the growth expression.

Tree establishment routines are also very suitable to use in gap models, because they depend on specific site factors that must be present or absent for seedling growth. The addition of stems from layering or root sprouting is more important in cold regions than elsewhere. Unmanaged establishment also may be bypassed and replaced with a specific planting regime of species, seedling sizes, and densities. Again, the integration of episodic events, especially of very occasional conditions conducive to establishment, provides realism to the models.

Mortality routines, too, are relevant. The increasing likelihood of death from some cause as the tree ages is reasonable in cold regions. Age-independent mortality, requiring some growth in successive years, is easily adjusted to information supplied in tree rings from current afforestation efforts (Ødum, 1990, Fig. 2). The array of available catastrophic mortality routines (fire, disease, pests, windthrow) provides choices needed to fit the model to local conditions.

Given these strengths, can a single model variant be recommended for use in predicting success of afforestation efforts in cold regions? Probably not, because each region has significantly different limits to tree growth and reproduction now and in response to expected climate change. Clearly some models will be less useful than others. In fact, most models contain major disadvantages for use in cold regions. All except FORSKA and FORSKA-2 treat trees as two dimensional disks of leaf area. This produces strong distortion and inaccuracy in growth simulation. JABOWA, FORTNITE, and SILVA do not include the effects of annual or monthly weather variation on tree growth. Also, they contain few of the physical limits to growth that control afforestation success in cold regions. FORET has no moisture or nutrient routines, and no winter cold effects. FORENA is general enough to work throughout eastern North America, but at the expense of the specificity needed to function in localized areas. The use of provenances instead of species described in the Coweeta experiments (Solomon, 1988) indicates that the FORET and JABOWA models are simply incapable of valid simulations for individual sites without prohibitively detailed site history.

Two model versions contain only minor disadvantages. By virtue of its superb treatment of the physical environment, the boreal model of Bonan is among the best for assessing afforestation in cold regions. It includes critical processes of permafrost dynamics and sun angle effects, and differentiates between direct and diffuse sunlight components. It also defines air temperature from solar radiation and surface albedo, which is very important at high latitudes. Weaknesses of the model are few, but possibly significant. It is superficially and unnecessarily empirical; in many cases its dynamics depend on interactions between a few arbitrary classes of effects and a few of plant response, rather than on continuous functions and relationships that can be derived from the ecophysiological literature. In addition, the model requires a region with a fairly short fire rotation and a lichen-mat growth and effect cycle in order to function (Bonan et al., 1990). Neither of these may be present at coastal sites where the model is to be implemented.

A second model of great suitability is FORSKA-2, the environmentally-driven version of FORSKA. Its treatment of tree morphology in relation to available light is far more realistic than other model versions with respect to the time-ordered development of stand structure, that is, the distribution of size classes and species composition of trees. The inclusion of sunlight-mediated photosynthetic equations, the self-pruning, and the relation of leaf area to height and productivity are all great improvements on earlier JABOWA and FORET-related models, and are of particular importance in high latitudes. The disadvantages of FORSKA-2 include its replicate model runs with differing site variables to generate a landscape-level composite, although the model can be run replicating simulations with the same site variables. FORSKA-2 also requires specialized biological and environmental data that may be unavailable or difficult to obtain. The model requires data for all species simulated on the photosynthetic compensation point (light level at which production by photosynthesis equals energy use by respiration). It also requires a measure of the sapwood area as a constant, although this may vary throughout the life of trees and, in any case, is not routinely measured in the field. If these problems can be overcome, however, FORSKA-2 is also an excellent choice for application in cold regions.

A combination of the environmental strengths of LOKI and the biological-ecological strengths of FORSKA would produce an exceedingly accurate and valid model for predictive purposes. This was the initial plan that led to development of the two models by Ph.D. students of H. H. Shugart (Bonan's model) and of I. C. Prentice (Leemans' model). Although the collaboration is well documented (Shugart et al., 1991), technical problems have kept the two models from being successfully integrated and applied. In addition to other motivations, the lack of a unified boreal model led Prentice and his coworkers to develop FORSKA-2. FORSKA-2 does not, however, include many of the physical phenomena and processes in LOKI that are so important in shrub-tundra and treeline situations. With increasing interest in afforestation of cold regions, it seems very likely that a functional, unified stand model for shrub-tundra environments will be developed soon.

REFERENCES

Aber, J. D., Botkin, D. B., and Melillo, J. M., 1978, Predicting the effects of different harvesting regimes on forest floor dynamics in northern hardwoods, Can. J. For. Res., 8:306-315.

Aber, J. D., and Melillo, J. M., 1982, FORTNITE: A Computer Model of Organic Matter and Nitrogen Dynamics in Forest Ecosystems, Research Bulletin R3130, University of Wisconsin, Madison, Wisconsin, 48 p.

Bartlein, P. J., Prentice, I. C., and Webb, T., III, 1986, Climatic response surfaces from pollen data for some eastern North American taxa, J. Biogeogr., 13:35-57.

Belcher, D. M., Holdaway, M. R., and Brand, G. J., 1982, A Description of STEMS: The Stand and Tree Evaluation and Modelling System, General Technical Report NC-94, U.S. Department of Agriculture, Forest Service, North Central Experiment Station, Minneapolis, Minnesota, 18 p.

Bennett, K. D., 1984, The post-glacial history of Pinus sylvestris in the British Isles, Quat. Sci. Rev., 3:133-155.

Billings, W. D., 1987, Constraints to plant growth, reproduction and establishment in Arctic environments, Arctic and Alpine Res., 19:357-365.

Blöndal, S., 1987, Afforestation and reforestation in Iceland, Arctic and Alpine Res., 19:526-529.

Bonan, G. B., 1988, Environmental Processes and Vegetation Patterns in Boreal Forests, Ph.D. Dissertation, University of Virginia, Charlottesville, Virginia, 286 p.

Bonan, G. B., 1990, Carbon and nitrogen cycling in North American boreal forests, II. Biogeographic patterns, Can. J. For. Res., 20:1077-1088.

Bonan, G. B., and Shugart, H. H., Jr., 1989, Environmental factors and ecological processes in boreal forests, Ann. Rev. Ecol. and Syst., 20:1-29.

Bonan, G. B., Shugart, H. H., and Urban, D. L., 1990, The sensitivity of some high-latitude boreal forests to climatic parameters, Climatic Change, 16:9-26.

Botkin, D. B., Janak, J. F., and Wallis, J. R., 1972, Some ecological consequences of a computer model of forest growth, J. Ecol., 60:849-872.

Botkin, D. B., Janak, J. F., and Wallis, J. R., 1973, Estimating the effects of carbon fertilization on forest composition by ecosystem simulation. In: Woodwell, G. M., and Pecan, E. V., eds., Carbon and the Biosphere, CONF-720510, NTIS, Springfield, Virginia, 328-344.

Bowling, S. A., [In press], Climate change as perceived by trees and climate modelers. In: Alden, J. N., Ødum, S., and Mastrantonio, J. L., eds., Forest Development in Cold Climates, Proceedings of an International Symposium, 1991 June 18-24 [Laugarvatn, Iceland], NATO, Plenum, New York, New York.

Burns, R. M., and Honkala, B. H., eds., 1990a, Silvics of North America: Volume 1, Conifers, Agriculture Handbook No. 654, Forest Service, U.S. Department of Agriculture, Washington D.C., 675 p.

Burns, R. M., and Honkala, B. H., eds., 1990b, Silvics of North America: Volume 2, Hardwoods, Agriculture Handbook No. 654, Forest Service, U.S. Department of Agriculture, Washington D.C., 877 p.

Cramer, W., and Prentice, I. C., 1988, Simulation of regional soil moisture deficits on a European scale, Norsk. Geogr. Tiddskr., 42:149-151.

Dale, V. H., and Hemstrom, M., 1984, CLIMACS: A Computer Model of Forest Stand Development for Western Oregon and Washington, Research Paper PNW-327, Forest Service, U.S. Department of Agriculture, Pacific Northwest Forest and Range Experiment Station, Portland, Oregon, 60 p.

Dale, V. H., Hemstrom, M., and Franklin, J., 1986, Modeling the long term effects of disturbances on forest succession, Olympic Peninsula, Washington, Can. J. For. Res., 16:56-67.

DeBell, D. S., 1990, Populus trichocarpa Torr. & Gray., black cottonwood. In: Silvics of North America: Volume 2, Hardwoods, Burns, R. M., and Honkala, B. H., eds., Agriculture Handbook No. 654, Forest Service, U.S. Department of Agriculture, Washington D.C., 570-576.

Ebell, L. F., and Schmidt, R. L., 1964, Meteorological Factors Affecting Conifer Pollen Dispersal on Vancouver Island, Publication No. 1036, Canadian Department of Forestry, Victoria, British Columbia, 28 p.

Fritts, H. C., 1976, Tree Rings and Climate, Academic Press, Inc., New York, New York, 567 p.

Fulton, M. R., 1991, A computationally efficient forest succession model: Design and initial tests, For. Ecol. and Manage., 42:23-34.

Fulton, M. R., [In press], A rapid simulation model for vegetation stand dynamics including mixed life forms. In: Solomon, A. M., and Shugart, H. H., Jr., eds., Vegetation Dynamics and Global Change, Chapman and Hall, Inc., New York, New York.

Grace, J., and James, J., [In press], The physiology of trees at treeline. In: Alden, J. N., Ødum, S., and Mastrantonio, J. L., eds., Forest Development in Cold Climates, Proceedings of an International Symposium, 1991 June 18-24, [Laugarvatn, Iceland], NATO, Plenum, New York, New York.

Harris, A. S., 1990, Picea sitchensis (Bong.) Carr., Sitka spruce. In: Burns, R. M., and Honkala, B. H., eds., Silvics of North America: Volume 1, Conifers, Agriculture

Handbook No. 654, Forest Service, U.S. Department of Agriculture, Washington D.C., 260-267.

Jacoby, G. C., and Ulan, L. D., 1981, Review of dendroclimatology in the forest-tundra ecotone of Alaska and Canada, Syllogeus, 33:97-128.

Jaeger, J. W., 1988, Developing Policies for Responding to Climatic Change, Report WCIP-1, WMO/TD-No. 225, World Meteorological Organization, Geneva, Switzerland, 53 p.

Junttila, O. and Nilsen, J., [In press], Physiological response of forest trees to low temperature and extreme light conditions. In: Alden, J. N., Ødum, S., and Mastrantonio, J. L., eds., Forest Development in Cold Climates, Proceedings of an International Symposium, 1991 June 18-24, [Laugarvatn, Iceland], NATO, Plenum, New York.

Kasanaga, H., and Monsi, M., 1954, On the light transmission of leaves and its meaning for the production of matter in plant communities, Jap. J. Bot., 14:302-324.

Kauppi, P., and Posch, M., 1985, Sensitivity of boreal forests to possible climatic warming, Clim. Change, 7:45-54.

Ker, J. W., and Smith, J. H. G., 1955, Advantages of the parabolic expression of height-diameter relationships, For. Chron., 31:235-246.

Kercher, J. R., and Axelrod, M. C., 1984, A process model of fire ecology and succession in a mixed-conifer forest, Ecology, 65:1725-1742.

Kimmins, J. P., and Scoullar, K. A., 1981, FORCYTE—a computer simulation approach to evaluating the effect of whole tree harvesting on nutrient budgets and future forest productivity, Mitt. D. Fortstl, Bundesversuchsanstalt, Wein, 140:189-205.

Kimmins, J. P., and Scoullar, K. A., 1983, FORCYTE-10: A User's Manual, Faculty of Forestry, University of British Columbia, Vancouver, British Columbia, 112 p.

Kramer, P. J., and Kozlowski, T. T., 1960, Physiology of Trees, McGraw-Hill Book Co., New York, New York, 642 p.

Larcher, W., 1983, Physiological Plant Ecology, 2nd Edition, Springer-Verlag, Berlin, 303 p.

Larsen, J.B., [In press], Testing winter desiccation resistance for species and provenance selection at timberlines. In: Alden, J. N., Ødum, S., and Mastrantonio, J. L., eds., Forest Development in Cold Climates, Proceedings of an International Symposium, 1991 June 18-24, [Laugarvatn, Iceland], NATO, Plenum, New York, New York.

Leemans, R., and Prentice, I. C., 1987, Description and simulation of tree-layer composition and size distributions in a primaeval Picea-Pinus forest, Vegetatio, 69:147-156.

Leemans, R., and Prentice, I. C., 1989, FORSKA: A General Forest Succession Model, Report 1989:2, Medd. f. Vaxtbiologiska Institutionen, University of Uppsala, Uppsala, Sweden, 45 p.

Leivsson, T.G., [In press], Altitude treeline in the Faroe Islands. In: Alden, J. N., Ødum, S., and Mastrantonio, J. L., eds., Forest Development in Cold Climates, Proceedings of an International Symposium, 1991 June 18-24, [Laugarvatn, Iceland], NATO, Plenum, New York, New York.

Lindgren, D., [In press], Provenance selection, testing, and breeding to accelerate adaptation and growth of trees and shrubs at tree limits. In: Alden, J. N., Ødum, S., and Mastrantonio, J. L., eds., Forest Development in Cold Climates, Proceedings of an International Symposium, 1991 June 18-24, [Laugarvatn, Iceland], NATO, Plenum, New York, New York.

Lines, R., [In press], Aspects of species and provenance choice at northern tree limits in maritime climates. In: Alden, J. N., Ødum, S., and Mastrantonio, J. L., eds., Forest Development in Cold Climates, Proceedings of an International Symposium, 1991 June 18-24, [Laugarvatn, Iceland], NATO, Plenum, New York, New York.

Loomis, R. S., Williams, W. A., and Duncan, W. G., 1967, Community architecture and the productivity of terrestrial plant communities. In: Greer, F. A., and Army, T. J., eds., Harvesting the Sun, Academic Press, New York, New York, 291-308.

MacLean, D. A., 1980, Vulnerability of fir-spruce stand during uncontrolled spruce budworm outbreaks: A review and discussion, For. Chron., 56:213-221.

Mielke, D. L., Shugart, H. H., and West, D. C., 1979, User's Manual for FORAR, a Stand Model for Composition and Growth of Upland Forests of Southern Arkansas, Report ORNL/TM-5767, Oak Ridge National Laboratory, Oak Ridge, Tennessee, 45 p.

Mikola, J., [In press], Provenance and individual variation in climatic hardiness of Scots pine, *Pinus sylvestris* (L.), in Northern Finland. In: Alden, J. N., Ødum, S., and Mastrantonio, J. L., eds., Forest Development in Cold Climates, Proceedings of an International Symposium, 1991 June 18-24, [Laugarvatn, Iceland], NATO, Plenum, New York, New York.

Mitchell, H. L., and Chandler, R. F., 1939, The Nitrogen Nutrition and Growth of Certain Deciduous Trees of Northeastern United States, Bulletin No. 11, Black Rock Forest Experiment Station, Cornwall-on-the-Hudson, New York, 94 p.

Mitchell, J. F. B., Manabe, S., Tokioka, T., and Meleshko, V., 1990, Equilibrium climate change. In: Houghton, J. T., Jenkins, G. J., and Ephraums, J. J., eds., Climate Change: The IPCC Scientific Assessment, Cambridge University Press, New York, New York, 131-172.

Monsi, M., Uchijima, Z., and Oikawa, T., 1973, Structure of foliage canopies and photosynthesis, Ann. Rev. Ecol. and Syst., 4:301-327.

Nikolov, N., and Helmisaari, H., 1991, Silvics of the circumpolar boreal forest tree species. In: Shugart, H. H., Bonan, G. B., and Leemans, R., eds., A Systems Analysis of the Global Boreal Forest, Cambridge University Press, Cambridge, United Kingdom, 13-84.

Noble, I. R., Shugart, H. H., and Schauer, J. S., 1980, A Description of BRIND, a Computer Model of Succession and Fire Response of the High Altitude *Eucalyptus* Forests of the Brindabella Range, Australian Capital Territory, Report ORNL/TM-7041, Oak Ridge National Laboratory, Oak Ridge, Tennessee, 96 p.

Ødum, S., 1990, Afforestation experiments reflecting the treeline conditions in Southwest Greenland, Medd. om Groenland, Biosci., 33:43-61.

Pastor, J., and Post, W. M., 1985, Development of a Linked Forest Productivity-Soil Process Model, Report ORNL/TM-9519, Oak Ridge National Laboratory, Oak Ridge, Tennessee, 162 p.

Pastor, J., and Post, W. M., 1986, Influence of climate, soil moisture, and succession on forest carbon and nitrogen cycles, Biogeochemistry, 2:3-27.

Perry, T. O., Sellers, H. E., and Blanchard, C. O., 1969, Estimation of photosynthetically active radiation under a forest canopy with chlorophyll extracts and from basal area measurements, Ecology, 50:39-44.

Prentice, I. C., 1986, The design of a forest succession model. In: Forest Dynamics Research in Western and Central Europe, PUDOC, Wageningen, the Netherlands, 253-256.

Prentice, I. C., Sykes, M. T., and Cramer, W., 1991, The possible dynamic response of northern forests to global warming, Global Ecology and Biogeography Letters, 1:129-135.

Prentice, I. C., Sykes, M. T., and Cramer, W., [In press], A simulation model for the transient effects of climate change on forest landscapes, Ecol. Model.

Ritchie, J. C., 1984, Past and Present Vegetation of the Far Northwest of Canada, University of Toronto Press, Toronto, Canada, 251 p.

Robertson, A., [In press], Impact of wind on northern forests. In: Alden, J. N., Ødum, S., and Mastrantonio, J. L., eds., Forest Development in Cold Climates, Proceedings of an International Symposium, 1991 June 18-24, [Laugarvatn, Iceland], NATO, Plenum, New York, New York.

Shugart, H. H., Jr., 1984, A Theory of Forest Dynamics, Springer-Verlag, New York, New York, 278 p.

Shugart, H. H., Jr., Bonan, G. B., and Leemans, R., eds, 1991, A Systems Analysis of the Global Boreal Forest, Cambridge University Press, Cambridge, United Kingdom, 565 p.

Shugart, H. H., Jr., and West, D. C., 1977, Development of an Appalachian deciduous forest succession model and its application to assessment of the impact of the chestnut blight, J. Environ. Manage., 5:161-179.

Shugart, H. H., Jr., and West, D. C., 1979, Size and pattern of simulated forest stands, For. Sci., 25:120-122.

Shugart, H. H., Jr., Mortlock, A. T., Hopkins, M. S., and Burgess, I. P., 1980, A Computer Simulation Model of Ecological Succession in Australian Subtropical Rainforest, Report ORNL/TM-7029, Oak Ridge National Laboratory, Oak Ridge, Tennessee, 48 p.

Skre, O., [In press], Growth of mountain birch, Betula pubescens Ehrh., in response to variable temperatures. In: Alden, J. N., Ødum, S., and Mastrantonio, J. L., eds., Forest Development in Cold Climates, Proceedings of an International Symposium, 1991 June 18-24, [Laugarvatn, Iceland], NATO, Plenum, New York, New York.

Sollins, P., Reichle, D. E., and Olson, J. S., 1973, Organic matter budget and a model for a southern Appalachian Liriodendron forest, Report EDFB/IBP-73/2, Oak Ridge National Laboratory, Oak Ridge, Tennessee, 150 p.

Solomon, A. M., 1986, Transient response of forests to CO_2-induced climate change: Simulation experiments in eastern North America, Oecologia, 68:567-79.

Solomon, A. M. 1988, Use of stand models at varying spatial scales to simulate forest responses to environmental changes. In: Seymour, R. S., and Leak, W. B., eds., Proceedings of the New England Growth and Yield Workshop, Miscellaneous Report 325, Maine Agricultural Experiment Station, 46-58.

Solomon, A. M., and Bartlein, P. J., [In press], Past and future climate change: Response by mixed deciduous-coniferous forest ecosystems in northern Michigan, Can. J. For. Res.

Solomon, A. M., and Shugart, H. H., Jr., 1984, Integrating forest stand simulations with paleoecological records to examine long-term forest dynamics. In: Agren, G. I., ed., State and Change of Forest Ecosystems, Swedish University of Agricultural Science, Uppsala, Sweden, 333-356.

Solomon, A. M., Tharp, M. L., West, D. C., Taylor, G. E., Webb, J. M. and Trimble, J. L., 1984, Response of Unmanaged Forests to Carbon Dioxide-Induced Climate Change: Available Information, Initial Tests, and Data Requirements, Report TR-009, U.S. Department of Energy, Washington D.C., 93 p.

Solomon, A. M., and Webb, T., III, 1985, Computer-aided reconstruction of late-Quaternary landscape dynamics, Ann. Rev. Ecol. and Syst., 16:63-84.

Solomon, A. M. and West, D. C., 1986, Atmospheric carbon dioxide change: Agent of future forest growth or decline? In: Titus, J. G., ed., Effects of Changes in Stratospheric Ozone and Global Climate, Volume 3: Climate Change, U.S. Environmental Protection Agency, Washington D.C., 23-38.

Thornthwaite, C. W., and Mather, J. R., 1957, Instructions and tables for computing potential evapotranspiration and the water balance, Public. in Climatol., 10:183-311.

Tigerstedt, P. M.A., [In press], Genetic diversity of tree populations at their arctic limits. In: Alden, J. N., Ødum, S., and Mastrantonio, J. L., eds., Forest Development in Cold Climates, Proceedings of an International Symposium, 1991 June 18-24, [Laugarvatn, Iceland], NATO, Plenum, New York, New York.

Tuhkanen, S., 1980, Climatic parameters and indices in plant geography, Acta Phytogeogr. Suec., Almqvist & Wiksell International, Stockholm, Sweden, 67:1-110.

Tuhkanen, S., 1984, A circumboreal system of climatic-phytogeographical regions, Acta Bot. Fennica, 127:1-50.

Urban, D. L., 1990, A Versatile Model to Simulate Forest Pattern: A User's Guide to ZELIG Version 1.0, Environmental Sciences Department, University of Virginia, Charlottesville, Virginia, 108 p.

Urban, D. L., Bonan, G. B., Smith, T. M., and Shugart, H. H., Jr., 1991, Spatial applications of gap models, For. Ecol. and Manage., 42:95-110.

Wardle, P., [In press], The causes of alpine timberlines: A review of the hypothesis. In: Alden, J. N., Ødum, S., and Mastrantonio, J. L., eds., Forest Development in Cold Climates, Proceedings of an International Symposium, 1991 June 18-24, [Laugarvatn, Iceland], NATO, Plenum, New York, New York.

Washington, W. M., 1990, Where's the heat? The ocean may be the missing sink, Nat. Hist., 90(3):66-72.

Washington, W. M., and Meehl, G. A., 1989, Climate sensitivity due to increased CO_2: Experiments with a coupled atmosphere and ocean general circulation model, Climate Dynamics, 4:1-38.

Watt, A. S., 1947, Pattern and process in the plant community, J. Ecol., 35:1–22.

Weinstein, D. A., Shugart, H. H., Jr., and West, D. C., 1982, The Long-Term Nutrient Retention Properties of Forest Ecosystems: A Simulation Investigation, Report ORNL/TM-8472, Oak Ridge National Laboratory, Oak Ridge, Tennessee, 142 p.

Yarie, J., 1986, FORCYTE-10: Extension of a stand-level growth and yield model utilizing nitrogen dynamics to taiga white spruce forests. In: Van Cleve, K., Chapin III, F. S., Flanagan, P. W., Viereck, L. A., and Dyrness, C. T., eds., Forest Ecosystems in the Alaskan Taiga: A Synthesis of Structure and Function, Springer-Verlag, New York, New York, 190-204.

CLIMATE CHANGE AS SEEN BY TREES AND BY CLIMATE MODELERS

Sue Ann Bowling

Geophysical Institute
University of Alaska Fairbanks
Fairbanks, Alaska

INTRODUCTION

Over the last few years, several attempts have been made to model how climatic warming due to carbon dioxide and other greenhouse gasses will affect forest distribution. One of the questions raised about such studies, that of whether forests can migrate fast enough to keep up with anticipated climate change, is probably not as serious as it seems. In our human-dominated world, if forests are considered desirable, trees will be planted in tundra that has warmed enough for them to survive, even if the nearest seed source is several thousand kilometers away. The best species and provenance of trees is addressed in a number of the other papers in this Proceedings. The question examined in this paper is more basic: How trustworthy are the results of climate change models? In particular, can the results of the models be used as a guide for planting seedings that will mature 50 to 100 years in the future?

The path the models take from the observed and continuing increase in the carbon dioxide content of the atmosphere to the geographical distribution of forests 100 years in the future is a tortuous one with a number of pitfalls. Further, no single person is normally responsible for or even familiar with all of the steps. Because of this, it seems appropriate to review these steps, with particular emphasis on those known to be poor approximations of reality.

THE PHYSICAL BASIS OF CLIMATE

Climate, in the broad sense and as it influences tree growth and survival, is not simply a matter of mean temperatures, precipitation, and wind, but is rather the sum total or ensemble (in the thermodynamic sense) of all of the weather experienced in a region. Statistics such as means and extremes are merely convenient partial descriptions of climate. This being the case, climate change needs to be looked at in terms of changes in the weather.

Forest Development in Cold Climates, Edited by J. Alden *et al.*
Plenum Press, New York, 1993

Weather, in turn, is nothing more than a by-product of the atmospheric transport of energy from regions of radiative surplus to those of radiative deficit, subject to conservation of mass, moisture, and momentum on a rotating planet. As human activities influence the distribution of the radiative surpluses and deficits in the atmosphere, they are influencing the driving mechanism of the whole weather (and ocean current) system.

Radiative surpluses and deficits are emphasized because electromagnetic radiation is the only way in which the Earth is able to gain or lose significant amounts of energy. Energy is transferred from the sun to the Earth primarily as electromagnetic waves with wavelengths shorter than four micrometers. The Earth in turn radiates energy to space at an intensity depending on temperature (via the blackbody function) and emissivity, with most of this thermal radiation being at wavelengths longer than four micrometers (Brasseur and Solomon, 1984). In a steady-state climate, the energy absorbed from solar radiation just balances that emitted to space from the Earth as a whole. Locally, however, there are major imbalances. At night and at high latitudes in the winter, for instance, the incoming solar radiation is zero, which for balance would require temperatures of -273 °C. At noon near the equator, on the other hand, radiative balance would require a temperature well above the boiling point of water.

Different wavelengths of electromagnetic radiation have different energies per photon, and react in different ways with the Earth's atmosphere and surface. The shortest wavelengths (X-rays and "hard" ultraviolet) are able to ionize atoms and molecules, that is, knock electrons free with a wide range of energies for the electrons. These wavelengths can be absorbed by a wide range of atmospheric gasses, and are almost completely absorbed above 100 kilometers in the atmosphere, producing the ionosphere (defined on the basis of ionization) and thermosphere (roughly the same region, but defined on the basis of temperature). Somewhat less energetic radiation, but still in the hard ultraviolet range, is able to knock apart stable molecules such as O_2 and N_2, leading to free atoms of oxygen and nitrogen which can then combine with the remaining molecules to produce relatively unstable products such as ozone, O_3. These unstable molecules can be knocked apart by somewhat softer (longer wavelength) ultraviolet. Ozone, in particular, forms a layer at around 20 to 50 kilometers elevation that provides a shield against the more dangerous part of the ultraviolet radiation that is not absorbed by O_2. The absorption of ultraviolet at these elevations produces a relatively warm layer in the middle and upper stratosphere.

Longer and less energetic solar wavelengths are usually not well absorbed by atmospheric gasses. For the most part, the longest ultraviolet, visible, and very near infrared radiation emitted by the sun reaches either clouds or the surface of the Earth. Absorption or reflection there depends on the nature of the surface. Dark coniferous forest, for instance, absorbs more than 90 percent of the incoming solar radiation reaching the surface, reflecting 10 percent or less back upward. Fresh snow, on the other hand, may reflect as much as 80 percent of the incoming solar radiation, leaving only 20 percent to be absorbed. Clouds also reflect a high percentage (around 60 to 70 percent) of the incident sunlight (List, 1951), and transmit much of the remainder. The fraction of sunlight reflected is called the "albedo." Because the albedo itself depends on climate and season, it is apparent even at this stage that climate is a non-linear process with powerful feedback mechanisms.

As we continue to even longer wavelengths (greater than four micrometers), several things happen. In the first place, radiation emitted by objects at normal terrestrial temperatures begins to be more important than that received from the sun. Albedo becomes very low for almost all normal terrestrial surfaces (polished metal is an exception), so most surfaces are emitting near-blackbody radiation. Finally, a new type of absorption and emission by gasses becomes important—that associated with rotational or vibrational energy levels of molecules. Although this process is not important for simple diatomic molecules such as O_2 or N_2, it is very important for some atmospheric components present in the natural atmosphere in smaller amounts, notably water vapor and carbon dioxide, and to a lesser extent, because of its lower concentration, methane. Each of these compounds absorbs at specific wavelengths and is capable of emitting at the same wavelengths. A photon emitted at the surface at a wavelength absorbable by carbon dioxide will probably be absorbed by a carbon dioxide molecule at some height in the atmosphere. This molecule will usually transfer its excess energy to the air by a collision before it reemits a photon at the same wavelength, but the warmth of the atmosphere will assure that a fraction of the carbon dioxide molecules are kept in the excited state and thus the layer will emit at the same wavelengths it absorbs. The intensity of the emission (which is both upward and downward) depends on the wavelength, the concentration of the gas, and the temperature of the atmospheric layer.

Actual escape of a photon to space from a specific layer is a probabilistic event that depends on the absolute column number of molecules capable of absorbing the photon above the layer. But we can define a "level of escape" for any given wavelength corresponding to any specific probability that a photon emitted with an upward inclination will escape the Earth's atmosphere before it strikes a molecule capable of absorbing it. If the amount of an active gas in the atmosphere is increased, the amount above any pressure height will be increased by the same proportion, and the emission to space will occur from higher in the atmosphere. So far as the surface balance is concerned, the downward radiation reaching the surface from radiatively active gasses will come from lower and warmer layers of the atmosphere, and thus the total longwave radiation absorbed by the surface will increase. The overall direct result is a temperature increase near the surface of the Earth.

Minor disturbances of both atmospheric gasses and surface albedo probably date back to the domestication of fire, and accelerated with the development of agriculture. The disturbances, however, were small compared with natural processes. During the last century, three things have combined to make human influences on the global radiation balance significant: (1) the explosive growth of human population has spread agricultural modification of albedo into the remaining "natural" areas of the earth's surface, and incidentally released carbon stored in standing forest biomass to the atmosphere as carbon dioxide; (2) the increasing use of fossil fuels has returned carbon buried over geological time to the atmosphere as carbon dioxide (recognized by Arrhenius, 1908, and further publicized by Callendar, 1938, 1949); and (3) the growth of the chemical industry has resulted in the release of compounds which do not occur in nature into the atmospheric system. The group which is best known are the chlorofluorocarbons (CFCs), which are implicated in the destruction of ozone. The three are, of course, interrelated; it is unlikely that the global population could be fed for long without fossil fuels and chemicals. The overall result, however, has been significant changes in the amounts and composition of minor atmospheric gasses.

The two changes most significant for high-latitude forest growth are (1) large fractional increases in several compounds that actively absorb thermal infrared radiation, most notably carbon dioxide and methane, and (2) the possible reduction of the ozone layer, especially in spring at high latitudes, due to the presence of CFCs in conjunction with natural dynamic and cloud physical processes. Both carbon dioxide and methane have increased measurably and significantly over the last half century, and carbon dioxide, which is produced by the burning of fossil fuels, seems likely to double its pre-industrial level by the middle of the

21st Century. The source of the increase in methane is still controversial, which makes its continued increase harder to forecast. Because both compounds affect the distribution of energy sources and sinks, which drives the whole weather and climate system, we can reasonably expect that some climate changes will occur as a result. But what changes are important?

CLIMATE AS SEEN BY A TREE

Trees do not read thermometers. Neither do they measure precipitation, win speed, or snowfall. In particular, they have no knowledge or interest in these parameters as measured at the nearest climatological station. This is not to imply either that trees are sentient or that climate is unimportant to a tree, but only to emphasize that a tree responds to the environment of its own living tissues, and not to what is measured a few kilometers or even a few meters away from its branches or roots. While the internal environment is related to the external environment, the relationship is not one-to-one (Fig. 1).

From the tree's point of view, then, the critical factors are those that affect the life processes of the tree (Fig. 2). These certainly include:

- Temperature of tree tissues. This includes the effect of temperature on the biochemical reactions taking place in metabolically active cells, cold damage to unhardened living tissues, and physical damage to the supporting structure due to expansion when water freezes. In treeline species adapted to the local climate, the second and third effects are generally avoided by the trees physiological response to changing day length.

- Water balance. The tree can be affected by any serious imbalance between the water absorbed by roots and that transpired through the leaves or needles or lost through injury or to sap-sucking insects.

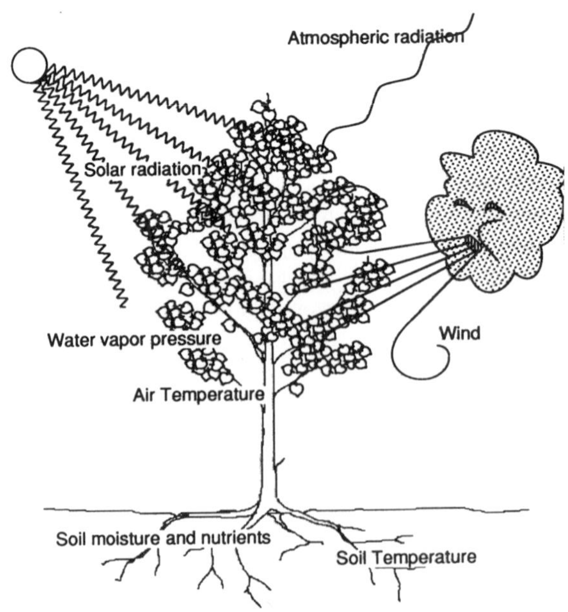

Fig. 1. A tree's climate as seen by a climatologist.

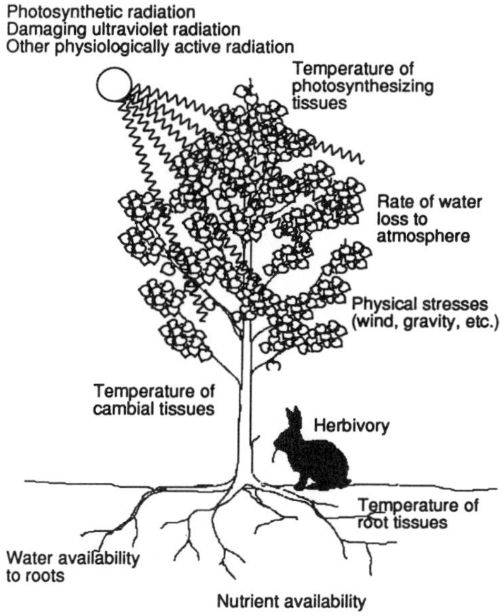

Fig. 2. A tree's climate as seen by the tree.

- Internal chemical environment. The tree needs small amounts of elements not present in water and carbon dioxide (most importantly but not exclusively compounds of nitrogen, potassium, and phosphorous), and it can be poisoned by other chemicals found in the soil or released into the tree by pathogenic microorganisms.

- Radiation. From the tree's point of view, at least three wavelength regions are of direct importance: (1) the chlorophyll absorption bands that lead to photosynthesis, (2) the red and far-red bands absorbed by phytochrome, which controls photoperiodic responses and dormancy, and (3) damaging ultraviolet wavelengths. The latter must be considered in any discussion of anthropogenic climatic change, as an increase in damaging ultraviolet wavelengths at high latitudes is a definite possibility if the ozone layer continues to weaken. In general, high latitude plant populations would be expected to have little resistance to ultraviolet damage but good day-length adaptation; high-elevation populations should have good ultraviolet resistance but would lack adaptation to the day-length cycles typical of high latitudes. Perhaps controlled hybridization experiments between populations from high elevations and those from high latitudes should be considered as a way of dealing with increased ultraviolet radiation at high latitudes.

- Physical damage. Physical damage includes both herbivory (with herbivores ranging from viruses through sheep to moose) and damage from fire, wind, or snow.

All parts of a tree's environment depend on climate as we normally think of it, but usually in complex ways. The temperature of a leaf or needle, for instance, is normally determined by:

- Incoming solar radiation (itself a function of season, latitude, elevation, cloudiness, surroundings, and leaf albedo, the last three of which may themselves vary with climatic factors).

- Incoming longwave radiation (a function of air temperature throughout the atmosphere, relative humidity, the carbon dioxide content of the air, ground temperature, the temperature of other leaves and trees, and the leaf's sphere of view).

- Latent heat loss caused by evaporation or evapotranspiration (dependent on leaf temperature, atmospheric relative humidity, and the availability of water to the leaf/needle).

- Air temperature.

- Wind speed and turbulence at the leaf, which has a strong effect on the efficiency of energy transfer between leaf and atmosphere and hence on whether leaf temperature is dominated by radiative balance (as it is in calm conditions) or air temperature (as it is in strong winds).

Leaf temperature is probably dominated by air temperature most of the time, but almost every other aspect of what we normally consider climate comes into the equation as well (Fig. 3).

The same is true of most parts of the environment as seen by the tree. Water balance is a function of precipitation, soil type (which itself is partly due to history), relative humidity, wind speed, and temperature, the geometry and health of the trees root system, and the population of sap-sucking insects.

The availability of nutrients and the uptake of toxic substances from the soil depends on soil history, precipitation (including pH), and microorganisms which may be sensitive to soil temperature and moisture.

The physical state of soil almost always depends on temperature-sensitive weathering processes as well as the past vegetation history. In marginal continental environments, mean annual temperature and surface vegetation may combine to produce permafrost, which in

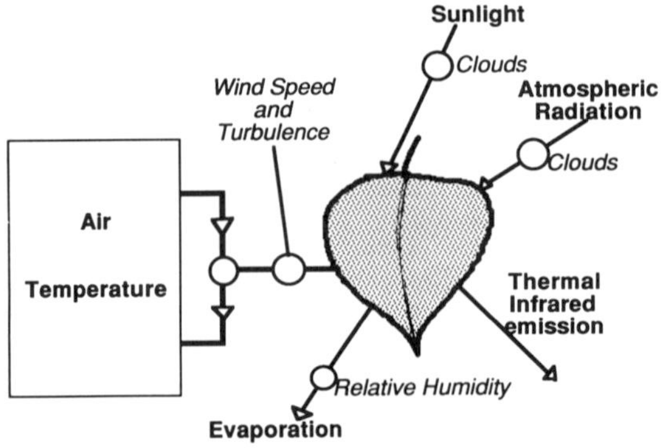

Fig. 3. Primary factors affecting the temperature of a leaf.

turn influences moisture transport and space for root growth as well as soil temperature. Shading of the soil by trees or tall grass may in turn influence the permafrost table.

Physiologically active radiation can be affected by aerosols and cloud cover as well as ozone.

Physical damage may be caused directly by wind or snow load, or indirectly by drought which leads to forest fire.

Damage by insect herbivores is often dependent on weather conditions which allow pests to survive. Improved weather could allow greater insect damage, for instance.

Numerous attempts have been made to link altitudinal/latitudinal treelines to the mean temperature of the warmest month, growing degree days, or wind speeds. Several of the papers in this volume deal with these efforts. In fact, the limiting factors for tree growth will vary with microsite and may or may not relate well to the values measured at the nearest climatological station. One promising approach is to relate growth to actual tissue temperatures, and tissue temperatures to a combination of radiation temperature, air temperature, and wind speed (Grace and James, in press). The relationship between climate and tree growth, however, is probably the lesser part of the problem.

CLIMATE AS SEEN BY A CLIMATE MODELER

Some simple effects of increases in the so-called "greenhouse" gasses due to human activities are relatively easy to calculate. Fourier (1827) recognized over a century ago that increased carbon dioxide with no other atmospheric changes would lead to warmer temperatures. Arrhenius (1908) estimated at the beginning of this century that a doubling of carbon dioxide would result in warming the earth by about 4 °C. He also recognized that burning fossil fuels could lead to such a result in the course of a few centuries. A somewhat more sophisticated analysis took explicit account of the fact that on a planet with as much free liquid water as ours, an increase in temperature would increase the amount of water vapor in the air. (Arrhenius recognized this fact, but it is unclear whether it was included in his calculations.) Because water is also a greenhouse gas, this would provide a positive feedback and increase the temperature still more. Manabe and Wetherald (1967) modeled a one-dimensional case in which carbon dioxide levels were twice the current

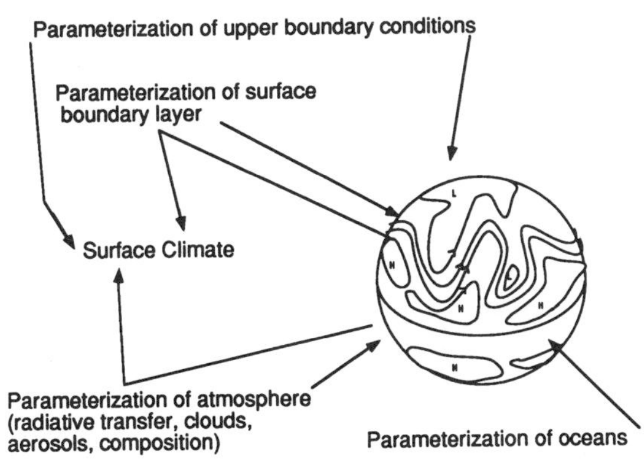

Fig. 4. Climate as seen by a climate modeler.

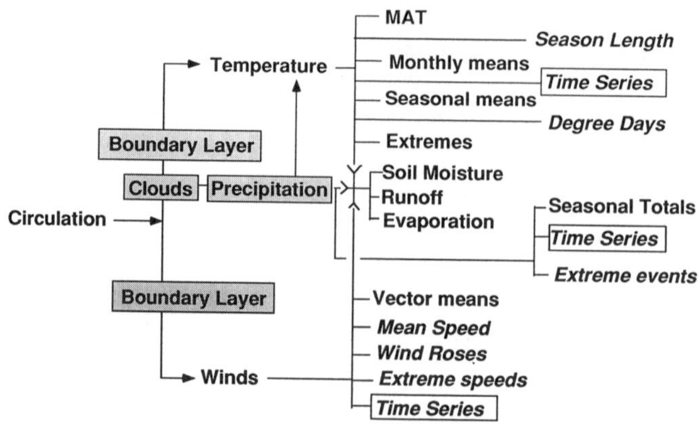

Fig. 5. Steps from modeled circulation to climatological outputs. Italicized outputs represent calculations which are not usually given as model outputs, but which could readily be obtained from the models if desired. Note that all climatological outputs must pass through the gray areas of the boundary layer and/or cloud physics.

value, solar input was equal to the current global average, energy was redistributed vertically through convection, and the relative humidity was held constant. Their model predicted a 2 °C warming. Very simple models allow us to say that even if the temperature of the air column stayed the same, an increase in carbon dioxide would decrease the likelihood of radiation frosts. Because the latest spring frosts and the earliest ones in autumn are usually of the radiation frost type, this would lead to slight increases in the length of the growing season, regardless of how the average temperature changed. Schneider (1975) gives a summary of these early energy balance models and their evolution.

When the problem is examined on a global scale, an increase in radiatively active gasses will undoubtedly produce changes in the location and intensity of the regions of positive and negative radiative balance in the atmosphere. Because the distribution of these regions drives the entire weather/climate/ocean current system, it is almost certain that the climate will change. But the climate system is extremely complex, and contains both negative and positive feedbacks. As an example, clouds play a major role in the radiation balance at both short and long wavelengths. A small increase in cloudiness, a change in cloud brightness, or even a small change in the height of cloud tops could produce a cooling large enough to balance the calculated warming due to greenhouse gasses (e.g., Ramanathan et al., 1989). It is by no means certain, however, that an increase in greenhouse gasses and other pollutants would result in increased cloud cover. Even if the cloud cover changed, this in itself would still represent a change in the distribution of the energy sources and sinks in the atmosphere. This, in turn, would lead to changes in the circulation of the atmosphere which would affect regional climates.

Most modern climate modeling is carried out using general circulation models (GCMs). A widely-publicized recent example of the use of a GCM is given in Hansen et al. (1988). The goal of the climate modeler using GCMs is:

- To input certain boundary conditions, notably atmospheric composition, surface conditions, and solar radiation.

- From these inputs and the physical laws governing the atmosphere, to calculate the wind, temperature, and moisture fields over some extended time period, the length of the period being at least several months and preferably tens of years (though simulations over this length of time are rare).

- From the calculated fields, to calculate or deduce local and global surface climates.

The physical laws governing the atmosphere are well known. They consist of:

- Newton's laws of motion as applied to a rotating sphere (essentially conservation of momentum).

- The perfect gas law.

- Two continuity equations, one for dry atmospheric gasses and one for water.

- The first law of thermodynamics, coupled with energy source and sink terms. These source and sink terms are computed internally from a radiative transfer equation and from latent heat derived from the water continuity equation.

In order to solve these equations, we need to know atmospheric composition, initial conditions, and boundary conditions. The atmospheric composition is important primarily for radiative transfer, and must include carbon dioxide, ozone, methane, some aerosols and exotics such as the CFCs, as well as the major components of the atmosphere. Parts of the atmospheric composition, notably water vapor, liquid water (clouds), and some aerosols (wind-lifted dust, for instance) should be and sometimes are calculated within the model. The model calculations of aerosols and clouds, which in turn affect the calculations of water vapor, precipitation, and radiative transfer, are at present generally agreed to be very poor (e.g., Cess et al., 1989; Cess et al., 1991).

The initial state of the atmosphere is of great concern in short-term forecasting, which starts from the same equations. In practice, this is a major limitation to forecasting, as the equation set is extremely non-linear, and subject to chaotic behavior—that is, two simulations with undetectable differences in initial conditions, which are in any case only roughly and sparsely measured, may diverge to totally different states (Lorenz, 1964). In climate modeling, the usual approach is to use initial conditions that approximate a present climate and assume that if the equations are integrated long enough the initial conditions will be lost as a transient, at least so far as the statistics of the simulation are concerned. The sampling of model results starts after the model appears to have "settled down" and lost the influence of initial conditions. Initial surface conditions, such as the presence or absence of sea ice, may well be important, but these would be included as boundary conditions. Independence of long-range climate statistics on initial state (within reasonable limits) is, however, an unproven assumption, especially if the role of the oceans is taken into account.

Ignoring for the moment the possibility that chaos-like conditions may occur on climatic time scales, let us look at how these equations are solved in practice. It is obviously impossible to specify initial conditions and solve the equations at an infinite number of points in the atmosphere. In practice, initial conditions are specified and the equations are solved at a finite number of points in some sort of three-dimensional grid over the surface of the earth. The horizontal and vertical spacing of gridpoints, and whether the equations are solved as waves with a finite number of wave numbers or directly at the gridpoints, differ among the models. In every case, however, the topographic and thermal characteristics of the lower boundary must be smooth on the scale of the grid used. Because to a first approximation the number of points varies inversely as the cube of the distance between points (this assumes that the vertical separation between layers is reduced in accord with the horizontal separation between points), and the model time step must also be reduced as the distance between points is reduced, the computer time needed to simulate a given amount of model time varies approximately as the fourth power of the spatial resolution. As a climate model must be run for a long model time to obtain accurate statistics, climate models are run on a much coarser grid (usually on the order of 5° of latitude for horizontal point spacing) than are weather forecasting models.

This has two implications. The first is that the topography is smoothed to the point that the Himalayas and the American Cordillera are essentially the only mountain ranges entering the simulations, and these two major ranges are greatly simplified. This means that regional climates with strong orographic control cannot be simulated well with these models. The second is that meteorological processes taking place at spatial scales less than several grid points must be parameterized, or approximated by empirical equations linking the small-scale processes to what the model says is going on at larger scales. Each model is "tuned" by adjusting the constants in the parameterizations until the model more or less reproduces the present climate when run with present boundary conditions. These parameterized processes include all convective processes, with or without cloud formation and precipitation, on scales up to hurricane or typhoon size. They also include all cloud microphysics, including precipitation processes, boundary layer processes involving transfer of energy and water vapor into and out of the atmosphere at the surface (Garratt, 1990), and indirectly any radiative process involving surfaces or clouds. Thus the energy sources and sinks in the models all enter through parameterized processes.

The solution of the equations results in matrices of vector winds, vertical motion, temperature, and moisture content at each time step and grid point. The wind and temperature fields away from the lower boundary are probably the most accurate output of the models.

The moisture field has an additional source of uncertainty in the very crude modeling of cloud formation. Although details vary among models, the basic principle is that if the mixing ratio goes above saturation, any excess water vapor condenses and precipitates out. In fact, condensed water may reach the ground as precipitation, fall to a lower and drier elevation of the atmosphere and evaporate there (which some models allow), or remain as droplets or ice crystals too small to fall. These eventually evaporate as drier air mixes with the cloud. Cloud density, optical properties (albedo), and lifespan, as well as the formation of precipitation, depend on nucleation processes for both water droplets and ice crystals; current models neglect nucleation entirely. The fractional cloud cover during convection (vertical overturning driven by heating at the ground surface), which in the real world may vary from near 0 to almost 100 percent, is often specified arbitrarily in the models.

As a result of this poor parameterization of clouds, both precipitation and atmospheric moisture content predicted by any model are likely to be much less accurate than the fields of motion and temperature in the free air. In addition, since the condensation process is a source of heat, and re-evaporation of clouds or precipitation an energy sink, poor parameterization of clouds distorts the energy sources and sinks driving the system and may have some effect on the motion and temperature fields.

Surface temperatures and winds are not given directly by the calculations, but require another step back through the parameterized boundary layer. This may be particularly critical at high latitudes in winter, as the parameterizations are at their worst when the surface radiative budget is negative and energy has to be transported downward. Under these circumstances both wind and temperature at the surface are likely to be very poorly modeled. Precipitation may be even worse. Soil moisture is calculated from temperature, precipitation, and surface winds, and is thus even more of a problem to model (Manabe and Wetherald, 1986). The problems with cloudiness, which affects both surface air temperature and radiative temperature, will also influence the climate as a tree perceives it.

Parameterizations generally differ among models, and a number of recent papers have compared models using different parameterization schemes and methods of calculation. There are difficulties with such comparisons. Differences in how well the model simulates present climate will carry over into differences in response to perturbations such as doubling carbon dioxide (Mitchell et al., 1987). Attempts to evaluate differences in treatment of specific problem areas are hindered by differences in the way clouds are coupled into other processes in different models. Mitchell et al. (1989) have presented results of a single model with different assumptions of cloud parameterizations, showing very high sensitivity of climate model outputs to how clouds are handled.

Several recent papers have addressed the problem. Cess et al. (1989, 1990, 1991) compare the effects of the cloud-climate feedbacks, snow-climate feedbacks, and climate feedbacks in general in as many as 19 different climate models. Mearns et al. (1990) compare variability of modeled and observed climates for three versions of the National Center for Atmospheric Research Community Climate Model. Cunnington and Mitchell (1990) have discussed the effects of convective parameterizations, and Grotch and MacCracken (1991) and Gutowski et al. (1991) have looked at regional variations among models. Kellogg and Zhao (1988) and Zhao (1988) have examined the areas of agreement and disagreement among models with respect to predictions of soil moisture.

Some of the models have been tested by using them to simulate past climates, such as ice age climates (Kutzbach, 1987; Rind, 1988). Unfortunately, no model has as yet been able to predict the onset of glaciation or deglaciation, though several can maintain ice age conditions as a steady stte.

In addition to the problems with parameterization of cloud and boundary layer processes, almost all models are hampered on long time scales by the slow feedbacks between boundary conditions and climate. On land, these feedbacks include such factors as desertification or the replacement of grassland with forest (or forest with grassland), as well as the melting or advance of ice sheets. These processes produce notable changes in albedo and in the availability of water for evaporation at the land surface.

Ocean surface temperatures and the extent of sea ice represent a problem of even more concern. While some models do have "swamp" oceans that respond to local energy balances, the real ocean has both surface and deep water energy exchanges and in fact transports a significant fraction of the total energy necessary to balance the horizontal radiative energy imbalances (Vonder Haar and Oort, 1973). Because the time scale for the deep circulation is so long—on the order of hundreds of years—there are serious mathematical problems with coupling oceanic and atmospheric models. Bryan et al. (1975) and Bryan (1984) began attacking the problem. Recent studies by Rind and Chandler (1991) using oceanic and atmospheric models that were run separately suggest that feedbacks between the major deep oceanic circulation and the atmosphere are such as to enhance climate change—that is, the feedback loop, which is not accurately handled in any existing model, is positive. Several researchers have pointed out the potential for ocean-atmosphere linking to produce major steps in climate (Broeker, 1987; Broeker et al., 1985; Covey, 1991; Covey and Barron, 1988). Comparison among models cannot even evaluate the magnitude of errors of this type, as all of the current models have the same weakness.

CONCLUSIONS

It is clear from the above that predicting future forest distribution from the output of a single climate model is a questionable enterprise. This, however, does not mean that the models are of no use.

So far as the present models are concerned, a key factor is consensus among models. Most or all the current models seem to agree on some results. A global temperature rise in response to increasing greenhouse gasses, for instance, is predicted by most models and also indicated by past correlations between carbon dioxide content and global temperature. There is considerably more disagreement among models about predicted changes in precipitation and regional climate change, and none of the present models can claim to predict precipitation with any accuracy in areas of strong relief. In general, current models can reasonably be used to predict qualitative forest boundary changes in regions where most of the state-of-the-art models are in agreement.

Beyond this, foresters need to push for improvement of the climate models in three ways.

First, models are needed that do a better job of parameterizing the processes that are related to both the energetics of the models and the relationship of the modeled surface climate to the modeled circulation. Probably the most important of these processes at the moment are

- Clouds and condensation processes. This includes not only clouds and their effect on the shortwave and longwave radiation balance, but also precipitation processes and the role of latent heat.

- Boundary layer processes involving transport of energy and moisture through the boundary layer. These are the processes that control both the climate derived from the modeled circulation and much of the energy flow driving the circulation.

The second major change needed is to include in the climate models processes currently treated as boundary processes, such as the circulation of the oceans. Although this is a formidable challenge mathematically, it may well be essential if we are going to be able to predict climate change.

Finally, foresters need to interact with modelers in determining what are the most useful outputs for their particular applications. Models work by creating their own internal weather, and almost any aspect of that internal weather can be captured and recorded if the modeler so desires. If a forester needs degree days above 30 °C, frost days in the summer months, number of hours with surface winds above a given threshold, consecutive days without precipitation, or a similar output, this information is generated by general circulation models. The accuracy of such non-standard data is subject to the same caveats is are the standard climate statistics, but getting the information out of a model is more a matter of communicating its importance to the modelers than of any real programming difficulties.

Ideally, future generations of climate models will be improved not only in the accuracy of their depiction of the complex, chaotic system we call climate, but in the availability of a broad range of kinds of climate data. It is up to the users of non-standard measures of climate, however, to work with modelers in developing these outputs.

REFERENCES

Arrhenius, S., 1908, The celestial bodies, in particular the Earth, as abodes of organisms. In: Worlds in the Making, Harper Brothers, 39-63.

Brasseur, G., and Solomon, S., 1984, Aeronomy of the Middle Atmosphere, D. Reidel, Dordecht, The Netherlands, 441 p.

Broecker, W. S., 1987, Unpleasant surprises in the greenhouse? Nature, 328:123-126.

Broecker, W. S., Peteet, D. M., and Rind, D., 1985, Does the ocean-atmosphere system have more than one stable mode of operation? Nature, 315:21-25.

Bryan, K., 1984, Accelerating the convergence to equilibrium of ocean-climate models, Journal of Physical Oceanography, 14:666-673.

Bryan, K., Manabe, S., and Pacanowski, R. C., 1975, A global ocean-atmosphere climate model. Part II: The oceanic circulation, Journal of Physical Oceanography, 5:30-46.

Callendar, G. S., 1938, The artificial production of carbon dioxide and its influence on temperature, Quarterly Journal of the Royal Meteorological Society, 64:223-237.

Callendar, G. S., 1949, Can carbon dioxide influence climate? Weather, 4:310-314.

Cess, R. D., Potter, G. L., Blanchet, J. P., Boer, G. J., Ghan, S. J., Kiehl, J. T., Le Treut, Li, H. Z.-X., Liang, Mitchell, F. B., Morcrette, J.-J., Randall, D. A., Riches, M. R.,Roeckner, E., Schlese, U., Slingo, A., Taylor, K. E., Washington, W. M., Wetherald, R. T., and Yagai, I., 1989, Interpretation of cloud-climate feedback as produced by 14 atmospheric general circulation models, Science, 245:513-516.

Cess, R. D., Potter, G. L., Blanchet, J. P., Boer, G. J., Del Genio, A. D., Déqué, M., Dymnikov, V., Galin, V., Gates, W. L., Ghan, S. J., Kiehl, J. T., Lacis, A. A., Le Treut, H., Li, Z.-X., Liang, X.-Z., McAvaney, B. J., Meleshko, V. P., Mitchell, J. F. B., Morcrette, J.-J., Randall, D. A., Rikus, L., Roeckner, E., Royer, J. F., Schlese, U., Sheinin, D. A., Slingo, A., Sokolov, A. P., Taylor, K. E., Washington, W. M., Wetherald, R. T., Yagai I., and Zhang, M.-H., 1990, Intercomparison and interpretation of climate feedback processes in 19 atmospheric general circulation models, Journal of Geophysical Research, 95:16601-16615.

Cess, R. D., Potter, G. L., Zhang, M.-H., Blanchet, J. P., Chalita, S., Colman, R., Dazlich, D. A., Del Genio, A. D., Dymnikov, V., Galin, V., Jerrett, D., Keup, E., Lacis, A. A., Le Treut, H., Liang, X.-Z., Mahfouf, J.-F., McAvaney, B. J., Meleshko, V. P., Mitchell, J. F. B., Morcrette, J.-J., Norris, P. M., Randall, D. A., Rikus, L., Roeckner, E., J. F. Royer, Schlese, U., Sheinin, D. A., Slingo, A., Sokolov, A. P., Taylor, K. E., Washington, W. M., Wetherald, R. T., and Yagai, I., 1991, Interpretation of snow-climate feedback as produced by 17 general circulation models, Science, 253:888-892.

Covey, C., 1991, Credit the oceans? Nature, 352:196-197.

Covey, C., and Barron, E., 1988, The role of ocean heat transport in climate change, Earth-Science Reviews, 24:429-445.

Cunnington, W. M., and Mitchell, J. F. B., 1990, On the dependence of climate sensitivity on convective parametrization, Climate Dynamics, 4:85-93.

Fourier, M., 1827, Mémoire sur les températures du globe terrestre et des espaces planétaires, Mem. de l'Académie Royale des Sciences de l'Institut de France, 7:569-604.

Garratt, J. R., 1990, The sensitivity of large-scale models to PBL and land-surface parameterizations, Planetary Boundary Layer Model Evaluation Workshop, 1989, August 14-15, [Reading, United Kingdom], published as WCRP-42, WMO/TD No. 378, 24-28.

Grace, J., and James, J., 1992, [In press], Physiology of Trees at Treeline. In: Alden, J. N., Odum, S., and Mastrantonio, J. L., eds., Forest Development in Cold Climates, Proceedings of an International Symposium, 1991, June 18-24, [Laugarvatn, Iceland], NATO, Plenum, New York, New York.

Grotch, S. L., and MacCracken, M. C., 1991, The use of general circulation models to predict regional climatic change, Journal of Climate , 4:286-303.

Gutowski, W. J., Jr., Gutzler D. S., and Wang, Wei-Chyung, 1991, Surface energy balances of three general circulation models, implications for simulating regional climate change, Journal of Climate, 4:121-134.

Hansen, J., Fung, I., Lacis, A., Rind, D., Lebbedeff, S., Ruedy R., and Russell, G., 1988, Global climate changes as forecast by Goddard Institute for Space Studies three-dimensional model, Journal of Geophysical Research, 93:9341-9364.

Kellogg, W. W., and Zhao, Zong-ci, 1988, Sensitivity of soil moisture to doubling of carbon dioxide in climate model experiments, Part I: North America, Journal of Climate, 1:348-366.

Kutzbach, J. E., 1987, Model simulations of the climatic patterns during the deglaciation of North America, The Geology of North America K-3, 425-446.

List, 1951, Smithsonian Meteorological Tables, sixth revised edition, Smithsonian Institution Press, Washington, D.C., 442-443.

Lorenz, E. N., 1964, The problem of deducing the climate from the governing equations, Tellus, 16:1-11.

Manabe, S., and Wetherald, R. T., 1967, Thermal equilibrium of the atmosphere with a given distribution of relative humidity, Journal of the Atmospheric Sciences, 24:241-259.

Manabe, S., and Wetherald, R. T., 1986, Reduction in summer soil wetness induced by an increase in atmospheric carbon dioxide, Science, 232:626-628.

Mearns, L. O., Schneider, S. H., Thompson S. L., and McDaniel, L. R., 1990, Analysis of climate variability in general circulation models: Comparison with observations and changes in variability in 2x CO_2 experiments, Journal of Geophysical Research, 95:20469-20490.

Mitchell, J. F. B., Wilson, C. A., and Cunnington, W. M., 1987, On CO_2 climate sensitivity and model dependence of results, Quart. J. Roy. Met. Soc., 113:293-322.

Mitchell, J. F. B., Senior, C. A., and Ingram, W. J., 1989, CO_2 and climate: A missing feedback? Nature, 341:132-134.

Ramanathan, V., Cess, R. D., Harrison, E. F., Minnis, P., Barkstrom, B. R., Ahmad, E., and Hartmann, D., 1989, Cloud-radiative forcing and climate: Results from the Earth radiation budget experiment, Science, 243:57-63.

Rind, D., 1988, Dependence of warm and cold climate depiction on climate model resolution, Journal of Climate, 1:965-997.

Rind, D., and Chandler, M., 1991, Increased ocean heat transports and warmer climate, Journal of Geophysical Research, 96:7437-7461.

Schneider, S. H., 1975, On the carbon dioxide-climate confusion, Journal of the Atmospheric Sciences, 32:2060-2066.

Vonder Haar, T. H., and Oort, A. H., 1973, New estimate of annual poleward transport by Northern Hemisphere oceans, Journal of Phys. Oceanography, 3:169-172.

Zhao, Zong-Ci, 1988, Sensitivity of soil moisture to doubling of carbon dioxide in climate experiments, Part II: The Asian monsoon region, Journal of Climate, 4:367-378.

EFFECTS OF WIND ON BOREAL FORESTS

Alexander Robertson

Forestry Canada
Newfoundland and Labrador Region
St. John's, Newfoundland, Canada

INTRODUCTION

Bonan and Shugart (1989) provide a conceptual model of many complex environmental interactions that control growth of trees in northern forests. Typical of such process models, the influence of wind on ecosystem dynamics is rather implicit. Wherever wind is strong and persistent, its effect on trees and stands is the most easily recognized of all the physical factors. Unfortunately, the critical wind speeds that control tree growth in northern boreal forests are largely unknown. It is, consequently, difficult to predict how managed forests and plantations will respond to wind and what the most appropriate silviculture options should be.

Wind influences every facet of forest growth and yield, including floristic diversity, dispersal, composition, productivity, and longevity (Hytteborn et al., 1987; White and Picket, 1985). Wind also affects distribution of insects and pathogens (DeMars and Hain, 1980; Veblen et al., 1991), the flammability of fuel and spread of forest fires (Flannigan et al., 1989), and the dispersion and deposition patterns of moisture and air pollutants (Fowler et al., 1989).

Literature on the meteorological aspects of wind and its role in ecophysiological processes is extensive. Readers are referred to articles elsewhere in these proceedings and also to a special series on these aspects in the Philosophical Transactions of The Royal Society of London (1989) entitled "Forests, Weather and Cimate." This overview highlights a few of the principle ecological aspects of wind on development of northern forests in cold climates, including the effects of changing surface wind patterns, forest migration, wind-shaped forests, disturbance in landscape ecology, and problems related to windthrow and silviculture. Some parts of a brief paper pertaining to effects of wind on forests in a changing climate (Robertson, 1991a) were modified and included in this paper.

Forest Development in Cold Climates, Edited by J. Alden *et al.*
Plenum Press, New York, 1993

SURFACE WIND PATTERNS AND CHANGING CLIMATE

Over a geological time scale of 10 million years, uplift of plateaus and mountains affects surface wind patterns sufficiently to cause large-scale changes in biota (Ruddiman, 1990). In the late Tertiary, for example, a diversion of westerly winds to a more northerly flow produced drier summers and winters and resulted in a shift from forest to grassland in central Asia and the American plains (Wolfe, 1985). During the same period, a deep subtropical high pressure zone along the Pacific coast of North America caused a shift from an arid to a semi-arid vegetation (Axelrod, 1937).

General circulation models constructed at the Goddard Institute for Space Studies (GISS-GCM) in New York show that many zonally-averaged atmospheric circulation patterns (Hadley cell, jet stream, total atmospheric transport, and others) remain much the same as they were 9,000 years ago (Rind, 1986). Because of poor resolution, prediction of bioclimatic distributions in general, and surface wind patterns in particular (Rowntree, 1990), by general circulation models is inaccurate (Prentice, 1990). Nevertheless, GISS-GCM models indicate that increased carbon dioxide concentrations in the atmosphere will warm the troposphere, cool the stratosphere, and increase the potential energy of the lower atmosphere (Rind et al., 1990). Apparently, the atmospheric energy budget might already have increased by 10 to 20 watts per square meter, with a concomitant increase in surface wind speeds over oceans and frequency of wind storms, heavy rainfall, prolonged droughts, and other severe meteorological events over continents (Dessens and Van Dihn, 1990; Flohn et al., 1990). Whether or not the rate of climatic change is unprecedented, changes in surface wind patterns will continue to exert a wide range of ecological effects on northern boreal forests.

EFFECTS OF WIND AT THE "ARCTIC FRONT"

From a synoptic climate standpoint, meteorologists often dispute the existence of a so-called "Arctic Front." Nevertheless, the "Arctic Front" represents the long-term average seasonal position of the zonal boundary between polar and southern air masses. Generally, the summer position of the Arctic Front coincides with the northern boundary and its winter position coincides with the southern boundary of the boreal forest in North America (Bryson, 1966) and Eurasia (Krebs and Barry, 1970).

Winds associated with shifts in the Arctic Front are known to affect the altitudinal and latitudinal position of treelines. In summer, the Arctic Front imposes alternating periods of cold and warmth and is reflected not only as a distinctive ebb and flow of species migration and tree growth, but also in the ability to grow tree and vegetable crops in boreal countries. Malmstrom (1960), for example, provides an excellent study of the difficulties imposed by the Arctic Front on crop production in Iceland.

The growth and distribution of larch (Larix Mill. spp.) in north-central Russia indicate that altitudinal treelines increase in 140- to 160-year cycles compared to 60- to 80-year cycles for latitudinal expansion of the boreal forest zone in response to short-term climatic changes (Gorchakovsky and Shiyatov, 1978). Presumably, the short-term climatic change is associated with changes in the mean position of the Arctic Front. In Lapland, winds associated with outbreaks of polar air masses exert a greater influence on growth variation in Scots pine (Pinus sylvestris L.) than either precipitation or minimum winter temperature (Pohtila, 1980). Likewise, patterns of tree growth at Churchill, Manitoba, reveal intermittent periods of arctic and temperate conditions depending on the position of the "Arctic Front" (Scott et al., 1988).

Palynological evidence suggests that shifts in the relative position of winter and summer storm tracks is one of the main factors influencing the movement and species composition of the forest-tundra ecotone in Labrador (MacPherson, 1985). With the passing of the oceanic climatic optimum and cooler ocean temperatures during the past 3,000 years, a general deterioration of the climate has perpetuated the coastal tundra in southern Labrador and Newfoundland (MacPherson, 1981). The vegetation gradient is somewhat reversed in Newfoundland and southern Labrador where tundra progresses northward into forest because of the cooling effect of moist, on-shore winds blowing over the cold Labrador Current. By comparison with other regions in the boreal zone, winters are mild in Newfoundland. Occasional above-freezing periods in winter melt the protective snow cover and expose soils to wind and freeze-thaw erosion. In fact, the tundra continues to expand northward in Newfoundland as a result of timber cutting and extensive fires in the post-European settlement period (Fig. 1). And while the treeline in parts of eastern Newfoundland has decreased in elevation (Meades, 1989), there is no evidence linking microclimatic deterioration to either shifts in the mean zonal position of the Arctic Front or a climatic cooling trend. In fact, in northern Manitoba and Labrador-Ungava, tree populations have increased at the treeline, although the treeline itself has not moved. Climatic warming favoring sexual regeneration has been suggested as the primary cause for increased population at the treeline (Payette et al., 1985; Scott et al., 1987).

In contrast to eastern North America, the alpine treeline in the Scandes Mountains of Fennoscandia has decreased 200 meters in response to a cooling trend of about 2 °C during the past 8,000 years (Kullman, 1988). An observed cooling trend since the 1950s was sufficient to cause a retrogression of the birch forest—alpine tundra ecotone (Kullman, 1989). Because the forest decline is most pronounced on windswept crests, Sveinbjörnsson (1983) suggests that the susceptibility of treeline tree species to mechanical damage—wind, glaze, and snow loading—contributes much more to decline than does the growth-limiting effect of cooler temperatures.

WIND AND FOREST MIGRATION

During the early post-glacial period (circa 11,000 years BP), white spruce (Picea glauca (Moench) Voss) migrated across 2,000 kilometers from south-central Alberta to the MacKenzie Delta region in a thousand year period (Ritchie and MacDonald, 1986). This period of particularly rapid spread of white spruce is correlated with a strong southerly component of surface winds 9,000 to 11,000 years ago (Kutzbach and Guetter, 1986).

General circulation models imply that the northern boundary of the boreal forest will move northwards by as much as 700 kilometers during the next century (Wheaton et al., 1987). Others maintain that certain plant communities associated with wet and dry sites will simply move as the rate of climate change dictates (Harrington et al., in press). Fast rates of treeline migration and dynamic replication of plant communities are improbable for many reasons. First, biomes do not migrate en masse; they are merely part of the shifting mosaic that reflects prevailing environmental conditions (Huntley and Webb, 1989). They do not "move" as a unit in response to climatic warming or cooling (Ritchie, 1989). For example, shrubs and herbs may move independently of tree populations (Cwynar and Spear, 1991).

The ability of most subarctic forest tree species to migrate in response to a warming trend is measured in terms of a few hundred meters per century on mineral soils. On mineral soil at the southern edge of the boreal forest, the annual northward range extension of a few species, such as beech (Fagus L.) and oak (Quercus L.), is at rates of several kilometers (Huntley et al., 1989). Paleoecological data suggest that during a 100- to 200-year warming period in Fennoscandia approximately 4,000 years ago, the rate of change in the range boundary of Scots pine (Pinus sylvestris L.) on drying peatlands was 375 to 800 meters per year over a maximum distance of 70 to 80 kilometers (Gear and Huntly, 1991). In northern regions, the greatest response of forests to climatic warming is expected to be in maritime areas of Labrador, southern Greenland, Iceland, northern Fennoscandia, and around the Bering Strait (Kauppi and Posch, 1988). Even the natural replacement of entire ecosystems in response to climatic warming—for example, replacement of red spruce forests by hardwood forest—will be delayed by a century or more at best (Davis and Botkin, 1985). Treelines may expand at much slower rates because of limited seed dispersal and delay in soil development (Pennington, 1986).

Fig. 1. Human-induced forest degeneration: (a) a shrub barren resulting from indiscriminate cutting in coniferous forests at Springdale, north-central Newfoundland; (b) kalmia heath following forest fire, Shawnadithit Lake, central Newfoundland; (c) Balsam fir (Abies balsamea (L.) Mill.) krummholz following generations of indiscriminate cutting, L'Anse-au-Clair, southern Labrador.

Regarding dispersal rates, Johnson and Webb (1989) point out that tree genera with light, wind-dispersed seed do not necessarily migrate faster than genera with heavy, animal-dispersed seeds. For example, Davis (1989) found that blue jays (Cyanocita cristata L.) were responsible for the northward spread of oak (Quercus species) and beech (Fagus grandifolia Ehrh.) at rates as rapid as those of wind-dispersed seeds of eastern hemlock (Tsuga canadensis (L.) Carr.).

Concerning the influence of geomorphic processes on patterns of tundra vegetation, Peterson and Billings (1980) emphasize the importance of wind on soil erosion and snow redistribution. Although large eolian soil complexes can be common in the forest-tundra ecotone, they are often unstable and prone to discontinuous colonization by forest vegetation (Filion and Morisset, 1983), especially if livestock grazing is involved (Arnalds, 1987; Runolfsson, 1987) (Fig. 2). Outliers of tree stands in the forest-tundra ecotone generally persist because of protection from wind (Nichols, 1976) and survive as krummholz mats in snowbed communities (Gimbarzevsky, 1977). Outliers in the forest-tundra ecotone are usually found on sites with adequate soil accumulation and are protected by a deep snow blanket (Hadley and Smith, 1987).

Wind is a primary factor in the initiation and spread of wildfires which, since the 1950s, have substantially expanded the tundra southward in northern Quebec. Furthermore, in the aftermath of wildfire, exposure of soil and plants to wind during both summer and winter retards the recolonization of uplands by trees. Many climate change models predict expansion and increased productivity of the boreal forest in response to climate warming (Wheaton et al., 1987), but less snow cover and increased exposure of plants and soil to wind may increase environmental stress and cause treelines to recede (Sirois and Payette, 1991). Given the current state of knowledge, it is impossible to predict how the forest-tundra ecotone will react to accelerated global warming (Kauppi and Posch, 1988).

The scale of expected human disturbance in the future will override any natural responses to climatic change and will, by natural regeneration, transform slow growing, shade-tolerant coniferous forests to light-demanding pioneer species. By necessity, "weed" species may become crop species, as indicated by the recent interest in aspen (Populus

Fig. 2. Remnant birch (Betula pubescens Ehrh.) in south Iceland. Birch forest once covered much of Iceland but was destroyed by cutting and sheep grazing.

tremuloides Michx.) (Fig. 3) as a major source of raw material for pulpwood in northern Alberta. Also, large-scale human disturbance may eliminate advanced disjunct colonies, restrict range extensions, and cause some species to become extinct. Multiple uses of forest lands and artificial changes to forest patterns, including the introduction of exotic plantations and intensive silviculture, have already substantially masked natural responses of indigenous species to climatic change in many northern regions.

The impact of wind in changing environments will become an important consideration in any forest management policy. The need to consider wind effects may be perceived by economists as negative but by ecologists as environmentally beneficial. Hypothetically, managers might have to choose between allowing the gradual blowdown of weakened suppressed trees to proceed naturally, or carry out thinning programs to harvest immature timber for an early cash crop. The first choice may entail a rather benign nutrient cycling process characterized by gradual changes in biological diversity. The second choice invokes abrupt and extensive changes in biological diversity, especially if it involves underplanting with other crop species. In windy climates, the decision to harvest or thin may be tempered

Fig. 3. Different paths of vegetation succession from cutting of pulpwood fir forests in Newfoundland: (a) healthy regeneration of balsam fir near St. George's; (b) mountain paper birch (Betula papyrifera var. cordifolia (Reg.) Fern.) regeneration near St. Anthony; (c) aspen (Populus tremuloides Michx.), with an understory of balsam fir and black spruce (Picea mariana (Mill.) B.S.P.), near South Brook (Hall's Bay).

by the prospects of negative physiognomic changes and species interactions resulting from deeper penetration of wind, radiation, and precipitation into managed stands, as well as increased windthrow and forest decline.

FORESTS SHAPED BY WIND

Wind acts on trees in many ways. At the microscale, the mechanical effects of shaking by wind reduce shoot growth (Rees and Grace, 1984). At the other extreme, severe exposure to persistent, strong winds creates krummholz (Holtmeier, 1985) and reduces longevity of subalpine forests (Hadley et al., 1990; Harrington, 1985) and coastal coniferous forests (Robertson, 1987a, 1989).

In summer, the cooling effect of wind lowers net photosynthesis (Grace, 1989) and reduces maturation of needles, making them more susceptible to winter desiccation (Hornbeck et al., 1986). Wind-induced damage causes deformation of crowns on the windward side (Allen, 1985). In tropical and temperate coastal areas, the mechanical effects of wind on trees are compounded by the deposition of wind-borne surfactants, sand, and sea-spray (Gellini et al., 1983; Edwards and Holmes, 1968). In winter, cuticular abrasion of needles and shoots by wind-borne ice crystals, and glaze and rime ice deposition, cause crown deformation (Baig and Tranquillini, 1980; Foster, 1988; Hadley and Smith, 1983, 1989; Robertson, 1987a) (Fig. 4).

According to the frost-drought theory, needles subjected to rapid freeze-thaw cycles caused by periods of solar radiation heating in calm periods, followed by increases in wind speed, are prone to desiccation (Hadley et al., 1990). To compensate for these environmental stresses, krummholz mat canopies create a special aerodynamic profile that includes production of a very high number of needles (Fig. 5a) to increase the tree's capacity for absorbing solar radiation and trapping a protective blanket of snow (Hadley and Smith, 1987).

Throughout the year, wind-induced root damage in dense forest stands contributes to decline (Rizzo and Harrington, 1988), lowers the rotation age, and induces compression-wood formation near the base of boles (Robertson, 1990) (Fig. 5b).

From a physiognomic standpoint, wind-induced deformation is biotic and abiotic streamlining that reduces the drag coefficient (Mayhead, 1973) and enables trees to grow in exposed locations (Figs. 5c, 5d). Trees adapt to wind stress with a downward allocation of biomass, including development of short, tapered boles, high wood density, compression (or tension) wood, short, thick, streamlined branches (bent to leeward), compact foliage, a root plate with strong flaring at the butt, and a thick, copious root system (Fig. 5e). The effectiveness of physiognomic adaptations to windy climate is illustrated by tree form in northern cities where many cultivated trees native to temperate climates survive persistent winds and frequent storms. Canopies of plantations and natural forests often exhibit strong curvature rising from the windward edges (Fig. 6a). This feature is common in spinneys and copses (forest patches) interspersed throughout the agricultural landscape of Britain (Robertson, 1987b) (Fig. 6b).

Some striking effects of wind on the physiognomy of forests include the characteristic smooth canopy of continuous forest or krummholz (Woolley, 1980) (Fig. 6c), the complex system of elongated, wedge-like "hedges" aligned in the same direction as the prevailing wind, and the snow barrier "ribbons" perpendicular to the prevailing wind in the forest-

Fig. 4. Adverse effects of wind on tree form: (a) Scots pine (*Pinus sylvestris* L.) shelterwood trees deformed by persistent strong winds, Tyn Rhos Farm, Bethel, North Wales; (b) bending of branches of mountain paper birch by ice glaze, St. John's, Newfoundland; (c) Scots pine plantation deformed by recurring ice loading, Bonavista, Newfoundland; (d) ice glaze loaded on the windward side of balsam fir trees in Torbay, Newfoundland, that stripped terminal shoots; (e) skeletal remnants of a balsam fir forest destroyed by sand dunes at Western Brook Pond, Gros Morne National Park, Newfoundland, that opened up the forest canopy and exposed trees to wind loaded with sea salt, sand, surfactants and ice-crystals from the frozen sea, and snow pack.

Fig. 5. Adaptation of canopies and trees to wind in Newfoundland: (a) compact, curved canopy, high foliage density, and cone production on the leeward side enables balsam fir krummholz to survive harsh seashore conditions at Daniel's Harbour; (b) fire-killed balsam fir krummholz reveals tree physiognomic streamlining and strong root plate for wind firmness; Port Saunders; (c) a balsam fir wave forest dead-tree strip followed by dense regeneration in response to disturbance by cutting, Ship Cove, Placentia Bay; (d) a strongly flagged sycamore maple (*Acer platanoides* L.) is streamlined to survive a particularly windy site on the campus of Memorial University of Newfoundland, St John's; (e) westward flagging of black spruce shaped by persistent northeast winter winds, Pistolet Bay.

tundra ecotone of Finland and the Rocky Mountains (Holtmeier, 1985), the wave forest with its cyclical strips of standing dead trees on the coastal plains of northern Newfoundland and Labrador (Robertson, 1987a), and the curved profiles of copses and spinneys in Britain (Robertson, 1987b).

Other graphic examples of wind-shaped krummholz occur worldwide and are known colloquially as "elfinwood" in the tropical highlands of South America and "tuckamore" in Newfoundland and Labrador. In eastern Canada, the "tuckamore" varies from the stunted forest "high tuckamore" of balsam fir, which occupies the snow-bed communities and ranges from one to four meters in height, to "low tuckamore" of dwarf white spruce (Picea glauca forma parviflora (Vict.) Fern. and Weath.) and dwarf black spruce (Picea mariana forma semiprostrata (Peck.) Blake), which form dense carpets five to 15 centimeters high on very exposed crests (Robertson and Roberts, 1982) (Fig. 6d). In southern Newfoundland and southern Labrador, where the climate is particularly windy, wet, and cold as a result of the onshore airmass passing over the cold Labrador current, both high and low tuckamore regenerate mainly by layering, and hence they are composed of genets and clones, although sheltered locations and occasional warm summers may be conducive to sexual reproduction. Clonal krummholz, in which layering predominates over sexual regeneration, also occurs in the less windy but cooler and drier boreal-subarctic transition zone (Henttonen et al., 1986; Sirois and Payette, 1991).

Apart from physical adaptations, Lawrence (1982) considers elfin forest in the tropical highlands to be a genetic adaptation to steep slopes and windy environments. Elfinwood includes phenotypic plasticity and possibly genetic differentiation. Genetic differentiation includes post-germination selection of tree forms best suited to windy sites. The high

Fig. 6. Wind-shaped canopy profiles: (a) strong curvature of a balsam fir stand, Sally's Cove, Newfoundland; (b) strong curvature of oak spinney, Prestwick, Scotland; (c) carpet of white spruce (Picea glauca forma parva (Vict.) Fern. and Weath.), Red Bay, Labrador; (d) fire-killed balsam fir revealing a smooth, streamlined canopy, Port Saunders, Newfoundland.

proportion of multiple-stemmed tamarack (Larix laricina (Du Roi) K. Koch) regenerated sexually on windy crests and knolls in Newfoundland may indicate a strategy for genetic adaptation, but there is no scientific evidence to confirm that.

Because the degree of crown and stem deformation is proportional to mean summer wind speed, several graphical grading systems and quasi-empirical indices have been developed for mapping surface winds using trees as biological indicators of mean wind speed (Robertson, 1987b).

Tree deformation has been used as a "biological indicator" of surface wind patterns worldwide: trade winds and westerlies in the Caribbean Islands (Jefferson, 1904); sea breezes along the Italian Riviera (Runge, 1957); and mountain and valley winds in the Allgauer Alps (Runge, 1959), the Tatra Mountains in central Europe (Plesnik, 1957), the Colorado Rocky Mountains (Plesnik, 1975), Hawaii (Noguchi, 1979) and Wales (Thomas, 1973), Newfoundland (Robertson, 1986), Japan and Eastern Europe (Yoshino, 1975), and throughout the United States (Wade and Hewson, 1979).

A study using trees to estimate mean winds in hilly terrain in eastern Newfoundland (Robertson, 1986) (Figs. 7a, 7b) indicated that indices based on variations in tamarack crown deformation are sensitive enough to detect prevailing surface wind features such as:

- convergent zones due to blocking and constricting wind flow near entrances to valleys,
- divergent zones and venturi effects where valleys become narrower at the upwind end, and
- rapid acceleration of up-slope wind due to convergence between two airstreams (one flowing uphill and the other blowing over the lee slope) to create a "jet" effect on the crest of hills.

Other techniques using trees and continuous forest canopies as biological indicators of wind flow include remote sensing of canopy roughness (Woolley, 1980), asymmetric cross-sections of bole wood (Fig. 7c) (Robertson, 1986, 1990, 1991b), and crown deformation in subalpine fir-spruce forests in New England, United States (Boyce, 1987; Holroyd, 1970). Sciecz et al. (1979) provide equations for predicting real time wind speed in subarctic black spruce-lichen forests in Labrador (Fig. 7d) from relative stand height weighted with the vertically-projected stand density and interpolation of observations at the nearest regional weather station.

WIND AS A NATURAL DISTURBANCE

The focus of research on wind effects in the northern boreal forest has largely been on abrupt, discontinuous wind disturbance in ecosystem dynamics, that is, mainly on windthrow (Smith et al., 1987). Likewise, ecological studies that adopt climax theories view wind disturbance as a discrete event that invokes homeostatic processes within a stable physical environment. For example, discontinuous disturbance releases understory vegetation from suppression and thereby accelerates succession towards a climax (steady-state) forest (Lorimer, 1980). An even-aged stand, with little net accumulation of biomass, is an example of a steady-state forest (Halle et al., 1978).

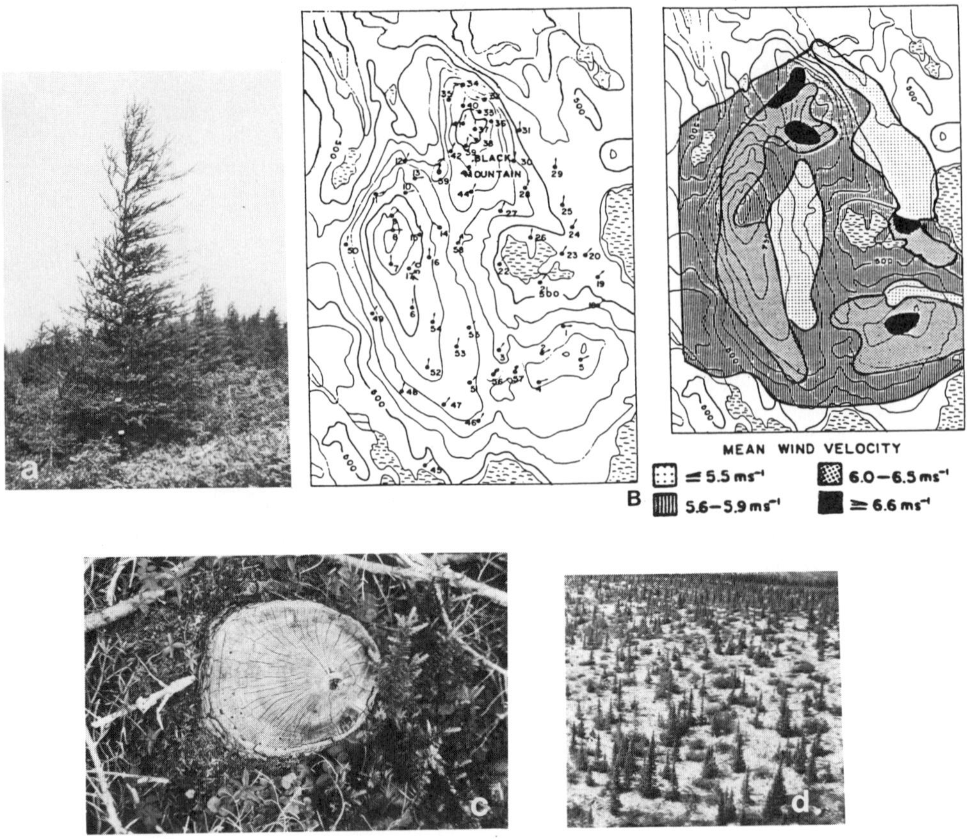

Fig. 7. Trees as biological indicators of wind speed and direction: (a) the sensitivity of larch makes it a particularly good genus for mapping wind flow; (b) map of wind direction (left) and wind speed (right) derived from tamarack (Larix laricina (Du Roi) K. Koch) deformation; (c) asymmetric cross-section of stumps can also be used to estimate mean wind flow; (d) black spruce-lichen forest of the type used by Sciecz et al. (1979) to estimate surface wind speeds.

By contrast, from a landscape ecology perspective, disturbance is a continuous process that controls landscape patterns (Turner, 1989). For example, heterogenous metapopulations are dependent on the rate and scale of constant biotic and abiotic disturbances, both in terms of patch dynamics (competition within and dispersal between patches) and across the landscape (Caswell and Cohen, 1991), which, in essence, is a shifting-mosaic steady state (Bormann and Likens, 1979). Likewise, wind-shaped patterns on many time and spatial scales affect physical processes such as carbon and nitrogen cycling (Bonan and Shugart, 1989), evapotranspiration (Black and Kelliher, 1989), energy budget (McCaughey and Saxton, 1988), turbulent transfer (Amiro, 1990; Raupach, 1989), pedogenesis (Brown and Martel, 1981; Norton, 1988), and, ultimately, microsite recolonization (Webb, 1988).

FORESTRY AND WINDTHROW

In an extensive review on silviculture in windy climates, Savill (1983) summarized the pervasive influence of wind on production forestry:

Wind damage to forests is a recurring hazard in many parts of the world. High winds can uproot trees, break stems, cause malformations and retard growth. They can also cause major disruptions to planned management and economic losses from higher logging costs, damage during storage, unharvested wood, lowered increment in residual stands, shortened rotations, etc.

Recurring wind storms associated with tropical hurricanes, tornadoes, and the passage of cold fronts cause considerable blowdown in forests and plantations (Fig. 8a). Strong surface winds, thought to be associated with low-level jet streams, cause extensive blowdown along a 160-kilometer corridor in the boreal pulpwood forest near Lac Suel, northwestern Ontario (Flannigan et al., 1989).

As an economic factor, windthrow is mainly a problem in temperate region plantations and pulpwood stands at the southern fringes of the boreal region where storms cause major blowdowns of valuable timber. Many additional problems associated with the aftermath of windstorms, such as fire hazards, cost of cleaning up, glutting of timber markets, and wood replacement, affect the competitive position of regional, national, and even global forest industries (Robertson, 1991a).

Research to understand and predict the causes and consequences of storm damage has concentrated on the aeromechanics of windthrow (Mayer, 1989), stability of forest canopies (Smith et al., 1987), and silvicultural techniques (Mathews, 1989; Savill and Evans, 1986).

The role of natural endemic windthrow of dead and weakened trees as an important component in forest turnover in subalpine fir forests is well known (Lang, 1985; Sprugel,

Fig. 8. Windthrow: (a) windthrow of insect-damaged balsam fir stands, Crabbes River, western Newfoundland; (b) windthrow on the edge of a residual stand of balsam fir in a cutover forest, Hawkes Bay, northwest Newfoundland; (c) endemic windthrow in a natural black spruce forest, Badger, central Newfoundland; (d) storm damage in a Sitka spruce plantation, Beaworthy, Devon, south England.

1976) (Fig. 8b). In terms of timber losses, however, unnatural endemic windthrow caused by inappropriate silvicultural practices is just as serious as infrequent storm damage. Among the more obvious examples of unnatural endemic windthrow is the progressive decline along the edges of semi-mature coniferous plantations and forests opened by patch- and strip-cutting. Within-canopy blowdown caused by poor choice of species, site preparation, and silvicultural intervention (Fig. 8c) is also among man-made causes of windthrow.

The rate of human-induced windthrow in residual forest stands ranges from two to 10 meters from the edge of the stand per year in coniferous pulpwood forests in Newfoundland (Robertson, 1987a) (Fig. 8d). Indeed, an early experiment in Newfoundland (Gilmour, 1926) showed that the annual volume of windthrow in one- to two-hectare forest blocks and strips can temporarily exceed the mean annual increment. The rate of windthrow ranged from 7 to 15 cubic meters per hectare per year in the blocks and three to six cubic meters per hectare per year along edges of strips, compared to an average growth rate of 2 to 2.5 cubic meters per hectare per year for commercial stands as a whole. Similarly, windthrow in residual stands of upland black spruce forests in northwestern Ontario can account for as much as 2 to 10 percent of the merchantable volume over a two- to four-year period following strip cutting (Fleming and Crossfield, 1983). Even greater losses have been recorded in the Abitibi region of Quebec, where windthrow over a three-year period accounted for as much as 20.2 percent of the initial volume of leave strips, most of which occurred during the second year after logging (Ruel, 1989). Also, residual pulpwood stands of black spruce on good sites suffer most from windthrow because tree form makes them susceptible to uprooting (Heinselman, 1957). Stand characteristics and cutting treatments affecting pattern of residual stands (width, orientation, and shape) are among the key variables involved in human-induced windthrow in black spruce forest on peatlands. Elling and Verry (1978) provide a simple equation to predict wind mortality in relation to basal area, height, age, site index, and type and depth of peat, and Smith et al. (1987) determined that the critical height at which black spruce pulpwood stands become susceptible to windthrow by periodic storms is 20 to 21 meters. They also showed that dense stands had an interlocking root mass and high wind resistance. Naturally, wind resistance is reduced and the stand declines proportionally to the additional effects of wind stress when timber harvesting, insects and disease, and aging weaken the root mass.

In windy places, where open-grown trees and contiguous forest stands are usually windfirm, opening of the forest by cutting leads to a rapid decline of trees on the exposed edges of residual blocks and strips. Trees that die standing remain upright for many years before they are blown down (Figs. 9a, 9b). The process is much the same as the wave forest (Robertson, in press). In a natural wave forest, canopy debris gradually decomposes before the surface is disturbed by windthrow and the soils undergo a cyclical taxonomic transformation. By contrast, in artificially-induced wave forests, changes in soil taxonomy

Fig. 9. Wave dieback: (a) wave dieback initiated by cutting in a balsam fir stand, Northwest Arm, Hare Bay, northern Newfoundland; (b) wave dieback in a larch plantation, Bethel, North Wales.

are not cyclical, and variations in the rate of physical and chemical transformation lead to unpredictable paths of vegetation succession. In most cases, weed species, particularly wind-dispersed pioneer herbs, shrubs, and trees, are more likely to replace the desirable crop species where reseeding on clearcuts and management options are restricted.

Not all windthrow should be considered harmful. From an ecological standpoint, windthrow is an integral part of the natural disturbance process (Foster, 1988). From Oregon to Alaska, for example, storms that cause substantial windthrow recur every decade or so (Harcombe, 1986) and create and enlarge gaps in old-growth forest that provide niches for shade-tolerant species such as Sitka spruce (Picea sitchensis (Bong.) Carr.) (Taylor, 1990). Some windthrow along stream edges is desirable for fish habitat because it dissipates stream energy, collects gravel, and forms pools (Swanson et al., 1976). In essence, the chain reaction initiated by windthrow maintains and rejuvenates biological diversity in riparian zones.

SILVICULTURAL STRATEGIES

Silvicultural strategies aimed mainly at reducing storm damage and increasing wind resistance include forest designs that manipulate the aerodynamic characteristics of stand architecture while retaining desirable stem form and wood quality characteristics. To achieve a wind-resistant forest, some advanced silvicultural programs advocate a dynamic approach starting with site preparation, species and/or clonal mixtures, planting densities, thinning schedules or selective topping, cutting patterns at economical rotation ages, and, finally, regeneration (Alexander, 1964; Jeglum, 1990; Mathews, 1989; Savill and Evans, 1986; Spittlehouse and Strathers, 1990). But, given long rotations of 50 to 80 years, or even short rotations of 15 to 20 years for tree crops, there are few places where silviculture has been practiced long enough to provide reliable guidelines for designing storm-resistant plantations and managed natural stands throughout much of the northern forest.

Nevertheless, several well-known silvicultural techniques, such as the mixed species shelterwood technique for long rotations, compensate for the more obvious effects of wind (Fig. 10) and improve site quality (Leibundgut, 1984). Another technique especially suited to barren sites in cold climates is the cultivation of multipurpose, short-rotation pioneer crop species such as Salix and Larix to provide quick shelter and site improvement (Fig. 11a). Longer-rotation crop species are under-planted after final thinning of the fast growing pioneer crop. Successful examples of this approach can be seen in the transition from poor sheep pastures to productive larch forests on the Duke of Atholl Estates in central Scotland established in the 18th Century by "Planter John" Murray (4th Duke of Atholl) (Anonymous, 1832) and also at Hallormstaður in east Iceland in this century (Blöndal, 1982) (Fig. 11b). Similarly, Robertson (1984) advocated short-rotation Salix for multiple bioenergy and agroforestry use as the initial step in site improvement for productive forestry in cold climates.

CONCLUSIONS

Wind is a major factor controlling forest development in the north. Variability in surface wind patterns is subject to climatic change on many time scales. Apparently, in response to global warming, the mean annual surface wind speed will somewhat decrease over land, but the intensity and frequency of storms will increase and be more destructive to the boreal forests.

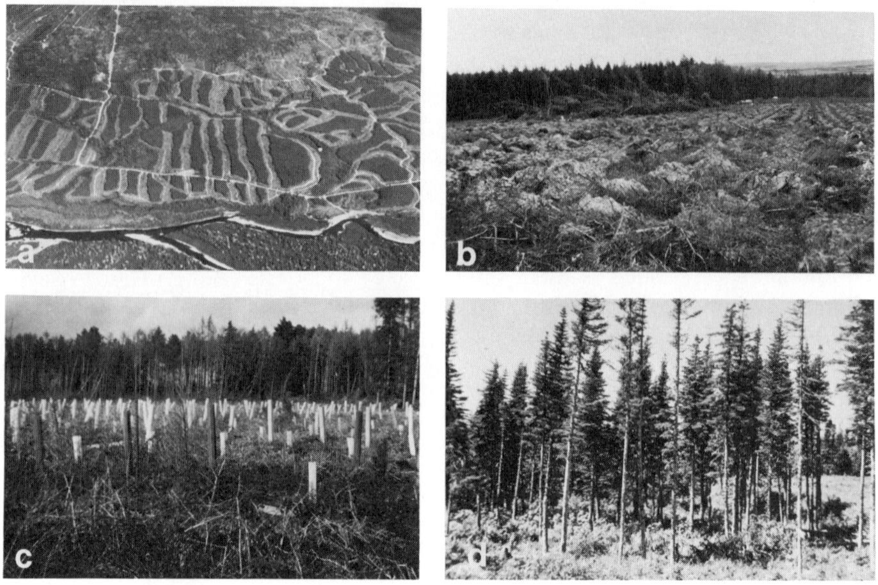

Fig. 10. **Silvicultural techniques for a windy climate: (a) strip cutting in black spruce stands to promote regeneration, northwest Gander River, central Newfoundland; (b) draining and mounding (dolloping) of clay soils to promote tree growth and increase wind-firmness (note windthrow on forest edge), Beaworthy, Devon, south England; (c) Tuley tubes used to provide early shelter for replanting in clearcut (note windthrow on forest edge) Baggely Wood, Oxford, England; (d) precommercial thinning of balsam fir stands (note dieback of edge trees), River-of-Ponds, northwest Newfoundland.**

The influence of an "Arctic Front" on forest development and growth is evident in some parts of the circumpolar subarctic forest and obscured in others. Nevertheless, winds associated with outbreaks of polar air exert a greater influence on the development of subarctic forests than do temperature and precipitation in parts of the northern circumpolar region.

Although wind may aid the rate of forest migration onto the tundra, development of suitable niches is the overriding factor controlling the dispersal efficiency of boreal tree species. Also, in many cases, tree species that have animal-dispersed seeds may migrate as efficiently as those that have wind-dispersed seeds. Some climate-change models predict higher productivity associated with forest migration onto tundra in warm periods. But increases in wind exposure of soils following wildfires have reduced forest productivity and extended tundra southward in northern and eastern Canada. Although tree populations in the forest-tundra ecotone fluctuate, the latitudinal position of the treeline does not shift appreciably. Because of a cooling trend, the alpine treeline has dropped in Fennoscandia. In many parts of Newfoundland and southern Labrador, the tundra is expanding into forested regions as a result of the combined effect of the Labrador Current and human disturbances.

Wind-shaped forests involve streamlined adaptations; notably, smooth curvature of canopies, interlocking root plates, high stand density, asymmetric crowns, compression (or tension) wood, high wood density, and a general downward allocation of biomass to enable forests and open-grown trees to survive windy climates. Because the relative asymmetric growth of trees and stands is proportional to the prevailing wind flow, trees can be used as

Fig. 11. (a) A one-year-old experimental plot of willows (Bowle's Hybrid) grown in a sheltered location has produced up to 28 tonnes per hectare per year; Pasadena, western Newfoundland; (b) Sigurður Blöndal, former Director of the Iceland Forest Service, hosting visitors in a magnificent 38-year-old Siberian larch (Larix sibirica Ledeb.) plantation at Hallormstaður, east Iceland.

biological indicators of wind speed and direction and of special features such as vortices and divergent and convergent zones that could cause problems for forestry.

Wind is a continuous disturbance feature in heterogenous landscape ecology patterns that influence ecophysiological processes of forest trees on many time and spatial scales. In essence, wind plays a major role in maintaining diversity in shifting-mosaic steady-state ecosystems.

Wind sets an upper limit to forest growth. However, the critical wind speeds that control growth processes in subarctic and boreal forests are virtually unknown except for pulpwood forests on peatland and for some plantations. Opening of the contiguous forest cover by cutting and fire induces serious decline and windthrow along forest margins. The many beneficial aspects of windthrow are often overlooked, especially with respect to nutrient cycling, wildlife habitat, and biological diversity.

The complex influence of wind needs to be incorporated into silvicultural strategies at all stages of forest development, including site classification, site preparation, forest and plantation design, harvesting and regeneration. Because very little research has been done on the impact of wind on northern boreal forests and plantations, however, a considerable knowledge gap needs to be addressed before appropriate silvicultural strategies can be formulated and implemented.

REFERENCES

Alexander, R. R., 1964, Minimizing windfall around clear cuttings in spruce-fir forests, For. Sci., 19(2):130-140.

Allen, S., 1985, Some processes involved in wind shaping of conifer trees, University of Edinburgh, Edinburgh, Scotland. Ph.D. thesis.

Amiro, B., 1990, Drag coefficients and turbulence spectra within three boreal forest canopies, Bound. Layer Meteor., 52:227-246.

Anonymous, 1832, Account of larch plantations on the estates of Atholl and Dunkeld, executed by the late John, Duke of Atholl, Trans. Highland Society of Scotland, New Series, 3:165-219.

Arnalds, A., 1987, Ecosystem disturbance in Iceland, Arctic and Alpine Res., 19(4):508-513.

Axelrod, D., 1937, A pliocene flora from the Mount Eden beds, southern California, Stanford University, Geological Science, 60:1-79.

Baig, M., and Tranquillini, W., 1980, The effects of wind and temperature on cuticular transpiration of Picea abies and Pinus cembra and their significance in desiccation damage at the alpine treeline, Oecologia, 47:252-256.

Black, T., and Kelliher, F., 1989, Processes controlling understory evapotranspiration, Phil. Trans. Roy. Soc. Lond., B 324:207-231.

Blöndal, S., 1982, Fremmede treslag i Hallormstad skogomrade, Ost-Island, Saertrykk fra Tiddskrift for Skogbruk, 82(1):1-11.

Bonan, G. B., and Shugart, H. H., 1989, Environmental factors and ecological processes in boreal forests, Ann. Rev. Ecol. Syst., 20:1-28.

Bormann, F., and Likens, G., 1979, Pattern and process in a forested ecosystem, Springer Verlag, New York.

Boyce, R., 1987, Wind direction and fir wave motion, Can. J. For. Res., 18:461-466.

Brown, J-L., and Martel, Y., 1981, Interets des chablis dans l'etude des paleosols, Geogr. Phys. Quatern., 35:87-92.

Bryson, R. A., 1966, Air mass, streamlines, and the boreal forest, Geogr. Bull., 8:228-269.

Caswell, H., and Cohen J., 1991, Disturbance, interspecific interaction and diversity in metapopulations, Biol. J. Linn. Soc., 42:193-218.

Cwynar, L. C., and Spear, R. W., 1991, Reversion of the forest tundra in the central Yukon, Ecology, 72(1):202-212.

Davis, M., 1989, Lags in the vegetation response to greenhouse warming, Climatic Change, 15:75-82.

Davis, M., and Botkin, D., 1985, Sensitivity of cool-temperate forests and their fossil pollen record to rapid temperature change, Quaternary Research, 23:327-340.

DeMars, C. J., and Hain, F. P., 1980, Bark beetle dispersal related to patterns of tree mortality and wind. In: Berryman, A. A., and Sopranyik, L., eds., Dispersal of forest insects: evaluation, theory and management implications, Proceedings of the 2nd IUFRO Conference, 1979 August 27-31, [Sandpoint Idaho], p. 66-78.

Dessens, J., and Van Dinh, P., 1990, Frequent Saharan dust outbreaks north of the Pyrenees: a sign of climatic change? Weather, 45(9):327-333.

Edwards, R., and Holmes, G., 1968, Studies of airborne salt deposition in some North Wales forests, Forestry, 41(2):155-174.

Elling, A., and Verry, E., 1978, Predicting wind-caused mortality in strip-cut stands of peatland black spruce, For. Chron., 54:249-252.

Filion, L., and Morisset, P., 1983, Eolian landforms along the eastern coast of Hudson Bay, Northern Quebec. In: Morisset, P., and Payette, S., eds., Tree-line ecology, Proc. of a Northern Quebec Tree-line Conference, Centre d'etudes Nordique, 1981 June 22-July 1, Université Laval, Québec City, Québec, p. 73-94.

Flannigan, M., Lynham, T., and Ward, P., 1989, An extensive blowdown occurrence in Northwestern Ontario, Proceedings of the 10th Conference, Fire and Forest Meteorology, 1989 April 17-21 [Ottawa]: Atmospheric Environment Service, Downsview, p. 65-71.

Fleming, R., and Crossfield, R., 1983, Strip cutting in shallow-soil upland black spruce near Nipigon, Ontario, III. Windfall and mortality in the leave strips: preliminary results, Info. Report O-X-354, Great Lakes Forest Research Centre, Saulte Ste. Marie, Ontario, Canada, 27 p.

Flohn, H., Kapala, A., Knoche, H., and Machel, E., 1990, Recent changes of the tropical water and energy budget and of mid-latitude circulations, Climate Dynamics, 4:237-252.

Foster, D., 1988, Species and stand response to catastrophic wind in central New England, United States, J. Ecology, 76:135-151.

Fowler, D., Cape, J., and Unsworth, M., 1989, Deposition of atmospheric pollutants on forests, Phil. Trans. Roy. Soc. London, B 324:247-265.

Gear, A., and Huntley, B., 1991, Rapid changes in the range limits of Scots pine 4000 years ago, Science, 251:544-547.

Gellini, R., Pantana, F., Grossoni, P., Bussotti, F., Barbaloni, E., and Rinallo, C., 1983, Further investigations of the disorder of the coastal vegetation in the Park of San Rossore (central Italy), European. J. For. Path., 15(3):145-157.

Gilmour, J., 1926, Clear-cutting of pulpwood lands, For. Chron., 2:1-2.

Gimbarzevsky, P., 1977, L'anse aux Meadows National Historic Park integrated survey of biophysical resources, Forest Management Institute, Information Rept., FMR-X-99, Ottawa, Ontario, 193 p.

Gorchakovsky, P. L., and Shiyatov, S., 1978, The upper forest limit in the mountains of the boreal zone of the USSR, Arctic and Alpine Research, 10: 349-363.

Grace, J., 1989, Tree lines, Phil. Trans. Roy. Soc. London, B 324:233-245.

Hadley, J., and Smith, W., 1983, Influence of wind exposure on needle desiccation and mortality for timberline conifers in Wyoming, U.S.A., Arctic and Alpine Research, 15:127-135.

Hadley, J. L., and Smith, W., 1987, Influence of wind on krummholz mat microclimate on needle physiology and survival, Oecologia (Berlin), 73:82-90.

Hadley, J., and Smith, W., 1989, Wind erosion of leaf surface wax in alpine timberline conifers, Arctic and Alpine Research, 21(4):392-398.

Hadley, J., Friedland, A., Herrik, G., and Amundson, R., 1990, Winter desiccation and solar radiation in relation to red spruce decline in the northern Appalachians, Can. J. For. Res., 21:269-272.

Halle, F., Oldeman, A., and Tomlinson, P., 1978, Tropical trees and forests: an architectural analysis, Springer Verlag, Berlin.

Harcombe, P., 1986, Stand development in a 130 year old spruce-hemlock forest based on age structure and 50 years of mortality data, Forest Ecology and Management, 14:41-58.

Harrington, J., Kimmins, J., Lavender, D., Zoltai, S., and Payette, S., [In press]. The effect of climatic change on forest ecology: Canada as an illustrative case, Proc. 10th. World Forestry Congress, 1991 October [Paris]: FAO Rome, Volume 2, p. 49-58.

Harrington, T., 1985, Growth and decline of wind-exposed red spruce and balsam fir in the White Mountains, Can. J. For. Res., 16:232-238.

Heinselman, M., 1957, Wind-caused mortality in Minnesota black spruce in relation to cutting methods and stand conditions, Proc. Amer. For., 1957:74-77.

Henttonen, H., Kanninen, M., Nygren, M., and Ojansuu, R., 1986, The maturation of Pinus sylvestris seeds in relation to temperature climate in northern Finland, Scandinavian J. For. Res., 1:243-249.

Holroyd, E., 1970, Prevailing winds in the White Mountains as indicated by flag trees, For. Sci., 16:222-229.

Holtmeier, F.-K., 1985, Climatic stress influencing the physiognomy of trees at the polar and mountain timberline. In: Turner, H., and Tranquillini, W. J., eds., Establishment and tending of subalpine forest: Research and Management, Proc. 3rd. Intern. Workshop, Reideralp, Switzerland: Berichte, Eidgenossische Anstalt für das Forstiche Versuchswesen, Switzerland, 270:31-40.

Hornbeck, J., Smith, R., and Federer, C., 1986, Growth decline in red spruce and balsam fir relative to natural processes, Water Air and Soil Poll., 31:425-430.

Huntley, B., and Webb, T., 1989, Migration: Species' response to climatic variations caused by changes in Earth's orbit, J. Biogeography, 16:5-19.

Huntley, B., Bartlein, P., and Prentice, I., 1989, Climatic control of the distribution and abundance of beech (Fagus L.) in Europe and North America, J. Biogeography, 16:551-560.

Hytteborn, H., Packham, J., and Verwijst, T., 1987, Tree population dynamics, stand structure and species composition in montane virgin forest of Vallibacken, northern Sweden, Vegetatio, 72:3-19.

Jefferson, M., 1904, Wind effects, J. Geogr., 3:2-20.

Jeglum, J., 1990, Modified harvesting to promote natural regeneration of black spruce. In: Tirus, B. D., Lavign, M. B., Newton, P., and Meades, W. J., eds., The Silvics and ecology of boreal spruces, Proceedings of an IUFRO Working Party Symposium, 1989 August 12-17, Newfoundland: Forestry Canada, St. John's, Inf. Rpt. N-X-271, p. 123-129.

Johnson, W., and Webb, T., 1989, The role of blue jays (Cyanocitta cristata L.) in the post-glacial dispersal of fagaceous trees in eastern North America, J. Biogeography, 16:561-571.

Kauppi, P., and Posch, P., 1988, A case study of the effects of CO_2-induced climatic warming on forest growth and the forest sector: A. Productivity reactions of the northern boreal forests. In: Parry, M., Carter, T., and Koijn, N., eds., The Impact of Climate Variations on Agriculture, Kluwer Academic Publishers, Dordrecht, The Netherlands, p. 183-196.

Krebs, J. S., and Barry, R. G., 1970, The Arctic Front and the tundra-taiga boundary in Eurasia, Geogr. Rev., 60:548-554.

Kullman, L., 1988, Holocene history of the forest-alpine tundra ecotone in the Scandes Mountains (central Sweden), New Phytol., 108:101-110.

Kullman, L., 1989, Recent retrogression of the forest-alpine tundra ecotone (Betula pubescens Ehrh. ssp. tortuosa (Ledeb.) Nyman) in the Scandes Mountains, Sweden, J. Biogeography, 16:83-90.

Kutzbach, J., and Guetter, P. J., 1986, The influence of changing orbital parameters and surface boundary conditions on climate simulations for the past 18,000 years, J. Atmos. Sci., 43(16):1726-1759.

Lang, G., 1985, Forest turnover and the dynamics of bole wood litter in subalpine balsam fir forests, Can. J. For. Res., 15:262-268.

Lawrence, R., 1982, Wind stress and elfin stature in rain forest trees: an adaptive explanation, Amer. J. Bot., 69:1224-1230.

Leibundgut, H., 1984, Die waldpflege (Care of Woods) (3rd. edition), Paul Haupt, Bern.

Lorimer, C., 1980, Age structure and disturbance history of a southern Appalachian virgin forest, Ecology, 61:1169-1184.

MacPherson, J. B., 1981, Development of the vegetation of Newfoundland and climatic change during the Holocene. In: MacPherson, A., and MacPherson, J., eds., The Natural Environment of Newfoundland, Past and Present, Department of Geography, Memorial University of Newfoundland, St. John's, Newfoundland, 295 p.

MacPherson, J. B., 1985, The post-glacial development of vegetation in Newfoundland and eastern Labrador-Ungava: synthesis and climatic implications, Syllogeous, 35:267-280.

Malmstrom, V., 1960, Influence of the Arctic Front on the climate and crops of Iceland, Proceedings of the 19th International Geographical Congress, dates, [Stockholm].

Mathews, J. D., 1989, Silvicultural systems, Oxford University Press, Oxford, 284 p.

Mayer, H., 1989, Windthrow, Phil. Trans. Roy. Soc., London, B 324: 267-281.

Mayhead, G., 1973, Some drag coefficients for British trees derived from wind tunnel studies, Agric. For. Meteor., 12:128-130.

McCaughey, J., and Saxton, W., 1988, Energy balance of storage terms in a mixed forest, Agric. For. Meteor., 44:1-18.

Meades, W. J., 1989, Recent tree-line drop in eastern Newfoundland. In: McIver, D., Street, R. B., and Auclair, A, N., eds., Climate Applications in Forest Renewal and Forest Production, Forestry Canada, Ottawa, Ontario, p. 264-267.

Nichols, H., 1976, Historical aspects of the northern Canadian treeline, Arctic, 29:38-47.

Noguchi, Y., 1979, Deformation of trees in Hawaii and its relation to wind, J. Ecology, 67:611-652.

Norton, D., 1988, Tree windthrow and forest soil turnover, Can. J. For. Res., 19:386-389.

Payette, S., Filion, I., Gauthier, L., and Boutin, Y., 1985, Secular climatic change in old-growth tree-line vegetation of northern Quebec, Nature, 315:135-138.

Pennington, W., 1986, Lags in adjustment of vegetation to climatic change caused by the pace of soil development, Vegetatio, 67:105-118.

Peterson, K., and Billings, W., 1980, Tundra vegetational patterns and succession in relation to microtopography near Athasook [*sic*], Alaska, Arctic and Alpine Research, 12(4):473-482.

Plesnik, P., 1957, Influence of wind on the shape of crowns and stem of spruce in the region of the upper tree-line in the Mountain Krwanska., Mala Tatra, Geograf. Casopis, Bratislava, 9(4):197-224. English Translation.

Plesnik, P., 1975, Some problems of the timberline in the Rocky Mountains compared with Central Europe, Arctic and Alpine Research, 5:77-84.

Pohtila, E., 1980, Climatic fluctuations and forestry in Lapland, Holarctic Ecology, 3:91-98.

Prentice, K. C., 1990, Bioclimatic distribution of vegetation for General Circulation Model studies, J. Geophys. Res., 95(D8):11,811-11,830.

Raupach, M., 1989, Stand overstory processes, Phil. Trans. Roy. Soc. London, B324:175-190.

Rees, D., and Grace, J., 1984, Effect of wind on the extension growth of Pinus contorta Douglas, Forestry, 53:145-153.

Rind, D., 1986, The dynamics of warm and cold climates, J. Atmos. Sci., 43:3-24.

Rind, D., Suozzo, N., Balachandra, N., and Prather, M., 1990, Climate change and the middle atmosphere, Part 1: The double CO_2 climate, J. Atmos. Sci., 47:475-494.

Ritchie, J., 1989, History of the Boreal forest in Canada. In: Fulton, R. J., ed., Geology of Canada, No.1, Quaternary Geology of Canada and Greenland, Geology of North America, Vol. K-1., Geological Society of America, Washington, D. C.

Ritchie, J., and MacDonald, G., 1986, The patterns of post-glacial spread of white spruce, J. Biogeography, 13:527-540.

Rizzo, D., and Harrington, T., 1988, Root movement and root damage of red spruce and balsam fir on subalpine sites in the White Mountains, New Hampshire, Can. J. For. Res., 18:991-1001.

Robertson, A., 1984, Willow plantations in agroforestry, Span, 27(1):32-34.

Robertson, A., 1986, Estimating mean windflow in hilly terrain from tamarack (Larix laricina (Du Roi) K. Koch) deformation, Int. J. Biomet., 30(4):333-349.

Robertson, A., 1987a, The centroid of tree crowns as an indicator of abiotic processes in a balsam fir wave forest, Can. J. For. Res., 17(7):746-755.

Robertson, A., 1987b, Use of trees to study wind, Arbor. J., 11:127-143.

Robertson, A., 1989, Tre og vindur, Arsriti Skograekterfelags Islands, p. 33-44.

Robertson, A., 1989, Tre og vindur, Arsriti Skograekterfelags Islands, p. 33-44.

Robertson, A., 1990, Directionality of compression wood in balsam fir wave forest trees, Can. J. For. Res., 20(8):1143-1148.

Robertson, A., 1991a, Some effects of wind on northern forests in a changing climate, Comm. For. Rev., 70(1/2):47-55.

Robertson, A., 1991b, Centroid of wood density, bole eccentricity and tree ring width in relation to vector winds in wave forests, Can. J. For. Res., 18(1):73-82.

Robertson, A., [In press], Wind and wave forests: a case study and implications for sivicul-ture. In: Alden, J. N., Mastrantonio, L. J., and Ødum, S., eds., Proceedings of a NATO-ARW Forest Development in a Cold Climate, 1991 June 18-23, [Laugarvatn, Iceland], Plenum, New York.

Robertson, A., and Roberts, B., 1982, Checklist of the alpine flora of Western Brook Pond and Deer Pond areas, Gros Morne National Park, Newfoundland, Rhodora, 84(873):101-115.

Rowntree, P., 1990, Estimates of future climatic change over Britain, Part 2: Results, Weather, 45(3):79-89.

Ruddiman, W., 1990, Changes in climate and biota on geologic time scales, Trends in Ecology and Evolution, 5(9):285-288.

Ruel, J-C., 1989, Mortalite du bois laisse sur pied a la suite d'une coupe par bandes dans trois regions du Quebec, For. Chron., 62(2):107-113.

Runge, F., 1957, Windgeformte Bäume in der Italienischen Riveria (Wind-formed trees in the Italian Riviera), Meteor. Rundsch., 10:47-48.

Runge, F., 1959, Windgeformte Bäume in der Talern der Allgauer Alpen (Wind-formed trees in valleys of the Allgauer Alps), Meteor. Rundsch., 12:98-99.

Runolfsson, S., 1987, Land reclamation in Iceland, Arctic and Alpine Research, 19(4):514-517.

Savill, P., 1983, Silviculture in windy climates: Review article, Forestry Abstracts, 44(8):473-488.

Savill, P., and Evans, J., 1986, Plantation silviculture in temperate regions, Oxford University Press, Oxford, 246 p.

Sciecz, G., Petzold, D., and Wilson, R., 1979, Wind in the subarctic forest, J. Appl. Meteor., 18:1268-1274.

Scott, P., Hansell, R., and Fayle, D., 1987, Establishment of white spruce populations and responses to climatic change at the treeline, Churchill, Manitoba, Canada, Arctic and Alpine Research, 19(1):45-51.

Scott, P., Fayle, D., Bentley, C., and Hansell, R., 1988, Large-scale changes in atmospheric circulation interpreted from patterns of tree growth at Churchill, Manitoba, Canada, Arctic and Alpine Research, 20(2):199-211.

Sirois, L., and Payette, S., 1991, Reduced postfire tree regeneration along a boreal forest-forest-tundra transect in northern Quebec, Ecology, 72(2):619-627.

Smith, V., Watts, M., and James, D., 1987, Mechanical stability of black spruce in a clay belt region of northern Ontario, Can. J. For. Res., 17:1080-1091.

Spittlehouse, D., and Strathers, R., 1990, Seedling microclimate, Land Management Rept., 65, British Columbia Ministry of Forests, Victoria, British Columbia, 28 p.

Sprugel, D., 1976, Dynamic structure of wave-regenerated Abies balsamea in northeastern United States, J. Ecology, 64:889-911.

Sveinbjörnsson, B., 1983, Bioclimate and effect on the carbon dioxide flux of mountain birch (Betula pubescens Ehrh.) at its altitudinal tree-line in the Tornetrask area, northern Sweden, Nordicana, 47:111-122.

Swanson, F., Lienkaemper, G., and Sedell, J., 1976, History, physical effects and management implications of large organic debris in western Oregon streams, Pacific

Northwest Forest and Range Experiment Station, General Technica Report PNW-56, Portland, Oregon, U. S. Department of Agriculture, Forest Service, 15 p.

Taylor, A., 1990, Disturbance and persistence of Sitka spruce (Picea sitchensis (Bong) Carr.) in coastal forests of the Pacific Northwest, North America, J. Biogeography, 17:47-58.

Thomas, J., 1973, Tree deformation by wind in Wales, Weather, 28:46-58.

Turner, M., 1989, Landscape ecology: The effect of pattern on process, Ann. Rev. Ecol. Syst., 20:171-197.

Veblen, T., Hadley, K., Reid, M., and Rebertus, A., 1991, The response of subalpine forests to spruce beetle outbreak in Colorado, Ecology, 72(1):213-231.

Wade, J., and Hewson, F., 1979, Trees as a local climate indicator, J. Appl. Meteor., 19:1182-1187.

Webb, S., 1988, Windstorm damage and microsite colonization in two Minnesota forests, Can. J. For. Res., 18:1186-1195.

Wheaton, E., Singh, T., Dempster, R., and Higginbotham, K., 1987, An exploration and assessment of the implications of climatic change for the boreal forest and forest economics of the Prairie Provinces and Northwest Territories: Phase 1. SRC Technical Report No. 211, Saskatchewan Research Council, Saskatoon, Saskatchewan, Canada, 282 p.

White, P., and Picket, S., 1985, Natural disturbance and patch dynamics: an introduction. In: Picket, S. T. A., and White, P. S., eds., The Ecology of Natural Disturbance and Patch Dynamics, Academic Press, New York, p. 9-13.

Wolfe, J., 1985, Distribution of major vegetation types during the tertiary. In: Sudquist E. T., and Broeker, W. S., eds., The Carbon Cycle and Atmospheric CO_2: Natural Variations in Archean to Present, American Geophysical Union, Geophysics Monograph, 32:357-375.

Woolley, S., 1980, Wind field mapping utilizing flagged trees as indicators of speed and direction. In: Wade, J. E., and Hewson, F. W., eds., A guide to biological wind prospecting, Department of Atmospheric Sciences, Oregon State University, Corvallis, Report DOE:EY-76-306-2227.

Yoshino, M., 1975, Climate in a small area: an introduction to local meteorology, Tokyo University Press, Tokyo, 549 p.

LONG-TERM BIOMETEOROLOGICAL MONITORING AT TWO FOREST SITES IN ICELAND AND NEWFOUNDLAND: INITIAL RESULTS

H. Thorgeirsson

Agricultural Research Institute
Reykjavík, Iceland

J. H. McCaughey

Department of Geography
Queen's University
Kingston, Ontario, Canada

A. Robertson and C. T. French

Forestry Canada
Newfoundland Forestry Centre
St. John's, Newfoundland, Canada

INTRODUCTION

Climatic variables such as temperature, precipitation, and wind determine tree survival and growth and can be related to the distribution of species or populations. Prediction of potential production of particular tree crops at the landscape level, however, requires a knowledge of environmental physics, particularly of the relative efficiency of various species in intercepting and transforming solar radiation to chemical energy as biomass (Cannell, 1989; Monteith, 1977). An important aspect of environmental physics is the site's energy balance, that is, the partitioning of net energy into soil heat flux, latent heat flux, sensible heat flux, and energy stored as biomass.

Several climatic variables have been used successfully to predict tree growth. Typically, the entire growing season is considered, and monthly mean and mean annual values of climatic variables are used as input. More intensive silvicultural management of fast-growing, short rotation tree crops, however, requires detailed knowledge of environmental influences that control growth rates and reliable models to predict potential productivity (Pereira and Landsberg, 1989; Perttu and Kowalik, 1989).

Forest Development in Cold Climates, Edited by J. Alden *et al.*
Plenum Press, New York, 1993

The lack of detailed knowledge of plant-atmosphere interactions has limited our ability to link climate with forest productivity. Standard meteorological stations are sparse and generally not located in forest areas. Also, the climatic variables usually measured at standard weather stations do not meet the requirements of biometeorology. Thus, they do not provide an adequate database from which to develop models for forest and plantation productivity. The development of economical and reliable automatic data logging and retrieval systems and high-resolution electronic sensors for quantifying plant climate variables has recently extended the capabilities of research and modeling (Brand, 1991).

We describe a field recording system to monitor environmental variables of importance for forestry and hydrology, specifically, parameters required for the partitioning of energy in plantations into principal components. We also report preliminary results from the initial stage of long-term biometeorological experiments in a black spruce (Picea mariana (Mill.) B.S.P.) plantation in western Newfoundland and a black cottonwood (Populus trichocarpa Torr. & Gray) plantation in Iceland. The experiment is especially relevant to forest development in Iceland, where considerable resources are invested in afforestation of treeless landscapes, and in reforestation of cutover land in Newfoundland. Integration of comparative studies of plantation productivity between Iceland and Newfoundland will benefit resource management in both countries.

THEORETICAL BACKGROUND

This study uses the Bowen ratio method to analyze the energy balance of forest sites. A brief synopsis of the theoretical background of this method follows. A detailed description of the methods and instrumentation pertaining to this study is found in McCaughey (1985a, 1986).

Radiation and Energy Balance

The radiation balance of a surface is given as:

$$Q^* = (K\downarrow - K\uparrow) + (L\downarrow - L\uparrow),$$

where Q^* is net radiation, $K\downarrow$ is global shortwave (solar) radiation, $K\uparrow$ is the reflected shortwave radiation, $L\downarrow$ is the incoming longwave (terrestrial) radiation, and $L\uparrow$ is the outgoing longwave radiation.

Energy balance analysis involves the partitioning of net radiation into smaller components using the principle of the conservation of energy, such that

$$Q^* = Q_e + Q_h + Q_g + Q_s + Q_p ,$$

where Q_e is latent heat flux density (evapotranspiration), Q_h is sensible heat flux density, Q_g is soil heat flux density, Q_s is the net energy storage in the biomass and air below the level of measurement of net radiation, and Q_p is the energy stored by photosynthesis (normally ignored as it is a small term and difficult to measure). All the terms of this equation have units of watts per square meter (W m^{-2}) for short-term (half-hourly or hourly) average values and either megajoules per square meter (MJ m^{-2}) or kilojoules per square meter (kJ m^{-2}) for longer periods.

Soil heat flux density can be calculated from measurements of soil heat flux at a given depth with a soil heat flux plate, coupled with measurements of soil temperature change in the layer above the plate and information on the soil's heat capacity and moisture status.

The ratio of sensible heat flux to latent heat flux is termed the Bowen ratio (β), where $\beta = Q_h/Q_e$. It can be calculated from accurate measurements of wet- and dry-bulb temperature differences in the boundary layer above the canopy. Q_e can be converted to evapotranspiration in terms of millimeters day^{-1} using the latent heat of vaporization. The Bowen ratio method for the estimation of evapotranspiration has been widely tested (Angus and Watts, 1984; Fuchs and Tanner, 1970; Lindroth and Halldin, 1990; McCaughey and Brintnell, 1984; Spittlehouse and Black, 1980).

METHODS AND STUDY SITES

Measuring System

The system used to measure the variables needed for the calculation of β, including the differences in wet- and dry-bulb temperature in the canopy boundary layer, is a **R**eversing **T**emperature **D**ifference **M**easurement **S**ystem (RTDMS) (McCaughey, 1986). The vertical position of the psychrometers is reversed between measurement periods to cancel systematic errors. The method will not eliminate height-dependent, systematic errors caused by differences in radiative heating (Spittlehouse and Black, 1980).

The present system design replaces an older rotary design used at the Petawawa National Forest Institute, Ontario, Canada (McCaughey, 1985a). The present system, installed at two sites in Newfoundland and one in Iceland, is an elevator design that interchanges the positions of the two psychrometers by slowly moving them up and down a pair of rails. The activation period and synchronous movement of the psychrometers along the rails is controlled by the data logger. The elevator design is more durable and has withstood frequent windstorms.

The wet- and dry-bulb sensors are mounted inside a radiation shield made from PVC tubing covered in reflective aluminized tape. Aspiration of the psychrometers at five meters per second is provided by a small DC-powered fan. The temperature sensors are 5-junction copper-constantan thermopiles coated with a high thermal conductivity resin for fast response (time constant of one minute). A single-junction thermocouple was also included in the thermopile for absolute temperature. Each thermopile was calibrated using a platinum resistance thermometer as a standard.

In addition to the RTDMS, each recording station is equipped with a net radiation sensor,[1] soil temperature thermocouples, soil heat flux plates, soil moisture blocks, tipping bucket rain gage, anemometers, and wind vane. The Newfoundland site is also equipped with a global radiation sensor.[2] The sensor signals are digitized using a Campbell[3] datalogger (either 21X or 7X models). The Iceland station is connected to line power and a

[1] Middleton Instruments.

[2] A global radiation sensor was installed at the site in Iceland on July 10, 1991.

[3] Campbell Scientific Corporation, Logan, Utah, United States.

telephone line for monitoring, reprogramming if necessary, and automatic data retrieval; the Newfoundland station is powered by an automobile battery.

<u>Analytical Software</u>

Analytical software has been developed for analysis of data from the RTDMS (McCaughey, 1990). The software consists of three basic programs: RADBAL (radiation balance), STORAG (storage terms), and ENBAL (energy balance). RADBAL solves the radiation balance equation. STORAG calculates the heat storage for soil and the vegetation canopy (including the latent and sensible heat storage in the canopy air space), and stratifies the canopy into layers, including a boundary layer according to the height of measurements of wet- and dry-bulb temperature. The output from STORAG is used as one of the inputs to ENBAL, which solves the energy balance using the data from the RTDMS. As well as providing surface convective fluxes, the ENBAL program also solves for the aerodynamic and canopy resistance and the surface-atmosphere decoupling coefficient. Ancillary software is available to find the zero plane displacement and the surface roughness.

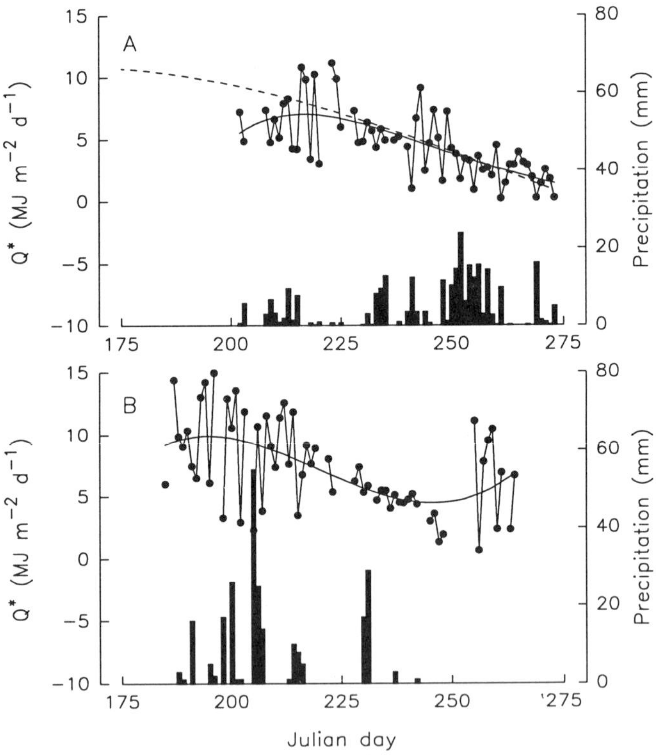

Fig. 1. Daily integrated net radiation (MJ m^{-2} day^{-1}; dots) and precipitation (millimeters; bars) for the Gunnarsholt site in Iceland (a) and St. George's site in Newfoundland (b). The daily integrated net radiation for a willow plantation in Uppsala, Sweden, is also shown (a) (dashed line, a trend line obtained by polynomial regression) (A. Lindroth, Swedish University of Agricultural Sciences, personal communication). Julian day 175 corresponds to June 24, and 275 to October 2.

Study Sites

The RTDMS system was installed at Salmon Pond in central Newfoundland from 1986–1988 (McCaughey and Robertson, 1989), then moved to St. George's in western Newfoundland in 1989. Another was installed at Gunnarsholt in southern Iceland in 1989. An earlier version of the system was installed at Petawawa, Ontario, Canada (McCaughey, 1985a, 1985b; McCaughey and Saxton, 1988).

The St. George's site (latitude 48°50′ N.) was a balsam fir forest, cut for pulpwood and prescribed burned in 1988 to remove logging debris. The site was planted in 1989 with container-grown black spruce seedlings, the same year the RTDMS was installed. The soil is shallow sandy loam on glacial till, with an intact 5- to 10-centimeter humus layer. Rotting stumps penetrate to a depth of 15 to 20 centimeters, and remnants of branches litter the surface. An herb and shrub layer, 30 centimeters in height, is developing on the site along with the planted black spruce.

The Gunnarsholt site was established in 1989. That year the RTDMS system was installed in a hayfield at Gunnarsholt (latitude 63°50′ N.) in South Iceland, the headquarters of the Icelandic Soil Conservation Service. In 1990, black cottonwood plants propagated from cuttings were planted with one meter spacing in a 16-hectare area surrounding the recording station. The previous fall, the sod had been stripped off, and prior to planting, the site was treated with herbicide. In 1990, the surface was essentially bare soil, with narrow strips of grass that had been left in the process of stripping the sod. Regular operation of the recording station started in July 1990. The eolian soils of the area are Andisols formed from a mixture of volcanic tephra, glass, and crystalline basalt. The age of the sediments is about 100 years. The soils are coarse (sandy loam) with high infiltration rates. Short range order clay (ferrihydrite and allophane) gives rise to favorable water-holding characteristics. The homogenous treeless surroundings of the Gunnarsholt site provide excellent fetch, which meets the most stringent criteria of energy balance models.

RESULTS

The Radiation Environment

The daily integrated net radiation is a measure of the total energy input to a site for a given day. Figure 1a shows this value for the Gunnarsholt site during July, August, and September, 1990. Also shown is the daily total precipitation. Monitoring began on July 21, 1990. The maximum daily integrated net radiation of 11.2 MJ m^{-2} day^{-1} occurred on August 11. Figure 1b shows daily integrated net radiation and precipitation for the St. George's site in 1990. Regular monitoring began on that site in early July. The maximum value there was 15.0 MJ m^{-2} day^{-1} on July 15, a practically clear day. Figure 1a shows, for comparison purposes, the daily integrated net radiation over a short rotation willow (*Salix viminalis* L.) plantation at Uppsala, Sweden (latitude 59°49′ N.) (A. Lindroth, Swedish University of Agricultural Sciences, personal communication) for the 1988 growing season. The maximum daily integrated net radiation there was 15.6 MJ m^{-2} day^{-1} on June 26.

As expected, the diurnal patterns of net radiation were quite variable. Figure 2 shows 30-minute averages of net radiation for a rainy day (12 millimeters precipitation) and a dry day (no precipitation) at the Gunnarsholt site. The negative values are a result of the energy emitted by the surface at night. The dry day was not cloud-free, however. There is evidence of clouds in the morning because the efflux values (negative net radiation) from

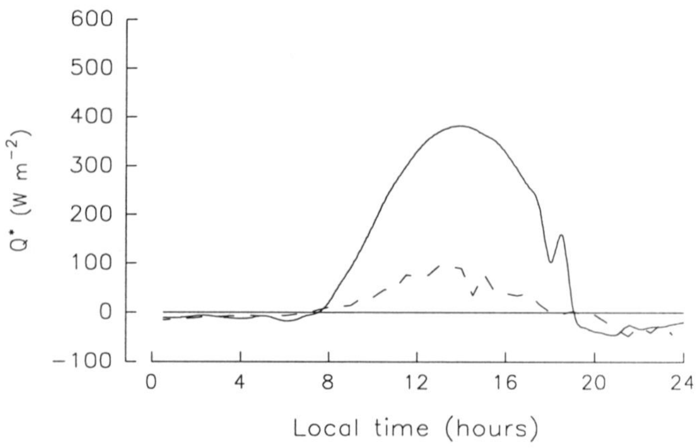

Fig. 2. Thirty-minute averages of net radiation (W m^{-2}) for a rainy day (August 29, 1990, dashed line, 12 millimeters precipitation) and a dry day (August 31, solid line, no precipitation) for the Gunnarsholt site, Iceland.

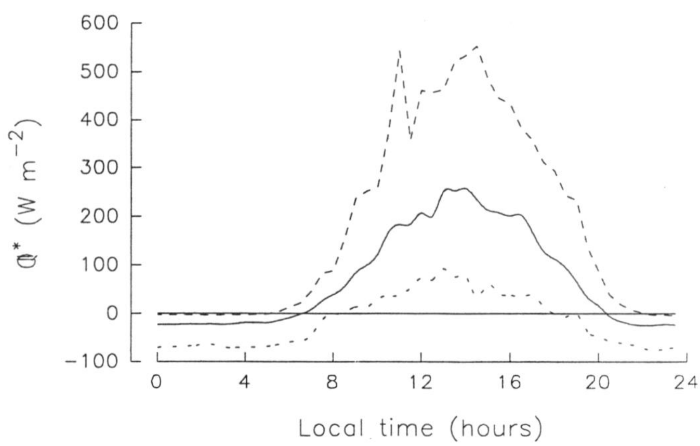

Fig. 3. Monthly minimum, average, and maximum 30-minute net radiation (W m^{-2}) values for August 1990 at the Gunnarsholt site, Iceland.

midnight until 6 a.m. are quite low and the values before the peak at 2 p.m. are lower than expected. The daily integrated net radiation on the rainy day was 1.0 MJ m^{-2} day^{-1}, and 9.1 MJ m^{-2} day^{-1} on the cloudy day. Figure 3 shows the monthly minimum, average, and maximum 30-minute net radiation values during August at the Gunnarsholt site.

Similar diurnal patterns were observed at the St. George's site. Figure 4 shows 30-minute averages of net radiation for two days at the St. George's site, a clear day (July 15, 1990) and a cloudy day (July 17, 1990), using the same scale as for Figure 2. The precipitation on July 17 was 17 millimeters and 2 millimeters on July 15, shortly after midnight. The daily integrated net radiation on July 15 was 15.0 MJ m^{-2} day^{-1}, and 3.3 MJ m^{-2} day^{-1} on July 17.

Net radiation is the balance between incoming and outgoing shortwave and longwave radiation. These components were analyzed for the St. George's site. Reflected shortwave was fairly constant at 3.3 MJ m^{-2} d^{-1} on average for the monitoring period July 7 through September 21, 1990. The incoming shortwave and longwave radiation was much more variable, as one would expect, and resulted in quite variable longwave and shortwave balance for individual days (Fig. 5).

Energy Balance

The components of the energy balance have been partitioned for the St. George's site using data from the RTDMS and the analytical software. Table 1 shows the components of the energy balance for two sample days. Net energy storage in biomass was not estimated and set equal to soil heat flux. Figure 6 shows net radiation, as well as latent, sensible, and soil heat fluxes, for the daytime period on July 15.

The Bowen ratio (β) indicates the balance of sensible (Q_h) to latent (Q_e) heat flux, when $\beta = 1$ and $Q_h = Q_e$. If water is limiting, Q_e is reduced, giving high β. Figure 7 shows β for the daytime periods at St. George's on the clear and cloudy days shown earlier. β is <1.0 for most of the day.

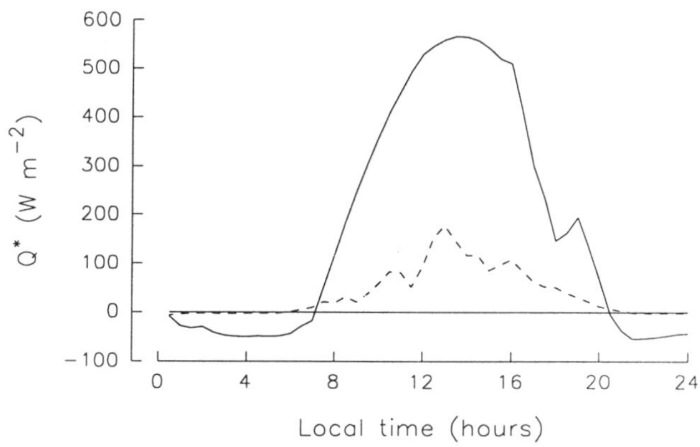

Fig. 4. Thirty-minute averages of net radiation (W m^{-2}) for a clear day (July 15, 1990, solid line, 2 millimeters precipitation shortly after midnight) and a rainy day (July 17, dashed line, 17 millimeters) for the St. George's site, Newfoundland.

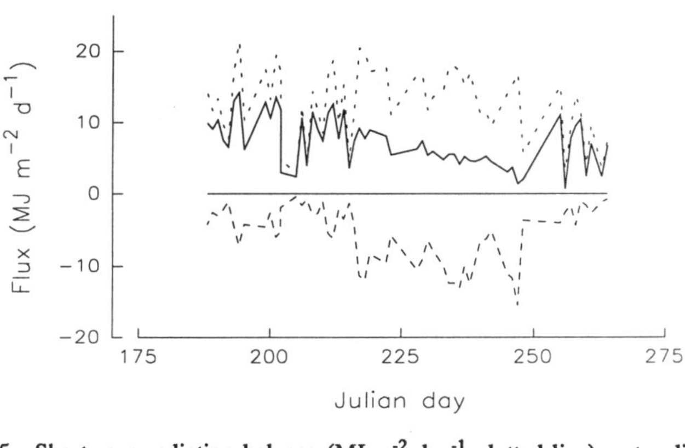

Fig. 5. Shortwave radiation balance (MJ m^{-2} day^{-1}; dotted line), net radiation balance (MJ m^{-2} day^{-1}; solid line), and longwave radiation balance (MJ m^{-2} day^{-1}; dashed line) for the St. George's site July 7 (Julian day 188) through September 21 (Julian day 264), 1990.

Table 1. Daily Integrated Net Radiation, Latent Heat Flux, Sensible Heat Flux, and Surface Soil Heat Flux (in MJ m^{-2} day^{-1}), and the Bowen Ratio for the St. George's Site on a Clear (July 15, 1990) and Cloudy (July 17, 1990) Day

	Clear day	Cloudy day
Daily net radiation	15.0	3.3
Latent heat flux	9.7	2.5
Sensible heat flux	4.4	0.4
Soil heat flux	0.9	0.4
Bowen ratio	0.45	0.16

Soil and Air Temperature

Figure 8a shows daily mean soil temperature at five centimeters depth for the two study sites and Figure 8b shows the daily mean air temperature. Soil temperature followed air temperature closely, with somewhat damped oscillations. As can be seen from Table 1, soil heat flux accounts for 6 percent and 12 percent of the daily energy input at the St. George's site on the clear and cloudy day, respectively. During the period July through September, soil heat flux averaged 0.46 MJ m^{-2} day^{-1}, or 6.3 percent of daily net radiation at the St. George's site. Soil heat flux was significantly lower at the Gunnarsholt site and was, on the average, slightly negative (-0.11 MJ m^{-2} day^{-1}) for the period.

Incoming radiation strongly affected the amplitude of the diurnal temperature variation. Figure 9a shows the 30-minute averages of air and soil (one centimeter depth) temperature

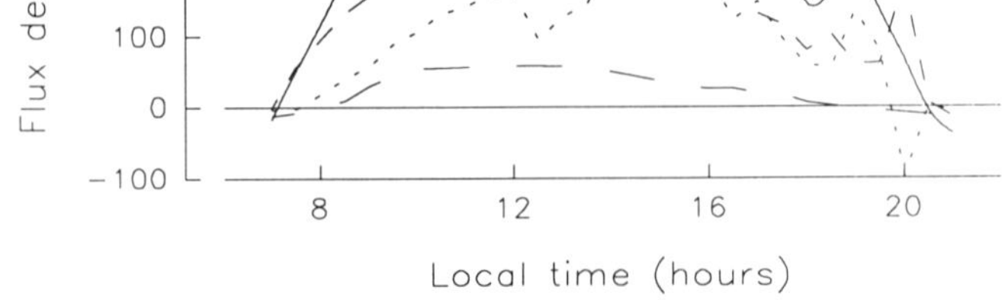

Fig. 6. Components of the energy balance for the daytime period of a clear day (July 15, 1990) at the St. George's site, Newfoundland. Net radiation (solid line), latent heat flux (medium dash), sensible heat flux (dotted line), and soil heat flux (long dash).

DISCUSSION

Results presented in this paper are preliminary and cover only a part of the first monitoring season. As for other long-term projects, the value of the data will increase with time. Growth measurements on the trees will begin during the 1991 growing season, adding another dimension to the data set. The purpose of reporting at this early stage is to draw attention to the experiment and to encourage cooperation at the Iceland and Newfoundland sites. Such cooperation could greatly enhance the value of these sites.

The data obtained so far indicate that the monitoring systems are functioning properly. The differences in the radiation environment between the two sites (Fig. 1) reflect differences in latitude between Iceland and Newfoundland. It should be noted, however, that monitoring on both sites began after the summer solstice (Julian day 171), when maximum values of net radiation can be expected.

The rainfall patterns at the two sites are quite different. Precipitation is less frequent at the St. George's site, but each rainfall event brings more precipitation (Fig. 1). This is consistent with differences in climate at the two sites.

The extreme day-to-day variation in net radiation can be seen in Figures 1-4. The difference between minimum and maximum daily integrated net radiation is more than tenfold for the month of August at Gunnarsholt. This variation is much greater than the variation in other factors such as air temperature (Fig. 8b).

The Bowen ratio was less than one for both the cloudy and the clear day at the St. George's site (Table 1). Proportionally less of the incoming energy was associated with latent heat loss on the clear day, giving a higher Bowen ratio compared to the cloudy day. Latent heat flux was markedly reduced in the afternoon on the clear day (Fig. 6), resulting in a higher Bowen ratio (Fig. 7). This might indicate slight water stress but could also be associated with the reduction in radiation that occurred about the same time (Fig. 6).

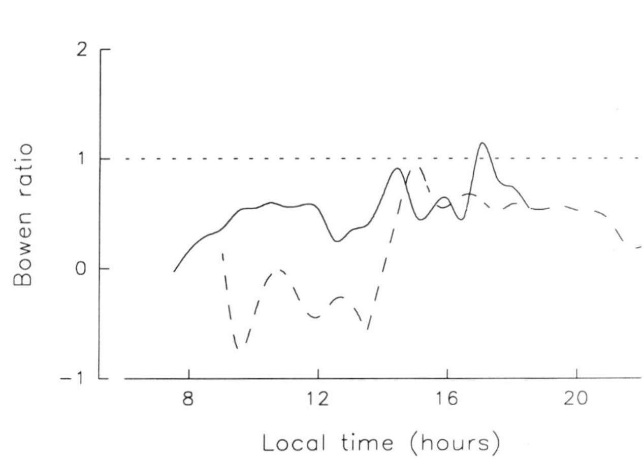

Fig. 7. The Bowen ratio (β) for a clear day (July 15, 1990, solid line) and a cloudy day (July 17, dashed line) for the St. George's site, Newfoundland.

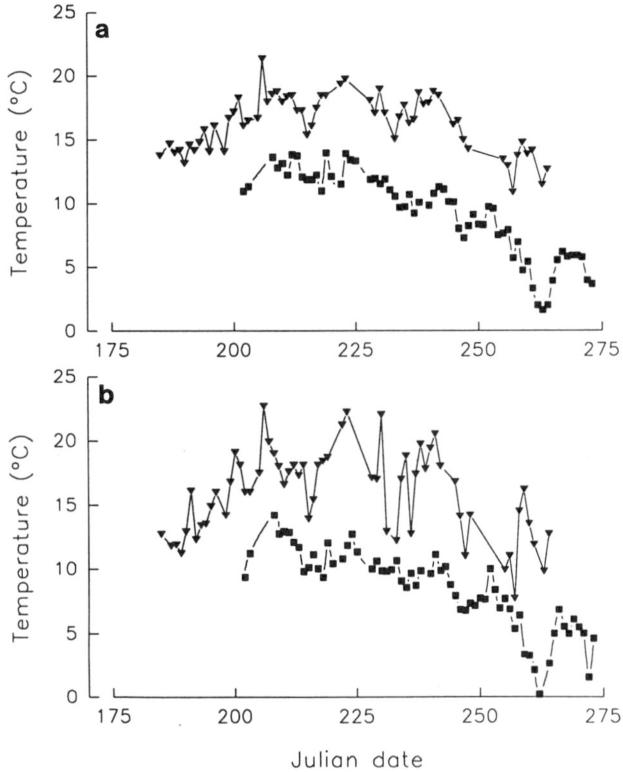

Fig. 8. Daily averages of soil temperature (°C) at 5 centimeters depth (a) and air temperature (b) for the St. George's site, Newfoundland (triangles), and the Gunnarsholt site, Iceland (squares).

Both soil and air temperatures are higher at the St. George's site (Fig. 8). More data are needed to characterize the seasonal dynamics of soil and air temperature at the two sites. The diurnal dynamics of soil and air temperature can be extracted from this data set, however. Soil temperature at one centimeter depth tracked air temperature on cloudy days when radiative cooling at night and heating from solar radiation during the day were minimal (Fig. 9). During clear days, however, soil temperature at one centimeter depth exceeded air temperature by as much as 5 °C during the day (Fig. 9). This was reversed by radiative cooling during a clear night. The lower soil heat flux at the Gunnarsholt site is most likely a result of site preparation that left the soil surface completely bare. The St. George's site, by contrast, has an intact humus layer and is covered by herbs and shrubs.

Literature on the energy balance of plant canopies is extensive but pertains mostly to low plant ecosystems, such as agricultural crops and alpine and arctic tundra, that are relatively easy to measure. By comparison, there are fewer measurements of energy flux in forests and plantations (Lindroth, 1985; Perttu and Kowalik, 1989; Price and Black, 1989). Tree canopies are more complex than agricultural crops because of height and roughness of canopy surface. These complexities cause technical and physical problems in terms of experimental design, logistics, and cost. Only within the last few years, as the result of rapid advancement in environmental physics and technology, has it been feasible to measure the energy balance of tree crops over the course of several growing seasons.

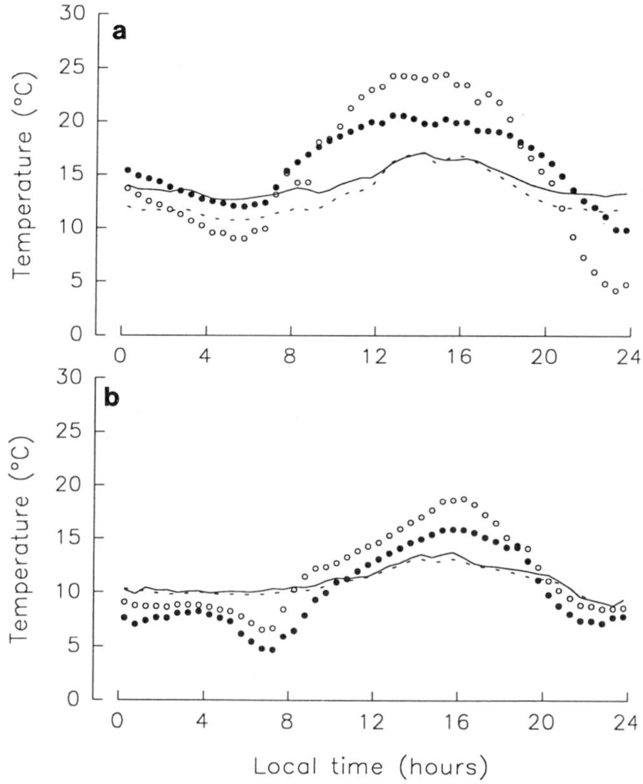

Fig. 9. Thirty-minute averages of air (filled circle and solid line) and soil (open circles and dashed line) temperature (°C) at 1 centimeter depth, for a clear day (July 15, 1990, symbols) and a rainy day (July 17, lines) at the St. George's site, Newfoundland (a) and for a rainy day (August 29, 1990, lines) and a dry day (August 31, symbols) at the Gunnarsholt site, Iceland (b).

Environmental data, such as the data obtained from the St. George's and the Gunnarsholt sites, can be analyzed in various ways to meet different objectives. Examples are given of an analysis of data from the St. George's site based on the energy balance/Bowen ratio method (Table 1; Figs. 6 and 7). The radiation balance has also been analyzed (Fig. 5). This kind of analysis is vital for understanding site energy balance, evapotranspiration, and crop responses.

Additional parameters are needed to predict forest productivity and will be obtained for both sites. These parameters relate to the characteristics of the plants growing on the sites and to soil factors such as water and nutrients. To quantify the energy captured by the tree crops, the radiation interception by the canopy will be estimated from direct measurements of leaf area index (LAI). After the canopies have developed, the radiation interception will be estimated by measuring directly the radiation penetration through the canopy. Radiation interception, coupled with measurements of above- and below-ground production, will be used to calculate light-use efficiency (Cannell, 1989; Goudriaan and Monteith, 1990; Jarvis et al., 1985; Monteith, 1977). Soil parameters will include soil water potential and concentration of key nutrients.

REFERENCES

Angus, D. E., and Watts, P. J., 1984, Evapotranspiration — How good is the Bowen ratio method?, Agricultural Water Management, 8:133-150.

Brand, D. G., 1991, The establishment of boreal and sub-boreal conifer plantations: An integrated analysis of environmental conditions and seedling growth, Forest Science, 37:68-100.

Cannell, M. G. R., 1989, Physiological basis of wood production: a review, Scandinavian Journal of Forest Research, 4:459-490.

Fuchs, M., and Tanner, C. B., 1970, Error analysis of Bowen ratios measured by differential psychrometry, Agricultural Meteorology, 7:329-334.

Goudriaan, J., and Monteith, J. L., 1990, A mathematical function for crop growth based on light interception and leaf area expansion, Annals of Botany, 66:695-701.

Jarvis, P. G., Miranda, H. S., and Muetzelfeldt, R. I., 1985, Modelling canopy exchanges of water vapour and carbon dioxide in coniferous forest plantations. In: Hutchison, B. A., and Hicks, B. B., eds., The Forest-Atmosphere Interaction, D. Reidel Publishing Company, Dordrecht, The Netherlands, 684 p.

Lindroth, A., 1985, Seasonal and diurnal variation of energy budget components in coniferous forests, Journal of Hydrology, 82:1-15.

Lindroth, A., and Halldin, S., 1990, Gradient measurements with fixed and reversing temperature and humidity sensors above a thin forest, Agricultural and Forest Meteorology, 53:81-103.

McCaughey, J. H., 1985a, A radiation and energy balance study of mature forest and clear-cut sites, Boundary-Layer Meteorology, 32:1-24.

McCaughey, J. H., 1985b, Energy balance storage terms in a mature mixed forest at Petawawa, Ontario — A case study, Boundary-Layer Meteorology, 31:89-101.

McCaughey, J. H., 1986, Energy balance measurement system for agroforestry applications in Newfoundland, Newfoundland Forestry Centre Information Report N-X-251. Forestry Canada, St. John's, Newfoundland, 26 p.

McCaughey, J. H. 1990, RTDMS program manual, Delta T Documents, Yarker, Ontario, Canada, 52 p.

McCaughey. J. H., and Brintnell, D. M., 1984, Evaluation of a Bowen ratio measurement system over forest and clear-cut sites at Petawawa, Ontario, Journal of Atmospheric and Ocean Technology, 1:276-282.

McCaughey, J. H., and Robertson, A. W., 1989, An energy balance measuring system for hydrological studies in remote areas. In: Proceedings Annual Conference and the 9th Canadian Hydrological Conference, 1989 June 8-18, St. John's, Newfoundland, Vol. IIA, Canadian Society of Civil Engineering, Ottawa, Canada, 317-331.

McCaughey, J. H., and Saxton, W. L., 1988, Energy balance storage terms in a mixed forest, Agricultural and Forest Meteorology, 44:1-18.

Monteith, J. L., 1977, Climate and the efficiency of crop production in Britain, Philosophical Transactions of the Royal Society of London, B Series, 281:277-294.

Pereira, J. S., and Landsberg, J. J., eds., 1989, Biomass production by fast-growing trees, Proceedings of the NATO Advanced Research Workshop held in Obidos, Portugal, 1987 October 6-10, NATO ASI Series, Series E, Applied Science, Vol. 166, Kluwer Academic Publishers, Dordrecht, The Netherlands, 288 p.

Perttu, K. L., and Kowalik, P. J., eds., 1989, Simulation Monographs, 30: Modelling of energy forestry: growth, water relations and economics, Pudoc, Wageningen, The Netherlands, 199 p.

Price, D. T., and Black, T. A., 1989, Estimation of forest transpiration and CO_2 uptake using the Penman-Monteith equation and a physiological photosynthesis model. In: Black, T. A., Spittlehouse, D. L., Novak, M. D., and Price, D. T., eds., Estimation of areal evapotranspiration, IAHS Publication No. 177, Institute of Hydrology, Wallingford, England, 213-227.

Spittlehouse, D. L., and Black, T. A. 1980, Evaluation of the Bowen ratio/energy balance method for determining forest evapotranspiration, Atmosphere-Ocean, 18:98-116.

GENETIC DIVERSITY OF TREE POPULATIONS AT THEIR ARCTIC LIMITS

Peter M.A. Tigerstedt

University of Helsinki
Department of Plant Breeding
Helsinki, Finland

INTRODUCTION

Adaptation and fitness of populations at their marginal limits have been of special interest to geneticists (Carson, 1955), physiologists (Levitt, 1980), evolutionists (Grant, 1963), and ecologists (Dunbar, 1968) for a long time. Marginal populations have been of interest to geneticists involved in studying the effect of strong natural selection on the genetic structure of populations. Does strong selection pressure in marginal populations result in genetic uniformity or do balanced systems retain diversity even at species margins? Evolutionists observe speciation at species margins and study overlapping marginal populations and hybrid swarms. Marginal populations have been investigated by physiologists to determine how adaptive mechanisms function, for example, in cold, heat, and drought resistance. Ecologists have looked at the simplicity of marginal ecosystems, and at the environmental fluctuations that are typical at species limits. One has questioned the validity of the "diversity-stability" rule under marginal conditions (Dunbar, 1968).

In forestry and agronomy, research has focused on how to breed trees and plants for marginal environments. In recent years, breeding for stability and sustainability of yields has been the main goal. In plant breeding programs, adaptation and yield stability have triggered more research on the importance of natural versus artificial selection. A fundamental question for the breeder is whether yield and adaptation are commensurate. One has studied the performance of wild plants, land races, open pollinated cultivars, and hybrids to find the answer. It has become clear that marginal populations of plants often have unique genetic structures that must be preserved as valuable and endangered genetic resources.

Preserving a green belt is important to all human activities at the margins of plant species. Loss of the green belt in the semi-arid tropics means desertification. In the subarctic, it means a decrease in forests and an increase in the treeless tundra. At the margins of species' distributions in both semi-arid tropic and subarctic climates, such calamities are the result of minute changes in microclimate; a critical change in either water, temperature, or both, may preclude forest and other plant regeneration.

Forest Development in Cold Climates, Edited by J. Alden *et al.*
Plenum Press, New York, 1993

Trees are probably the most important living component of the marginal, perennial green belt. The green belt of forest trees checks the spread of both desert and tundra. Globally, the green belt of trees is gradually being accepted as a "green lung" that absorbs and stabilizes carbon in the atmosphere, counteracts global warming, and stabilizes precipitation.

INDICATORS OF CLIMATIC ADAPTATION

Native trees are attuned precisely to local seasonal variation of photoperiod, temperature, and precipitation. Morphological, phenological, and physiological responses of trees cloned in several environments are probably the most sensitive indicators of climatic change now available. The timing of growth and development in tree species is based on many additive genes which interact to adjust the biological clock of the tree.

The Vegetative Cycle

Growth initiation and cessation in trees follow temperature and daylength gradients in subarctic climates and drought gradients in the tropical climates. Temperature, daylength, and drought interact to determine rhythm and length of the vegetative phase. The complicated interaction of temperature and light appears to be species specific at the arctic tree limits. Some evergreen conifers and deciduous hardwoods require dormancy to continue growth; others do not (Heide, 1985; Stern and Roche, 1974). At northern tree limits, larch (*Larix* Mill.), a deciduous conifer, forms the coniferous treeline over large areas (Fig. 1). Larch appears to be among the hardiest species in the northern boreal forest zone.

Many provenance trials and progeny tests of provenance hybrids have shown that vegetative phenology is highly heritable. Hybrids generally have growth profiles intermediate between those of their parents. The vegetative cycle of Scots pine (*Pinus sylvestris* L.), for example, facilitates genetic adaptation along a clinal transect from the south of Finland to the northern treeline (Mikola, in press).

The tree growth profile is also influenced by prevailing oceanic, continental, and alpine climates (Fig. 2). Thus, the growth profile of a tree provenance depends on climate as well as on its proximity to the treeline. Growth profiles of hybrids between provenances are intermediate between their parents. The genetic manipulation of growth profiles is an efficient way to tailor trees for best possible yields. For example, the growth profile of the hybrid between maritime Japanese larch (*Larix kaempferi* Sarg.) and continental Siberian larch (*L. sibirica* Ledeb.) is adjusted for maximum yield in a continental-maritime transition climate.

The Generative Cycle

Flower primordia are formed in the year preceding flowering. At conclusion of reproductive bud development in late spring, proper temperature, light, and time are required to complete meiosis and form pollen and egg cells. The seasonal course of embryology, including meiosis, is species specific (Luomajoki, 1986). Formation of the female gametophyte, development of pollen and egg cells, fertilization, and maturation of seed is often disturbed under marginal conditions. Low seasonal temperatures at northern treelines frequently prevent flowering and seed maturation. Thus, in marginal habitats, the generative cycle is less flexible than the vegetative cycle. In Scots pine, for example, seed fails to mature north of a 950 °C growing-degree-day isotherm (Sarvas, 1967). The complicated

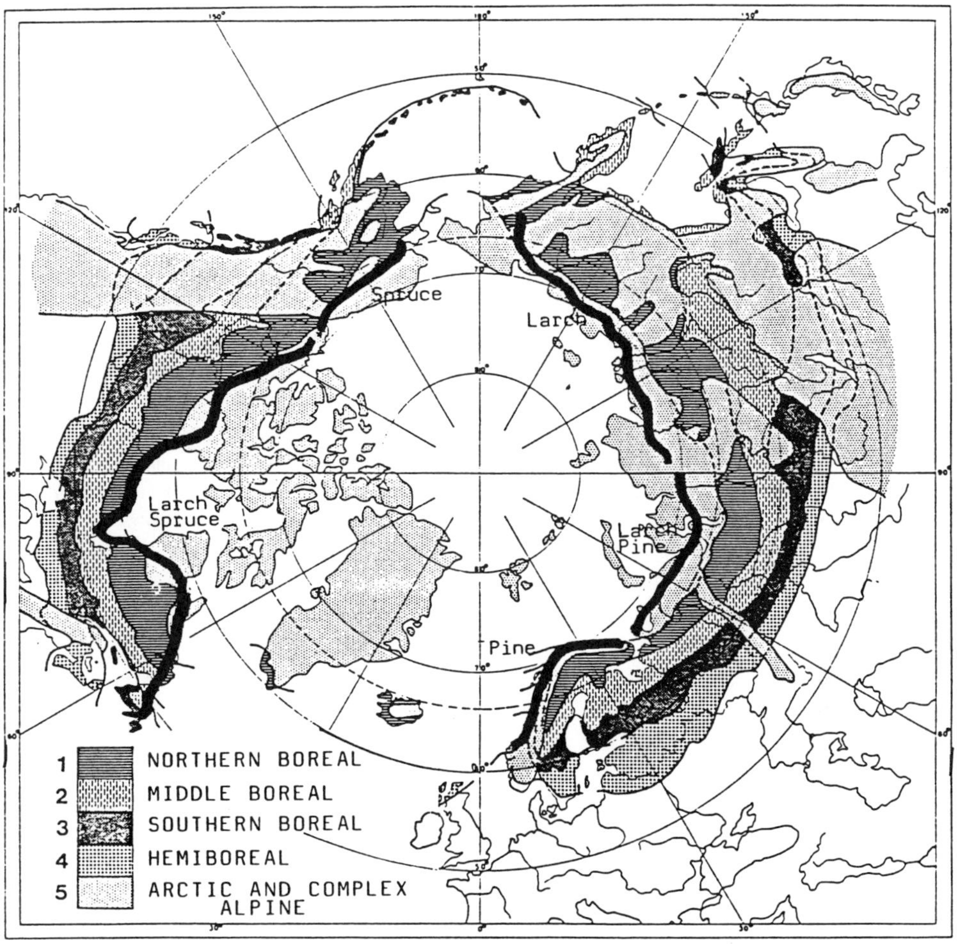

Fig. 1. The coniferous treeline (heavy black line) superimposed on the boreal zone subdivisions.

meiotic cycle of flowering is probably the most temperature-sensitive mechanism in trees, and a threshold temperature sum is required to complete seed maturation.

<u>The Natural Regeneration Cycle</u>

Adaptation means fitness, and fitness means the ability of a genotype to produce offspring to the next generation. In simple genetic terminology:

$$\mathbf{W = 1 - s},$$

where **W** is fitness and **s** is the selection coefficient. Selection coefficients are strong at tree limits where cumulative mortality in natural populations can be more than 50 percent (Fig. 3). Complete mortality is equivalent to a selection coefficient of unity, that is, the genotype is not leaving offspring for the next generation. Natural forest regeneration close to treeline is thus under high natural selection pressure. Only an exceptional tree can complete its generative cycle, and only a few offspring survive, grow, and reproduce.

Several lessons are apparent from the discussion so far. First, the genetic resources of stands close to treeline are of special adaptive value and must be preserved in situ for future natural regeneration and artificial tree breeding work. Precious natural genetic resources are threatened around the globe, including those of subarctic and subantarctic regions. Second, the high mortality recorded in experimental plantations close to treeline calls for special methodology in reforestation programs. Attention must be given to the importance of natural selection in every generation. Natural regeneration under marginal conditions, wherever possible, will conserve locally-adapted populations. Sowing of seed mimics natural regeneration and is preferable to tree planting. Sowing allows for natural selection among many plants in young stands. Third, populations near treelines are precisely adapted to local environments. To avoid loss from maladaptation, seed must not be transferred to foreign habitats. The only exception is the transfer of seed to slightly better climates, that is, from the north to south, or from high to lower altitudes. But even transfer of seed from one site to another may cause maladaptation near treeline, as shown for two provenances (2 and 8) in Figure 3. In this case, the two northern provenances came from more continental (8) or maritime (2) climates. Fourth, priority must be put on adaptation in forest-tree breeding close to the treeline. It is better to sacrifice yield and have slow-growing trees than no forest at all.

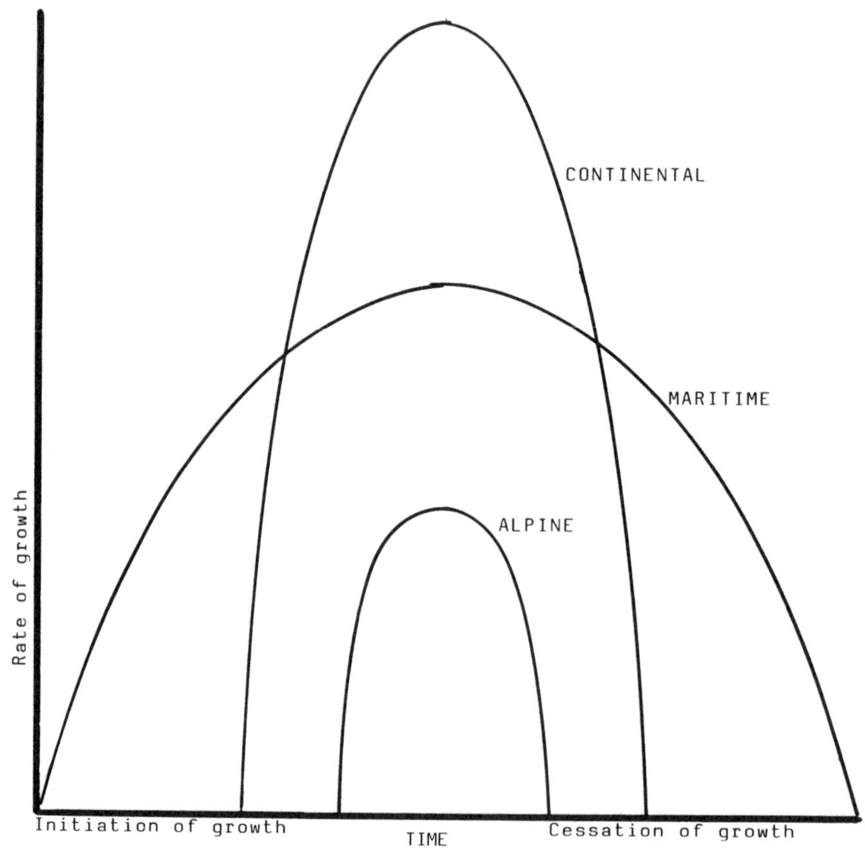

Fig. 2. Theoretical growth profiles of maritime, continental, and alpine tree species.

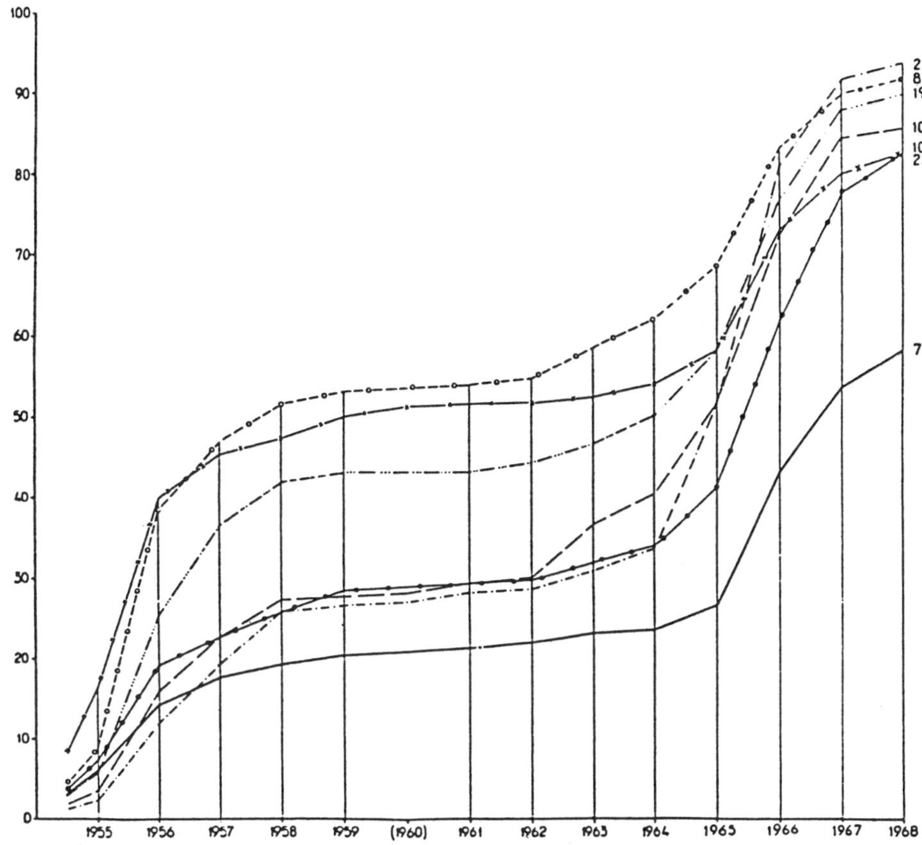

Fig. 3. Cumulative mortality of Scots pine (Pinus sylvestris L.), provenances growing close to treeline (Eiche and Gustafsson, 1970).

Maladapted trees that fail to produce viable seed in marginal environments may resort to vegetative regeneration at tree limits. Clonal regeneration is well known in many of the hardwoods, for example, in cottonwood (Populus L.) and birch (Betula L). In conifers, alpine fir (Abies lasiocarpa (Hook.) Nutt.), white spruce (Picea glauca (Moench) Voss.), Norway spruce (P. abies (L.) Karst.) and black spruce (P. mariana (Mill.) B.S.P.) may, under marginal treeline conditions, coppice from their roots. Vegetative regeneration helps to conserve well-adapted genotypes in marginal conditions, and could be used by tree breeders to propagate selected genotypes beyond tree limits. Plant physiologists may learn how mature tissues of vegetatively-reproducing conifers revert to a juvenile phase of development during the regeneration process. Clonal forestry at the treeline preserves the best genotypes and avoids the toll of natural selection in each new generation. Heterozygosity of individuals that reproduce vegetatively can be determined from isozyme markers. Recently, experiments with cross-pollinating crop plants (sorghum and pearl millet) in the marginal zone of the semi-arid tropics have shown better yield and stability of heterotic hybrids than have open-pollinated varieties with less heterozygosity. Marginal populations of conifers seem to retain high heterozygosity. Generally, heterozygosity seems to indicate higher tolerance and superior adaptation.

THE ISOZYME EVIDENCE

Allozyme variation of forest-tree populations has been analyzed for more than 20 years. Results have generally shown that wind-pollinated tree species retain large amounts of genetic variation in their populations and that loci are generally occupied by several allozymes each (Hamrick and Godt, 1990). Individual heterozygosity is high and often conforms to expected heterozygosity under random mating. Allozyme (gene-allele) variation within stands (populations) is usually large. Clines in allozyme frequencies can sometimes be distinguished, but just as often, they are absent. Linkage disequilibriums are rare. Seed embryos and young stands have an excess of homozygotes, but old stands appear to be in Hardy-Weinberg equilibrium or have excess heterozygotes (Tigerstedt et al., 1982).

Apparently the high degree of inbreeding depression, particularly after selfing, leads to selection through competition at all stages of the zygotic cycle. Hundreds of studies have been reported and the genetic structure of tree populations, largely unknown 20 years ago, are now known almost as well as in fruit fly (Drosophila) populations.

Variation at Species' Margins

We are interested here in marginal tree populations. The best examples come from the study of conifers, either at alpine or subarctic tree limits. Norway spruce and Scots pine are dominant tree species from 60° to almost 70° N. latitude in Finland. The northern limits of both species are in Lapland, roughly at latitude 69° north. Isozyme variation in Norway spruce is as extensive at northern tree limits as it is further south, and differences in allozyme frequencies between northern and southern populations can be detected in some enzyme systems (Tigerstedt 1973, 1974, 1979). Some allozyme systems show gene frequency shifts correlated to latitude; others show no such trends. Natural Scots pine stands near northern tree limits in Finland retain large allozyme variation in several enzyme systems (Kärkkäinen 1990, 1991). These studies showed large allozyme variation and absence of homozygotic excess in spite of partial population isolation and low stand densities. Strong selection against inbred genotypes may have counteracted the effects of selfing expected in low-density populations. In spite of failures in flowering and seed maturation, marginal tree populations, during their sexual life span of more than a hundred years, can produce thousands of recombinant progenies. Thus, natural selection of adapted genotypes in marginal habitats may have an exceptionally strong effect on the population structure.

Natural selection against inbred genotypes appears to be roughly identical in marginal and central populations of white spruce in Alaska (Alden and Loopstra, 1987). There, 97 percent of the total genetic diversity was among trees within stands, and 92 percent of the loci analyzed were polymorphic. Marginal populations were as genetically diverse as populations 17 kilometers away from the margin. Some evidence of inbreeding was found in marginal white spruce populations in Nouveau-Quebec, Canada (Tremblay and Simon, 1989). The populations in Nouveau-Quebec had roughly the same structure as populations studied by Alden and Loopstra. Marginal populations of sugar maple in northern Ontario, Canada, were also polymorphic at 38 percent of loci analyzed and genetically different as a result of isolation (Perry and Knowles, 1989).

More examples are not necessary. Marginal tree populations are genetically valuable resources for colonizing non-forested areas at treelines. But genetic variation along the forest margin and strong cumulative selection that may discard as much as 50 to 90 percent of the planted seedlings must be considered when planning afforestation programs. Special methods must be developed for sowing seeds or mass planting on marginal sites. Breeding for better adaptation is the single most important step in establishing forests beyond present tree limits.

CLIMATIC TYPES AND ADAPTATION

The boreal zone and its biotic subdivision, particularly its longitudinal subdivisions (Hämet-Ahti, 1981) constitute a rigid framework for describing gross climatic types. The arctic and complex mountain oroarctic (alpine) areas constitute the arctic limits of tree populations. These areas are the transcontinental subzones of interest. The complexity of zoning has resulted in several studies and classifications that cannot be described in detail here. Both subarctic and oroarctic treelines are greatly influenced by the degree of humidity along the maritime-continental gradient.

The Maritime Growth Profile

The maritime growth profile (Fig. 2) has a long and cool growing season. The growing season starts and stops slowly, and maximum growth is achieved at comparatively low day temperatures. Typical profiles of this kind are found on the west and northwest coasts of Norway, in coastal areas of British Columbia, Canada, in southern Alaska, the north and northwest coasts of Hokkaido, Japan, and in the Sakhalin and Kurile Islands of far eastern Russia. Sitka spruce (*Picea sitchensis* (Bong.) Carr.) from the Gulf of Alaska and Scots pine from coastal regions of Norway represent tree species of this growth profile. Species of these origins are sensitive to frost. They survive well in mild winters but sustain frost damage when transferred to more continental climates. The growth profile is genetically determined, and variation follows latitude and altitude clines. Note that no time scale is given in Figure 2 because of this clinal variation. The same kind of variation is also present in the two other growth profiles described.

The Continental Growth Profile

A typical growth profile of continental origin is shown in Figure 2. Forests of the continental growth profile have a relatively short, intense growing season that starts and stops abruptly. The fast start and abrupt stop is an advantage for species adapted to continental climates where winter turns into summer in a few weeks and summer ends with night frosts in early August. This general profile exists in Siberia, east of the Ural Mountains, far into continental eastern Siberia, in central Alaska, the Yukon and away from the large lakes in Northwest Territories, and southeast to central and northern Ontario. Typical subarctic treeline species of the continental growth profile are white spruce and balsam poplar (*Populus balsamifera* L.), in North America, and Siberian spruce (*Picea abies* subsp. *obovata* (Ledeb.) Domin), Siberian larch, and Scots pine in Asia. On transfer to maritime areas, trees of such origins grow slowly and are susceptible to injury from late spring frosts.

The Alpine Growth Profile

Subdivision of the Arctic zone by altitude brings us to oroarctic areas. Some of the slowest growing and hardiest trees come from these locations. Species with the alpine growth profile start growth late in the spring, and stop in early summer. Growth is slow even at the peak of the growth period (Fig. 2). Typical species of the alpine growth profile are alpine fir, alpine larch (*Larix lyallii* Parl.), and Engelmann spruce (*Picea engelmannii* Parry ex. Engelm.) at high-altitude treelines in the northern Rocky Mountains of North America.

The Transitional Growth Profile

The transitional growth profile is typical of semi-maritime climate types; it is intermediate between maritime and continental. This growth profile is encountered west of the Ural Mountains in the White Sea region of Russia, and in Finland, Sweden, and parts of eastern Norway. It is also found in central British Columbia, on the plateau between the Coast Range and the Rockies, in south central Alaska, and in the far east of Russia around the Sea of Okhotsk. This profile area has a sharp geographic gradient (in space) between continental and maritime climates. Within the area, climate changes unpredictably (in time) depending on large weather front movements. Such rapid fluctuations in climate probably cause disruptive natural selection (Mather, 1955) for high climatic tolerance. Examples of this growth profile include western Siberian larch (*Larix sukaczewii* N. Dyl.) west of the Urals, white birch (*Betula pendula* Roth), Scots pine, and Norway spruce in large parts of Scandinavia, and white spruce, lodgepole pine (*Pinus contorta* Dougl. ex Loud.), and other species of central British Columbia. Origins of this kind are relatively tolerant to transfer and some are exceptionally adaptable to marginal sites in both maritime and continental climates.

GROWING TREES NORTH OF TREELINE

Cultivation of trees north of the present treeline is possible with adapted seed or plant material. Tree species introduced to non-forested areas should have a wide genetic base for natural selection to be effective. Low population density at the margin and isolated pockets of trees in stands along the treeline may have caused genetic differences between populations in the same marginal zone (Alden and Loopstra, 1987; Perry and Knowles, 1989; Tremblay and Simon, 1989). Genetic variation, which increases the chance of success beyond the present treeline, can be maximized with mixtures of seed from isolated pockets of different origin. Seed mixtures from isolated pockets may be impractical for large-scale afforestation because flowering and seed set in stands at northern treelines are infrequent. For example, good seed years for treeline populations of Scots pine may occur only once or twice in a hundred years. This problem was eliminated by establishing seed orchards of northern Scots pine in better climates in central Finland. Flowering and seed set occur at frequent and regular intervals in central Finland, but contamination from background pollen produced by local populations fertilized from 30 percent (Muona and Harju, 1989) to 70 percent (Pakkanen et al., 1991) of the ovules produced in the orchard. In summary, seed crops are genetic mixtures of marginal and central provenance hybrids, and the pure marginal provenances. The mixture depends on weather conditions during flowering and varies greatly from year to year. For this reason, seeds collected from orchards in central Finland are not suitable for harsh climates at Scots pine treelines. Pollen contamination can be prevented if seed orchards are located in exotic environments where local species are sexually incompatible with the seed-producing populations in the orchards.

Black spruce of treeline origins in Alaska and Canada appears to grow better than local Norway spruce on organic soils in northern Scandinavia. Alpine treeline provenances of alpine fir from British Columbia exceed the coniferous treeline in Finnish Lapland. On productive soils in Lapland, Siberian larch of proper origins has lower cumulative mortality and grows faster than local Scots pine (Tigerstedt et al., 1983). The same larch provenance, which originates from locations west of the Ural Mountains, has an unusually wide adaptation, and grows well from 60° to 68° N. latitude in Finland and in Iceland. Exceptionally wide adaptation may have resulted from disruptive selection in a transitional climatic zone where maritime and continental climates meet. Should this be the case, it may be advantageous to look for species from transition climates in other parts of the world, such as Douglas-fir (Pseudotsuga menziesii (Mirb.) Franco), lodgepole pine, and Engelmann spruce from the transition zone in British Columbia. Douglas-fir and lodgepole pine are competitive in parts of Scandinavia and produce a considerably higher yield than Norway spruce and Scots pine on proper sites.

Preliminary experiments in Lapland show that exotic species are sometimes more productive than the native species at the treeline. Why are exotics sometimes superior to indigenous species? Superiority of introduced species has also been recorded in temperate and subtropical climates, for instance, of Monterey pine (Pinus radiata D. Don) from California in New Zealand and Australia, and species of eucalypts (Eucalyptus) introduced to Africa and South America from Australia. In general, exotic species and transferred provenances of indigenous species often yield more wood than native local populations. In both cases, foreign seed origins may produce more biomass at the expense of fitness achieved from generations of natural selection. Trees are naturally selected for maximum fitness rather than for maximum yield. This knowledge should be of special concern when cultivating trees for wood production beyond the present polar treelines.

REFERENCES

Alden, J., and Loopstra, C., 1987, Genetic diversity and population structure of Picea glauca on an altitudinal gradient in interior Alaska, Can. J. For. Res., 17:1519-1526.

Carson, H. L., 1955, The genetic characteristics of marginal populations of Drosophila, Cold Spring Harbor Symp. Quant. Biol., 20:276-287.

Dunbar, M. J., 1968, Ecological Development in Polar Regions, New Jersey, Prentice-Hall.

Eiche, V., and Gustafsson, Å., 1970, Population research in the Scandinavian Scots pine (Pinus sylvestris L.), Recent experimentation. In: Hecht, M. K., and Steere, W. C., eds., Essays in Evolution and Genetics in Honor of Theodosius Dobzhansky, Amsterdam, Holland, North Holland.

Grant, V., 1963, The Origin of Adaptations, New York and London, Columbia University Press.

Hämet-Ahti, L., 1981, The boreal zone and its biotic subdivision, Fennia, 159:69-75.

Hamrick, J. L., and Godt, M. J. W., 1990, Allozyme diversity in plant species. In: Brown, A. H. D., Clegg, M. T., Kahler, A. L., and Weir, B. S., eds., Plant Population Genetics, Breeding, and Genetic Resources, Sunderland, Massachusetts, Sinauer Associates Inc.

Heide, O., 1985, Physiological aspects of climatic adaptation in plants with special reference to high-latitude environments. In: Kaurin, Å., Junttila, O., and Nilsen, J., eds., Plant Production in the North, Tromsø, Norway, Norwegian University Press.

Kärkkäinen, K., 1990, The reproduction of Scots pine at the northern tree line, Oulu, Finland, University of Oulu, 50 p. M.S. thesis. In Finnish.

Kärkkäinen, K., 1991, Selfing, variation in flowering and inbreeding depression in northern Scots pine populations, Oulu, Finland, University of Oulu, 81 p. Lic. thesis. In Finnish.

Levitt, J., 1980, Responses of Plants to Environmental Stress, Vol. 1. New York, NY, Academic Press.

Luomajoki, A., 1986, Timing of microsporogenesis in trees with reference to climatic adaptation, a review, Acta For.Fenn., 196:1-32.

Mather, K., 1955, Polymorphism as an outcome of disruptive selection, Evolution, 9:52-61.

Mikola, J., [In press], Provenance and individual variation in climatic hardiness of Scots pine in northern Finland. In: Alden, J. N., Ødum, S., and Mastrontonio, J. L., eds.,

Forest Development in Cold Climates, Proceedings of an international symposium, 1991 June 18-24, [Laugarvatn, Iceland], NATO, Plenum, New York, New York.

Muona, O., and Harju, A., 1989, Effective population sizes, genetic variability, and mating system in natural stands and seed orchards of Pinus sylvestris, Silvae Genetica, 38:221-228.

Pakkanen, A., and Pulkkinen, P., 1991, Pollen production and background pollination levels in Scots pine seed orchards of northern Finnish origin. In: Pollen contamination in seed orchards, Proceedings of the meeting of the Nordic group for tree breeding, Umeå, Sweden, SUAS, For. Genetics and Plant Phys. Rep. 10.

Perry, D. J., and Knowles, P., 1989, Allozyme in sugar maple at the northern limit of its range in Ontario, Canada, Can. J. For. Res., 19:509-514.

Sarvas, R., 1967, Climatological control of flowering in trees. In: Proceedings of XIV International Union of Forest Research Organization Congress, Munich, Germany, DVFFA, Munich.

Stern, K., and Roche, L., 1974, Genetics of Forest Ecosystems, Berlin, Heidelberg, New York, Springer-Verlag.

Tigerstedt, P.M.A., 1973, Studies of isozyme variation in marginal and central populations of Picea abies, Hereditas, 75:47-60.

Tigerstedt, P.M.A., 1974, Genetic structure of Picea abies populations as determined by the isozyme approach. In: Proceedings of International Union of Forest Research Organization Congress, Stockholm, Sweden, SUAS, Stockholm.

Tigerstedt, P.M.A., 1979, Genetic adaptation of plants in the subarctic environment, Holarctic Ecology, 2:264-268.

Tigerstedt, P.M.A., Rudin, D., Niemelä, T., and Tammisola, J., 1982, Competition and neighbouring effect in a naturally regenerating population of Scots pine, Silv. Fenn. 16:122-129.

Tigerstedt, P.M.A., Pohtila, E., and Nevala, S., 1983, Larch, a potential tree for Lapland? Oulu, Finland, Res. Inst. of Northern Finland, C47:47-55. In Finnish.

Tremblay, M., and Simon, J-P., 1989, Genetic structure of marginal populations of white spruce (Picea glauca) at its northern limit of distribution in Nouveau-Quebec, Can. J. For. Res., 19:1371-1379.

POTENTIAL SPECIES AND PROVENANCES FOR FOREST DEVELOPMENT IN COLD CLIMATES

Maximilian Hagman

Forest Research Institute
Forest Tree Breeding Station
Maisala, Finland

INTRODUCTION

The cultivation of plants outside their natural range is as old as human culture, and in many cases their cultivation has been going on so long that we cannot be sure of their original source. In the beginning, exotic trees were cultivated as ornamentals, but forestry soon became the major objective of introducing exotic trees. Investigations through the years have shown that exotic tree species are often more vigorous and productive than indigenous species in northern climates. Trees of foreign origin were cultivated in cold regions where indigenous species were difficult to regenerate and the yield and quality of forest products were low. The endemic flora of cold regions are usually impoverished in commercial tree species. More research is needed to improve the management of exotic species, however, and to secure the most productive provenances for future tree breeding.

Procedures for introducing potential species and provenances will be reviewed for the continental parts of the Nordic countries. Maritime species, provenances, and conditions are reviewed in these proceedings by Lines (in press). Cold climates at high latitudes and high altitudes, however, present quite different conditions. For very interesting tree-limit problems in the tropics, see the review by Franz (1979) and the paper on the Austrian Alps by Hempel (1961).

FINDING THE RIGHT SOURCES

Geographical Distribution

The potential range of every tree species is governed by the climate where it evolved. Within the potential range, each species has the optimum habitat where growth is best and it can compete successfully with other species.

Forest Development in Cold Climates, Edited by J. Alden *et al.*
Plenum Press, New York, 1993

As a general rule, tree species require moist sites when transferred from their natural habitat toward their heat limit and dry sites when transferred towards their cold limit. Many examples from mild climates have demonstrated that the existing natural range of a species by no means represents its optimum region.

Cold and frost affect all life functions and the distribution of plants. The annual cycle of growth and dormancy of cold-climate tree species must be attuned to large variations in annual temperature. Species in cold climates must be adapted to late spring frosts, limited heat sums during the growing period, early frosts in the autumn, the length of the autumn and growing seasons, thaw periods during winter, and very low temperatures during midwinter.

Most widely distributed plants evolve local races that are well adapted to local climates. In addition, every species seems to have a cold tolerance limit beyond which it is damaged or killed by freezing. The limit cannot always be judged from the geographical origin of the species and must be determined by experiment.

By studying the climatic data and plant geography at high latitudes and altitudes in different parts of the world, we may find regions where potential species and provenances match the environment of the planting site. This approach was followed by the founder of the Mustila Arboretum and many other tree growers in Finland and elsewhere (Tigerstedt, 1986). In searching for adapted seed sources, it is particularly useful for researchers familiar with the plant geography and forest ecology of their homeland to investigate the climate and geography in the source region. The investigations of the climate and species in British Columbia on behalf of the Finnish Forest Research Institute before lodgepole pine (*Pinus contorta* Dougl. ex Loud.) and Douglas fir (*Pseudotsuga menziesii* (Mirb.) Franco) were introduced in the 1930s are good examples (Kujala, 1945).

When the climate and plant geography cannot be investigated, valuable information can still be found in old references such as Mayr (1906) and Schenk (1939, part I).

Because Pleistocene glaciation reduced the Nordic flora to only a few tree species, the historical development of the vegetation must be studied through plant geography (Wulff, 1950). The geography and climate of potential seed sources must be studied in conjunction with the environment of regions where the trees are to be planted. Fortunately, the boreal regions are well mapped in this respect (Ball, 1986; Bryson, 1966; Hansen and Brown, 1980; Hämet-Ahti, 1981; Kallio, 1975; Rouse, 1991).

A general description of the vegetation at the seed source is not enough to guarantee success of introduced species. Many introductions have failed because the silvicultural requirements of the species were not known. Personal experiences are again valuable, but in absence of empirical data, works such as Fowells (1965) and Schenk (1939, parts II and III) are useful sources of information.

Climate

The distribution of northern species depends on the combined effects of many different climatic factors and their complex interactions. Knowledge about the adaptation of species and provenances to the frost-free growing period, minimum and maximum temperatures, precipitation, wind, and the annual distribution of these and factors that regulate the rhythm of growth and dormancy, such as day and night length, are important.

Decisions about what seed source to use for afforestation at tree limit must be based on climatic factors observed for a long time in large areas within the natural range of the species. With the exception of air temperature, however, few climatic factors have been measured in this way. Air temperature is influenced by several factors that affect the survival and growth of trees: radiation, cloudiness, wind, site conditions, and altitude.

Since ancient times, plant geographers have tried to relate the distribution of plants to temperature (Gams, 1931). Several authors have constructed air temperature indexes, such as temperature sums above certain threshold values and temperature-based growth units, to predict plant distributions and identify hardy populations. Distribution limits based on air temperature indexes are reliable only in main features because they do not account for the effect of local topography or temperatures below the threshold value. Nor do they recognize that variation is more extreme on plant surfaces than in meteorological observation shelters. Mork (1941) defined his growth unit as the effect of temperature on growth when the mean temperature for the six warmest hours of the day is 8 °C. For Norway spruce (*Picea abies* (L.) Karst.), the height growth is one percent of the total growth intensity at 8 °C. For northern Norway, Bergan (1989) found Mork's units to be the best general expression for the effects of temperature on forest growth at timber line. Sarvas (1972) established fundamental relationships between growing season temperature sum and the growth of Scots pine (*Pinus sylvestris* L.). His method has applications in studies of tree growth of other species as well.

Plants respond directly to the energy absorbed, so that a measure of energy that reaches the plant would be more useful than the sum of ambient temperatures. Records of irradiation and energy balances are rarely available, however.

Temperature is not the only important factor. Moisture is another major feature limiting plant growth. Continental and maritime vegetation can be estimated from temperature and precipitation records. Temperature and precipitation indexes that can be used for model constructions were suggested by Gams (1931) and used successfully for mapping mountain vegetation (Gams, 1932).

Until recently, the compilation of numerous and complex meteorological data has been restricted by the scarcity of station information and by the scale of numerical work. Modern computers have solved many of these problems. According to a very recent review (Booth, 1991), new methods of climatic analysis to assist species and provenance research have worked well in the tropics and could be used to search for potential combinations of seed sources and planting sites in cold climates.

The Experimental Approach

Despite their usefulness, the plant geographical and climatological approaches provide only a general indication of the potential performance of introduced species. Cold and drought tolerance, growth and yield at new sites, and the reproductive behavior of exotics must be proven by experimentation.

Because trees are long-lived, sometime during their lifetime they are likely to be exposed to extreme conditions. During these "bottleneck years," many promising sources may fail to survive environmental extremes. Long-term field experiments are necessary to identify adapted provenances, but short-term laboratory tests of frost hardiness will reduce time and costs.

Fortunately, our knowledge of the processes that lead to cold and frost damage and hardiness of plants has increased considerably since tree introductions began in the last century (Hänninen and Pelkonen, 1988; Larcher et al., 1985; Sakai and Eiga, 1985). Methods that have been developed and tested to determine the hardiness of indigenous tree species (Nilsson and Eriksson, 1986) could be applied to introduced species.

It is necessary, however, to develop and apply methods for determining the fitness of species to spring, summer, and autumn as well as winter phases of plant development. For example, a simple and reliable method is needed to determine the onset and degree of dormancy in indigenous as well as introduced tree species.

Special habitats may require special methods for measuring adaptability. It is not certain that arctic plants are adapted to alpine conditions and vice versa. The degree of ultraviolet light absorption by the leaves might be an indicator of better adaptation to high altitudes. Perhaps a test for high or low flavonoid content could also be used. These substances are well known to protect cells from short wavelength, high-energy radiation.

LEARNING FROM EARLIER CULTIVATIONS

Many countries have a tradition of planting exotic species for dendrological studies. When dendrological plantations are located in cold areas, especially when they are represented by stands rather than single trees, they can provide useful information for locating adapted provenances. In Finland, plantations at the Mustila Arboretum and experiments of the Forest Research Institute were established for dendrological studies and have provided invaluable information about the exotic species in cold habitats (Heikinheimo, 1956; Lähde et al., 1984). Cultivations of the Forest Research Institute are located as far north as the Arctic Circle, where they have grown for more than 50 years.

In the Nordic countries, much data has accumulated from provenance experiments with Scots pine, Norway spruce, and lodgepole pine (Pinus contorta Dougl. ex Loud.). Results of these experiments can be found in the comprehensive literature elsewhere and will not be reviewed here. Neither will I go into the rules for provenance transfers that are available from many countries. Many of these rule books also contain the background literature on past provenance transfer decisions (Anonymous, 1982). Instead, I will review a few of the more recent provenance transfer experiences, mainly from Finnish Lapland. The results are applicable to northern parts of Sweden as well.

Larch species have been widely planted and studied more extensively than other genera in northern Finland and are the subject of a recent review (Hagman, in press). Only a few general conclusions will be made here.

Of the larches, western Siberian larch (Larix sukaczewii N. Dyl.) is a possible alternative to the indigenous Scots pine and Norway spruce on advantageous sites. The most promising seed sources of Larix sukaczewii are the well known Raivola plantation near St. Petersburg (Ilvessalo, 1923) and the indigenous populations in the Arkhangelsk region of Russia. The total stem wood yield of Arkhangelsk larch after 52 years at the Arctic Circle in Finland has been 162 cubic meters, bark included, per hectare. One plot of the Raivola larch produced 235 cubic meters, bark included, in 55 years.

The larches from central Siberia, Siberian larch (Larix sibirica Ledeb.), Dahurian larch (Larix dahurica Turcz.), and their natural hybrid, Larix czekanowskii Szaf., are not adapted to the climate of Finnish Lapland. They are continental species and probably suffer from change in daylength and cold summers. Similar experiences from larch transfers are reported from the Soviet Union by Iroshnikov (in press).

Dahurian larch, from central Siberia, has generally failed in Finland even at southern latitudes, but a sample of Larix gmelini Rupr. from Magadan (latitude 59°38′ N., longitude 151°50′ E.) has 92.9 percent survival after seven years in the field in a treeline experiment at Utsjoki, Finland (latitude 69°49′ N., longitude 27°08′ E.) (Table 1). More sources from northeastern Siberia should be sampled when political conditions permit.

Tamarack (Larix laricina (Du Roi) K. Koch) provenances from Alberta and Saskatchewan are hardy in Utsjoki, and an Alaskan provenance has performed well on swamp soil at the Teuravuoma Experimental Forest, western Finnish Lapland (Numminen, 1981). From the recent paper by Packee (in press), it is obvious that more research on this species is necessary.

Pinus contorta is another North American species that has promise in Scandinavia. In numerous provenance studies, survival and juvenile height growth equal or even exceed the survival and height growth of native Scots pine. Provenances from the northern limit of Pinus contorta in the Yukon of Canada have been best, but many experiments have shown large variations between stands within a provenance.

Because of difficulties in sampling material from very northern sources, only a few experimental plantings have been made with other high latitude species, and those are still young. Superior seed sources might be selected from these plantations, however.

The survival of the different species and seed sources after twelve years in central Lapland is summarized in Table 2. The trees were grown at the Pakatti Nursery (latitude 67° N.) and some plantations were made in 1977. This trial demonstrates that provenances from British Columbia, Ontario, and Alberta are not suitable for Lapland. Potential provenances are from Yukon and Alaska. The survival of black spruce (Picea mariana (Mill.) B.S.P.) was nearly the same as the survival of local Norway spruce. The survival of lodgepole pine from Yukon was slightly better than the survival of local Scots pine, but Scots pine provenances transferred from northern sources to Pakatti survived better than the local source.

Performance of the true firs, subalpine fir (Abies lasiocarpa (Hook.) Nutt.) and Siberian fir (Abies sibirica Ledeb.), both from southern Finland plantations, and the subalpine fir of British Columbia origin, was outstanding.

Survival of Siberian larch declined sharply over the years, but the reason is unknown. The site might not be the best for this species. Survival of silver birch (Betula pendula Roth) from Northern Finland is remarkable good. The hardwoods are important species in cold areas of Lapland.

On the initiative of Professor Paavo Kallio and the Kevo Subarctic Research Station of the University of Turku, Turku, Finland, three "treeline parks" were started in cooperation with the Finnish Forest Research Institute. In these plantations located in Utsjoki, Lapland at latitude 69°49′ N. (285 meters above sea level), 69°15′ N. (201 meters), and 69°46′ N. (100 meters), respectively, various species and origins of Abies, Larix, Picea, and Pinus from North America and Larix from Eastern Siberia have grown since 1977. Because of difficulties in obtaining enough material for these plantations, replications in space and time

Table 1. Survival of Different Species, Origins, and Provenances at "Treeline Parks" in Utsjoki, Lapland, Latitude 69°46′ N., Longitude 27°01′ E., Altitude 100–285 meters. Origin and Provenance According to OECD Terminology (1974)

Species	Origin/ latitude N.		Field Age (years)	Survival (percent)
Picea mariana	Alaska	65°10′	10	92.9
	Alaska	65°10′	76	8.8
	Yukon	62°02′	10	64.2
Picea glauca	Alaska	63°27′	7	70.5
	Alaska	64°54′	7	65.8
	Yukon	63°40′	7	90.7
	Manitoba	unknown	3	57.5
Picea abies	Finland	67°38′	10	99.0
Pinus contorta	Yukon	63°21′	10	89.2
Pinus sylvestris	Finland	63°00′	7	2.0
	Finland	66°30	7	18.9
	Finland	69°20′	7	74.0
Larix sukaczewii*	Prov. Kittilä, Finland	67°10′	10	88.0
Larix gmelini	Magadan	59°38′	7	92.9
Larix laricina	Alberta	54°07′	7	68.4
Abies balsamea	Alberta	57°25′	7	100.0

* Origin unknown

are inadequate and the results are preliminary. In addition, sites in the treeline parks are poor and may be inferior for some of the species tried.

Survival of species at the Utsjoki treeline is shown in Table 1. Overall results support the conclusion that North American spruces are promising in northern Finland. Survival of black spruce and white spruce was almost equal to the survival of local Norway spruce. The North American spruces were slightly taller at 10 years of age. As in the Pakatti experiments, the best provenances were from Alaska and the Yukon. Unfortunately, the Utsjoki experiments did not have enough Siberian larch to compare it with the Magadan larch. Parallel experiments should be made with these species in Lapland.

The subalpine and Siberian firs should be investigated at other treeline sites in northern Finland, particularly because the survival and growth of older practical plantations show the same success. Species from the northern Far East of Asia could provide valuable material at least for maritime conditions in Scandinavia. So far, seed collections have not been made in that region.

Although provenances of the potential softwood species are fairly clear, very few provenance investigations have been initiated for the hardwoods. Silver birch and downy birch (Betula pubescens Ehrh.) are local species and good seed sources are available. The common aspen (Populus tremula L.) grows to the treeline in Scandinavia, but other poplars have not been tried.

Table 2. Twelve-Year Survival of Species, Origins, and Provenances Grown in the Pakatti Nursery, Central Lapland, Latitude 67°40′ N., Longitude 24°56′ E. Origin and Provenance Are According to Organization for Economic Development and Cooperation (OECD) Terminology (1974)

Species	Origin	latitude N.	Altitude (meters)	Survival (percent)
Picea glauca	B.C.	55°40′	670	1.5
	B.C.	50°15′	930	4.1
	B.C.	55°20′	unknown	16.8
	Yukon	60°02′	"	54.0
	Alaska	64°53′	200	38.2
Picea mariana	Ontario	49°25′	unknown	1.8
	Alberta	53°36′	"	6.9
	Yukon	60°07′	"	80.8
	Alaska	64°46′	"	83.3
	Alaska	64°46′	"	90.0
Pinus contorta	B.C.	56°46′	"	3.3
	Yukon	63°19′	"	80.0
Pinus silvestris	Kittilä	67°38′	"	34.0
	Inari	69°32′	"	53.6
	Utsjoki	69°45′	"	63.6
Picea abies	Kittilä	67°38′	"	92.8
	Kittilä	67°57′	"	90.4
	Kittilä	67°38′	"	78.8
Larix sukaczewii*	Prov. Kittilä, Finland	67°38′	"	28.8
Abies lasiocarpa*	Prov. Punkaharju, Finland	61°43′	"	70.0
Abies sibirica*	Prov. Punkaharju, Finland	61°43′	"	86.0
Picea engelmannii	B.C.	56°14′	"	6.1
Betula pendula	Kittilä	67°38′	200	78.0

* Origin unknown

Silvicultural benefits from a birch shelterwood are many, and where afforestation is difficult, the application of a "Vorwald" of a deciduous species would help in establishing conifers. Betula pubescens, and in some cases Alnus species, should be tried. Birch shelterwoods were reported to be of great value in afforestation of conifers in northern Norway (Bergan, 1989). Even when conditions do not permit the production of industrial timber, coppicing of birch, for example, could provide fuelwood of economic value (Ferm et al., 1982). Perhaps Populus species could be managed for the same purpose if adapted material could be found and animal damage eliminated. A hybrid aspen (Populus tremula L., South Finland x P. tremuloides Michx., Alaska) is the fastest growing tree on swamp soil in the Teuravuoma Experimental Forest, western Lapland.

Clonal material of polyploid or hybrid origins must be considered for afforestation of difficult tree-limit sites in the future. Clonal forestry is very open for further experiments.

SEED PROCUREMENT

Procurement of seeds or vegetative propagules of northern origins is a major problem for many tree-limit species. The desired forests are often located far from forestry operations. Roads are often absent or difficult to travel, and seed collection can rarely be handled by the local administrations. Seed collections must be organized separately at great costs for each forest development program and species. If trees must be felled for seed, as for example in the larch seed collection in northwestern Russia (Blöndal and Benedikz, 1990; Snorrason and Blöndal, 1990), the original parents will be lost. Flowering and seed ripening are irregular in cold climates and seed quality is usually low (Mork, 1957).

Variation in seed quality among individual trees within the same stand may be large, as was observed during a 1987 and 1988 seed collection expedition to North America (Table 3). Variation in seed quality may be caused either by variation in the microclimate in the stand, or by genetic variation among trees for adaptation in seed ripening. If the variation in seed quality has a genetic basis, reproductive fitness of the populations can be improved by establishing seed stands locally from the best individuals, and in the country of seed use with seedlings from the most vigorous collections.

Seed orchards established with the best individuals from progeny tests may also improve the growth and reproductive fitness of the populations. For climatic reasons, seed orchards are often located in regions favorable for flowering and seed production where background pollen and physiological influences must not be allowed to interfere with the hardiness of the progenies. Background pollen is less important for rare exotics, but so-called "aftereffects" have been reported for Norway spruce seed produced in southern orchards (Björnstad, 1981). The possibility of aftereffects in other species must be investigated.

The population structure of seed stands and seed crops produced in mild climates must be followed and compared with the structure of the indigenous populations and progenies growing in harsh conditions. The performance of populations reproduced in mild climates must be checked in laboratory and field trials, and changes in isozyme and monoterpene patterns and other gene frequency markers must be related to changes in climate and performance. Laboratory and field investigations of all material obtained for provenance research and of the original natural populations must be initiated as soon as possible.

RESEARCH PLANS FOR EXOTIC SPECIES

A comprehensive review of the need for further experiments with exotic species has recently been prepared in Sweden (Martinsson, 1988; Rosvall, 1988). On the basis of the reviews, earlier discussions, and correlation of results from northern Finland and Sweden (Rosvall et al., 1984; Rosvall, 1985), a decision was made in the late 1980s to initiate further investigations of the species and provenances that show promise in northern Scandinavia. Because species from northwestern North America were of common interest to the Nordic countries, plans were made to fill in the provenance and species network with new collections from Alaska and Canada. A tour to collect species from northwestern Russia was also desired but was prevented by organizational difficulties.

North American collections were carried out as a cooperative effort among the Nordic countries in 1987–1988. Unique reproductive material was collected from 138 stands and

Table 3. Range in Percent of Fully-Developed Seed Recovered from about 18 Single-Tree Samples per Origin of Several North American Species Collected in 1987 and 1988

Species	Origin/	N. latitude	Altitude (meters)	Range (percent)
Picea mariana	B.C.	54°57′	800	96.0-9.0
	B.C.	60°35′	850	78.2-4.5
	Alaska	62°50′	700	11.3-0.3
	Alaska	62°10′		34.7-0.5
	Alaska	62°45′	750	54.6-7.3
	Alaska	65°00′	710	59.2-10.2
	Alaska	65°31′	1000	0.0
	Alaska	65°54′	1205	1.3-1.0
	Alaska	66°48′	550	8.6-0.0
	Alaska	67°19′	350	35.1-1.3
	NWT	60°25′	330	43.5-6.8
	NWT	61°02′	150	67.7-17.0
	NWT	61°54′	170	60.7-1.3
	NWT	61°50′	1000	34.6-1.0
	NWT	62°00′	230	83.5-18.5
	NWT	62°30′	160	62.0-10.2
	NWT	63°55′	170	83.5-5.5
	NWT	67°29′	100	54.4-12.7
	NWT	68°20′	90	24.0-0.0
	Yukon	63°17′	1080	63.2-1.5
	Yukon	66°23′	600	18.8-0.0
Picea glauca	Alaska	63°08′	800	1.2-0.0
	Alaska	63°45′	950	5.8-0.0
	Alaska	65°15′	450	51.7-1.8
	Alaska	66°59′	200	23.5-0.0
	Alaska	67°35′	20	36.7-2.8
	Alaska	68°03′	1000	4.6-0.0
	Alaska	68°07′	750	47.5-1.0
	Yukon	60°46′	1000	7.5-0.0
	Yukon	61°00′	850	30.7-0.8
Abies lasiocarpa	B.C.	54°50′	800	56.5-7.8
	B.C.	56°08′	300	34.7-0.0
	B.C.	55°35′	730	60.6-1.0
	B.C.	59°20′	100	2.8-0.5
	B.C.	60°34′	1170	0.0
	Alaska	56°12′	500	0.3-0.0
	Yukon	62°17′	1370	0.0
	Yukon	62°45′	1000	65.7-5.0
	Yukon	63°18′	1150	20.5-0.3
	Yukon	64°00′	1300	5.5-0.0
Pinus contorta	Yukon	60°06′	1000	65.1-14.8
	Yukon	60°08′	800	72.2-29.8
Larix laricina	Alaska	65°12′	100	39.1-0.5
	Alaska	65°17′	400	26.3-0.0
	NWT	67°57′	754	8.0-7.0
Pinus banksiana	NWT	61°05′	200	96.0-70.6
	NWT	61°55′	170	96.2-84.5
	NWT	62°38′	250	91.0-42.0

2,559 single trees of jack pine (Pinus banksiana Lambert), lodgepole pine, subalpine fir, tamarack, white spruce, black spruce, mountain hemlock (Tsuga mertensiana (Bong.) Carr.), and balsam poplar (Populus balsamifera L.) (Table 3). In addition, seven bulk seed lots and some living plants were obtained for each of the eight species.

With this material at hand, the Nordic Provenance Research Group under the leadership of Ola Rosvall, Institute of Forest Tree Improvement, Sävar, Sweden, established the following research objectives for the northern marginal regions of Finland and Scandinavia:

1. Insure that the species and provenances selected for research are optimal for all marginal regions where climate limits forestry.

2. Prepare for genetic improvement and seed production, that is, classify and plant the material as base populations for selection and breeding.

3. Establish reference materials for species and provenances at different localities in the Nordic countries.

The reference materials serve as a standard for breeding indigenous species, and for other forest research including experiments on silvicultural methods, technique of stand management, and diseases and insect control. The delineation of Nordic seed procurement areas, and testing of provenances and families in laboratory, nursery, and field conditions, form an important part of the research work.

Research procedures include:

1. Development and testing of climatic models against phenological data to determine which provenance-specific parameters can be related to local meteorological information.

2. Growth rhythm studies in vitro.

3. Nursery experiments in controlled environments.

4. Provenance experiments in the field.

5. Family testing in the field.

The work will be divided among the Nordic countries on the basis of climate and geographical conditions and local interests. The provenance and family tests will require large resources, including personnel and financial support. Results will benefit northern forestry and the people living in harsh and remote cold regions.

Whether atmospheric carbon dioxide accumulations will cause global warming and other climate changes at northern tree limits is not clear at the present. Kullman (1985) showed that climatic changes have always been large in northern regions. The mountainous forests of Scandinavia are growing in a region of great climatic variability and uncertainty. If the climate changes, eastern and northeastern flora might, in such circumstances, have genetic adaptation problems.

The most important feature of each genetic system is the level of compromise it establishes between fitness in order to secure immediate survival, and flexibility in adaptive response to future changes in the environment.

The most efficient population structure for promoting evolutionary diversification is the division of a large population into several sub-populations that are at least partially isolated in distinct habitats. Only the future will show if we have been able to manipulate evolutionary change in this respect.

REFERENCES

Anonymous, 1982, Lesosemennoe rayonirovanie osnevnyh lesoobrazujuscih porod bSSSR, Ljesnaja promys-lennost, Moskva, 368 p.

Ball, T. T., 1986, Historical evidence and climatic implications of a shift in the boreal forest tundra transition in central Canada, Climatic Change, 8:121-134.

Bergan, J., 1989, Reforestation and afforestation at high altitudes in North Norway. In: Forest regeneration at northern latitudes close to timber line, General Technical Report, PNW-GTR-247, Portland, Oregon, U. S. Department of Agriculture, Pacific Northwest Research Station, 21-24.

Björnstad, A., 1981, Photoperiodical after-effects of parent plant environment in Norway spruce (Picea abies (L.) Karst.) seedlings, Meddelelser fra Norsk Institutt for Skogforskning, 36(6):1-30.

Blöndal, S., and Benedikz, T., 1990, Öflun lerkifræs fráArkhangelskhéradi (Procurement of larch seeds from Archangel district), Arsrit Skógræktarfélags Islands 1990, 117-125.

Booth, T. H., 1991, Where is the world? New climatic analysis methods to assist species and provenance selection for trials, Unasylva, 42(165):51-57.

Bryson, R. A., 1966, Air masses, streamlines and the boreal forest, Geographical Bulletin, 8:228-269.

Ferm, A., Kauppi, A., and Rinne, P., 1982, The potential of birch for biomass production, International Energy Agency Forestry Planning Group B., Report JAB, 34:1-67.

Fowells, H. A., 1965, Silvics of forest trees of the United States, Agric. Handb. 271, U. S. Department of Agriculture, Washington, D.C., 762 p.

Franz, H., 1979, Ökologie der Hochgebirge, Verlag Eugen Ulmer, Stuttgart, 495 p.

Gams, H., 1931, Die klimatische Begrenzungen von Pflanzenarealen und die Verteilung der hygrischen Kontinentalität in den Alpen, I, Zeitschrift der Gesellschaft für Erdkunde zu Berlin, 321-346.

Gams, H., 1932, Die klimatische Begrenzungen von Pflanzenarealen und die Verteilung der hygrischen Kontinentalität in den Alpen, II, Zeitschrift der Gesellschaft für Erdkunde zu Berlin, 52-68; 178-198.

Hagman, M., [In press], Experiences with Larix species in Northern Finland. In: Packee, E. C., ed., Proceedings Symposium Northern Forest Silviculture and Management, 1957, August 16-22, IUFRO S1.05-12, Lapland, Finland.

Hämet-Ahti, L., 1981, The boreal zone and its biotic subdivision, Fennia, 159:69-75.

Hänninen, H., and Pelkonen, P., eds., 1988, Frost hardiness and over-wintering of forest trees, Silva Fennica, 22:179-255.

Hansen, R. K., and Brown, J., 1980, Coastal-inland distribution of summer air temperatures and precipitation in northern Alaska, Arctic and Alpine Research, 12:403-412.

Heikinheimo, O., 1956, Tuloksia ulkomaisten puulajien viljelystä Suomessa, Communicationes Instituti Forestalis Fenniae, 46(3):1-129.

Hempel, R., ed., 1961, Ökologische Untersuchungen in der subalpinen Stufe I, Mitteilungen der forstlichen Bundes-Versuchsanstalt Mariabrunn, 59:1-431.

Iroshnikov, A.I., [In press], Larix in the USSR. In: Packee, E. C., ed., Proceedings Symposium Northern Forest Silviculture and Management, 1957 August 16-22, Lapland, Finland.

Ilvessalo, L., 1923, Raivolan lehtikuusimetsä, Communicationes ex Instituto Questionum Forestalium Finlandiae Editae 5, 1-101.

Kallio, P., 1975, Reflections on the adaptation of organisms to the northern forest limit in Fennoscandia, Proceedings, Circumpolar Conference on Northern Ecology, 1975, September 15-18, Ottawa, National Research Council of Canada, Ottawa, 21-37.

Kujala, V., 1945, Waldvegetationsuntersuchungen in Kanada mit besonderer Berücksichtigung der Anbaumöglichkeiten kanadischer Holzarten auf natürlichen Waldböden in Finnland, Suomen Tiedeakatemian Toimituksia Sarja A. IV, Biologica 7.

Kullman, L., 1985, Ekologiska ramar för skogsodling i klimatiskt marginella höjdlägen (Ecological frames for artificial regeneration in climatically marginal high-altitude areas). In: Nilsson, J. E., ed., Reforestation material for harsh northern sites, Report 4, Department of Forest Genetics and Plant physiology, Swedish University of Agriculture, Umeå, 188-200.

Larcher, W., Häckel, H., and Sakai, A., 1985, Meteorologische Pflanzenpathologie, Witterung und Klima als Umweltfaktoren, Kälte und Frost, Handbuch der Pflanzenkrankheiten, Verlag Paul Parey, Berlin, Hamburg, 326 p.

Larsen, J. A., 1989, The Northern Forest Border in Canada and Alaska, Biotic Communities and Ecological Relationships, Ecological Studies 70, Springer-Verlag, New York, 255 p.

Lines, R., [In press], Species and provenance choice at northern tree limits in maritime climates. In: Alden, J. N., Odum, S., and Mastrantonio, J. L., eds., Forest Development in Cold Climates: Proceedings of an international workshop, 1991 June 18-24, [Laugarvatn, Iceland], NATO, Plenum, New York.

Lähde, E., Werren, M., Etholén, K., and Silander, V., 1984, Ulkomaisten havupuulajien varttuneista viljelmistä Suomessa, Communicationes Instituti Forestalis Fenniae, 125:1-87.

Martinsson, O., 1988, Främmande trädslag i praktiskt svenskt skogsbruk, I. Allmän översikt över några av de mest intressanta trädslagen samt behov av forskningsinsatser, Kungliga Skogs-och Lantbruksakademiens Tidskrift, 127:85-89.

Mayr, H., 1906, Fremdländische Wald- und Parkbäume für Europa, Paul Parey Verlag, Berlin, 568 p.

Mork, E., 1941, Om sambandet mellom temperatur og vekst, Meddelelser fra det Norske Skogforsöksvesen, 27:1-89.

Mork, E., 1957, Om frökvalitet og fröproduktion hos fura i Hirkjölen, Meddelelser fra det Norske Skogforsöksvesen, 14(48):353-379.

Nilsson, J.-E., and Eriksson G., 1986, Freeze testing and field mortality of Pinus sylvestris L. in northern Sweden, Scandinavian Journal for Forest Research, 1:205-218.

Numminen, E., 1981, Ulkomaisten puulajien viljely Pohjois Suomen soilla, Metsäntutkimuslaitoksen tiedonantoja, 6:53-60.

OECD, 1974, OECD Scheme for the control of forest reproductive material moving in international trade, C(74) 29 final of 5th March, 1974, OECD, Paris, 24 p.

Packee, E. C., [In press], Larch in North America. In: Packee, E.C., ed., Proceedings Symposium Northern Forest Silviculture and Management, 1957 August 16-22, IUFRO S1.05-12, Lapland, Finland.

Rosvall, O., 1985, Val av svartgransprovenienser i NorraSverige, Institutet för Skogsförbättring, Skogsträdsförädlings-information 1985/86, 2:1-6.

Rosvall, O., 1988, Främmande trädslag i praktiskt svenskt skogsbruk, III. Kunskapsläge och försöksverksamhet i norra Sverige, Kungliga Skogs-och Lantbruksakademiens Tidskrift, 127:95-105.

Rosvall, O., Strömberg, S., and Andersson, B., 1984, Skogsförbättrings proveniensförsök med contortatall i norra Sverige, Institutet för Skogsförbättring, Årsbok, 1984:117-159.

Rouse, W. R., 1991, Impacts of Hudson Bay on the terrestial climate of the Hudson Bay lowlands, Arctic and Alpine Research, 23(1):24-30.

Sakai, A., and Eiga, S., 1985, Physiological and ecological aspects of cold adaptation of boreal conifers. In: Kaurin, Å., Junttila, O., and Nilsen, J., eds., Plant Production in the North, Norwegian University Press, Tromsö, 157-170.

Sarvas, R., 1972, Investigation on the annual cycle of development of forest trees, Active period, Communicationes Institut Forestalis Fenniae, 76(3):1-110.

Schenk, C.A., 1939, Fremdländische Wald-und Parkbäume I-III, Paul Parey Verlag, Berlin, 615 p; 645 p; 640 p.

Snorrason, A., and Blöndal, S., 1990, Skógur og skógabúskapur i Arkhangelskhéradi (Forests and Forestry in Arkhangel oblast), Arsriti Skógraektarfélags Island 1990, 99-115.

Tigerstedt, P. M.A., 1986, Arboretum Mustila, Publications of the Finnish Dendrological Society, 2:1-28.

Wulff, E., 1950, An introduction to historical plant geography, Plant Sciences Books X, Chronica Botanica Company, Waltham, Mass., 223 p.

SPECIES AND PROVENANCE CHOICE AT NORTHERN TREE LIMITS IN MARITIME CLIMATES

Roger Lines

Forestry Commission
Northern Research Station
Roslin, Midlothian, United Kingdom

INTRODUCTION

This chapter is based on information gathered over 34 years as a silviculturist for the British Forestry Commission and on related work by other researchers both in Great Britain and in other northern countries. The main object of the Forestry Commission in 1919 was to grow productive timber crops for strategic purposes on land of low agricultural value. Over the years, as improved silvicultural techniques became available, trial plantations of many species and provenances were extended to sites with ever more limiting conditions for tree growth.

Great Britain lies mainly in the Temperate Mixed Broadleaved Woodland Zone from latitude 50° N. to nearly 61° N. Small remnants of the Northern Coniferous Forest of Scots pine (Pinus sylvestris L.) occur in north Scotland, and patches of tundra-like vegetation are found above 1,000 meters elevation. Similarities between the distribution of the native pine forest and birch scrub in the Highlands and distribution of these species in Norwegian and Finnish Lapland were noted by McVean (1964). The warming influence of the North Atlantic Drift conceals the fact that Sullom Voe in the Shetland Islands is further north than Cape Farewell at the southern tip of Greenland. Aberdeen (latitude 57°15′ N.) is on the same latitude as Kodiak Island, Alaska.

The British climate varies regionally from semi-continental in southeast England to extreme Atlantic maritime in the Shetland Islands and the outer islands of the west coast of Scotland (Tansley, 1949). Oceanic climates show a small range of annual temperature variation; thus, the difference between the mean temperature of the warmest and coldest months is only 8 °C in the Shetland Islands, while at Cambridge it is 13.6 °C. By comparison, the continental climate of Berlin has a range of 19 °C, and Cordova, Alaska a range of 15.1 °C. The mean daily minimum temperature of the coldest month is above 1 °C at both Stornoway, Outer Hebrides, and Lerwick, Shetland, although the lowest temperatures on record at both sites are -11.6 °C and -8.9 °C, respectively. Precipitation is well-distributed

Forest Development in Cold Climates, Edited by J. Alden *et al.*
Plenum Press, New York, 1993

throughout the year in northwest Britain; thus, rain falls on 263 days per annum at Stornoway, where the annual precipitation is 1,266 millimeters. Snow seldom lies long on the islands. With infertile soils, high atmospheric humidity, and high precipitation, the resulting natural vegetation on these extreme Atlantic sites is mainly blanket bog (Tansley, 1949).

Strong persistent winds are a major factor affecting tree growth (Lines and Howell, 1963; Miller et al., 1987). At the Butt of Lewis, the northernmost point of the Outer Hebrides, long-term records show that 29 percent of all winds exceed 39 kilometers per hour (24 miles per hour) and nearly half the hourly winds were above this speed in January. Another feature of the maritime climate of the Scottish Highlands is the frequent alternation of freezing and thawing of the upper soil horizons in winter. Green (1964) claimed that this, together with high winds, made the climate of this area much more severe for plants than the climate of the more continental Norwegian fells (upland plateau).

Exposure increases rapidly with increasing elevation in maritime climates and this reduces tree productivity. Worrell and Malcolm (1990) showed that productivity of Sitka spruce (Picea sitchensis (Bong.) Carr.) is strongly related to elevation at any one location, but that for the same elevation it is lower at northern or coastal sites compared with spruce on inland and southern sites. This pattern of variation could be explained in terms of known patterns of windiness (Miller, 1985) and growing-season temperature. For purposes of forest management, Britain has been divided into seven zones of varying wind exposure (Fig. 1).

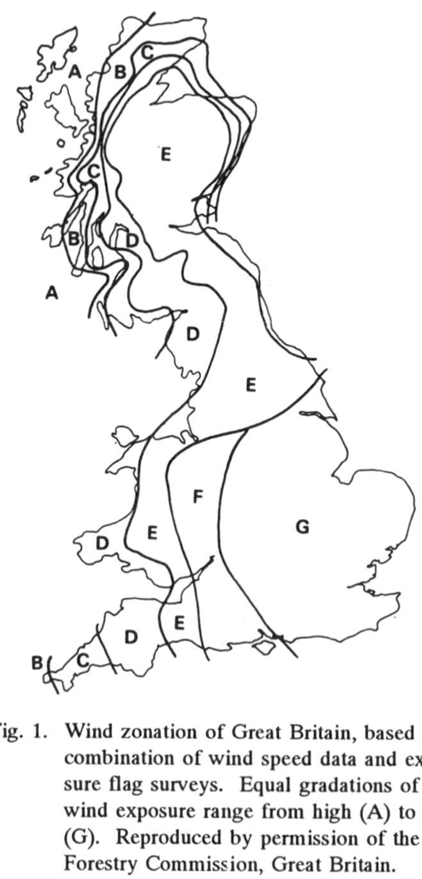

Fig. 1. Wind zonation of Great Britain, based on a combination of wind speed data and exposure flag surveys. Equal gradations of wind exposure range from high (A) to low (G). Reproduced by permission of the Forestry Commission, Great Britain.

The elevational limits for commercial plantability range from 150 meters in the northern islands to 600 meters in the Central Highlands of Scotland (Miller et al., 1987).

SPECIES TRIALS

Most of the tree flora of Great Britain was eliminated during the Quaternary glaciation, leaving Scots pine as the only native conifer with commercial potential. The dearth of tree species in Britain stimulated more interest in the introduction of exotic species than in countries with a wide range of native species. Mitchell (1980) lists only 28 species as truly native. By contrast, Hunt (1972) lists 341 conifers that currently grow in the British Isles. If broadleaved species are included, Hillier (1981) lists 8,000 trees and shrubs in Britain.

The choice of species available, combined now with knowledge of their silvicultural and physiological characteristics, appears to offer wide possibilities for introducing exotics. Few are sufficiently hardy and vigorous, however, to grow at northern tree limits in a maritime climate. In addition, an introduced species may be from a seed origin that is well adapted to growing on a sheltered arboretum site in southern Britain but may be quite unsuitable for severe northern conditions. Only 14 species have been tested by planting many seed origins on a wide range of British sites (Lines, 1987a).

Selection Criteria

Ideally, a research program of species and provenance trials should form an integrated part of any afforestation scheme. The sequence should be (1) arboretum phase, (2) species elimination phase, (3) species testing phase, (4) range-wide provenance trials, (5) restricted provenance phase, and (6) a provenance proving phase, planting the most promising origins in larger plots to give reliable estimates of volume production. Factors to consider when choosing a species include these:

Climatic matching
Soil characteristics in its native range and on the proposed sites
Knowledge of a species' performance in its native range and as an exotic
Genetic variation within the species
Safety in relation to climate, pests, and diseases
Timber properties and yield

Climatic matching can only be a general guide, as many forest trees with a wide range have become adapted to different climates. For example, lodgepole pine (*Pinus contorta* Doug. ex Loud.) growing in the Yukon is adapted to a very different climate than a population of this species growing on the coast of California. Many trees are inherently adaptable. In Iceland, both the outstandingly oceanic Sitka spruce and the strongly continental Siberian larch (*Larix sibirica* Ledeb.) are very successful. Wood (1955) gave data for detailed climate matching between Britain and regions of northwest America to facilitate better choice of species and provenances from that area. Bjarnason (1967) also used climatic matching as a guide for species introduction to Iceland from Alaska.

Soil preferences for many exotics are described in the literature, for example, Fowells (1965) and Wood (1955). It would be unwise to attempt to grow species that show clear soil preferences in their native range on soils of a very different character with respect to fertility, structure, or acidity.

Performance of a species in its native range can be found in the literature (Fowells, 1965). This kind of information must be interpreted with care, however. The performance of Monterey pine (Pinus radiata D. Don) in California would scarcely justify its widespread use in New Zealand and South America. The same could be said for lodgepole pine from the northern parts of its range, which is now planted on more than one million hectares in Sweden. Macdonald et al. (1957) have assembled a mass of data for a large number of species grown as exotics in Britain, including details of performance in their country of origin and provenance variation.

Genetic variation within a species is considered later in a separate section on provenance.

Safety is of supreme importance in long-term afforestation investments. Introduced species must be able to tolerate climatic events that may occur only once in 50 to 100 years. Testing in long-term arboreta is of particular value in avoiding losses from rare climatic extremes (Edwards and Howell, 1962). Silen (1962) noted that a high proportion of the 164 exotic conifers at the Wind River Arboretum in Washington State failed over a period of 50 years. In 1981, much damage to Sitka spruce in Britain was caused by an exceptional combination of a very warm spell followed by frost, which initiated cambial activity before the trees had flushed (Redfern, 1982). A similar incident occurred during April 1963 in Iceland and caused such severe damage to Sitka spruce that it curtailed the use of this species for several years (Bjarnason, 1967). Species from warm southern latitudes tend to continue active growth into the autumn and then become injured by frost. In northern Scotland, Sitka spruce from Oregon frequently suffer die-back; roble (Nothofagus obliqua (Mirb.) Bl.) and rauli (Nothofagus procera (Poepp. & Endl.) Oerst.) show both top die-back and cambial damage to stems on exposed sites, whereas nirre (Nothofagus antarctica (Forst.) Oerst.) is largely undamaged.

Insect attacks on exotic species are hard to predict in advance. The pine shoot moth (Rhyacionia buoliana Denis & Schiff.) devastates stands of lodgepole pine in the Netherlands, yet it is relatively unimportant at high latitudes. By contrast, a sudden outbreak of the pine beauty moth (Panolis flammea Denis & Schiff.) in 1967 severely damaged lodgepole pine, which had grown in Scotland for forty years (Stoakley, 1979). In Sweden, lodgepole pine is more vulnerable to small rodents than native Scots pine (Hagner and Fahlroth, 1974). Fungal attacks are seldom more serious on exotic species than on native species in the northern maritime region. The most serious potential threat comes from attacks of rust fungi on lodgepole pine, such as the western gall rust (Endocronartium harknessii Moore), which needs no alternate host (Martinsson, 1980).

Wood properties of tree species and provenances growing at northern limits may be very different from those of native trees growing at lower latitudes. Annual rings are likely to be narrower and the amount of late wood in relation to early wood may be reduced (Harding, 1972).

Establishment Problems

Before species trials can be established at northern tree limits, satisfactory techniques must be developed for raising hardy planting stock. Standard nursery practices may be inappropriate because species vary greatly in their silvicultural characteristics. For example, species such as subalpine fir (Abies lasiocarpa (Hook.) Nutt.) grow slowly as seedlings and may require four years to reach the minimum size for forest planting, while hybrid larch (Larix xeurolepis Henry) can attain plantable size in one year. Methods of raising plants in Britain are described by Aldhous (1972). Raising plants as containerized seedlings in greenhouses has gained favor, especially in Iceland (Svanbergsson, 1986), and Blondal (1989) describes similar methods for producing containerized stock in Norway at 66° N.

Sitka spruce and lodgepole pine are easier to establish in forest plantations than Abies or Tsuga species, and are better able to withstand frost lift and low nitrogen levels. Frost lift is severe on soils stripped of vegetation in Iceland. The problem of frost lift can be alleviated by sowing Nootka lupine (Lupinus nootkatensis Donn) to establish vegetation cover before planting. In Scotland, frost lift is rare and competing vegetation is usually removed by ploughing. Frost lift can occur on some mineral soils, however.

On British moorlands, growth of Picea and Abies species is reduced by competition for nitrogen from dominant heather (Calluna vulgaris (L.) Hull.). Pinus and Larix are less susceptible to nitrogen deficiency and are used in "nursing mixtures" to aid establishment of the more productive Picea and Abies (Taylor, 1991). Alternatively, herbicide control of Calluna or fertilization can be used to alleviate nitrogen deficiency (McIntosh, 1983).

MAJOR SPECIES FOR NORTHERN TREE LIMITS

Sitka spruce. Sitka spruce is the only spruce with a natural range almost entirely in the maritime zone. For more than 3,000 kilometers along the Pacific coast of North America, between 39° N. and 61° N., it grows in a narrow belt, mainly at elevations below 300 meters. Productivity is high, with up to 24 cubic meters of stemwood per hectare per annum in Britain (Hamilton and Christie, 1971). Although Sitka spruce prefers mesic sites in North America (Roche and Haddock, 1987), it grows well on a wide range of sites, including infertile peatland in Scotland (Low, 1987b). In Britain, Sitka spruce is superior to all other common species when grown on extremely exposed sites (Macdonald et al., 1957). The foliage is resistant to mechanical injuries and the physiological effect of wind in cooling the tissues during the growing season seems to affect this species less than others. Exposure to wind remains an important factor, however, in explaining site-to-site variation in growth (Worrell and Malcolm, 1990).

Roots of Sitka spruce cannot tolerate prolonged anaerobic conditions (Coutts and Philipson, 1987). Winter waterlogging, common on many sites in Britain, results in shallow rooting with accompanying risk from windthrow. Arnoy (1986) considered Sitka spruce to be the best species for peat soils in the coastal areas of northern Norway (Kaasen and Skaret, in press). The tallest trees in the Faroe Islands are Sitka spruce, about 60 years old and 20 meters tall (Odum et al., 1989). This species appears to grow less well on mild oceanic sites, where high winter temperatures fail to control infestation of the spruce aphid (Elatobium abietinum Walker), which elsewhere is controlled by severe winter frosts (Lines, 1987b; Carter, 1972).

Norway spruce (Picea abies (L.) Karst.). Norway spruce also has a wide range, from north of the Arctic Circle to 41° N., and from 200 to 2,000 meters above sea level. At the boreal fringe of its range, the growing season (characterized by mean daily temperatures above 6 °C) may average only six weeks per year. Norway spruce grows on a wide variety of sites, although it suffers from Calluna competition on moorlands. It is far less resistant to exposure than Sitka spruce, and its roots do not tolerate anaerobic conditions or soil compaction. Productivity varies greatly with site. Christie and Lines (1979) compared yield tables for Norway spruce in 27 European regions. Volume increment varies between six and 20 cubic meters per hectare per annum in Britain, between six and 15 cubic meters in West Norway, and between four and eight cubic meters in north Sweden.

Lodgepole pine. Lodgepole pine ranges as far north as 64° N. in the Yukon and as far south as 31° N. in California, and from sea level to 3,900 meters in elevation (Critchfield, 1980). Productivity varies widely among sites (Christie and Lines, 1979) and among provenances in exotic plantations. Maximum production in Britain never reaches that of Sitka spruce; on northern sites, growth averages six to eight cubic meters per annum (Stevenson and Mackay, 1991). In Iceland, maximum growth on the best sites is five cubic meters per hectare (Sigurgeirsson, 1988), while in West Norway, lodgepole pine grows faster than Scots pine (Magnesen, 1980). Frost damage is seldom a problem unless southern provenances are used, but damage from small rodents and deer can be severe and stem form can be poor (Arnoy, 1986). Nutrient requirements are so low that it can be grown on a wider range of sites than most other species. Lodgepole pine roots are better adapted to anaerobic conditions than are those of many other species (Coutts and Philipson, 1978).

Scots pine. Scots pine is the most widely distributed of the pines. It ranges from Spain in the west to almost the Sea of Okhotsk in the east, and from Greece and Turkey in the south to the Arctic Ocean; it is found from sea level to 2,400 meters in elevation. Primarily a continental species, it colonized Scotland and Norway centuries before Norway spruce expanded after the last glacial period. Scots pine tolerates a wide range of soils and climates, though in Scotland climatic browning is frequent. Nutritional requirements are low and frost damage is unusual. Scots pine was used in all of the early British trials on northern sites, but other species grew faster and suffered less climatic damage on almost all types of peat (Zehetmayr, 1954). In Iceland, experience was similar and wide-scale use of Scots pine in 1950-1960 was abandoned after serious infestations by the wooly aphid (*Pineus pini* Macq.). In West Norway, climate damage and infestation problems are less important, although Arnoy (1986) found that even there, Scots pine was much less productive than Sitka spruce.

Japanese larch (*Larix kaempferi* (Lambert) Carr.). The natural range of Japanese larch extends over several mountainous areas of central Honshu around latitude 36° N. and elevations from 1,200 to 2,400 meters, but it has readily adapted to much more northerly attitudes as well. It is a pioneer species on volcanic ash beds and grows well on loose-textured soils that are low in nitrogen. Japanese larch has been widely used as a nurse species in northern Britain (Edwards, 1957), and it forms a component of shelterbelts on exposed sites such as the Shetland Islands (Low, 1987a; Stewart, 1962). Damage from autumn frosts sometimes occurs because the shoots harden late in the season. Stem and crown form are poorest on exposed infertile sites. Productivity on exposed northern sites in Britain is seldom more than four to six cubic meters per hectare per annum. In West Norway, Arnoy (1986) reported very fast growth of Japanese larch, but with such poor stem form that it was unsuitable for timber.

Noble fir (*Abies procera* Rehd.). Noble fir is the most widely used true fir (*Abies* spp.) in British trial plantations. Its natural range is on the upper slopes from 1,000 to 1,800 meters in elevation of the Cascade Range of Washington and Oregon, where the climate is highly continental and the growing season is short. A natural wax coating on the needles helps to restrict injury on exposed sites. Early growth is slow, which makes establishment difficult (Aldhous and Low, 1974). After canopy closure, shoot growth increases. The maximum yield in Britain is 22 cubic meters per hectare per annum, but even on infertile exposed sites, yield may exceed 12 cubic meters. Nutrient requirements are not high, but it is very susceptible to *Calluna* competition. This species was widely used in pioneer planting at Loch Ossian at latitude 57° N. and over 400 meters in elevation (Keenleyside, 1985).

MINOR SPECIES

The list of possible species for afforestation of northern maritime sites is nearly endless. In Iceland, for example, 100 species are listed by Hansson (1989), although current planting is almost entirely devoted to five species: Sitka spruce, the natural hybrid Picea xlutzii Little, lodgepole pine, black cottonwood (Populus trichocarpa Torr. & Gray), and western Siberian larch (Larix sukaczewi Dyl.) (Lines, 1990). In the Faroe Islands, a similar long list is given by Odum et al. (1989), but most planting is now Sitka spruce and lodgepole pine. Many other species have started well in sheltered gaps within older plantations (Lines, 1987b), including noble fir, western hemlock (Tsuga heterophylla (Raf.) Sarg.), Sitka alder (Alnus sinuata (Reg.) Rydb.), grand fir (Abies grandis Lindl.), and even Japanese cryptomeria (Cryptomeria japonica (L.F.) D.). New trials with nirre (Nothofagus antarctica (Forst.) Oerst.), coique (N. betuloides Mirb.), and lenga (N. pumilio (Poepp. & Endl.) Krass.) are promising (Odum, 1989). In West Norway, Arnoy (1986) mentions twelve species in trials on six peatland sites, none of which compared favorably with Sitka and Norway spruces.

In Britain, Zehetmayr gave detailed results of species trials on peatlands (1954) and heathlands (1960). Of 50 minor species tried on peat, only Serbian spruce (Picea omorika (Pancic) Purkyne), western hemlock, noble fir, and balsam fir (Abies balsamea (L.) Mill.) showed promise. Lines (1963) reviewed the choice of species for northern Britain and suggested 10 that merited investigation. Of these, only mountain hemlock (Tsuga mertensiana Sarg.) and Macedonian or Balkan pine (Pinus peuce Grise.) showed promise for use on northern exposed sites. The former proved so slow-growing that it was abandoned. Macedonian pine, although starting more slowly than Scots pine, later produced 30 percent more basal area on infertile exposed sites (Lines, 1985).

PROVENANCES

Sitka spruce. Of 48 provenance trials with Sitka spruce in Britain, 25 are in Scotland. Many of the trials on exposed high-elevation sites in England and Wales are also relevant. Results have been summarized by Lines (1980, 1987a), and a more detailed report is being edited. In general, there was a cline for increasing vigor with decreasing latitude of provenance as far south as 46° N. There were some exceptions; provenances from Queen Charlotte Island, for example, grew faster than expected. Interaction of provenance and site was significant, with northern provenances growing relatively better on northern and exposed sites than southern provenances. Within the Alaska group, the tallest provenances were from south of latitude 57° N. and the shortest from Kodiak Island and Prince William Sound. Provenances from the white spruce (Picea glauca (Moench) Voss) introgression zones near Lawing, Alaska, and on the Skeena River, British Columbia, also grew slowly. However, Faulkner et al. (1985) showed that controlled crosses with white spruce from eastern Canada grew 20 percent taller than Sitka spruce in north Scotland. Time of growth cessation varied widely among provenances (Kraus and Lines, 1976), which helps to explain differences in autumn frosting (Cannell and Sheppard, 1982). All provenances were subject to defoliation by the green spruce aphid (Elatobium abietinum Walker), although growth recovery was slower for an Alaska provenance than for southern provenances (Carter and Nichols, 1988).

In west Norway, the tallest provenances in an 11-year trial were from 54° to 55° N.

latitude (Magnesen, 1986). Those from below 50° N. suffered severe frosting in the nursery. In Denmark, Alaska and British Columbian provenances proved most frost hardy but grew more slowly than those from the Queen Charlotte Islands and Forks, Washington (Madsen, 1984). In Iceland, only Alaska provenances survived in early trials. The preferred provenances for afforestation were Cordova and Homer, although trees from Sitka grew best in a 1970 experiment (Benedikz, 1976). Trees of more southerly origin are growing well at Hagavik in southwest Iceland (Lines, 1990). The Norwegian results at latitude 60° N. and those of Illingworth (1980) on sites with a fairly severe climate in northern British Columbia suggest further Icelandic trials of provenances from 54° to 55° N. may be desirable.

Norway spruce. There are many experiments with Norway spruce in Britain, but few on northern exposed sites. In an experiment very close to the sea at Minard (in County Argyll), Scotland, the Polish, Romanian, and Czech provenances were tallest and Scandinavian provenances were among the shortest. Dietrichson (1980) summarized the results of nearly 300 provenance trials with Norway spruce in Nordic countries, including many in west and north Norway. He noted that Carpathian provenances usually outgrew local native provenances in Scandinavia, but some suffered severely from autumn and winter frosts. In Iceland, inherently slow-growing provenances from northern Norway are preferred. Hardier and faster-growing provenances could perhaps be found in Poland or in western Russia than in Scandinavia.

Lodgepole pine. Most of the 83 experiments with lodgepole pine in Britain are established on northern or upland sites (Lines, 1987a). Of the 340 seedlots tested, the fast-growing southern coastal provenances were unsuitable for most sites because of early instability and snow damage, whereas those of north coastal origins were much less affected. Provenances from Alaska and the north coast of British Columbia are the best choice for use in mixture with Sitka spruce, while those from the lower Skeena River, British Columbia are recommended for pure plantations. For the most exposed northern coastal sites in Scotland, a land race has been developed from a Hollis, Alaska provenance. A technical bulletin on this species will be published soon (Lines, in press).

In west Norway, experiments have shown that coastal provenances from as far south as Oregon survive and grow fast, but with very poor form (Magnesen, 1980). On exposed northern sites, Alaska provenances later outgrow southern provenances. In maritime climates, provenances of inland origins decline in the pole stage. The most widely used provenance in Iceland is from Skagway, Alaska, although Sigurgeirsson (1988) concluded from a survey of 15 provenances in most of the older plantations that the Skagway provenance is climatically suited only to parts of the country. Here again, a local land race could be developed profitably.

Scots pine. In Britain, 28 experiments with about 150 seedlots have been established (Lines, 1987a). Results from experiments planted from 1928 to 1935 in Scotland showed that high elevation and high latitude provenances from continental climates grew slowly, while British provenances grew best (Lines and Mitchell, 1965). Later experiments to find suitable provenances were only partly successful and pointed to the superiority of trees from the northwest area near Loch Maree. Forrest (1980) showed that this strain is genetically distinct from five other Scottish strains. The tree improvement program is sufficiently advanced to provide all seed for afforestation from seed orchards and seed stands in western Scotland.

Japanese larch. British experiments with Japanese larch were described by Lines and Mitchell (1969). The fastest growing origins were from the Suwa region of Nagano and the lower slopes of Mt. Fuji in Japan; for exposed northern regions in the British Isles, however, high elevation provenances may be preferable because they become dormant long before low elevation sources. Tree form is so highly varied that seed from second or third generation plantations on the island of Hokkaido, north of the natural range of Japanese larch, could be superior in cold, maritime climates.

Noble fir. Seven experiments were planted in Britain from 1982 to 1984 with up to 27 provenances of noble fir. Preliminary results show little early height variation among provenances and suggest that the introgression zone of noble fir and Shasta red fir (Abies magnifica var. shastensis Lemm.) south of latitude 44° N. should be avoided in future seed collections for high latitudes (Lines, 1987a).

CONCLUSIONS

Widespread species and provenance trials show that with a few exceptions the most successful species at northern tree limits in maritime climates are Sitka spruce, its hybrid with white spruce, and lodgepole pine, all from northwest America. Norway spruce, larch species, and Scots pine are more variable than lodgepole pine and Sitka spruce, and grow well in some areas but fail in others. The importance of selecting suitable seed origins cannot be over-emphasized. Testing young plants in growth chambers and in freezing trials may help to reduce the number of provenances used in field trials but cannot substitute for such trials, which should last for many years. Successful pioneer plantations may act as gene banks and produce local land races that are well adapted to the new environment.

REFERENCES

Aldhous, J.R., 1972, Nursery Practice, Forestry Commission Bulletin 43, HMSO, London, 184 p.

Aldhous, J.R., and Low, A.J., 1974, The potential of western hemlock, western red cedar, grand fir and noble fir in Britain, Forestry Commission Bulletin 49, HMSO, London, 105 p.

Arnoy, B., 1986, Tree species on peatland in West Norway, Norsk Institutt for Skogforskning, Rep. 10/86, 24 p.

Benedikz, T., 1976, Progress report on the international ten provenance experiment with Sitka spruce in Iceland. In: O'Driscoll, J., ed., IUFRO Sitka spruce International ten provenance Experiment—Nursery stage results, Department of Lands, Forest and Wildlife Service, Dublin, 124-139.

Bjarnason, H., 1967, Skogsaken i Island og dens utvikling, Medd. Det Norske Skogforsoksvesen, 84:74.

Blöndal, S., 1989, Skogur undir heimskautsbaug, Arsrit Skograektarfelags Islands, 59-76.

Cannell, M. G. R., and Sheppard, L. J., 1982, Seasonal changes in frost hardiness of provenances of Picea sitchensis in Scotland, Forestry, 55(2):137-153.

Carter, C. I., 1972, Winter temperatures and survival of the Green Spruce Aphid, Forestry Commission Forest Record 84, HMSO, London, 10 p.

Carter, C. I., and Nichols, J. F. A., 1988, The green spruce aphid and Sitka spruce provenances in Britain, Forestry Commission Occasional Paper 19, Edinburgh, 7 p.

Christie, J. M., and Lines, R., 1979, A comparison of forest productivity in Britain and Europe in relation to climatic factors, For. Ecology and Manage., 2:75-102.

Coutts, M. P., and Philipson, J. J., 1978, Tolerance of tree roots to waterlogging, I. Survival of Sitka spruce and lodgepole pine, New Phytol., 80:63-69.

Coutts, M. P., and Philipson, J. J., 1987, Structure and physiology of Sitka spruce roots, Proc. Roy. Soc. Edinburgh, 93B, 131-144.

Critchfield, W. B., 1980, The distribution, genetics and silvics of lodgepole pine, Proc. IUFRO Joint Meeting of Working Parties, Vancouver, Canada, 1978, Vol. I, Ministry of Forests, British Columbia, Canada, 65-94.

Dietrichson, J., 1980, Norway spruce provenance trials in Nordic countries, Proc. I.U.F.R.O. Joint Meeting of Working Parties on Norway spruce provenances and Norway spruce breeding, Bucharest, 1979, Lower Saxony Forest Research Institute, Escherode, W. Germany, 3-14.

Edwards, M. V., 1957, Larix leptolepis. In: Exotic Forest Trees in Great Britain, Forestry Commission Bulletin 30, HMSO, London, 69-73.

Edwards, M. V., and Howell, R. S., 1962, Planning an experimental programme for species trials, Paper to 8th Commonwealth Forest Conference, East Africa, Forestry Commission, London, 11 p.

Faulkner, R. F., Fletcher, A. M., and Samuel, C. J. A., 1985, Genetics—Forest progeny tests, Forestry Commission Report on Forest Research, HMSO, London, 28 p.

Forrest, G. I., 1980, Genotypic variation among native Scots pine populations in Scotland based on monoterpene analysis, Forestry, 53(2):101-128.

Fowells, H. A., 1965, Silvics of forest trees in the United States, USDA Forest Service, Agriculture Handbook 271, Washington, D.C., 762 p.

Green, F. H. W., 1964, The Climate of Scotland. In: Burnett, J. H., ed., The Vegetation of Scotland, Oliver and Boyd, Edinburgh, 15-35.

Hagner, S., and Fahlroth, S., 1974, On the prospects of cultivating Pinus contorta in North Sweden, Sveriges Skogsvardsforbunds Tidskrift, 4:47.

Hamilton, G. J., and Christie, J. M., 1971, Forest Management Tables, Forestry Commission Booklet 34, HMSO, London, 201 p.

Hansson, O. V., 1989, Islensk trjaheiti, Arsrit Skograektarfelags Islands, 121-128.

Harding, T., 1972, Some observations on the wood of young Sitka spruce of different provenances gorwn at Bush (Midlothian) and Wykeham (Yorkshire), Forestry Commission Report on Forest Research, HMSO, London, 135-136.

Hillier, H. G., 1981, Hillier's Manual of Trees and Shrubs, David and Charles, Newton Abbot, 576 p.

Hunt, D. R., 1972, Reference list of conifers and conifer allies grown out of doors in the British Isles. In: Napier, E., ed., Conifers in the British Isles, Proc. 3rd Conifer Conference, The Royal Hort. Soc., London, 109-122.

Illingworth, K., 1980, Sitka spruce provenance trials three years after planting in British Columbia, Proc. I.U.F.R.O. Joint Meeting of Working Parties, Vancouver, Canada, 1978, Vol. 2:311-326, Ministry of Forests, Province of British Columbia, Canada, 65-94.

Kaasen, N. O., Skaret, G., and Kaasen, T., [In press], Afforestation of Lutz spruce, Sitka spruce, and Norway spruce in the Vesterallen Islands, Norway. In: Alden, J. N., Mastrantonio, J. L., and Ødum, S., eds., Forest Development in Cold Climates: Proceedings of an international symposium; June 18-24, 1991, [Laugarvatn, Iceland] NATO, Plenum, New York.

Keenleyside, J. C., 1985, Loch Ossian plantations Corrour, Inverness-shire, a fresh look at species and growth, August 1982, Scottish Forestry, 39(4):275-281.

Kraus, J., and Lines, R., 1976, Patterns of shoot growth, growth cessation and bud set in a nursery test of Sitka spruce provenances, Scottish Forestry, 30(1):12-24.

Lines, R., 1963, The choice of species, Forestry Commission Journal, No.32, 45–51.

Lines, R., 1980, The I.U.F.R.O. Experiments with Sitka spruce in Britain, Proc. I.U.F.R.O. Joint Meeting of Working Parties, Vancouver, Canada, 1978, Vol. 2:211-225, Ministry of Forests, Province of British Columbia, Canada.

Lines, R., 1985, The Macedonian pine in the Balkans and Great Britain, Forestry, 58(1):27-40.

Lines, R., 1987a, Choice of seed origins for the main forest species in Britain, Forestry Commission Bulletin, HMSO, London, 61 p.
Lines, R., 1987b, Forestry in the Faroe Islands, Scottish Forestry, 41(2):102-114.
Lines, R., 1990, Forestry in Iceland, Scottish Forestry, 44(2):85-94.
Lines, R., 1992, [In press], Experiments on lodgepole pine seed origins in Britain, Forestry Commission Bulletin 66, HMSO, London.
Lines, R., and Howell, R. S., 1963, The use of flags to estimate the relative exposure of trial plantations, Forestry Commission Forest Record 51, HMSO, London, 31 p.
Lines, R., and Mitchell, A. F., 1965, Results of some older Scots pine provenance experiments, Forestry Commission Report on Forest Research for 1964, HMSO, London, 172-194.
Lines, R., and Mitchell, A. F., 1969, Provenance: Japanese larch, Forestry Commission Report on Forest Research, HMSO, London, 43-44.
Low, A. J., 1987a, Tree planting prospects in Shetland and Orkney, Scottish Forestry 41(4):282-296.
Low, A. J., 1987b, Sitka spruce silviculture in Scottish forests. In: Henderson, D. M., and Faulkner, R., eds., Sitka spruce, Proc. Roy. Soc. Edinburgh, 93B:93-106.
Macdonald, J., Wood, R. F., Edwards, M. V., and Aldhous, J., 1957, Exotic Forest Trees in Great Britain, Forestry Commission Bulletin 30, HMSO, London, 167 p.
Madsen, S. F., 1984, Increment, frost damage and death in the I.U.F.R.O. international ten provenance experiment with Sitka spruce, established in spring 1975 in Denmark, Paper to IUFRO Working Party Meeting in Edinburgh, Sept. 1984.
Magnesen, S., 1980, A short note on Pinus contorta in West Norway, Proc. I.U.F.R.O. Joint Meeting of Working Parties, Vancouver, Canada, 1978, Vol. I:83-84, Ministry of Forests, Province of British Columbia, Canada.
Magnesen, S., 1986, The international Sitka spruce provenance experiment in West Norway, Norsk institutt for Skogsforsksning Rep., 1/86, Aas, Norway, 12 p.
Martinsson, O., 1980, How dangerous are the North American pathogens to the Scandinavian lodgepole pine. In: Pinus contorta as an exotic species, Research Note 30, Department of Genetics, Swedish University of Agricultural Sciences, Garpenberg, Sweden, 257-262.
McIntosh, R., 1983, Nitrogen deficiency in establishment phase Sitka spruce in upland Britain, Scottish Forestry, 37(3):185-193.
McVean, D. N., 1964, Woodland and Scrub. In: Burnett, J. H., ed., The Vegetation of Scotland, Oliver and Boyd, Edinburgh, 144-167.
Miller, K. F., 1985, Windthrow Hazard Classification, Forestry Commission Leaflet 85, HMSO, London, 14 p.
Miller, K. F., Quine, C. P., and Hunt, J., 1987, The assessment of wind exposure for forestry in upland Britain, Forestry, 60:179.
Mitchell, A. F., 1980, Native British trees, Research Information Note, 53/80, Forestry Commission, Farnham, Surrey, 4 p.
Ødum, S., 1989, A 10 year old trial with South American trees and shrubs with special regard to the Nothofagus spp. In: Hojgaard, A., Johansen, J., and Ødum, S., eds., A century of tree-planting in the Faroe Islands, Annales Societatis Scientarium Faroensis, Suppl. XIV, 125-156.
Ødum, S., Hansen, L., and Rasmussen, S., 1989, Annotated inventory of trees and shrubs in gardens and plantations. In: Hojgaard, A., Johansen J., and Odum, S., eds., A century of tree-planting in the Faroe Islands, Annales Societatis Scientarium Faroensis, Suppl. XIV, 80-124.
Redfern, D. B., 1982, Spring frost damage in Sitka spruce, Forestry Commission Report on Forest Research, HMSO, London, 27 p.
Roche, L., and Haddock, P. G., 1987, Sitka spruce in North America with special reference to its role in British forestry, Proc. Roy. Soc. Edinburgh, 93B, 1-12.

Sigurgeirsson, A., 1988, Stafafura a Islandi, Arsriti Skograektarfelags Islands, 3-36.
Silen, R. R., 1962, A discussion of forest trees introduced into the Pacific Northwest, Journ. of Forestry, 60(6):407.
Stevenson, A. W., and Mackay, A. F., 1991, Forestry in the Highland Region, Forestry, 64(2):103-140.
Stewart, G. G., 1962, Kergord plantations, Shetland, Forestry, 35(1):35-56.
Stoakley, J. T., 1979, Pine beauty moth, Forestry Commission Forest Record 120, HMSO, London, 11 p.
Svanbergsson, A., 1986, Skograektarfelag Reykjavikur 40 ara, Arsriti Skograektarfelags Islands, 5-18.
Tansley, A. G., 1949, The British Islands and their vegetation, Cambridge University Press, United Kingdom, 930 p.
Taylor, C. M. A., 1991, Forest Fertilization in Britain, Forestry Commission Bulletin 95, HMSO, London, 45 p.
Wood, R. F., 1955, Studies of northwest American forests in relation to silviculture in Great Britain, Forestry Commission Bulletin 25, HMSO, London, 43 p.
Worrell, R., and Malcolm, D. C., 1990, Productivity of Sitka spruce in Northern Britain, Forestry, 63(2):106.
Zehetmayr, J. W. L., 1954, Experiments in tree planting on peat, Forestry Commission Bulletin 22, HMSO, London, 110 p.
Zehetmayr, J. W. L., 1960, Afforestation of upland heath, Forestry Commission Bulletin 32, HMSO, London, 145 p.

SEED COLLECTIONS FROM NORTH AMERICAN TREES FOR MARGINAL SITES IN THE NORDIC COUNTRIES

Gisle Skaret
The County Forest Service of Nordland
Mosjøen, Norway

Ola Rosvall
The Institute for Forest Improvement
Savar, Sweden

INTRODUCTION

All Nordic countries have marginal areas where forestry cannot be practiced due to the harsh climate, a lack of native commercial forests, slow growth, poor regeneration of the indigenous species, or low quality of the forest products. Harsh climate and absence of commercial forests are especially pronounced on Greenland, the Faroe Islands, Iceland, and northern Norway.

Large areas of the boreal forests of Sweden and Finland are also slow-growing and difficult to regenerate. A paucity of native flora and a lack of tree species with desirable growth and wood quality characteristics are common features. Research and experience from commercial silviculture have shown that use of exotic tree species can improve forest yield. The successful introduction of lodgepole pine (Pinus contorta var. latifolia Engelm.), Sitka spruce (Picea sitchensis (Bong.) Carr.), and Lutz spruce (Picea xlutzii Little), the natural Sitka x white spruce (Picea glauca (Moench) Voss) hybrid, to the Nordic countries is well known. Research to optimize provenance selections and make future tree breeding possible for these and other species has intensified. In addition, large seed supplies are needed for future afforestation and reforestation programs. These objectives can be achieved with an extensive research program designed to test provenances and families from the extreme northern limits of their ranges. Ultimately, such research will improve afforestation success and increase the yield of forestry products in marginal areas.

To accomplish these objectives, a joint Nordic Research Program, financed by the Nordic Forest Research Cooperation Committee of the Nordic Council of Ministers, was formed in 1987. A summary of seed and other reproductive material collected from species in western North America, and plans for testing the materials in the Nordic countries, follows.

Forest Development in Cold Climates, Edited by J. Alden *et al.*
Plenum Press, New York, 1993

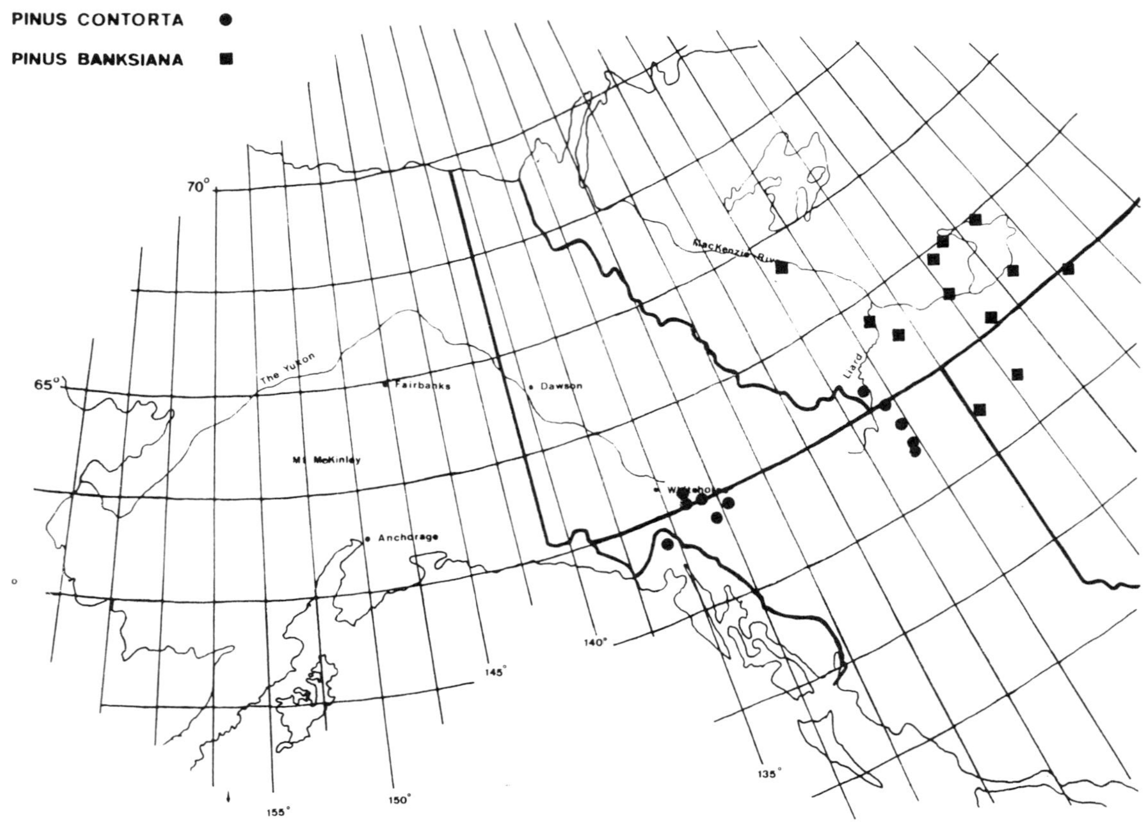

Fig. 1. Provenances of 1987 Pinus banksiana and P. contorta seed collections in Alaska, Yukon, and Northwest Territories.

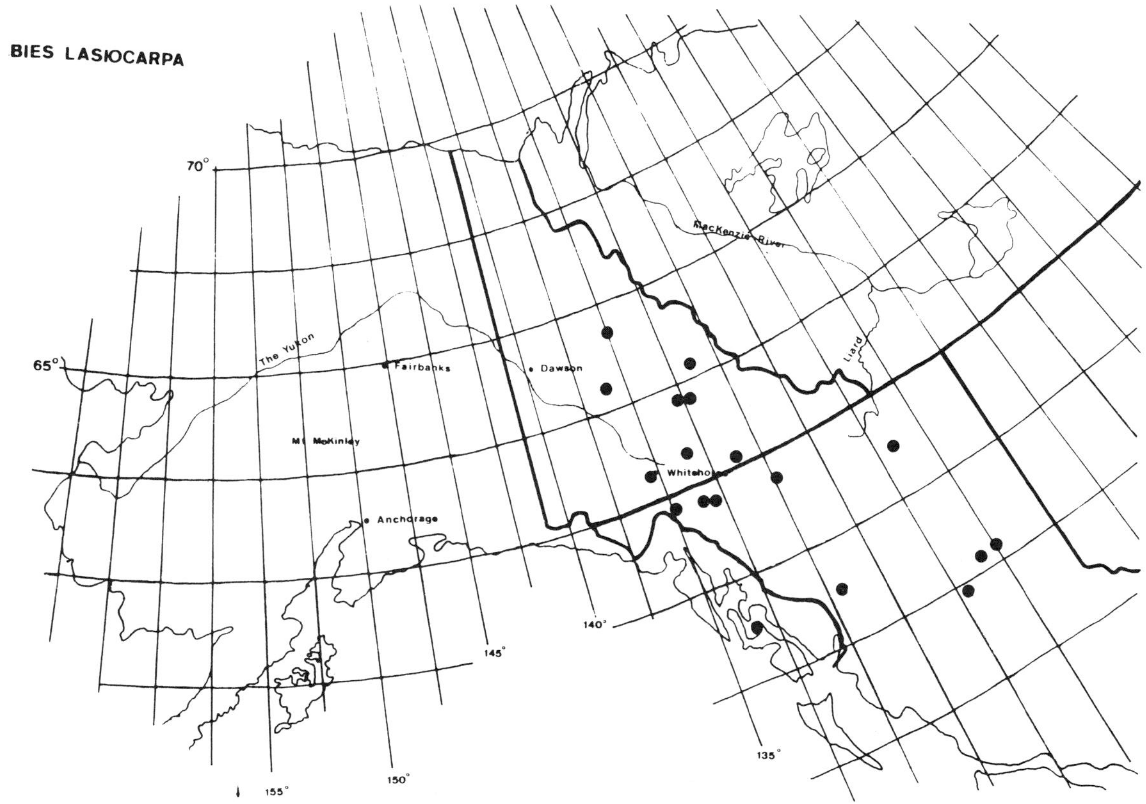

Fig. 2. Provenances of 1987 and 1988 Abies lasiocarpa in Alaska, British Columbia, and the Yukon.

MATERIAL AND METHODS

Species and Provenance Origin

The following species were of major interest and identified for extensive seed collection: lodgepole pine (Pinus contorta var. latifolia), shore pine (Pinus contorta Doug. ex Loud. var. contorta), jack pine (Pinus banksiana Lamb.), subalpine fir (Abies lasiocarpa (Hook.) Nutt.), white spruce (Picea glauca (Moench) Voss), black spruce (Picea mariana (Mill.) B.S.P.), and tamarack (Larix laricina (Du Roi) K. Koch).

Only limited seed collections were possible from Picea sitchensis, Picea xlutzii, western hemlock (Tsuga heterophylla (Raf.) Sarg.), mountain hemlock (Tsuga mertensiana (Bong.) Carr.), western red cedar (Thuja plicata Donn ex. D. Don), Alaska cedar (Chamaecyparis nootkatensis (D. Don) Spach), Pacific silver fir (Abies amabilis (Dougl.) ex Forbes), black cottonwood (Populus trichocarpa Torr. & Gray), willow (Salix L.) spp., alder (Alnus Mill.) spp., and birch (Betula L.) spp.

The area of primary interest for the seed collection was between 65° and 70° N. latitude. The phenology of provenances suitable for research in the Nordic countries must be synchronized with the growing season of the planting site. Hardiness and the timing of growth termination are especially important for adaptation and survival of the introduced species. Timing is regulated by photoperiod and temperature in spring and late summer. Latitude determines the photoperiod, which means that provenances adapted to marginal areas in the Nordic countries must come from the same latitudes, that is, northern British Columbia, Alberta, the Yukon and Northwest Territories, and Alaska. This conclusion is supported by many provenance experiments.

Further east in North America, preferred forests are limited to latitudes less than 55° N. The collection area also has the same southern border as the commercial seed zone for Pinus contorta and Picea sitchensis in the Nordic countries. Both species have been tested extensively and used for afforestation in the Nordic countries.

Plans were to collect seed (and pollen for Picea glauca) for producing 1,100 seedlings per tree from 15 to 20 trees per provenance. The number of provenances varied with species and ranged from 10 to 30. Altogether, plans were to collect seed from 106 different provenances.

Stands selected for seed collection were easily accessible from roads and produced good cone crops. Remote stands, however, were also included in the collection plans. Stand age, quality, growth, and size were not major selection criteria. Stands were sampled at high elevations when they were easily accessible from roads, but populations from northern latitudes were sampled in preference to populations from high elevations.

A very quick selection procedure was adopted for sampling trees within stands. Trees selected for seed collection were from the dominant crown class, average or better in stem form or quality, and were growing at least 100 meters apart. The distance between stands sampled for each species averaged 220 kilometers.

Fieldwork

Seed collections were initiated for the serotinous species Pinus contorta, P. banksiana, and Picea mariana. Vigor and germination of seed stored in serotinous cones decrease

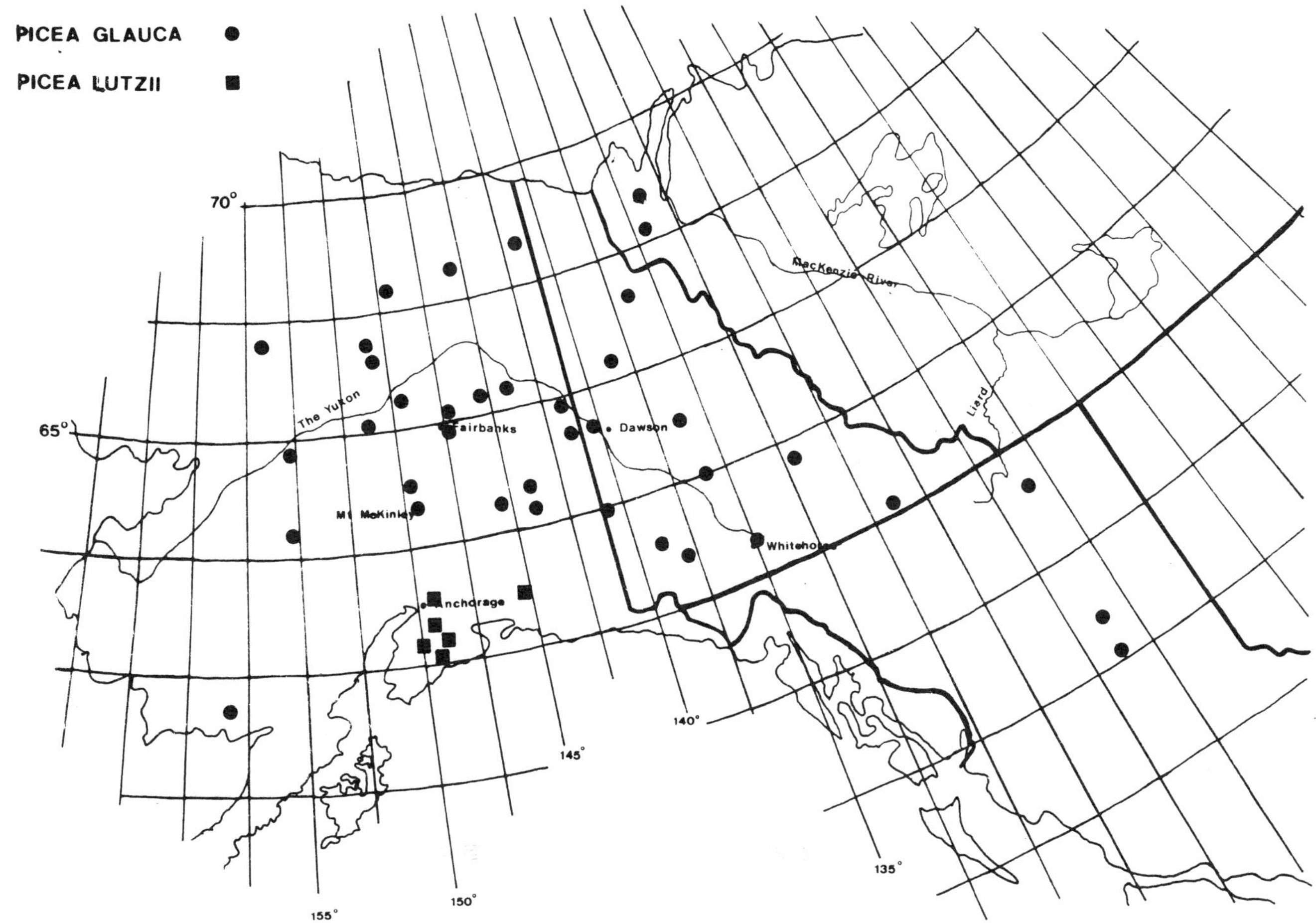

Fig. 3. Provenances of 1987 and 1988 Picea glauca and P. xlutzii seed collections in Northwest Territories.

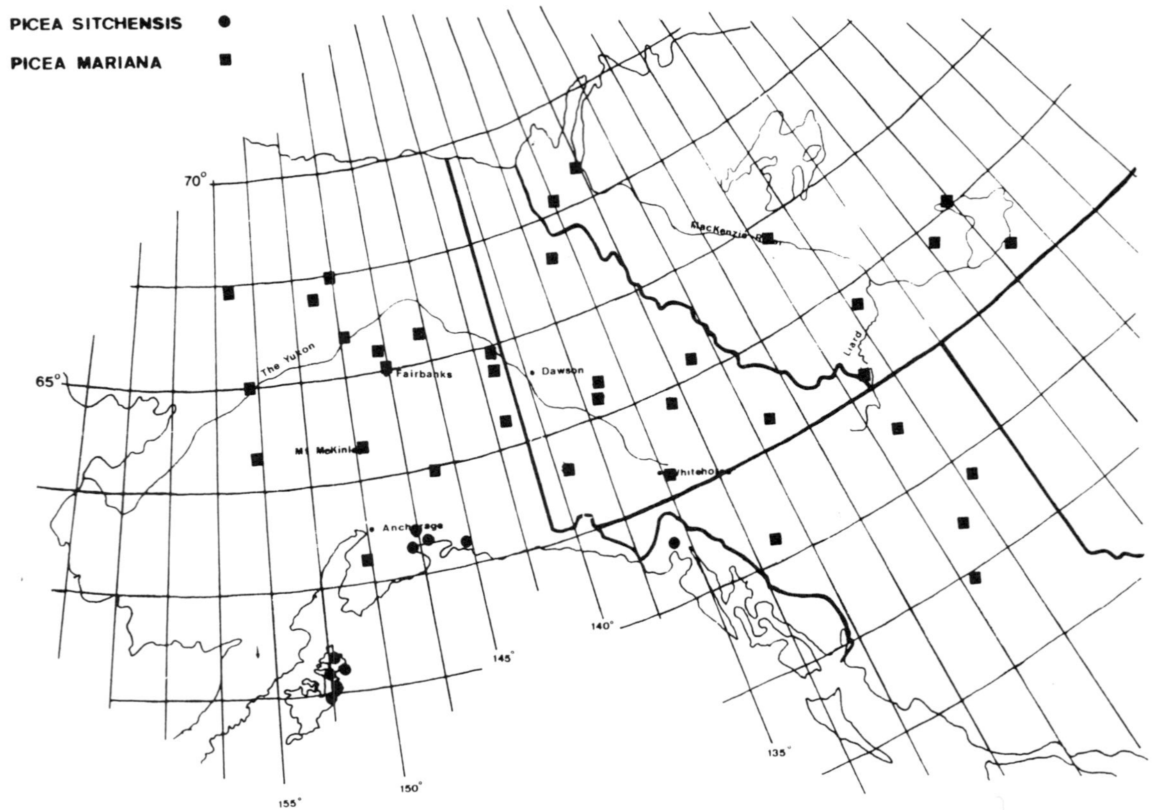

Fig. 4. Provenances of 1987 and 1988 Picea mariana and P. sitchensis seed collections in Northwest Territories.

slowly with time, so that cones several years old can yield seed that germinates as well as seed from fresh cones. Cones on the non-serotinous species had to be collected during the two-week period after seed maturation and before seed dispersal in late summer and early autumn.

Cones were collected from either felled trees or tops cut from standing trees. Branches with staminate cones, 8 to 10 meters from the ground, were pruned for Picea glauca pollen collections. The location, species composition, height, age, density, and soil conditions were documented for each stand.

The staminate cones were dried for dehiscence and the pollen was cleaned of foreign material and dried for storage. Seed cones were also dried for seed extraction at Yellowpoint Propagation Facilities in British Columbia, the University of Alaska-Fairbanks, and the State Forest Plant Breeding Station in Denmark. The Forest Research Station in Ruotsinkyla, Finland, was responsible for final seed cleaning, for quality examination, and inventory.

A Nordic seed bank for storing the collections was established at the Kolari Research Station in northern Finland. In a number of stands, seedlings were dug, packaged in moist moss from the forest understory, and delivered to the closest airport for transport to the Arboretum Nursery at Hørsholm, Denmark. Most collections were made in areas that were accessible by vehicle. In Alaska, however, roads were limited and a small aircraft was used for transportation.

Many people and organizations were involved in the field work. Two collection teams worked during each of two field seasons. In the Northwest Territories, Alberta, British Columbia, and the Yukon, cones were collected from P. banksiana and P. contorta during June and July 1987. In Alaska, the Yukon, and British Columbia, stands and trees were selected and cones were collected in September 1987 and 1988. The Picea glauca pollen was collected by one team during a three-week period from May 15 to June 5, 1990.

RESULTS

During 1987 and 1988, seeds were collected from 10 conifers and 141 stands (Table 1). In addition to the seed collections listed in Table 1, seedlings of Chamaecyparis nootkatensis, Picea glauca, and six additional conifers (Table 2), and seed and seedlings of 34 broadleaf deciduous trees and shrubs (Table 3), were collected for shipment to Hørsholm, Denmark.

Provenances of the major conifers in Tables 1 and 2 are shown in Figures 1 to 5. Provenances of Betula papyrifera Marsh. and the broadleaf trees and shrubs are shown in Figure 6. Fifty-three pollen samples were collected from thirteen Picea glauca stands in southcentral and interior Alaska (Fig. 7).

DISCUSSION

Tree Species and Distribution of Stands

The aim of the collection project was to procure seed from as many trees as possible in a limited time from a very large geographic area at high latitudes in North America. Rapid

Table 1. Seed Collections in 1987 and 1988 from Coniferous Species

Tree Species	Number of stands sampled	Number of trees sampled
Abies lasiocarpa	40	352
Chamaecyparis nootkatensis	1	1
Juniperus horizontalis	1	1
Larix laricina	7	99
Picea glauca	39	737
Picea xlutzii	5	108
Picea mariana	36	743
Picea sitchensis	10	137
Pinus banksiana	12	242
Pinus contorta	10	129

Table 2. Plant Collections in 1988

Tree species	Number of stands	Number of plants
Abies amabilis	2	75
Abies lasiocarpa	5	1849
Chamaecyparis nootkatensis	1	6
Picea glauca	2	1031
Picea sitchensis	4	314
Thuja plicata	1	46
Tsuga heterophylla	2	292
Tsuga mertensiana	4	482

tree selection and cooperation from provincial, state, and federal forestry organizations of North America enabled the Nordic Forest Research Council collection teams to achieve these goals. For economic and genetic reasons, field time was devoted to tree felling, cone collection, and cone transportation rather than to evaluating phenotypic differences between stands and trees and collecting descriptive field information. Intensive phenotypic selection is not efficient in wild populations, and genes of the unknown pollen parents are of the same interest as genes of the phenotypically variable seed parents.

Table 3. Bulk Seed and Plant Collections for Broadleaf Trees and Shrubs

Species	Number of seed samples	Number of plants
Alnus crispa (Ait.) Purch, American green alder	1	
Alnus sinuata (Reg.) Rydb., Sitka alder	7	
Alnus tenuifolia Nutt., thinleaf alder	2	
Amelanchier alnifolia Nutt., Saskatoon service-berry	1	
Amelanchier florida Lindl., Pacific serviceberry	1	
Betula glandulosa Michx., resin birch	1	
Betula glandulosa x papyrifera, Yukon birch	1	
Betula papyrifera Marsh., paper birch	2	18
Cornus stolonifera Michx., red-osier dogwood	1	
Elaeagnus commutata Bernh., silverberry	2	
Oplopanax horridus (Sm.) Miq., Devils club	5	
Populus balsamifera L., balsam poplar	1	3
Poplus balsamifera x trichocarpa	1	86
Populus tremuloides Michx., quaking aspen	1	42
Populus trichocarpa Torr. & Gray., black cotton-wood	3	161
Potentilla fruticosa L., bush cinquefoil	3	
Ribes bracteosum Dougl., stink currant	2	
Ribes lacustre (Pers.) Poir., swamp gooseberry	1	
Ribes laxiflorum Purch., trailing black currant	1	
Rosa acicularis Lindl., prickly rose	5	
Rosa nutkana Presl., Nootka rose	1	
Rubus parviflorus Nutt., western thimbleberry	1	
Rubus pedatus Sm., five-leaf bramble	1	
Rubus spectabilis Pursh., salmonberry	1	
Salix alaxensis (Anderss.) Cov., feltleaf willow	2	75
Salix arbusculoides Anderss., littletree willow	1	33
Salix sitchensis Sanson., Sitka willow	1	46
Salix sphenophylla A. Skvortz., wedgeleaf wil-low	1	20
Shepherdia canadensis (L.) Nutt., buffaloberry	6	

Table 3 (Continued). Bulk Seed and Plant Collections for Broadleaf Trees and Shrubs

Species	Number of seed samples	Number of plants
Sorbus sitchensis Roem., Sitka mountain-ash	4	-
Spirea beauverdiana Schneid., Beauverd spirea	2	-
Vaccinium alaskaense Howell., Alaska blueberry	2	-
Vaccinium ovalifolium Sm., early blueberry	2	-
Vaccinium parvifolium Sm., red huckleberry	2	-
Viburnum edule (Michx.) Raf., highbush cranberry	4	-

The major species were collected in the following order:

1. Pinus banksiana

Pinus banksiana seeds were collected in the Northwest Territories and Alberta during June and July 1987. The collections were well distributed, especially in the Northwest Territories, where 10 of the 12 stands sampled for this species were located. At many localities, stands were easy to find. The northernmost stand was near the Blackwater River, Northwest Territories, at latitude 63°58′ N., 300 kilometers north of Fort Simpson. The area is accessible only by helicopter.

The southernmost stand of P. banksiana from which collections were made was at Rainbow Lake, Alberta, latitude 58°21′ N. This locality is within the zone of hybridization with Pinus contorta.

2. Pinus contorta

Stands of Pinus contorta were sampled in the border area between Yukon and British Columbia for provenance research in Iceland. The stands were at Bennett Lake, Yukon, at latitude 60°06′ N. and between 134°42′ and 133°47′ W. longitude, and within the introgression zone with shore pine (Pinus contorta Doug. ex Loud. var. contorta). Only 50 percent of the trees sampled were serotinous. The northern range of P. contorta was sampled in 1978 in connection with the seed collection for the breeding program in the Nordic countries.

3. Picea mariana and Picea glauca

Picea glauca and P. mariana comprised 75 stands sampled and more than half of the total collections. The very northern limit of the spruce in the Mackenzie River Delta, Northwest Territories, could not be sampled because of bad weather. Consequently, the northernmost samples were from a Picea glauca stand south of Inuvik at latitude 68°20′ N.

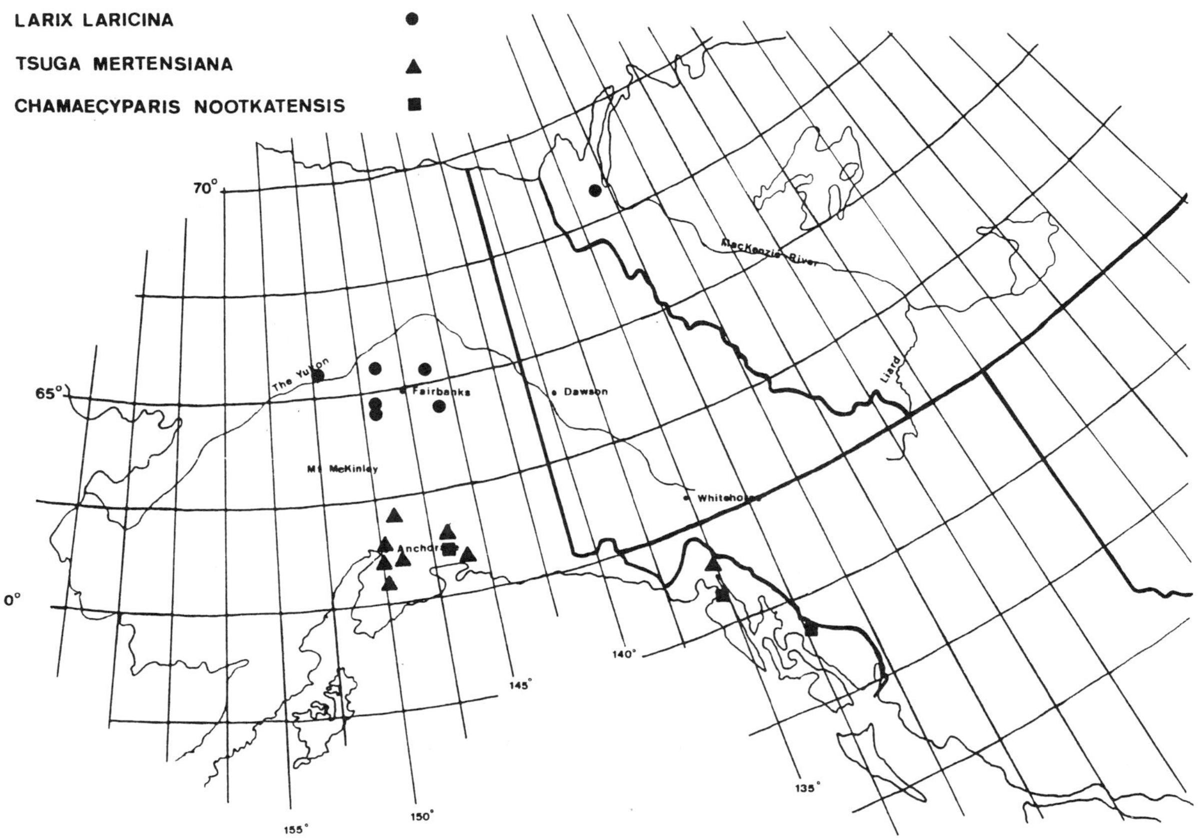

Fig. 5. Provenances of 1987 and 1988 Larix laricina, Tsuga mertensiana, and Chamaecyparis nootkatensis seed collections from Alaska and Northwest Territories.

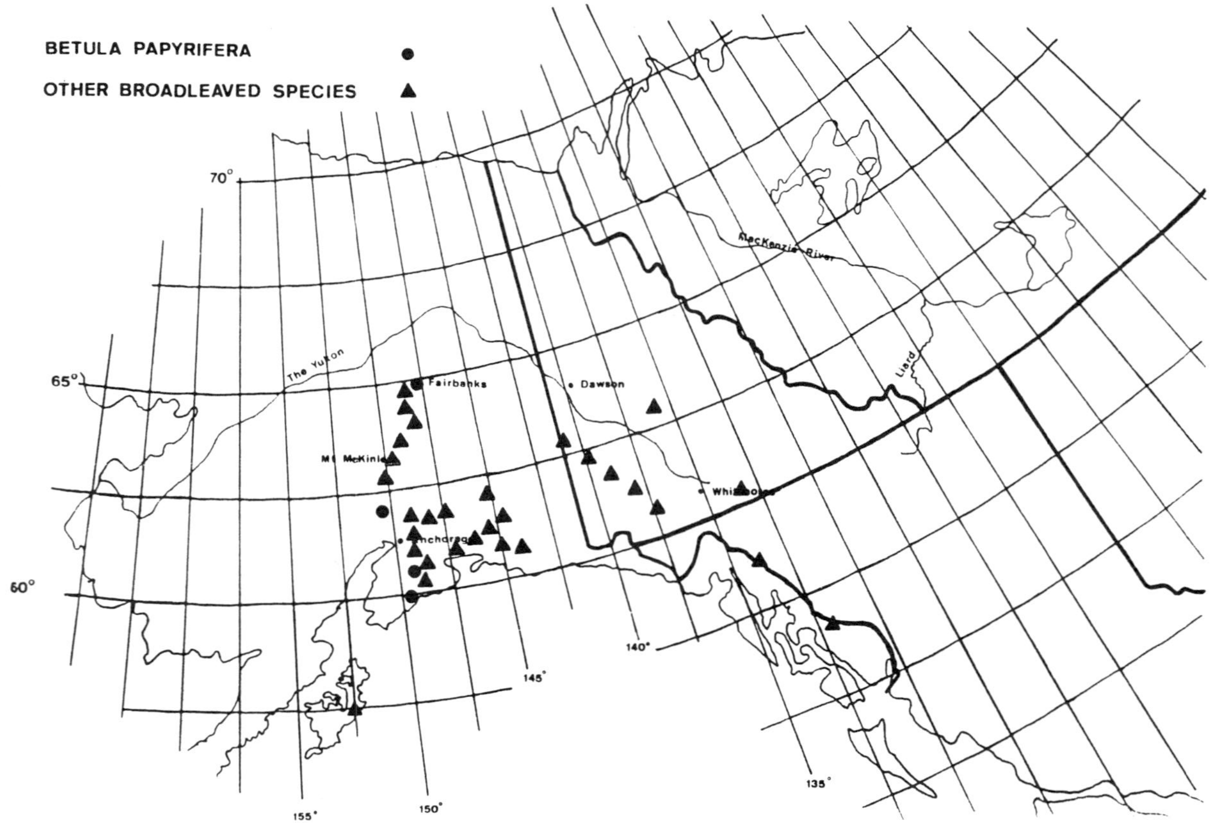

Fig. 6. Provenances of 1987 and 1988 seed and seedlings of the broadleaf deciduous species collected in Alaska and Yukon for provenance research in the Nordic countries.

The forests were large and growth rates were surprisingly fast at these latitudes and conditions. Some of the P. glauca trees sampled were 20 meters tall.

Air travel was necessary to sample the stands in Alaska. The northernmost stand sampled was at Mancha Creek, latitude 68°12′ N. This population of P. glauca continues along the Firth River into Canada and into the side valleys of the Firth, approaching to within 32 to 48 kilometers (20 to 30 miles) of the Beaufort Sea. In 1987, the spruce seed moth (Laspeyresia youngana Kearfott) and the inland cone rust (Chrysomyxa pirolata Wint.) caused severe loss of white spruce seed in western Alaska; consequently, adequate samples do not exist for this area.

4. Picea sitchensis and Picea xlutzii Little

Picea sitchensis was introduced to northern regions of Scandinavia about 70 years ago. It was used extensively for afforestation and established in many provenance trials. For this reason, collections were limited to those stands that had not been sampled in Alaska and tested in Scandinavia.

Picea xlutzii was sampled from hybrid populations on the Kenai Peninsula, Alaska. The degree of P. sitchensis introgression decreased with the distance from the sea and with increasing altitude.

5. Abies lasiocarpa

Abies lasiocarpa seed was collected in southeast Alaska, the Yukon, and British Columbia in 1987 and 1988. The species is a typical subalpine tree that forms the altitudinal timberline in mountainous areas. In the Yukon, stands are found 1,250 meters above sea level as far north as Keno Hill at 64° N. latitude. In northern British Columbia, stands were found at both high and low elevations. The most extensive stand of A. lasiocarpa was sampled at Meziadin Lake, British Columbia, at latitude 56°08′ N., longitude 129°25′ W.

6. Larix laricina

In the northern portions of its range, Larix laricina is usually associated with the broadleaf species in open stands and with Picea mariana on muskeg habitat. Larix laricina was sampled from six stands in interior Alaska, but the northernmost collection was made south of Inuvik, Northwest Territories, at latitude 67°57′ N.

PROJECT COSTS

The total cost of the project, exclusive of salaries, was 900,000 Norwegian kroners, approximately $150,000 U.S. Each seed sample, on the average, cost about 360 Norwegian kr. ($60 U.S.). This cost included expenses for both collecting and extracting the seed.

Salaries for the Nordic expedition were paid by their respective countries. The final personnel costs for seed viability analysis were financed by Finland.

FURTHER PLANS

The four objectives of future experimental work are (1) to provide the best provenances

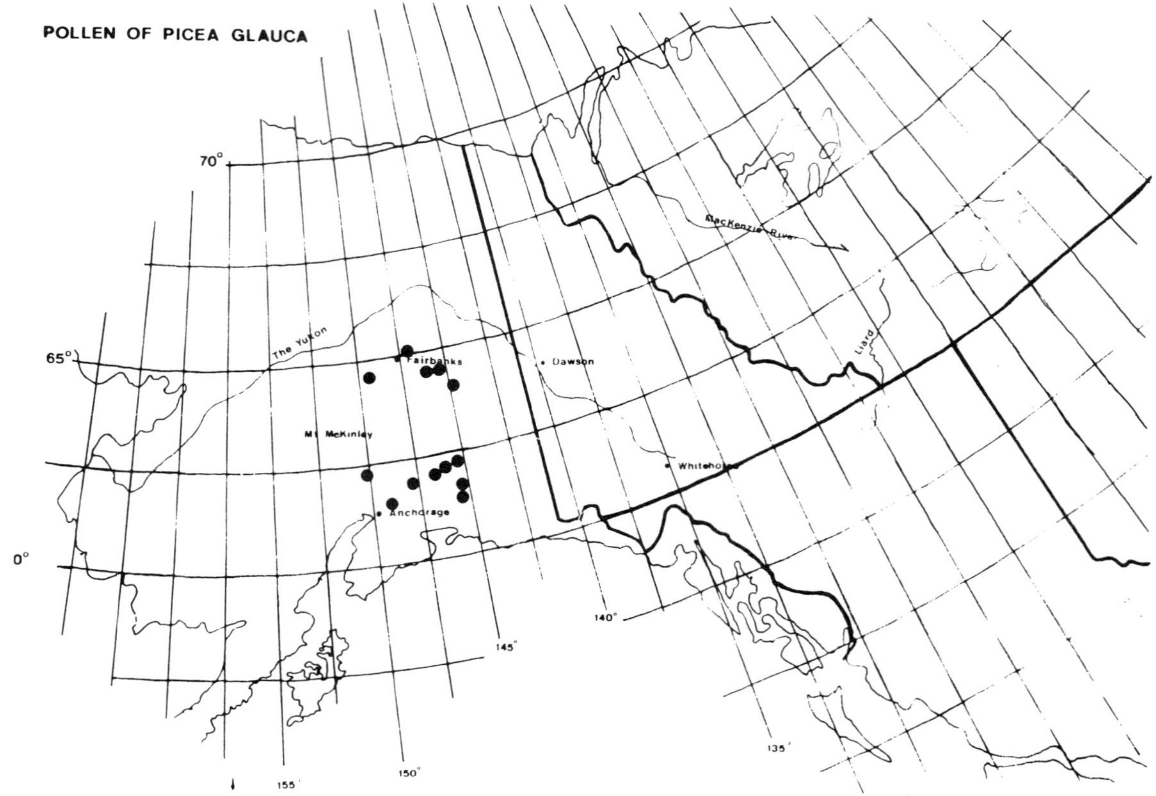

Fig. 7. Provenances of the 1990 P. glauca pollen collections in Alaska for tree breeding in the Nordic countries.

possible for afforestation of harsh climates that limit forestry in the Nordic countries; (2) to insure future genetic improvement of the introduced species by planting base populations for different selections; (3) to determine the genetic variation in environmental fitness and growth traits; and (4) to create reference material of species and provenances on contrasting sites. Reference material would be used, for example, to facilitate exchange of breeding material for reforestation of the exotic and native tree species and for other scientific purposes.

Growth rhythm and hardiness responses of families to temperature and photoperiod will be studied in controlled environments. Growth and timber quality and tolerance of environmental extremes will be studied in long-term field experiments. Provenance test designs will include small experiments on many sites (many provenances, 20 seedlings per provenance, and 10 to 20 sites). Family tests of variation within provenances will total 20 to 50 seedlings per family on at least four sites in a factorial subdivision of harsh and mild sites and north and south locations. In some cases, a third factor is the cline in the continental–maritime climate in Scandinavia. As knowledge accumulates, seed sources can be selected and planted in large plot experiments to assess silvicultural treatments and long-term yields.

The collection of P. glauca pollen is only the first step in breeding P. xlutzii. The next step is to complete the controlled crosses and establish systematic progeny tests of the hybrids. For the next generation of improvement, scions from outstanding phenotypes in the best crosses will be collected for vegetative propagation (grafting) and future breeding.

ACKNOWLEDGEMENTS

The seed collection project was successful for many reasons. The field work was planned for three seasons and was estimated to require five to ten years, but it was accomplished in two years due to unusually warm summers. In addition to the skillful work of the Nordic collaborators, the project coordinators acknowledge the help of various United States and Canadian organizations, ranging from the federal labor and forest authorities to local officers at remote customs stations and forest headquarters. The cooperation of research colleagues in North America contributed significantly to the success of the project. Dr. John Alden of the Institute of Northern Forestry, Fairbanks, Alaska, Tom Malone of the University of Alaska–Fairbanks, and Don Piggot of Yellowpoint Propagation, Ladysmith, British Columbia, were particularly helpful.

THE USES OF LIGNOSES IN HORTICULTURE IN ICELAND

Jóhann Pálsson

Municipal Park Commissioner
Reykjavík, Iceland

The history of continuous horticulture in Iceland extends over little more than 100 years. The Icelandic flora is relatively lacking in indigenous lignoses (Fig. 1), so it was necessary for the pioneers in horticulture to obtain trees and shrubs from abroad. A century ago, overseas contacts were almost entirely limited to Denmark, which was the sole overseas source for such material. The material available was the same species and cultivars found in northwest Europe.

The Icelandic climate and soil differ considerably from those of neighboring countries. The cold, maritime climate with low summer temperatures and variable winters is particularly unfavorable to most lignoses. In addition, Iceland is very windy, a result of the low-pressure zones that frequently pass over Iceland on their west-to-east route across the North Atlantic Ocean.

Many of the first experiments in gardening failed and did little to increase the expectations of the inhabitants. Icelanders had for generations fought against major soil erosion in a treeless, overgrazed country, and they had little faith in the possibilities of horticulture.

Trees like Norway spruce (Picea abies (L.) Karst.), white spruce (Picea glauca (Moench) Voss), common horse-chestnut (Aesculus hippocastanus L.), and European beech (Fagus sylvatica L.) lingered on as krummholz or died. Other species, such as wyche elm (Ulnus glabra Huds.), sycamore (Acer psuedoplantanus L.), Swedish whitebeam (Sorbus intermedia (Ehrh.) Pers.), Finnish whitebeam (Sorbus x hybrida L.), and the indigenous mountain ash (Sorbus aucuparia L.) grew over time, and now many attractive trees about ten to thirteen meters high are found in the old gardens (Benedikz, 1991).

Horticulture in Iceland progressed very slowly until after World War II, when the situation started to change (Fig. 2). Three circumstances are mainly responsible for the resurgent interest in horticulture. First, the oldest gardens had by that time become quite

Forest Development in Cold Climates, Edited by J. Alden *et al.*
Plenum Press, New York, 1993

attractive and stood out as examples of what was possible to accomplish, and the pioneers had by then gained valuable experience. Second, the economy of the country and the demographic distribution of the inhabitants had changed considerably. More people lived in villages or towns and had a small piece of land around their houses to cultivate. Some even had summerhouses in the country and thus had an opportunity to do some landscaping.

Finally, after gaining political independence in 1944, Icelanders established direct contacts with other countries. These contacts gave individuals and organizations involved in forestation efforts in Iceland the opportunity to send tree-seed-collecting expeditions to geographically remote places where the prevailing climate was similar to that of Iceland. Alaska has been the most important source of tree seed since 1943 and 1944, when the first collections were carried out. Since 1945, numerous excursions have been made to various regions in Alaska and other parts of North America, as well as to Scandinavia and the USSR (Blöndal, 1977).

The importation of trees has first and foremost been handled by the Icelandic Forestry Service, which initially concentrated mostly on species suitable for timber production. Nevertheless, this new material was also appreciated for gardening and landscaping. The majority of trees and taller shrubs available in nurseries today were commercially introduced in Iceland after 1950. The most popular trees, apart from the indigenous mountain ash and birch (Betula pubescens Ehrh.), are Sitka spruce (Picea sitchensis (Bong.) Carr.), Engelmann spruce (Picea engelmannii Parry ex. Engelm.), lodgepole pine (Pinus contorta Dougl. ex. Loud.), Siberian larch (Larix sibirica Ledeb.), black cottonwood (Populus trichocarpa Torr. & Gray), and goat willow (Salix caprea L.).

Fig. 1. Reykjavík in about 1872, before the initiation of horticulture.

Fig. 2. Reykjavík today. Public parks and gardens are established features in the environment.

A distinctive feature of older gardens in Iceland is that the trees stand too close together. The amateur gardeners did not realize how tall the young plants would grow, and they wanted to obtain tight and high vegetation as soon as possible to provide shelter for people and sensitive plants. The result was shady gardens.

It was only later that the usefulness of shrubs for providing shelter was generally understood. For shrubs as well as trees, it was necessary to look for material suitable for Icelandic conditions. It was also important to find species and forms that would suit all the different uses to which shrubs can be put in horticulture and landscaping: shelter, hedging, decoration, borders, and ground cover.

The variety of shrubs in Icelandic gardens today is quite extensive. Many species within the genera Lonicera, Sorbus, Spirea, and Ribes have proved particularly hardy. Similarly, some wild roses (Rosa L.) and lilacs (Syringa L.) have done very well (Figs. 3 and 4). Several species that are by now common in Icelandic gardens are unknown outside botanical gardens in neighboring countries. Among these are Úlfareynir (Sorbus x hostii (Jacq. f.) K. Koch.), a large shrub or small tree bearing pink flowers in spring and large scarlet berries in autumn (Fig. 5); honeysuckle (Lonicera deflexicalyx Batal.), a large shrub that can reach a height of four to five meters and has a profusion of golden flowers in spring (Fig. 6); Lonicera hispida Pall., a one and a half meter high shrub with large, funnelform flowers in spring and oblong bright-red fruit about two centimeters long (Fig. 7); and the indigenous wooly willow (Salix lanata L.), which is in great demand as a low hedge or border.

Fig. 3. Taeckholm's rose (Rosa pimpinellifolia L.), a seedling from the Botanical Garden of Reykjavík. The abundant semi-double white flowers have shades of pale pink at the edges of the petals. This clone is now frequently planted in the municipal parks of Reykjavík.

Fig. 4. Lilacs (Syringia L.) from the series Villosae Schneid. Many species and hybrids from this series are popular in Icelandic gardens.

Fig. 5. Úlfareynir (Sorbus x hostii (Jacq. f.) K. Koch.), a very popular little tree or shrub with pink flowers and scarlet fruits.

Fig. 6. Honeysuckle (Lonicera deflexicalyx Batal.), a hardy and profusely flowering shrub about four meters high.

The willows are perhaps the most characteristic feature of modern Icelandic gardens. Goat willow (Salix caprea L.), dark-leaved willow (S. myrsinifolia var. borealis (Fr.) Hyl.),and bay willow (S. pentandra L.) have the ability to turn into trees 10 to 12 meters high, although the willows are primarily used to create fast-growing hedges to provide shelter. The above-mentioned types are popular for that purpose, together with felt-leafwillow (S. alaxensis (Anderss.) Cov.) and an indigenous hybrid Salix cv. "Brekkuvíðir," which can grow to form most attractive one-to-two meter high hedges in even the windiest locations. The botanical gardens founded in Reykjavík and Akureyri about thirty years ago have provided a large part of the various shrubs available on the market today (Fig. 2).

Today, when expeditions are sent out to collect seeds from trees, the shrubs are not neglected. Two recent expeditions, the first led by Óli Valur Hansson in 1985 (Árnason et al., 1986), brought among other things valuable collections of alder (Alnus Mill.), poplars (Populus L.), and willows from Alaska and the Yukon. The second expedition, led by Vilhjálmur Lúdvíksson in 1989, travelled to the vicinity of Magadan and the Kolyma River in northeast Siberia and brought back many interesting species such as Chosenia bracteosa (Trautv.) Nakai and Salix berberfolia Pall. This material is at present being cultivated in

Fig. 7. Lonicera hispida Roem. & Schult., a compact shrub about 1.5 meters high with white funnelform flowers and bright red oblong fruits up to two centimeters long.

experimental fields. All new species of lignoses that prove hardy enough to tolerate the conditions in Iceland are warmly received, and, with a little imagination, the right niche for them is usually found.

REFERENCES

Árnason, Á., Gudmundsson, B., and Hansson, Ó. V., 1986, Plant Collecting in Alaska and Yukon, 1985, Icelandic Forestry Association Reports, 59-60.

Benedikz, Th., 1991, Stærðýmissa trjátegunda á Íslandi (The height of different species of trees in Iceland), 1991, Icelandic Forestry Association Reports, 65-69.

Blöndal, S., 1977, Innflutningrur trjátegunda til Íslands (The import of tree species to Iceland), Skógarmál, Reykjavík, 173-223.

ACCELERATING ADAPTATION OF TREES AT TREE LIMITS BY SELECTIVE BREEDING

Dag Lindgren

Department of Forest Genetics and Plant Physiology
Swedish University of Agricultural Sciences
Umeå, Sweden

INTRODUCTION

This paper focuses more on theoretical deductions than on empirical results, and deals more with long-term strategies than with short-term tactics. Speculation about breeding for tree-limit conditions is an interesting challenge and intellectual exercise, but under many circumstances it is hardly realistic. I reflect about what the tree limit environment and adaptation should be, although some of these reflections might be dubious or oversimplified. In addition, I identify patterns of adaptation to the tree-limit environment, with the caveat that several of these patterns can be challenged by exceptions.

Research on Breeding Trees for Cold Climates

The Nordic countries constitute a majority of the eight nations north of latitude 60° N.: the United States (Alaska), Canada, Russia, Iceland, Denmark (Greenland), Norway, Sweden, and Finland.

Several factors make northern Scandinavia suitable for commercial forestry. The climate is warmer on the northern Scandinavian peninsula than at comparable latitudes elsewhere because of the Gulf Stream. In addition, because no place is far from the coast, the cost of transporting resources is acceptably low. Forests are accessible, and the population and infrastructure are sufficient to support forest operations. On the other hand, the population is small enough so demands on the land for purposes other than forestry are limited.

The forest industry is important in the Nordic countries. Research has focused on species, provenances, and breeding for reforestation, and much work on breeding trees for reforestation of cold climates has been accomplished.

In contrast, North America has been regarded as a land of nearly unlimited forest resources and has had little need to study reforestation problems near tree limits. The

Forest Development in Cold Climates, Edited by J. Alden *et al.*
Plenum Press, New York, 1993

driving forces for forest research in North America have been harvesting, products, marketing, and recreation, rather than afforestation of marginal areas.

In central and southern Europe, forest limits occur in alpine environments, where forested areas are small and the terrain often too steep for production forestry. Reforestation might be undertaken for protection from avalanches or erosion, to enhance scenic values, or for other environmental reasons. In Iceland, the major purpose of forestry is land reclamation. In China and New Zealand, high-altitude mountain slopes are reforested for protection against erosion. The natural vegetation in New Zealand is not well adapted to high altitudes, and thus exotics have been successful above timberlines (Ledgard, 1980). Little has been published on the genetics of trees at tree limits, however.

THE TREE-LIMIT ENVIRONMENT

The tree-limit environment and its interactions with trees have been the subject of numerous books, symposia, and investigations (Arno and Hammerly, 1984; Benecke and Davis, 1980; Bäckström, 1984; Kaurin et al., 1985; Lindgren, 1986; Nilsson, 1985; Salzburg et al., 1986). I use tree limit to mean the limit of the range of a species as a forest tree in an undisturbed situation. "Tree limit," "timberline," or "forestline" is the area beyond existing closed forests. Above the closed forest, trees often exist as bushes, sometimes known as elfinwood or "krummholz," although krummholz often has genetic implications. Treeline is where trees disappear from the forest-tundra ecotone, even if the species extends further in bush form (the "Kampfzone").

Tree-limit conditions depend on latitude. At high latitudes, the differences in temperature and night length between the seasons are larger and more important than at low latitudes in initiating the onset and termination of growth during the annual dormancy and growth cycle. Thus, low summer temperature and critical night length are more important at high latitudes than at low latitudes.

Low Summer Temperature

From a practical point of view, low temperature during the growing season combined with a short growing period is the most limiting factor for treeline species. Tree limit often coincides with the 10 °C isotherm of average temperature of the warmest month, which is usually July (Arno and Hammerly, 1984; Tuhkanen, 1982). Other factors might appear important locally, but are generally speaking not as important.

Variation around a polynomial regression of tree limits for Scots pine (Pinus sylvestris L.) and Norway spruce (Picea abies (L.) Karst.) on latitude for the entire Swedish mountain chain was almost ±200 meters (Kullman and Hofgaard, 1987), corresponding to a difference of about ±1 °C in mean July temperature. Part of the deviation was probably associated with temperature, because summer temperature is a function of the degree of continentality as well as of latitude and altitude. Thus, factors unrelated to summer temperature could explain about as much of the difference in Swedish tree limits as is explained by a temperature difference of ±1 °C.

Even if summer temperature is a key factor in the position of treeline, its influence is mainly indirect. Air temperature in the range 5 to 10 °C is, in itself, not an obstacle for growth, but low temperatures cause secondary effects, such as slow mobilization of nitrogen and late snow melt. Low temperature probably limits survival more than growth.

Several factors other than low summer temperature make life difficult for trees at tree limits. Summer frosts, deep and heavy snow, wind, intense light in spring, desiccation, insufficient soil drainage, and drought are discussed in greater detail in other papers in this volume. Low summer temperatures can amplify the effects of other factors; cool summers often result in immature needles and shoots that are more susceptible to frost, sun, and desiccation injury (Wardle, in press).

A stand of trees can affect its own environment. Cutting mature trees can lower the treeline. It is often wise to leave some existing plants and trees to stabilize the local environment, rather than create a situation where forest culture is required. Another reason to save existing vegetation is that man-made stands of uniform age might be less robust and stable than multi-aged stands.

Figure 1 illustrates that growing-season temperature at tree limits varies widely from year to year. Climatic variation over different time scales has several consequences:

1. Natural selection acting for a long time might have adapted local populations to a historic average climate, but that does not mean that local populations are fine-tuned to the actual weather at a site.

2. Yearly variations in summer temperature on a single site are larger than temperature differences between tree limit sites at the same latitude; year-to-year variations in summer temperature are larger than variations in the edaphic environment for the same area, and local adaptations are unlikely to determine tree limits.

3. In documented experiences and empirical results, the good years are probably over-represented. This reduces the general value of documented empirical experience to an unknown degree.

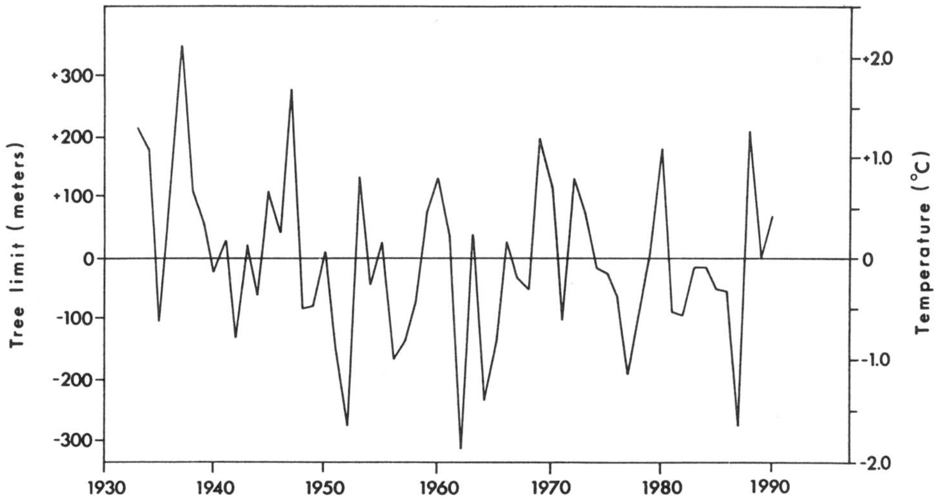

Fig. 1. Annual variation in average temperature during the growing season at forest areas in the interior of northern Sweden. If the temperature of individual summers limits growth and survival, annual variation is interpreted as a hypothetical change in tree limit of 167 meters per 1 °C. Theoretical tree limits are based on mean summer temperatures for six weather stations from 1931 to 1960 (Adapted from Lindgren et al., 1989).

4. Results from unreplicated experiments might not be representative of average conditions over time.

5. Weather is unpredictable and results of field experiments are not reproducible. Thus, to predict tree survival, theoretical deductions are necessary to complement empirical results from field experiments. At tree limits, temporal variation in weather cannot be ignored in selecting adapted populations.

GENETIC ADAPTATION

Nature's Strategies

Survival characteristics and mechanisms are probably of primary adaptive significance in the natural selection processes of extreme environments (Heide, 1985). Natural selection in extreme climates may operate more between the organism and the abiotic environment than through biological competition. Genotypes from extreme environments can survive in mild environments if they are protected from competition with locally-adapted species and genotypes.

Perenniality is a characteristic of high-latitude flora. A large below-ground biomass in roots and rhizomes and early and rapid establishment of leaf canopy are common. Vegetative reproduction, polyploidy, and apomixis are more common in the extreme conditions at arctic tree limits than at the alpine tree limits of lower latitudes. The trend is to sacrifice reproductive flexibility for higher survival.

Nature seems to create new subspecies for tree-limit conditions. About 12 tree–species complexes have evolved subspecies adapted to the extremely cold climates from latitudes 35° to 70° N. (Clausen, 1963). These complexes are found among the larches (Larix), pines (Pinus L.), spruces (Picea A. Pietr.), firs (Abies Mill.), hemlocks (Tsuga (Endl.) Carr.), junipers (Juniperus L.), poplars (Populus L.), willows (Salix L.), birches (Betula L.), alders (Alnus Mill.), and mountain ashes (Sorbus L.). Some common characteristics of tree species that form northern hemisphere tree limits are described by Clausen (1963):

1. Tree species from extremely cold habitats in one part of the earth are likely to have one or several close relatives in similar habitats in other parts. Thus variant forms from the same complexes are found in Europe, Asia, and North America.

2. Tree-limit species in cold habitats have considerable latitudinal extension.

3. Transitional forms often exist between related species. Species of most complexes intercross freely, so that each has a large gene pool.

Characteristics of tree-limit species support the hypothesis that the evolutionary solution to the "unstable niche" at tree limits is to develop species complexes that form new species, subspecies, and races when and where a tree-limit niche appears. The approach is similar to a natural process in a single population with a flexible and rich gene-pool. Hybridization also appears to have a place in breeding for niches close to the tree limit.

In contrast to central ecotypes, tree-limit ecotypes sometimes have distinct characteristics such as narrow crowns or bushy growth. To prove that tree-limit populations are genetically different, central and tree-limit populations must be experimentally compared in common environments. The more radical morphological differences among populations of forest trees often seem to be caused by environmental factors.

When origins from an elevational gradient are compared in the same experiment, those from higher elevations often harden earlier and thus grow more slowly than those from low elevations (Holzer, 1967).

Cold Acclimation

The most important survival trait at northern tree limits is sufficient early cold acclimation, generally believed to be triggered by decreasing day and increasing night length in late summer and early autumn (Ekberg et al., 1979). A physiological response is induced when the trees "perceive" a certain critical night length (Dormling, 1979). Hänninen et al. (1990) demonstrated that a model based only on night length was more efficient than models including temperature sum for timing of growth cessation in relation to climatic adaptation of northern woody plants. Perennial plants often seem to rely on night length as a calendar for growth cessation and hardening, especially at high latitudes, where night length varies greatly over time. Night length at a given latitude is calendar-related and is a more reliable predictor of low temperatures in the weeks ahead than is the current temperature, which may be colder or warmer than normal. Other factors, such as light quality, minimum temperature, heat sum, age, and preconditioning, also influence growth cessation.

Cold acclimation is the gradual process, occurring in several distinct steps, whereby plant tissues become resistant to injury from low temperatures. The end of the active growth period lacks clearcut visible point events, which makes objective measurement of the cold hardening process difficult (Koski, 1985). Nevertheless, it is characterized by the ability of cells to withstand subfreezing temperatures without lethal intracellular ice formation. Cold hardening begins after completion of stem elongation and diameter growth, formation of winter buds, lignification of late summer xylem and phloem cells, and the induction of dormancy. In deciduous species, it also follows leaf senescence and abscission.

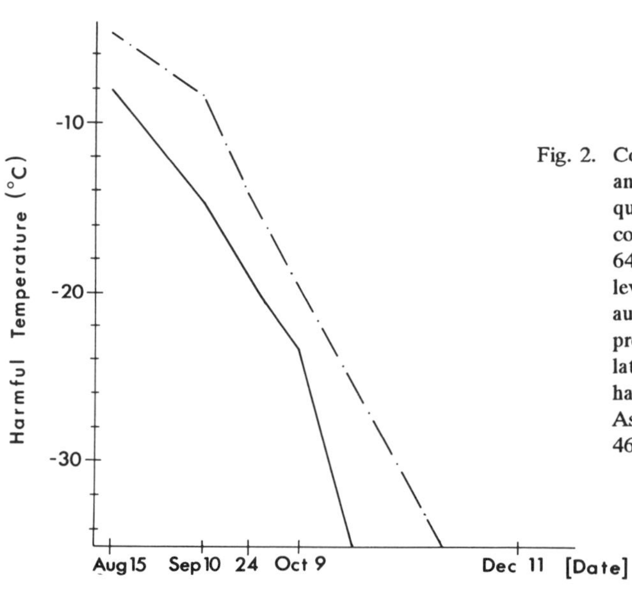

Fig. 2. Cold acclimation of two Scots pine provenances as measured by the temperature required to induce severe damage to needles collected in a trial at Renberget, Sweden, at 64° N. latitude and 280 meters above sea level (masl) on different dates in the autumn. The northern early-hardening provenance (solid line) is from 67°47′ N. latitude and 432 masl and the southern late-hardening provenance (broken line) is from Askerud, Sweden, 61°06′ N. latitude and 460 masl (Lindgren and Nilsson, 1992).

If new needles and shoots fail to mature fully during the brief, cool summer, they will be susceptible to both frost damage in autumn and desiccation in winter. Early frost damage and winter desiccation are probably associated with inadequate cuticle formation in the new needles and shoots (Tranquillini, 1979). The maturation of needles and shoots requires cessation of growth and the loss of succulence. Tree limits occur where the summer temperatures barely allow for completion of growth and ripening of shoots and needles.

A search for provenances that stop growing early in autumn should be made in order to find breeding stock for afforestation at arctic tree limits. Increased night length late in the growing season triggers cold acclimatization at high latitudes.

The end of the active growth period often is measured as bud set or cessation of terminal shoot elongation. Empirical correlations between the date of bud set and initial hardiness have been high in field experiments. Cessation of growth and onset of dormancy and cold hardiness begin in late summer, before actual winter hardening has developed. These are early phenomena, however, that have little causal relation to what is really necessary for full winter hardiness. Their importance may be overemphasized because they are easy to observe.

Balance Between Hardiness and Growth

Tree growth is severely reduced at tree limits because trees are active for only a small part of the year. Plants at tree limits use much of their resources to attain the "awake" state necessary to initiate growth and maturation. Thus, few resources remain for stem growth, compared to those available to trees in more favorable environments.

Genotypes from tree-limit populations must sacrifice growth for hardiness. This is an optimum survival strategy only at the tree limit. In mild environments, genotypes that invest more in growth will outcompete tree-limit genotypes. Because trees need a minimum amount of energy fixed by photosynthesis to live, the degree to which growth can be sacrificed by reducing the active period is limited.

The striking variation in growth among stands at the same distance from tree limits indicates that environmental factors other than summer temperature are more important for growth and less important for survival in cold climates.

GENETIC ADAPTATION AT TREE LIMITS

Tree Breeding Versus Evolution

Natural selection has acted on trees for millions of years. Are trees then perfectly adapted to local conditions? Probably not. In particular, genotypes of forest-tree populations are probably not well adapted for planting, especially at tree limits.

People demand much higher survival rates for trees in plantations than nature provides for trees in the wild. Natural regeneration will maintain a population if only a single seed among the millions from each tree produces a similar tree in the lifetime of the parent tree. Humans, however, want most planted seedlings to grow into timber trees. Natural regeneration is successful if it produces a new population once in a century; people want successful plantations every year. By maintaining high genetic variation, nature is able to find an optimum balance between survival and growth through trial and error at the population level. People, however, do not want to plant the bottom fraction of the gene pool, which is almost certain to die. In nature, populations often regenerate under cover of other vegetation, which gives shelter. Reforestation with selected populations requires hardier genotypes to withstand planting (or seeding) in less protected environments. Thus, the selected populations produced by tree breeders must outperform nature by several magnitudes, a pretty tough task.

There are examples of successful plantations close to the current timber limit, such as Scots pine (Pinus sylvestris L.) in northern Finland (Timonen, 1989). These plantings are still not close to the absolute tree limit, which is formed by mountain birch (Betula tortuosa Ledeb.).

Natural adaptation has two components: survival and reproduction. The tree-limit environment is unfavorable for reproduction. Natural regeneration is not a problem for the tree breeder, who has an advantage over nature in marginal areas. From the forester's point of view, resources spent producing cones and pollen are wasteful.

Adaptation of Tree-Limit Populations

Tree species adapted only to tree-limit conditions can have great difficulty surviving, because tree-limit conditions are variable and unpredictable. Such species may vanish when, for example, the tree limit in a warm period exceeds the top of a mountain (Ford, 1982).

The likelihood of finding desirable genetic materials for tree limits in northwest Europe is particularly low. The number of tree-limit species is larger in Siberia and North America than in Scandinavia, because the east-west European mountain chains blocked the southward escape routes of the Scandinavian flora during repeated Pleistocene glaciations. Thus, the Scandinavian species are probably not as well adapted to tree-limit conditions as North American or Siberian species (Martinsson, 1986).

In harsh climates, natural regeneration was probably most successful in the warmest years. Thus, the natural populations at tree limit might not regenerate at average tree-limit conditions, but instead might be adapted to average conditions several hundred meters below the tree limit (Fig. 1).

Tree species with wide distribution and long life cycles must adapt to average conditions in space and time. Populations at tree limits are growing as far from typical conditions for the species as is compatible with their survival, and therefore are as maladapted as is compatible with their existence.

Because survival is the basic criterion for evolution, and because every ancestor has survived, it can be argued that survival capacity of the offspring must be high. We may then ask the question: If survival is inherited, and favored by natural selection, why is it not higher than it is? The probable answer is that growth is in conflict with survival in the harshest conditions. Genotypes surviving under the harshest conditions do not survive under mild conditions, where the capacity to withstand competition for limited resources is more important than the ability to survive environmental extremes.

Natural selection for survival is intense at the tree limit. We may assume that this will lead to high genetic survival and low genetic diversity. This is true only if heritability (h^2), the proportion of the total (phenotypic) variance that is attributed to average gene effects, for survival ability is high. It is difficult to define heritability for survival. First, survival is an average, but usually we are dealing with unique genotypes. With time, survival tends towards zero. Because all parents survived, there is no phenotypic variation in the parental generation with respect to survival capacity. Still, meaningful observations can be made about heritability of survival ability. Based on my experience with heritability for other characteristics, it might be reasonable to assume heritabilities in the range of 0.1 to 0.3. Bengt Andersson, SkogForsk, Sävar, Sweden (personal communication, 1991) found that among a number of Scots pine progeny tests in harsh conditions, heritability of survival never exceeded 0.3. Heritability of survival under natural conditions is probably much lower than in planted field trials, because the plantation environment is relatively uniform.

Calculations of probable survival, if the survivors become parents of a new generation, are given in Table 1. The calculations assume that individual phenotypic toughness (i.e., survival ability) is the sum of genetic and environmental components. According to the model, the phenotype will survive if the phenotypic toughness exceeds a certain threshold (Nilsson, 1991). A standardized normal distribution is assumed.

Expected survival of progeny is calculated as follows:

$$Q\frac{(ih^2-x_t)}{\sqrt{1-ih^2(i-x_t)}},$$

where

Q	=	the cumulative normal distribution (upper tail),
x_t	=	the truncation point for toughness corresponding to the parental survival,
i	=	the corresponding intensity of selection, and
h^2	=	the narrow-sense heritability.

If heritability for survival ability is reasonably low, surviving trees established in environments where survival is low will not produce offspring with high survival ability (Table 1). Many generations may be required in the same environment for natural selection to increase survival percentage. The environment is not constant over time and gene flow will counteract multi-generation processes. Thus, the survival of tree-limit progeny will often be unsatisfactory in plantations at tree limit.

High Gene Flow Threatens Adaptation

Because ancestors of tree-limit populations are from milder environments, all gene migration is negative for adaptation at the most extreme range of tree distribution. Because pollen migrates long distances and adaptation is inherited, it is unlikely that trees at the tree limit are well-adapted.

By making assumptions about how air moves, it is possible to calculate an average distance of pollen dispersal as a function of pollen sedimentation in motionless air (Table 2).

If pollen survives air conditions for several hours, gene flow by pollen is possible over considerable distances. Andersson (1963) reports that about 75 percent of Scots pine pollen germinated after it was wind-transported from Germany over the Baltic Sea to the south tip of Sweden, a distance of more than 70 kilometers. If local climate varies with altitude or aspect, the distance between mild and harsh habitats need not be far. In mild conditions,

Table 1. Expected Survival (percent) of Progeny from Surviving Genotypes for Different Heritabilities (h^2) of Survival Ability in the Parent Population

Survival (percent)			
$h^2 = 0.0$[1]	$h^2 = 0.1$	$h^2 = 0.3$	$h^2 = 0.5$
1	2	4	9
5	7	12	21
10	12	19	30
20	23	31	43
30	34	42	53
50	53	60	69

[1] **The first column (h^2=0) is identical to survival in the parent population.**

regions, pollen transported from mild environments will pollinate female strobili at tree limits. In northern Scandinavia, many female strobili of Norway spruce and Scots pine are receptive before dehiscence of the pollen. Harsh conditions at tree limits also reduce pollen production. For widely distributed wind-pollinated trees, large amounts of pollen will be shed in mild environments and transported considerable distances. Thus, in harsh environments, pollen transported from milder environments will be more plentiful than local pollen. Such pollen is therefore very likely to be the father of many Norway spruce and Scots pine seeds produced near tree limits.

In nature, a number of factors limit pollen migration and reduce gene flow towards the tree line. These factors include

- Pollen weight. Heavy pollen has fast sedimentation rates and short expected flight distances.

- Species distribution. Uncommon species do not produce enough distant pollen to dilute the local pollen. Populations confined to tree limits are usually small and do not produce large amounts of pollen.

- Topography. In flat and uniform landscapes, even long-distance pollen will be from tree-limit populations.

- Transport mode. Pollen vectors other than wind may be less efficient in long distance transport.

- Viability. Short survival time of pollen after shedding will limit gene flow over long distances.

For Scots pine in Sweden, variation in hardiness is clinal and seems to depend more on latitude than on altitude (Prescher, 1986). These observations are compatible with long pollen flights.

Table 2. Theoretical Pollen Flight at Wind Speeds of 10 meters/second (Stanley and Linskens, 1974)

Species	Sedimentation (centimeters/ second)	Dispersion (kilometers)
Norway spruce (*Picea abies* (L.) Karst.)	6.8	22
Scots pine (*Pinus sylvestris* L.)	3.7	68
English oak (*Quercus robur* L.)	4.0	199
Alder (*Alnus* Mill. spp.)	2.8	547
European white birch (*Betula verrucosa* Ehrh.)	2.4	

Seeds also play a major role in gene flow. Animals also can transport some seeds considerable distances. In alpine conditions, updrafts are capable of carrying seeds several hundred meters upslope.

Species that form bands along altitudinal and arctic tree limits usually do not compete successfully below tree limits. Gene flow of species confined to tree-limit environments will not prevent adaptation of tree-limit populations. For example, the tree-limit species subalpine fir (*Abies lasiocarpa* (Hook.) Nutt.) does not exist at milder sites and cannot dilute the pollen of treeline populations with clouds of maladapted pollen from mild sites.

ADAPTING TREES TO TREE LIMITS

Environmental Change

Environment may be more important than genetics for long survival and growth of trees at their limits. On marginal sites, improvement of both the environment and genetics is required. Tree improvement cannot replace environmental improvement, however. Several methods exist to improve the tree-limit environment. Severe competition both above and below ground, and chemical war in the ground, can be reduced with scarification. Drainage, physical support, shelter from solar radiation and desiccation, reduction of animal populations, and nurse crops are techniques that can improve survival at tree limits. Environmental improvement should start with the best microsites. Once trees are established, they provide shelter for their neighbors and offspring.

Conditioned Plants

Healthy planting stock has an advantage over natural seedlings. Nursery-grown seedlings reach transplant size under nursery conditions in one or two growing seasons, thereby avoiding many years of exposure to harsh natural conditions to achieve the same stage of development. Plants raised in a nursery can be physiologically adapted (by manipulating the light regime, for example) to the physical environment where they will be transplanted.

Even the environment during seed maturation may affect seedling vigor (Andersson, 1989; Bjørnstad, 1981; Dormling and Johnsen, 1992; Lindgren and Wang, 1986). This opens new possibilities for increasing hardiness. Introduction of vegetative propagation techniques introduces additional options for manipulating the physiological state of planting stock.

Genetic Variation

A genetic strategy for adapting trees to arctic and alpine tree limits is meaningful only if survival and growth traits are genetically variable and if the magnitude of the variation is known. Genetic variation in survival close to tree limits can be demonstrated at four different levels: between species and between populations within a species (Fig. 3), between different offspring within a population (Fig. 4), and among clones within a population (Fig. 5).

Survival of species near northern tree limits depends on latitude of seed origin (Fig. 3). Adaptation of northern provenances to harsh sites is superior to that of southern provenances, demonstrating that it is meaningless to compare species without testing provenances from a wide range in latitude.

Test for Early Cold Acclimation

A large fraction of the genotypes in natural populations ceases growth too late for planting on harsh open areas. Several short-term tests for early cold acclimation of genotypes are available (Andersson, 1992; Mikola, in press; Nilsson, in press). Growth cessation can be measured by bud set, by color of stem tops, and by resistance of tissues to injury after freezing of whole potted plants or parts of plants.

Latitude and Night Length

Nights exceeding a minimum critical length occur later in the autumn with increasing latitude (Fig. 6). A provenance adapted to cease active growth at a particular night length will enter dormancy and develop hardiness earlier if it is propagated at a latitude south of its natural origin. Thus, early cold acclimatization is obtained by using provenances adapted to latitudes north of the planting site. In many situations, local provenances will not finish

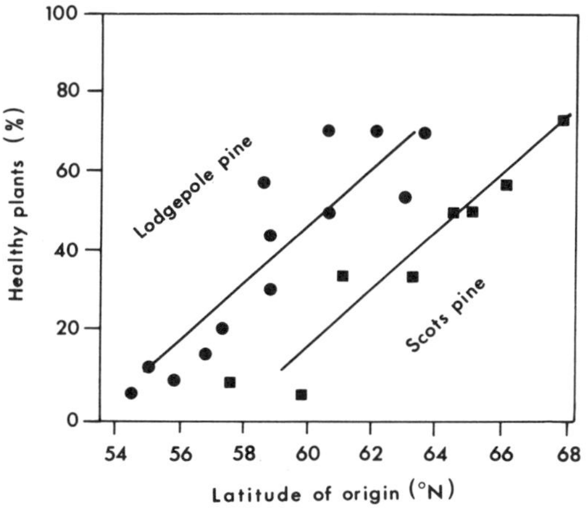

Fig. 3. Genetic variation at the species level. Percent of healthy plants three growing seasons after planting as a function of provenance origin for Scots pine and lodgepole pine. The experimental plantation is at Whitehorse, Yukon Territory (60°40′ N. latitude and 660 meters above sea level) (Source: Lindgren and Lindgren, 1990).

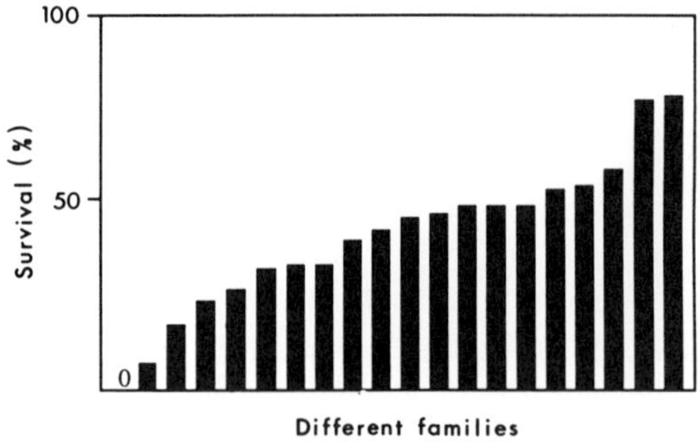

Fig. 4. Genetic variation in survival at the family level, as demonstrated by survival of open-pollinated Scots pine families. The families were sampled at random from a natural stand at latitude 66°50′ N., 180 meters above sea level, and tested in a provenance experiment at Kåbdalis, Sweden, 66°27′ N., 180 meters above sea level (Source: Eriksson and Lundkvist, 1986).

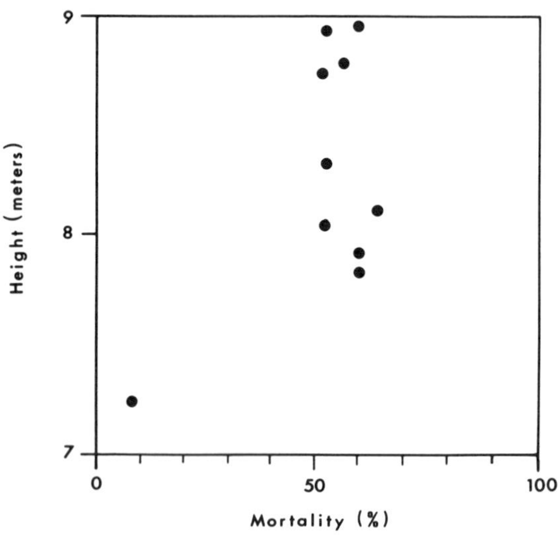

Fig. 5. Genetic variation in height and survival at the genotype level: survival and height growth of Scots pine clones in an experiment with grafts. The experimental site was at 65°54′ N. latitude and 37 meters above sea level (Lindgren, 1984a). The clones were from approximately the same latitude and altitude. One clone survived much better than the others, but grew slowly.

growth early enough to escape frost injury after planting on harsh exposures. Northern provenances, if available, could solve this problem.

Creating New Land Races

Local populations adapt to harsh conditions by natural processes. Adaptation and growth improve with each generation that imported crops are grown and harvested in their new environment. Natural selection seems to work surprisingly fast. For instance, Ledgard (1980) states that for provenance trials at high altitudes, "trees from seed collected in New Zealand from parents of unselected origin will invariably rank high." It may be cheaper and faster to let natural selection do the work than to rely on sophisticated breeding techniques.

For many species, sexual reproduction at tree limits is a problem. Thus, land races must be developed from tests and selection in planted trials at relevant tree-limit conditions. If pollen migration is a problem, it may also be necessary to control pollination.

Stopping Gene Flow

Gene flow as a result of pollen and seed migration from mild climates is a major reason populations do not adapt to altitudinal tree limits. For seed production and breeding,

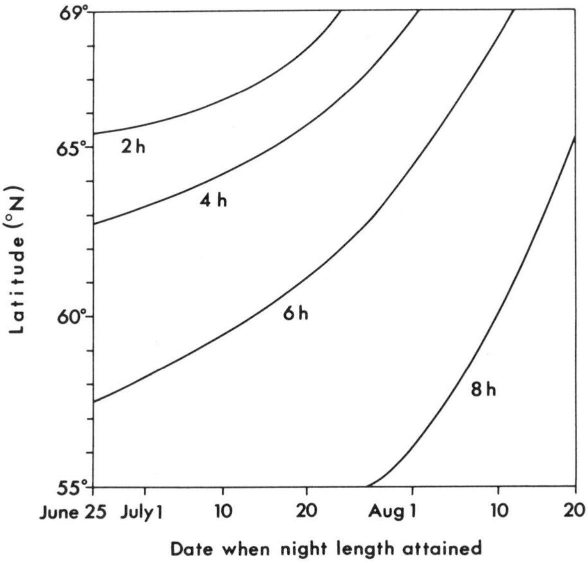

Fig. 6. Dates for night lengths from two to eight hours at latitudes from 55° to 69° N. (Based on calculations by Lindgren, 1984; see Appendix 2).

gene flow from mild climates can be limited by choosing a species confined to tree limits, by controlled pollination, or by isolating breeding and seed production areas from natural pollen sources. The most efficient method is controlled pollination. Controlled pollinations, however, are complicated, time consuming, administratively demanding, expensive, and not always reliable.

For breeding minor exotics in Scandinavia, stands can be isolated from extraneous pollen sources as an alternative to controlled pollination crosses. It is thus possible to arrange sub-lines of the breeding population as isolated groups of trees. A place isolated from wild pollen of native Scots pine and Norway spruce cannot be found in Scandinavia, however. But places outside Scandinavia are isolated from extraneous pollen sources. For instance, sub-lines of breeding material of Norway spruce can perhaps be propagated in Iceland for use at latitudes from 62° to 68° N. in the other Nordic countries. Iceland may be less ideal for propagating and breeding Scots pine because it suffers from the insect problems of subarctic maritime climates.

Redundancy

To compensate for low survival at tree limits, seedlings can be planted at intervals of less than two meters. More genotypes for the same price can be obtained by artificial seeding with shelters than by planting. To obtain full stocking, areas that fail within a plantation must be reseeded or replanted. Natural regeneration from trees at their limits should be encouraged and protected.

Vegetative Propagation

In sexual propagation, progeny means for adaptive traits are expected to regress toward the population means. This mechanism is likely to prevent the progeny of the few survivors in a harsh environment from being as successful as the parents. Probably the only way to produce plants with a reasonably high likelihood of survival is by cloning genotypes tested for transgressive hardiness.

To test a discontinuous character such as survival capacity, it is necessary to replicate the genotype in a common environment. Seedlings are unreplicated genotypes and can, strictly speaking, not be tested for survival. Survival ability cannot be increased to any great extent by selecting survivors alone, because heritability of survival is low (Table 1). Heritability of survival ability can be increased by testing clones, so that substantial improvement in survival is possible.

A way to produce seed with high survival in tree-limit environments is to repeat superior controlled crosses that have been tested in replicated experiments. Controlled crosses are so expensive and difficult that breeding programs may be acceptable only if combined with vegetative propagation of the few seeds obtained.

Clonal propagation provides a method of testing the same genotypes in different environments and selecting clones that can survive under variable conditions in time and space at tree limit environments. Clones that perform well in harsh environments can be sexually propagated for seed production and breeding in milder environments.

Vegetative propagation may be the simplest technique for producing plants adapted to tree limits because seed set is generally poor in harsh environments close to tree limits. Genetic improvement from clonal selection and vegetative propagation is at a dead end, however. Sexual recombination between selected genotypes is necessary for long term breeding to accelerate adaptation of trees at environmental limits.

In the absence of sexual reproduction close to tree limits, nature often relies on vegetative propagation for species such as Norway spruce (Tiefenbacher, 1989). By turning to vegetative propagation, genotypes can be reproduced without wasting energy on sexual reproduction. Energy conservation is particularly important at the tree limit, where resources are scarcely sufficient for trees to survive.

Genetic Engineering

Can we find a gene therapy for the problem of adaptation at tree limits? In a time perspective spanning centuries, the answer is "maybe." Over the shorter time, I am less optimistic. Changing single genetic factors is probably not enough to surpass the natural variation available for classic breeding. Evolution of adaptive traits is efficient in modifying numerous single factors. Tampering with adaptation means changing fundamental processes of the organism. Improving adaptation requires much more basic knowledge than trivial tasks such as genetic engineering to induce trees to produce a compound that is toxic or unattractive to predators. If a genetic cure for poor adaptation is attempted in an entirely new, untested, and artificial way, at least a full rotation (at least a century for the major conifers at arctic tree limits) of field tests would be required before it could be applied on a large scale.

Tests or Deductions?

Forest trees in harsh climates have rotation times that span centuries. Foresters are not willing to wait for results of long-term experiments, and results of short-term field trials are unreliable. In addition, the future climate is likely to be different from the climate during both short- and long-term test periods. Pathogens and plant-eating insects will also evolve. Sites are variable and experimental conditions may not conform to future forest practices. For these reasons, theoretical deductions may be just as important as actual test results.

The introduction of lodgepole pine (Pinus contorta var. latifolia Engelm.) at high altitudes in northern Sweden demonstrates the value of both deductive reasoning and empirical results. Since the early 1980s, Sweden has planted large areas with lodgepole pine close to tree limits to provide jobs for citizens in remote areas and to provide a supply of raw material for the forest industry. Old forests were not to be cut if they could not be regenerated, but indigenous species were difficult to regenerate. Thus, use of lodgepole pine was encouraged for afforestation of harsh areas. We analyzed the possible adaptation of lodgepole pine to these environments from the distribution of indigenous species, climate comparisons, climate chamber experiments, and analogies with experience from other species (Lindgren, 1981; Lindgren, 1985; Lindgren and Lindgren, 1985). We concluded for several reasons that was probably not well-adapted to the far north, to high altitudes, or to the region near timberline close to the Scandes Mountains. Therefore, we considered that the use of lodgepole pine for commercial plantations was justified only on the basis of an extensive positive experience.

At a meeting at Narvik, Norway, in 1983, I called it a paradox that although lodgepole pine plantations ought not to function well close to tree limits, many young plantations still seemed to survive satisfactorily. Someone from the auditorium remarked that the paradox may vanish. Since 1986, lodgepole plantations close to the forest limit have suffered extensive damage from infection with Gremmeniella abietina (Lagerb.) Morolet, a pathogen that girdles the stems of pine species (Karlman, 1990), because they were maladapted to the severe weather conditions of northern Sweden that climaxed with the cold summer of 1987 (Fig. 1). As a result, planting lodgepole pine close to the forest limit was forbidden by the Swedish National Board of Forestry in 1987. It is too early to make a final economic and biological evaluation of the damage. Probably a final evaluation will find that compared to Scots pine, lodgepole pine has been successful even in areas close to tree limits. From a psychological and political point of view, however, it would have been better if less area had been planted near forest limits. The early warnings had only limited effect, but they might offer an example of the value of arguments other than results from field tests and empirical experience.

CONCLUSIONS

From my experience, here is a geneticist's recipe for successful plantations at the tree limits:

- Make certain you have solid and reliable funding for decades ahead. Your benefactor should not be concerned about economic returns, at least not from the sale of wood.

- Give preference to species that are naturally confined to tree limits. Geographic and climatic comparisons between seed origins and plantation sites might be relevant when choosing species and populations for afforestation of harsh environments.

- Choose provenances from latitudes north of the plantation site.

- Test for early winter acclimation.

- Reduce or eliminate pollen flow from milder environments.

- Conduct clonal tests and propagate the most successful clones vegetatively.

- Start with the best microsites for tree survival.

- Follow the best available advice concerning optimum environments, particularly with respect to drainage, physical support, protection against sun and desiccation, and shelterwood. Breeding cannot replace good silvicultural practice.

- Cooperate with natural forces whenever possible.

- Consider actions, such as alternative silvicultural methods, species, and provenances, to reduce the consequences of failure.

- Keep animals away.

Be forewarned: Failure is more likely than success! Some luck in timing afforestation attempts at tree level is necessary, but have faith—global warming may solve the problem for you!

ACKNOWLEDGEMENTS

Constructive comments were received from many persons. Special thanks are extended to Bo Axelsson, Bengt Andersson, Lars Christersson, Ingegerd Dormling, Björn Elfving, Inger Ekberg, Gösta Eriksson, Anders Fries, Claes Fries, Björn Hånell, Leif Kullman, Katarina Lindgren, Mats Lindgren, Jan-Erik Nilsson, and Aðalsteinn Sigurgeirsson for their reviews and contributions. The presentation of this paper at the NATO Workshop "Forest Development in Cold Climates" in June 1991 was supported by the Swedish Council for Forestry and Agricultural Research. The editors, in particular John Alden, have been extremely helpful.

APPENDIX I: POTENTIAL EFFECTS OF GLOBAL WARMING ON FOREST PRODUCTION AND CARBON FIXED BY FORESTS IN SWEDEN

For this conference, I was encouraged to estimate the effect of global warming on Swedish forests and the carbon sequestered by them. The results of this endeavor are most likely very unreliable, but the effort is important to prepare mankind for an uncertain future.

Land can be classified into many categories. Swedish terminology for classifying domestic forest land on the basis of harsh climate is complex and confusing (Skogsstyrelsen, 1989), so I will use a simplified scheme. On maps of Sweden, lines have been drawn dividing the country into the following three "harshness" categories (Nilsson, 1990):

1. Mountain areas, where forests are sparse and afforestation is seldom attempted.
2. Land not classifed as Category 1, where forest regeneration is difficult because of a cool climate.
3. Land where the cool climate offers no particular problems for forest regeneration.

The borders between categories 1 and 2, and 2 and 3, are approximately 200 meters and 300 meters below the altitude tree limits, respectively.

Area in each category in Sweden is as follows (in million hectares):

	Category			
	1	2	3	Total
Forest land	4.0	3.2	21.8	28.9
Marsh and peatland	0.7	0.7	2.0	3.4
Above tree limit (alpine)	3.4	0	0	3.4
Other land	0.1			0.1
Total land area	8.2	4.0	29.0	40.8

Productive forest land has the capacity to produce ≥ one cubic meter of stemwood per hectare per year. Of the four million hectares of forest land in Category 1, 1.5 million hectares are productive forest land. Of this, about one third has been set aside as natural reserves. Since 1986, the trend in Swedish forestry has been to back down from the tree limits because of environmental and economic reasons, and also because adapted genetic material is not available for reforestation.

Marsh and peatlands that support forest are classified as forest land. Much of the marsh and peatland now considered beyond the tree limit could be drained and turned into productive forests. This could increase Sweden's annual forest production by around 12 million cubic meters (Hånell, 1985). Because of administrative, economic, and environmental reasons, few marshes and peatlands will be developed in the foreseeable future. Wetlands might be a major natural carbon dioxide sink. Drainage will release some carbon dioxide, but the net loss is expected to be small and is not included in the calculations. Forestation of abandoned farm land, forest tree breeding, use of exotics, and other silvicultural activities will also increase forest production. An increase in atmospheric carbon dioxide as well as other human-related changes in the soil and atmosphere might change Swedish forest production even in a constant climate. In the short term, an increase in production is likely.

Generally, temperature increases with decreasing altitude. The effect on forest production of a 1 °C-warmer growing season might be equivalent to sinking the land 167 meters (Lindgren et al., 1989). That change would probably turn about two million hectares of nonproductive Category 1 forest land into productive Category 2 forest land at an average annual growth rate of two cubic meters per hectare. On some sites, the increase in growth might require a few decades (Tuhkanen, 1982). In other circumstances, the increase might be delayed for a millennium or longer until tree limits have reached higher elevations and are in equilibrium with growing-season temperature. For most of the productive forest land (around 20 million hectares), the annual production would increase approximately 0.7 cubic meters per hectare (Hägglund and Svensson, 1982). Probably the delay in response would be short. Thus, the total increase in annual growth would be about 18 million cubic meters.

For a limited area, warm summers will cause water deficits and reduce production. The standard deviation between the temperatures of growing periods in different years is 0.8 °C (Fig. 1, main paper). Actual temperatures vary more than 1 °C over small geographic areas. Thus, a 1 °C increase in growing season temperature will not lead immediately to dramatic unpredictable consequences.

Vegetation will easily acclimatize to the new climate. Winter temperatures are expected to increase twice as much as summer temperatures in Sweden. Thus, my estimate for the effect of a 1 °C increase in the average annual temperature corresponds to about a 0.67 °C increase during the vegetation period and an increase of about 12 million cubic meters of wood annually, or approximately 15 percent. Each cubic meter of stem wood consists of approximately 0.2 tons of carbon. Roots, needles, twigs, branches and bark may bind similar amounts of carbon, but much of it will be withdrawn for less time than carbon bound in the stem wood. Rising temperatures thus could cause an additional two to five million tons of carbon per year to be withdrawn from the atmosphere, an amount far short of the 17 million tons carbon per year that Sweden currently adds to the atmosphere. The possible temporary counteraction of global warming by increased forest growth still would not compensate for Sweden's total carbon dioxide production. The accumulation of wood in Swedish forests, however, will significantly decrease Sweden's current net contribution to global warming.

A quantitative estimate of the effects of a 3 °C increase on forest productivity would be too unreliable to be meaningful, but a massive increase in growing season temperature certainly would be associated with many environmental changes that would interact and accumulate in unpredictable patterns and chain events. Linear extrapolation will not work as a prediction method. Consequences are unpredictable, but they are likely to be dramatic. When consequences do become evident, it probably will be much too late to start counter-measures. I suppose we now enter a millennium of unstable and variable climate before a new equilibrium is attained.

Under conditions of a warmer climate, a possible but hypothetical scenario is that new species will invade the Arctic tundra regions of northern Siberia and northern Canada from the south and out-compete the present tree-limit populations. There are large areas without mountain ranges where forests could establish new high-altitude treelines. Even where mountains exist, new niches at tree limit will be narrow and edaphically variable. Even if humans transfer tree-limit species to suitable habitats that form in the north, populations might not migrate fast enough to prevent extinction. In addition, fast-migrating populations will not find a refuge beyond the shores of the Arctic Ocean. Thus, when the climate cools again after the next glaciation, the gene resources for northern tree-limit species might not be available to repopulate present tree limits.

APPENDIX II: AN ALGORITHM FOR DAY LENGTH AS A FUNCTION OF DATE AND LATITUDE[1]

Many algorithms have been derived and utilized for day length. To our knowledge, however, simple daylength algorithms are not readily available for biological work.

To avoid unimportant details, daylength is calculated for the period when the sun is above the horizon at longitude 0° W. A program for a more complex version of the algorithm is available from Dag Lindgren. The program ("DAGL") is written in QBASIC and will run on any modern DOS-based computer.

[1] Prepared by Mats Lindgren, Department of Astronomy, The University of Uppsala, Uppsala, Sweden

Step 1. Calculate the equatorial coordinate δ as follows:

$$\delta = \arcsin [\sin (23.439-0.0000004n) \times \sin (L + 1.915\sin g + 0.020\sin 2g)],$$

where

L = mean longitude of the sun, 280°.460 + 0.9856474n,
g = mean anomaly of sun, 357°.528 + 0.9856003n, and
n = Julian day, starting at 0 hour Greenwich Mean Time, January 1, 2000

Step 2. Determine if the sun is circumpolar, that is, if the day is 24 or 0 hours. For latitude "Φ," the test is whether

$$\delta + 0.833 > 90° - \Phi$$

Step 3. Calculate daylength for day "n" as a function of latitude "Φ" as follows:

$$\textit{Daylength (radians)} = 0.1337 \arccos \left[\frac{\cos 90°.833 - (\sin\delta \bullet \sin\Phi)}{\cos\delta \bullet \cos\Phi}\right]$$

The value 90°.833 corresponds to the angle from the zenith to the central point of the sun's disk. The angle must be exceeded for the upper edge of the sun's disk to be invisible on a regular and unobstructed horizon.

REFERENCES

Andersson, B., 1992, Forecasting Pinus sylvestris field mortality by freezing tests—methods and applications, Umeå, The Swedish University of Agricultural Sciences, Department of Forest Genetics and Plant Physiology, 24 p. Ph.D. thesis.

Andersson, B., 1989, Effect of maternal soil treatment on first year growth and autumn hardiness of Pinus sylvestris full sib families, Scand. J. For. Res., 4:163-173.

Andersson, E., 1963, Seed stands and seed orchards in the breeding of conifers, Proceedings of the World Consultation on Forest Genetics and Tree Improvement, FAO/FORGEN, Paper 8/1, 18 p.

Arno, S. F., and Hammerly, R. P. H., 1984, Timberline: Mountain and arctic forest frontiers, The Mountaineers, Seattle, Washington, 304 p.

Benecke, U., and Davis, M. R., eds., 1980, Mountain environments and subalpine tree growth, Forest Research Institute, New Zealand Forest Service, Christchurch, New Zealand, Tech. Paper No. 70.

Bjørnstad, Å., 1981, Photoperiodical after-effects of parent plant environment in Norway spruce seedlings, Medd. Norsk Institutt for skogforskning, 36:6, 1-30.

Bäckström, P.-O., 1984, ed., Skogsföryngring i fjällnära skogar (Forest regeneration in mountainous forests), Forskarrapport (Report of a research team), Umeå, The Faculty of Forestry, The Swedish University of Agricultural Sciences, 86 p. In Swedish.

Clausen, J., 1963, Tree lines and germ plasm—a study in evolutionary limitations, Proceedings of the National Academy of Sciences, United States, 50:860-868.

Dormling, I., 1979, Influence of light intensity and temperature on photoperiodic response of Norway spruce provenances, Proceedings of the IUFRO working parties on Norway spruce provenances and Norway spruce breeding, Bucharest, 1979, Lower Saxony Forest Research Institute, Escherode, Germany, p. 398-408.

Dormling, I., and Johnsen, Ø., 1992, Effects of parental environment on full-sib families of Pinus sylvestris, Can. J. For. Res., 22:88-100.

Ekberg, I., Eriksson, G., and Dormling, I., 1979, Photoperiodic reactions in conifer species, Holarctic Ecology, 2:255-263.

Eriksson, G., and Lundkvist, K., 1986, Adaptation and breeding of forest trees in boreal areas, Umeå, The Swedish University of Agricultural Sciences, Department of Forest Genetics and Plant Physiology, Report 6, 67-80.

Ford, M. J., 1982, The changing climate: Responses of natural fauna and flora, George Allen and Unwin, London.

Heide, O. M., 1985, Physiological aspects of climatic adaptation in plants with special reference to high-latitude environments. In: Kaurin, Å., Junttila, O., and Nilsen, J., eds., Plant Production in the North, Norwegian University Press, Oslo, p. 1-22.

Holzer, K., 1967, Das Wachstum des Baumes in seiner Anpassung an zunehmende Seehöhe. In: Egger, J., ed., Ökologie der Alpinen Waldgrenze, Mitteilungen der forstlichen bundes-Versuchsanstalt, 75:427-456.

Hånell, B., 1985, Skogliga våtmarker–en resurs för ökad skogsproduktion, Skogsfakta-inventering och ekonomi, No 8. In Swedish.

Hägglund, B., and Svensson L.-G., 1982, Ståndortsförhållanden på skogsmark och myr i Sverige, Umeå, Swedish University of Agricultural Sciences, Department of Forest Survey, Report 35, 1-122. In Swedish with English summary.

Hänninen, H., Häkkinen, R., Hari, P., and Koski, V., 1990, Timing of growth cessation in relation to climatic adaptation of northern woody plants, Tree Physiology, 6:29-39.

Karlman, M., 1990, Gremmeniella infection on lodgepole pine in northern Sweden. In: Proceedings of the Joint Meeting of Western Forest Genetics Association and IUFRO Working Parties Douglas-fir, Contorta Pine, Sitka Spruce and Abies Breeding and Genetic Resources, Olympia, Washington, United States, August 1990, Paper 2.156, 10 p.

Kaurin, Å., Junttila, O., and Nilsen, J., eds., 1985, Plant production in the north, Norwegian University Press, ISBN 82-00-07385-8, 356 p.

Kullman, L., and Hofgaard, A., 1987, Klimatisk hasardgräns i fjällnära skogar (Climatic risk border in mountainous forests), Sveriges Lantbruksuniversitet-Svenska Naturskyddsföreningen (Swedish Univercity of Agricultural Sciences, Swedish Association of Environmental Protection), ISBN 91-576-3276-6. In Swedish.

Ledgard, N. J., 1980, Specific and intraspecific growth differences in tree trials up to timberline in the Craigieburn range, New Zealand. In: Benecke, U., and Davis, M. R., eds., Mountain environments and subalpine tree growth, Forest Research Institute, New Zealand Forest Service, Christchurch, New Zealand, Tech. Paper 70, 137-153.

Lindgren, D., 1981, Var försiktig med contortan i norr! (Careful with lodgepole in the north!), Skogen, 11:64-65. In Swedish.

Lindgren, D., 1984, A Scots pine trial in northern Sweden at latitude 66°, Umeå, Swedish University of Agricultural Sciences, Department of Forest Genetics and Plant Physiology, Arbetsrapport, 13:1-9.

Lindgren, D., 1984, Latitude and night length, Umeå, Swedish University of Agricultural Sciences, Department of Forest Genetics and Plant Physiology, Arbetsrapport, 11:1-4.

Lindgren, D., 1985, Contortatall—varifrån och vart (Lodgepole pine—where and from where?), Umeå, Swedish University of Agricultural Sciences, Department of Forest Genetics and Plant Physiology, Report 4, 159-168. In Swedish with English summary.

Lindgren, D., ed., 1986, Provenances and forest tree breeding for high latitudes, Umeå, Swedish University of Agricultural Sciences, Department of Forest Genetics and Plant Physiology, Report 6, 287 p.

Lindgren, D., and Lindgren, K., 1985, Pinus contorta in the midnight sun—expectations and experiences. In: Kaurin, Å., Junttila, O., and Nilsen, J., eds., 1985, Plant production in the north, Norwegian University Press, Oslo, p. 181-190.

Lindgren, D., and Lindgren, K., 1990, A Canadian-Swedish species genotype environment interaction study. In: Proceedings of the Joint Meeting of Western Forest Genetics Association and IUFRO Working Parties on Douglas-fir, Contorta Pine, Sitka Spruce and Abies Breeding and Genetic Resources, 1990 August 20–24, Olympia, Washington, United States, Paper 2.209, 10 p.

Lindgren, D., Lindgren, K., Odin, H., and Eriksson, B., 1989, Temperaturen under vegetationsperioden i norra Sveriges skogsland 1933-1988, Umeå, Swedish University of Agricultural Sciences, Department of Forest Genetics and Plant Physiology, Arbetsrapport, 30:1-18. In Swedish.

Lindgren, D., and Wang, Q., 1986, Are genetic results influenced by environment during seed maturation, Conference proceedings, A joint meeting of IUFRO Working Parties on Breeding Theory, Progeny Testing and Seed Orchards, 1986 October 13-17, Williamsburg, Virginia, United States, 192-199.

Lindgren, K., and Nilsson, J.-E., 1992, Cold acclimation of Pinus sylvestris and Pinus contorta as measured by freezing tolerance of detached needles, Scandinavian Journal of Forest Research, 7:309-315.

Martinsson, O., 1986, Historic background to the natural distribution of tree species in Scandinavia, Umeå, Swedish University of Agricultural Sciences, Department of Forest Genetics and Plant Physiology, Report 6, 281-287.

Mikola, J., [In press], Provenance and individual variation in climatic hardiness of Scots pine in northern Finland. In: Alden, J. N., Ødum, S., and Mastrantonio, J. L., eds., Forest Development in Cold Climates, Proceedings of an International Symposium, 1991 June 18-24, [Laugarvatn, Iceland], NATO, Plenum, New York, New York.

Nilsson, J.-E., ed., 1985, Skogsodlingsmaterial för hårda nordliga lägen (Reforestation materials for harsh northern sites), Proceedings of a seminar at Umeå, 1985 February 19-20. Umeå, Swedish University of Agricultural Sciences, Department of Forest Genetics and Plant Physiology, Report 4, 1-200. In Swedish, with English summaries.

Nilsson, J.-E., 1991, Estimating the effect of pollen contamination on the seed orchard progeny, Umeå, Swedish University of Agricultural Sciences, Department of Forest Genetics and Plant Physiology, Report 10, 66-85.

Nilsson, J.-E., [In press], Selection and Breeding of Scots pine for northern Sweden. In: Alden, J. N., Ødum, S., and Mastrantonio, J. L., eds., Forest Development in Cold Climates, Proceedings of an International Symposium, 1991 June 18-24, [Laugarvatn, Iceland], NATO, Plenum, New York, New York.

Nilsson, N.-E., ed., 1990, Skogen (The Forests), Volume 2 of the National Atlas of Sweden, ISBN 91-87760-06-1, SNA Publishers, Stockholm, 144 p. In Swedish and English.

Prescher, F., 1986, Transfer effects on volume production of Pinus sylvestris L.: A response surface model, Scandinavian Journal of Forest Research, 1:285-292.

Salzberg, K. A., Fredriksson, S., and Webber, P. J., eds., 1986, Restoration and vegetation succession in circumpolar lands, Proceedings of the seventh conference of the Comité Arctique International, 1986 September 7–13, [Reykjavík, Iceland], Arctic and Alpine Research, 19(4), 586 p.

Skogsstyrelsen, 1989, Fjällnära skogar (Montaneous forests), Utredningsrapport till Jordbruksdepartementet (Investigation Report to the Department of Agriculture). In Swedish.

Stanley, R. G., and Linskens, H. F., 1974, Pollen, Biology Biochemistry Management, Springer Verlag, Berlin, 307 p.

Tiefenbacher, H., 1989, Natürliche und künstliche vegetative Vermehrung von Fichten der subalpinen Kampfzone (Picea abies Karst.) (Natural and artificial vegetative propagation of spruces in the subalpine "Kampfzone"), Universität für Bodenkultur in Wien. Ph.D. Dissertation.

Timonen, M., 1989, Artificial regeneration on the polar timber line in Finnish Lapland. In: Martinsson, O., Packee, E. C., Gasbarro, A., and Lawson, T., eds., Forest regeneration at northern latitudes close to timberline, Gen. Tech. Rep. PNW-GTR-247, Portland, Oregon, U. S. Department of Agriculture, Forest Service, Pacific Northwest Research Station, 17-21.

Tuhkanen, S., 1982, Trädgränsen och klimatet (Treeline and climate), Naturen, p. 85-94. In Swedish.

Tranquillini, W., 1979, Physiological ecology of the alpine timber line, Ecological Studies 31, Springer Verlag, New York, 137 p.

Wardle, P., [In press], Causes of alpine timberline: A review of the hypotheses. In: Alden, J. N., Ødum, S., and Mastrantonio, J. L., eds., Forest Development in Cold Climates, Proceedings of an International Symposium, 1991 June 18-24, [Laugarvatn, Iceland], NATO, Plenum, New York, New York.

SELECTION AND BREEDING OF SCOTS PINE FOR NORTHERN SWEDEN

Jan-Erik Nilsson

Department of Forest Genetics and Plant Physiology
Swedish University of Agricultural Sciences
Umeå, Sweden

INTRODUCTION

Sweden lies between latitudes 55° and 69° N. This means there are big differences in the environment and in the prerequisites for growth and survival of plants, including forest trees. In southern Sweden, mortality from weather and climate stress is not a major problem for native forest species although, in some places, frost causes damage during the growth period. To the north and at higher elevations, however, regeneration problems and mortality increase because of poor adaptation to climate. Above tree limit, adaptation is so poor that stress factors such as long-lasting snow cover, short and cold growth periods, short summer nights, hard winds, and low winter temperatures prevent natural regeneration. Tree mortality usually cannot be attributed to a specific event because decline is slow and damage is not visible until long after the injury. Often, the first visible signs of poor adaptation and future mortality are insect and fungi attacks on trees already weakened by climate stress.

The strong influence of geography on survival of Scots pine (Pinus sylvestris L.) in northern Sweden is well documented. The severity index (Eriksson et al., 1980) is commonly used to characterize northern Swedish localities with respect to Scots pine regeneration. The index was derived from the largest series of Scots pine provenance trials in Sweden (Eiche, 1955), and is the expected mortality (percent) of the local provenance 20 years after planting. At tree limits, the severity index is 100, and decreases by approximately six units per degree towards southern latitudes and by 10 units per 100 meter decline in elevation (Eriksson et al., 1980). Mortality caused by climate and weather stresses, however, continues after the age of 20 years (Ståhl and Andersson, 1985).

Survival has a major influence on the volume of wood produced by Scots pine in north Sweden (Eriksson et al., 1980; Persson and Ståhl, 1990; Remröd, 1976). Survival of Scots pine is clinally related to the latitude of origin. Expected survival on a site 20 years after planting increases by approximately 10 percent for every degree of latitude the seed source is north of the planting site; that is, survival increases from 20 percent for the local provenance to 30 percent on a site located at latitude 65° N. and 550 meters in elevation by using seeds from latitude 66° N. and the same elevation. A transfer of approximately two

Forest Development in Cold Climates, Edited by J. Alden *et al.*
Plenum Press, New York, 1993

or three degrees latitudes to the south will result in considerable gain in survival and volume production without severe loss in height (Eriksson et al., 1980; Remröd, 1976). Compared to latitude, seed transfer from high altitudes to low altitudes within the commercial forest area has little or no effect on field survival (Eriksson et al., 1980; Persson and Ståhl, 1990; Remröd, 1976). It should be noted, however, that the effect of altitude on hardiness has not been as thoroughly investigated as the effect of latitude.

The need for hardy seeds in the interior of northern Sweden cannot be met by north to south transfers because the northern limit of Scots pine is exceeded and seed production is poor at tree limits. This paper presents the breeding of Scots pine for northern Sweden and research on possibilities and problems concerning selection and seed orchard management.

THE BREEDING PROGRAM

A lack of suitable regeneration material for northern Sweden prompted development of the first grafted seed orchards of Scots pine in the late 1950s. The trees were phenotypically selected for growth and quality in mature natural stands in northern Sweden. With clinal variation in mind, the plus trees were separated into different seed orchards on the basis of homogenous geographical origins. This allows assignment of seed orchard crops to suitable regions or seed orchard zones, each with relatively homogenous climate (Fig. 1). As indicated, the purpose of breeding in northern Sweden is to produce large amounts of improved genetic material for reforestation of commercial forest sites and not to produce seeds for tree limits which are usually 100 to 200 meters above the breeding zones. By establishing northern seed orchards along the coast and up to five degrees of latitude south of the clone origin, inadequate flowering and seed maturation at northern and high-altitude localities was circumvented.

To obtain information on breeding values of the selected plus trees, progeny trials were planted with full-sib or poly-cross families after most clones reached a fertile age, usually 10 to 15 years after seed orchard establishment; that is, in the late 1970s for most northern seed orchards.

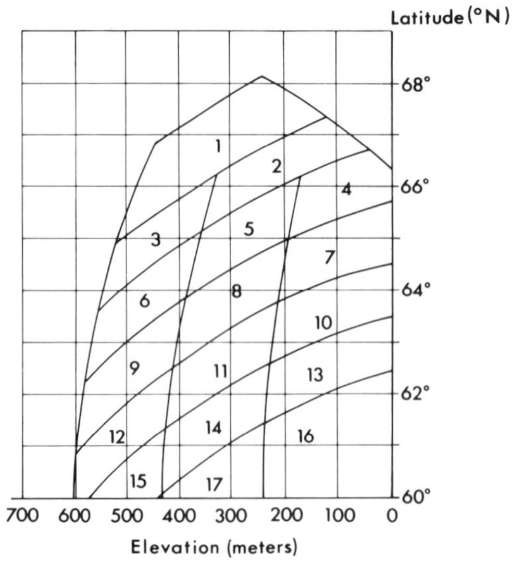

Fig. 1. The seventeen Swedish seed orchard zones (Numbers 1-17) for Scots pine north of latitude 60° N.

After the late 1970s, a new series of second-generation multi-clone seed orchards in Sweden was established to cover the seed demands of all seed orchard zones for Scots pine. Tree selection for the series was based on the results of early progeny freezing tests. To facilitate comparisons with neighboring trees, plus trees tested for the new seed orchards were selected in 20- to 40-year-old plantations. In addition, plus trees from the first-generation seed orchards, selected on the basis of their performance in progeny field and freezing tests, were included in the new orchards. Tested Finnish plus trees were also included in orchards for the northernmost seed orchard zones. The new seed orchards were designed for future genetic thinning, and were, like the first-generation orchards, established in milder climates than those of the clone origins.

The progeny testing of the plus tree clones guarantees that trees in the second-generation seed orchards are hardier than trees in the first-generation orchards. Seeds from the second-generation orchards are expected to be available for commercial plant production around 2000.

The Swedish tree improvement program, based on material collection and retrospective testing, is being revised to incorporate long-term multiple generation breeding (Danell, 1991). The objectives of the long-term multiple generation breeding are to sustain genetic stability, to conserve genetic variation, and to increase the efficiency of short-term breeding programs and improve genetic gain. Multiple closed breeding populations, each with genetic profiles adapted to different light and temperature regimes, will be selected for breeding.

The temperature regimes will cover both present and all possible future temperature conditions. Adaptation profiles of each population will be delineated from measurements of survival and growth characteristics rather than from geographic origins. For medium- and short-term breeding programs, elite open lines of limited size and longevity will be selected as branches from closed breeding populations and improved by intensive recurrent selection.

Open-pollination seed orchards will probably be inefficient in advanced generations, because genetic gain realized from closed seed-producing populations will be greater than the gain from the open seed-producing populations of today. New techniques to control pollination will be necessary to maximize gains. Improvement in individual traits will be less from multi-trait than from individual trait selection, but future economic gains will be greater.

The forest tree breeding programs described here, along with much of the applied research, are managed by Forestry Research Institute of Sweden (formerly known as the Institute for Forest Improvement). Basic research on breeding and seed orchard maintenance is conducted mainly in the genetic departments of the Swedish University of Agricultural Sciences, Umeå, Sweden. Cooperation and exchange of experience between the two organizations increases the prospect of successful tree breeding.

SELECTION FOR NORTHERN LATITUDES

Because low survival is a major problem at northern latitudes, increased hardiness is a main objective of commercial breeding of Scots pine for northern Sweden. To guarantee fast progress in breeding, reliable methods are needed for progeny testing. Further increase in hardiness between successive generations can be obtained if reliable and easily applied methods are available for phenotypic screening.

Selection in Field Experiments

At northern localities, the growth of forest trees is slow, and many years pass before trees extend above the snow where they are exposed to winter stresses such as strong winds, large temperature fluctuations, and desiccation. Selection for survival before the trees spend at least a few years above the winter snow may be unreliable. As a result, trees in progeny tests are not usually evaluated until they are 10 to 15 years old. In addition, winter damage and mortality must be visible before survival can be evaluated. It is impossible to predict the age of injury and mortality when field experiments are planned. All field trials are, therefore, unlikely to be practical for hardiness selection. Selection among and within families is also planned in full-sib family field tests.

Progeny Freeze Testing

Although progeny field tests are necessary for estimating breeding values and plus tree selection for the next seed orchard generation, they are not reliable for early breeding decisions to reduce generation times. Short-term experiments are needed for large-scale and reliable early testing.

Adaptation of Scots pine to climate gradients in northern Sweden has caused similar clinal relationships between latitude of origin and a number of phenological traits such as proportion of dry matter in needles or shoots (Hagner, 1970; Langlet, 1936, 1943), autumn frost hardiness (Andersson, 1985; Andersson and Westin, 1990; Dormling et al., 1977; Nilsson and Eriksson, 1986), and first-year bud set (Mikola, 1982). Such clinal adaptations are related to field mortality and autumn cold acclimation, and are useful as selection criteria for early cold acclimation.

Artificial freeze testing of young plants under controlled conditions in early autumn is often used in Sweden to study genetic variation in cold acclimation and hardiness, and in commercial breeding of Scots pine. The method can be used on plants grown in climate chambers, greenhouses, and other controlled environments, or outdoors. In controlled environments, the seedlings are first exposed to growth conditions with constant photoperiod, usually 18-20 hours of light depending on the provenance (Fig. 2). After 8 to 10 weeks of growth at approximately 20 °C day and 15 °C night temperatures, the plants are exposed to cold acclimation conditions by gradually increasing night length at a rate of one hour per

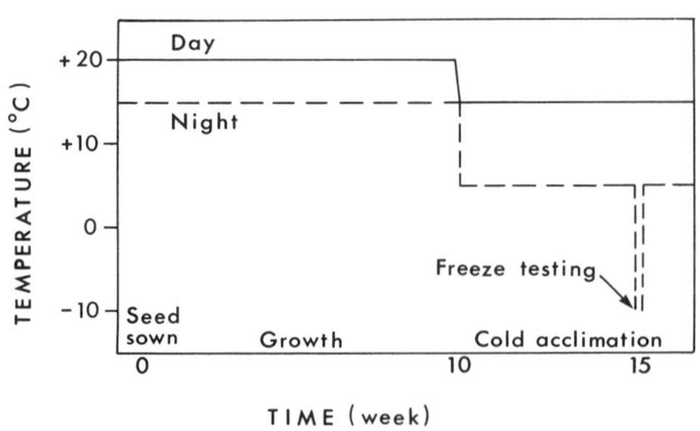

Fig. 2. A typical temperature regime for an artificial freezing experiment of one-year-old Scots pine.

week and reducing the day-night temperatures by 5 to 10 °C. After a few weeks of cold acclimation, the seedlings are gradually cooled in a freezing chamber to a predetermined temperature (often -10 °C). Damage to the seedlings is related to the level of cold acclimation at the time of testing, and can be determined about two weeks after freezing by visual assessment of the discoloration of needles and stems. For maximum resolution of cold acclimation between progenies, it is important that the date of freeze testing be adjusted to the median hardiness level of the test material. Progeny ranking is unreliable if freezing damage is too slight or too severe. Freezing damage can be assessed before leaf and stem injury is apparent, for example by measuring leaf chlorophyll fluorescence and freeze-induced electrolyte leakage, but these techniques have not been used for selecting hardy Scots pine in the Swedish breeding program.

Cold acclimation of one-year-old Scots pine seedlings, as assessed by artificial autumn freezing tests, is closely related to the latitude of origin and field mortality at 18 years of age (Fig. 3). Field survival is also related to family differences in early cold acclimation tests (Nilsson and Andersson, 1987; Nilsson et al., 1991). Thus, trials at both provenance and family levels indicate that Scots pine can be selected for survival at northern latitudes from progeny freeze tests in early autumn.

The level of cold hardiness in early autumn is inherited equally from both parents (Norell et al., 1986; Nilsson and Walfridsson, 1990), and plus trees can be tested as either mothers or fathers. The high general combining ability of cold hardiness facilitates progeny testing of parents with flowers of only one sex. Low specific combining ability justifies the use of multi-clone seed orchards in breeding for early cold acclimation.

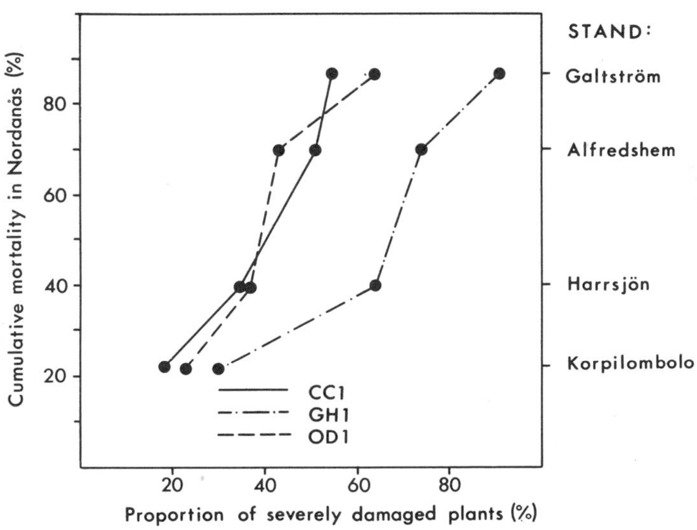

Fig. 3. Relationship between cumulative mortality of Scots pine after 18 years in a progeny field trial at Nordanås, latitude 64° N., longitude 18° E., 400 meters, and freezing damage to one-year-old seedlings grown in a climate chamber (CC1), greenhouse (GH1), and outdoors (OD1). The four populations tested are from latitude 67° N., 185 meters (Korpilombolo), latitude 64° N., 360 meters (Harrsjön), latitude 63° N., 8 meters (Alfredshem) and latitude 62° N., 5 meters altitude (Galtström) (Nilsson and Eriksson, 1986).

Interaction of genotype with environment was not found for freezing damage when Scots pine progenies were grown and cold acclimated during the same year at geographically widespread locations (Nilsson et al., 1991). Also, when grown in artificial conditions (climate chamber or greenhouse), progeny ranking is usually similar to the ranking obtained under natural conditions. This indicates that early autumn freeze testing is a reliable selection method for early cold acclimation in "normal" years. Ranking of progenies can be altered, however, by strongly manipulating the environment during growing and cold acclimation periods (Nilsson, 1988; Nilsson et al., 1991). For safety reasons, Scots pine progenies should be selected for cold acclimation under abnormal weather conditions, especially those unfavorable for cold hardening.

Repeated freeze tests of new plants on different occasions during the autumn gives more information on the variation in cold acclimation rhythm than does a single freezing test. Repeated freeze testing can also be used to study cold acclimation responses to a range of environmental conditions. For large-scale selections, a series of freeze tests may be impractical and uneconomical. If the number of plants available for testing is limited, repeated freeze tests of detached shoots or needles from a single plant may provide more information than a single whole-plant freeze test.

Repeated freeze testing of plants, shoots, or needles during a major part of the cold acclimation period, has not been used in commercial breeding of Scots pine. Future studies will indicate whether repeated freeze testing is a suitable alternative to whole-plant freeze testing in early autumn.

Phenotypic Selection

The cold climate of northern and especially, high altitude sites, slows growth and advances the age at which timberline species become fertile. Progeny testing is delayed until trees are sexually mature, and the majority of young trees on severe localities are excluded from progeny selection. Reliable methods for phenotypic selection at early ages would, in combination with flower stimulating treatments, considerably shorten generation times in a breeding program.

Phenotypic selection is often strongly influenced by variations in the environment. If environmental variation is large, the genetic gain from phenotypic selection for traits with low or moderate heritability is limited. Therefore, phenotypic selection is not recommended for ranking genotypes growing in different environments. For advanced generation breeding, however, selection among even-aged phenotypes at homogenous locations, such as in progeny field trials, has more potential for identifying the best genotypes than first-generation phenotypic selections from heterogenous locations. But even within sites, phenotypic selection is a relatively rough screening method for finding the most promising genotypes.

Phenological traits similar to those evaluated in progeny tests such as the proportion of needle dry matter in the autumn can also be used for phenotypic selection. Also, resistance of detached current-year shoots to autumn frost (Nilsson et al., 1991) or needles[1] (Lindgren and Nilsson, 1992), and the attachment of needles to the current-year shoots (Nilsson and

[1] Nilsson, J. E., and Walfridsson, E. A., 1989, Dokumentation av klonarkiv och avkommeförsök med tallindivider från extrema lokaler vid trädgränsen (Documentation of clone archives and progeny trials of Scots pine from extreme sites at the timberline), Department of Forest Genetics and Plant Physiology, Swedish University of Agricultural Sciences, Umeå, unpublished report.

Åman, 1986; Rummukainen, 1982) have similar clinal relationships as field survival to the latitude of origin. The merits of phenotypic selection within provenances for increasing survival in north Sweden, however, need further investigation before it is utilized in commercial selection and breeding of Scots pine.

The length of the annual growth period for Scots pine positively affects total shoot elongation. Because late growth cessation delays initiation of cold hardening, phenotypic selection for growth is probably not a very profitable way to increase field survival. This hypothesis is supported by negative correlations between parent height and rate of cold acclimation of the offspring (Nilsson, 1990; Nilsson and Åman, 1986). On the other hand, selection among the smallest trees to achieve early growth cessation and cold acclimation is not profitable from the growth aspect. Phenotypic selection for height must be supplemented by selection that counteracts possible negative selection for hardiness. Supplementary selection for tree quality and saw timber, as used in the commercial breeding of Scots pine, probably has a positive effect on hardiness and survival. Progeny testing, however, is a safe way to prevent clones of low hardiness from entering northern seed orchards.

Nilsson et al. (1991) found that a combination of first-year progeny freeze testing and freeze testing of detached shoots in autumn was the best selection criterion for vigorous progeny at ten years of age. The study also indicated that this selection method could increase survival without reducing height growth.

HARDINESS OF SEED ORCHARD TREES

Variation in breeding values for field survival and growth[2] is large between parent clones within the first generation seed orchards. Artificial freezing of progeny shows that genetic variation in the rate of cold acclimation is also large (Nilsson and Andersson, 1987; Nilsson et al., 1991). The genetic diversity within seed orchards mandates progeny testing to identify the hardiest seed orchard clones from which to collect cones for reforestation of the harshest sites.

Scots pine seeds produced in first generation seed orchards are not sufficiently hardy for reforestation of large areas in northern Sweden. In fact, the seed orchard progenies are less hardy than progenies from native stands of the same source. Studies of Andersson (1985), Ericsson (1988), and Andersson and Westin (1990) showed that the hardiness of the progeny of young seed orchards is similar to, and growth is better than, open-pollinated progenies from natural stands approximately halfway between the clone origin and the seed orchard location. A number of explanations have been proposed for reduced hardiness of the seed orchard progenies, including the method of plus tree selection and limited pollen production in combination with contamination of seed orchards with outside pollen.

Plus tree selection for the first generation seed orchards was based on growth and wood quality. As mentioned previously, phenotypic selection for rapid growth probably reduces hardiness, whereas selection for wood quality counteracts the decrease in hardiness. The hypothesis of null selection for hardiness in the first-generation seed orchards is supported by the same ten-year field survival of controlled plus tree crosses and progenies from natural stands of comparable geographical origin (Anonymous, 1984). The hardiness of seed

---

[2] Personal communication, Bengt Andersson, Research Leader, the Institute for Forest Improvement, Sävar, Sweden.

orchard crops can be increased by genetic thinning of inferior clones, or by collecting seeds from clones that produced the most vigorous progeny in field tests.

Variation in flowering time and intensity also affects the genetic composition of the seed orchard offspring. Even if flowering time and intensity are not directly related to hardiness, reduction in flowering decreases the effective population size. This is especially serious at young ages when the internal pollen production of the seed orchard is small and outside pollen contributes most of the male gametes. Between 21 and 55 percent of the pollen in young Swedish Scots pine seed orchards is from background sources (El-Kassaby et al., 1989; Nagasaka and Szmidt, 1985; Wang et al., 1991).

Because the climate is cooler, local pollen dispersal is later at northern than at southern latitudes. Northern populations, however, initiate female receptivity and pollen dispersal at lower heat sums than more southern ones. This means that little local pollen is available at the time of female receptivity in south transferred seed orchards. The contaminating pollen might, therefore, come from more southern localities where pollen dispersal is at the same time or even earlier than in the seed orchard. In this case, the contaminating pollen is less hardy than the pollen from the orchard clones, and the hardiness of the seed orchard progeny is reduced.

Research on seed orchard crops indicates that the origin and hardiness of outside pollen varies among years. Fluctuations in the hardiness of pollen can be obtained from yearly variations in wind direction and weather conditions that influence the time of local pollen dispersal. As an example, northern winds during the receptive period of a south transferred seed orchard might delay pollen dehiscence and decrease the availability of outside pollen for fertilization. Also, long rainy periods might reduce the amount of background pollen in the air.

Artificial mass pollination (Yazdani et al., 1986) with hardy pollen, and delay of female receptivity by evaporative cooling of the orchard clones until the proportion of background pollen is low, reduces the contamination from southern pollen and increases the hardiness of the seed orchard progeny. In fact, female receptivity in coastal seed orchards is somewhat delayed by cool air in early summer. Promotion of male flowering by hormone treatment may also reduce the proportion of background pollen from unselected trees of southern origin.

To understand more about the nature of pollen migration and its effect on progeny hardiness, 25 archives of Scots pine clones are planned in central and northern Sweden. Beginning in 1992, the same ten clones, each selected for abundant female and minimum male flowering, will be planted in each archive. Because the archives will not be located near pollen dispersing Scots pine stands, the clones will mainly be fertilized by background pollen. By testing the background pollinated progenies for hardiness-related phenological traits such as artificial freeze testing or bud set, yearly and geographical variation in the fertilizing background pollen can be studied. Utilizing the very strong relationship between autumn phenology and latitude of origin, progeny variation in phenology can be interpreted in terms of geographical variation in pollen origin.

It is sometimes proposed that hybridization effects may be positive when mating geographically distant populations. In a recent study on spontaneous hybridization of Scots pine in northern Sweden, however, survival and growth of the hybrids were not superior to natural populations halfway between seed and pollen sources; the reverse was true (Nilsson, 1992). Therefore, it is unlikely that the seed orchard progeny will perform better than the breeding values of the orchard male and female gamete populations, regardless of the origin of the outside pollen.

USE OF MARGINAL POPULATIONS FOR BREEDING

Scots pine plus trees for northern Sweden have so far been selected in natural stands, cultivated plantations, and in approximately 20-year-old field experiments on commercial forest land in northern Sweden. For the latest seed orchard generation, plus trees from northern Finland are also included. Plus tree selections have been restricted to vigorous high-quality phenotypes.

Because of inferior phenotypes and absence of genetic information, Scots pine from extreme localities in the forest-tundra ecotone and near the barren mountains are not included in the Swedish breeding program. Lack of flowering and seed maturation precludes genetic studies of timberline pine and genetic knowledge is inadequate in marginal areas. With financial support from The Swedish National Board of Forestry, we established two archives of Scots pine near Umeå with clones from the tree limit and tundra transition zone, latitude 68° to 69° N., and a series of five full-sib progeny field trials in northern Sweden to provide a basis for future research on timberline pine[1].

The clone archives were planted in 1985-86 with approximately 200 clones representing low, middle, upper, and tree limit population on four mountain slopes, and a geographical wide-spread forest-tundra ecotone in northwest Sweden. Trees were selected from extreme environments without regard for phenotype. The grafts are now established and studies on phenotype and phenology can begin. When the grafts are fertile, mating programs will be initiated for careful genetic studies of the timberline Scots pine. In a preliminary study, Yazdani and Nilsson (1986) found that monoterpene compositions of two mountain-slope populations in northern Sweden, were different from monoterpene compositions of southern populations. Differences in monoterpenes indicate that timberline populations in northern Sweden may have differentiated genetically from the southern populations.

At the extreme limits of growth near timberline, female cones were found on a few trees from 1982 to 1984, but the seeds were not viable and progeny tests could not be established. Small amounts of viable pollen were collected and a mating program was carried out on mother clones in northern seed orchards and in clone archives. Pollen from northern plus tree clones were used for reference crosses. The full-sib families were planted in 1987 and 1988 in a series of five progeny field trials at latitudes from 66° to 68° N. in interior Sweden. The series of experiments will provide the first field results on genetic variation in survival and growth of Swedish timberline Scots pine. The first evaluation is planned in 1992.

In a preliminary freezing experiment of surplus one-year-old timberline progenies from the field series, we found large genetic variation in the rate of cold acclimation of Scots pine (Fig. 4). Results also indicate that timberline populations acquire cold hardiness earlier than plus trees in today's breeding program for northern Sweden. Thus, timberline popula-

---

[1] Nilsson, J. E., and Walfridsson, E. A., 1990, Repeated freeze testing of detached needles of Scots pine during cold acclimation, Umeå, Swed. Univ. Agr. Sciences, Dept. of Forest Genetics and Plant Physiology, unpublished report.

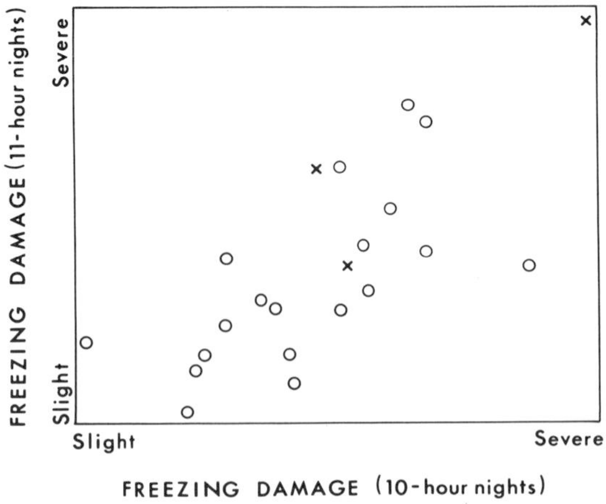

Fig. 4. Freezing damage of one-year-old full-sib families from the tree limit and tundra transition zone in northern Scandinavia (o) and from plus trees in northernmost first-generation Swedish seed orchards (x). Freezing tests were performed during cold acclimation at 10- and 11-hour nights. Half of the plants were tested on each occasion.

tions may be of interest in future selection and breeding of Scots pine for early cold acclimation at localities with short and cold summers.

SUMMARY

The goal of long-term breeding of Scots pine in northern Sweden is to gradually improve adaptation and growth, and to meet the demands of economic forestry. Today, high timber production and quality are given priority. For harsh sites in the interior of northern Sweden, wood production depends more on survival than on rapid growth. Thus, the object of tree breeding in northern Sweden is to improve hardiness while in south and central Sweden it is to improve growth.

The clinal adaptation of Scots pine to latitude means that selection for survival is limited to trees of northern origin. Plus tree selection for various geographical regions is based on progeny testing for survival in field trials and early cold acclimation in artificial freezing tests. For practical, economic, and experimental reasons, short-term progeny experiments, such as artificial whole-plant freeze testing during early cold acclimation, are very effective in breeding programs aimed at short generation times. Development of new methods to further increase the reliability of early testing and selection for important traits will make early selection even more effective than it is today. Phenotypic and phenologic screening for hardiness traits before trees are fertile will also shorten generation times.

Although breeding of Scots pine is not designed for extreme sites close to tree limits, selections from tree-limit localities may provide interesting alternatives in the breeding of Scots pine for harsh sites within forest limits. Even with very effective plus tree selection criteria, pollen contamination from outside sources reduces the quality of seed orchard crops. Understanding natural pollen migration and how to reduce the effects of pollen contamination in seed orchards is important for the success of forestry at high latitudes and elevations.

REFERENCES

Andersson, B., 1985, Frystest av norrländska tallplantage- skördar, Freezing tests of northern Scots pine seedlings from seed orchards. In: Nilsson, J. E., ed., Reforestation material for harsh northern sites. Proceedings of a seminar at Umeå, 1985, February 19-20, Department of Forest Genetics and Plant Physiology, Swedish University of Agricultural Sciences, Umeå, Report 4:42-56.

Andersson, B., and Westin, J., 1990, Frystestresultat för material (Pinus sylvestris L.) från svenska och finska tallfröplantager och naturbestånd (Results of freezing tests on Scots pine (Pinus sylvestris L.) seedlings from Swedish and Finnish seed orchards and natural stands), The Institute for Forest Improvement, Uppsala, Report 17:1-31.

Anonymous, 1984, Lovande resultat från testning av norrländska tallfröplantager, Stockholm, Skogen, 5/6:86.

Danell, Ö., 1991, Survey of past, current, and future Swedish forest tree breeding, Silva Fennica, 25:241-247.

Dormling, I., Eriksson, G., and Jonsson, A., 1977, Foto- och termoperiodiska reaktioner hos tall och contortatall, Photo- and thermoperiodic reactions of Pinus sylvestris and Pinus contorta, Royal College of Forestry, Department of Forest Genetics, Stockholm, Research Notes 27, 48-57.

Eiche, W., 1955, Cold damage and plant mortality in experimental plantations with Scots pine in northern Sweden, Studia For. Suecica, 36:1-218.

El-Kassaby, Y. A., Rudin, D., and Yazdani, R., 1989, Levels of outcrossing and contamination in two Pinus sylvestris L. seed orchards in northern Sweden, Scand. J. For. Res., 4:41-49.

Eriksson, G., Andersson, A., Eiche, V., Ifver, J., and Persson, A., 1980, Severity index and transfer effects on survival and volume production of Pinus sylvestris in northern Sweden, Studia For. Suecica, 156:1-32.

Ericsson, T., 1988, Odlingstest av tallplantagefrö i Norrland (Scots pine seed orchard tests in northern Sweden, Results from assessments in summer 1984 of field trials planted 1973-75), The Institute for Forest Improvment, Uppsala, Report 1, 1-46

Hagner, M., 1970, A genecological investigation of the annual rhythm of Pinus silvestris L., Studia For. Suecica, 80:1-40.

Langlet, O., 1936, Studier över tallens fysiologiska variabilitet och dess samband med klimatet, Stockholm, Meddelanden från Statens Skogsförsöksanstalt, 29:219-470.

Langlet, O., 1943, Photoperiodismus und Provenienz bei der gemeinen Kiefer (Pinus silvestris L.), Stockholm, Meddelanden från statens skogsförsöksanstalt, 33:295-330.

Lindgren, K., and Nilsson, J. E., 1992, Cold acclimation of Pinus sylvestris and Pinus contorta provenances as measured by freezing tolerance of detached needles, Scand. J. For. Res., 7:309-315.

Mikola, J., 1982, Bud-set phenology as an indicator of climatic adaptation of Scots pine in Finland, Silva Fennica, 16:178-184.

Nagasaka, K., and Szmidt, A. E., 1985, Multilocus analysis of external pollination of a Scots pine (Pinus sylvestris L.) seed orchard. In: Gregorious, H. R., ed., Population genetics in forestry, Lecture Notes in Biomathematics 60:134-138.

Nilsson, J. E., 1988, Variation in the rate of winter hardening of one-year-old plus tree families of Scots pine raised in different environments, Silva Fennica, 22:213-224.

Nilsson, J. E., 1990, Variation in early winter hardening within families of Pinus sylvestris L. from northern Sweden, Scand. J. For. Res., 5:17-27.

Nilsson, J. E., 1992, Growth and survival of spontanous provenance hybrids of Pinus sylvestris, Scand. J. For. Res., 7:193-203.

Nilsson, J. E., and Andersson, B., 1987, Performance in freeze testing and field experiments of full-sib families of Pinus sylvestris L., Can. J. For. Res., 17:1340-1347.

Nilsson, J. E., and Åman, M. L., 1986, The phenological rhythm of Pinus sylvestris clones and its relationship with damage of one-year-old offsprings after freeze testing. In: Lindgren, D., ed., Provenances and Forest Tree Breeding for High Latitudes, Proceedings of the Franz Kempe symposium at Umeå, 1986 June 10-12, Department of Forest Genetics and Plant Physiology, Swedish University of Agricultural Sciences, Umeå, Report 6, 127-147.

Nilsson, J. E., Andersson, B., and Walfridsson, E. A., 1991, Progeny freeze testing, progeny field testing and parental phenology of Pinus sylvestris L. clones in northern Sweden, Scand. J. For. Res., 6:177-195.

Nilsson, J. E., and Eriksson, G., 1986, Freeze testing and field mortality of Pinus sylvestris L. in northern Sweden, Scand. J. For. Res., 1:205-218.

Nilsson, J. E., and Walfridsson, E. A., 1990, Autumn freeze testing of one-year reciprocal families of Pinus sylvestris L., Scand. J. For. Res., 5:501-511.

Norell, L., Eriksson, G., Ekberg, I., and Dormling, I., 1986, Inheritance of autumn frost hardiness in Pinus sylvestris L. seedlings, Theor. Appl. Genet., 72:440-448.

Persson, B., and Ståhl, E. G., 1990, Survival and Yield of Pinus sylvestris L. as related to provenance transfer and spacing at high altitudes in northern Sweden, Scand. J. For. Res., 5:381-395.

Remröd, J., 1976, Val av tallproveniensed i norra Sverige (Choosing Scots pine (Pinus sylvestris L.) provenances in northern Sweden), Department of Forest Genetics, The Royal College of Forestry, Stockholm, Research Notes 19, 1-132.

Rummukainen, von U., 1982, Die Haftfähigkeit von Kiefernnadeln, Allgemeine Forstzeitschrift, 14:416-418.

Ståhl, E. G., and Andersson, S., 1985, Avgång och skadefrekvens i 30-åriga tallproveniensförsök i hårda klimatlägen (Mortality and frequency of damage in 30-year-old Scots pine trials in severe climatic conditions). In: Nilsson, J. E., ed., Reforestation material for harsh northern sites, Proceedings of a seminar at Umeå, 1985 February 19-20, Department of Forest Genetics and Plant Physiology, Swedish University of Agricultural Sciences, Umeå, Report 4, 66-74.

Wang, X. R., Lindgren, D., Szmidt, A. E., and Yazdani, R., 1991, Pollen migration into a seed orchard of Pinus sylvestris and the methods of its estimation using allozyme markers, Scand. J. For. Res., 6:378-385.

Yazdani, R., Hadders, G., and Szmidt, A. E., 1986, Supplemental mass pollination in a seed orchard of Pinus sylvestris L. investigated by isozyme analysis, Scand. J. For. Res., 1:309-315.

Yazdani, R., and Nilsson, J. E., 1986, Cortial monoterpene variation in natural populations of Pinus sylvestris in Sweden, Scand. J. For. Res., 1:85-93.

PROVENANCE AND INDIVIDUAL VARIATION IN CLIMATIC HARDINESS OF SCOTS PINE IN NORTHERN FINLAND

Jouni Mikola

Foundation for Forest Tree Breeding
Helsinki, Finland

INTRODUCTION

In the optimum zone of Scots pine (Pinus sylvestris L.) in northern Finland, seed matures when 77 percent of the local average annual heat sum is reached (Sarvas, 1967, 1970). In absolute units, the temperature sum required for seed maturation is much higher for southern provenances than for northern ones. In the northern marginal areas, 950 growing degree days (d.d.) at a 5 °C threshold is the lowest average heat sum to which the annual development cycle of Scots pine is genetically adapted. In places where the average annual heat sum falls below this limit, for example, north 65° to 66° N. latitude in Finland (Fig. 1), all provenances and individuals have roughly the same minimum heat requirement of 950 degree days for complete ripening of the seed (Ryynänen, 1982; Sarvas, 1967, 1970).

Sarvas (1969) suggested that in marginal areas, natural selection for short and cool growing seasons exhausted the genetic variation in the annual development of Scots pine. This hypothesis seems true for generative processes, but not for annual vegetative development. Extensive field studies on the growth and survival of provenances in northern Sweden show that a climatic gradient in the genetic adaptation of Scots pine extends to the northern edge of the marginal area (Eiche and Andersson, 1977; Langlet, 1959). In Finland, the timing of first-year bud set is linearly related to latitude and average heat sum of the seed origins throughout the country, and to northern timberline (Mikola, 1982).

Scots pine appears to be poorly adapted to the prevailing climate throughout the marginal area defined by Sarvas (1967, 1970), at least for commercial forestry. Because of generative maladaptation, good seed crops are infrequent for most of the northern half of Finland. Moreover, high seedling mortality hinders reforestation of Scots pine in the north, even when local seed is used.

To ensure the availability of mature seed, about 1,800 hectares of seed orchards of more than 4,000 northern Scots pine "plus" trees have been established in southern and

Forest Development in Cold Climates, Edited by J. Alden *et al.*
Plenum Press, New York, 1993

central Finland. Pollen contamination from surrounding forests, however, is a serious problem. The hybrid seed resulting from background pollination is not hardy enough for reforestation of northern habitats (Mikola, 1986; Nikkanen, 1982).

Extensive genetic testing of seed orchard progenies of northern Scots pine was started in Finland in the 1970s. In recent years, much data has accumulated on the survival of these trees in the field. This paper summarizes the results of representative experiments for improving the hardiness of reforestation material by individual tree selection and breeding. The suitability of field and other tests for assessing individual differences in climatic hardiness and the use of natural provenances as reference materials are emphasized.

MATERIAL AND METHODS

The data presented here are based on progeny tests of north Finnish Scots pine plus trees selected for tree breeding and seed production. Provenance or population samples from natural stands have been included as reference materials in all progeny tests. All experiments were planned and measured by the Finnish Forest Research Institute, Department of Forest Genetics, in Vantaa, Finland. The studies were established and maintained by the National Board of Forestry (Finnish Forest Service). Only a few representative experiments were chosen for analysis. Many other experiments, however, support the findings and conclusions in this paper.

All field tests reported here were composed of six to 10 complete randomized blocks of 25-tree plots (150 to 250 plants per entry) in each test site. The original spacing of plants was two by two meters or, on a few sites, one by one meter. For this paper, the data analysis was limited to the comparison of entry means, that is, the percentage of average survival of families and provenances at each test site.

Linear correlation analysis was used to illustrate the adaptive significance of provenance variation and the usefulness of natural population samples as reference materials in progeny testing for climatic hardiness and adaptation. Strong correlations between geographic origin and survival are evidence of intense selection for climatic hardiness. Weak correlations, on the other hand, indicate that the plant mortality and climate relationship is negligible, and mortality is caused by stress factors other than harsh climate. The idea of evaluating the resolving power of genetic experiments with the aid of suitable reference materials has been discussed in an earlier paper by Mikola (1986).

The study of individual variation in hardiness was based on the consistency of progeny performance on different sites. Strong correlations in family survival between two or more sites demonstrate real genetic differences in hardiness. Together with the performance of reference provenances, these correlations indicate which test sites give reliable and realistic information for determining the climatic hardiness of mother trees.

RESULTS AND DISCUSSION

Seed Orchard Material in Testing of Hardiness

When large-scale tests of breeding values of northern Scots pine plus trees were started in Finland, there was good reason to wonder whether open-pollinated provenance hybrid seed from south-transferred seed orchards could be used. Results of a series of experiments

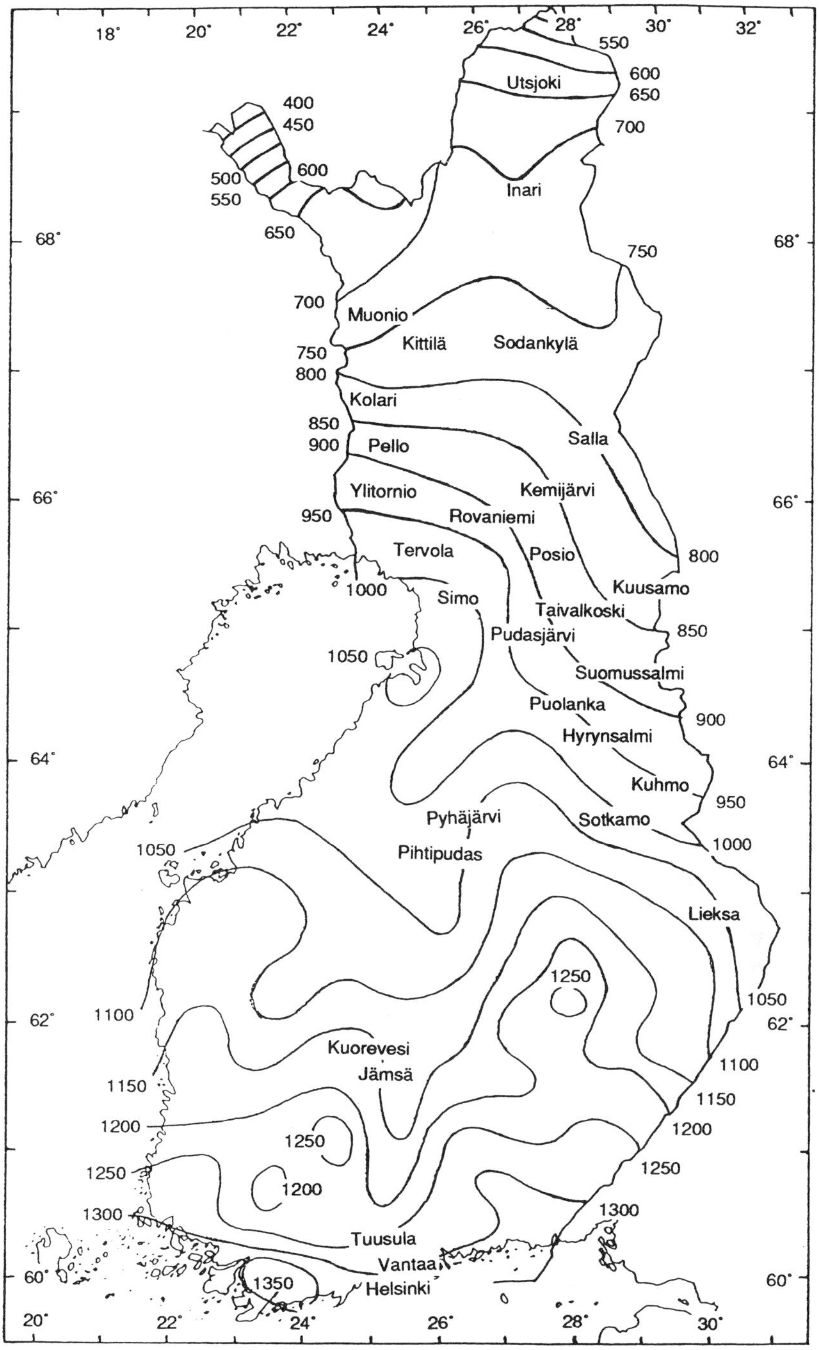

Fig. 1. Isograms of average annual heat sum (growing degree days at a 5 °C threshold) in Finland. Means of years 1931-1960, at 50 d.d. intervals. Modified from Kolkki (1966). Geographic locations mentioned in the text or in Figures 2 to 5 are also shown.

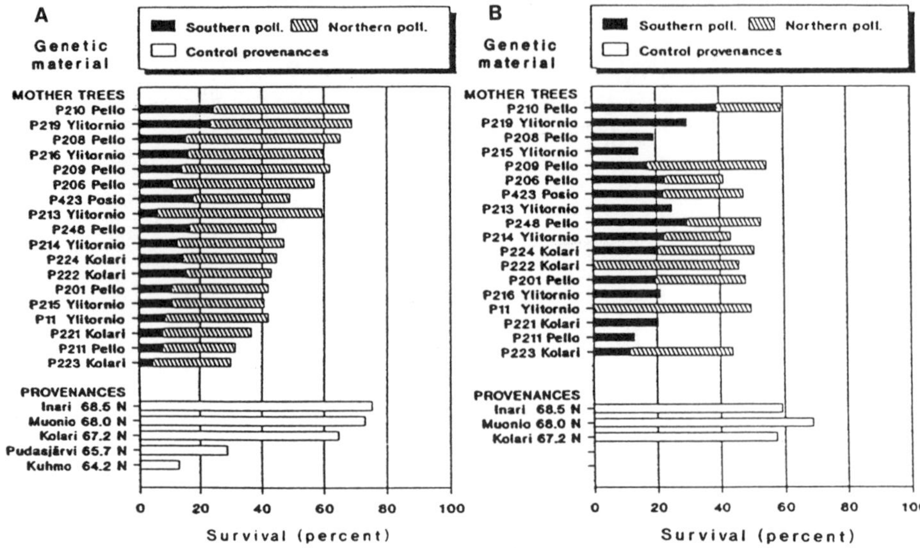

Fig. 2. Variation in 10-year survival (percent) among southern-pollinated and northern-pollinated seed orchard progenies of 18 mother trees at two harsh fieldsites in northern Finland. (A) Site 1 = KOLARI, latitude 67° N.; (B) Site 3 = POSIO, latitude 66° N. (Source: Finnish Forest Research Institute, Department of Forest Genetics, National Register, Vantaa, Finland; data of experiment series No. 519/1-3.)

to answer this question are shown in Figure 2. The family materials consist of both open-pollinated northern x southern progenies and control-pollinated northern x northern progenies of 18 mother clones originating from latitudes 66°30′ to 67°18′ N. in western Lapland. A pollen mixture of clones originating from the same area was used in controlled crosses. The open-pollinated seed was harvested in a seed orchard at Jämsä, southern Finland (latitude 62°00′ N.) in 1969. At that time, the orchard did not produce pollen, and all seed must have resulted from background pollination from southern pine forests. The control-pollinated seed was produced in the same seed orchard a few years later.

On Site 1 at Kolari (Fig. 2a), the variation in survival among the mother trees is very consistent between northern and southern family groups (r = 0.675***).[1] This suggests that individual genetic effects in hardiness are additive, that is, about one half of the average hardiness of progenies is determined by the mother tree, regardless of the origin or genetic adaptation of the father population. Therefore, open-pollinated seed orchard material appears to be suitable for screening mother trees for hardiness, assuming that the pollen pool is the same for all progenies. This condition is fulfilled only in young orchards where all pollen is from background sources.

Survival variation among provenances is large on Site 1 (Fig. 2a), and closely associated with geographic origin (r = 0.922*). The significant correlation of survival and geographic origin indicates that test conditions are effective in showing genetic variation in climatic hardiness.

[1] Significance levels of the correlation coefficients (r):*** $p<0.001$, ** $p<0.01$ * $p<0.05$, n.s. = $p\geq 0.05$.

The open-pollinated families survived, on the average, about as well as the control provenance from latitude 64°00′ N. Survival of the control provenance corresponded closely to the arithmetic origin of seed orchard materials halfway between the origin of mother trees (latitude 66°30′ to 67°18′ N.) and the father population (62°00′ N.). Compared to the reference provenances, most of the control-pollinated northern x northern families, however, had lower survival than expected on the basis of their geographic origin (66°30′ to 67°18′ N.). The low survival of northern families produced in southern orchards may be due to: (1) below average hardiness of plus trees in their native populations, (2) southern pollen contamination in the seed orchard crosses, and (3) physiological after-effects resulting from the southern locality of seed development (Björnstadt, 1981).

On Site 3 at Posio (Fig. 2b), experimental material was inadequate because of a shortage of plants at the time of establishment. Although Site 3 is less informative than Site 1, the results are in agreement. Mother tree effects on 10-year survival on Site 3 are fairly consistent in the two progeny groups (northern x southern and northern x northern, r = 0.606 n.s., degrees of freedom = 7). Also, correlations in family survival between Sites 1 and 3 are strong: 0.725*** and 0.649* for southern-pollinated and northern-pollinated progenies, respectively. Site 2 of this series of experiments was excluded from the analysis in Figure 2 because it was located in the mild climate of Simo at latitude 65°48′ N. (altitude 100 meters), where overall survival was high and variation among the test materials was low.

Variation in Hardiness Within Provenance

Figure 3 shows the survival of open-pollinated seed orchard progenies in a series of field tests for screening northern plus trees cloned in three seed orchards at Jämsä in southern Finland (latitude 62°00′ N.). All clones originated from between 67 and 69° N. latitude in northern Finland. Most clones were from a few stands in Inari at latitude 68°36′ to 68°42′ N. The mother trees were vigorous individuals from several natural populations or provenances in climatically harsh sites where annual heat sum averaged only 700 to 800 growing degree days. A set of eight natural provenances, sampled from the northern edge of the species' range at 69° N. to the seed orchards at 62° N. latitude, was included in the tests as reference material. The family and provenance entries were the same on all experimental sites shown in Figure 3.

The variation in survival between families is strikingly large, and fairly consistent, for all four sites. The survival of seed orchard progenies averaged slightly less than the survival of natural provenances from latitudes around 65° N., about halfway between the maternal and paternal origins of the seed orchard material. The best families on each site, however, survived as well as the northernmost control provenance from latitude 69°42′ N. Survival of the weakest families was as poor as survival of the southernmost provenance from latitude 62°00′ N. This provenance represents the natural population of the seed orchard locality, and the average climatic adaptation of the father population of background-pollinated seed orchard progenies.

The similarity of survival ranking of families on different sites clearly demonstrates that a great part of the observed variation is due to genetics. Site-to-site correlations in the average survival percentages of the families ranged from 0.596*** to 0.378* (d.f. = 27). The variation between control provenances indicates that all four sites were effective in revealing genetic differences in climatic adaptation; correlations between the survival percentages and origin latitudes of provenances on Sites 1, 2, 3, and 4 are 0.914***, 0.922***, 0.724*, and 0.864**, respectively.

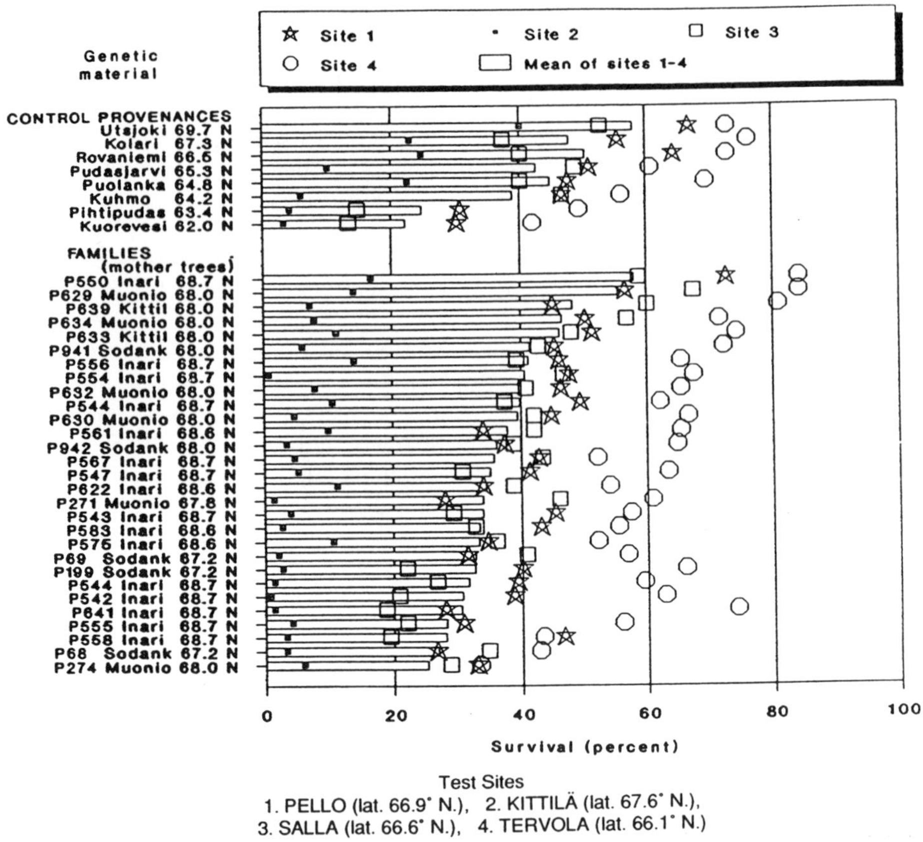

Fig. 3. Variation in 5-year field survival of open-pollinated seed orchard progenies of high-latitude plus trees on four sites in northern Finland. Test sites are as follows: 1. PELLO (lat. 66°54′ N.), 2. KITTILÄ (lat. 67°36′ N.), 3. SALLA (lat. 66°36′ N.), 4. TERVOLA (lat. 66°06′ N.) (Source: Finnish Forest Research Institute, Department of Forest Genetics, National Register, Vantaa, Finland; data of experiment series No. 792/1-6.)

This series of experiments included two other test sites, located further south in a milder climate: Site 5 at Hyrynsalmi, latitude 64°36′ N., and Site 6 at Sotkamo, latitude 63°48′ N. These sites were omitted from the analysis because they provided little information about hardiness variation. Correlation coefficients in family survival between these sites and Sites 1 through 4 were all positive, but not significant. Correlations between the survival and latitude of the control provenances were not significant (0.689 and 0.379, respectively; d.f. = 6) on Sites 5 and 6.

Effect of Geographic Origin on Family Hardiness

Figure 4 summarizes results from a series of field tests designed to study the significance of geographic origin in the search for exceptionally hardy individuals. The family materials consist of open-pollinated seed orchard progenies of five mother plus tree clones from each of six regions in northern Finland. Eight natural provenance samples were included as controls, but a shortage of plants precluded establishment of adequate control material on most sites. With one exception, however, the family material is the same on all sites.

As in previous studies, family variation in survival was very similar from site to site in this experiment. Site-to-site correlations in survival percentages of all families ranged from 0.582*** to 0.491**. Thus, all sites provided consistent resolution of real genetic differences. Even the deficient control material of natural provenances suggests that all three sites shown in Figure 4 have been efficient in resolving genetic differences in climatic adaptation. Correlation coefficients between survival and latitude of origin of reference provenances were high (0.689, 0.909*, and 0.923 on Sites 1, 2, and 4, respectively), but except for Site 2, a survival-origin relationship could not be established due to the small number of observations.

Site 3 of this series was excluded from the report for the same reasons that some sites were excluded in the two previous series of experiments. Site 3 is located in a mild climate at Pyhäjärvi (latitude 63°54′ N.), and survival of both families and provenances diverges from the results at Sites 1, 2, and 4.

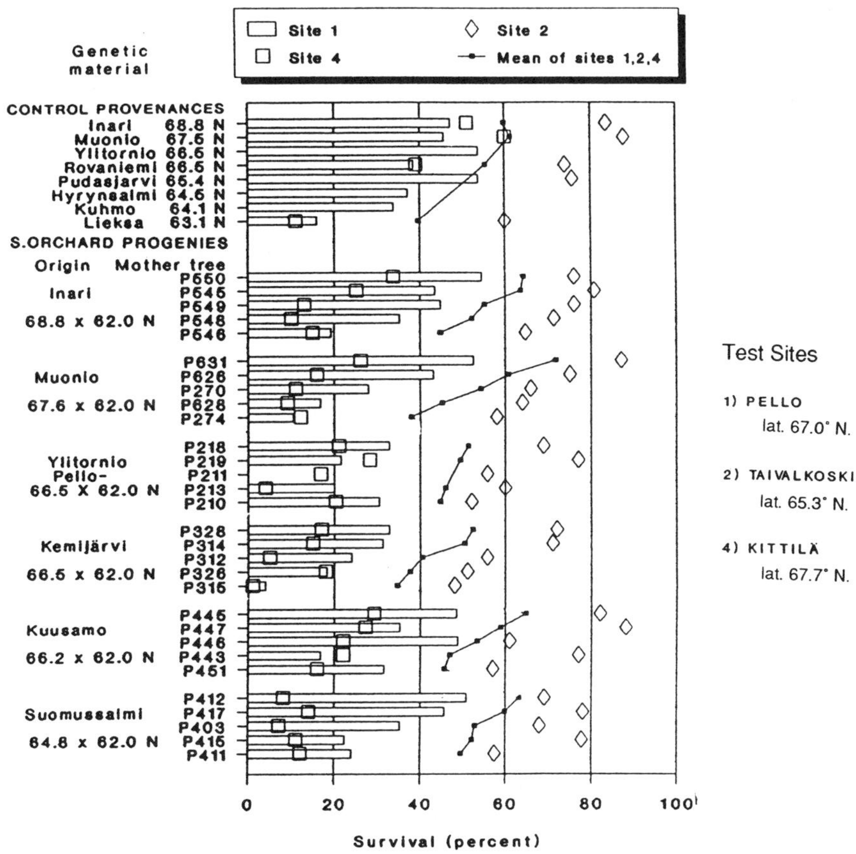

Fig. 4. Variation in 5-year field survival of open-pollinated seed orchard progenies of 30 plus tree clones, representing six geographic origins, at three sites in northern Finland. Test sites are as follows: 1. PELLO (lat. 67°00′ N.), 2. TAIVALKOSKI (lat. 65°18′ N.), 4. KITTILÄ (lat. 67°42′ N.) (Source: Finnish Forest Research Institute, Department of Forest Genetics, National Register, Vantaa, Finland; data of experiment series No. 979/1-4.)

The geographic origin of trees in North Finland appears not to have any conspicuous effect on the hardiness of individual mother trees or their progenies. Average survival of progeny groups was slightly different, but this variation was not related to geographic location or climatic severity. Absence of a relationship between survival and geographic origin may be due to the small sample size (five mother clones per region), or to bias in sampling. All mother clones for the study were chosen only on the basis of availability of seed in very young seed orchards, but this may have caused unintentional selection for climatic adaptation and hardiness.

In any case, it appears that both hardy and non-hardy Scots pine can be found throughout northern Finland. The individual variation in hardiness within provenances seems to cover the average differences between natural provenances separated from each other by four to five degrees in latitude (Figs. 2, 3, and 4), or by 400 to 500 kilometers in north-south distance.

Indirect Testing Methods

In the 1970s, discovery of relationships between the geographic origin of Finnish Scots pine and the timing of growth cessation in one-year-old seedlings (Mikola, 1982) prompted interest in developing a handy early method of screening breeding materials for climatic adaptation and hardiness. Results have been disappointing, however, for progeny testing of individual trees.

In Figure 5, the results of a greenhouse test are compared with field survival of the same material. The survival data is the same as shown in Figure 3, that is, the average survival of families on four field sites. As demonstrated previously, field survival in this series of experiments provides a very accurate resolution of the real hardiness differences among families. The greenhouse test was carried out at Tuusula in southern Finland (latitude 60°18′ N.). It was started several years after the establishment of the field tests, with 17 of the 29 open-pollinated seed orchard families of northern mother clones which were still available as stored seed. The control provenance materials in Figure 4, however, are only roughly comparable in the two experiments; they originate from the same regions in both experiments, but not from the same stands.

A very strong relationship appears to exist at the provenance level between the first-year-growth period in the greenhouse and field survival after 5 years ($r = -0.913^{***}$). Both field survival and the duration of first year growth are closely associated with the latitude of origin; for growth duration the correlation with latitude is even stronger ($r =-0.995^{***}$) than for survival ($r = 0.938^{**}$).

At the family level, the correlation between survival and the length of the first-year growth period is weaker ($r = -0.494^{*}$) than at the provenance level. The range of variation among families for both field survival and the first-year growth period is, however, almost the same as the range of variation among provenances for these traits. Similar relative variation of growth rhythm characteristics and correlations with field survival have been found for family and provenance materials in other greenhouse experiments, and even in freezing tests of northern Scots pine.[2] In general, physiological growth rhythm characteristics of young plants seem to be very strongly associated with subsequent hardiness and field survival at the provenance level but not at the family level.

---

[2] Author's unpublished data in the Finnish Forest Research Institute, Department of Forest Genetics, Vantaa, Finland, and unpublished data from B. Andersson, the Institute for Forest Improvement, Sävar, Sweden.

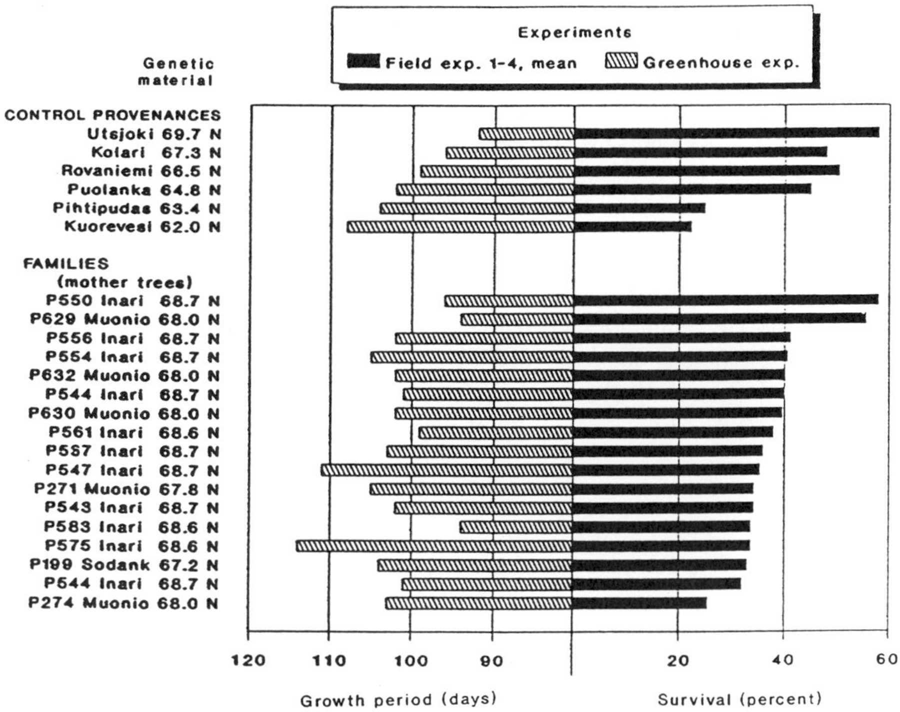

Fig. 5. Variation in 5-year field survival and first-year growth period among 17 open-pollinated seed orchard progenies and 6 control provenances of Scots pine. Survival data is based on four test sites, and is shown in more detail in Figure 2. The growth period is the time interval from sowing to 50 percent budset. (Source: Finnish Forest Research Institute, Department of Forest Genetics, National Register, Vantaa, Finland; data of experiments No. 792/1-4 and 1159.)

CONCLUSIONS

Although field testing is expensive and time consuming, observation of survival in harsh field conditions is the most realistic and reliable method of testing climatic adaptation and hardiness of tree breeding materials at present. Resolution of genetic differences can be achieved rapidly even in field tests, simply by locating the test sites in sufficiently severe conditions near northern and high altitude tree limits. The necessity of conducting survival tests may be very frustrating for the practical forester who needs to allocate land and labor for such tests. Most of the test sites reported in this paper, for instance, have shown clear-cut genetic variation in survival of Scots pine at 5 or 10 years of age. Many of these sites, however, must have been abandoned in subsequent years as the average survival percentages of background-pollinated seed orchard progenies have fallen to near zero.

Indirect evaluation of hardiness of Scots pine from annual growth rhythm characteristics works well for provenance or population materials, but not for individual trees or family materials. The strong correlations observed at the provenance level between geographic origin, field survival, and the timing of growth cessation were expected and are explainable. At the family level, however, the associations between the above characteristics are not as clear, although the same causal relations should prevail.

The experiments reported here demonstrate that the climatic hardiness of individual Scots pine trees is genetically determined and that the variation within populations of northern Finland is large. The large variation within provenances suggests that the hardiness of northern reforestation material can be improved by progeny testing and breeding of individual tree selections. The potential for improving hardiness of northern Scots pine seems to be better than the potential for improving growth rate or timber quality of more southern Scots pine in Finland.

Problems in the present method of producing improved Scots pine seed (background pollination in southern orchards established with northern clones) limits the effectiveness of hardiness breeding. Geographic transfer of seed orchards to sites beyond the range of Scots pine may eliminate pollen contamination (Koski, 1987). The development of effective vegetative propagation methods for Scots pine may finally offer a solution to all pollen contamination problems.

REFERENCES

Björnstadt, Å., 1981, Photoperiodical after-effects of parent plant environment in Norway spruce (Picea abies (L.) Karst.) seedlings, Communic. Norwegian Forest Res. Inst., 36(6):1-30.

Eiche, W., and Andersson, E., 1977, Survival and growth in Scots pine (Pinus sylvestris L.), provenance experiments in northern Sweden, Theor. and Appl. Gen., 26(5-6):168-175.

Kolkki, O., 1966, Tables and maps of temperature in Finland during 1931-1960, Meteorol. Yearbook of Finland, Vol. 65, Part 1a:1-42.

Koski, V., 1987, Long geographic transfers, a possible way of eliminating pollen contamination in advanced-generation seed orchards, Forest Ecol. and Manage., 19:267-271.

Langlet, O., 1959, A cline or not a cline—a question of Scots pine, Silvae Genetica, 8(1):13-22.

Mikola, J., 1982, Bud-set phenology as an indicator of climatic adaptation of Scots pine in Finland, Silva Fennica, 16(2):178-184.

Mikola, J., 1986, Reference material as an aid in interpreting progeny test results, Proceedings of a IUFRO Conference, A Joint Meeting of Working Parties on Breeding Theory, Progeny Testing and Seed Orchards, 1986 October 13-17, Williamsburg, Virginia, The North Carolina State University-Industry Cooperative Tree Improvement Program, North Carolina State University, Raleigh, North Carolina, United States, 482-491.

Nikkanen, T., 1982, Survival and height growth of North Finland x South Finland hybrid progenies of Scots pine in intermediate areas, Folia Forestalia, 527:1-31.

Ryynänen, M., 1982, Individual variation in seed maturation in marginal populations of Scots pine, Silva Fennica, 16(2):185-187.

Sarvas, R., 1967, The annual period of development of forest trees, Proceedings of the Finnish Academy of Science and Letters 1965, Helsinki, 211-231.

Sarvas, R., 1969, Genetical adaptation of forest trees to the heat factor of the climate, Second World Consultation on Forest Tree Breeding, 1969 August 7-16, Washington, D.C., FAO and IUFRO, Paper FO-FTB-69-2/15, 1-11.

Sarvas, R., 1970, Establishment and registration of seed orchards, Folia Forestalia, 89:1-24.

HYBRIDIZATION AMONG PROVENANCES OF LODGEPOLE PINE

Anders Fries

Department of Forest Genetics and Plant Physiology
Swedish University of Agricultural Sciences
Umeå, Sweden

INTRODUCTION

This paper describes a hybridization experiment among provenances of lodgepole pine (Pinus contorta var. latifolia Engelm.). Lodgepole pine is one of four varieties of Pinus contorta Dougl. ex Loud. from North America (Critchfield, 1957). It is characterized by wide geographical distribution and broad ecological amplitude (Critchfield, 1980), which has resulted in large genetic variation. This variation and the population structure must be known when introducing the species for commercial use.

Lodgepole pine has been planted as a commercial species in Sweden since the 1970s. Based on a number of provenance tests (Lindgren, 1983; Lindgren et al., 1988; Rosvall et al., 1984), provenance transfer recommendations are now available. Provenance testing should continue, however, because lodgepole pine has only recently been introduced in Sweden.

Differences in the performance of the provenances in this study are influenced by the population history of the species. Factors such as the distribution before and during the latest glaciation, and routes of migration, influence survival and growth of the provenances in the study area. The latest glaciation ended approximately 16,000 years ago in the areas of the most southern provenances, and approximately 10,000 years ago in the areas of the most northern provenances (Reeves, 1973; Wheeler and Guries, 1982). Following glacial retreat, the southern populations migrated north. Populations of lodgepole pine may, however, have survived in ice-free refugia in Alberta, the Yukon, and Alaska (Critchfield, 1980; Hultén, 1937; Nyland, 1980). During the glaciation, these isolated populations should have evolved independent of the southern populations. Genetic isolation in combination with the discontinuous distribution of lodgepole pine in the Yukon today could cause large differentiation among the populations. Fossil pollen records of MacDonald and Cwynar (1985) contradict the northern refugia theory. The fossil pollen records, however, indicate that all lodgepole pine in the Yukon originated from populations that arrived in British Columbia 12,000 years ago and reached the Yukon approximately 3,000 ± 1,000 years ago. In

Forest Development in Cold Climates, Edited by J. Alden *et al.*
Plenum Press, New York, 1993

addition, allelic frequencies indicate that the species is still, or possibly just stopped, migrating northward (Cwynar & MacDonald, 1987). Their conclusion is that mechanisms other than northern refugia must explain the genetic differentiation among northern populations.

Provenance hybrids that diverge from mid-parent breeding values to impart heterosis may be useful in tree breeding. Studies of heterosis are important because plus trees of different origins are planted together in Swedish seed orchards, and provenance hybrids are often produced. Furthermore, in active breeding, selections from provenances or plus trees that perform well after acclimatization to Swedish conditions are crossed and provenance hybrids are created. Hybrids between coastal and inland provenances are used in British breeding programs to combine rapid growth and good stem form (Aldhous, 1976; Lines, 1976).

This study aims to determine which provenances hybridize and if heterosis can be utilized to improve growth and stem form. The study was made by evaluating field tests of the oldest full-sib families of lodgepole pine in Sweden. The families originated from unselected parent trees, but full-sib progenies of plus trees are now available in field tests. The unselected parents originated from areas where the plus trees were selected. Growth capacity, growth rhythm, weather damage and flowering were investigated. As a measure of growth rhythm and hardiness, proleptic shoots that formed late in the growing season were also observed. Most investigations have demonstrated that the inheritance of growth rhythm is under additive gene control (Mikola, 1982; Norell et al., 1986; Rehfeldt, 1977).

MATERIAL AND METHODS

The Plant Material

The plant material in the present study is full-sib families of lodgepole pine in field tests at two locations: Bergbäck and Uvberget (Table 1). The full-sib families were produced by controlled cross pollinations in a provenance trial which was established by Svenska Cellulosa Aktiebolaget (Swedish Cellulose Company) at Ånge (Nilsson, 1981). The origins of the provenances are given in Table 2, and illustrated together with the test sites on maps in Figure 1. Within each provenance combination (Fig. 2), individual trees were crossed to create full-sib families.

The controlled crosses were made from 1977 to 1981, and in April 1983, seeds were sown in containers at the Institute for Forest Improvement in Sävar (latitude 63°54′ N., longitude 20°33′ E., and altitude 35 meters). Seedlings were grown one year in the greenhouse before planting at Bergbäck and Uvberget in May and June 1984.

The field tests incorporated single tree plots without blocks at 2x2 meter spacing. The test sites were clearcut in 1982, and soil was harrowed in the autumn of 1983. Both trials were planted in mesic sandy tills on gentle soils and south-facing slopes. Site index at Bergbäck varied from T17 to T19 (T17=a dominant height of 17 meters for Scots pine (Pinus sylvestris L.) at 100 years of age (Hägglund and Lundmark, 1982), and the site index at Uvberget ranged from T21 to T22. The test plot at Uvberget was cleaned of competing vegetation in 1986 and 1987, and damage by moose, although slight, had reduced growth. About 75 percent of the measurable saplings were bushy from moose browsing. Six years after planting, 22 percent of the saplings at Uvberget and 27 percent at Bergbäck had been severely damaged or were dead, and were not measured.

Table 1. Location and Size of the Field Tests, and Location of the Provenance Trial for the Controlled Crosses

Site (Sweden)	Latitude (N.)	Longitude (E.)	Altitude (Meters)	Saplings (Number)	Full-sib families (Number)
Bergbäck	64°03′	19°51′	185	3735	362
Uvberget	61°03′	14°52′	270	3659	332
Ånge	62°32′	15°42′	200		

Table 2. Provenances of the Parent Trees for the Controlled Crosses

Provenance No[1]	Name	Latitude (N.)	Longitude (W.)	Altitude (Meters)
581	Salmon Arm, B.C.[2]	50°39′	119°24′	1,400
583	Quesnel, B.C.	53°04′	122°26′	670
586	Bear Lake, B.C.	54°30′	122°40′	730
589	Wonowon, B.C.	56°46′	121°53′	910
590	Mile 181[3], B.C.	57°31′	122°54′	1,230
591	Fort Nelson, B.C.	58°50′	122°42′	450
592	Toad River, B.C.	58°52′	125°21′	760
593	Mile 545[3], B.C.	59°43′	127°13′	610
594	Mile 630[3], B.C.	60°03′	128°35′	910
595	Teslin, Yukon	60°12′	132°45′	910
598	Whitehorse, Yukon	60°42′	134°48′	850
605	Stewart Crossing, Yukon	63°22′	136°42′	480
606	Kananaskis, Alberta	51°02′	115°01′	1,400

[1] ID number, Institute for Forest Tree Improvement, Sweden. The prefix "Is" is omitted in this paper

[2] B.C. = British Columbia

[3] On the Alaska Highway

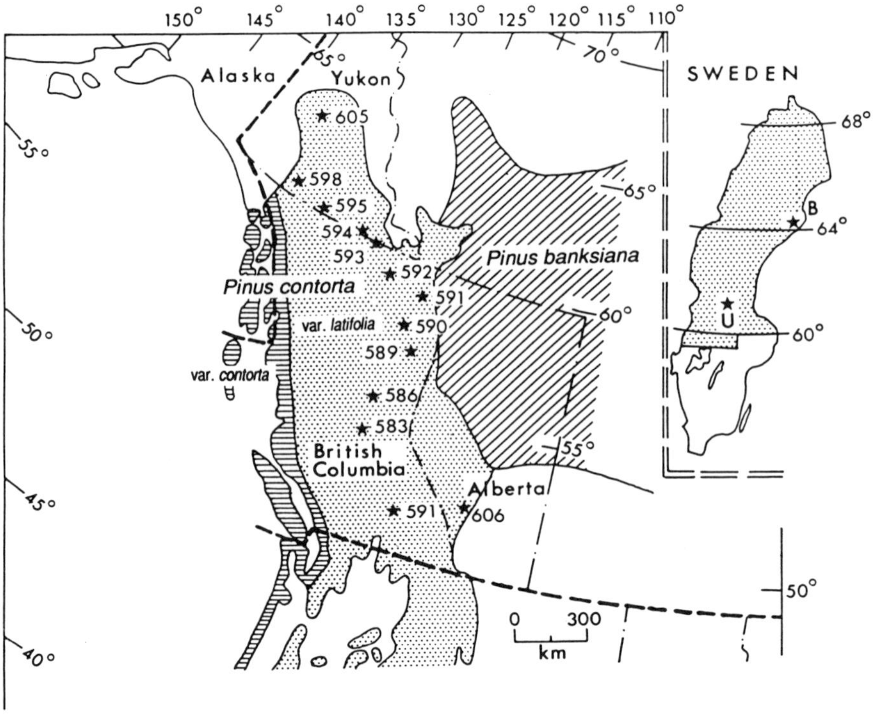

Fig. 1. Lodgepole pine provenances in Canada (left) and field tests in Sweden (right). The natural distribution of Pinus contorta var. contorta and var. latifolia is indicated by dots and of Jack pine (Pinus banksiana Lamb.) by lines. Dots on the map of Sweden show the area designated for commercial use of lodgepole pine. "B" is Bergbäck and "U" is Uvberget.

Measurements

In August 1989, when the saplings were seven years old (six years in the field), the following measurements were recorded:

1. *Vigor (V89)* in four classes: 0=undamaged, 1=slightly damaged but vigorous, 2=severely damaged, and 3=dead. Only saplings in classes 0 and 1 were included in the analyses of growth data.
2. *Growth pattern (VF89)* in four classes: 0=normal growth, 1=crooked stem, 2=forked stem, or stem with spiked knots, and 3=bushy growth.
3. *Height 1989 (H89):* Total height including growth in 1989.
4. *Height 1987 (H87):* Height from the soil surface to the second highest whorl when measured in 1989.
5. *Primary damage (SK1):* major damage.
6. *Secondary damage (SK2):* other damage with small effect on growth.
7. *Type of damage:* the following types of damage were registered: moose, field voles, insects, fungi, temperature, climate, unsuitable planting spot (shallow planting, planting in high moisture, etc.).
8. *Flowering:* presence of female and male cones.

a) Bergbäck

Male provenance \ Female provenance	581	583	586	589	590	591	592	593	594	595	598	605
581				102,0	100,2							109,0
583					119,6						111,0	
586					111,0 > xx					115,3	124,2 > xx	101,0
589				108,2	99,8 > x				102,5		107,2	98,8
590	72,9				93,6						96,1	
591						112,9				< x 96,3		115,8 > x
592					> xxx 106,4		106,5				106,2 > xxx	115,0 > xx
593												122,0
594				141,0	102,2				97,6			118,1
595					> x 97,7					100,5	104,1	98,4
598				96,6	96,4						97,2	97,4
605					> xx 106,1							105,4
606					96,5							111,8

b) Uvberget

Male provenance \ Female provenance	581	583	586	589	590	591	592	593	594	595	598	605
581				131,3	124,7							128,7 > xxx
583					132,5						145,0	
586					128,4 > x					121,7	113,7	119,0
589				126,5	134,8 > x						127,0	121,9
590	102,1				119,2						125,6	
591						131,4				136,2 > xx		134,2 > xxx
592					> xx 124,8		124,2				137,6 > x	129,5 > x
593												
594					>x 122,1				126,2			117,5
595					>xxx 126,9					124,2	121,2	124,2 > xx
598				127,7 > x	>xxx 124,7						122,5	> x 124,8 > xx
605					> x 128,3 > x							121,5
606					> xxx 137,8							121,9

Fig. 2. Average family height in 1989 in centimeters, and results from comparisons between the provenance hybrids (above and below the diagonal) and the non-hybrids (on the diagonal). The index for the height relation and level of significance (>xx, etc.) refers to the comparison between the hybrid and non-hybrid closest to the index: xxx, p<0.001; xx, p<0.01; x, p<0.05; no index: not significant at the 5% level; >, taller than the non-hybrid; <, shorter than the non-hybrid.

Most of the weather damage was to the buds and shoots and was usually caused by low temperatures at the beginning or end of the growing season. Growth increment from 1987 to 1989 was calculated using single tree data. A preliminary evaluation showed strong correlation between height in 1989, height in 1987, and growth from 1987 to 1989, and the results from the analyses on these three traits were similar (Table 3). Therefore, only results for height in 1989 (H89) are presented.

The majority of the saplings at Uvberget, 97.5 percent of the vital or slightly damaged, or 75 percent of the total outplanted, were bushy as a result of damage by moose. This was independent of provenance and family. Furthermore, the separate analyses on all saplings, and on saplings with and without moose damage, were nearly identical (Table 3). Therefore, to base inferences on the largest sample of homogeneous material possible, the analyses of the Uvberget trial were made only on saplings classified as bushy.

Statistical Analyses

The test material consists of about 350 full-sib families from 33 between- and 8 within-provenance combinations (Table 1 and Fig. 2). Each provenance combination contributed between one and 19 full-sib families, and each full-sib family contributed between one and 29 saplings. In the analyses of qualitative traits (vigor, damage, proleptic shoots, and flowering), the full-sib families were grouped into half-sib families, and calculations were made on arcsine-transformed frequencies of the number of saplings within half-sib families. Standard t-tests for difference between provenance combinations were then performed. The restrictions on minimal numbers for pairwise t-tests were 50 observations from more than one half-sib family. For quantitative parameters, t-tests on absolute values were employed for the analyses. Variance components were used for quantifying the relative importance of the breeding values, and were calculated for male and female provenances separately, and for male and female provenances combined. The calculations were made for the two sites separately and combined using the DIALL program of Schaffer and Usanis (1969). The formula included the effects of general combining ability (GCA), specific combining ability (SCA), selfing (Self) and reciprocal crosses (Rec), and was as follows:

$$y_{fmp} = \mu_{fmp} + GCA_f + GCA_m + SCA_{fm} + Rec_{fm} + Self_p + Error_{fmp} ,$$

where y_{fmp} is the observed value, μ_{fmp} is the average for all saplings in the study, f is the female provenance, m is the male provenance, and p represents the parent provenance when it is the same for the male and female parents.

Analyses of variance for estimating variance components and significance tests of differences between female and male provenances, and female and male saplings within provenances, were based on full-sib family averages. For these analyses, the SAS VARCOMP and SAS GLM procedures were used.

Regression analyses were made on growth traits as linear regressions using averages for full-sib families.

RESULTS

Survival, Weather Damage, and Prolepsis

Table 4 gives the frequencies of undamaged or slightly-damaged saplings, saplings with

Table 3. Height of Provenance Hybrids and Within-Provenance Crosses, and Significance of Difference between the Groups: *N*, number of observations; *M*, mean; *S*, standard deviation (Significances: ***, $p<0.001$; *, $p<0.05$; *NS*, not significant)

Site	**BERGBÄCK, Sweden**					**UVBERGET, Sweden**						
Genetic level for estimate	**Full-sib family**			**Proven. combi-nation**	**Single tree data**	**Full-sib family**			**Proven. combi-nation**	**Single tree data**		
VF89[a]	**All**			**All**	**All**	**3**	**3**	**3**	**3**	**3**	**< 3**	**All**
Trait	**H87**	**H89-H87**	**H89**	**H89**	**H89**	**H87**	**H89-H87**	**H89**	**H89**	**H89**	**H89**	**H89**
Provenance hybrids												
N	268	268	268	33	1780	244	244	244	30	1310	342	1652
M (cm)	49.2	54.7	103.9	106.0	100.0	62.8	65.1	127.9	126.5	126.5	143.3	130.0
S (cm)	11.2	10.4	18.1	11.8	26.4	10.7	11.4	16.0	8.2	19.5	21.6	21.0
S/M (%)	22.7	19.0	17.4	11.1	26.3	17.0	17.5	12.5	6.4	26.3	15.1	16.2
Within-provenance crosses												
N	94	94	94	8	935	88	88	88	8	662	232	894
M (cm)	47.7	54.0	101.7	102.7	98.6	60.3	63.3	123.6	124.4	121.8	140.7	126.8
S (cm)	10.1	7.7	13.7	6.5	25.2	6.4	8.8	11.1	3.7	19.6	23.0	22.1
S/M (%)	21.2	14.2	13.5	6.4	25.5	10.6	13.8	9.0	3.0	25.5	16.3	17.5
Sign.	NS	NS	NS	NS	NS	*	NS	*	NS	***	NS	***

[a] VF89 = growth pattern 3: Bushy form, in most cases due to moose damage, VF89 = growth pattern 0, 1 or 2: normal growth, crooked, forked or with spiked knots, respectively.

weather damage on buds and shoots, and of proleptic saplings. Frequencies are given for each of the two crosses, and for progenies grouped according to parental origin: one group where one or both parents originate south of latitude 60° N. and one group where both parents originate north of this latitude. Differences between provenance hybrids and within-provenance crosses were not significant, and difference due to parental origin occurred only for prolepsis at Bergbäck where proleptic shoots were more common in the northern material ($p<0.05$).

With respect to weather damage, three hybrids differed significantly from the corresponding within-provenance crosses (Fig. 3). In these three hybrids, t-tests were used for comparing frequencies of vital and slightly-damaged saplings within half-sib families with the frequencies within the corresponding non-hybrids. In three half-sib comparisons, the hybrids were significantly more damaged, and in three other comparisons they were less damaged (see the numbers in the brackets). In 14 out of the 20 family comparisons, however, differences were not significant.

Male provenance \ Female provenance	592	595	598	605
592	82,2%		NS (0 of 3) 74,8% x > (1 of 3)	
595		75,0%	x < (1 of 4) 72,9% x > (2 of 4)	x < (1 of 3) 58,6% x < (1 of 3)
598			58,4%	
605				81,4%

Fig. 3. A comparison of the percent of vital or slightly-damaged saplings among provenance hybrids and corresponding non-hybrid half-sib families. Differences were considered significant at $p<0.05$. The number of comparisons between provenance hybrid and non-hybrid families is enclosed in brackets. For example, x > (2 of 4) indicates that in two of four half-sib comparisons, the hybrid sustained significantly more damage than the nonhybrid.

Growth Traits

Mean values and standard deviations for growth traits are given in Table 3. Damage by moose was extensive at Uvberget, and in Table 3 the single-tree data in the three columns to the right are given for all vigorous saplings, for the saplings not damaged by moose (<3), and for the saplings damaged by moose (3). Except for total height among the undamaged saplings, the relationships between the three groups are the same. For the full-sib families at Uvberget, provenance hybrids were significantly taller than within-provenance crosses (non-hybrids) in both 1987 and 1989 ($p<0.05$). The relative variation, estimated as the coefficient of variation (S/M), was also larger for the provenance hybrids. Increment for the provenance and within-provenance crosses from 1987 to 1989 were not significantly different but followed the same pattern as H87 and H89. At Bergbäck in 1987, provenance hybrids were also taller than the within-provenance crosses, and their advantage increased until 1989. The differences were, however, not significant.

Table 4. Results of *t*-tests[a] for Percent Vigor, Weather Damage, and Prolepsis between Provenance Hybrids and Non-hybrids, and between Progenies with One or Both Parents from South of Latitude 60° N. and Both Parents from North of Latitude 60° N.

	TYPE OF CROSSING			PARENTAL ORIGIN			
	Hybrid	**Non-hybrid**	**Sign.**	**<60°N.**	**>60°N.**	**Sign.**	**Average**
Undamaged or slightly damaged							
BERGBÄCK	72.1	74.8	NS	73.5	70.6	NS	72.8
UVBERGET	78.6	74.9	NS	79.0	73.5	NS	77.7
Climate damage							
BERGBÄCK	6.3	3.1	NS	6.3	3.3	NS	5.7
UVBERGET	<0.1	<0.1	-	<0.1	<0.1	-	<0.1
Prolepsis							
BERGBÄCK	12.1	16.0	NS	10.2	21.7	*	13.1
UVBERGET	2.2	1.6	NS	1.6	3.4	NS	3.0

[a] For significances, see Table 3.

Mean height of the provenance combinations, and results of the t-tests for significance of the differences between the provenance hybrids and the within-provenance crosses are shown in Figure 2. Table 5 shows that provenance differences in 1989 height were not significant. At Bergbäck, male and female parents within provenance differed significantly ($p<0.001$), and the variance components were around 20 percent. Differences were not significant at Uvberget.

At Bergbäck, the GCA variance in 1989 height was 5.15 percent of the total variance ($p<0.05$) (Table 6). In the combined data for both trials, GCA contributed 1.16 percent ($p<0.001$) of the total variance, but GCA was not significant at Uvberget. SCA contributed only 0.7 percent and selfs 1.3 percent of the total variance for both sites. SCA and self effects were not significant. Reciprocal crosses and site effects were significant and contributed 88.5 percent of the total variance for the combined data. There was no correlation between the actual value of GCA (i.e., the breeding value) for a parent clone and the variance of its SCA-values.

Flowering

Frequency of families producing male or female seed cones was generally lower than three percent. Female flowers (cones) were produced on five percent of the hybrids and 11 percent of the non-hybrids at Uvberget. Differences in cone production among hybrids and within-provenance crosses were not significant.

DISCUSSION

The results of this study indicate that heterosis does not contribute to the variation in height growth among provenance hybrids of lodgepole pine. The study was conducted in practical field tests and only a few traits could be investigated in detail. The provenance test in which the crosses were made was not designed for breeding purposes. For this reason, selected parents were difficult to cross, and the mating design became unbalanced. Although the qualitative data were transformed, accuracy was low because the frequencies were either very low (survival, flowering, and damage by weather and other agents), or very high (moose damage at Uvberget).

No weather damage occurred at Uvberget, while 6.3 percent of the hybrid and 3.1 percent of the non-hybrid crosses were damaged at Bergbäck. The difference between hybrid and non-hybrid crosses was not significant, however. The three provenance hybrids (598x592, 598x595 and 605x595) that differed significantly from their respective non-hybrid crosses, and for which hybrid vigor was investigated in detail (Fig. 3), were not more susceptible to weather injury than the non-hybrids. Differences in weather injury among half-sib families pooled from the crosses were usually not significant, and the few provenance hybrids with significantly less injury than the within-provenance crosses were as common as the provenance hybrids with more injury. Also, in contrast to most reports, southern provenances did not sustain more damage than northern provenances.

The exposure to disease-causing organisms was also too low to demonstrate inherent resistance against fungi or other pathogens. However, genetic control of disease resistance has been demonstrated at both the provenance and the family-within-provenance levels of genetic variation (van der Kamp, 1988; Yanchuk et al., 1988; Ying and Hunt, 1987). Van

Table 5. Variance Components in Percent of the Total Variance *(V)* for Provenance of Female Parent *(FP)*, Provenance of Male Parent *(MP)*, Female Parent Tree within Provenance *(F w FP)* and Male Parent Tree within Provenance *(M w MP)*, and Results from Significance Tests[1] (Height of Saplings in 1989)

Source	Bergbäck			Uvberget		
	V (%)	df	Sign	V (%)	df	Sign[1]
FP	3.7	8	NS	4.9	8	NS
MP	0	12	NS	0	11	NS
F w FP	23.4	117	***	0	101	NS
M w MP	19.8	48	***	3.4	43	NS
Error	53.1	176		91.7	158	

[1] For significances, see Table 3.

der Kamp (1988) saw significant interaction in the infection rate between lodgepole pine provenances and the geographic source of spores. Furthermore, infection scores increased with altitude of plant origin (Yanchuk et al., 1988), and heritabilities of infection resistance at the individual, family, and provenance levels were high (Ying and Hunt, 1987). Thus, resistance to certain pathogens, and growth and fitness traits may be combined in hybrids between provenances or between families within provenances. Inoculation tests are, however, difficult to perform both indoors and in the field, because pathogens and the host interact with environmental factors that are difficult to control and reproduce.

For height growth, GCA was significant and more important than SCA (Table 6). High GCA in comparison with SCA indicates that heterosis is uncommon. On the other hand, the provenance hybrids were taller than the within-provenance crosses (Table 3). Hybrids between provenances 590 and 605 were significantly taller than the crosses within the corresponding parent provenances (Fig. 2), and they were largely responsible for the differences in height between the provenance hybrid and non-hybrid crosses. Faster growth of the hybrids was more evident at the milder site (Uvberget) than at the severe site (Bergbäck), where differences between the hybrid and the non-hybrid crosses were not significant. In addition, the variance component for SCA, although not significant, was larger at Uvberget (Table 6). Thus, heterosis for growth rate may be better expressed under favorable than unfavorable growing conditions. The correct interpretation for the significantly taller hybrids of provenances 590 and 605 may, however, be that the hybridization released inbreeding depression from matings among relatives within the provenances. In the natural range of lodgepole pine, northern populations are generally smaller and more scattered than southern populations. Matings among relatives are more common within small scattered populations. Provenance 590 is from high altitude where infrequent seed crops and isolation may increase matings among relatives and increase inbreeding depression.

Table 6. Variance Components in Percent of the Total Variance *(V)* for General Con bining Ability *(GCA)*, Specific Combining Ability (*SCA*), Selfing *(Self)*, Reciprocal Crosses *(Rec)* and Site Effects for Sapling Height in 1989

Source	Bergbäck			Uvberget			Bergbäck & Uvberget		
	V(%)	df	Sign[1]	V(%)	df	Sign[1]	V(%)	df	Sign[1]
GCA	5.15	12	*	4.4	12	NS	1.2	12	***
SCA	0	16	NS	23.4	15	NS	0.7	16	NS
Self	23.4	8	NS	0	8	NS	1.3	8	NS
Rec	71.5	4	***	72.2	3	***	27.5	4	***
Site	-	-	-	-	-	-	61.0	1	***
Error	-	-	-	-	-	-	8.3	26	

[1]For significances, see Table 3.

Significant correlation between the breeding values of the clones and the variation in the SCA of their families was lacking. Thus, when selecting clones or parents with a high GCA, the possibility of producing fast-growing full-sib families from SCA is remote. Furthermore, possible hybrid effects are unlikely to increase general adaptability and growth.

On the whole, heterosis was not expressed, although fast-growing hybrids might occasionally be found. These results are supported by studies of hybrids between Scandinavian and European provenances of Norway spruce (Picea abies (L.) Karst.) (Eriksson and Ilstedt, 1986; Ilstedt and Eriksson, 1986). The variance component for GCA was larger than the variance component for SCA. The within-family variation could be large, however. On harsh sites, non-hybrids of Norway spruce produced more volume than hybrids, while on mild sites, the provenance hybrids grew taller than the non-hybrids. Wood quality traits (branching and stem straightness) were statistically superior for the provenance hybrids on mild sites, and for the non-hybrids on harsh sites.

The effects of reciprocal crosses were large, especially between provenances 581 and 589. Provenance 581 was from a high altitude site (1,400 meters) where seed crops are infrequent and crossing among close relatives is common. The estimates were, however, based on only a few observations and degrees of freedom. Thus reciprocal effects were not conclusive.

Proleptic shoots were more frequent at Bergbäck than at Uvberget, and for progeny from northern provenances. Progenies of northern origins are expected to be proleptic because they cease growth early and initiate growth again when transferred to areas with longer vegetation periods. Plants with proleptic growth are thus hardier than plants without prolepsis. Differences between hybrids and non-hybrids in frequency of prolepsis were not apparent, and heterosis for growth rhythm traits was not demonstrated. Therefore, models of intermediate inheritance of growth rhythm traits (Eriksson et al., 1978; Mikola, 1982) are plausible.

This study demonstrated that hybrid effects among lodgepole pine provenances are manifested in the mild environments. In contrast, prolepsis, and thus growth rhythm and hardiness, follow a latitudinal trend and are intermediately inherited. Evidence for heterotic height growth may result from release of inbreeding depression.

ACKNOWLEDGEMENTS

The Swedish Council for Forestry and Agricultural Research funded the crossing program, and Svenska Cellulosa Aktiebolaget provided the provenance trial for the crosses. The Institute for Forest Improvement in Sävar produced the plants and Mo och Domsjö AB and Korsnäs AB own the land for the field trials. At the Department of Forest Genetics and Plant Physiology, Swedish University of Agricultural Sciences in Umeå, Jan-Erik Nilsson and Dag Rudin initiated the crossing program. Contributions were made by Dag Lindgren, Karin Ljung, Kjell Olofsson, and their co-workers. Sven Andersson, Nils Ganered, Lars-Göran Lejdebro, and Stefan Löfmark maintained and measured the trials.

REFERENCES

Aldhous, J. R., 1976, Lodgepole pine seed zones with reference to British requirements, For. Comm. Res. and Dev. Pap. 114, Great Britain, 6-39.

Critchfield, W. B., 1957, Geographic variation in Pinus contorta, Maria Moors Cabot Found. Publ. 3, Harvard Univ., Cambridge, Mass., 118 p.

Critchfield, W. B., 1980, Genetics of lodgepole pine, Res. Pap. WO-37, U.S. Department of Agriculture, Forest Service, 57 p.

Cwynar, L. C., and MacDonald, G. M., 1987, Geographical variation of lodgepole pine in relation to population history, American Naturalist, 129:463-469.

Eriksson, G., and Ilstedt, B., 1986, Stem volume of intra- and interprovenance families of Picea abies (L.) Karst., Scand. J. For. Res., 1:141-152.

Eriksson, G., Ekberg, I., Dormling, I., and Matérn, B., 1978, Inheritance of bud-set and bud-flushing in Picea abies (L.) Karst., Theor. Appl. Genet., 53:3-19.

Hägglund, B., and Lundmark, J. E., 1982, Handledning i bonitering med Skogshögskolans boniteringssystem, Del 1-3, National Board of Forestry, Jönköping, Sweden, 53, 70, and 121 p.

Hultén, E., 1937, Outline of the history of arctic and boreal biota during the quaternary period, Bokförlaget Thule, Stockholm, Sweden, 168 p.

Ilstedt, B., and Eriksson, G., 1986, Quality of intra- and interprovenance families of Picea abies (L.) Karst., Scand. J. For. Res., 1:153-166.

Lindgren, K., 1983, Provenances of Pinus contorta in northern Sweden, Swedish University of Agricultural Sciences, Department of Forest Genetics and Plant Physiology, Umeå, Sweden, 91 p. Ph.D. thesis.

Lindgren, K., Lindgren, D., and Rosvall, O., 1988, Förflyttningsrekommendationer för provenienser av contortatall i Sverige, Arbetsrapport nr 27, Dept. For. Gen. and Plant Physiol., Swedish University of Agricultural Sciences, Umeå, Sweden, 43 p.

Lines, R., 1976, The development of forestry in Scotland in relation to the use of Pinus contorta, For. Comm. Res. and Dev. Pap. 114, Great Britain, 2-5.

MacDonald, G. M., and Cwynar, L. C., 1985, A fossil pollen based reconstruction of the late Quaternary history of lodgepole pine (Pinus contorta ssp. latifolia) in the western interior of Canada, Can. J. For. Res., 15:1039-1044.

Mikola, J., 1982, Bud-set phenology as an indicator of climatic adaptation of Scots pine in Finland, Silva Fenn., 16:178-184.

Nilsson, J. E., 1981, Flowering in Pinus contorta, Swedish Univ. of Agric. Sci. Rep. No. 2, 128 p.

Norell, L., Eriksson, G., Ekberg, I., and Dormling, I., 1986, Inheritance of autumn frost hardiness in Pinus sylvestris L. seedlings, Theor. Appl. Genet., 72:440-448.

Nyland, E., 1980, Lodgepole pine in the Yukon. In: Proceedings of the IUFRO working party meeting 1980 on Pinus contorta provenances (S2-02-06) in Norway and Sweden, Pinus contorta as an exotic species, Swedish Univ. of Agric. Sci., Res. Notes No. 30, 27-43.

Reeves, B. O. K., 1973, The nature and age of the contact between the Laurentide and Cordilleran ice sheets in the western interior of North America, Arct. Alp. Res., 5:1-16.

Rehfeldt, G. E., 1977, Growth and cold hardiness of intervarietal hybrids of Douglas-fir, Theor. Appl. Genet., 50:3-15.

Rosvall, O., Strömberg, S., and Andersson, B., 1984, Skogsförbättrings proveniensförsök med contortatall i norra Sverige (Lodgepole pine provenance trials in northern Sweden), Institute for Forest Improvement, Årsbok, Sweden, 117-159.

Schaffer, H. E., and Usanis, R. A., 1969, General least squares analysis of diallel experiments, A computer program-DIALL, North Carolina State University, Genet. Dep. Res. Rep. No. 1, Raleigh, North Carolina, United States, 57 p.

van der Kamp, B. J., 1988, Susceptibility of lodgepole pine provenances to geographically separate western gall rust spore sources, Can. J. For. Res., 18:1203-1205.

Wheeler, N. C., and Guries, R. P., 1982, Biogeography of lodgepole pine, Can. J. Bot., 60:1805-1814.

Yanchuk, A. D., Yeh, F. C., and Dancik, B. P., 1988, Variation of stem rust resistance in a lodgepole pine provenance-family plantation, For. Sci., 34:1067-1075.

Ying, C. C., and Hunt, R. S., 1987, Stability of resistance among Pinus contorta provenances to Lophodermella concolor needle cast, Can. J. For. Res., 17:1596-1601.

GENETICS, CYTOGENETICS, AND MOLECULAR GENETICS OF ICELANDIC BIRCH: IMPLICATIONS FOR BREEDING AND REFORESTATION

Kesara Anamthawat-Jónsson
Agricultural Research Institute
Keldnaholt, Reykjavik, Iceland

J. S. Heslop-Harrison
Department of Cell Biology
JI Centre for Plant Science Research
Norwich, United Kingdom

Thorsteinn Tómasson
Agricultural Research Institute
Keldnaholt, Reykjavik, Iceland

INTRODUCTION

Icelandic birch (Betula pubescens Ehrh.) is the major indigenous woody species in Iceland. Conservation of the remaining native birch forests is a recognized priority. The species has little commercial value because of its slow growth and irregular form; however, it is important for land reclamation, environmental protection, amenity planting, and for nature reserves.

A long-term goal is to gain a better understanding of the genecology of the species, thereby increasing its ecological and economic value in the subarctic environment, particularly in Iceland. Icelandic birch is especially suitable for forestation purposes. It has an extensive pool of genetic variation, and it is indigenous, widespread, and well-adapted to subarctic conditions. Results of genecology studies of Icelandic birch can also be used as a model for improvement and utilization of other tree species in cold climates.

Little is known about the genetic diversity of Icelandic birch. Our research program is designed to characterize the genetic base of the species and its relatives. Morphological and physiological features, growth and development, forms and habit, tolerance to environmental stress, and adaptability are potentially useful traits that can be improved through genetics. Using available biotechnology, we will study the genetic variation at DNA and chromosome levels as well as the morphological and physiological variation where genotype and environment interactions are important. Such studies will be integrated and applied in selection and breeding programs.

Forest Development in Cold Climates, Edited by J. Alden *et al.*
Plenum Press, New York, 1993

Methods of studying genetic diversity in Icelandic birch include (1) observations of plant morphology in various locations in Iceland, (2) cytological analysis and molecular cytogenetic studies of plants from natural populations and from experimental hybrids, and (3) detection of molecular polymorphisms in the plant populations and characterization of the genetic resources. Preliminary results are reported here.

GENETIC VARIATION AND ADAPTABILITY

Icelandic birch is a morphologically variable species. Löve and Löve (1956) placed it in one collective species, B. pubescens. The Icelandic birch woodlands are part of the mountain birch zone that extends north of the coniferous forest line from the forest tundra of the northern Kola Peninsula, through the mountainous region of northwest Fennoscandia, to Iceland and south Greenland (Sulkinoja, 1990). For convenience, we refer to all mountain birch as B. pubescens Ehrh.. Some authors refer to the mountain birch of Fennoscandia as B. pubescens subsp. tortuosa (Ledeb.) Nyman (Sulkinoja and Valanne, 1980), and others regard it as the separate species B. tortuosa Ledeb. (Vaarama and Valanne, 1973).

Variation in the B. pubescens complex has long been recognized. Since Linné, the taxonomic status of this birch has been uncertain. Hagman (1971) reviewed data about the lack of effective reproduction barriers in the genus Betula. Interfertility and wide geographic distribution in isolated and harsh environments has lead to variability which may or may not be regarded as speciation. Many extant birch populations in northern climes may have descended from relict Ice Age populations, and hence may have evolved from many different and isolated gene pools. For example, the birch in northern Sweden and Norway is entirely different from that in Greenland. Even birch populations in north or east Iceland are different from those of the central areas. When seedlings from these diverse sources are grown together in an experimental garden, their phenotypic differences are maintained.

Genetic Diversity

The study of morphological variation requires a systematic method of selecting appropriate characters and evaluating the selected characteristics from an adequate sample size. Information from existing morphological indices such as those used by Elkington (1968) is applicable. Leaf characters are critical and include leaf length and width ratio, angle of lamina apex, angle of leaf base, and serration of the leaf margin. Quantitative methods of measurement, such as the discriminant function of leaf shape measurement of birch (Atkinson and Codling, 1986), are needed to counteract environmental influences. Reliable methods of analysis are important because of the high degree of variation within the species.

Morphological and physiological variation often reflect the ability of plants to adapt to the environment. Many physical pressures affect the survival of plants in Iceland: cold climate, short growing season, long day-length at high latitude, oceanic and windy weather, volcanic soils lacking nutrient and clay component, and snow-ice packing in winter. Biological factors such as sheep grazing and wood cutting also reduce plant survival. It is difficult to assess the capacity of a species to adapt to the environment; experiments must be designed to study the effect of selected individual factors. For example, photosynthesis and respiration have been extensively studied and are thought to be important factors influencing growth and survival of plants under cold conditions (Skre, 1991). Genecology, the study of intraspecific variation of plants in relation to environment, must also be considered because such variation indicates physiological and evolutionary aspects of differentiation (Heslop-Harrison, 1964).

Genetic Variation and Adaptation

Superficial observations of Icelandic birch show substantial phenotypic variation in nearly every character examined: growth habit and form, leaf shape and size, color of leaf senescence in autumn (yellow to red), time of flowering, maturation of seed, morphology of seed and seed bract, and seed production and germination. Large variations are apparent both within populations and between populations at different geographical locations. Birch trees are variable in growth habit, ranging from monocormic (single stem) to polycormic (several stems arising from buds at the base of the plant). The height of mature trees varies from less than one meter to 14 meters. A common shrubby type of Icelandic birch is shown in Figure 1. This tree is bushy, polycormic, and less than one meter high. Stems are crooked and almost horizontal. Leaves are small and rounded. Leaf length and width are similar, and the angle of the leaf apex is wide. The tree has several characteristics similar to those of dwarf arctic birch (*B. nana* L.).

Growth of Icelandic birch seedlings in a controlled environment (growth cabinet) was compared with seedlings of lowland birch (*B. pendula* Roth) and dwarf arctic birch under the same conditions. The Icelandic birch seedlings grew much more slowly. Growth and elongation of internodes appeared to be suppressed, while the number of leaves and nodes were similar to that of lowland birch. This could be a sign of early dormancy in response to the 16-hour daylength in the growth cabinet. Photoperiodic adaptation was demonstrated in tree species that span wide ranges of latitudes (Vaartaja, 1959). Under a controlled environment, apical growth of seedlings of paper birch (*B. papyrifera* Marsh.) from the north (69° N.) and south (42° N.) latitudes was measured at different photoperiods. Photoperiods of less than 16 hours led to early dormancy in the northern races, but southern races continued to grow at 12-hour photoperiods. Glasshouse experiments (Vaarama and

Fig. 1. Icelandic birch from Bifröst, Borgarnes.

Valanne, 1973) showed that the mountain birch of Finnish Lapland had a short, genetically-controlled growth period and a fixed dormancy difficult to break without cold treatment.

The morphology and biology of mountain birch in northern Finland have been described extensively by the research group at Kevo Subarctic Research Institute at Utsjoki, Finnish Lapland (Kallio et al., 1983; Sulkinoja and Valanne, 1987; Vaarama and Valanne, 1973). The birch of Finnish Lapland was also compared with mountain birch in southwest Greenland (Sulkinoja, 1990). Several morphological and physiological characteristics have been closely correlated to the adaptability of the species to subarctic climate, including short growing season and low temperature in the summer and extreme conditions in the winter. Polycormic growth, for example, is a trait that enables trees to survive grazing and winter damage. This trait also ensures vegetative propagation and compensates for low seed production and poor germination. Many of these adaptive characters not only occur in Icelandic birch but also in the dwarf arctic birch, suggesting hybrid introgression between the two species.

CYTOGENETICS AND INTROGRESSION

Hybrid Introgression

Hybrid introgression has been suggested as one cause of the wide genetic variation in Icelandic birch (Anamthawat-Jónsson and Tómasson, 1990; Elkington, 1968). Löve and Löve (1956), however, concluded that introgressive hybridization between Icelandic and dwarf arctic birches did not occur because of the low frequency and sterility of natural hybrids. Elkington (1968) studied Icelandic birch populations in northwest Iceland using histograms of hybrid indices based on morphological characteristics compared with birch of the same species from north Scotland. The analysis showed that Icelandic birch had a broad range of variation that was closer to the index of dwarf arctic birch than to the index of the birch from north Scotland, indicating hybrid introgression in Icelandic birch. Cytogenetic studies of Icelandic birch from natural populations and controlled crosses (Anamthawat-Jónsson and Tómasson, 1990) suggested that the gene flow from dwarf arctic birch to Icelandic birch occurred through hybridization, followed by backcrossing of the hybrids with Icelandic birch. Based on its chromosome complement and phenotype, the backcross progeny was Icelandic birch, but it carried some characteristics of dwarf arctic birch.

The two birch species can form F_1 hybrids that are triploid (2n=3x=42) because dwarf arctic birch is diploid (2n=2x=28) and Icelandic birch is tetraploid (2n=4x=56). Natural triploid hybrids were found in Iceland (Löve and Löve, 1956), north Scotland (Kenworthy et al., 1972) and north Finland (Sulkinoja and Valanne, 1980). They were intermediate in morphology, as were the F_1 hybrids from our crossing experiments. If morphology reliably reflects hybridization, the natural F_1 hybrids would be quite common in Iceland. This can be verified only by determining the chromosome number of the hybrid trees directly. All plants grown from seeds collected in nature had the Icelandic birch chromosome number of 56 (Fig. 2A), although some of the parental plants resembled hybrids. Crossing experiments suggested that selection was so strong that stabilization of chromosome number occurred in the first generation of backcrossing (Anamthawat-Jónsson and Tómasson, 1990). Apparently, either the chromosomally unbalanced gametes were inviable or the unbalanced embryos did not survive. Sulkinoja and Valanne (1980) tried to germinate seeds collected from natural triploid hybrids in north Finland. Germination percentage was very low, averaging only 2.4 percent in good years. In addition, many seedlings were deformed and did not survive when transplanted into soil.

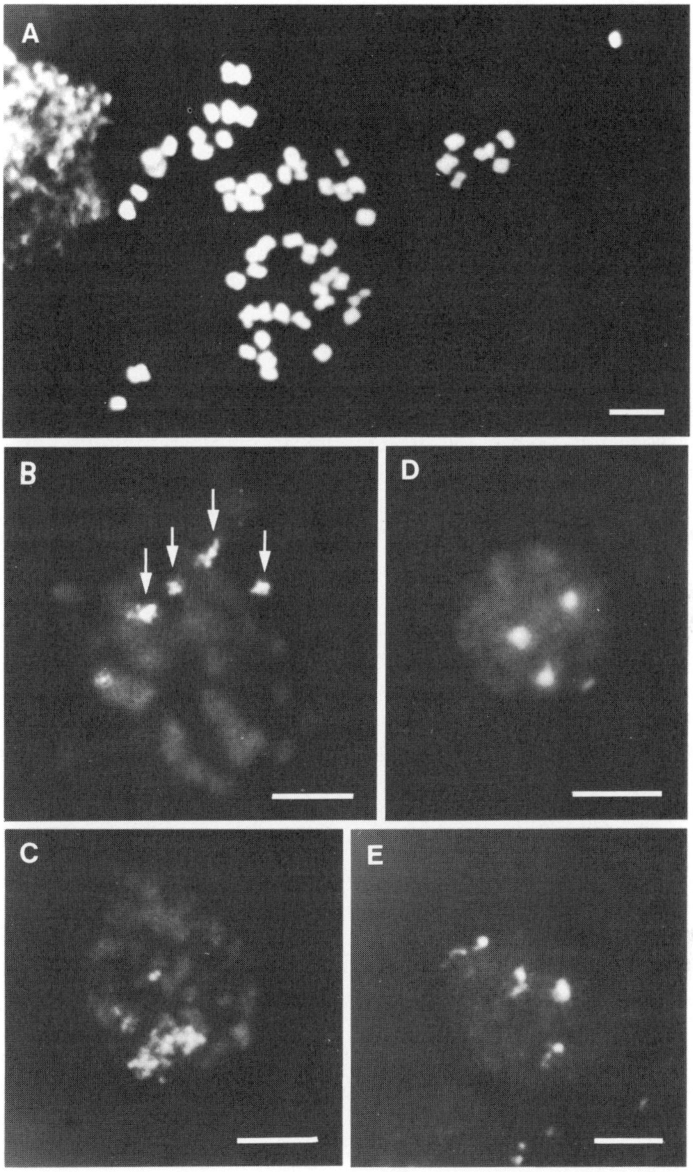

Fig. 2. Chromosome spreads of root-tip cells (scale bar is three micrometers). A: Metaphase spread of Icelandic birch from Bifröst, Borgarnes, showing 56 chromosomes. B: In situ hybridization of ribosomal genes to metaphase chromosomes of tetraploid lowland birch from east England, showing sites of ribosomal genes on two pairs of chromosomes (indicated with arrows). C: The same in situ hybridization to an interphase nucleus of the tetraploid lowland birch showing a dispersed pattern of probe hybridization. D: Interphase nucleus of Icelandic birch from Bæjarstaðir showing condensed sites of ribosomal genes. E: Interphase nucleus of dwarf arctic birch showing partly decondensed pattern of in situ hybridization with the same ribosomal probe.

Morphology of progenies (with 2n=56) from the experimental backcrosses of the F_1 hybrids with Icelandic birch was highly variable, similar to that found in nature. Many characteristics of dwarf arctic birch were observed: small plant size, orbicular (round) leaf shape, horizontal growth habit, and red color of autumn leaves. It is interesting, therefore, to investigate whether such transfers of characters can be identified using cytogenetic techniques, molecular markers, and phenotype analysis.

Molecular Cytogenetics

The transfer of traits through introgression is likely to occur by chromosome recombination between two different parental genomes at meiosis in the F_1 hybrids. The recombination could result in a replacement of chromosome segments of Icelandic birch with those of dwarf arctic birch. Such recombinant chromosomes may carry genes of adaptive value or genes responsible for desirable traits. Transfer of traits can be detected using the technique of DNA hybridization to chromosomes (in situ hybridization). We have identified the parental origin of chromosomes in hybrids between cereal crops and their related species (Anamthawat-Jónsson et al., 1990) and detected the recombination with alien chromosome segments in cereal crops (Heslop-Harrison et al., 1990). Alternatively, selected DNA sequences that are species- or genome-specific can be isolated, cloned, and used to identify chromosome recombination with in situ hybridization. For example, a DNA sequence specific to dwarf arctic birch would be able to identify chromosome segments from its own species in the backcross or introgressant plants. Advantages of detecting recombination with DNA hybridization include speed, directness, and independence of environmental factors. Detection can be done at early seedling stages and hence save generation time.

In situ hybridization of birch chromosomes has not been reported before, probably because the technique is new and the cytology of small chromosomes is demanding. Our preliminary results on birch in situ hybridization are shown in Figure 2. Seedling root tips were pretreated in ice water for 24 hours, fixed in 1:3 acetic acid:ethanol, digested with cellulase and pectinase, and squashed. Chromosome preparations were stained with fluorochrome DAPI (4′,6-diamino-2-phenylindole) for examining chromosome morphology (Fig. 2A). In situ hybridization of ribosomal genes (clone pTa71 isolated from wheat by Gerlach and Bedbrook, 1979) to chromosomes was carried out using a non-radioactive fluorescence, the fluorochrome FITC (fluorescein isothiocyanate), and shown as bright areas, while the grey areas represented total DNA (Fig. 2B-E).

Ribosomal genes (rDNA) transcribe ribosomal RNA, which forms a major component of the protein synthesizing machinery in the cell. In birch, the in situ localization of these

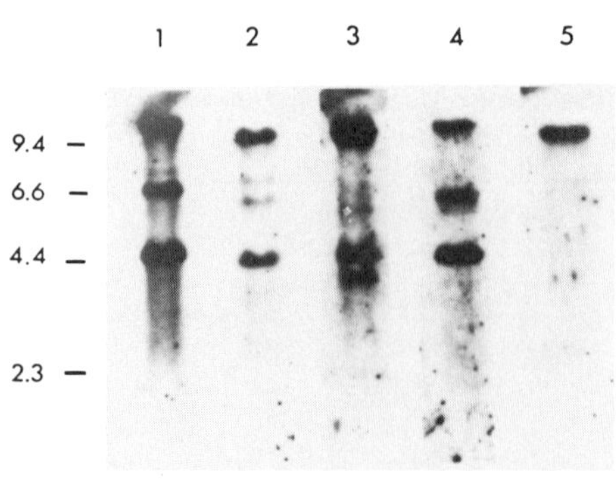

Fig. 3. Chemiluminescent Southern hybridization of ribosomal genes (clone pTa71) to *Dra*I restriction enzyme digested genomic DNA of: (1) dwarf arctic birch, (2-4) Icelandic birch from different populations, and (5) lowland birch from east England. The numbers on the left indicate fragment size in kilobases (kb). The major ribosomal DNA *Dra*I fragments are about 11 kb in size in all birch species used in this study. Some of the fragments contain extra *Dra*I site(s), and hence are cleaved into smaller fragments.

genes revealed two key features. Firstly, the ribosomal gene loci in the genome of tetraploid birch were localized on two pairs of metaphase chromosomes (Fig. 2B). Secondly, the pattern of localization in interphase nuclei varied from cell to cell and from one species to another. For example, the sites of hybridization were dispersed in some nuclei (Fig. 2C) and more condensed in others (Fig. 2D-E). Similar differences were observed in wheat and rye (Leitch et al., 1992) and were thought to relate to gene expression or structural differences in the genes between species.

MOLECULAR GENETICS AND POLYMORPHISMS

Marker Technology

Although molecular analysis of genomes is becoming applicable for selection of annual crop plants, such as wheat and maize (corn), the technology is even more advantageous for tree species. Molecular methods (using DNA or isozyme markers) overcome the species drawbacks normally associated with conventional tree breeding programs: slow growth, large size, longevity, and long time before fruiting or seed set.

A saturated map of molecular markers would be extremely useful for analysis of birch genetics. The small size of the genome means that only a small proportion consists of highly repetitive DNA. Unlike wheat and other cereals (Sharp et al., 1989), many clones can be used to give restriction fragment length polymorphisms (RFLP). The chance of finding segregating genes is high because inbreeding has been much less than in most crops, and birch has substantial genetic variability. The intercrossability between the species will also help in generating a molecular map. It is also likely that the new random amplified polymorphic DNA (RAPD) marker technology (Williams et al., 1990) will give a good molecular genome map for birch.

At the same time a molecular map is being generated, it is important to link the markers with phenotypic or enzyme characters of the plants. Most characters of interest to plant breeders show phenotypic variability within a particular crop species (Snape et al., 1990). Some of these, such as disease resistance, may be clearly segregating single gene effects (McIntosh, 1988). Most characteristics of interest to breeders, however, are controlled by many genes and show continuous genetic variation. Considerable progress is being made in analyzing such quantitative trait loci (Snape et al., 1990). Complete coverage of a genome with DNA markers will, with relatively little effort, account for the total variation for quantitative characters.

Molecular Polymorphisms

The analysis of polymorphic markers at the DNA level will allow the development of complete marker maps of a genome. We found that polymorphisms in DNA sequence were common in the genome of Icelandic birch. Methods used for detection of molecular variation include: (1) isolated DNA sequences, such as ribosomal genes, that give different RFLP pattern among samples from natural populations or segregating progenies in crosses, and (2) polymerase chain reaction technology such as RAPD analysis that is not dependent on restriction sites or knowledge of the whole genome. For both methods, DNA was isolated from leaves of birch growing in nature and from seedlings grown in a controlled environment.

For detecting variation in the ribosomal genes in birch, we used the non-radioactive chemiluminescence method described by Anamthawat-Jónsson et al. (1990). Isolated birch

DNA was digested with a number of restriction enzymes that cut the DNA at various specific sites. The restricted DNA fragments in each enzyme digest were separated by gel electrophoresis. The size-fractionated DNA was then transferred to a nylon membrane and chemically denatured before hybridization with a single-stranded DNA probe. The probe used in this study was the clone pTa71 again. Preliminary results (Fig. 3) showed that restriction enzyme *Dra*I (the enzyme that cleaves DNA at AAA/TTT site) produced variation in restriction fragment length (RFLP) in the ribosomal genes, particularly among different samples of Icelandic birch. The same DNA probe detected RFLP in cereal species and identified a rye DNA transfer in wheat (Heslop-Harrison et al., 1990). The highly repetitive ribosomal genes are useful as probes because coding regions in the genes are conserved in a wide range of species, but the non-coding regions evolve rapidly and reflect phylogeny (Knaak et al., 1990). The genes were used to show the differences between Brassica species (radish group) and between some accessions within a species (Delseny et al., 1990). Because genetic variation in Icelandic birch is unusually high, the ribosomal genes would also be a good marker for detecting polymorphisms at the population level. In addition, the construction of restriction maps of these genes in birch will be useful in understanding how the DNA sequences undergo evolutionary change or whether introgression can be detected at the DNA level.

We have also begun to detect polymorphisms in the genome of Icelandic birch using RAPD analysis following the method described by Howland et al. (1991). The method uses a heat-stable DNA polymerase to amplify genomic DNA segments targeted by a single DNA primer of arbitrary sequence, followed by gel electrophoresis and ethidium bromide fluorescent staining of the amplified products. The results shown in Figure 4 were obtained from birch genomic DNA treated with Amplitaq DNA polymerase, using a synthesized DNA primer of 10 nucleotides arranged in the following sequence, GTAGTACAGG. The reaction was performed in a thermocycler for 45 cycles, each cycle involved heating the samples at 91 °C for 1 minute, followed by 35 °C for 1 minute and 72 °C for 2 minutes. The amplified DNA was electrophoresed in an 0.8 percent agarose gel.

Icelandic birch appears to be highly variable in its DNA sequence, resulting in different band patterns (Fig. 4). Genetic relationships can be worked out by analyzing the

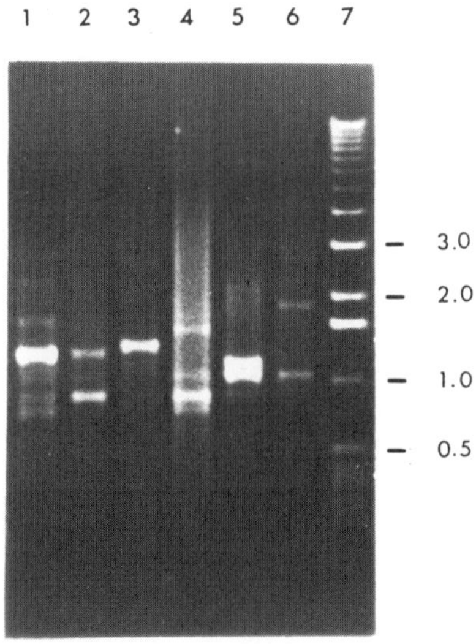

Fig. 4. Random amplified polymorphic DNA (RAPD) of: (1) lowland birch from east England, (2) dwarf arctic birch, and (3-6) Icelandic birch from different populations. Lane (7) shows DNA size marker in kilobases.

bands that individuals have in common and those that are different. The technique is simple, effective, and readily applicable to studying Icelandic birch because it does not depend on knowledge of the whole genome. The technique may also be used to identify specific segregations from crosses in breeding programs. In addition, RAPD markers are found more frequently than RFLPs, at least in corn and soybean. As a result, the construction of RAPD mapping may be faster than RFLP mapping. Some RAPD markers are polymorphic in the DNA sequences not accessible with RFLP method (Chang and Meyerowitz, 1991).

Genome Size in Trees

In common with many other angiosperm trees, and in contrast with many gymnosperms, birch has a very small amount of DNA in its nucleus (Bennett and Smith, 1976; Mowforth, 1986). The amount of DNA in a haploid and unreplicated genome of a diploid birch species (B. pendula) was about 0.4 picogram (Dr. J. Maluszynska, University of Vienna, personal communication) and 0.8 picogram in the tetraploid B. pubescens (Mowforth, 1986). Angiosperms have DNA that encodes about 50,000 genes in their nuclei, or some 100 million base pairs (megabase pairs) of DNA. The nuclear DNA of diploid birch species, including genes and non-coding DNA sequences, is only a few times greater than this amount. Other angiosperms have DNA up to 2,500 times the amounts required for gene coding.

There are many advantages in working with a species with low nuclear DNA (Meyerowitz, 1989). Genes can be cloned and complete "libraries" of the genome can be made and screened for useful genes. One of the largest international biological programs now underway is for the analysis of Arabidopsis thaliana (L.) Heynh., a species in the Brassica family which has a very small genome. The molecular analysis of such species has served as a basis for understanding and manipulating species with larger genomes.

Although work on the molecular analysis of birch genomes, or indeed most tree genomes, is limited, preliminary experiments already show that genome organization can be studied both by molecular genetics and molecular cytogenetics. We hope that this data can be linked with selection and breeding programs to increase the usefulness of native birch, particularly for land reclamation and forestry in Iceland.

SUMMARY

Icelandic birch is highly polymorphic, probably because of evolutionary divergence from various gene pools and also because of hybrid introgression with the coexisting dwarf arctic birch. Through natural selection, many adaptive characteristics associated with morphology, growth, and development have ensured the survival of Icelandic birch in extreme climatic conditions. Growth and form of the species can be improved through breeding both by controlled introgressive hybridization with introduced gene pools, and by directional selection from the existing genetic variation.

The transfer of traits through introgression normally occurs at gene and chromosome levels. Interspecific hybridization between Icelandic birch (tetraploid, 2n=4x=56) and dwarf arctic birch (diploid, 2n=2x=28) gives rise to F_1 triploid hybrids (2n=3x=42). Cytogenetic analysis has shown that surviving plants from the first generation backcross and plants derived from seeds collected in nature have a stable tetraploid chromosome number of 56. Chromosome recombination may have resulted in replacement of segments or whole chromosomes of dwarf arctic birch in Icelandic birch chromosome complement. Detection of such transfers should be possible using the technique of in situ hybridization.

Segments or whole chromosomes from dwarf arctic birch acquired through chromosome recombination often carry genes of adaptive value or genes responsible for desirable traits. To identify such genes without environmental influences, and to be able to select in early stages of tree growth, markers are required at the molecular level. Hybridization of ribosomal genes to restriction enzyme-digested and size-fractionated DNA revealed molecular variation within the species. While transcribed regions of this gene family are conserved, the spacers are known to undergo evolutionary change. Mutation is likely to cause restriction fragment length polymorphisms (RFLP), and some fragments may become species-specific which are useful in detecting introgression. Random amplified polymorphic DNA (RAPD) shows characteristics specific to species, populations, or individuals. RAPD markers may be used to identify polymorphisms, for DNA fingerprinting, and to follow genetic segregations through crosses in breeding programs.

ACKNOWLEDGEMENTS

We thank Dr. D. Howland of the University of East Anglia for advice on RAPD experiments, Prof. T. Skroppa at the Agricultural University of Norway for birch seed stocks, and BP Venture Research and Icelandic Science Council for support.

REFERENCES

Anamthawat-Jónsson, K., and Tómasson, T., 1990, Cytogenetics of hybrid introgression in Icelandic birch, Hereditas (Lund, Sweden), 112:65-70.

Anamthawat-Jónsson, K., Schwarzacher, T., Leitch, A. R., Bennett, M. D., and Heslop-Harrison, J. S., 1990, Discrimination between closely related Triticeae species using genomic DNA as a probe, Theor. Appl. Genet. (Germany), 79:721-728.

Atkinson, M. D. and Codling, A. N., 1986, A reliable method for distinguishing between Betula pendula and B. pubescens, Watsonia (London, UK), 16:75-76.

Bennett, M. D., and Smith, J. B., 1976, Nuclear DNA amounts in angiosperms, Phil. Trans. R. Soc. Lond. B (London, UK), 274:227-274.

Chang, C., and Meyerowitz, E. M., 1991, Plant genome study: restriction fragment length polymorphism and chromosome mapping information, Current Opinion Genet. Dev. (UK), 1:112-118.

Delseny, M., McGrath, J. M., This, P., Chevre, A. M., and Quiros, C. F., 1990, Ribosomal RNA genes in diploid and amphidiploid Brassica and related species: organization, polymorphism, and evolution, Genome (Canada), 33:733-744.

Elkington, T. T., 1968, Introgressive hybridization between Betula nana L. and B. pubescens Ehrh. in north-west Iceland, New Phytol. (Cambridge, UK), 67:109-118

Gerlach, W. L., and Bedbrook, J. R., 1979, Cloning and characterization of ribosomal RNA genes from wheat and barley, Nucl. Acids Res. (Oxford, UK), 7:1869-1885.

Hagman, M., 1971, On self- and cross-incompatibility shown by Betula verrucosa Ehrh. and Betula pubescens Ehrh., Commun. Inst. Forest. Fenn. (Finland), 73:1-125.

Heslop-Harrison, J., 1964, Forty years of genecology. In: Cragg, J. B., ed., Advances in Ecological Research 2. London, UK, Academic Press, 159-247.

Heslop-Harrison, J. S., Leitch, A. R., Schwarzacher, T., and Anamthawat-Jónsson, K., 1990, Detection and characterization of 1B/1R translocations in hexaploid wheat, Heredity (London, UK), 65:385-392.

Howland, D. E., Oliver, R. P., and Davy, A. J., 1991, RAPDs -- Random amplified polymorphic DNA, poster at the meeting "RFLP and QTL in Breeding," University of East Anglia, Norwich, UK.

Kallio, P., Niemi, S., Sulkinoya, M., and Valanne, T. 1983, The Fennoscandian birch and its evolution in the marginal forest zone, Nordicana (Finland), 47:101-110.

Kenworthy, J. B., Aston, D., and Bucknall, S. A., 1972, A study of hybrids between Betula pubescens Ehrh. and B. nana L. from Sutherland -- an integrated approach, Trans. Bot. Soc. Edinb. (Scotland), 42:517-539.

Knaak, C., Hamby, R. K., Arnold, M. L., Leblanc, M. D., Chapman, R. L., and Zimmer, E. A., 1990, Ribosomal DNA variation and its use in plant biosystematics. In: Kawano, S., ed., Biological Approaches and Evolutionary Trends in Plants, London, UK, Academic Press, 135-158.

Leitch, A. R., Mosgöller, W., Shi, M., and Heslop-Harrison, J. S., 1992, Different patterns of rDNA organization at interphase in nuclei of wheat and rye, J. Cell Sci. (UK), 101.

Löve, A., and Löve, D., 1956, Cytotaxonomical conspectus of the Icelandic Flora, Acta Horti Gotoburgensis, 20:135-140.

McIntosh, R. A., 1988, Catalogue of gene symbols for wheat. In: Miller, T. E., and Koebner, R. M. D., eds., Proc. 7th International Wheat Genetics Symposium, UK, Bath Press, 1225-1323.

Meyerowitz, E. M., 1989, Arabidopsis, a useful weed, Cell (Massachusetts, USA), 56:263-269.

Mowforth, M. A. G., 1986, Variation in nuclear DNA amounts in flowering plants: an ecological analysis, Sheffield, UK, University of Sheffield. Ph.D. Thesis.

Schwarzacher, T., Leitch, A. R., Bennett, M. D., and Heslop-Harrison, J. S., 1989, In situ localization of parental genomes in wide hybrid, Ann. Bot. (London, UK), 64:315-324.

Sharp, P. J., Chao, S., Desai, S., and Gale, M. D., 1989, The isolation, characterization and application in the Triticeae of a set of wheat RFLP probes identifying each homoeologous chromosome arm, Theor. Appl. Genet. (Germany), 78:342-348.

Skre, O., 1991, Physiology of plant survival under cold conditions, with particular reference to dark respiration as a factor limiting growth at timberline (a literature review), Medd. Nor. Inst. Skogforsk. (Norway), 44.1:1-34.

Snape, J. W., Law, C. N., Worland, A. J., and Parker, B. B., 1990, Targeting genes for genetic manipulation in crop species. In: Gustafson, J. P., ed., Gene Manipulation in Plant Improvement II, 19th Stadler Genetics Symp., New York, USA, Plenum Press, 55-76.

Sulkinoja, M., 1990, Hybridization, introgression and taxonomy of the Mountain birch in SW Greenland compared with related results from Iceland and Finnish Lapland, Meddelelser om Gronland, Bioscience, 33:21-29.

Sulkinoja, M., and Valanne, T., 1980, Polyembryony and abnormal germination in Betula pubescens subsp. tortuosa, Rep. Kevo Subarctic Res. Stat. (Finland), 16:31-37.

Sulkinoja, M., and Valanne, T., 1987, Leafing and bud size in Betula provenances of different latitudes and altitudes, Rep. Kevo Subarctic Res. Stat. (Finland), 20:27-33.

Vaarama, A., and Valanne, T., 1973, On the taxonomy, biology and origin of Betula tortuosa Ledeb., Rep. Kevo Subarctic Res. Stat. (Finland), 10:70-84.

Vaartaja, O., 1959, Evidence of photoperiodic ecotype in trees, Ecol. Monogr. (USA), 29:91-111.

Williams, J. G. K., Kubelik, A. R., Livak, K. J., Rafalski, J. A., and Tingey, S. V., 1990, DNA polymorphisms amplified by arbitrary primers are useful as genetic markers, Nucl. Acids Res. (Oxford, UK), 18:6531-6535.

CHLOROPLAST DNA DIVERSITY, PHYLOGENETICS, AND HYBRIDIZATION IN PICEA

Aðalsteinn Sigurgeirsson[1] and Alfred E. Szmidt[2]

[1]Iceland Forest Service
Mosfellsbær, Iceland

[2]Swedish University of Agricultural Sciences
Umeå, Sweden

INTRODUCTION

The spruce genus (Picea A. Dietrich) is holarctically distributed and represents a prominent ecological component of northern hemisphere forests, especially those close to the low-temperature limits of forest growth (Hustich, 1953; Schmidt-Vogt, 1977). Ecologically, the genus is tolerant of acidic soils, prolonged cold, and heavy rain and snowfall in winter, factors that might explain its prominence in boreal forests (Page and Hollands, 1987). The genus is one of only eleven genera of tree species that have evolved tolerance to extremely cold climates (Clausen, 1963; Hustich, 1953). Some Picea species are of major commercial value for timber production and are the objects of intensive breeding efforts and extensive population-genetic investigations (Furnier et al., 1991; Lagercrantz and Ryman, 1990; O'Reilly et al., 1985; Yeh and El-Kassaby, 1980).

However, speciation patterns, genetic interrelationships among species, and other questions about the evolutionary genetics of the genus are inadequately resolved (Aldén, 1987; Schmidt-Vogt, 1977).

In recent years, we have studied the use of restriction fragment analysis for revealing chloroplast DNA (cpDNA) diversity within and among Picea species. The goal was to improve our understanding of the evolutionary history of this genus. This paper reviews recent findings and, based on them, suggests different hypotheses for the evolutionary patterns observed.

Forest Development in Cold Climates, Edited by J. Alden *et al.*
Plenum Press, New York, 1993

THE PROBLEM

The genus Picea is noted for its complex and amorphous phylogenetic structure (Page and Hollands, 1987; Wright, 1955). The current understanding of species interrelationships is based mainly on morphological, anatomical, and chemical comparisons, and on the degree of cross-compatibility among species (Schmidt-Vogt, 1977). The biological validity of this knowledge is, however, clouded by inconsistency and controversy among authorities concerning circumscriptions of species and species groups (sections) derived on the basis of these diverse criteria (Aldén, 1987; Schmidt-Vogt, 1977). Not uncommonly with Picea, we are faced with the thorny problem of species definition and delimitation. In this genus, the terms "taxonomic" and "biological" species cannot be applied interchangeably. In fact, morphologically very dissimilar species may be highly interfertile (Fowler, 1987; Mikkola, 1969; Wright, 1955), while similar species may be incompatible or nearly so (Ernst et al., 1989; Gordon, 1976). Many Picea species are difficult to classify as species or ecotypes; hence, current circumscriptions may be colored by artifacts of a taxonomist's inclination to "split" or to "lump" (Schmidt-Vogt, 1977). Moreover, several morphologically distinct species are known to cross and intergrade where geographic boundaries between them are indistinct or nearly absent (Bobrow, 1972). In regions where species are suspected to intergrade, it is difficult to tell whether the variation among localities represents one variable species or two or more species. Consequently, studies of genetic variation within Picea species are sometimes thought to be confounded by variation resulting from natural interspecific hybridization (Daubenmire, 1968; El-Kassaby et al., 1988; Fowler et al., 1988).

DNA MARKERS

If the goal is to obtain genetic and evolutionary information, direct examination of genetic variation at the DNA level allows inference of genotypes from phenotypes, and is one major advantage over analyses of gene products, such as morphological or biochemical characters. The development of DNA markers, especially restriction fragment length polymorphisms (RFLPs), provides virtually unlimited qualitative genetic markers that are easily characterized. RFLPs serve as ideal genetic markers for studying selectively neutral genetic variation and unravelling evolutionary patterns. They generally have little effect on morphology or fitness, do not exhibit ontogenetic or phenotypic plasticity, and are ubiquitously distributed throughout the plant and animal kingdoms (Birky, 1988). Their use depends on the ability of restriction enzymes to cleave DNA at specific four, five, or six nucleotide recognition sites, allowing the detection of polymorphisms created by point mutations, insertions, deletions, and inversions on the genome.

THE CHLOROPLAST GENOME

The development of higher plants is dependent on a coordinated expression of three distinct genomes: chloroplast (cp), mitochondrial (mt), and nuclear. Relative to the other genomes, cpDNA is small (120,000 to 160,000 base-pairs). The cp genome harbors ribosomal- and transfer-RNA genes, and many protein-coding genes necessary for the photosynthetic apparatus (Zurawski and Clegg, 1987). A commonly accepted theory for the origin of chloroplasts is that they represent descendants of ancient unicellular organisms that lost their biological autonomy and became permanently ensconced as intracellular symbionts of higher plants (Cavalier-Smith, 1982; Gray and Doolittle, 1982). Because changes in size, gene content, and sequence during 400 million years of evolution have been small, the plant cpDNA has been described as "a genome frozen in time" (Birky, 1988).

Applying RFLPs for the purpose of analyzing cpDNA variation elicits several advantages for evolutionary studies on plants, owing to unique features of the genome: uniparental inheritance, compact size, and slow rate of sequence and structural evolution. In recent years, the availability of cpDNA RFLPs has brought about a surge of comparative studies aimed at clarifying evolutionary relationships among plant groups (Palmer, 1987; Palmer et al., 1988). Most of these studies concerned angiosperm genera of agronomic importance; few have focused on gymnosperms. The conservative evolution of cpDNA tends to limit the amount of useful sequence variation that can be found intraspecifically, or among very closely related species (Birky, 1988; Palmer et al., 1988). Nevertheless, given sufficient analysis, such variation can be detected and employed in genetic studies of population structure of plants (Govindaraju et al., 1989; White, 1990).

Most genetic markers are nuclear-DNA-encoded morphological or biochemical characters. Because these markers are biparentally inherited, their interpretation may be complicated by recombination. The inheritance of cpDNA is uniparental in plants and follows a line over generations, changing only by occasional mutations. Its mode of genetic transmission is mainly maternal among angiosperms, but strictly paternal among all gymnosperms investigated to date (Neale et al., 1989). The "linear" inheritance of cpDNA provides unique opportunities for its use as a genetic marker to follow evolutionary lineages, especially for insights into the origin of species-hybrid complexes (Rieseberg et al., 1990; Sigurgeirsson et al., 1990, 1991; Wang and Szmidt, 1990).

DIVERSITY IN CHLOROPLAST DNA: *PICEA* VS. *PINUS*

When compared to the genus *Pinus*, for example, the genus *Picea* has little morphological variation (Wright, 1955). Ecological differentiation is also low in *Picea*; the genus is mainly confined to cool-temperate or boreal forest zones (Page and Hollands, 1987), whereas species of *Pinus* occupy a wide spectrum of biotypes, ranging from tropical to subarctic (Axelrod, 1986; Li, 1953; Mirov, 1967). Recently, we carried out a restriction fragment analysis to contrast levels of cpDNA differentiation among these two genera (Sigurgeirsson and Szmidt, in press). We examined 31 species of *Picea*, representing the bulk of all taxonomically described species, and included *Pinus sylvestris* L., from the subgenus *Pinus*, and *P. cembra* L., from the subgenus *Strobus*.

Levels of cpDNA differentiation among the *Picea* species were low relative to the cpDNA differentiation between the two *Pinus* species (Fig. 1a). The highest divergence between *Picea* species (measured as $\hat{d}$; the estimated number of nucleotide substitutions per site; see Nei and Miller, 1990) was $\hat{d} = 0.0137$ between *P. breweriana* S. Wats. and *P. glauca* (Moench) Voss, whereas the divergence between *Pinus sylvestris* and *P. cembra* was $\hat{d} = 0.0418$. Provided that cpDNA evolves at similar rates in the two genera, the low level of cpDNA divergence between the *Picea* species relative to the *Pinus* species suggests that all *Picea* species radiated from their immediate, common ancestor long after the separation of *Pinus* into two subgenera. This evidence is congruent with fossil evidence, which suggests that *Pinus* differentiated from other Pinaceae genera in the Mesozoic era, about 140 to 135 million years (Ma) before present (Alvin, 1960; Axelrod, 1986; Miller, 1976). The oldest confirmed fossil of an ancient *Picea* species dates only to the Oligocene epoch, 38-25 Ma (Miller, 1989). This suggests that the recent phylogenetic radiation and age of *Picea* species are responsible for their weak differentiation in morphology and ecological preference, rather than because "there is something inherent in the spruce germ plasm that resists change," as postulated by Wright (1955).

PHYLOGENETIC INFERENCE FROM CHLOROPLAST DNA

More than 60 species have been described for the genus Picea. Picea has been revised several times during the past century, mainly on the basis of morphological criteria (Schmidt-Vogt, 1977; Schmidt, 1989). The genus has proven recalcitrant to phylogenetic analysis, however, and systematists do not agree on a phylogenetic hypothesis that properly summarizes evolutionary genetic relationships (Aldén, 1987). Phylogenetic groupings from artificial crossing experiments lack concordance with groupings based on morphology, and do not form a tenable basis for subdivisions of the genus (Mikkola, 1969; Wright, 1955).

Molecular genetic information can be of considerable value for appraising genetic discontinuities within and among species (Avise, 1989) and for reconstructing phylogenies

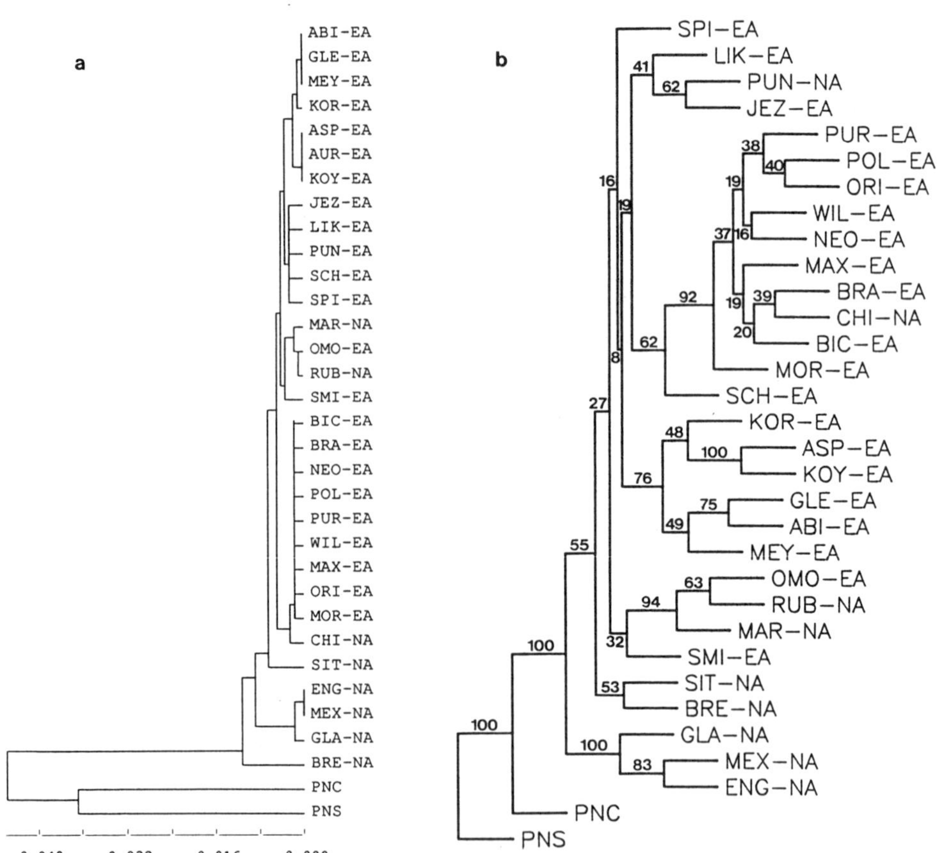

Fig. 1. (a) UPGMA (Unweighted pair-group average) dendrogram for Picea and Pinus species, derived from $\hat{d}$ values, estimated number of nucleotide substitutions per site (Nei and Miller, 1990), in the RESTSITE program (v. 1.1) of the RESTSITE computer package (v. 1.1.; developed by J. C. Miller, University of Wisconsin). For abbreviations of species names, see Table 1. Origin (by continent) denoted by "EA" (Eurasia) and "NA" (North America). (b) Majority-rule Wagner parsimony tree, constructed with the use of bootstrap (Felsenstein, 1985), and using the MIX program of PHYLIP (Phylogeny Inference Package), v. 3.1. (J. Felsenstein, University of Washington, Seattle, Washington), with 100 replicate runs. Numbers at nodes refer to the percentage of bootstrap samples where this branching was noted.

Table 1. Species Included in Phylogenetic Studies

	Species	Distribution
1.	Picea abies (L.) Karst. (ABI)	Northern Eurasia
2.	P. asperata Mast. (ASP)	Central China
3.	P. aurantiaca Mast.[1] (AUR)	Central China
4.	P. bicolor (Maxim.) Mayr (BIC)	Japan
5.	P. brachytyla (Franch.) Pritz. (BRA)	Central China
6.	P. breweriana S. Wats. (BRE)	N.W. United States
7.	P. chihuahuana Martinez (CHI)	Mexico
8.	P. engelmanni (Parry) Engelm. (ENG)	Western N. America
9.	P. glauca (Moench) Voss (GLA)	North America
10.	P. glehnii (Fr. Schmidt) Mast. (GLE)	East Asian islands
11.	P. jezoensis (Sieb. et Zucc.) Carr. (JEZ)	East Asian islands
12.	P. koraiensis Nakai (KOR)	Korea/Manchuria
13.	P. koyamai Shirasawa (KOY)	Japan
14.	P. likiangensis (Franch.) Pritz. (LIK)	Central China
15.	P. mariana (Mill.) Britt. (MAR)	North America
16.	P. maximowiczii Regel ex Mast. (MAX)	Japan
17.	P. mexicana Martinez (MEX)	Mexico
18.	P. meyeri Rehd. et Wils. (MEY)	Central China
19.	P. morrisonicola Hayata (MOR)	Taiwan
20.	P. neoveitchii Mast. (NEO)	Central China
21.	P. omorika (Pancic) Purkyne (OMO)	Serbia
22.	P. orientalis (L.) Link (ORI)	Caucasus
23.	P. polita (Sieb. et Zucc.) Carr. (POL)	Japan
24.	P. pungens Engelm. (PUN)	W. United States
25.	P. purpurea Mast.[2] (PUR)	Central China
26.	P. rubens Sarg. (RUB)	E. North America
27.	P. schrenkiana Sarg. (SCH)	Central Asia
28.	P. sitchensis (Bong.) Carr. (SIT)	W. North America
29.	P. smithiana (Wall.) Boiss. (SMI)	Central Asia
30.	P. spinulosa (Griff.) Henry (SPI)	Himalayas
31.	P. wilsonii Mast. (WIL)	Central China
32.	Genus Pinus, Subgenus Strobus: P. cembra L. (PNC)	N. Asia, C. Europe
33.	Genus Pinus, Subgenus Pinus: P. sylvestris L. (PNS)	Europe, N. Asia

[1] = P. asperata var. aurantiaca (Mast.) Boom
[2] = P. likiangensis var. purpurea (Mast.) Dallim. & Jacks.

(Palmer, 1987). With these objectives in mind, we set out to reconstruct a cpDNA-based Picea phylogeny, by comparing cpDNA restriction fragments among 31 Picea species, using eight restriction enzymes (Sigurgeirsson and Szmidt, in press). As noted before, we found the overall level of cpDNA differentiation in the genus to be low. Nevertheless, based on cpDNA affinities, a well-defined and statistically supported structure with species groups was discerned within the genus. Results for different approaches to phylogeny reconstruction among 31 species were largely in agreement and suggest several cpDNA groups (Figs. 1 and 2).

Monophyletic groups (clades) within some species groups could not be resolved because of small differences, indicating recent evolutionary divergence. An example of this poor resolution concerned Picea chihuahuana Martinez and a number of Asian species which exhibited very near affinities at the cpDNA level. These are referred to as the P. brachytyla complex, including P. bicolor (Maxim.) Mayr, P. brachytyla (Franch.) Pritz, P. maximowiczii Regel ex Mast., P. morrisonicola Hayata, P. neoveitchii Mast., P. orientalis

(L.) Link, P. polita (Sieb. et Zucc.) Carr., P. purpurea Mast., and P. wilsonii Mast. (Figs. 1 and 2). Most species in the P. brachytyla complex harbored virtually identical cpDNAs, distinguishable only on the basis of one or very few RFLPs. These findings agree with the observations of Wright (1955) and Gordon (1968), who suggested close affinities between P. chihuahuana and the P. brachytyla complex, which comprised many of the above species, and which is centered in China. We found P. purpurea (P. likiangensis var. purpurea) nested within this same cpDNA group. This result did not agree with its common taxonomic placement as a subspecies of P. likiangensis (Franch.) Pritz. (Table 1). It is worth noting, however, that Wright (1955) considered P. purpurea to be morphologically closer to the P. brachytyla complex than to P. likiangensis.

Other examples of close agreement between cpDNA affinities and earlier systematic schemes include P. abies (L.) Karst., P. asperata Mast., P. aurantiaca Mast., P. glehnii (Fr. Schmidt) Mast., P. koyamai Shirasawa, and P. koraiensis Nakai. These species share several morphological features (Rushforth, 1987) and exhibit high interfertility in artificial crossing experiments (Mikkola, 1969). Moreover, the almost identical cpDNAs of P. asperata and P. aurantiaca agree with the usual placement of P. aurantiaca as a variety of P. asperata within the genus (Schmidt-Vogt, 1977).

It is generally recognized that genetic differences between species often reflect adaptation and ecological preference to climatic or edaphic factors. Hence, different congeners are frequently visualized as occupying different peaks on an "adaptive landscape" (Wright, 1932). Such adaptations may arise at different stages during the evolution of a genus. Some may be modifications resulting from recent environmental selection after an extant species diverged from another species. Other adaptations may be modifications within extinct, common ancestors of extant species which manifest themselves in particular suites of adaptations shared among closely related members of the same genus. In this context, similarities in ecological preference between Picea species with close cpDNA affinities may indicate joint retention of adaptive characters. Among these is the P. brachytyla complex. These species, including P. chihuahuana, share a quasi-oceanic climate with cool temperatures throughout the year and daily fluctuations exceeding seasonal fluctuations (Ohsawa, 1990; Schmidt, 1989). A cpDNA species group comprising P. abies and allied species (Figs. 1 and 2) occupies the more seasonal, boreal forest zones north of the P. brachytyla complex's range in Eurasia. According to Kuan (1981), the southward distribution of the P. abies species group in China is limited by their high demand for light during the summer.

BIOGEOGRAPHY OF PICEA

Of the 34 extant Picea species recognized as distinct by Aldén (1987), nine are found in North America, two in Europe, and the remainder (23) in various parts of Asia. Mainly on the basis of the large number of species, Asia has long been thought to represent the origin for the genus, from where species later migrated into Europe and North America (Li, 1953; Wright, 1955).

Reconstructions of migrational histories in the absence of complete fossil and historical evidence depend on a reliable phylogenetic framework of reference (Futuyma, 1986). In the past, hypotheses on interrelationships of Picea species were poorly supported and did not provide an adequate basis for biogeographic inferences. Many species were not easily delimited, as already explained. In fact, the morphological uniformity observed within the genus led some botanists and foresters to infer that all Picea species represent either hybrids or subspecies (Gordon, 1976). As an example, Hustich (1953) postulated that the taxonomic distinction between P. abies in Eurasia and P. glauca in North America might be due merely to exaggeration of the Atlantic Ocean and Bering Strait as phytogeographic barriers.

We have used the cpDNA-based phylogeny as a framework for testing hypotheses concerning the biogeography of Picea (Sigurgeirsson and Szmidt, in press). We found the degree of between-species cpDNA divergence highest among North-American species. Furthermore, North American species branch closest to the hypothesized phylogenetic root of the genus (Fig. 1b). From the evidence, we are unable to find support for the oft-cited postulate of Wright (1955), that Asia was the center of origin for the genus. The present apportionment of cpDNA variation among continents is more easily reconcilable with the notion that North America is the center of diversity in the genus. This would also suggest that North America is the center of origin.

The apparent discordance between the partitioning of taxonomic diversity and cpDNA diversity among continents could be most easily explained in terms of the unsatisfactory taxonomic status of several species, especially those from China (Schmidt-Vogt, 1977; Wright, 1955). Hence, given a more rigorous taxonomic treatment, Schmidt-Vogt (1977) suggested that the 22 Chinese Picea species could be reduced to three.

Occasionally, cpDNA affinities are closer between species native to different continents than between species with adjacent or neighboring ranges on the same continent. For example, the North American P. mariana and P. rubens show closer affinities to P. omorika from Europe than to other North American species. Together, these three species form a well-supported clade (Fig. 1b), which is compatible with both morphology and crossability data (Fowler, 1983; Rushforth, 1987; Wright, 1955). The cpDNA evidence suggests that recent intercontinental migrations have occurred since the original radiation of Picea. Picea species commonly grow near arctic or alpine treelines where the climatic history has been turbulent, with rapid and recurrent cycles of range expansions and contractions (Huntley and Webb, 1989). Indeed, Picea did not contract from the high arctic to more southern latitudes until the end of the Pliocene (2 million years ago), at a late stage of Picea evolution (Florin, 1963). At this time, the main route for the migration of temperate conifer and hardwood species between Eurasia and North America was apparently across the North Atlantic, with Iceland providing an interrupted pathway into the early Quaternary (Raven and Axelrod, 1974). Bearing this in mind, it is not surprising to observe close genetic affinities between species with highly disjunct ranges, and distant relationships between species with adjacent or sympatric ranges.

ANALYSES OF PICEA HYBRID ZONES

Studies of interspecific crossability between different Picea species suggest, with exceptions, that internal barriers to hybridization are weak (Mikkola, 1969; Wright, 1955). Hence, opportunities for interspecific gene exchange among a number of Picea species are ample, once they come into contact. Indeed, several cases have been reported where two or more apparently distinct but interfertile Picea species meet, mate, and produce hybrids (Bobrow, 1972). The purported existence of hybrid zones involving numerous taxa has led to the suggestion that interspecific hybridization, followed by introgression, has been a factor of profound evolutionary significance in Picea (Bobrow, 1972; Wright, 1955).

CpDNA polymorphisms offer powerful and penetrant genetic markers for establishing or refuting ambiguous cases of interspecific gene flow in Picea and other species exhibiting paternal cpDNA inheritance. CpDNA RFLPs in Picea usually appear to be either species-diagnostic, that is, intraspecifically invariant and unique for the given species, or common to and diagnostic for certain species groups (El-Kassaby et al., 1988; Sigurgeirsson et al., 1990; Stine et al., 1989). Owing to the paternal mode of inheritance, a RFLP on the chloroplast genome can serve as a species-diagnostic pollen marker for following gene flow from one species to another. Thus, the species siring the progeny of a given seed parent can be inferred from the species-diagnostic cpDNA-types present among the progeny (Sigurgeirsson et al., 1990). In population samples dominated by one species, it is even possible to detect trace levels of alien cpDNA (Sigurgeirsson et al., 1990, 1991; Stine et al., 1989).

We have carried out a series of studies aimed at elucidating the dynamics and spatial distribution of a spruce complex found on and around the Kenai peninsula in Alaska. In this region, P. glauca, P. mariana, and P. sitchensis supposedly migrated from separate Pleistocene refugia and are sympatric. Natural hybrid zones of P. xlutzii Little (P. glauca x P. sitchensis) were first described in the region by Little (1953) on the basis of morphology. Hybridization or introgression appears strongly corroborated from other lines of genetic evidence (Copes and Beckwith, 1977). There is also some morphological evidence for the localized existence of natural hybrids between P. mariana and P. sitchensis, as well as trihybrids (P. xlutzii x P. mariana) occurring in the region (J. Alden, Institute of Northern Forestry, USDA Forest Service, Fairbanks, Alaska, personal communication).

The ranges of P. glauca and P. sitchensis are for the most part allopatric or parapatric due to distinctly different ecological preferences (Daubenmire, 1968; Roche, 1969). P. sitchensis is dependent on an oceanic climate for proper development, concomitant with abundant humidity and rainfall, and small variation in annual temperature (Malcolm, 1987). The wide transcontinental distribution of P. glauca suggests that it is better adapted to dry continental climates with wide seasonal amplitude. The Kenai peninsula is, however, an "intermingling and mixing of interior continental forests of P. glauca and rainy coastal forests of P. sitchensis" (Little, 1953), and has an abundance of complex transitional zones between continental and oceanic climatic regimes (Bonan and Shugart, 1989). The abundance of intermediate habitats may provide niches for genotypes adaptively intermediate between the norm for P. sitchensis and P. glauca.

We have identified a number of RFLP markers for distinguishing the cpDNAs of P. glauca, P. mariana, and P. sitchensis (El-Kassaby et al., 1988; Sigurgeirsson et al., 1990, 1991; Szmidt et al., 1988). These markers have been tested on a variety of standard individuals and have been shown to be intraspecifically invariant. In one study (Sigurgeirsson et al., 1990), the markers were used to confirm earlier morphological and allozyme evidence that P. xlutzii represents a hybrid product of P. glauca and P. sitchensis. In each composite progeny sample, we observed only markers diagnostic for P. sitchensis and P. glauca, concurrently. This suggested that the fathers of the resulting open-pollinated progeny harbored divergent cpDNAs. Two separate populations differed, however, in the proportions of seed sired by each species. The abundance of seed sired by each respective parent species appeared to correlate with the tendency of the maternal parent to resemble either P. glauca or P. sitchensis morphologically. Later, closer examinations have failed to confirm P. mariana-diagnostic cpDNA in the progenies (authors' data). Thus, the seed sired by P. mariana is apparently negligible.

In a separate study (Sigurgeirsson et al., 1991), we analyzed composite population samples to determine the extent of P. glauca penetration into P. sitchensis populations in the region. The frequent occurrence of P. sitchensis and P. glauca sympatry and, by inference, interspecific gene flow on the Kenai peninsula, were confirmed. Furthermore, recent analyses of 63 populations indicate the presence (albeit often in very low amounts) of P. glauca cpDNA in most allopatric populations of P. sitchensis in Alaska, but the lack of P. sitchensis cpDNA in allopatric populations of P. glauca (authors' data).

Relative to the genus as a whole, Picea glauca and P. sitchensis are morphologically distinct species (Page and Hollands, 1987; Wright, 1955), and are always separated rather widely in the system of the genus (Schmidt-Vogt, 1977). In terms of pairwise comparisons between Picea species, we find genetic divergence between them high at the cpDNA level ($\hat{d}$ = 0.0104) (Sigurgeirsson and Szmidt, in press). Despite their apparently high evolutionary distance, they cross as well as conspecifics (Fowler, 1987; Wright, 1955) and apparently produce hybrids wherever they come into contact, naturally or under cultivation (Roche and Fowler, 1975). Hence, all lines of evidence, apart from high crossability, would suggest that the hybrid between P. glauca and P. sitchensis is a cross between genotypes separated by genetic divergence nearly as high as any found within the genus.

A potential object of further investigation is whether the interpecific hybridization observed in the area of contact has produced an aggressive, self-perpetuating hybrid (P. xlutzii) that is better adapted to the climatic transition zone than the parent species, or whether the hybrid is simply a fortuitous byproduct of interfertility and sympatry without any hybrid advantage per se. Among other things, the loss of vigor, resistance, viability, and other ecologically important traits has been associated with hybrid zones and is thought to be an important barrier between species (Whitham, 1989). Hence, hybrids may exhibit outbreeding depression due to the disruption of coadapted gene complexes, but nevertheless be numerous in transition zones due to the inability of parent species to isolate themselves from each other's genes.

CHLOROPLAST DNA POLYMORPHISM IN NORWAY SPRUCE

Norway spruce (Picea abies (L.) Karst.) is native to montane and boreal forests of Europe and northern Asia and occupies the largest geographic range of any Picea species. Picea abies has long been known to exhibit high levels of morphological variation and a geographic structure that can be partitioned on the basis of morphological, physiological, and biochemical variation (Lagercrantz and Ryman, 1990; Schmidt-Vogt, 1977). The significance of this variation has been controversial from the systematic point of view. The controversy has centered on the question of whether to delineate a separate taxon, P. obovata Ledeb., within P. abies, or whether to treat it as an ecotype with a northern and eastern distribution.

Employing restriction analysis for the purpose of assaying intraspecific cpDNA variation, we detected two distinct cpDNA types (plastotypes) among European populations of P. abies on the basis of two, coinherited RFLPs (authors' data). A tentative examination of their distribution in Europe revealed that one plastotype is predominantly distributed in Russia and northern Fennoscandia, while the other is distributed in central Europe. The distribution of plastotypes is geographically reminiscent of the subdivision of the P. abies taxon into P. abies and P. obovata. Although the pattern observed suggests a parapatric distribution for the two plastotypes in Europe, at least one wide zone of contact and intergradation is present in central and southern Sweden. The pattern of intergradation in Sweden is congruent with earlier reports of morphological, physiological, and allozyme variation within the species (Krutzsch, 1986; Lindquist, 1948; Lundkvist, 1979; Lundkvist and Rudin, 1977).

An analysis of cpDNA differentiation in P. abies and numerous closely allied species in the P. abies alliance supports the treatment of P. abies as a strongly differentiated, monophyletic assemblage. These plastotypes appear to represent two lineages within P. abies that diverge more than P. asperata and P. koyamai or P. meyeri and P. koraiensis, for example.

IMPLICATIONS FOR SPECIES AND GENE CONSERVATION

Concern is currently growing over the imminent loss of global biodiversity resulting from both direct and indirect human pressures on the environment and its genetic resources (Anonymous, 1990; Falk, 1990; Peters, 1990). The consensus among authorities is that a major extinction spasm is imminent, with dire biological and economical consequences (Vane-Wright et al., 1991). Even woody species from widespread and economically significant genera, including those of temperate climates, are thought to be threatened or disappearing (Falk, 1990).

The cosmopolitan distribution of Picea throughout much of the northern hemisphere might buffer the genus against the threat of extinctions or severe losses of diversity. Much of the range of Picea is occupied, however, by a small number of taxa at northerly latitudes, for example P. abies, P. glauca, and P. mariana, while most species in the genus are rare endemics at southern latitudes. At least one of these, P. chihuahuana, is considered endangered (Gordon, 1968).

Rare endemics in the genus Picea typically exhibit narrow ecological tolerance and are relicts from past ice ages found in isolated belts or pockets in cool, marginal, mountainous habitats of warm climates (Rushforth, 1986; Wright, 1955). Hence, a rapid rise in global temperatures and the subsequent habitat displacement predicted by several workers (Peters, 1990) may leave many endemic species vulnerable to extinction. Aside from the obvious moral and aesthetic arguments for preserving the present array of Picea species, economic arguments for conserving these species are foremost. Some species may have value for cultivation within and beyond their present range, or may hold valuable genes for breeding programs of widespread congeners (Rushforth, 1986). A lineage represented by a single endemic species may represent many million years of divergent evolution, and may have accumulated genes for rare adaptations which would be lost on extinction. To illustrate, P. omorika, endemic to a tiny range in the Drina Mountains of Serbia, has only a few thousand trees. Nevertheless, its tiny gene pool exhibits allozyme variation comparable to that of the widespread P. abies (Kuittenen et al., 1991). Moreover, P. omorika is remarkably successful at many sites outside its native range, and is more tolerant of atmospheric pollution than any of its congeners (Rushforth, 1987).

Questions may arise about how to maximize biotic diversity within the genus Picea. A straightforward strategy is to define priorities for safeguarding viable populations of each species. There is a problem, however, connected to the question of taxonomic artifact and the confusing taxonomy of Picea, as described above. In conservation biology, the term biodiversity is commonly used as a measure of the number of species that inhabit the planet. Hence, the problem of measuring biodiversity is commensurate with the problem of defining what constitutes a species. The reality of existing taxonomies as a reference for distinguishing different species or populations as evolutionarily distinct entities that are worthy of conservation efforts may sometimes have a weak basis (Avise, 1989; O'Brien and Mayr, 1991). Many descriptions of species trace to meager morphological evidence derived from cursory surveys in the last century. When subjected to molecular tests, morphology-based distinctions of "species" are often poorly corroborated (Avise, 1989). Hence, some groups showing little evolutionary differentiation at the DNA level have been recognized as taxonomically distinct, while phylogenetically distinct forms have not been recognized as distinct taxa. In light of this, the adequacy of biotic descriptions should be considered before embarking on costly conservation programs.

Our studies suggest the feasibility of identifying evolutionarily distinct members of the genus with cpDNA-based tests. Results of such studies, especially if complemented with other molecular analyses of genetic variation within and among taxa, could resolve phylogenetic breaks within the genus more clearly than morphology-based taxonomies and thus weigh evolution in conservation priorities (Vane-Wright et al., 1991). So far, our cpDNA-based results indicate (1) the evolutionary distinctness of certain species, for example, P. breweriana, P. chihuahuana, and P. omorika; (2) the near absence of evolutionary distinctness of some species, suggestive of their ecotype status, for example, P. asperata and P. aurantiaca; (3) the resolution of certain putative hybrid species or introgressed populations, such as P. xlutzii; and (4) the resolution of major genetic breaks within species with divergent evolutionary histories, for example, P. abies.

CONCLUSIONS

Several new insights into the evolutionary status of the genus Picea and its individual species have been obtained from cpDNA restriction fragment studies. We have only begun, however, to discover the wealth of evolutionary information that can be obtained from molecular genetic methods. Analyses of cpDNA, mtDNA, or nuclear DNA sequence variation, either through restriction mapping (Doyle et al., 1990) or polymerase chain reaction technology (Mullis and Faloona, 1987), hold promise for learning more about patterns and rates of evolutionary divergence within Picea. Given time, patience, and resources, the genetic structure of this ecologically and economically important genus may eventually be resolved.

ACKNOWLEDGEMENTS

This work was supported by grants from The Cellulose Industries Council for Technology and Forest Research, The Swedish Council for Forestry and Agricultural Research, the Alaska Science and Technology Foundation, the Nordic Council for Agricultural Research, and the Iceland Forest Service.

REFERENCES

Aldén, B., 1987, Taxonomy and geography of the genus Picea, Intern. Dendrol. Soc. Yearb. 1986, 85-96.

Alvin, K. L., 1960, Further conifers of the Pinaceae from the Wealden formation of Belgium, Bull. Inst. Roy. Sci. Natur. Belgique, 146:1-39.

Anonymous, 1990, Biological diversity, its conservation and use for sustainable agricultural, forestry and fisheries development, FAO Working Paper, IDWG Sub-Group on Biological Diversity, United Nations, FAO, Rome, 41 p.

Avise, J. C., 1989, A role for molecular genetics in the recognition and conservation of endangered species, TREE, 4:279-281.

Axelrod, D. I., 1986, Cenozoic history of some western American pines, Ann. Missouri Bot. Gard., 73:565-641.

Birky, C. W., 1988, Evolution and variation in plant chloroplast and mitochondrial genomes. In: Gottlieb, L. D., and Jain, S. K., eds., Plant Evolutionary Biology, Chapman and Hall, New York, N.Y., United States, 23-53.

Bobrow, E. G., 1972, Die Introgressive Hybridisation in der Gattung Picea A. Dietr. (Introgressive Hybridization in the genus Picea A. Dietr.), Symp. Biol. Hung., 12:141-148.

Bonan, G. B., and Shugart, H. H., 1989, Environmental factors and ecological processes in boreal forests, Annu. Rev. Ecol. Syst., 20:1-28.

Cavalier-Smith, T., 1982, The origin of plastids, Biol. J. Linnean Soc., 17:289-306.

Clausen, J., 1963, Tree lines and germ plasm; a study in evolutionary limitations, Proc. Natl. Acad. USA, 50:860-868.

Copes, D. L., and Beckwith, R. C., 1977, Isoenzyme identification of Picea glauca, P. sitchensis and P. xlutzii populations, Bot. Gaz., 138:512-521.

Daubenmire, R., 1968, Some geographic variations in Picea sitchensis and their ecologic interpretation, Can. J. Bot., 46:787-798.

Doyle, J. L., Doyle, J. L., and Brown, A. H. D., 1990, A chloroplast-DNA phylogeny of the wild perennial relatives of soybean (Glycine subgenus Glycine), Congruence with morphological and crossing groups, Evolution, 44:371-389.

El-Kassaby, Y. A., Sigurgeirsson, A., and Szmidt, A. E., 1988, The use of restriction analysis of chloroplast DNA in classifying hybrid spruce seedlots. In: Hällgren, J.-E., ed., Proceedings of the Frans Kempe Symposium on Molecular Genetics of Forest Trees, 1988 June 14-16, Swedish Univ. Agric. Sciences, Umeå, Sweden, 67-88.

Ernst, S. G., Hanover, J. W., and Keathley, D. E., 1989, Assessment of natural interspecific hybridization of blue and Engelmann spruce in southwestern Colorado, Can. J. Bot., 68:1489-1496.

Falk, D. A., 1990, Endangered forest resources in the U.S., integrated strategies for conservation of rare species and genetic diversity, For. Ecol. Manage., 35:91-107.

Felsenstein, J., 1985, Confidence limits on phylogenies, an approach using the bootstrap, Evolution, 39:783-791.

Florin, R., 1963, The distribution of conifer and taxad genera in time and space, Acta Horti Bergiani, 20:121-326.

Fowler, D. P., 1983, The hybrid black x Sitka spruce, implications to the phylogeny of the genus Picea, Can. J. For. Res., 13:108-115.

Fowler, D. P., 1987, The hybrid white x Sitka spruce: species crossability, Can. J. For. Res., 17:413-417.

Fowler, D. P., Park, Y. S., and Gordon, A. G., 1988, Genetic variation of red spruce in the Maritimes, Can. J. For. Res., 18:703-709.

Furnier, G. R., Stine, M., Mohn, C. A., and Clyde, M. A., 1991, Geographic patterns of variation in allozymes and height growth in white spruce, Can. J. For. Res., 21:707-712.

Futuyma, D. J., 1986, Evolutionary Biology, 2nd Ed, Sinauer Assoc., Sunderland, Mass., United States, 600 p.

Gordon, A. G., 1968, Ecology of Picea chihuahuana Martinez, Ecol., 49:880-896.

Gordon, A. G., 1976, The taxonomy and genetics of Picea rubens and its relationship to Picea mariana, Can. J. Bot., 54:781-813.

Govindaraju, D. R, Dancik, B. P., and Wagner, D. B., 1989, Novel chloroplast DNA polymorphism in a sympatric region of two pines, J. Evol. Biol., 2:49-59.

Gray, M. W., and Doolittle, W. F., 1982, Has the endosymbiont hypothesis been proven?, Microbiol. Revs., 46:1-42.

Huntley, B., and Webb, T. III, 1989, Migration: species' response to climatic variations caused by changes in the earth's orbit, J. Biogeography, 16:5-19.

Hustich, I., 1953, The boreal limits of conifers, Arctic, 6:149-160.

Krutzsch, P., 1986, An investigation on bud set in Norway spruce (Picea abies). In: Lindgren, D., ed., Proceedings of the Frans Kempe Symposium on Provenances and

Forest Tree Breeding for High Latitudes, 1986 June 10-11, Swedish Univ. Agric. Sciences, Umeå, Sweden, 21-31.

Kuan, C-T., 1981, (Fundamental features of the distribution of Coniferae in Sichuan), Acta phytotax. sin., 19:393-407. In Chinese.

Kuittinen, H., Muona, O., Kärkäinen, K., and Borzan, Z., 1991, Serbian spruce, a narrow endemic, contains much genetic variation, Can. J. For. Res., 21:363-367.

Lagercrantz, U., and Ryman, N., 1990, Genetic structure of Norway spruce (Picea abies), concordance of morphological and allozymic variation, Evolution, 44:38-53.

Li, H. L., 1953, Present distribution and habitats of the conifers and taxads, Evolution, 7:245-261.

Lindquist, B., 1948, The main varieties of Picea abies (L.) Karst. in Europe, with a contribution to the theory of forest vegetation in Scandinavia during the last Pleistocene glaciation, Acta Horti Bergiani, 14:249-342.

Little, E. L., Jr., 1953, A natural hybrid in Alaska, J. Forest., 51:745-747.

Lundkvist, K., 1979, Allozyme frequency distributions in four Swedish populations of Picea abies as determined by isozyme analysis, Hereditas, 90:127-143.

Lundkvist, K., and Rudin, D., 1977, Genetic variation in eleven populations of Picea abies as determined by isozyme analysis, Hereditas, 85:67-74.

Malcolm, D.C., 1987, Some ecological aspects of Sitka spruce, Proc. Royal Soc. Edinburgh, 93B, 85-92.

Mikkola, L., 1969, Observations on interspecific sterility in Picea, Ann. Bot. Fenn., 6:285-339.

Miller, C. N., 1976, Early Evolution in the Pinaceae, Rev. Palaeobot. Palynol., 21:101-117.

Miller, C. N., 1989, A new species of Picea based on silicified seed cones from the Oligocene of Washington, Amer. J. Bot., 76:747-754.

Mirov, N. T., 1967, The Genus Pinus, The Ronald Press Company, New York, N.Y., United States, 602 p.

Mullis, K. B., and Faloona, F. A., 1987, Specific synthesis of DNA in vitro via a polymerase-catalyzed chain reaction, Methods Enzymol., 155:335-350.

Neale, D. B., Marshall, K. A., and Sederoff, R. R., 1989, Chloroplast and mitochondrial DNA are paternally inherited in Sequoia sempervirens D. Don Endl., Proc. Natl. Acad. Sci. USA, 86:9347-9349.

Nei, M., and Miller, J. C., 1990, A simple method for estimating average number of nucleotide substitutions within and between populations from restriction data, Genetics, 125:873-879.

O'Brien, S. J., and Mayr, E., 1991, Bureaucratic mischief: recognizing endangered species and subspecies, Science, 251:1187-1188.

Ohsawa, M., 1990, An interpretation of latitudinal patterns of forest limits in South and East Asian Mountains, J. Ecol., 78:326-339.

O'Reilly, G. J., Parker, W. H., and Cheliak, W. M., 1985, Isozyme differentiation of upland and lowland Picea mariana stands in northern Ohio, Silvae Genet., 34:214-221.

Page, C. N., and Hollands, R. C., 1987, The taxonomic and biogeographic position of Sitka spruce, Proc. Royal Soc. Edinburgh, 93B:13-24.

Palmer, J. D., 1987, Chloroplast DNA evolution and biosystematic uses of cpDNA variation, Am. Nat., 130:S6-S29.

Palmer, J. D., Jansen, R. K., Michaels, H. J., Chase, M. W., and Manhart, J. R., 1988, Chloroplast DNA variation and plant phylogeny, Ann. Missouri Bot. Garden, 75:1180-1206.

Peters, R. L., 1990, Effects of global warming on forests, For. Ecol. Manage., 35:13-33.

Raven, P. H., and Axelrod, D. I., 1974, Angiosperm biogeography and past continental movements, Ann. Missouri Bot. Garden, 61:539-673.

Rieseberg, L. H., Carter, R., and Zona, S., 1990, Molecular tests of the hypothesized hybrid origin of two diploid Helianthus species (Asteraceae), Evolution, 44:1498-1511.

Roche, L., 1969, A genecological study of the genus *Picea* in British Columbia, *New Phytol.*, 68:504-554.

Roche, L., and Fowler, D. P., 1975, Genetics of Sitka spruce, Res. Pap. WO-26, Washington, D.C., U. S. Department of Agriculture, Forest Service, 15 p.

Rushforth, K., 1986, Mexico's spruces, rare members of an important genus, *Kew Magazine*, 3:119-124.

Rushforth, K., 1987, Conifers, Christopher Helm, London, 232 p.

Schmidt, P. A., 1989, Beitrag zur Systematik und Evolution der Gattung *Picea* A. Dietr., *Flora* (Jena), 182:435-461.

Schmidt-Vogt, H., 1977, Die Fichte, Band I. Verlag Paul Parey, Hamburg, 629 p.

Sigurgeirsson, A., and Szmidt, A. E., [In press], Phylogenetic and biogeographic implications of chloroplast DNA diversity in *Picea*, *Nord. J. Bot.*

Sigurgeirsson, A., Szmidt, A. E., and Alden, J. N. 1990, A molecular study of interspecific hybridization in the spruce complex of Alaska. In: Talbert, S., ed., Proceedings of Joint Meeting of Western Forest Genetics Association and IUFRO Working Parties S2.02-05, 06, 12 and 14, Douglas-fir, Contorta Pine, Sitka spruce and Abies Breeding and Genetic Resources, Olympia, Washington, United States, 1990 August 20-24, 6.14-6.23.

Sigurgeirsson, A., Szmidt, A. E., and Karpinska, B. 1991, Alaskan *Picea sitchensis* populations infiltrated with *Picea glauca* genes: a study using DNA markers. In: Fineschi, S., Malvolti, M. E., Cannata, F., and Hattemer, H. H., eds., Biochemical markers in the populations genetics of forest trees, SPB Academic Publishing, The Hague, The Netherlands, 197-207.

Stine, M., Sears, B. B., and Keathley, D. E., 1989, Inheritance of plastids in interspecific hybrids of blue spruce and white spruce, *Theor. Appl. Genet.*, 78:768-774.

Szmidt, A. E., El-Kassaby, Y. A., Sigurgeirsson, A., Aldén, T., Lindgren, D., and Hällgren, J-E, 1988, Classifying seedlots of *Picea sitchensis* and *P. glauca* in zones of introgression using restriction analysis of chloroplast DNA, *Theor. Appl. Genet.*, 76:841-845.

Vane-Wright, R. I., Humphries, C. J., and Williams, P. H., 1991, What to protect? Systematics and the agony of choice, *Biol. Conserv.*, 55:235-254.

Wang, X.-R., and Szmidt, A. E, 1990, Evolutionary analysis of *Pinus densata* (Masters), a putative Tertiary hybrid. 2. A study using species-specific chloroplast DNA markers, *Theor. Appl. Genet.*, 80:641-647.

White, E. E., 1990, Chloroplast DNA in *Pinus monticola*. 2. Survey of within-species variability and detection of heteroplasmic individuals, *Theor. Appl. Genet.*, 79:251-255.

Whitham, T. G., 1989, Plant hybrid zones as sinks for pests, *Science*, 244:1490-1493.

Wright, J. W., 1955, Species crossability in spruce in relation to distribution and taxonomy, *Forest Sci.*, 1:319-350.

Wright, S., 1932, The roles of mutation, inbreeding, crossbreeding, and selection in evolution, Proceedings of XI Internat. Congr. Genetics, 1:356-366.

Yeh, F. C., and El-Kassaby, Y. A., 1980, Enzyme variation in natural populations of Sitka spruce (*Picea sitchensis*). I. Gentic variation patterns among trees from 10 IUFRO provenances, *Can. J. For. Res.*, 21:487-422.

Zurawski, G., and Clegg, M. T., 1987, Evolution of higher-plant chloroplast DNA-encoded genes: Implications for structure-function and phylogenetic studies, *Ann. Rev. Plant Physiol.*, 38:391-418.

GENETIC VARIATION IN EARLY FROST TOLERANCE OF SPRUCE FROM NORTHWESTERN NORTH AMERICA

Jon Dietrichson

Department of Forestry
Agricultural University of Norway
Aas, Norway

INTRODUCTION

On the north and west coast of Norway, both Sitka spruce (Picea sitchensis (Bong.) Carr.) and Lutz spruce (Picea xlutzii Little), the spontaneous hybrid of Sitka and white spruce (Picea glauca (Moench) Voss) backcrossed to the parent species, are suitable for reforestation (Little, 1953). Sitka spruce is better adapted to the wind and ocean salts than is Norway spruce (Picea abies (L.) Karst). Sitka spruce, however, has coarse branching, fast growth, and wide annual rings. As a result, the wood quality at rotation ages of 50 to 60 years has become a subject of discussion and study. In contrast, trees in the early 1960 Lutz spruce plantations in the Vesteraalens at latitudes 68° to 69° N. in northern Norway have slender stems, narrow crowns, and fine branches (Pettersen, 1981). Because Lutz spruce has sufficient frost tolerance for the coastal conditions of northern Norway, the trees have good growth and form. In the late 1950s, the Forest Research Institute in western Norway imported seeds from Lawing, Alaska, which is in a major introgression zone of Sitka and white spruce in northwest America (Little, 1953) and is assumed to be the origin of the successful plantations in northern Norway. Genetic variation of Sitka spruce and white spruce from the introgression regions in Alaska must be tested before new seed and breeding materials are introduced from northern Norway.

Frost Resistance

Frost resistance is one of the most important criteria for introducing forest trees to cold environments. Growth and wood quality in trees depend on avoidance of frost damage (Dietrichson, 1964; Eiche, 1966). Because early frost tolerance of the first year's growth in Scots pine (Pinus sylvestris L.) correlates with field survival up to at least 18 years of age, early testing for frost resistance is a step forward in modern forest tree breeding (Nilsson and Erikkson, 1986). This chapter summarizes the early results of a freezing test of 57 genetic entries for Sitka, white, and Lutz spruce from northwest America in Norway, with Norway spruce serving as a control.

Forest Development in Cold Climates, Edited by J. Alden *et al.*
Plenum Press, New York, 1993

Genetic material for the freezing tests came from trees from six different geographic sources (Table 1):

1. Norway spruce from four stands of four to nine single-tree families, each from central Norway.

2. White spruce from three stands in the Yukon Territory, Canada.

3. Lutz spruce from five stands on the Kenai Peninsula, Alaska.

4. White spruce with possible Sitka spruce introgression, taken from seven stands in the region north and east of Anchorage, Alaska.

5. White spruce from four stands in interior Alaska.

6. Sitka and Lutz spruce from five Alaska stands in which the identity of progenies from different families was maintained during the study.

The number of families tested from each stand is shown in Table 1. Stand and family samples overlap for the Tustumena Lake, Seward to Moose Pass, Valdez, and Thompson Pass provenances; that is, the separated families originate from each of the four stand samples.

SEED SOWING AND SEEDLING GROWTH

Seeds were sown January 22-23, 1986, in multipot containers, each holding 95 40-milliliter pots containing a standard peat and 30 percent perlite growing medium. For each progeny and stand sample, three seeds were sown in each of approximately 120 pots. All samples were randomly subdivided into four blocks of 30 pots each. After sowing, the containers were kept cool until March 17, when they were transferred to a phytotron. Two replicates were set in each of two growth chambers. To promote germination after sowing, the pots were maintained under continuous light at 20 °C for the first three weeks. After a few weeks, the pots containing more than one seedling were thinned.

The temperature was maintained at a constant 20 °C for the first six weeks, after which the night temperature was reduced to 15 °C for the seventh week, 10 °C for the eighth week, and then to 5 °C until freezing was carried out from the sixteenth to nineteenth weeks. The day temperature was maintained at 20 °C throughout the growth period.

From the first through the third week, light was continuous. The fourth week the night length was three hours and then increased one hour weekly to a maximum of 14 hours, a period well beyond the critical photoperiod for growth cessation of all seed sources (Junttila and Skaret, 1990). A constant night length of 14 hours was maintained until the freezing tests were complete after the nineteenth week. Seedlings were grown under a fertilization program described by Johnsen (1989) and were 50 to 100 millimeters tall when the freezing tests began.

FREEZING TESTS

The freezing tests were conducted on July 3, 10, 13, and 21. Seedlings for each of the four tests were selected randomly from blocks within one or the other of the two growth

Table 1. Provenances and Families in the Freezing Study

No.	Provenance (Stand)	Latitude	Longitude	Elevation (meters)	Number of families
		STAND SAMPLES			
		Norway			
1	Hemnes[1]	66°05′ N.	13°51′ E.	125	9
2	Selbu	63°06′ "	11°17′ "	260	8
3	Snåsa	64°12′ "	12°20′ "	90	9
4	Grane	65°37′ "	13°15′ "	60	4
		Yukon Territory			
6	Jakes Corner	60°23′ N.	133°47′ W.	860	unknown
7	Little Fox	61°22′ "	135°38′ "	890	"
8	W. of Dawson City	64°05′ "	139°27′ "	370	8
		Kenai Peninsula			
9	Portage-Girdwood	60°52′ N.	148°58′ W.	10	8
10	Cooper Landing	60°28′ "	149°30′ "	200	7
11	Tustumena Lake	60°15′ "	151°11′ "	100	9
12	Summit Lake	60°40′ "	149°25′ "	400	9
13	Seward-Moose Pass	60°15′ "	149°20′ "	170	10
		Anchorage Region			
14	Glenn Highway	61°48′ N.	148°50′ W.	220	9
15	Copper Center	61°57′ "	145°15′ "	450	8
16	Valdez	61°03′ "	145°45′ "	0-200	6
17	Thompson Pass	61°10′ "	145°45′ "	500	4
18	Chitina	61°30′ "	144°28′ "	300	8
19	Palmer	61°31′ "	149°02′ "	30	10
50	Talkeetna	62°10′ "	150°06′ "	130	10
		Interior Alaska			
49	Central/Circle City	65°34′ N.	144°49′ W.	230	8
51	Tetlin Junction	63°19′ "	142°38′ "	500	10
52	Jarvis Creek	64°00′ "	145°53′ "	380	5
53	Bonanza Creek	64°41′ "	148°12′ "	130	11
		PROGENY SAMPLES			
20-28	Tustumena Lake	60°15′ N.	151°11′ W.	100	9
29-38	Seward-Moose Pass	60°15′ "	149°20′ "	170-240	10
39-44	Valdez	61°03′ "	145°45′ "	0-200	6
45-48	Thompson Pass	61°10′ "	145°45′ "	500	4
54-58	Kruzof Island	57°00′ "	135°40′ "	50	5

[1] The Hemnes material is from controlled crosses in Lyngdal Seed Orchard at latitude 58° N.

chambers, but always from the same block for each test. Twenty-four plants of each genetic entry were lifted for each test and randomly distributed as three-seedling plots in new multipot containers. Before freezing, the roots were insulated by putting the multipot containers into larger containers of polystyrene foam, and the surface of each pot was insulated with one to two centimeters of perlite. Three tests were incomplete for five of the samples because the seedlings died before freezing. The freezing tests started at +10 °C and lasted for 16 hours. After remaining at 10 °C for three hours, the temperature was reduced by 6 °C per hour until the minimum test temperature was reached. This was maintained for two hours, and then the temperature was raised at a rate of 6 °C per hour to 10 °C. The pots were kept at 10 °C for three hours and then transferred to the greenhouse.

Minimum temperatures for the four freezing tests were as follows: Test 1, -10.0 °C; Test 2, -14.5 °C; Test 3, -9.0 °C; and Test 4, -9.5 °C. Seedlings in each test were selected from the same growth chamber.

RATING OF FREEZING INJURY

Freezing injury was scored as browning of needles four to five weeks after freezing. The rating scale used follows Johnsen (1989):

0	=	No discoloration
1	=	0.1 to 9.9 percent brown or discolored needles
2-10	=	Brown or discolored needles at 10 percent intervals from 10 to 99.9 percent
11	=	All needles completely brown (100 percent)

STATISTICAL ANALYSIS

The mean freezing injury was calculated for all genetic entries in each of the four freezing tests. Analyses of variances were made on arcsine-transformed data, groupwise for stands within regions and for families in each of the five provenances (stands). The four Norway spruce control provenances were not analyzed. Transformation of discrete data by non-normal distribution before statistical analysis is a safe standard procedure (Steel and Torrie, 1960). In this study, needle damage, a continuous variable, was classified into eleven discrete classes that were not normally distributed (Johnsen, 1989). A Student-Newman-Keuls test of statistical difference among each group of provenances and families was made to rank the mean values.

RESULTS

The means of the four freezing tests are shown in Figures 1 and 2. Tables 2 and 3 group the genetic entries statistically.

Region Samples

Needle damage was not statistically different among stands sampled from the Yukon, but among sources sampled from interior Alaska, only number 53, from Willow Island on the Tanana River near Fairbanks, sustained more injury than those from upland sites at Central/Circle, Tetlin Junction, and Jarvis Creek (Fig. 1, Table 2). Of the sources east and north of Anchorage, the Sitka spruce provenance from Valdez was injured more than the white spruce provenances from the Glenn Highway, Copper Center, Chitina, and Palmer, and the Lutz spruce provenance from Thompson Pass. Among the sources from the Kenai region, the Sitka spruce-dominated provenances (Portage to Girdwood and Seward to Moose Pass) sustained more injury than the white spruce-dominated sources from Cooper Landing, Tustumena Lake, and Summit Lake.

Table 2. Statistical Grouping of the Differences between Stands within Each of the Four Regions for Needle Injury at Five Percent or Less Probability of Error. Stands with the Same Letter Belong to the Same Group (See Table 1 and Fig. 2).

Yukon		Kenai		Anchorage		Interior Alaska	
No.	Group	Group		No.	Group	Group	No.
8	A	9	A	16	A	53	A
7	A	13	A	15	B	51	B
6	A	10	B	19	B	49	B
-	-	12	B	14	B	52	B
-	-	11	B	50	B	-	-
-	-	-	-	18	B	-	-
-	-	-		17	B	-	-

Family Samples

Only Lutz spruce families from Tustumena Lake and the pure Sitka spruce families from Kruzof Island were not statistically different in freezing resistance (Fig. 2 and Table 3). Family samples from Seward to Moose Pass and from Valdez were grouped into three statistically different injury classes. From the Valdez provenance, family 39 sustained least injury and families 40, 41, 42, and 44 sustained the most injury. Needle injury to families 42 and 44 overlapped injury to an intermediate class including family 43. Family samples

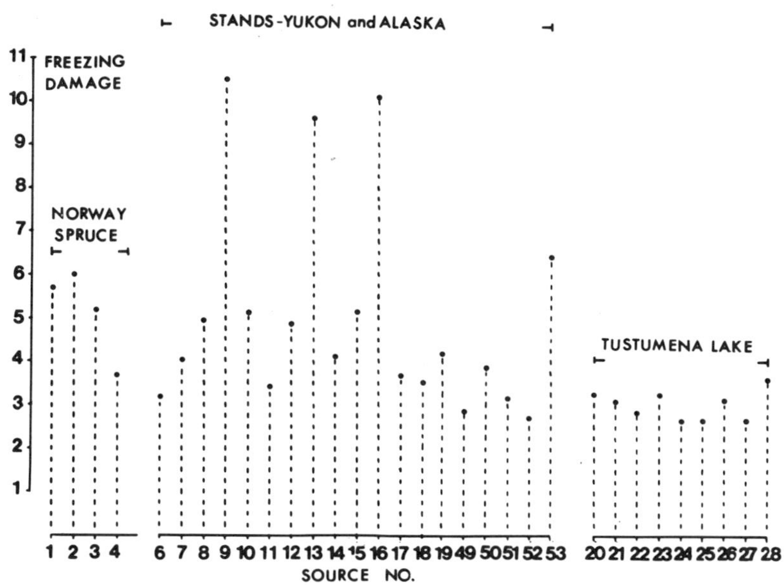

Fig. 1. Mean freezing damage of seedlings in four tests: Norway spruce controls, stand (provenance) samples from Yukon and Alaska, and family progenies from Tustumena Lake.

from Thompson Pass were statistically grouped into two injury classes. Families 46 and 48 sustained the most injury, families 45 and 47 the least, and family 46 overlapped both the high and low injury classes.

DISCUSSION

Early frost tolerance in trees is related to the timing of growth cessation. Growth cessation and dormancy are controlled by genetics, environment, and their probable interaction. Results of any artificial freezing test must be interpreted with great care and always in reference to the particular test environment.

In Tromsø, at latitude 69°39′ N., Junttila and Skaret (1990) found that white spruce from interior Alaska enters dormancy and develops winter hardiness earlier than Sitka and Lutz spruce from the Gulf of Alaska. They studied outdoor winter injury on two-year-old white spruce seedlings of six white spruce provenances from interior Alaska, and two assumed Lutz spruce provenances from Seward and Tyonek (at Cook Inlet west of Anchorage). Three Sitka spruce stands located between latitude 58° and 60° N. and five crosses of Norway spruce of central Norway origin were used as controls.

At the provenance level of genetic variation, the early frost damage demonstrated in this study supports the two-year winter survival result of Junttila and Skaret (1990). White spruce from interior Alaska is adapted to a sub-arctic climate and under stable climatic conditions enters dormancy faster and acclimates earlier to both early frost and winter stress than Sitka spruce, Lutz spruce, and Norway spruce.

The most interesting result is the genetic variation in freeze damage detected between the twelve stands from the Kenai Peninsula and Anchorage regions (Fig. 1, Table 2). Stand

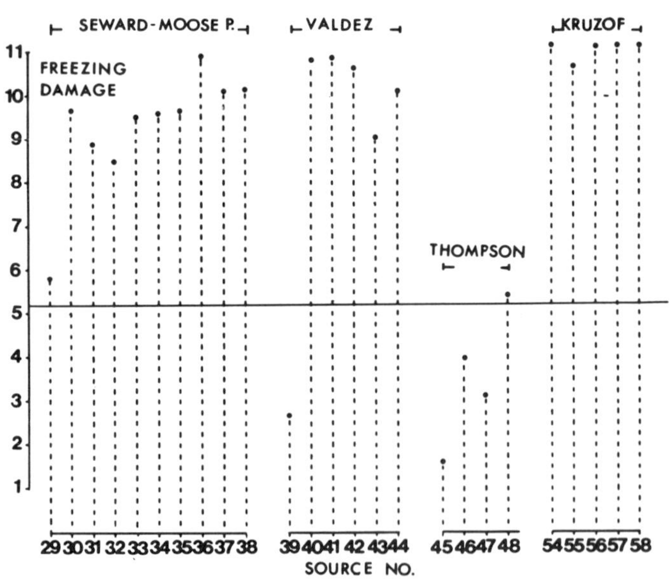

Fig. 2. Mean freezing damage among different families from four stands (provenances) (Table 1). The horizontal line is the mean of the four Norway spruce controls.

Table 3. Statistical Grouping of Families for Needle Injury in Five Stands at Five Percent or Less Probability of Error. Numbers with the Same Letter Belong to the Same Group and Are Listed in Order of Decreasing Freezing Injury. Numbers Are Sorted according to the Means (see Figs. 1 and 2).

STAND NAME																
Tustumena Lake		Seward-Moose Pass				Valdez				Thompson Pass			Kruzof Island	
No.	Group	No.	Group			No.	Group			No.	Group		No.	Group
20	A	36	A	-		41	A	-		48	A	-	58	A
21	A	38	A	B		40	A	-		46	A	B	57	A
22	A	30	A	B		42	A	B		47	-	B	54	A
23	A	37	A	B		44	A	B		45	-	B	56	A
24	A	35	A	B		43	-	B		-	-	-	55	A
25	A	34	A	B		39	-	-	C	-	-	-	-	-
26	A	33	A	B		-	-	-	-	-	-	-	-	-
27	A	31	A	B		-	-	-	-	-	-	-	-	-
28	A	32	-	B		-	-	-	-	-	-	-	-	-
-	-	29	-	-	C	-	-	-	-	-	-	-	-	-

9, from Portage to Girdwood, had the least freezing tolerance, and stand 11 from Tustumena Lake had the highest freezing tolerance in the Kenai Peninsula region. The frost damage to stand 9 was the same as frost damage to the Sitka spruce families 54 to 58 from Kruzof Island (Fig. 2), approximately 800 kilometers further southeast. Hardiness of the Tustumena Lake stand, on the other hand, only 60 to 70 kilometers southwest of the Portage-to-Girdwood stand at 100 meters above sea level, were intermediate in freezing injury, indicating that variation within stands is large. The ten families from the Seward to Moose Pass region were highly variable (Fig. 2, Table 3), ranging from interior white spruce to coastal Sitka spruce in frost susceptibility. In contrast, the nine families from the Tustumena Lake stand showed no variation in frost injury (Fig. 1, Table 3).

Of the seven stands from the Anchorage region, the Valdez source, a Sitka spruce provenance, was the most severely damaged. The hardiness level of the six other provenances in the Anchorage region indicates that the populations are highly influenced by continental white spruce and that some variation occurs within stands. The within-stand analyses of both the Valdez stands (39-44) and the Thompson Pass stands (45-48) showed that family differences were statistically significant.

CONCLUSION

The main conclusion of this study is that spruce on the Kenai Peninsula and on the nearby coastal mainland has very large genetic variation in early freeze tolerance among, as well as within, stands. It is likely that this variation resulted from hybridization and introgression between white spruce and Sitka spruce. The hybridization first described by Harold J. Lutz on the Kenai Peninsula (Little, 1953), and later by and later by Daubenmire (1968) in the Skeena River Valley of British Columbia, Canada, was based on morphological variation in cone characteristics. The more recent and modern biochemical studies by Sigurgeirsson et al. (1990), using cpDNA restriction fragments as genetic markers, have shown that genes from several of the Sitka spruce populations on the Kenai Peninsula have infiltrated the continental white spruce in very unequal proportions between stands where the species are sympatric. Thus the large genetic variation in freezing damage demonstrated among the Lutz spruce provenances and families in this study is not surprising.

Although white spruce from interior Alaska flushes very early in the spring (author's unpublished data), ceases growth early in the summer, and develops high early frost tolerance (Junttila and Skaret, 1990), it is not suited to the cool and climatically unstable coastal environment of northern Norway. Furthermore, late-flushing Sitka spruce from coastal Alaska might fail to enter dormancy and develop early frost tolerance in the coastal environment of northern Norway. Lutz spruce from the central Kenai Peninsula (Pettersen, 1981), however, is surprisingly well-adapted to northern Norway. This natural hybrid deserves more attention from Norwegian tree breeders and seed collectors.

One method of introducing Lutz spruce to northern Norway would be to test families for growth rhythm, hardiness, critical photoperiods, and cpDNA fragment lengths, in the outstanding plantations described by Pettersen (1981). The best parents could be cloned into breeding archives or into first-generation seed orchards, similar to the Sitka spruce and Lutz spruce orchards established in Norway. But with variable gene frequencies from either the Sitka or white spruce, the long-term outcome over generations of selections would be difficult to predict.

A second method of adapting pure Lutz spruce to north coastal Norway would be to hybridize pure Sitka and white spruce from disjunct populations. Such work has been started in Scotland (Faulkner, 1987). In Norway, the first two small seed orchards were established in the 1960s with phenotypic selections from Sitka spruce plantations. Selections were of provenances from latitude 57° N. and are outside the Kenai Peninsula introgression area. Flowering is regular and abundant, and the selections can be crossed with white spruce from interior Alaska. But again, the outcome of breeding over generations is difficult to predict. Segregations back to the parent species would lead to large variation in the F_2 as well as successive generations.

The third method of adapting Lutz spruce to the north coastal climate of Norway would be to simply procure seed from the best trees in the best natural stands and plantations of Lutz spruce to fill the immediate demand for reforestation, and thereafter subject the plantations to one or more generations of natural mass selections.

ACKNOWLEDGEMENTS

I am grateful to Dr. John Alden, Institute of Northern Forestry, Fairbanks, Alaska, United States, who collected most of the seeds and gave valuable comments to the manuscript. The Norwegian Ministry of Agriculture provided financial support for the tests as part of their program to breed trees suited to the harsh marginal climates in northern Norway. Bjarne Frøystad, of Stavanger, Norway, and Skjalg Slettjord, of Skjomen, Norway, members of the committee responsible for this task, recommended the study. Inger Apeland, Geir Østereng, Reidar Tutturen, and Øystein Johnsen, all at the Norwegian Forest Research Institute, provided technical assistance. As well as being familiar with growth chambers and the freezing technique, Øystein Johnsen analyzed the data and provided valuable comments on the manuscript. I am also grateful to comments from Aage Langhammer and Linda Hjeljord.

REFERENCES

Daubenmire, R., 1968, Some geographical variations in Picea sitchensis and their ecologic interpretations, Can. J. Bot., 46:787-798.

Dietrichson, J., 1964, Proveniensproblemet belyst ved studier av vekstrytme og klima (The provenance problem illustrated by studies of growth-rhythm and climate), Meddr Norske SkogforsVes, 19:507-656.

Eiche, V., 1966, Cold damage and plant mortality in experimental provenance plantations with Scots pine in northern Sweden, Stud. For. Suec., 36:1-219.

Faulkner, R., 1987, Tour Notes, European Tree Breeders' Association visit to Scotland and North England, September 3-10, 1987, Forestry Commission, Roslin, Midlothian, Scotland, 67 p.

Johnsen, Ø., 1989, Freezing-testing young Picea abies plants, Scand. J. For. Res., 4:351-367.

Junttila, O., and Skaret, G., 1990, Growth and survival of seedlings of various Picea species under northern climatic conditions: Results from phytotron and field experiments, Scand. J. For. Res., 5:69-81.

Little, E. L., Jr., 1953, A natural hybrid in Alaska, J. For., 51:745-747.

Nilsson, J-E., and Eriksson, G., 1986, Freeze-testing and field mortality of Pinus sylvestris L. in northern Sweden, Scand. J. For. Res., 1:205-218.

Pettersen, G., 1981, En vekstundersøkelse av aktuelle bartreslag ved skogreising i Lofoten og Vesterålen (Investigation of growth of potential conifer species at a plantation on the Lofoton and Vester Islands), Hovedoppgave ved Institutt ved Skogskjøtsel (Institute of Forest Management), NLH, Norway, 73 p. M.S. Thesis.

Sigurgeirsson, A., Szmidt, A. E., and Karpinska, B., 1990, Alaskan Picea sitchensis populations infiltrated with Picea glauca genes: A study using DNA markers. In: Hattemer, H. H., and Fineschi, S., eds., Proc. of a Conference for Biochemical markers in the population genetics of forest trees, SPB Academic Publishing, The Hague, The Netherlands, 197-207.

Steel, R. G. D., and Torrie, J. H., 1960, Principles and procedures of statistics with special reference to the biological sciences, McGraw-Hill Inc., New York, 481 p.

HISTORY OF TREE PLANTING ON THE ALEUTIAN ISLANDS

David Bruce

Pacific Northwest Research Station
Forest Service, U. S. Department of Agriculture
Portland, Oregon, United States

INTRODUCTION

Before the voyages of Christopher Columbus to the New World, native Americans planted fruit and nut trees near their settlements, but the first trees planted in North America for the purpose of wood production appear to have been Sitka spruce (Picea sitchensis (Bong.) Carr.) planted by the Russians at Unalaska Bay in the first decade of the 19th Century. Hardwoods planted in Rhode Island in 1820 on an old field were profitably harvested in the next 57 years (Anonymous, 1992). The live oak (Quercus virginiana Mill.) plantings on Santa Rosa Island in Florida were made after 1830. A few successful plantations on the Aleutian Islands and some natural regeneration at Unalaska suggest that getting trees to grow there is easy, but recent experience indicates this is not true. Interest continues in tree planting near Unalaska Bay where experience has shown that small-wood production is feasible. Farther west there is interest in plantings for amenity purposes.

Many misleading errors exist in the rather obscure historical record of early tree planting. This paper recounts what really happened in the Aleutians and to draw conclusions about what can be learned from these and from other plantings at high latitudes.

The Alaska Peninsula, the Aleutian Islands, and the Commander Islands are volcanic mountains that form an arc nearly 2,600 kilometers long and separate the frigid Bering Sea from the warmer Japanese current in the north Pacific Ocean (Fig. 1). The Islands lie between latitudes 51° N. and 55° N.; the eastern end of the Alaska Peninsula is about 58° N. The climate is cool, wet, windy, foggy, and cloudy. Mean monthly temperature in summer (July through September) averages about 10.0 °C on most islands and 0 °C in January. Annual variation in the growing season mean is about 2.5 °C. Summer temperatures on the island of Shemya average 8.5 °C or 1 °C lower than on Attu or Adak. In summer, these two islands are about 1 °C cooler than Unalaska and 3 °C cooler than Kodiak.

Forest Development in Cold Climates, Edited by J. Alden *et al.*
Plenum Press, New York, 1993

Coniferous trees grow at the eastern end of the Alaska Peninsula, on the Kamchatka Peninsula, and formerly on the islands between. Fossil wood from Amchitka, Atka, Unga, and other islands attests to the presence several million years ago of trees related to modern Metasequoia and hardwoods. In 1817, petrified wood seen at Unalaska was said to be from the bed of a lake on Umnak (Kotzebue, 1821). Carbonized tree stumps found offshore from Amchitka in 23 meters of water, were identified as conifers and carbon-dated at more than 40,000 years of age (Barr et al., 1973). In 1944, I heard reports of buried tree stumps unearthed by bulldozers during construction work on Amchitka and Ummak. I was unable to locate samples of wood from these stumps in Anchorage. Because they were described as stumps, it seems likely they were from trees growing on the islands rather than from driftwood, although driftwood could easily have been preserved in cold water-logged soils. In any event, if these trees once grew in the Aleutians, they were not planted.

Pollen samples from Umnak and Adak suggest that climatic conditions 3,500 to 8,500 years ago may have been more favorable for tree growth than they are today (Heusser, 1973, 1978). A much higher percentage of willow pollen is found in layers of soil laid down in that period than in more recent years.

Vitus Bering in 1741, and later Russian explorers, found no trees on the islands, only shrublike willow (Salix L. sp.), Sitka alder (Alnus sinuata (Reg.) Rydb.), and Siberian mountain-ash (Sorbus sambucifolia (Chalm. & Schlect.) Roem.) The Aleuts used driftwood to build boats and houses and occasionally for fuel. A shortage of driftwood would have suggested the need to plant trees to European settlers accustomed to harvesting wood from forests.

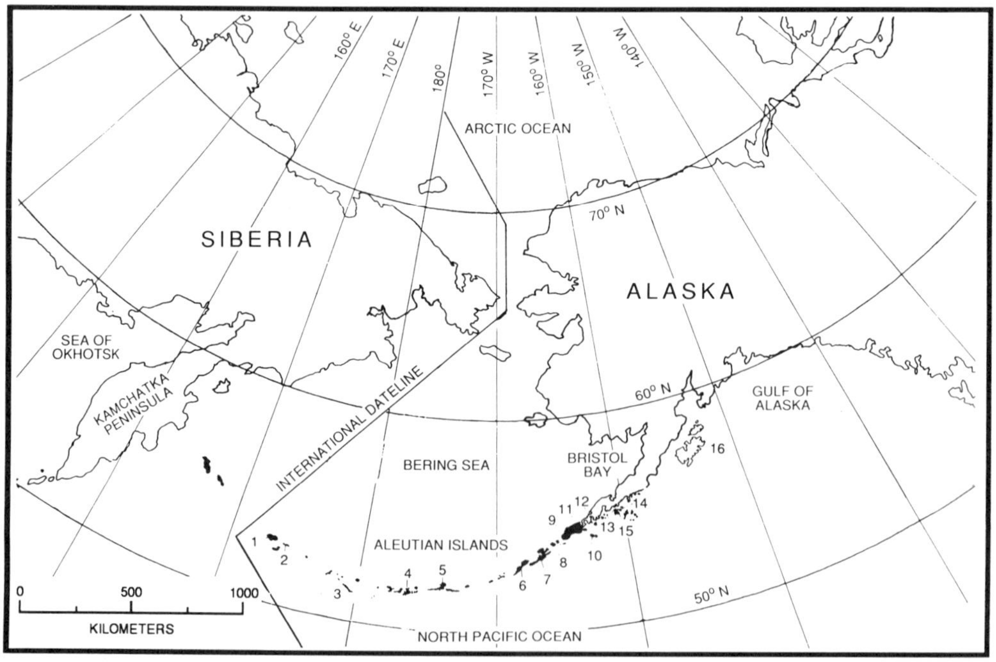

Fig. 1. Aleutian Islands: 1. Attu, 2. Shemya, 3. Amchitka, 4. Adak, 5. Atka, 6. Umnak, 7. Unalaska, 8. Akutan, 9. Unimak, 10. Sanak, 11. False Pass on Unimak I. Places along the Alaska Peninsula: 12. Cold Bay, 13. Unga, 14. Sand Point, 15. Squaw Harbor.

Three 19th Century plantings of Sitka spruce on islands in Unalaska Bay survived for 100 years or more (Fig. 2). A few old trees had been planted near buildings and protected by residents. There are many possible causes for the failure of small spruce that survived transplanting and winter winds. The two greatest hazards in the 19th Century were fire and animal damage. Earliest hazards were rooting by hogs, trampling and browsing by cattle, and after 1825, browsing by goats or sheep. The three old groups of planted spruce were on small islands that undoubtedly were free of browsing animals while the trees were young. Four old spruce trees are now growing in the city park at Unalaska, and single old trees grow at Nateekin Bay and Broad Bay on the western shore of Unalaska Bay. Most of the trees growing now at Unalaska were planted during World War II or regenerated naturally since 1952.

The three islands with old plantations of Sitka spruce and the most living trees or stumps are:

1. Amaknak Island, 24 trees growing near a small lake (Veniaminov, [1840] 1985).

2. Expedition Island, 58 trees growing on a southeast-facing slope near the center of the island (Author's notes, 1944).

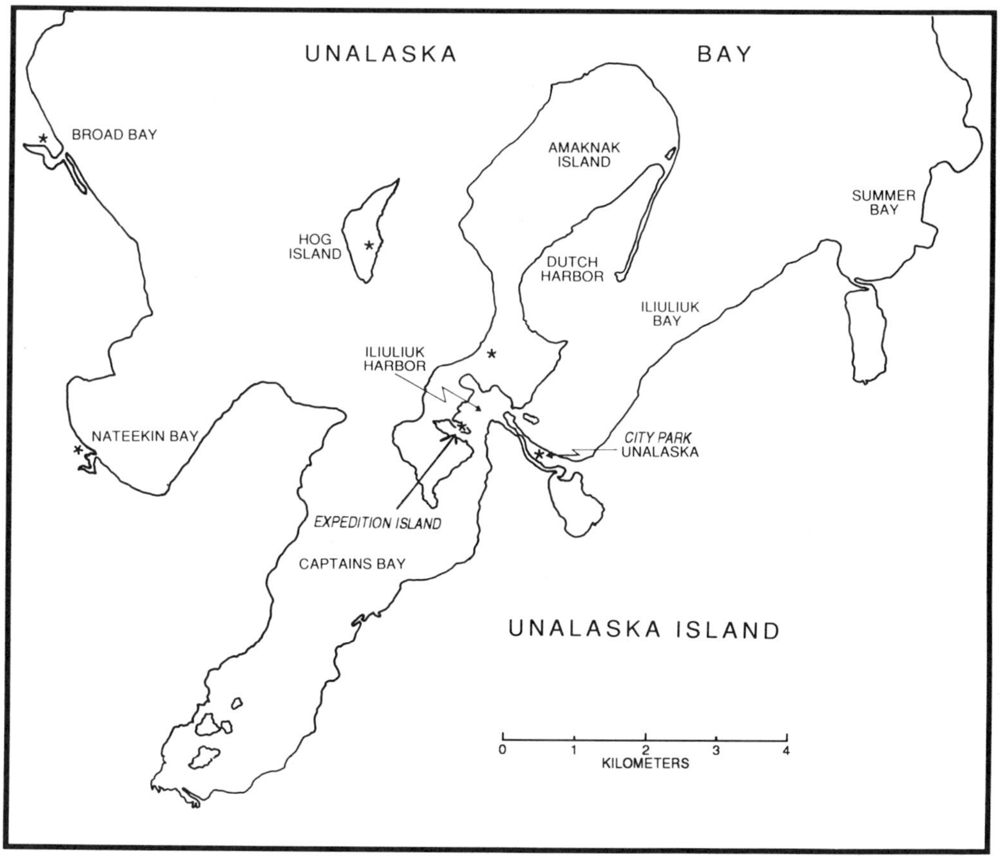

Fig. 2. Unalaska Bay. Tree locations: 1. Amaknak Island, 2. Hog Island, 3. Expedition Island, 4. City Park, Unalaska, 5. Nateekin Bay, 6. Broad Bay.

3. Hog Island, one clump of 21 trees near a spring, probably including four mother trees and 17 sprouts, and 60 meters away a group of seven old stumps (Alden, personal communication, 1989). (See Acknowledgements for full names and affiliations of most authors of personal communications.)

No first-hand account of any early planting in the Aleutians has been found. Some reports are obviously second hand; in some reports, the planting date and tree size do not agree.

THE FIRST RUSSIAN TREE PLANTING

The earliest record I found of a proposal to plant trees in the Aleutians is in a February 1790 report from Grigorii I. Shelikhov, the merchant who initiated the founding of the Russian-American Company, to the Governor of Irkutsk and Kolyvan Provinces (Bashkina et al., 1980). He wrote, "The good sites near every settlement have been explored . . . One place has been designated as a proper site for a harbor, another place for tree growing, and some places... for carrying on trade..." Many of Shelikhov's plans were never carried out, but copies of this report or others were probably kept in company files in both America and Russia.

Chamberlain Nikolai P. Rezanov was Shelikhov's son-in-law, a shareholder in the Russian American Company, and an official in the Czar's court. In 1803, before Rezanov went to Japan as an ambassasor and to Russian America to inspect company posts, he received instructions from Russian Foreign Minister Nikolai P. Rumiantsev, approved by Czar Alexander I (Bashkina et al., 1980), to emphasize "the spread of the Christian faith, the enlightenment of that region, the establishment of trade, shipbuilding, the strengthening of agriculture, animal husbandry, and other economic matters." The complete Russian version of these instructions was examined and no direct mention of trees or tree planting was found (Russian Ministry of Foreign Affairs, n.d.).

This leads to the question of when and by whom the trees on Amaknak Island in Unalaska Harbor were planted. Many people have accepted both the date 1805, and Captain Frederik Litke's statement that the plantation was established by order of Rezanov (Litke, [1835] 1987). Rezanov gave two sets of written instructions to Aleksandr A. Baranov concerning the conduct of business in America. In 1805, Baranov was both general manager of the Russian American Company and governor of Russian America. The first set of instructions was given to Baranov on September 9, 1805, soon after Rezanov's arrival at Sitka; the second was dated July 20, 1806, shortly before his departure to Russia. Both contained 17 numbered articles. Kyril T. Khlebnikov ([1835] 1963, 1973) (Tikhmenev, [1861-63] 1978), in Baranov's biography, mentions only a few of the first set of instructions. The first six articles of the second set of instructions concerned problems confronting the administration and the inhabitants of Russian America (Bashkina et al., 1980). The other articles recommended improvements in administration of the settlements but there was no direct mention of tree planting (Yudin, n.d.).

Failure to find direct reference to tree planting in these instructions suggests that it may have been discussed during one of the several councils convened by Rezanov at Sitka in the fall and winter of 1805-1806. These meetings were attended by Ivan A. Kuskov who was Baranov's assistant and who was in command at Sitka from September 30, 1806, to August 30, 1808, while Baranov was at Kodiak.

Father Ivan Veniaminov, the priest at Unalaska from 1824 to 1834, said the planting date was about 1805 (Veniaminov, [1840] 1985). But Rezanov visited Unalaska in July 1805, so little time was left that year for tree planting. Also, the settlers at Sitka were near starvation in the winter of 1805-1806 and inhabitants of Unalaska suffered an epidemic the following winter. Emelian Larionov, the company agent at Unalaska, who had been decorated by Rezanov during his July 1805 visit, died June 3, 1806, and was replaced by Fedor Burenin, who had been his assistant (Dr. Lydia Black, personal communication, 1989). A new manager was likely to have received more definite and detailed instructions on how to improve conditions than Larionov, who had gone to Unalaska in 1798 as manager of Shelikhov's post there, and as a possible replacement for Baranov as general manager in Russian America. It appears likely the trees were shipped from Sitka in 1807 by Kuskov to Burenin for planting.

In any event, the planting must have been before July 1813, when agent Burenin was replaced by Ivan V. Kriukov. A planting date between 1807 and 1813 would be supported by both Adelbert von Chamisso's and Veniaminov's early descriptions of the trees on Amaknak. Chamisso, who was a scientist aboard Kotzebue's ship, probably saw the trees in 1817. Otto von Kotzebue visited Unalaska for one week in September 1816 and for 13 weeks in spring and summer of 1817 (Kotzebue, 1821). He lived with Kriukov, then the company agent, during each visit. Information about the source of seedlings must have come from Kriukov. Had Kriukov been in charge of the planting, Chamisso probably would have known this and reported it.

Chamisso (Kotzebue, 1821) reported that:

> Trees are still found on the peninsula of Alaska and on the neighboring island of Oonemak, which is separated from it by a narrow channel and which seems to partake of the nature of the continent. Oonalashka, and other islands of the chain, are entirely destitute of them. It has been attempted to plant pines, a kind of Abies brought from Sitka, at Oonalashka; most of them have perished, the others seem scarcely to thrive, but the plantation is still young, and it is well known how ill coniferous trees bear transplanting.

This statement does not include mistakes in tree identification. In those days, pine was often used as a synonym for conifer, and one of the Alaskan trees listed by Franklin P. Hough (1878) was Abies sitchensis, the scientific name Hough used for Sitka spruce.

Chamisso's statement about trees on Unimak was repeated by Baron Ferdinand von Wrangell ([1839], 1980), who added a comment about a tree climbed there. Chamisso's diary ([1856], 1986) gives a more complete version of this:

> A son of Kriukov, a cheerful lad, had gone from Unalaska to Unimak, for him this was the breadth of the world. He had seen trees there; he had even climbed up a tree and swung from its branches. He told us about that with great pride, but also with no small fear that he would be taken for a liar for this unusual report. He made every effort to explain to us in a credible manner what a tree was.

Veniaminov ([1840] 1985) challenges this and says that only shrubby willow and alder thickets are found on the peninsula but no other trees such as birch (Betula L. sp.), poplar (Populus L. sp.) and conifers. Recent observations confirm Veniaminov's statement, except that the range of balsam poplar (Populus balsamifera L.) extends about half way down the peninsula and shrublike dwarf arctic birch (Betula nana L.) grows at the western end of the peninsula (Viereck and Little, 1972). Veniaminov appears to have been correct, so the boy's tree most likely was on Kodiak Island, whether he climbed it himself, or repeated a story he had heard.

In January 1818, Captain L. A. Hagemeister, who temporarily had replaced Baranov as company manager, included a cryptic paragraph in a letter to G. G. Potorochin, the manager on Kodiak (Pierce, 1984): "Please send the enclosed seeds to Unalaska. The trees grow quickly and in two years can serve as firewood. Order the manager to plant some of them immediately." Had the seeds been planted and produced firewood in two years or even longer, Veniaminov undoubtedly would have heard about it and mentioned it in his book ([1840] 1985). Hagemeister apparently was trying to accomplish something he thought Baranov had neglected.

In 1823, Ivan V. Kriukov, after 40 years in the Aleutians, visited his family in Irkutsk, met Ivan Evseevich Popov Veniaminov, and interested him in going to the Aleutians. It is unlikely that tree planting was discussed at any length, although it may have been mentioned. Veniaminov arrived at Unalaska in July 1824. While at Unalaska from 1824 to 1834 he devised an Aleut alphabet. One of his helpers was Kriukov's son Stepan. Another of Kriukov's sons, Vasilii, painted icons for the church at Unalaska. Perhaps one of them had told the story about tree climbing.

Litke was at Unalaska August 22-31, 1827 (Litke, [1835] 1987). He wrote, "The commandant Petrovsky ... and Father Ivan Veniaminov did their best ... to relieve our boredom ..." He also stated that the plantation "on a little island adjacent to Unalaska" was established by order of Rezanov. Whether he got this information from Petrovsky or Veniaminov or brought it with him from Russia, they undoubtedly shared whatever they knew about the trees. Litke and Veniaminov remained friends and corresponded concerning publication of Veniaminov's book (Veniaminov, [1840] 1985).

Veniaminov ([1840] 1985) described the trees in 1834, although he probably visited them with Litke in 1827, and possibly at other times. Chamisso's statement implies either he saw dead seedlings in the plantation in 1817 or that he saw them elsewhere, but they were not mentioned by Veniaminov. Traces of dead seedlings might disappear in 10 to 17 years. Veniaminov wrote,

> Around 1805 there were brought to Unalaska from Sitkha' several very small, 2 to 3 years old, spruce seedlings, which were planted in different localities around the main settlement. They survived only on Amakhnak. There are 24 in number and some of them now (in 1834) are taller than 2.1 meters and about 46 centimeters in circumference at the root. All the others are from 0.6 to 1.5m tall.* They are in (locations) surrounded by hills alongside a lake and a swampy place on an elevated level place. In 1833 for the first time there appeared cones (fruit) on several trees. It is to be observed (from the branches) that in their first years these trees grew more sideways than in height. The lower branches are very long and very close to one another, but in the last six or seven years the trees began perceptibly to grow upward in height so that the distance along the tree trunk of the whorls of branches which grew within the last five years has increased, and is equal to the

* For uniformity, all measurements have been changed from original units to metric.

> distance between branches which grew during the first 10 or 12 years. It is worth noting that several of the trees, all planted at one and the same time, are so small and slender that it seems they came out of the earth no more than four years ago and their branches are very few. Why did they grow so slowly?

This statement neither confirms nor denies mortality of seedlings on Amaknak. Also, only one or two trees may have been as tall as two meters. The comment about growth in the last six or seven years suggests Veniaminov last visited the trees with Litke. A recent observation of tree growth in the Aleutians may help us interpret this statement. The tallest Sitka spruce in a 1965 photograph (Fig. 3) of the Adak National Forest was about 1.8 meters tall and had six whorls of branches above one meter (Gullickson, 1965). This photograph was taken 21 years after planting in the fall of 1944; tree heights would be similar to those seen on Amaknak in 1834.

The trees on Amaknak were on a wetter site and were more widely spaced than those on Adak, both of which could contribute to a larger basal diameter for a given height. Veniaminov reported that the trees were two or three years old when planted. One of the Adak trees was seven or eight years old when planted and the others were about that age. Other unknown differences in growing conditions make comparisons risky, but their greater basal diameter suggests that the Amaknak trees were older than those in the picture. The size difference together with earlier estimates of 1805 suggest that the planting date of 1807 is more likely than any later date.

The slow growth of the young trees on Amaknak Island probably discouraged other planting near Unalaska Bay and may explain Hagemeister's 1818 letter. The second planting probably was done some years after Veniaminov observed in 1834 that the trees finally had started to grow.

THE SECOND RUSSIAN TREE PLANTING

Father Veniaminov has already been mentioned because his book, "Notes on the Islands of the Unalaska District" (Veniaminov, [1840] 1985), contains the most complete

Fig. 3. Adak National Forest, October 1965, twenty-one years after planting. The tallest tree was then 1.8 meters. Photo by J. M. Gullickson.

early description of the first Russian tree planting. He also should be credited with the second Russian planting, although not in the way that is generally accepted.

In 1834, Governor Wrangell had Veniaminov transferred from Unalaska to Sitka. He arrived there in November 1834. In November 1838, Veniaminov left Sitka to go by ship to Saint Petersburg. He arrived in June 1839 and worked on his book, which was published in 1840. In November 1839, his wife died, and in November 1840 he took monastic vows and was renamed Innokentii.

In 1838, manager Ivan A. Kupreianov at Sitka sent some cones of the Califonia tree chaga (cedar or redwood), to Atka, stating that autumn is considered the best time for planting, preferably in places sheltered from wind, and requesting periodic reports on the results (Arndt, personal communication, 1992)

In August 1841, the article, "About Sowing and Planting trees on Unalaska Island," by O. B. Fisher (F. B. Fisher?) was published in the Russian Journal of Forestry (Fisher, [1841] 1968). The name O. B. Fisher that appears on the Russian typed copy of the article on file in Juneau may be an error by the typist (the Cyrillic F is the same as the Greek phi). Fedor Bogdanovich Fisher was director of the Imperial Saint Petersburg Botanical Garden from 1823 to 1850, and corresponded with Kupreianov about trees in America (Pierce, 1990).

Lutz (1963) speculates that because the introduction said at least a few trees transplanted by Innokentii on Unalaska Island adapted well, that Veniaminov may have planted trees between 1834 and 1841. Considering the slow initial growth of spruces planted in the Aleutians, it seems more likely that Veniaminov had described the older trees to Fisher while at the Academy of Sciences in Saint Petersburg, and that Fisher's introduction was written after Veniaminov had received the rest of the instructions and left Saint Petersburg. Veniamonov apparently requested the instructions in 1839 which suggests he planned to use them for future planting. If he had already had trees planted, he probably would have mentioned it in his 1840 book.

In January 1841, Veniaminov left Saint Petersburg, crossing Siberia to Okhotsk which he left August 19, 1841, and arrived in Sitka in September to serve as Bishop of Alaska. Chevigny's (1967) statement that he stopped at Unalaska on his way east to Sitka appears to be wrong (Arndt, personal communication, 1991). Veniaminov sailed west in May 1842 and arrived at Unalaska the end of the month. After a month at Unalaska, he went to Atka and in August to Kamchatka. He left there a year later and returned to Sitka in September 1843. This chronology suggests that his first chance to encourage tree planting at Unalaska, after receiving Fisher's written instructions, was in June 1842.

Veniaminov's published letters (Barsukov, 1897) were examined, searching specifically for some mention of tree planting and for letters to Litke (Arndt, personal communications, 1990, 1991). The letters in 1842 that describe Veniaminov's visit to Unalaska emphasize that he arrived on Holy Ascension Day, the name day of the church at Unalaska, and how happy the people were to see him again. This implies that they had not seen him since 1834. What he did while at Unalaska was not mentioned. Merely leaving instructions for tree planting would hardly merit mention. If he had brought trees to plant, they would have been from Sitka. Later letters said he again visited Unalaska in 1848, but gave no details. A biographical book about Veniaminov by Barsukov (1883) scarcely mentions the 1842 and 1848 trips to Unalaska and says nothing about tree planting.

The priest at Unalaska from 1834 to 1843, Father Grigori Golovin, appears to have left no day-to-day journals that might mention tree planting. Such journals are not in indexes of either the Russian church collections now in the Library of Congress or those at Kodiak (Arndt, personal communication, 1990). It seems most likely that Veniaminov left instructions for tree planting with Golovin and that it was done in 1843. In 1844, Andrei P. Sizoi replaced Golovin as priest. Innokentii Shaiashnikov, who was educated by Father Netsvetov at Atka, went with Netsvetov to the Yukon in 1844 to establish the first Orthodox mission there. In 1848, Shaiashnikov travelled to St. Michael, where Veniaminov ordained him as priest. They then went to Unalaska where Shaiashnikov became priest of the Church of the Holy Ascension (Black, personal communication, 1992). He therefore would have had no first-hand information about the second tree planting. Shaiashnikov is the most likely source of information about this second planting at the time of the American takeover in 1867. The trees undoubtedly came from the nearest source, Kodiak. This is consistent with the source stated by Captain P. N. Golovin ([1862] 1979), who visited Kodiak in 1861, but the date given by Captain Golovin is for the first planting. It is unlikely that Veniaminov suggested tree planting to Grigorii Golovin in 1834, and it was done before Veniaminov's return in 1842.

George Davidson (1868), who worked for the U. S. Coast Survey, is definite about where these trees were planted, but not about when. By saying, "an island in the western roadstead," he meant Hog Island, which lies west of Amaknak Island (Fig. 2). Davidson said the trees were planted some 30 years since (sic) and were not yet 76 centimeters tall. This suggests he based the time estimate on information obtained from someone who knew that Grigorii Golovin had the trees planted but did not know when they were planted. The size of the trees on Hog Island compared with those on Adak and Amaknak 20 to 30 years after planting suggests they were planted less than 30 years before, which is consistent with an estimated planting date of 1843.

Charles Bryant (1870) says Veniaminov had 20 trees planted 30 years since (sic); this time estimate coincides with Davidson's statement. Bryant probably visited the Amaknak planting because he reported the trees were 7.6 meters tall and were bearing cones. His tree count agrees with the approximate number of trees alive on Amaknak at that time. He may have mentioned the stream that was the outlet to the lake, and not the lake itself, because before his visit he had been told the trees had been planted near a small stream.

Franklin B. Hough in his report on forestry (1878) is specific about trees being planted on two islands in Iliuliuk Harbor. If Amaknak Island is <u>in</u> the harbor, part of the harbor must be west of the island, which is where Hog Island lies. In his "History of Alaska," Bancroft (1886) refers to all of Unalaska Bay as Iliuliuk Harbor, so this usage was not unusual at that time. The confusion in names may stem from a chart in the U. S. Coast Survey's 1869 "Harbors in Alaska" that was labelled in large letters "Illouliouk and Captain's Harbors". In smaller letters the chart indicated the main body of water was Unalaska Bay. Hough's comment on planting date agrees with Davidson's and so probably came directly or indirectly from the same source. I believe the mixup in names led to the later belief that the second planting was on Expedition Island, because this was the only island in what is now known as Iliuliuk Harbor. The island is now a peninsula.

John Muir's (1883) comments suggest he saw only the trees in the settlement and did not inquire about other plantings. John Cantwell (1887) may have seen the Amaknak trees.

Isabel S. Shepard (1898), whose husband was Captain of the U. S. Steamer Rush, did not mention the existence of a second plantation when describing her 1889 visit to the one on Amaknak Island; apparently she did not know about the old one on Hog Island. Possibly the planting of the trees on Expedition Island was no longer fresh in anyone's memory. The Bering Sea Patrol did not establish headquarters and have yearlong residents at Unalaska until 1895 (Denfeld, 1987).

The trees on Hog Island Davidson reported to be 76 centimeters tall probably are the same ones that Mrs. Pelagia McCurdy, Unalaska, saw on Hog Island about 50 years later when she was a girl. She mentioned the trees in a letter to R. W. Tindall (1970):

> We hope to preserve the trees on Expedition Island by Father Veniaminof. Did you know that a grove had also been planted on Hog Island? My grandfather, Nikifer Diakanoff had a camp, warehouse, bath house (steam, banya), smokehouse, barabara, above ground living quarters, on Hog Island and I was with my grand parents during my early childhood. Nearby the trees use to be a hot spring. Also Hog Island was a village site in the days before the Russians settled in Unalaska. As I child I remember the trees on Dutch Harbor [Amaknak], then the branches hung low to the ground and they were very beautiful specimens.

THE THIRD PLANTING AT UNALASKA BAY

This raises the question of who planted the trees on Expedition Island and when they were planted. I believe this was done after the United States purchased Alaska in 1867, and that it was done by the Bering Sea Patrol, the U. S. Coast Survey, or the Alaska Commercial Company. Had this third planting been much earlier than 1878, Hough might have reported three plantings rather than two. Had the third planting been done earlier by the Russians, Innokentii Shaiashnikov would have included it in his reports.

The first direct evidence I have found about the plantation on Expedition Island is a 1909 photograph that suggests trees fully occupy the area (Jochelson, 1925). This would have made the trees 30 or 40 years old then. A comment by archeologist Vladimir I. Iokhel'son (Waldemar Jochelson), who worked at Unalaska from January to June 1909,

Fig. 4. A 1910 view of Expedition Island and the 19th century grove of Sitka spruce from Haystack Hill, Unalaska Island. Photo courtesy of Alaska and Polar Regions Department, University of Alaska Fairbanks, Archives #78-4-106, Boaz Collection.

seems to bear that out: "In the strait between the last neck of land and Unalaska is a small island called Expedition Island, on which Weniaminoff has planted some fir trees brought from Sitka." There is a similar photograph taken in 1910 in the Boaz Collection in Fairbanks (Fig. 4). The third evidence is a visit to the island in 1917 or 1918 by a teacher from the Jesse Lee Home who described the trees as a "group of scrubby evergreen trees" (Winchell, 1951). A recent ring count (1989) of 101 on a core from one of the largest trees on Expedition Island (Alden, personal communication, 1989) suggests a planting date between 1867 and 1877.

OTHER OLD TREE RECORDS

Before the 1897 Gold Rush, any available small boat would have been used to get from Unalaska Island to Amaknak Island. In Gold Rush days, there was a path from Dutch Harbor to the old ferry dock that went past the plantation on Amaknak Island (Reed, 1969):

> The settlement of Dutch Harbor was on a flat facing north of the harbor and was surrounded by grassy hills dotted with wildflowers. I recall that a broad path led through a gap in these hills south through a little valley where, beside a small lake, was a grove of spruce trees planted by the Russians and said at that time to be the only trees on the whole Aleutian chain. The path continued to a boat landing opposite the pretty little village of Unalaska on Unalaska Island.

By 1909, Dutch Harbor was a ghost town; the manager of the North American Commercial and two clerks were the only white residents (Jochelson, 1925). After the radio station was built in 1911, the trees would not have been visible from the new path past the radio towers to the ferry dock.

Bernhard E. Fernow (1901) saw the trees on Amaknak in 1898 and gave a size range of 15 to 61 centimeters and heights of 7.6 to 9.1 meters. He reported one group of 12 trees and another of seven, all vigorous and in good health, except where fire had damaged some.

Fig. 5. Sitka spruce on Amaknak Island, planted in 1807. Fire has damaged the seven trees on the right. Photo July 1898, by E. S. Curtis, Library of Congress, USZ62-55417.

Photographs taken about this time suggest that three of the trees alive in 1898 died soon afterwards. A photograph (Adams, 1903) by C. H. Roesch apparently was taken before the fire occurred in 1897 or 1898 because no defoliation is evident. Fernow (1901) published a photograph by E. S. Curtis in July 1898 that shows one tree severely defoliated and two with some damage (Fig. 5). Cantwell (1904) published a photograph taken in 1899 that shows more deterioration. A photograph in the Bancroft Library collection at Berkeley, California, probably taken in 1900, shows one dead tree and two with only a few tufts of needles. The fire probably was started by some of the 5,000 people who camped near Unalaska Bay during the winter waiting to seek gold when the ice broke up in the spring of 1898. They may have damaged or felled trees at other locations near Unalaska Bay.

Iochel'son's 1909 excavations were within 100 meters of the trees on Amaknak and a like distance from the trees on Hog Island. It is hard to imagine he did not see the trees on these islands, but they were not mentioned in his report (Jochelson, 1925).

In July 1944, I interviewed two long-time residents of Unalaska before visiting the plantations on Amaknak and Expedition Islands. C. H. Hope, at the Post Office told me that a priest planted the trees on Expedition Island about 60 years before. He said the ones in the village were planted by the father of a native who was about 70 years old. Art Harris, at a store called the Arcade, reported that the trees on Expedition Island were planted 40 years before 1944 by the U. S. Coast and Geodetic Survey. He said there were four or five trees about six meters tall on Hog Island in a swale, near a stream, and that they were planted long before those on Expedition Island. In May 1986, Henry Swanson of Unalaska, age 91, told me there were a few big stumps on one side of a stream on Hog Island. He believed that the trees were cut shortly before World War II. He also reported there was a clump of small spruce on the other side of the stream; he did not know if they had been planted or not.

I believe Hope was using local wisdom, based either on Bryant's or Iochel'son's report or on their source, when he attributed the planting on Expedition Island to a priest. His planting date would have been 1884 or earlier, with the trees in the village being planted before those on the island. Although Harris said the U. S. Coast and Geodetic Survey, formerly the U. S. Coast Survey, did the planting, they did not build the four houses on Amaknak until 1922. The Coast Survey, however, dispatched reconnaissance teams along the Alaska coast almost annually between 1867 and 1880. Also, the Bering Sea Patrol easily could have transported from Kodiak the 80 or more seedlings that must have been planted on Expedition Island. It is hard to guess which agency should get credit for the planting, or if it was done by the Alaska Commercial Company.

Harris mentioned four or five trees on Hog Island that were about six meters tall. An oblique aerial photo taken in May 1942 suggests there were two clumps of trees on the island; a vertical in June 1951 shows only one. Thirty-seven Quonset huts were built on the island in the fall of 1942 and were occupied until about 1945. It seems likely the trees were cut during that period. The four trees near the spring may have escaped cutting because they were too small at the time or were surrounded by too many sprouts. Because no one in the Army mentioned trees on Hog Island in July 1944, the old trees may have been cut earlier. Not having examined a chart for Unalaska Bay, I assumed that Harris was using a second name for Amaknak, and neither visited Hog Island nor inquired specifically about it.

GROWTH OF TREES AT UNALASKA

The largest of the trees planted about 1807 on Amaknak Island was a little over 2.1

meters tall with a basal diameter of about 15 centimeters in 1834 (Veniaminov, [1840] 1985). In about 1870, the largest was about 7.6 meters tall and 51 centimeters in diameter (Bryant, 1870). In 1877, Cantwell reported that the trees were exactly the same in number as when planted fifty years before. He probably was using Bryant's count of 20, rather than Veniaminov's 1834 count of 24 or Chamisso's 1817 report that implies more than 24. Fernow's 1898 estimates of tree size are probably the most reliable until recent years.

After 1898, the first estimate of size I found was by Dr. Alberts [Elberts?], Director of Alaska U. S. Experiment Stations (Merritt, n.d.):

> I was at Unalaska in the summer of 1928 and saw the trees. At that time, they were about 10.7 meters in height and 46 centimeters in diameter at the base. The trunks are quite conical in form and the sets of branches starting from the terminal bud each spring are very close together. It appears that cones have formed in some years, but if seeds were produced, they failed to produce seedlings, because at the present time, there are no other trees in the region aside from those which were planted in 1805.

In 1943, Ernest P. Walker said "a clump of 13 spruces planted at Dutch Harbor in 1805 are now about 7.6 meters high" (Walker, 1943) (Fig. 6).

In July 1944, I found one group of 10 live trees and two old stumps near the pond and four other trees a short distance away near a small building (Fig. 7). The area near the trees had been covered with crushed rock, several vehicles were parked there, and I called it

Fig. 6. Sitka spruce on Amaknak Island, planted in 1807. Snow-covered lake is to the right. Photo April 1943, National Archives, 80-G-72716.

Fig. 7. Sitka spruce on Amaknak Island, planted in 1807. Trees are surrounded by U. S. Navy motor pool. Photo July 1944, by the author.

a motor pool in my notes. At about that time, a wooden culvert had been installed, perhaps to provide drainage under the parking area. When it clogged, it became part of a dam about 90 centimeters above the base of the trees, and caused the death of at least four trees

(Tindall, n.d.). I noted no defoliation in 1944, and my photo shows none. A June 1946 U.S. Navy map shows six trees as small circles on the edge of the lake and two about 30 meters away next to a small building. Whether this represents the living trees at that time, or the map scale was too small to easily draw more circles, is not certain. The small building apparently burned about 1950, killing all survivors of the four nearby trees. In 1944, my diameter measurements on 12 trees ranged from 30 to 56 centimeters and I estimated heights to be 7.6 to 9.1 meters. Two trees were forked at 1.4 meters and the larger of those was 89 centimeters at 0.9 meters.

In 1958, Von J. Johnson, Bureau of Land Management, Anchorage (Memorandum to Area Fire Control Officer, 1958), measured 10 standing trees, six alive and four dead. The living trees were from 28 to 69 centimeters in diameter and 7.3 to 8.2 meters in height. Ring width on an increment core suggested an average diameter growth of about 2.5 centimeters in six years, once tree growth was well underway. Johnson noted a bark beetle (Dendroctonus spp.) on a dead tree, and stated: "Numerous seedlings averaging 1.3 cm high were found 40 meters from planting on gravel fill on the lee of prevailing winds—northwest." He had copies of older photographs annotated as to site, but was unable to take similar exposures. Johnson submitted seven photographs and negatives and returned six old photographs to the Bureau of Land Management office in Anchorage (Fig. 8a). None of these could be located by Tindall in the Bureau's files in 1976 or later.

In 1976, Tindall measured the six living trees and average diameter was 59 centimeters and average height was 10.9 meters, with the tallest being 11.6 meters. In 1977, Dan Bishop, USDA Forest Service at Juneau, measured the diameters of all trees recorded by Johnson 19 years earlier. In about 1983, an infestation of spruce bud moth (Zeiraphera spp.) was noted, and new needles on shoots of several nearby younger trees were eaten in 1985 and 1986. In 1985, Jurate Mazeika, Director, Department of Parks, Culture, and Recreation, Unalaska, told me the trees would be sprayed to control these insects.

In 1985, only four of the old trees were still standing. The dead ones had been felled, including two that died after 1976 (Fig. 8b). Tindall and I measured the two tallest living trees in 1985 at 12.6 and 12.9 meters in height. The instruments used in 1976 and 1985 were a Relascope and tape, so these latest heights are fairly accurate. I did not use an instrument in 1944 and Johnson did not say how he estimated height. These measurements are summarized in Table 1. Many height measurements seem to be incorrect. For example, it is likely the trees grew in height between 1898 and 1944, yet my estimates were the same as Fernow's. Tindall's 1976 Relascope measurements make it apparent I was wrong.

Fig. 8. (a) Sitka spruce on Amaknak Island, planted in 1807. Six trees were still alive and seedlings grow in the foreground. Photo 1958, by V. J. Johnson. (b) Sitka spruce on Amaknak Island, planted in 1807. Four old trees were still standing, and the young seedlings have grown considerably. Photo 1985, by R. W. Tindall.

Table 1. Growth of Sitka Spruce on Amaknak Island

Date	Observer	Number of trees	Height (meters) Min.	Avg.	Max.	Diameter (centimeters) Min.	Avg.	Max.
1834	Veniaminov	24	-	-	2.1	-	-	15 (gl)[a]
1870	Bryant	20	-	-	7.6	-	-	51 (gl)
1898	Fernow	17	7.6	-	9.1	15	-	61 (bh)[b]
1901	(Photograph)	3 dead	-	-	-	-	-	-
1928	Alberts	-	-	10.7	-	-	46	- (gl)
1943	Walker	13	-	7.6	-	-	-	-
1944	Bruce	12	7.6	-	9.1	30	-	56 (bh)
		2 forked	-	-	-	57	-	89 (1m)[c]
		2 stumps	-	-	-	-	-	-
1958	Johnson	6 alive	7.3	-	9.2	28	-	69 (bh)
		4 dead	-	-	-	28	-	47 (bh)
1976	Tindall	6	-	10.9	11.6	-	59	- (bh)
1977	Bishop	6 alive	-	-	-	34	-	70 (bh)
		4 dead	-	-	-	26	-	41 (bh)
1985	Bruce and	4 alive	-	-		-	-	- -
	Tindall	2 tallest	-	12.7	-	-	-	-

[a] gl = groundline
[b] bh = breast high
[c] 1m = one meter

My notes indicate that there were 59 trees on Expedition Island in 1944, but the source of this estimate was not recorded. I measured 13 diameters (from 19 to 45 centimeters) among the larger trees. Estimated heights ranged from 6.1 to 9.2 meters. In 1958, Johnson measured diameters of six dead and 47 live trees. Living trees ranged from 11 to 49 centimeters and estimated average height was 8.5 meters. In 1977, Bishop found the largest tree to be 54 centimeters. In July 1977, Tindall found 44 live trees, nine dead trees, and five stumps. This verifies my 1944 rumor and Johnson's 1958 tally, if one lonesome small tree is included in the 1944 count. Tindall got an average diameter of 31.5 centimeters and height 10.7 meters. In 1985, Tindall and I measured two of the tallest trees as 13.6 and 14.2 meters (Fig. 9).

In 1985, a diesel power plant was installed in a cannery on Expedition Island. Heat and fumes from the cannery subsequently defoliated and killed some of the trees. After defoliation that started in 1985 and 1986, all tree tops became visible. The tallest two trees in June 1987 were 16.1 and 16.8 meters but were not the same trees measured in 1985 (Alden and Bruce, 1989). Forty-two trees were measured; 17 were dead and 25 were alive. Their average diameter was 32.9 centimeters and height was 12.1 meters. Alden used the 1987 measurements to estimate volume growth; if the trees were planted about 1872 instead of 1842, the growth rate increased from his estimated 2.7 cubic meters to 3.4 cubic meters per hectare per year. Table 2 summarizes these measurements. Alden's 1987 measurements are the best and Johnson's 1958 diameters seem reliable. Unfortunately, the date of planting is not known.

Table 2. Growth of Sitka Spruce on Expedition Island

Date	Observer	Number of trees	Height (meters)			Diameter at breast height (centimeters)		
			Min.	Avg.	Max.	Min.	Avg.	Max.
1944	Bruce	58	6.1	-	9.2	-	-	45
1958	Johnson	47 alive	-	8.5	-	11	-	49
		6 dead	-	-	-	9	-	18
1977	Bishop	53	-	-	-	-	-	54
1977	Tindall	44 alive	-	10.7	-	-	31.5	-
		9 dead	-	-	-	-	-	-
		5 stumps	-	-	-	-	-	-
1985	Bruce and Tindall	2 tallest	13.6	13.9	14.2	-	-	-
1987	Alden	25 alive	5.9	-	16.8	20	-	55
		17 dead	4.3	-	15.7	16	-	43
		42 all	-	12.1	-	32.9	-	-

In 1988 and 1989, the trees and stumps on Hog Island were examined. Alden counted seven old stumps near a small stream and 21 Sitka spruce stems about 60 meters upstream (Fig. 10). The largest stump was 46 centimeters in diameter and 41 centimeters tall. The largest stem in the upstream clump was 32 centimeters in diameter at breast height (1.4 meters) and 7 meters tall. In this group of trees, 17 small stems appeared to layer from four mother trees. One of the large stems had 93 rings about a meter above the ground. This

Fig. 9. Sitka spruce on Expedition Island. Photo 1985, by R. W. Tindall.

suggests that it was planted about 1870. The ring count on the stump was about 70, but it should have been about 80 or 90 if the tree was planted in 1843 and cut in 1943. Four sources of error contribute to underestimates of ring counts on old stumps: unknown age of first ring, unknown date of mortality, rings not formed due to unfavorable growing conditions, and rings not noticed because of wood deterioration.

SPRUCE REPRODUCTION

Veniaminov ([1840] 1985) wrote that cones first formed on some trees on Amaknak in 1833. Bryant (1870) said cones are formed but the seeds do not ripen. Shepard (1898) saw the trees in 1889, and reported, "the cone does not arrive at perfection." At about the same time, Réclus (1890) said the trees showed good growth but did not reproduce.

In 1898, Fernow reported two spruce seedlings on Amaknak Island. The two-rail fence in the photograph accompanying Fernow's article might have controlled cattle, but certainly not sheep. A 1904 photograph shows a small herd of sheep near the Dutch Harbor dock. In 1944, I photographed a lonesome, small spruce on the edge of a bank on Expedition Island about 15 meters from the plantation; I thought it was the lone survivor of a recent planting rather than a seedling. At the time it looked to me like wind-pruned krummholz, but now I believe it had been heavily browsed. No one else reported any reproduction on Amaknak until Johnson in 1958: "Numerous seedlings averaging 13 centimeters high were found 40 meters from planting on gravel fill on the lee of prevailing northwest winds." He did not mention the seedlings on Expedition Island noted by Admiral Harley D. Nygren (Letter to A. E. Helmers, USDA Forest Service, Juneau, 1970), who reported: "First noticed many seedlings in 1958 not sure if present earlier. In 1959 dug up several, shipped to Seattle, two planted..." There also is a report that 250 seedlings were shipped to Adak from Expedition

Fig. 10. Sitka spruce on Hog Island. Old stumps in the foreground and clump of 21 stems in the background. Photo 1989, by J. M. Alden.

Island in 1958 by Richard Haney, who was then in the Navy (Alden, personal communication, 1990). In downtown King Cove, there are six Sitka spruce two to three meters tall that were transplanted from Dutch Harbor in 1961.

The Navy did two things during War II that encouraged spruce regeneration. They added a crushed rock fill near the trees on Amaknak which provided a competition-free mineral soil for germination and early growth, and they got rid of all livestock. The summer of 1952 was the warmest in the period 1949 to 1958 at Adak and Cold Bay, and therefore is a likely spruce seed year for the Aleutians. The average summer temperature in the Aleutians since 1940 possibly is at least 1 °C warmer than it was from 1800 to 1890, when the northern hemisphere was still recovering from the Little Ice Age. This change makes production of viable seed more likely, because even current temperatures seem to be marginal for seed to ripen.

PREWAR PLANTINGS

About 1943, copies of a few letters dated 1932 to 1939 that mention seedlings sent to the Aleutians or the Aleutian peninsula were forwarded to the Engineer Office, Alaskan Department, U. S. Army, Anchorage, by the USDA Forest Service, Juneau. On April 4, 1932, bundles of about 15 Sitka spruce seedlings were shipped, one to Mrs. A. M. Ambur at Akutan and the other to G. B. Henderson at Kashega. Recently I saw a photograph taken in 1976 of two spruce trees on Akutan about two or three meters tall. Those at Kashega were planted April 18 in a fenced area near the school. On June 3, 29 silver fir (<u>Abies</u> <u>amabilis</u> Doug. ex Forbes) seedlings from Sand Point, Washington, were planted near the creek and lake on the school grounds and eight were planted in Kashega village.

In 1934, the crew of the Coast Guard cutter Chelan planted several hundred spruce and cedar (sic) nursery seedlings on Expedition and Unalaska Islands. In 1935, at least one-half of the spruce were doing well, but only one cedar was alive and thriving. In 1935, the crew planted a number of Colorado blue spruce (<u>Picea</u> <u>pungens</u> Engelm.), Norway spruce (<u>Picea</u> <u>abies</u> (L.) Karst.), golden arborvitae (<u>Thuja</u> <u>occidentalis</u> L.), and Cryptomeria (<u>Cryptomeria</u> <u>japonica</u> (L.f.) D. Don) at Unalaska, mostly on the grounds of the U. S. Coast Guard. Two sets of photographs were forwarded to Coast Guard headquarters at Seattle and San Francisco.

In May 1936, the cutter Haida received at Seward five boxes, each containing about 18 spruce and hemlock (<u>Tsuga</u> (Endl.) Carr.) seedlings that were delivered for planting to the village of Unalaska, the village of Dutch Harbor, the Aleutian Live Stock Company at Umnak, N. J. McMillan at Atka, and to Mike Hodikof at Chickagof on Attu. A few hemlock and spruce were planted on Atka Island in May 1936. Only six survived the first year, and in February 1938, the planter N. J. McMillan, then at Noatak, doubted if any of them were still alive.

In August 1937, the Coast Guard Commander at Unalaska reported that some seedlings did not survive because they were improperly planted or lacked protection. The context suggests that these were the ones planted at Unalaska in 1934; those planted in 1935, which were protected from October 15 to May 15 with canvas on wood frames, grew as much as 30 centimeters.

In December 1939, members of the Unga Women's Club reported on seedlings sent July 1938. Apparently none of the birch seedlings were still growing, only the pine or fir.

According to a report in the Alaska Geographic Quarterly (Anonymous, 1980):

> Just after the turn of the century, well-heeled citizens of Unga enjoyed high success in landscaping with Sitka spruce and, although today the site is a ghost town, these trees have grown on a magnificant scale and are reseeding to the point that a forest will soon replace the toppled buildings.

This is followed by a picture of a cottonwood (Populus sp.) that seems to be 12 meters tall, growing between a cannery building and a hillside at Squaw Harbor. The accompanying text says spruce has also been successfully planted there.

Other locations where trees were planted more than 50 years ago include False Pass, Sanak, and Sand Point. In a group of spruce near a house at False Pass there are four stems about six meters tall that were planted in the 1930s. There is a nearby group of large spruce including one forked tree 13.5 meters tall with the two stems 59 and 57 centimeters in diameter that was planted in 1932. In 1944, I was told there were six conifers about 1.8 meters tall near a cemetery on Sanak. I have seen a photo of a group of trees at Sand Point and heard of another photo that was accompanied by the statement that the trees were planted about 1900. In 1985, I was told that these trees were 12 meters tall.

There undoubtedly were many other trees transplanted to the islands in the 1930s and earlier. Fox farmers have been credited with some of the tree planting on the islands before World War II.

WORLD WAR II PLANTING

During World War II, more people lived in the Aleutian Islands than at any other time before or since. The terrible weather, desolate scenery, and isolation caused severe morale problems. Tree planting was one of the many activities undertaken to improve living conditions. Military construction began at Dutch Harbor in August 1940. Airfield construction at Cold Bay and on Umnak began in December 1941. Adak was first occupied by Army troops in August 1942, and Amchitka was occupied in January 1943. The military population on the islands peaked late that summer at about 150,000 and diminished in 1944 to 50,000. Before the Japanese surrender in August 1945, some Aleutian posts were put on a standby basis.

Officers of the 151st Engineers at Cold Bay wrote to the USDA Forest Service in Juneau in March 1942 saying they were interested in starting a tree planting program. The Forest Service, with the help of the 42nd Engineers, furnished the trees. The first shipment was 115 wild seedlings of European ash (Sorbus aucuparia L.) and 25 red alder (Alnus rubra Bong.). The only living trees I found near Cold Bay in 1944 were a few spruce.

In July 1942, 10,000 spruce seedlings probably supplied by the 42nd Engineers at Juneau were planted near Fort Glenn on Umnak. More conifers probably were planted in 1943 near Pustoi Bay on Umnak. In early 1944, a large number of seedlings were held in storage for several weeks before planting. Some of those planted were destroyed by road construction near Umnak Pass, and others north of the cemetery were soon dead. Some were called Douglas-fir but this identification is not positive. In July 1944, my check of about 1,000 of the 1942 spruce seedlings showed that 15 percent were in good or fair condition and 44 percent were dead. In 1985, I was told that quite a few trees were still growing on Umnak.

Some trees were probably planted at Dutch Harbor the year before the Japanese bombed in June 1942. When I visited in July 1944, my hosts were recent arrivals and dates of planting (1941–1943) were not known for the many young trees. In May and June 1944, five shipments of 50 to 100 seedlings each were received at Unalaska by Army troops, including 50 hardwoods from Fort Richardson. Some of the tree shipments undoubtedly were from Kodiak; others may have been from Juneau.

The earliest reported planting on Adak was 1,000 spruce by the 18th Engineers in February 1943 in Nurses Creek and Hospital Creek valleys. Bruce and Court (1945) incorrectly described their survival, saying that only six were still alive in 1944. Only six of the seedlings seen during a hike down one of the two drainages were alive. In 1984, 53 old Sitka spruce were found in these valleys.

In April 1944, about 25 spruce and 25 willow were planted near Landrum Creek. In 1984, only three misshapen spruce and more than a dozen Barclay willows (Salix barclayi Anderss.) were found there. Other spruce may have been cut long ago as Christmas trees.

Before June 9, trees had been planted in 1944 at four upland locations on Adak; more were on hand for planting later that month. Those trees were protected by burlap wind screens 60 centimeters high. None of the spruce planted in the uplands were found in 1984. The road system in the upland areas made them too easy to find, and apparently they had been transplanted or cut earlier. In 1984, we found four clumps of Barclay willows, two in valleys and two upland, and two clumps of salmonberry (Rubus spectabilis Pursh) that appeared to have been introduced 30 to 40 years previously.

GENERAL BUCKNER'S TREE PLANTING PROGRAM

Early in 1944, General Simon B. Buckner, Alaskan Department, U. S. Army, decided that, as a morale booster, a more systematic tree planting program in the Aleutians should be started. At that time, I administered a civilian construction contract at Whittier, Alaska, for the Engineer, Alaskan Department. In a March response to an inquiry from the Executive Officer of Engineer Military, Alaskan Department, I wrote a memo describing action useful in transplanting trees to the Aleutians. In April, 1944, I received a letter directing me to obtain soil samples from several islands and to prepare a general information bulletin on tree planting. I was assigned full time to the project in May, and from June 4 to August 3, I was in the Aleutians digging soil samples, observing older plantings, selecting potential planting sites, but mostly waiting for transportation west from Adak. After two weeks waiting for air transport, I went on the first available troop ship to Attu. The notes I took were cursory—to help prepare a planting bulletin and report.

On April 22, 1944, I wrote to Professor H. J. Lutz at the Yale School of Forestry describing the proposed tree planting program and requesting pertinent publications and suggestions. He responded promptly with two publications and a great deal of advice. Much of this was included in the planting instructions I prepared later.

Before leaving for the Aleutians from Anchorage, I requested weather summaries from XI Air Force Service Command and met Lieutenant Arnold Court, a climatologist, then assigned to Weather Control. Later that year, he prepared a summary of weather conditions in the Aleutians, insisted we write our article for Geographical Review (Bruce and Court, 1945), got it approved by the U. S. Army, Air Force, and Navy after I left Alaska, and wrote several letters to me in 1945. These are the main sources of information about the results of the 1944 planting program.

Soil samples were shipped to Professor S. C. Vandecaveye, State College of Washington in Pullman. He provided prompt analyses of both chemical and physical characteristics. A brief summary of his findings was included in his last letter dated September 25, 1944:

> ... under the prevailing climatic conditions of the various stations sampled, the plant nutrient status and reaction of the soils represented by the samples we have analyzed appears to be adequate for tree growth if suitable subsoil drainage is provided either naturally or artificially.

The Army posts scheduled to receive trees in the 1944 planting program were at Cold Bay, Umnak, Adak, Amchitka, Shemya, and Attu (Fig. 1). Dutch Harbor was omitted because many trees already were planted there. Cold Bay was put on housekeeping status in August 1944, although it was reactivated in 1945 to conduct a training program for the Russians.

Orders based on drafts I prepared at Adak covering choice of planting site, methods of planting, and protection after planting were sent to each of the six posts in July 1944. On July 15, 1944, orders were sent to Fort Richardson, Ladd Field, Seward, and Kodiak detailing the type of tree to ship, method of digging, storing, and shipping. Had the war lasted longer, much of this program might have been put into effect. After I drafted a report on the project, I was assigned to other duty on September 15, 1944, and heard little about tree planting before I left Alaska in January 1945.

On May 7, 1945, Court reported from Shemya:

> Yesterday I managed to visit the "Post Tree Nursery" on a steep hillside in the NE side of this island... The trees look fairly healthy despite the helter-skelter planting. About two-thirds of the conifers have green needles, and most of the deciduous have life in the branches, although there is no sign of leaves. There are also two flourishing conifers near the nurse's quarters at the hospital.

On June 30, 1944, I had reported to Engineer, Alaskan Department, that no trees had been received at Shemya to date.

On November 2, 1945, Court took two photographs of trees planted in the Palisades area on Adak (Fig. 11a). Comparison of one of these with a 1963 photograph of what was then named the Adak National Forest (Fig. 11b) shows these are the same trees. The 1945 photographs show about 50 burlap enclosures about 60 centimeters high protecting small spruce trees. These enclosures were the same kind as those used in other 1944 plantings. No trees had been planted in this area when I took soil samples nearby early in June 1944. The letter accompanying the 1945 pictures said there were about 40 conifers and 20 hardwoods in the planting.

Thomas B. Glazebrook of Portland, Oregon, a member of the 799th Engineer Forestry Company on Afognak Island, oversaw the digging of about 50 spruce seedlings and shipping them west on the ship's deck. This mode of shipping was specified in the July 1944 orders, as was the 1.2 meter spacing of the plantings. Because the shipment included no hardwoods, it is possible hardwoods already had been shipped from Kodiak. Evidence that the

Fig. 11. (a) Adak National Forest, one year after planting. Photo November 1945, by A. Court. (b) Adak National Forest. Adak's struggling 'Forest' even has a litter barrel. Photo May 1963, by H. L. Baila.

seedlings went to Adak is lacking, just that the post at Kodiak was following the July 1944 orders. The Forestry Company left Afognak in December 1944. Evidence that the other posts shipped seedlings has not been found. The rapidity with which the war was winding down in Alaska suggests that this program was put on hold late in 1944, particularly because General Buckner left Alaska in June 1944. The only trees shipped may have been those from Kodiak and Afognak. The only plantings from Buckner's program I am sure of are at Shemya and Adak. Plantings on Attu and Amchitka are likely, and possible at Umnak and Cold Bay.

ADAK AND ITS NATIONAL FOREST

In 1950, Captain Anderson of the U. S. Navy asked his supply source at Kodiak for trees to plant on Adak, and the skipper of the supply ship brought a consignment (W. Metcalf, Berkley, California, personal communication, 1964). Unfortunately, how many trees were planted and where were not permanently recorded. Some of the shipment may be the present Marine Memorial.

If or where the trees transferred from Unalaska to Adak in 1958 by Richard Haney were planted is unknown.

Early reports on the U. S. Army plantings of Sitka spruce on Adak echo the concern shown 120 years earlier about slow initial growth. Bank et al. (1950) suggested trying boreal birches and alders as well as black spruce (Picea mariana (Mill.) B.S.P.), indicating that little success had been had with Sitka spruce. Later it was reported (Bank, 1953):

> On Amaknak Island in Unalaska Bay in 1805, Russian settlers planted 24 seedlings of Picea sitchensis which came from Sitka ... Other Sitka spruces were later planted on Expedition Isle in Roose Bay and Unalaska Village. All these groves have been depleted by dying trees, but parts of the original plants are still alive, and, except for the stunted appearance and gnarled form of some of the trees, they seem healthy. Many regularly produce cones. However, none has successfully reproduced by seedlings. More recently, at Adak, several dozen Sitka spruces were planted on a hill and on a lowland. These plantings are failures; the trees that have survived are stunted, and there has been no reproduction.

Natural regeneration at Unalaska Bay and the Adak National Forest suggests that this assessment was premature.

On March 2, 1964, the *Seattle Times* published a short note accompanied by a picture taken in May 1963 (Fig. 11b). It describes the littlest National Forest on Adak, saying there was a group of spruce trees from 1.2 to 3.7 meters high covering an area of about 0.1 hectare. A sign at the plantation read: "You are now entering and leaving Adak National Forest." Next to it was a little barrel with the words "Keep Adak Green." This news release appeared in papers nationwide. The USDA Forest Service offered an award to the lonesome G. I. who planted the trees. When I saw the picture I matched it with one of Court's 1945 pictures, and then informed the Forest Service how and when the Adak National Forest had been planted.

In 1965, a forestry graduate from the University of Washington, J. M. Gullickson, who was in the Navy on Adak, wrote to the Forest Service Experiment Station in Juneau requesting information about tree species suitable for planting on Adak. The response included two publications about tree planting in the Aleutians and information about climate and the suggestion that he try planting Sitka spruce from Kodiak. It also inquired about the fate of 500 Sitka spruce wildings shipped from Juneau to the Boy Scouts on Adak in 1962. This correspondence ended with an illustrated report by Gullickson dated October 7, 1965 on trees planted on Adak (Gullickson, 1965).

The trees shipped to Adak in 1962 were watered for a while and the local radio station asked for volunteers to plant them. No volunteers appeared and the trees died.

Gullickson's report on living trees on Adak, mostly Sitka spruce, includes five areas: the National Forest, six spruce at the Marine Memorial, 13 spruce in Nurse's Creek drainage (actually Hospital Creek), a few individual spruce near different buildings, and about 60 trees of various species planted in August 1965 near the Naval Communications Station. His photographs and measurements are an important part of the Adak record.

Gullickson reported that the Sitka spruce in the National Forest were planted around 1950 and they ranged in height from 0.6 to 1.8 meters. In the spring of 1965, an unspecified amount of 10-20-10 fertilizer was applied. A 0.7 meter tall pine had been planted near the spruce trees in August 1965.

The spruce in the Marine Memorial were larger and ranged from 1.5 to 2.5 meters tall with the largest 8.9 centimeters at the base. Five trees were in a clump and one tree was about 30 meters south. A sign in one of his photographs reads, "Wilson, Carnahan and Borum Memorial Park."

The 13 Sitka spruce in Hospital Creek were from 0.3 to 1.9 meters tall. Most suffered from wind damage. The rather random location of the trees suggested to Gullickson that they were established by seeding. My present information suggests that they are the survivors of irregularly spaced wildings planted in the valley in 1943.

Few of the trees planted near buildings survived, and an attempt to identify survivors in 1984 with those in 1965 was not successful. The small pine near the Adak National Forest in 1965 had disappeared, although we found four very small two- or three-needle pine seedlings west of the National Forest in 1985.

In January 1966, about 200 Scots pine 2-0 seedlings were planted about 8 feet apart near a lake (probably Palisades Lake) by two Navy engineers. They also applied 23 kilograms of 20-9-7 fertilizer to the National Forest. Their note also says, "9 kg 20-5-5 (Scott's) applied by Miller 5-65." If this was also applied to the National Forest, it may have been fertilized three times at a total rate of 2.5 tons per hectare in 1965 and 1966. Over the years, there may have been other fertilizer applications.

In 1984, we found seven Sitka spruce near an old Quonset hut site on the upper slope of Landrum Creek. I estimated that they were planted about 1964. This was not one of the locations mentioned by Gullickson. Before the next summer, one of them had been removed. In 1984, we also found a small spruce of a similar vintage about 10 meters south of the Marine Memorial.

In 1975, there was a large afforestation effort on Adak. At least six trees had been cut the previous year in the National Forest either maliciously (Dr. Clifford C. Amundsen, ecologist, University of Tennessee, personal communication, 1988) or as Christmas trees (Spencer, personal communication, 1984). The Commanding Officer of the Naval Base issued orders that have been effective in protecting the National Forest ever since, and also started a program of afforestation. This program was assisted by biologists from the U. S. Fish and Wildlife Service at Adak and by Amundsen, who recommended eight planting sites and listed 17 tree species that should be tried. Lieutenant Commander Gerald J. Jacobson wrote to the governors of 50 states requesting that tree seedlings be shipped to Adak for planting (Jacobson, 1975). Nine states shipped more than 3,000 seedlings which were planted after October 24, 1975 (Amundsen, 1985). Survival of trees was checked at four sites annually for five years beginning in 1979. More than half the trees were still alive in 1983. Actual survival, however, would have been much lower because dead seedlings are hard to find and Adak residents moved others.

In September 1984, the Adak National Forest had 33 living Sitka spruce and two small stumps at about 1.2 meter spacing in a rectangle roughly 9 by 11 meters (Fig. 12). Immediately to the north were eight Barclay willows in an area about five meters square. No trees were found north of the spruce and west of the willows. Perhaps other hardwoods such as balsam poplar or black cottonwood (Populus trichocarpa Torr. and Gray) had been planted there, because in 1945 there were half as many hardwoods as conifers. Six of the spruce survived the cutting in 1974. The count of 27 undamaged spruce agrees closely with one of Jacobson's 1975 news releases which indicates that the tree population of Adak increased from 28 trees to 2,128 after part of the 1975 planting was completed.

The average height of trees in the Adak National Forest in 1984 was 3.2 meters and the average diameter was 16.3 centimeters at about 15 centimeters above the ground. The trees probably averaged 0.5 meters tall when planted in 1944, 1.2 meters tall in 1965, and grew about 2.0 meters in the last 19 years. The benefits of close planting are shown by the 1984 measurements. The 12 trees on the outer margin that had not been cut in 1974 averaged 2.7 meters in height. The 15 inner trees averaged 3.6 meters, with the tallest 4.3 meters.

One of the two stumps was cut near the ground and a section was later sanded and examined under magnification. It was 7.6 centimeters in diameter. The ring measurement suggested the tree was 7 or 8 years old when transplanted in 1944. Growth was minimal for 3 years after transplanting, and it was cut after the 1962 growing season. A low branch lived for two years and traces of early 1963 growth were seen higher on the stem. A closeup photograph taken in 1945 suggests that some seedlings lost most of their new needles in 1945, and that older needles kept the trees alive for 3 or more years until they recovered from transplanting. I believe the two stumps found in 1984 were the largest trees in the plantation in 1962.

Measurements at the Marine Memorial in 1984 of the four largest trees averaged 4.0 meters in height and 21.3 centimeters in diameter near the ground. In 1965, they averaged 2.0 meters in height. In 1984, they produced many more cones than any of the trees on the National Forest.

In 1986, trees at three of the 1975 planting sites were measured (Alden, personal communication, 1986). Half to three-fourths of the 174 Sitka spruce found were in good condition except for yellowing of the foliage. Heights of living Sitka spruce ranged from 0.2 to 1.4 meters. Previous tallies by Navy volunteers indicated that white spruce (Picea glauca (Moench) Voss) and black spruce were planted at these locations, but none was found in 1986. Jacobson's news release on November 13, 1975, stated that 1,000 trees arrived courtesy of Alaska Airlines, the Ketchikan Boy Scouts, and the Alaska Logger's Association. The trees were Sitka spruce and western hemlock (Tsuga heterophylla (Raf.) Sarg.) dug near Ketchikan. A later news release reported that trees would be planted at

Fig. 12. Adak National Forest. Reference pole is two meters tall. Photo by W. R. Briggs, September 1984.

eight afforestation sites and two nurseries in the future. This suggests that about 100 of the Ketchikan seedlings were planted at each site and that before 1986, they had been tallied as white and black spruce. Only one hemlock and seven Scots pine were found alive in 1986. With no planting site maps, other trees found alive in 1983 were not relocated in 1986. Another correction to Amundsen's 1985 report is his identification of the large trees at the Marine Memorial as Colorado blue spruce. In 1984, I noticed color variation in these trees and asked Dr. Leslie A. Viereck if any appeared to be hybrids. He examined them carefully and said they were Sitka spruce.

OTHER ALEUTIAN ISLANDS

In 1985, I received a letter from the commander of the Shemya Air Base Group which indicated a report I sent him on the Adak National Forest solved a Shemya riddle -- the origin of 10 "spruce bushes" that cuttings had shown to be 40 years old. He sent a picture of one of the four trees near a small lake and said the other six were on a hillside at the other end of the island. They were about 72 centimeters tall with stem diameters of 20 centimeters. The four near the lake were buried by snow in the winter but the others were exposed to the incessant wind. There also were about 10 willows on the island.

In 1945, seeds of five eastern conifers and 10 shrubs were planted on Amchitka near the base of some Aleut middens (Nelson, letter, 1945). In 1985, I heard two reports about three or four trees either 30 or 120 centimeters tall growing near the old officer's club on Amchitka. Recently I learned there are but two survivors.

In 1987, Tindall visited Attu and U. S. Coast Guardsmen at the Loran Station showed him some of the nearby Sitka spruce. He did not examine trees in more remote areas. Trees on exposed sites were about 70 centimeters tall if growing singly, and nearly twice that height if in the middle of a clump. Photographs indicate single trees on protected sites were two meters tall and about three meters tall in clumps. The tallest trees were planted near a chapel and were protected until the walls collapsed about 1984. Now the tops are dying. Tindall estimated the trees were 25 to 30 years old; I believe they were planted during World War II.

The text accompanying a photograph of a clump of spruce planted near the Atka airstrip during World War II reports that one is about 65 centimeters in diameter at the base (Anonymous, 1980). These trees apparently replaced 20-year-old spruce that grew near a church and were killed when it burned in 1942.

At least two reports document survival of Colorado blue spruce in the Aleutians. One group of trees appears to be the blue phase of Sitka spruce and the other tree was a 1.7-meter nursery seedling that possibly lived for a year or two.

In 1985, we found four noble fir (Abies procera Rehd.) and six Douglas-fir (Pseudotsuga menziesii (Mirb.) Franco) near a building in Unalaska. They had been grown in large containers and were taken indoors during the winter. Apparently they were planted outside about 1981 because the tip of the healthiest noble fir had one very short internode followed by two that were not as long as lower nodes. The other noble fir were smaller, but still growing, and the Douglas-fir were either dead or dying.

I have mentioned surviving Sitka spruce, hemlock, and pine from the 1975 planting on Adak. Amundsen's 1986 report claims 5-year survival for white, black, and blue spruce, white, Scots, and Japanese black pine (Pinus thunbergiana Franco), Siberian larch (Larix siberica Ledeb.), fir (sic), and paper birch (Betula papyrifera Marsh.), but for some of the species, identification seems questionable.

On Shemya in 1985 and 1987, there was poor survival of recently planted trees. In 1985, a large group of Norway spruce about one meter tall that had been planted in 1984 had died. Some of 25 recently planted 60- to 120-centimeter tall black and white spruce wildings from Elmendorf Air Base near Anchorage had also died in 1985 and were replaced by Sitka spruce we left when we visited Shemya. Five spruce, a willow, and a dogwood (Cornus L.) were alive in 1987 but they were in very poor condition.

PLANTINGS AFTER 1985

Since 1985, many nursery seedlings have been planted on Adak by the USDA Forest Service in cooperation with the U. S. Navy, and at Unalaska and other coastal locations north and east of Unalaska by Re-Tree International and their cooperators. These organizations are keeping tree planting and survival records. No reports are available that give more than an indication of initial survival. Survival often has been poor because of difficulty in obtaining dormant seedlings late in the season at the source nurseries and in providing adequate site preparation and seedling protection after planting.

On Adak, in May 1987, seedlings of 90 Sitka spruce provenances and other species were transplanted into four beds about 10 meters long with seedlings spaced at 10 by 15 centimeters. The transplant beds were in a small ravine near the Adak National Forest and were further protected by windscreens. Trees planted included Manitoba maple (Acer negundo L.), balsam poplar, black cottonwood, aspen (Populus tremuloides Michx.), paper birch, European white birch (Betula pendula Roth var. carelica (Merrklin)), Siberian larch, Dahurian larch (Larix dahurica Turcz.), lodgepole pine, the natural hybrid lodgepole x Jack pine (Pinus contorta x Pinus banksiana Lamb.), Scots pine, black spruce, white spruce, Norway spruce, and a putative P. mariana x P. sitchensis hybrid.

Early in 1987, two foresters from Plum Creek Timber Company planted several hundred Engelmann spruce, western larch (Larix occidentalis Nutt.), western white pine (Pinus monticola Dougl. ex D. Don), noble fir, and Douglas-fir near one of the housing units on Adak. None of these trees were found in 1988 or 1989.

OTHER ISLANDS AT HIGH LATITUDES

Sitka spruce grows well on other islands at high latitudes, but so do other species that should be considered candidates for planting in the Aleutian Islands.

Hultén (1960) reported that Japanese stone pine (Pinus pumila Regel), common in Kamchatka, occurs on Bering Island. One recent report indicates that this species is on Mednyy Island, another denies its presence on either of the Commander Islands (Critchfield and Little, 1966). Thus, its performance after planting is not known.

Iceland has an extensive and varied record of tree planting. Small groups of trees planted in 1899-1907 include some of the largest trees now in Iceland (Iceland Forest Service, 1986). Between 1935 and 1950, Scots pine and Norway spruce nursery stock from Norway and Sitka spruce seedlings from Alaskan seed were planted. About 1960, a woolly aphid (Pineus pini Maeq.) damaged Scots pine so severely that it has almost disappeared from plantations. In April 1963, the temperature fell from 22 °C to -10 °C, accompanied by a northerly gale, and wiped out most of the Sitka spruce plantations. Norway spruce was then planted on former Sitka spruce sites, but the climate deteriorated and this species did not thrive. Coastal lodgepole pine from Skagway and western Siberian larch (Larix sukaczewii Dyl.) and Siberian larch have since been planted. Siberian larches are now the most commonly planted species in Iceland. Alpine fir (Abies lasiocarpa (Hook.) Nutt.) and Lutz spruce are also planted today and northern black cottonwood has the fastest early height growth of all species. The tallest tree in Iceland in 1986 was an Engelmann spruce 16.7 meters tall, planted in East Iceland in 1905, but nearby, a 33-year-old cottonwood was 16.4 meters tall. The climate of Iceland is more continental than in the Aleutians, so it is not safe to assume that trees best-adapted for Iceland will be best for the Aleutians.

Tree planting on the Falkland Islands, which lie between latitudes 51° S. and 52°30′ S., has met limited success. (Low, 1986). The two large islands are larger than any of the Aleutians but are much smaller than Iceland. The first large-scale planting in the early 1890's included Nothofagus (N. betuloides (Mirb.) Oerst.), poplar, Austrian pine (Pinus nigra Arnold), and Scots pine. From 1920 to 1925, a forester had poor results with planting stock from Britain, but better success with seedlings raised near Port Stanley. In Port Stanley, the most common tree today is Monterey cypress (Cupressus macrocarpa Hartw.), followed by Austrian and lodgepole pines. Other conifers include Sitka spruce and Monterey pine (Pinus radiata D. Don). This is astonishing. Monterey pine is supposed to be killed by even light frost, but there is a frost-resistant hybrid with knobcone pine (Pinus attenuata Lemm.) that has foliage like Monterey pine (Anonymous, 1948). Many trees are six meters tall in the lee of buildings. Near farms, Monterey cypress is most common but Sitka spruce, white poplar, and various pines including Corsican (Pinus nigra var. maritima (Ait.) Melville) have been used for shelter planting. Planting instructions prepared for the Falklands (Low, 1986) appear quite applicable to the Aleutians, which have higher rainfall and more loamy subsoils.

The Shetland Islands are at latitude 60° N. The principal species in the 1913-1920 plantations are Sitka spruce and Japanese larch. Some noble fir have grown nearly as well as Sitka spruce, and a few sycamore-maple and Japanese larch have reached large sizes. The plantations are mainly shelterbelts, and mountain pines (Pinus mugo Turra) have contributed to their success for these purposes (Stewart, 1962). Trees in several 1954 plantings have grown well, with Sitka spruce reaching 9 to 10 meters and lodgepole pine 8 to 9 meters. Other species suggested for future planting are Japanese larch (Larix leptolepis (Sieb. and Zucc.) Gord.) and hybrid larches (Larix x eurolepis Henry), birch, sycamore-maple, and mountain ash [both rowan and whitebeam (Sorbus aria var. intermedia (L.) Crantz)] (Low, 1987).

The Faroe Islands are at about 62° N. latitude, but the North Atlantic Drift moderates their climate. Mean monthly summer temperatures are about the same as at Unalaska, but winter temperatures are about 2 °C warmer (Lines, 1983). Successful ornamental planting of trees in the middle of the 19th Century predates the first effective plantation by about 50 years (Hojgaard et al., 1989). By 1985, about 78 hectares had been planted, mostly near the capitol, Torshavn. Hundreds of species of trees and shrubs have been planted. Many survive but grow slowly, because of the cool summers. Several tree species grow well in the shelter of buildings or older planted trees. Foremost among the successful pioneer trees in plantations is shore pine, followed by Sitka spruce, noble fir, Lawson cypress, Alaska yellow cedar (Chamaecyparis nootkatensis (D. Don) Spach), Leyland cypress (x Cupressocyparis leylandii), Japanese larch, mountain pine, western red cedar, sycamore-maple, Sitka alder, Icelandic birch, rowan, whitebeam, and northern black cottonwood. Several South American and New Zealand species, planted 10 years previously, showed promise. Four insect pests, including a species of Zeiraphera, and two fungi are attacking some of the introduced trees.

I will add to these candidates a few species: from coastal Alaska, mountain hemlock (Tsuga mertensiana (Bong.) Carr.) and Pacific silver fir, and from eastern Asia, Japanese black pine, reputedly salt resistant, and Jeddo spruce (Picea jezoensis (Sieb. and Zucc.) Carr.), which has many of the characteristics of Sitka spruce. The hybrid Leyland cypress might do better than Monterey cypress, one of its parents, because its other parent is Alaska yellow cedar, a species that grows as far west as Prince William Sound.

DISCUSSION

Clearly Sitka spruce can survive and grow to tree size in the Aleutian Islands. The trees that have lived the longest, either the older plantings or those made during World War II, were all wildings from Southeast Alaska or Kodiak. Although they are more expensive and require special care in transplanting, wildings have several advantages over nursery stock. They are innoculated with mycorrhizae, and have older needles that reduce loss of current foliage after transplanting. They have stouter stems and are not as easily frost-heaved or flattened by heavy snow. And they are larger than nursery stock and compete better with native vegetation after they recover from transplanting shock.

Because most other conifers planted in the Aleutians have been nursery stock, it is not clear if their failure to survive is due to poor adaptation to the Aleutian environment (lack of salt tolerance, for example) or to poor survival of nursery stock, poor transplanting techniques, and inadequate subsequent care. In such an adverse site, survival for 20 years or more does not guarantee long-term success. For example, the 45-year-old Adak National Forest still may be toppled by a winter gale.

In the Aleutians, the growing season is later than at mainland nurseries, so refrigerated nursery stock should be acclimatized for two or more years in lath houses or protected transplant beds. After that, outplantings should be protected against wind, both by choice of planting site and by windscreens. Some sheltered sites may have excessively wet soils that require drainage ditches. Best success should result from plantings designed as shelterbelts with seven or more rows of trees at close spacing (1.0 to 1.5 meters). Before height growth starts, annual or periodic weeding may be needed, along with judicious thinning. Ornamental plantings are easier than shelterbelts; they require loamy soil and good drainage, avoidance of excessive fertilizer and other soil amendments, and protection against wind and trampling. The more intensive these measures, the greater the possibility of successful tree growth of species other than Sitka spruce.

Some people object to tree planting in the Aleutians because it introduces species that are not native. But trees will not overrun the islands like European dandelions invade American lawns. On most islands, tree survival is marginal. Aside from the potential of damage by introduced insects and disease, the major enemies are indifference, neglect, and people armed with axes, shovels, or off-road vehicles.

ACKNOWLEDGEMENTS

For 15 weeks in the summer of 1944, I was assigned to a tree-planting program of the

U. S. Army, Alaskan Department, on the Aleutian Islands. During this time I visited nine Army posts between Seward and Attu that were to be involved in the program. I also drafted orders about the tree planting and prepared reports. In 1964, I became involved in some publicity about the Adak National Forest, one of the few accomplishments of the program. In 1984, accompanied by Dr. Leslie A. Viereck, Institute of Northern Forestry, USDA Forest Service, Fairbanks, I revisited Adak at the invitation of Walter R. Briggs, Staff Forester of the Naval Facilities Engineering Command, Department of the Navy, Silverdale, Washington. In 1985, on the way home from tree planting on Adak, I went to Shemya by invitation of the commanding officer, Colonel Tom T. Walker of the U. S. Air Force, and saw trees that were more than 40 years old. In both 1985 and 1986, I was invited to Unalaska, and in 1986 to Kodiak, to help in tree planting sponsored by Frank H. Lockyear, head of Re-Tree International, Wilsonville, Oregon. I was accompanied on those trips by Dr. John N. Alden, Institute of Northern Forestry, USDA Forest Service, Fairbanks, and the late Richard W. Tindall, formerly District Manager, USDI Bureau of Land Management, Anchorage.

Dr. Harold J. Lutz described the history of Russian tree planting near Unalaska Bay in the early 1800s (Lutz, 1963). His report provides much detail I did not repeat. Since 1963, several translations of old Russian documents concerning Alaska have been published, including: "The United States and Russia, The Beginning of Relations, 1765-1815 (Bashkina et al., 1980), several edited by Dr. Richard A. Pierce, University of Alaska, Fairbanks, and those sponsored by the Oregon Historical Society, Portland. I recently have discussed Russian America with Dr. Lydia T. Black, Department of Anthropology, University of Alaska, Fairbanks, and Elizabeth A. P. Crownhart-Vaughan at the Oregon Historical Society; both contributed to my understanding of the history of the Russian American Company. Katherine L. Arndt, University of Alaska, Fairbanks, has read parts of several Russian books for me, searching for references to trees and tree planting, and sending me translations. In this report, I added recently published information to Lutz's references, repeated some things that Lutz reported, and used new evidence in my interpretation of history. I am not surprised that so little can be found about tree planting in the historical record; it probably did not seem important to the early settlers.

My sources for recent history are copies of a file of letters sent to the Alaskan Department, U. S. Army, by the USDA Forest Service, Juneau, in about 1943, and notes, letters, and reports I wrote or received while assigned to the Alaskan Department tree planting project and soon thereafter. In 1964 and 1965, I received some correspondence about the Adak National Forest. Since 1981, I have interviewed several people and collected copies of photographs, maps, reports, and publications. I thank all those I have named and many others who have helped me assemble the information reported here.

REFERENCES

Adams, B., 1903, The Unalaska spruce plantation, Forestry and Irrigation, 19:382-385.

Alden, J., and Bruce, D., 1989, Growth of historical Sitka spruce plantations at Unalaska Bay, Alaska, Portland, Oregon, U. S. Department of Agriculture, Forest Service, Gen. Tech. Rep. PNW-GTR-236, 18 p.

Amundsen, C. C., 1985, Afforestation on Adak Island, Alaska. In: Central Aleutian tundra, Ecological manifestations of maritime tundra landscapes in the central Aleutian Islands (Amchitka, Adak), Alaska, prepared for U. S. Department of Energy under contract No. DOE-A505-76EV04480 ORO-4180-10, Appendix II, p. a11-a17.

Anonymous, 1948, Tree breeding at the Institute of Forest Genetics, U. S. Department of Agriculture, Forest Service, USDA Misc. Publ. No. 659, Washington, DC, 14 p.

Anonymous, 1980, Aleutian trees. In: The Aleutians, Alaska Geographic Quarterly, Alaska Geographic Society, Anchorage, 7(3):67-69.

Anonymous, 1992. In: Yale Forest School News, 78(4):22, New Haven, 28 p.

Bancroft, H. H., 1886, History of Alaska, The Works of Hubert Howe Bancroft, Vol. 33, San Francisco, A. L. Bancroft & Co., 775 p., p. 18.

Bank, T. P., II, Spaulding, A. C., Miller, H. A., and Bank, J. F., 1950, The University of Michigan expedition to the Aleutian Islands, 1948-49, Botanical Gardens and Museum of Anthropology, University of Michigan, Ann Arbor, p. 114-115.

Bank, T. P. II, 1953, Biological succession in the Aleutians, Pacific Science, VII(4):493-503.

Barr, L., Ellis, R. J., and Helle, J. H., 1973, Fossil tree stumps found in situ on submerged ridge at Amchitka Island, Alaska, Fishery Bulletin, 71(4):1,099-1,103.

Barsukov, I. P., 1883, Innokentii, mitropolit Moskovskii i Kolemskii, po ego sochineniiam, pis'mam i razkazam sovremennikov; (Innokentii, Metropolitan of Moscow and Kolomna, in his writings, letters, and the stories of his contemporaries), Synodal'naia tipografiia (Synod Press), Moscow. In Russian.

Barsukov, I. P., 1897, Pis'ma Innokentiia, mitropolita Moskovskago i Kolomenskago, kniga pervaia, 1828-1855, (Letters of Innokentii, Metropolitan of Moscow and Kolomna, Book One, 1828-1855), Synodal'naia tipografiia, (Synod Press) St. Petersburg. In Russian.

Bashkina, N. N., Bolkhovitinov, N. N., Brown, J. H., and others, 1980, The United States and Russia, the beginning of relations, 1765-1815, Documents 134, 191, and 233, Washington, D.C., U. S. Govt. Printing Office, 1,184 p.

Bruce, D. and Court, A., 1945, Trees for the Aleutians, Geog. Rev., 35(3):418-423.

Bryant, C. 1870, Reports of Captain Charles Bryant. In: Letter of the Secretary of the Treasury communicating the reports of Captain Charles Bryant, late special agent of the Treasury Department for Alaska, and H. A. McIntyre, special agent of the Treasury Department, U. S. Cong., 41st, 2d sess., S. Ex. Doc. 32, 24 p.

Cantwell, J. C., 1904, Report of the operations of the U. S. Revenue Steamer Nunivak on the Yukon River Station, Alaska, 1899-1901, U. S. Cong., 58th, 2d sess., Sen. Doc. 155, 325 p.

Cantwell, J. C., 1887, Among the Aleuts, The little settlement of Oonalaska, San Francisco Sunday Chronicle, San Francisco, California, January 16.

Chamisso, Adelbert von, [1856] 1986, The Alaska diary of Adelbert von Chamisso, naturalist on the Kotzebue voyage 1815-1817, translated by Robert Fortwine from: Reise um die Welt, Tagebuch (Voyage Around the World, Diary), Cook Inlet Historical Society, Anchorage, 69 p., p. 44, 46.

Chevigny, Hector, 1967, Russian America, The great Alaskan venture 1741-1867, reprinted 1973, Ballantine Books Inc., New York, 242 p., p. 184.

Critchfield, W. B., and Little, E. L., Jr., 1966, Geographic distribution of the pines of the world, Washington, DC, U. S. Department of Agriculture, Forest Service, Misc. Pub. No. 991, 97 p., p. 5.

Davidson, G., 1868, K. Report of Assistant George Davidson relative to the coast, features, and resources of Alaska territory, Report of the Hon. Hugh McCulloch, Secretary of the Treasury, on Alaska. In: Russian America, U.S. Cong., 40th, 2d sess., H. Ex. Doc. 177 (Part 1), 361 p.

Denfeld, D. C., 1987, The defense of Dutch Harbor, Alaska, from military construction to base cleanup, Alaska District, U. S. Army Corps of Engineers, Anchorage, 1987, 282 p., p. 16.

Fernow, B. E., 1901, Forests of Alaska. In: Alaska, history, geography, resources, Harriman Alaska Expedition, Vol. 2, Doubleday, Page, and Co., New York, 283 p., p. 235-256.

Fisher, O. B., (Fisher, F. B.?) [1841] 1968, About sowing and planting trees on Unalaska Island: O posievie i posadkie derev na ostrovie Unalashkie, Journal of Forestry: Lesnoi Zhurnal, St. Petersburg, No. 8 August 1841, p. [149]-158, translated by A. E. Helmers from typewritten copy on file, USDA Forest Service, Forestry Sciences Laboratory, Juneau, Alaska, 7 p.

Golovin, Captain P. N., [1862] 1979, The end of Russian America, translated by Basil Dmytryshyn and E.A.P. Crownhart-Vaughan, Oregon Historical Society, Portland, 250 p.

Gullickson, J. M., 1965, Trees planted on Adak Island, Alaska, Report on file at USDA Forest Service, Forestry Sciences Laboratory, Juneau, Alaska.

Heusser, C. J., 1973, Postglacial vegetation on Umnak Island, Aleutian Islands, Alaska, Rev. of Paleobotany and Palynology, Elsevier Sci. Pub. Co., Amsterdam, 15:277-285.

Heusser, C. J., 1978, Postglacial vegetation on Adak Island, Aleutian Islands, Alaska, Bulletin of the Torrey Botanical Club, 105(1):18-28.

Hojgaard, A., Johansen, J., and Ødum, S., Eds., 1989, A century of tree planting in the Faroe Islands, Foroya Frothskaparfelag, Torshavn, 187 p.

Hough, F. P., 1878, Report upon forestry, Washington, DC, U .S. Government Printing Office, 650 p.

Hultén, E., 1960, Flora of the Aleutian Islands and westernmost Alaska peninsula, with notes on the flora of the Commander Islands, 2d Edition, revised, New York, New York, Hafner Pub. Co., p. 67.

Blöndal, S., Benedikz, Þ., Ottósson, J., 1986, Forestry in Iceland, Iceland Forest Service, Reykjavik, Iceland.

Jacobson, G., 1975, Adak afforestation project, Scrapbook on file at Adak Museum, Adak Island, Alaska.

Jochelson, W. (Iochel'son, V.), 1925, Archeological investigations in the Aleutian Islands, Washington, DC, Carnegie Institution of Washington, 145 p.

Khlebnikov, K.T., [1835] 1963, Life of Aleksandr Andreevich Baranov, 65 p.: Zhizneopisanie Aleksandra Andreevicha Baranova, St. Petersburg, 1835. In: The earliest history of Alaska, three translations by M. B. Ricks, 1963, Cook Inlet Historical Society, Anchorage.

Khlebnikov, K. T., [1835] 1973, Baranof, Chief Manager of the Russian colonies in America: Zhizneopisanie Aleksandra Andreevicha Baranova glavnovo pravitelya Russikh koloni in Amerikye, St. Petersburg, 1835, Kingston, Ontario, Limestone Press, 140 p.

Kotzebue, O. E., 1821, A voyage of discovery into the South Seas and Bering's Straits, Longman, Hurst, Rees, Orme, and Brown, London, Vol. III, 442 p., p. 287.

Lines, R., 1983, Forestry in the Faroe Islands, Scottish Forestry, 41(20):102-114.

Litke, F., [1835] 1987, A voyage around the world, 1826-1829, Vol. I, Translated from French edition Voyage autour du monde...Traduit du Russe...par...F. Boye (Paris, 1835), by R. Marshall...edited by R.A. Pierce, Limestone Press, Kingston, Ontario, 230 p.

Low, A. J., 1987, Tree planting prospects in Shetland and Orkney, Scottish Forestry, 41(4): 282-296.

Low, A. J., 1986, Tree planting in the Falkland Islands, Forestry, 59(1):59-84.

Lutz, H. J., 1963, History of Sitka spruce planted in 1805 at Unalaska Island, Alaska, by the Russians, Juneau, AK, U. S. Department of Agriculture Forest Service, Northern Forestry Experiment Station, 25 p.

Merritt, M. L., n.d., Notes on the western and northern distribution of tree growth in Alaska, USDA Forest Service, Juneau, mimeograph, 3 p.

Muir, J., 1883, Botanical notes on Alaska, Cruise of the Revenue-Steamer Corwin in Alaska and the N.W. Arctic Ocean in 1883, Notes and memoranda: medical and anthropological, botanical, ornithological, Doc. 429, U. S. Treasury Department, Washington, DC, p. 45-53.

Pierce, R. A., 1984, The Russian-American Company, Correspondence of the governors, communications sent; 1818, Kingston, Ontario, Limestone Press, 194 p., p. 23.

Pierce, R. A., 1990, Russian America: A biographical dictionary, Kingston, Ontario, Limestone Press, 555 p.

Réclus, E., 1890, Nouvelle geographie universelle, la terre et les hommes, Vol. 15, Amerique boreale (New universal geography of the earth and man, Vol. 15, Boreal America), Librairie Hachette et Cie., Paris, 721 p. In French.

Reed, I. M., 1969, Boyhood in Nome gold camp, Fairbanks, Alaska, University of Alaska, Fairbanks.

Russian Ministry of Foreign Affairs, (n.d.), Vneshniaia politika Rosii XIX i nachala XX veka, Dokumenty Rossiikogo Ministerstva Inostrannykh Del: (Foreign policy of Russia of the 19th and beginning of the 20th century, Documents of the Russian Ministry of Foreign Affairs) Vol. I, p. 491-497. In Russian.

Shepard, I. S., 1898, The cruise of the U.S. Steamer "Rush" in Behring Sea, summer of 1889, The Bancroft Co., San Francisco, 257 p.

Stewart, G. G., 1962, Kergord Plantations, Shetland, Forestry, 35(1):35-56.

Tikhmenev, P. A., 1978 [1861-1863], A history of the Russian-American Company, translation of Istoricheskoe obozrenie obrazovaniia Rosiisko-Amerikanskaia kompaniia, St. Petersburg, 1861-63, 2 volumes, translated and edited by R. A. Pierce and A. S. Donnelly, University of Washington Press, Seattle and London.

Tindall, R. W., n.d., The forest at Dutch Harbor, Anchorage, AK, typewritten, 5 p.

Veniaminov, I. E. P. (Innokentii, Metropolitan of Moscow), [1840] 1985, Notes on the Islands of the Unalaska District: Zapiski ob ostrovakh Unalashkinskago otdiela, published under auspices of the Russian American Co. 1840, St. Petersburg, translated by L. T. Black and R. H. Geoghegan, edited and introduction by R. A. Pierce, 1984, Limestone Press, Kingston, Ontario, 511 p.

Viereck, L. A., and Little, E. L., Jr., 1972, Alaska trees and shrubs, Agric. Handb. 410, Washington, DC, U. S. Department of Agriculture, 265 p.

Walker, E. P., 1943, Alaska: America's continental frontier post, Smithsonian Institute, War Background Studies, No. 13, 57 p.

Winchell, M. E., 1951, Home by the Bering Sea, Caldwell, ID, Caxton Printers, Ltd., 226 p.

Wrangell, F. P. von, [1839] 1980, Russian American Statistical and Ethnographic Information, translated from German edition 1839 by Mary Sadouski, 1980, Kingston, Ontario, Limestone Press, 204 p., p. 178-179.

Yudin, G. V., n.d., Documents relative to history of Alaska, Washington DC, Library of Congress (microfilm copy of three boxes, accession number 18292), Box I, Folder II.

AFFORESTATION OF LUTZ SPRUCE, SITKA SPRUCE, AND NORWAY SPRUCE IN THE VESTERAALEN ISLANDS, NORWAY

Nils O. Kaasen and Gisle Skaret

County Forest Service of Nordland
Mosjøen, Norway

Tore Kaasen

SAS Institute
Trollaasen, Norway

INTRODUCTION

Tree species from Alaska and Canada have been used for afforestation in northern Norway since the late 1920s. Afforestation projects were small in the beginning because the performance of foreign species far north of their natural range was unknown.

It has become apparent over time that several tree species, such as blue spruce (Picea pungens Engelm.) from the Rocky Mountains, and white spruce (Picea glauca (Moench) Voss) and eastern hemlock (Tsuga canadensis (L.) Carr.) from eastern Canada, are unsuccessful at high latitudes.

In contrast to species from low latitudes and eastern north America, Sitka spruce (Picea sitchensis (Bong.) Carr.) from Alaska seems well-adapted to the climate of northern Norway. From yields measured in production plots established in the first plantings, we decided in the 1960s that Sitka spruce had a future in the humid coastal regions of northern Norway (Fig. 1). The only evidence of maladaptation and reduced growth was susceptibility to the green spruce aphid (Elatobium abietinum Walker). The green spruce aphid caused mild defoliation of provenances from Petersburg and Kruzof Island, Alaska, indicating that provenances from southeast Alaska might be poorly adapted to northern Norway. Today, plantations are established with seed from Seward, Alaska, which is on the Kenai Peninsula at about 60° N. latitude.

The growth potential of Sitka spruce is suggested by a mean annual increment of 11 cubic meters per hectare for six 55-year-old growth and yield plots established at 68° to 69° N. latitude in the Vesteraalen Islands by the Norwegian Forest Research Institute (NFRI),

Forest Development in Cold Climates, Edited by J. Alden *et al.*
Plenum Press, New York, 1993

Ås-NLH. Extrapolation of these data to 80-year-old plantations indicates that the final mean annual increment will be 13 to 15 cubic meters per hectare, and the standing volume will be about 700 cubic meters per hectare. According to our records, the provenance is Baranof Island, Alaska, at 57° N. latitude. The mean annual increment for 55-year-old Norway spruce (Picea abies (L.) Karst.) in the Vesteraalens is only about seven cubic meters per hectare (Fig. 2).

In 1950, Professor Harold Lutz, Yale University School of Forestry, described a new hybrid from natural crossing between Sitka and white spruce; it was later named Lutz spruce (Picea xlutzii Little) (Little, 1953). At about the same time, two Norwegian professors, H. H. H. Heiberg and T. Ruden of the Agricultural University of Norway, Ås-NLH, observed the natural hybrid on a trip to study the conifers of Alaska and British Columbia and suggested that the hybrid might have a place in Norwegian afforestation, particularly in northern Norway (Heiberg, 1952).

Lutz spruce was introduced some years later, although we recently observed it and not Sitka spruce in a 1935 research plot at Gravdal, latitude 68° N., in the Lofoten Islands. Old plantation records indicated that the provenance was the Kenai Peninsula, Alaska. From this discovery and accurate records filed by those who received plants of the "Kenai" Lutz spruce provenance, we intend to establish additional yield plots in stands from the 1935 plantations. Lutz spruce was unknown in 1935 and therefore the provenance is recorded as Sitka spruce in our records.

Fig. 1. Sitka spruce planted at Brevik, Norway, at 67° N. latitude. The stand is 70 years old and the provenance is Kruzof Island, Alaska, at 57° N. latitude. Epicormic branching is common but the mean annual increment and current annual increment are about ten cubic meters per hectare and 20 cubic meters per hectare, respectively. The Kruzof Island provenance is the poorest of the Sitka provenances planted during the 1920s. Photo by N. O. Kaasen, June 27, 1991.

Fig. 2. Fifty-five-year-old Norway spruce (foreground) and Sitka spruce (background). Plantations with mean annual increments of 6.3 and 12.0 cubic meter per hectare at Ås, 69° N. latitude, in the Vesteraalens. The Sitka spruce provenance is Baranof Island, Alaska, at 57° N. latitude. Photo by N. O. Kaasen, July 2, 1991.

Fig. 3. After clear-cutting Norway spruce, this site near Stokmarknes, Norway, 68°35′ N. latitude, was reforested with Lutz spruce in 1976. Height growth of largest trees averaged about 50 centimeters per year. Photo by N. O. Kaasen, July 3, 1991.

GEOGRAPHY AND SHORT NIGHTS

The Lofoten and Vesteraalen Islands are the northernmost part of Nordland County, Norway. At latitude 68° to 69° N., the Vesteraalens are eight to nine degrees north of provenances from the Kenai Peninsula, Alaska. In previous years, nursery losses were severe due to failure of plants from Alaska to harden under the short nights and long days of summer at high latitudes. Today, nursery plants receive long-night treatments which, if continued for a sufficiently long period in late summer, seem to increase survival. In the field, however, the plants survive under natural daylengths without evidence of frost damage.

FOREST MENSURATION PROJECT

Based on observations of its superior performance, afforestation of Lutz spruce has increased in the years after the first successful plantations (Fig. 3). The decision to change from Norway to Lutz spruce was delayed for fear of maladaptation and possible loss of yields in later years. As a result, the Nordland and Troms County Forestry offices, in cooperation with the NFRI, decided to establish a series of growth and yield studies for Norway, Sitka, Lutz, and Engelmann spruce (Picea engelmannii Engelm.) in the Lofoten and Vesteraalen Islands. Norway spruce is not indigenous to the Vesteraalen Islands but has been used in afforestation there since 1900 (Fig. 4). It must be transferred from its natural range south of 66° N. latitude for afforestation north of the Arctic Circle. Engelmann spruce was later omitted from the study because there were too few stands in which to establish plots.

NFRI will assume part of the project at a later date. How much and which parts will depend upon the results of previous and present investigations. At present, forest mensuration beyond the natural limits of spruce in Norway has not attracted special scientific interest, and it was therefore necessary for the Forestry County Offices of northern Norway to establish a series of long-term growth and yield studies. To provide employment, the

Fig. 4. Norway spruce in a sheltered area at Skjoldehamn, Norway, 69° N. latitude. Mean annual increment is about seven cubic meters per hectare. The stand is about 80 years old and was thinned several times. It was originally planted at a density of 10,000 seedlings per hectare. Photo by N. O. Kaasen, July 2, 1991.

studies were initiated in 1984, a year of severe unemployment in northern Norway. The studies were called the Forest Mensuration Project. Plots established in 1984 were scheduled for measurement at five-year intervals for the duration of the plantations.

A thesis on the growth of Sitka, Lutz, and Norway spruce in the Lofoten and Vesteraalen Islands was the first effort to compare the growth of these species in Norway (Pettersen, 1981). The thesis provided the purpose and background for our investigation. As district forester for the area, Pettersen was available for selecting stands for the research plots in the Forest Mensuration Project.

FIELD WORK

During the summer of 1984, we established, measured, and registered 109 plots. Twenty-six plots were Sitka spruce, 43 were Lutz spruce, 34 were Norway spruce, and six were Engelmann spruce plantations. To avoid logging disturbances, a five-meter buffer zone was established around the plots. Each inner plot was 20 x 20 meters and contained approximately 100 trees for a total of about 11,000 trees in the project. Where plantations of each species already existed on the same site, plots of Sitka, Norway, and Lutz spruce were established adjacent to each other. In plantations dating from the same year, only one plot was established for each species per geographic area.

Based on Pettersen's 1981 thesis, each tree in the 1984 survey was measured for height and diameter at breast height (1.37 meters) outside bark, and for spike knots, crooked stems, top breakage, and other indications of mechanical damage. Damage from frost, grazing, insects, fungi, and snow also was assessed. In addition, we recorded diameter and length of the branches at the first branch whorl above breast height. Maps of the location and number of each tree were prepared to assist in selecting the best parents for tree breeding.

The second survey in the Forest Mensuration Project was completed in 1989. In this survey, only height and diameter were recorded. Differences among species in height and diameter growth are reported at 0.05 or less probability of error.

RESULTS

The oldest and youngest Lutz spruce plots were 22 and 12 years from seed when measured in 1984. Although 10 years is a very short period for regression analysis of growth, we still found that Lutz and Sitka spruce grew faster than Norway spruce in height (Fig. 5a) and diameter (Fig. 5b). As we suspected before the plots were measured, Lutz spruce grew slower than Sitka spruce and faster than Norway spruce in both diameter and height. The height of Lutz spruce was expected to be closer to Sitka spruce than the results show, however. Differences in growth between Sitka and Lutz spruce were not significant at this early stage.

Growth rates in 1989 were the same as in 1984 (Figs. 6a and 6b). Diameter and height growth of Norway spruce were significantly less than Sitka or Lutz spruce at $P \leq 0.05$ of error. For height growth, however, the lower confidence line of Sitka spruce and the upper confidence line of Lutz spruce were very close to the mean height growth of each species (Fig. 6a).

On the basis of the 1989 results and our general impression that Lutz spruce grew as fast as Sitka spruce on productive sites, we analyzed the measurements for both low- and high-quality sites. The sites were classified as low or high quality on the basis of growth rate for each tree species. On high-quality sites, each tree species exceeded its overall mean annual growth rate, but grew less than average on low-quality sites.

Linear regression analyses for the high-quality sites showed that both the mean annual height and diameter growth of Lutz spruce were inside the confidence limits for height and diameter growth of Sitka spruce (Figs. 6c and 6d). Mean annual height and diameter growth of Norway spruce, however, was clearly inferior to that of Sitka and Lutz spruce.

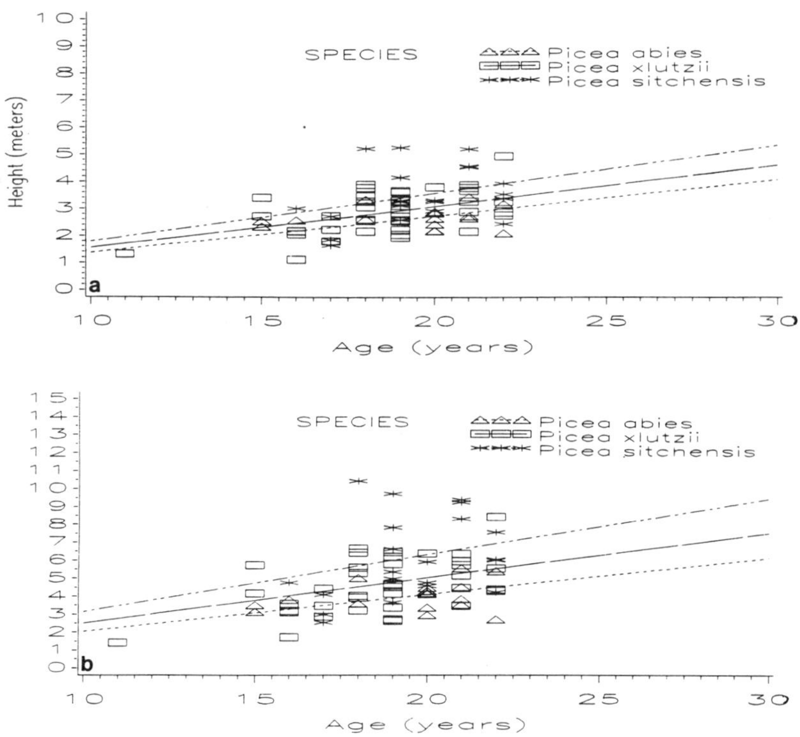

Fig. 5. Nordland Forest Mensuration Project. Plots were <23 years old in 1984. (a) Rate of height growth for all sites in 1984; (b) rate of diameter growth for all sites in 1984.

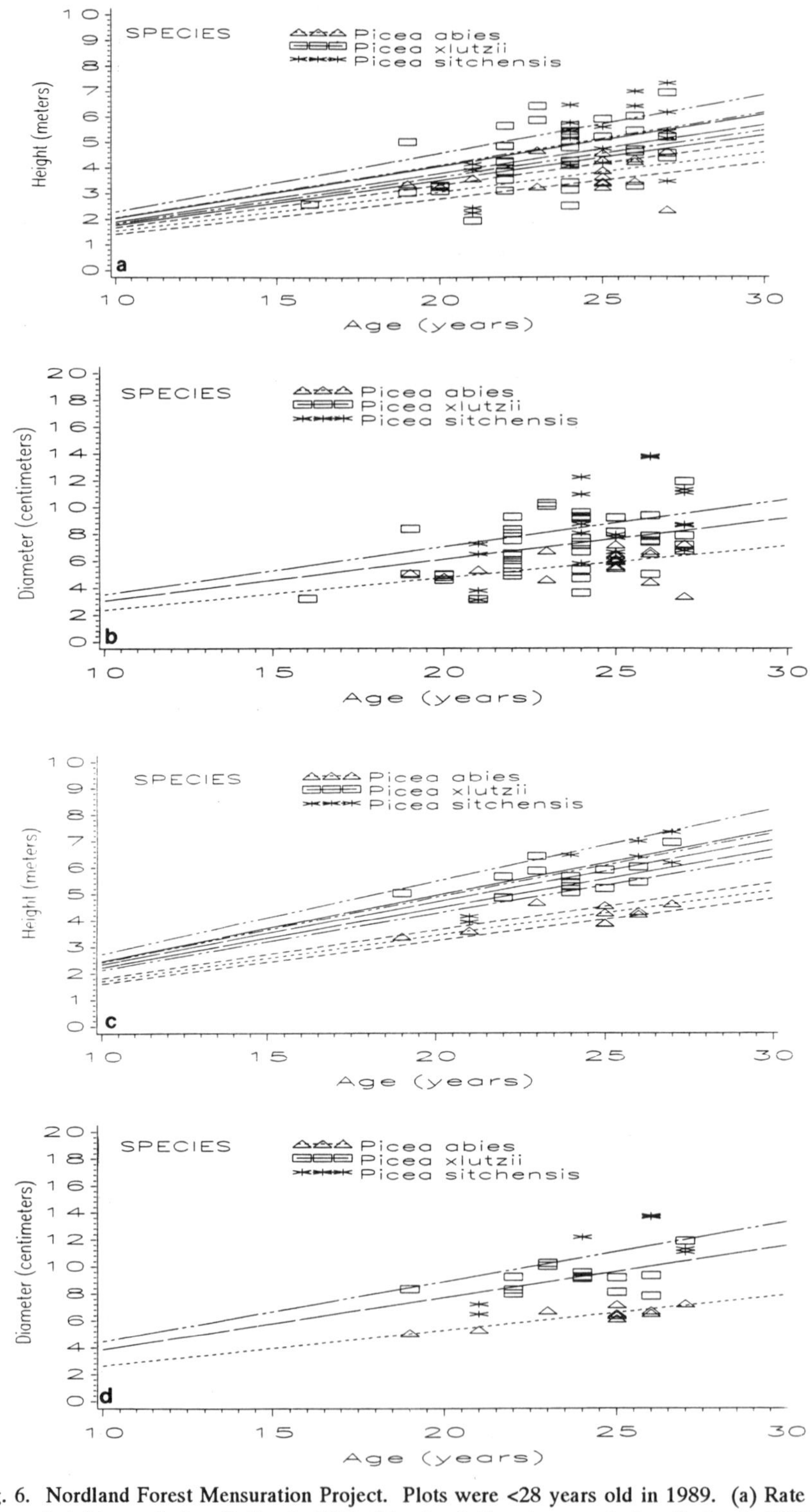

Fig. 6. Nordland Forest Mensuration Project. Plots were <28 years old in 1989. (a) Rate of height growth for all sites in 1989; (b) rate of diameter growth for all sites in 1989; (c) rate of height growth for high-quality sites in 1989; (d) rate of diameter growth for high-quality sites in 1989.

Differences among species in height and diameter growth on low-quality sites were the same as for high-quality sites (Figs. 7a and 7b). Growth rates for Lutz spruce were not statistically different from Sitka spruce, and growth of Norway spruce was statistically inferior to the growth of Sitka and Lutz spruce. Differences among the species were, however, less pronounced on low-quality sites than on high-quality sites.

A diameter-over-height analysis showed that the form of Norway spruce was superior to the form of Lutz and Sitka spruce. Differences in form among the tree species are similar to the differences in height growth on high- and low-quality sites.

DISCUSSION

Confidence intervals constructed from plantation data in the Vesteraalen and Lofoten Islands in northern Norway showed that Sitka spruce has wide variation in growth, whereas Norway and Lutz spruce demonstrate much less variation in growth rates among sites. Provenances for Sitka spruces ranged from Kruzof Island at 56° N. latitude in southeast Alaska to 61° N. latitude in Prince William Sound and on the Kenai Peninsula. Lutz spruce provenances, on the other hand, were confined to the Kenai Peninsula, and Norway spruce provenances were from 65° to 66° N. latitude in Norway. The wide variation in growth rates among provenances of Sitka spruce indicates that potential gains from provenance research in the Vesteraalen Islands and neighboring maritime regions are large.

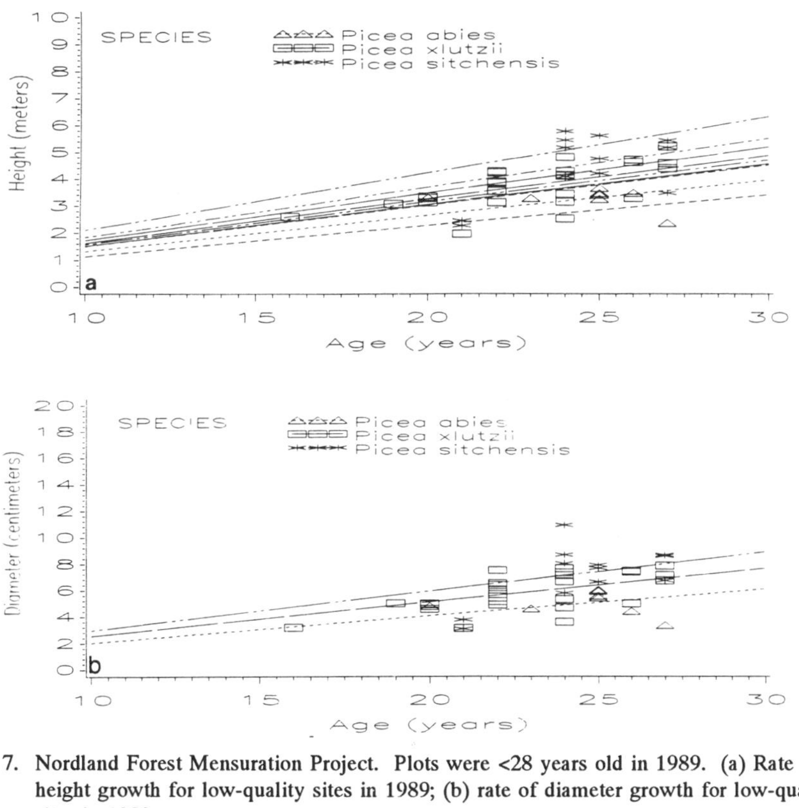

Fig. 7. Nordland Forest Mensuration Project. Plots were <28 years old in 1989. (a) Rate of height growth for low-quality sites in 1989; (b) rate of diameter growth for low-quality sites in 1989.

Fig. 8. Afforestation of Norway spruce at altitudinal tree limits (150 meters) on Amdoya Island, 69°05′ N. latitude. Wind and drifting snow cause needle loss. Lutz spruce is more hardy and sustains less damage than Norway spruce at this altitude. Note the typical chlorotic (yellow) condition of Norway spruce, known locally as "yellow spruce" in cold maritime climates. Photo by N. O. Kaasen, July 2, 1991.

Results of our survey indicate that Sitka spruce will grow more wood than Norway spruce and probably more wood than Lutz spruce in the cold maritime climate of northern Norway. These conclusions are partly confirmed by NFRI's oldest production plots in Sitka and Norway spruce plantations in the Vesteraalens (Bauger, 1970). The large branch diameter, stem taper, and epicormic branching of Sitka spruce (Herman, 1964), however, may detract from large volume production. The diameter and height growth of Lutz spruce are less than the diameter and height growth of Sitka spruce, and more than the diameter and height growth of Norway spruce at the same age. Lutz spruce may be superior to Norway and Sitka spruce in maritime-continental climates that experience frequent early frost. Bergan (1970) suggests that Sitka and Lutz spruce are preferable to Norway spruce in windy, maritime areas (Fig. 8). He also indicates that Lutz spruce should be preferred to Sitka spruce in the northern part of north Norway.

Heiberg (1952) wrote that differences in growth are not apparent between pure Sitka spruce and hybrid trees that are morphologically similar to Sitka spruce, and between pure white spruce and hybrids similar to white spruce. These relationships were not investigated in this project but could result in an interesting and rewarding study for northern Norway. Little (1953) noted that intermediate F_1 hybrids did not express heterosis for growth.

In 1983, the year before the Forest Mensuration Project, the Nordland County Forestry Office decided to increase afforestation of Lutz spruce in the Vesteraalen and Lofoten Islands. Results of the 1984 and 1989 surveys support this decision. In connection with the Forest Mensuration Project, the effects of stand density on limb size, natural pruning, and epicormic branching need to be investigated for each species. These traits determine wood quality and economic value. The Forest Mensuration Project will not be complete until the effects of stand density are analyzed and stocking guidelines are prepared for optimum growth and wood quality on high- and low-quality sites.

Next, it would be interesting to determine if each species requires different soil and vegetation conditions and how soil and vegetation influence tree performance. But the most interesting and rewarding results stem from the application of genetic principles to improve adaptation and growth of spruce beyond its natural range in northern Norway.

ARTIFICIAL CROSSING OF SITKA AND WHITE SPRUCE

The goal of the Forest Mensuration Project was to compare growth of the major maritime spruce species in the Vesteraalen Islands. But we soon found that detailed

Fig. 9. Paper pollination bags enclose the strobili (seed cones) of Sitka spruce at Stokmarnes, Norway, 68°35′ N. latitude. The pollination bags isolate the strobili from foreign pollen of Sitka and Norway spruce. When fully receptive, in early June 1991, the strobili were pollinated with white spruce from interior Alaska to produce the first F_1 hybrids of Sitka and white spruce (Lutz spruce) in northern Norway. Photo by N. O. Kaasen, July 3, 1991.

observations on growth, form, and vigor of every tree entered into the computer system could be used to select superior phenotypes for intra- and inter-specific breeding. For example, the best tree in every 10 Sitka spruce could be easily identified in each of the 26 Sitka spruce plots to provide a breeding population of about 250 parents.

When abundant flowering was observed for Sitka and Lutz spruce in the spring of 1985, we decided to initiate an applied improvement program in cooperation with the NFRI. Because white spruce pollen was unavailable, superior Sitka spruce phenotypes were crossed with superior Lutz spruce phenotypes. Seeds of the crosses were collected and planted with control material in research plots to estimate additive and nonadditive genetic variances, heritabilities, and gains from selection of growth traits and to provide populations for selecting parents of the next generation.

At the same time the crosses were made, scions were taken from the parents and grafted onto rootstock. The grafts were planted in clone banks that will probably be converted or transplanted into seed orchards.

In 1985, the quality of white spruce in northern Norway was not satisfactory for tree breeding and pollen could not be obtained for crossing experiments. We were aware that pollen could be collected from white spruce in Alaska but were not sure how to collect it. In 1987, the Nordic Forest Research Council granted funds to collect seeds from tree-limit species and sites in western North America. The seed was for provenance trials in marginal climates in the Nordic countries. When establishing the seed collection program, means for pollen collection from white spruce were also included. White spruce finally produced an abundant pollen crop in 1990, and several liters of pollen were collected from about 20 stands in interior Alaska.

We were eager to have abundant flowering in Sitka spruce as soon as possible to make sure that the pollen would not deteriorate. Fortunately, the weather in the Vesteraalen Islands initiated flower buds on Sitka spruce in 1990, and Professor Jon Dietrichson of the Agricultural University, Ås-NLH, Norway, and Gisle Skaret, of the County Forest Service of Nordland, Mosjøen, Norway, carried out a factorial breeding experiment with 32 parents of each species in early June 1991 (Fig. 9). It was the first artificial crossing between Sitka spruce and white spruce in Norway.

If the crossing experiment is successful, the first plants of artificial Lutz spruce will be outplanted in field trials before the 1994 Nordic Forestry Congress meeting is held in Norway. Pollen of white spruce from interior Alaska is expected to impart cold, drought, and early frost resistance. Sitka spruce, on the other hand, is expected to impart rapid growth and large stem volume. The crossing experiment and field trials are designed to reveal additive, heterotic, and epistatic gene effects in the hybrids.

REFERENCES

Bauger, E., 1970, Sammenligning mellom sitkagranens og granens høydeutvikling på Vestlandet og i Nord-Norge (Comparison between height development of Sitka spruce and Norway spruce in West Norway and North Norway), Reports of the Forest Research Institute of West Norway, Nr. 50, Vol. 14, 149-221.

Bergan, J., 1970, Skogreising i Nord-Norge (Afforestation in North Norway), Tiddskrift for Skogbruk (The Norwegian Journal of Forestry), No. 78, 251-261.

Heiberg, H. H. H., 1952, Inntrykk fra Nord-Amerikas skogbruk (Impressions of forestry in North America), Tiddskrift for skogbruk, bilag, utgitt ved utvalget for faglig hjelp i landbruket, Faglig hjelp under Marshall-planen, 21 p.

Herman, F. R., 1964, Epicormic branching of Sitka spruce, Research Paper PNW-18, Portland, Oregon, U. S. Department of Agriculture, Forest Service, Pacific Northwest Forest and Range Experiment Station, 18 p.

Little, E. L., Jr., 1953, A natural hybrid spruce in Alaska, Journal of Forestry, Vol. 51, 745-747.

Pettersen, G., 1981, En vekstundersøkelse av aktuelle bartreslag ved skogreising i Lofoten og Vesterålen (An investigation into growth of actual conifers in afforestation in Lofoten and Vesteraalen Islands), Agricultural University of Norway, Ås-NLH, Norway, 73 p. M.S. thesis.

DEVELOPMENT OF AN ENVIRONMENTAL FOREST AND ARBORETUM ON THE OUTER COAST OF FINNMARK COUNTY, NORWAY

Skjalg Slettjord

Forsheim Research Station
Klubbvik, Norway

INTRODUCTION

During the autumn of 1977, the Vardø City Council in Svartnes Bay (70°22′ N. lat., 31°05′ E. long.) in eastern Finnmark County, Norway, developed a plan for establishing a Nature Conservation Center in the vicinity of an undersea tunnel connecting Vardø Island to the mainland (Fig. 1). The objectives of the plan were to improve the local environment, provide amenities to homes and communities, increase local employment, and sustain development on the coast of Finnmark.

To accomplish these objectives, the City Council formed a city planning group from members of the City Council, the Gardening Society and the Agricultural Society of Finnmark and invited the Forsheim Research Station of Troms County to assist in adapting plant materials to the coast of Finnmark.

Based on knowledge of the climate, vegetation and soil on the coast of Finnmark, and availability of hardy plant material collected from field excursions to other cold-climate regions of the world, the City Planning Committee recommended that the Center include research on conservation of natural resources at Svartnes Bay. The aim of the Nature Conservation Center is to conduct research on the local vegetation and environment, and to educate the public on the conservation of local natural resources.

THE SVARTNES BAY RESEARCH AREA

The collaboration established between city planners and researchers during development of the Nature Conservation Center provides an example for others to follow in developing forests in areas with cold climates.

A climate protection zone of 300 hectares will surround the Center. The protection zone will provide sufficient experimental area for developing and testing an environmental forest concept. It will include wind shields and snowdrift collectors to protect an arctic

Forest Development in Cold Climates, Edited by J. Alden *et al.*
Plenum Press, New York, 1993

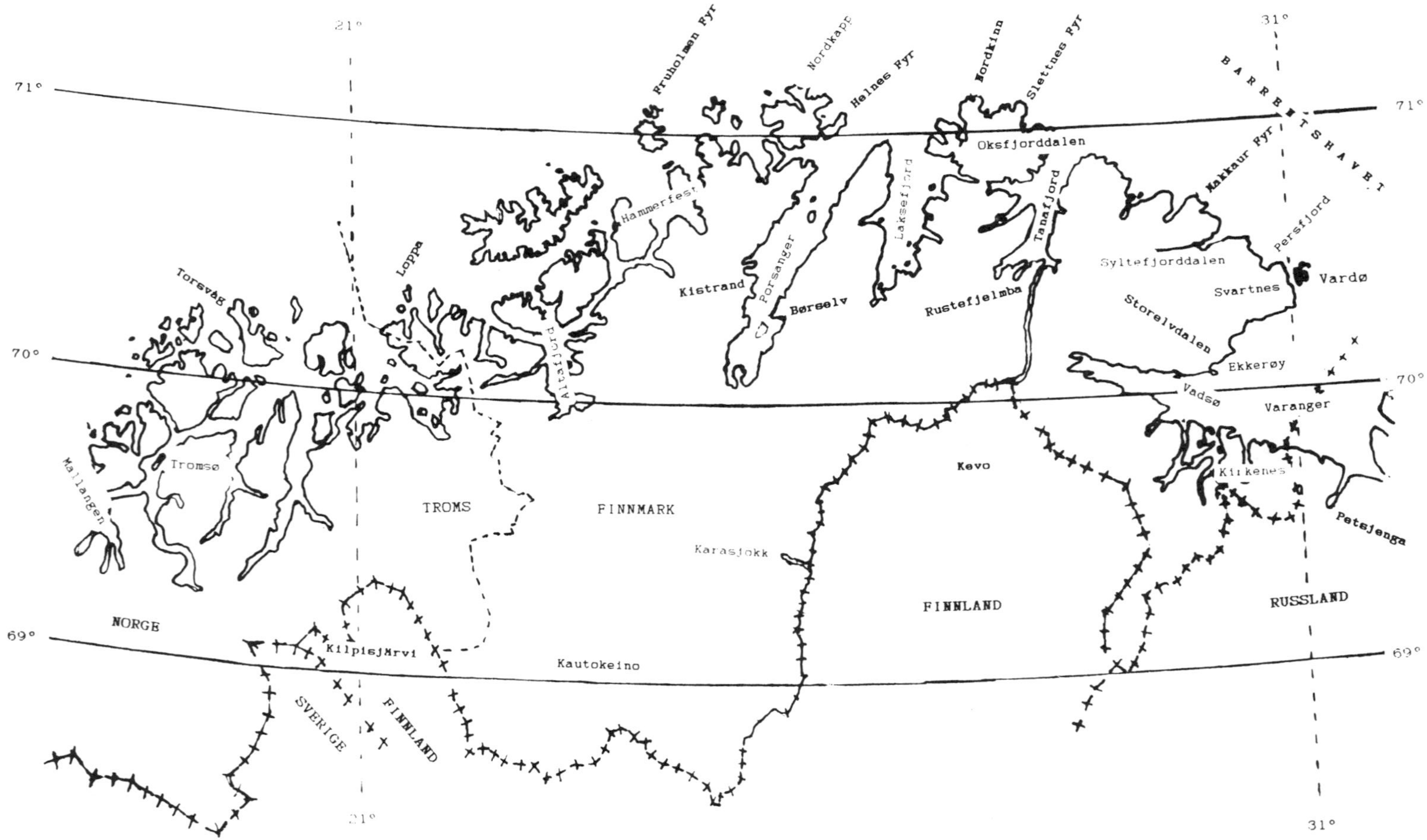

Fig. 1. The coast of Troms and Finnmark Counties, Norway.

arboretum, tree nursery, and plant breeding facilities. Buildings, for example, will be constructed in a pattern that will moderate the adverse effects of wind. The entire Center and wind protection zone will be fenced to eliminate grazing by reindeer and sheep. The climate-protection zone will be divided into several small areas for afforestation experiments such as the "Krokenfeltet," a 5-hectare area that has been fenced, drained, and planted in species trials.

The climate and soil of Svartnes Bay are representative of most coastal areas of Finnmark County. It is windswept and cold, but lies within the Nordic Birch Forest Zone. The Nordic Birch Forest Zone encompasses the current and potential distribution of mountain birch (Betula pubescens Ehrh. ssp. tortuosa (Ledeb.) Nyman) in Fennoscandia, Iceland, and Greenland. Phytogeographically, the Nordic Birch Zone links these countries ecologically and is the basis for the specialized bio-ecological research institutions located at northern tree limits in each country. Therefore, experimental results from the Svartnes area have wide application throughout the birch forest zone, and can be safely used anywhere human habitation is possible on the coast of Finnmark.

Never before has Finnmark County had a forest research station to meet the needs of its citizens. All other Norwegian research institutions are located in areas where the climate is much warmer than on the outer coast of Finnmark. For afforestation to succeed under the the extreme climatic conditions of the outer coast, both improved genetic plant material and new technology are necessary. An institution such as the Nature Resource Center, with its nursery and breeding arboretum, is therefore necessary to produce seed and seedlings adapted to the outer coast. We conclude that the people living in this northern area never have had the same opportunities for service and information from the public research institutions as the people in the more central parts of the country.

The Nature Conservation Center is managed by the Vardø City Council, the shelter planting consultant, and the author. Under the auspices of the Vardøhus Museum, the project has gradually grown with the establishment of a mountain birch provenance study and species trials. For it to continue, the nursery, buildings, arboretum, and other new facilities must be completed.

EARLY SETTLEMENT ON THE COAST OF FINNMARK

A general review of the settlement period on the coast of Fennocandia and vegetation change since the last glaciation is needed to assess the feasibility of environmental afforestation and an arctic arboretum at Svartnes Bay. Information about the climate and vegetation during the settlement period is available from Quaternary geology, palynology, archaeological records, and local history.

A map from Einar Niemi's book "Vadsøs Historie" (1983) shows the extent and retreat of the ice after the Würm Ice Age, and probable immigration routes of the first settlers to the coast of Finnmark in the Older Stone Age (Fig. 2). The discovery of early Stone Age settlements on the northern coasts is documented by Professor Povl Simonsen (1959). The settlements are from the early Komsa Period (7500 to 7000 B.C.) of the Older Stone Age and are among the oldest in Norway. The Komsa Period extended to 3000 B.C. (the end of Older Stone Age) and included several additional cultures that were followed by four cultures in the Younger Stone Age extending to the time of Christ. The settlements in those days were very scattered and had only a few hundred inhabitants each.

The Samish Iron Age, from the time of Christ to 1500 A.D., includes the Viking Period, from the 9th to the 11th Centuries, and the Middle Ages, from 500 to 1450 A.D.

About 890 A.D., the north Norwegian Earl Ottar of Malangen near Tromsø, Troms County, journeyed to England, where he told King Alfred about north Norway and his travels along the coasts of Finnmark and the Kola Peninsula to the White Sea (Fig. 3). This is the first written account of settlement in northern Norway and a key source of knowledge about the northernmost part of Norway in ancient times. It indicates that Malangen was the northernmost Norwegian settlement; land even further north was uninhabited except for a few localities where the "Finns" (Old Norwegian for "Samish") had settled during the early Stone age. The Samish hunted in winter and fished at sea in summer.

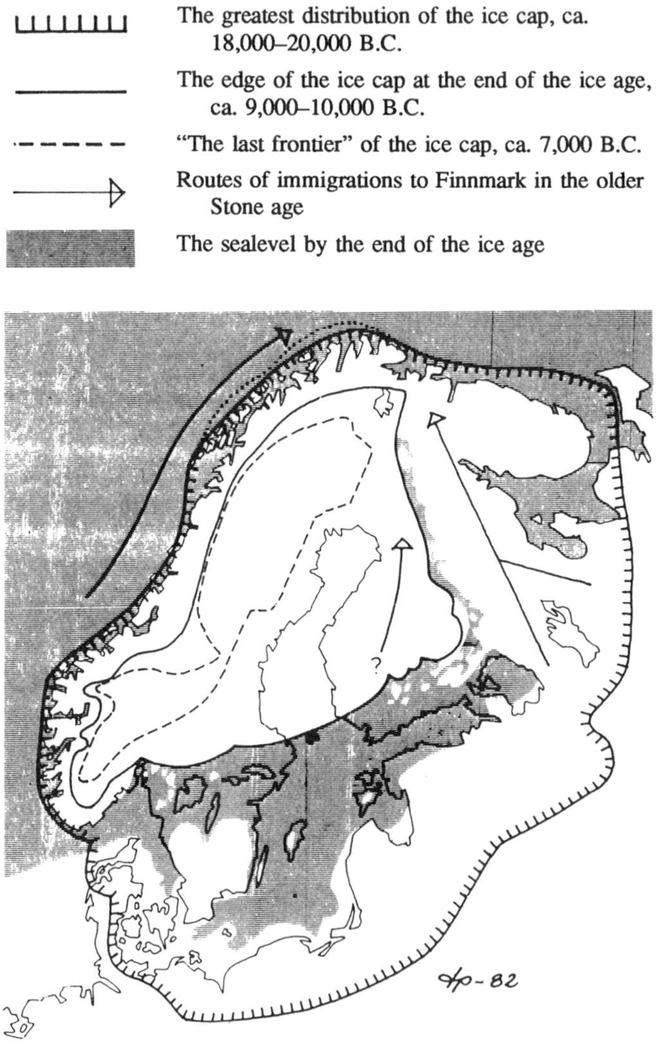

Fig. 2. Phases in the retreat of the Arctic ice cap at the conclusion of the last Ice Age (Niemi, 1983).

Finnmark has always been rich in resources of great importance to Norway. Exploitation of the natural resources of Finnmark started during the Viking period, when Ottar and his Viking successors occupied this large area and imposed taxes, known as "finneskatten," on the Samish population, and continued after the 10th Century, when the finneskatten was taken over by the Norwegian king. The valuable furs, by which the Samish paid the tax, initiated trade between Europe and Norwegian settlements on the north coast (Balsvik, 1989).

During the remainder of the Middle Ages, markets for stockfish grew as Catholicism spread throughout Europe and the population increased. The stockfish trade was probably the most important reason for Norwegian colonization of the Samish area. The fishing station settlements that suddenly appeared along the coast of Finnmark and the Kola Peninsula from 1250 to 1300 A.D. provided the stockfish for trading centers such as Bergen, on the west coast of Norway.

For public safety and to secure hegemony over the Finnmark and Kola districts, a church and fort were built at Vardø early in the 14th Century. Vardø thus became the center of administration and defense for the entire northern coast of Europe, which was known as "Vardøhus Len."

Cultural activities grew rapidly from 1400 to 1550 A.D. The population of Vardø in 1500 A.D. was about 300. For comparison, Oslo, the capital of Norway, had about 1,800 inhabitants in 1575. Both the fishstock trade and the population declined from 1600 to 1700 A. D., stabilized in the 18th Century, and grew again in the 19th Century. In the year 1801, the total number of inhabitants of all settlements on the coast of Finnmark was close to 8,000. As a result of new technology in trade and community development during the last 150 years, the population of Finnmark County has grown to 70,000.

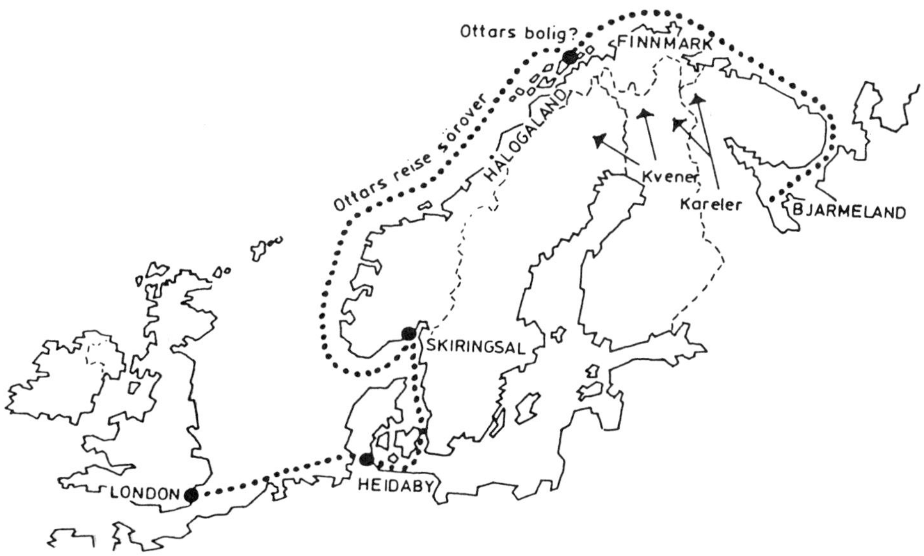

Ottars journeys to The White Sea and London, ca 890.
(Drawing: Tor Poppe)

Fig. 3. The journey of the Norwegian Earl Ottar of Malangen, who provided the first accounts of settlement on the coast of Finnmark County, Norway, in 890 A.D. (Niemi, 1983).

VEGETATION DEVELOPMENT

Recent pollen analyses during tree-limit investigations at Domsvatnet in Vardø showed that the Holocene climate was highly variable, and that light-demanding trees and shrubs such as mountain birch migrated to the outermost coast of Finnmark in the Komsa Period from 7500 to 3000 B.C. (Hyvärinen, 1976).

During the Younger Stone Age, the climate cooled again, and mountain birch retreated to its present range. But birch was not totally eliminated from its former range. Several vigorous forests have withstood both grazing and wood harvest since settlement and can still be found on the outer coast of Finnmark, a marginal area even for the most hardy species. Because the climate of the outermost coast is severe, the native vegetation is less productive and has fewer species than further inland.

In many ways, the outer coast flora resembles the alpine flora of the inland mountains. The present coastal vegetation results from a very long evolutionary response to many different factors, both natural—including climate, soils, other organisms, and the genetic quality of the flora—and cultural, and their interactions. The cultural factors include anthropogenic activities such as agroforestry, buildings and road construction, grazing, wood harvest, and so forth.

With alteration of one or more natural or cultural factors, the vegetation changes physiologically and genetically until a new balance is restored between it and the environment. The effects of negative cultural factors, such as overgrazing and excessive harvest, on the composition and vigor of the native flora are more serious in marginal than in central environments. Over-grazing and excessive clear cutting have permanently damaged the ecosystems of the outermost coast of Finnmark more than the ecosystems further inland.

As mentioned previously, northern settlements were widely scattered during the Younger Stone Age and the Samish Iron Age of the Viking period (3000 B.C. to 1200 A.D.). Human activities during this period had little effect on the vegetation, but after the 14th Century, the population increased very rapidly in the outer coast where the vegetation was most sensitive to grazing and clear cutting. Increased human pressure might explain why birch forests are not as extensive on the coasts as they are inland.

The extreme and harsh climate is the simplest explanation for the lack of forests on the coast of Finnmark. But the relatively strong impact of human activities must be a major factor in the deforestation of the coast as well. Well-established forest vegetation modifies the climate both within the forest and in the surrounding areas. Positive climatic effects of a fully-stocked forest will preserve both the forest and the most vulnerable species during periods of unfavorable climate.

As populations grew in the coastal area, the pattern of vegetation, which evolved from several thousand years of migration and selection, changed to the broken distribution of the present day. Local names in Persfjord, near Fort Vardøhus, 20 to 25 km from Vardø, indicate that introduced grasses were grown in an extensive woodland area where horses were first introduced in Finnmark. I have heard many old stories from Vardø about how the forests on both sides of Varangerfjorden were clear cut by soldiers and slaves from the fort. Fishermen from Troms County, in need of fuel, also clear cut woodlands along the coast of Finnmark when they returned from fishing.

Other evidence of widespread forests during the Komsa Period are large roots and trunks preserved in the bogs along the outer coast. According to Håkon Rasmussen, a fisherman from Smelror, local bog peat was not used for fuel, because large roots and trunks were so numerous that the peat fell to pieces when it dried. The Smelror bog, about 40 meters above sea level at Svartnes Bay, dates the forests to the early Komsa Period, about 7000 B.C.

In 1948, I found eight-centimeter diameter birch trunks at a depth of 30 to 40 centimeters in another bog at Svartnes Bay at 12 to 15 meters above sea level. The height above sea level indicates that the birch may date from the Younger Stone Age, 1500 to 1000 B.C.

On the basis of these observations, it appears that birch forests were widely distributed over most of the coastal lowlands of Finnmark during the Komsa Period. Improper farm management, agroforestry, and the onset of the little Ice Age in the 13th Century forced the birch forests to retreat from most of the marginal areas on the coast to present tree limits in the fjords. Here the human impact has been in balance with the natural environment since arrival of the first settlers.

As the coast was deforested by grazing and clear-cutting, wind protection was lost and the local climate and soil deteriorated. In addition, the climatic margins are so narrow that even small changes in the local environment are catastrophic for the most sensitive vegetation. The deterioration of the inland climate in Røros, at 62°30′ N. latitude in southern Norway, and loss of soil and vegetation during the 17th Century is a classic example of the adverse effects of deforestation.

Forest vegetation would probably recover from the impact of development throughout the coastal area of Finnmark, if negative human activities such as over-grazing and excessive clear cutting could be reduced for a very long time. For example, after electricity and oil replaced wood and bog as energy sources following World War II, birch became well-established in several new areas sheltered by <u>Salix</u> species along the fjord in the Vadsø-Krampenes commune of eastern Finnmark.

Not all cultural practices have negative effects on the vegetation and landscape, however. The success of beneficial plants depends on climate, soil, the associated biota, and their interactions. These factors can be manipulated by providing shelter from wind, irrigation during drought, draining wet soils, and fertilizing nutrient-poor soils. Plant materials can be adapted to the harsh environment of the north coast by recurrent plant selection and breeding.

Only those who live in this marginal and barren land know how much well-adapted plant materials mean for the well being of its citizens. The desire of people living beyond natural tree limits to grow flowers, shrubs, and trees, or vegetables is usually far more intense than the desire of those living on optimum sites within tree limits.

But factors that limit tree growth are not as restrictive as we might think. With new technology and adapted plant materials leading to afforestation success in marginal habitats, productive forests may be advanced well beyond present limits.

ENVIRONMENTAL FORESTS

Environmental forests are multi-purpose forest ecosystems managed for the well being of people. In contrast to traditional forestry, environmental forests are managed for all resources, including water, soil, recreation, esthetic values, and associated flora and fauna, as well as wood products. They provide shelter from the wind and remove carbon dioxide from the atmosphere. They can be developed by long-term culture of the natural and introduced vegetation as production forests are harvested for timber. Environmental forests also protect lakes and rivers from airborne and waterborne by products of industry and agriculture. Airborne industrial pollutants precipitate or are deposited directly on the ground and on vegetation in rain and snow. At low concentrations, many of the dissolved substances are plant nutrients, but at high concentrations, they are toxic to flora and fauna alike.

Environmental forests are probably the least expensive and most efficient method of removing low concentrations of nutrients from the air and water. Through photosynthesis, green plants transform carbon dioxide and water into organic carbon and oxygen. Other nutrients are absorbed through root systems and accumulated in the biomass. Trees with a deep root system, high transpiration and water uptake rates, and high pollution tolerance are superior to many other plants in absorbing and transforming nutrients in the runoff from farms, industry, and sewage-treatment plants, along with atmospheric carbon, into valuable lumber, and at the same time returning oxygen to the atmosphere.

In the barrens along the Arctic coast and in the coastal mountains of Finnmark Country, as well as in more climatically favorable areas, environmental forests are beneficial for their wood products and for the amenity they add to of homes and villages.

Environmental forests were first established at the Vardøhus Museum by the Nordic Arboretum Committee as part of the inter-Nordic project "Fjellbjørkeskogens Økosystem" and "Arktisk Arboretum."

Climate Protection

Climate is the average sum of the weather conditions for a region over a period of years. Weather, on the other hand, is the condition of the atmosphere with respect to temperature, humidity, wind speed, and barometric pressure at a given time and location. Weather and climate are determined by heat, radiation, precipitation, and wind. Each of these factors can be classified individually as heat climate, radiation climate, and so on. Climatic protection is, therefore, protection against one or more of these factors.

By means of houses, clothing, heating, and cooling, people have learned to regulate their microclimate, and by means of environmental forests, the local climate of their surroundings, especially wind. Wind increases evaporation and surface cooling. As shelterbelts, environmental forests reduce wind speeds, thus preventing damage to plants, homes, and villages; reduce snow drifting; and moderate chill factors, which in turn reduces heating costs. Artificial windbreaks provide shelter, but plantation forests are more efficient and less expensive.

Environmental forests are necessary to restore the local climate and soils on the coast and in the mountains of Finnmark. Both climate and soils have deteriorated from many centuries of intense and uncontrolled cutting and grazing. The situation is similar in Iceland, where over-grazing during the 19th and 20th Centuries destroyed large areas of natural birch woodlands and caused severe soil erosion and deterioration of the climate.

Once established in the northern areas, environmental forests will increase the well being of people by protecting soil, cultivated plants, and production forests from the wind.

THE CLIMATE AND SOILS OF COASTAL FINNMARK

Climate

With an area of about 42,000 square kilometers, Finnmark is the largest county of Norway. This coastal district lies between 70° and 71° N. latitude and between 21° and 31° E. longitude. The outermost coastline, which extends from Troms County in the southwest to the Russian border in the east, is about 500 kilometers. Of this distance, 477 kilometers lie north of 70 degrees.

The climate varies with topography and wind exposure, and is colder in the eastern part of the county than in the western. The following climatic data, including temperature, precipitation, and wind, are from meteorological observations at Vardø, Ekkerøy, and Kirkenes, all on the outer coast of eastern Finnmark (Fig. 1). Tetraterm temperature, the mean temperature of the four warmest months, was calculated for the June-to-September growing season from 1930 to 1960 (Table 1). Table 2 shows the mean monthly and annual precipitation for the same period, Table 3 gives the average dates when mean daily temperatures exceeded threshold values, and Table 4 shows the average length of the four seasons at each location. For wind and fog data (Table 5), Kirkenes was replaced by Hammerfest.

Used with caution, the tetraterm temperature is a reliable estimation of the growth potential for most tree-limit species in cold climates. Minimum tetraterm temperatures for completing the annual growth and development processes of potential tree species for introduction to the outer coast of Finnmark are given in Table 6.

Soils

Soils that developed from glacial deposits in the coastal areas of Finnmark can be divided into three general classes: "Koppmoll," bog, and aerial (eolian).

Koppmoll is washed moraine deposits from old beach ridges. Water easily penetrates the uppermost layer of Koppmoll deposits, but underlying fine-grained materials block the downward transport of water. The poor drainage maintains high water tables and forms bogs, especially between the ridges of old deposits.

Koppmoll is very common in the coastal areas of Finnmark. It is nutrient-poor, not suitable for higher forms of vegetation, and subject to severe erosion. Wind and grazing may create large areas of Koppmoll that support only crust lichens, or a sparse cover of *Empetrum* L. and a few grass species. On the other hand, I have observed large areas of mountain birch on Koppmoll deposits in eastern and western Finnmark.

Bog is the other dominant soil type in Finnmark. Bogs are found in areas where excess ground water and cool temperatures inhibit decomposition and mineralization of dead plants. As a result, soils are very acidic and nutrient-poor. The quality of bogs in Finnmark varies from those that support lush sedge (*Carex* L. species) to those that are nutrient-poor and support only the sphagnum (*Sphagnum* L. spp.) mosses. Shallow bogs are usually found on easily-drained and relatively shallow slopes, and are often only 0.1 to 1 meter deep.

Aerial or sand dune deposits, the third soil type, cover less area than bog and koppmoll

Table 1. Mean Monthly and Tetraterm (*TETR*) Temperatures (°C), 1931 to 1960

	JAN	FEB	MAR	APR	MAY	***JUN***	***JUL***	***AUG***	***SEP***	OCT	NOV	DEC	***TETR***
V[1]	-4.3	-5.2	-4.0	-0.8	2.6	***6.2***	***9.1***	***9.7***	***6.8***	2.5	-0.5	-2.7	***7.95***
E[2]	-5.6	-6.3	-4.9	-1.1	3.1	***7.3***	***10.4***	***10.5***	***6.8***	2.0	-1.3	-3.7	***8.75***
K[3]	-10.3	-11.1	-7.6	-2.2	3.1	***9.0***	***12.4***	***11.2***	***6.7***	0.8	-4.4	-7.8	***9.83***

[1] V = Vardø
[2] E = Ekkerøy
[3] K = Kirknes

Table 2. Mean Monthly and Annual Precipitation (in millimeters), 1931 to 1960

	JAN	FEB	MAR	APR	MAY	JUN	JUL	AUG	SEP	OCT	NOV	DEC	SUM
V[1]	45	46	47	36	36	37	41	52	63	56	43	43	545
E[2]	48	48	47	37	37	36	43	58	56	49	46	45	550
K[3]	29	22	21	21	24	37	61	64	51	33	31	30	424

[1] V = Vardø
[2] E = Ekkerøy
[3] K = Kirknes

Table 3. Average Date (shown as month/day) from 1931 to 1960 When Daily Mean Temperature Exceeded or Fell below Values from 0 to 11 °C

	Temperature Greater Than					Temperature Less Than				
	0	3°	6°	9°	10°	11°	10°	6°	3°	0°
V1	27/4	19/5	10/6	9/7				22/9	15/10	9/11
E[2]	27/4	16/5	3/6	24/6	5/7		18/8	21/9	11/10	2/11
K[3]	24/4	9/5	24/5	10/6	17/6	22/8	30/8	24/9	10/10	26/10

[1] V = Vardø
[2] E = Ekkerøy
[3] K = Kirknes

Table 4. Normal Length (in days) of Seasons When Daily Temperatures (T_d) were <0 °C (Winter), 0–10 °C (Spring), >10 °C (Summer), and 0–10 °C (Autumn) and When Growing Season Temperature Was ≥ 3 and 6 °C, Respectively, from 1931 to 1960

	Winter	Spring	Summer	Autumm	Vegetation period	
					$T_d \geq 3$ °C	$T_d \geq 6$ °C
V[1]	168	79	0	118	150	105
E[2]	175	69	45	45	149	111
K[3]	179	54	75	57	155	124

[1] V = Vardø; [2] E = Ekkerøy; [3] K = [illegible]s

Table 5. Average Number of Days per Year Within the Indicated Wind Speeds (meters/second) and Number of Days of Fog in June, July, and August. Data from 1931 to 1960.

	Wind Speed Ranges		
	10.8–13.8 meters/second	13.9–20.7 meters/second	20.8–24.4 meters/second
Vardø	131	33	12.4
Ekkerøy	95	9	1.5
Hammerfest	83	12	2.6

	Number of Days with Fog
Vardø	18.3
Ekkerøy	9.8
Hammerfest	5.0

Table 6. Tetraterm Temperature Requirements of Potential Tree-limit Species

Species	Tetraterm (°C)	Literature
Pinus sylvestris L.	8.4	
Picea abies L.	8.4	
Sorbus aucuparia L.	7.7	
Alnus incana L.	7.7	
Prunus padus L.	7.7	Skinnemoen/Helland
Populus tremula L.	7.6	(1969)
Betula pubescens Ehrh.	7.5	
Juniperus communis L.	5.3	
Betula nana L.	4.3	
Salix polaris Walenb.	1.4	
Pinus		
Tree-limit	5.5	
Forest-limit	7.7	Enquist (1933)
Seed ripening-limit	10.5	Hagem (1917)
Picea		
Tree limit	6.4	
Forest limit	8.4	Aas (1964), Lindquist (1948)
Seed ripening limit	10.3	Hagem (1917)
Betula		
Tree limit	4.7	
Forest-limit	6.5	Aas (1964, 1969)
Seed ripening-limit	<8.2[1]	Mork (1944)

[1] Dry seed was used for germination analysis. Fresh seed germinates at lower ripening temperatures than dry seed (Kielland-Lund, Institute of Nature Conservation, Agricultural University of Norway, Ås, personal communication).

soils. Aerial deposits also vary in composition. Mollusk shells in some sites increase the soil pH. Dryas L., a calciferous mountain plant, is often found among the flora that struggles for survival in the sand dune areas.

THE COASTAL MOUNTAIN BIRCH ZONE

The coastal zone between Vardø and Kirkenes encompasses the climatic extremes in eastern Finnmark (Tables 1–5). Kirkenes is among the most productive and favorable areas for plant growth, while Vardø, less than one degree of latitude north of Kirkenes, is among the most windswept and unfavorable areas for afforestation in Norway.

The northeastern limit of mountain birch in Fennoscandia is at Ekkerøy, about 11 kilometers east of Vadsø between Vardø and Kirkenes on the north side of the Varangerfjord (Fig. 1). The eastern outermost natural mountain birch stand of any significance in the Varanger Region of Finnmark is found in the Storelv Valley, only five kilometers east of Ekkerøy.

Only a few scattered and low-growing birch stands are found just north of Vardø, but further north and west, lush birch forests grow in combination with willow (Salix L. spp.) and European ash (Sorbus aucuparia L.) in small stands or niches in an otherwise treeless landscape.

Further north, astonishingly lush mountain birch, tall Salix thickets, and Sorbus aucuparia are found in Oksfjorddalen, on the island of Loppakalven off the coast of western Finnmark, and particularly in the Syltefjord valley about 45 kilometers northwest of Vardø (Fig. 1).

Differences in climate between treeless Vardø and Kirkenes are greater than expected, but differences in climate between Vardø and Ekkerøy near the eastern limits of mountain birch forests are less than expected (Tables 1–5). The presence of birch forests at Ekkerøy and their absence at Vardø support the hypothesis that human activities in early times nearly led to the extinction of the birch in the coastal districts of eastern Finnmark.

The climate of northwest Finnmark and the vigorous stands of birch found there indicate that mountain birch forests can be re-established throughout the coastal region of Finnmark.

SPECIES FOR ENVIRONMENTAL FORESTS

From the 30-year average tetraterm temperatures at Vardø, Ekkerøy, and Kirkenes (Table 1), and the tetraterm requirements of potential tree-limit species (Table 6) Sorbus aucuparia, Alnus incana, Prunus padus, Populus tremula, and Betula pubescens should grow at Vardø. Because of the high tetraterm requirements for growth and seed maturation, Pinus sylvestris and Picea abies are questionable tree-limit species on the coast of Finnmark.

Among the Salix species, only the hardy polar willow (Salix polaris Wahlenb.) is included in Table 6. Salix, however, has fewer requirements for heat, moisture, and light than mountain birch. The tetraterm requirements of species and provenances from Alaska are unknown and will be determined in future field trials in Finnmark County.

The indigenous Salix and downy birch (Betula pubescens Ehrh.) should provide adapted materials for establishing environmental forests in the coastal areas of Finnmark. Salix species are easy to propagate from either seeds or cuttings and are suitable as pioneer plants. Among more than 300 species, 30 are found in Norway. Most Salix flower before leaf emergence in early spring, and hybridization among species is common. Arctic species vary widely in height, ranging from the one-centimeter Salix polaris to the 10-meter goat willow (Salix caprea L.). Salix species may form 6- to 8-meter high thickets in sheltered areas near Vardø.

The two birch species, dwarf arctic birch (Betula nana L.) and B. pubescens, also hybridize in northern Finnoscandia. The first hybrids and their introgression products might have been the evolutionary pathway of B. ssp. tortuosa (Kallio et al., 1983). From research, we know that selection for adaptation of birch genotypes to climatic conditions at the Finnoscandia treeline is intense (Håbjørg, 1972).

TREE BREEDING

Different species and strategies of selection and breeding are required to meet key objectives of local societies for environmental forests. Selecting and propagating adapted plant materials are prerequisites for success in establishing environmental forests. New information about photoperiod and light requirements for the onset and termination of dormancy, and vegetative propagation shows that environmental forests should be established with local ecotypes.

The large genetic variation within populations of Salix and Betula species, and the ability of these species to hybridize, offers plant breeders ways to create new hybrids for tree-limit conditions. Seeds and cuttings have been collected from many different species and climatic ecotypes indigenous to the most exposed areas on the Finnmark coast for establishing an arctic breeding arboretum and an environmental forest at Svartnes Bay.

Arctic and subarctic origins preserved in the arboretums of neighboring countries are another source of plants suitable to northern Finnmark. Climatic ecotypes of conifer and broad-leaved hardwood species from Alaska are available from plant materials collected for research in Greenland and preserved in the Danish Arboretum in Hørsholm. Plant materials from areas with climates and growing conditions similar to eastern Finnmark are also available from Iceland and the Faeroe Islands. With assistance from the Nordic Arboretum Committee, plant materials compatible with the northern birch tree-limit zone can be collected from circumpolar regions for trials at the Vardø Arctic Arboretum. From 1987 to 1989, the Nordic Forest Research Council collected many seeds and plants from Alaska and the Yukon Territories (Skaret and Rosvall, in press). These collections might contain species and provenances suitable for preservation in the Vardø Arctic Arboretum. Species and provenances preserved in the Vardø Arboretum will be used to breed adapted land races for developing environmental forests, and pollution-tolerant shelterbelts, and for forestry in the mountain and coastal regions of Finnmark County.

Most of the plants used in the climate stress experiments at Svartnes were produced at the Forsheim Research Station in cooperation with the University of Tromsø, Troms County, and were produced from seeds and cuttings of forest vegetation collected from niches on the Finnmark coast. The Danish material preserved in Hørsholm, and the plant material from Finland and Iceland were tested at Forsheim before they were transplanted in Finnmark.

The Forsheim tests were preliminary. The test environment at the Forsheim nursery is two degrees of latitude south and 1,100 kilometers southwest of Vardø and is not representative of the environment in the Svartnes region. Therefore, a forest tree nursery at Vardø will be included in the facilities for the arctic arboretum and environmental forest in the Nature Conservation Center at Svartnes.

PLANTATION ESTABLISHMENT AND PRELIMINARY RESULTS

Re-vegetation of the coastal districts in Finnmark will require shelterbelts, protection of plantations from grazing, and species adapted to the local climate and soils. Indigenous willows, such as wooly willow (*Salix lanata* L.) and tea-leaved willow (*S. phylicifolia* L.), grow rapidly and form natural shelterbelts on the coast. As the willows grow older, hardy genotypes of larger and less wind-tolerant species can be transplanted into the natural shelterbelts formed by the willows.

At forest limits in the Storelvdal valley at Krampenes (Fig. 1), for example, mountain birch migrates to new sites sheltered by willow thickets. Birch seeds germinate in small clearings between the willow trees, and seedlings that survive critical establishment periods eventually outgrow the willow.

Using the succession strategies of the indigenous vegetation, environmental forests can be established in marginal areas as follows:

1. For wind protection and a favorable local climate, establish thickets with indigenous *Salix* species.
2. Prepare (scarify) the soil to improve growth.
3. Select and transplant superior provenances of the most hardy tree species into the shelter provided by the willow thickets.
4. Fence the plantation to prevent cutting and grazing.

The first plantation was established at the Svartnes Nature Conservation Center in 1982. Six hundred lodgepole pine (*Pinus contorta* Dougl. ex Loud.) seedlings of 19 Alaska provenances were planted in reversed grass turf on a slope without shelter from willow or other nurse crops. In 1983, using the above strategy, seedlings of *Pinus contorta*, subalpine fir (*Abies lasiocarpa* (Hook.) Nutt.), Engelmann spruce (*Picea engelmanni* Parry ex Engelm.), Sitka spruce (*P. sitchensis* (Bong.) Carr.), Lutz spruce (*P. xlutzii* Little), and Norway spruce (*P. abies* L. (Karst.) were transplanted to small openings among natural willow thickets in the plantation. The *Pinus contorta*, *A. lasiocarpa*, *Picea sitchensis*, and *P. xlutzi* were from Alaska, and the *P. abies* was from the natural limits of *Picea* in Rana, Nordland County, Norway. The *P. sitchensis* and *P. xlutzii* seedlings were supplied by Hørsholm Arboretum, and *P. abies* seedlings were from the Troms County nursery at Andselv, Norway.

Several ramets of a tree form of *Salix phylicifolia* L. from Persfjord, Vardø, were also transplanted to the Svartnes Environmental Forest Project. The ortet of this clone was about five- to six-meters tall. The diameter at the base of the stem was 20 centimeters. In addition to the *S. phylicifolia* clone, other outplantings include seedlings of seed-propagated *Salix caprea* L. from Skjomen, Norway, mountain birch provenances from Finnmark, and individuals of Siberian larch (*Larix siberica* Ledeb.) from an arboretum at Hallormsstadur, Iceland. The mountain birch provenances were part of a study by the Nordic Forest Birch Group on the adaptation and productivity of birch on a coast-to-inland gradient in northern Fennoscandia.

Both broadleaf and conifer seedlings were cultivated in plastic pots at Forsheim before they were transported to Svartnes. Mortality of unprotected Pinus contorta on reversed turf was severe. On the other hand, relatively few (10 to 15 percent) seedlings planted between the willow trees or on the lee of furrowed soil ridges in 1983 died. The oldest spruce were one meter tall after seven years, but birch from Finnmark sustained less mortality than all other species. At the present time, all trees are well-protected by snow. The critical test of survival will come after the trees grow above the snow cover and are subjected to potential wind dessication and cold injuries.

REFERENCES

Aas, B., 1964, Bjørke- og barskogsgrenser i Norge (Birch and treelines of conifer forests in Norway), University of Oslo. M.S. Thesis.

Aas, B., 1969, Climatically-raised birch lines in southeastern Norway, 1918–1968, Norsk Geolog. Tidsskr., 23:119-130.

Balsvik, R. R., 1989, Vardø: Grensepost og fiskevær (Vardø: Border town and fishing village), Vardø Kommune/Dreyer bok, Stavanger, p. 274.

Enquist, F., 1933, Trædgrænsundersøkningar (Studies on treeline), Svenska Skogsv. Før. Tidsskrift., 31:145-214.

Hagem, O., 1917, Furuens og granens frøsætning i Norge (Seed production in pine and spruce in Norway), Meld. Vestl. Forstl. Fors. Sta. (Reports of the Forest Research Institute of West Norway), 1(2):1-188.

Helland, A., 1912, Tregænser og sommervarmen (Treeline and summer heat), Tidesskrift for Skogbruk.

Hyvärinen, H., 1976, Flandrian pollen deposition rates and treeline history in northern Fennoscandia, Boreas, 5:163-175.

Håbjørg, A., 1972, Effects of photoperiod and temperature on growth and development three latitudinal and three altitudinal populations of Betula pubescens Ehrh., Norwegian Landbr. Høgskole, 51(2):1-27.

Kallio, P., Niemi, S., Sulkinoja, M., and Valanne, T., 1983, The Fennoscandian birch and its evolution in the marginal forest zone. In: Morisset, P., and Payett, S., eds., Tree-line ecology, Proceedings of the Northern Québec Tree-Line Conference, 1981 June 22–July 1, [Kuujjuarapik, Québec], Québec City, Québec, Centre d'etudes nordique, p. 101–110.

Lindquist, B., 1948, The main varieties of Picea abies in Europa, Acta Hortia Berganiani, 14(7):249-342.

Mork, E., 1944, Om bjørkefrutens modning, bygning og spiring (Concerning ripening, morphology, and germination of birch seed), Meld. Norsk Skogsfors., Vest. 8:421-470.

Niemi, E., 1983, Vadsøs Historie, Fra øyvær til kjøpstad (History of Vadsø: From an island community to a town), Vadsø Kommune/Helfjord Boktrykkeri, p. 26.

Simonsen, P., Jordfunn fra Vardø (Soil find[s?] from Vardø), Vardøhus Museum Årbok, 1959:8-56.

Skaret, G., and Rosvall, O., [In press], Seed collections from North American trees for marginal sites in the Nordic countries. In: Alden, J. N., Ødum, S., and Mastrantonio, J. L., eds., Forest Development in Cold Climates, Proceedings of an International Symposium, 1991 June 18-24, [Laugarvatn, Iceland], NATO, Plenum, New York, New York.

FOREST DEVELOPMENT IN ICELAND

Jón Loftsson

Iceland Forest Service
Reykjavík, Iceland

HISTORY

When Iceland was settled 1,100 years ago, at least 30 percent of the land area was covered by woodland. Icelandic birch (Betula pubescens Ehrh.) was the major tree species (Sigurðsson, 1977). Little is known about the dimensions of the trees, but they were utilized by the farming society for firewood, charcoal, and even as small building timber. The woodlands were also felled to make way for stock grazing. But grazing hindered woodland regeneration and deprived the soil of its protective cover, eventually causing severe erosion.

By the end of the 19th century, the woodlands had been reduced to a mere one percent of the total land area, and many farming districts suffered from severe soil erosion problems (Bjarnason, 1942). About the same time, between 1899 and 1907, the first serious tree-planting program was initiated by a group of Danish forestry enthusiasts. The trials were supervised by C. E. Flensborg, who later became director of the Danish Heath Society. This was a relatively large program for the period; 900 hectares of woodland were fenced at Hallormsstaður and Vaglir for protection from livestock. Small stands of exotic tree species were planted and three small nurseries were established (Bjarnason, 1957). The oldest groups of trees from this period are mountain pine (Pinus mugo Turra), Siberian cedar (Pinus sibirica Du Tour), bristlecone pine (Pinus aristata Engelm.), Norway spruce (Picea abies (L.) Karst.), Engelmann blue spruce (P. engelmannii Parry ex Engelm.) (Fig. 1), Siberian larch (Larix sibirica Ledeb.), European larch (L. decidua Mill.), and subalpine fir (Abies lasiocarpa (Hook.) Nutt.) (Fig. 2). All of the North American species probably originated from Colorado.

In 1907, the Icelandic parliament approved new forest legislation, both to protect the remaining woodlands and to create new forests. The Icelandic Forest Service (IFS) was created and A.F. Koefoed-Hansen, a Danish forestry graduate, was appointed Chief Forester. The country was divided into four regions, each supervised by a district forester. The Forestry Act was revised in 1928 and again in 1955 (Blöndal et al., 1986). The last revision is still in force with a few amendments concerning shelterbelts and farm forestry. In the

Forest Development in Cold Climates, Edited by J. Alden *et al.*
Plenum Press, New York, 1993

spring of 1991, Parliament enacted a further amendment to subsidize afforestation of 15,000 hectares of private lands in eastern Iceland.

From 1908 to 1935, the IFS, under Koefoed-Hansen's directorship, terminated trials with exotic species. Efforts were concentrated on enclosing the birch woodlands. During this period, the IFS acquired a large part of the property that it manages today.

The early trials with exotic species were rather disappointing, but the birch woodlands that had been protected by fencing regenerated rapidly. From about 1920 until 1960, the climate improved significantly and the country passed through one of the mildest climatic periods in its history. The mild climate undoubtedly accelerated natural regeneration of the enclosed birch woodland.

A series of events occurred around 1930 which were later to prove beneficial to Icelandic forestry. In 1930, the Icelandic Forestry Association was founded and forestry began to be supported by the general public. In 1935, Hákon Bjarnason, the first Icelandic forestry graduate, assumed directorship of the IFS. Under his directorship, afforestation of exotic conifers was emphasized and forestry began to change. Planting stock was purchased from Norway and efforts were made to procure seed from regions with similar length of vegetation period, including Alaska, Siberia, northwestern Russia, and northern Scandinavia.

Fig. 1. Thirty-year-old Engelmann spruce (Picea engelmanii (Parry) ex. Engelm.) near the town of Stykkishólmur, West Iceland. Provenance: Colorado, U.S.A.

During World War II, Iceland was occupied by United States Armed Forces. New contacts were established with Alaska to compensate for the loss of contacts in Europe. Bjarnason (1946) travelled to Alaska in 1945 and collected seed of several species and many origins of Sitka spruce (Picea sitchensis (Bong.) Carr.). In the post-war period the forestry effort expanded. Nursery capacity increased from a few thousand seedlings per year before World War II to more than a million seedlings by 1958 (Bjarnason, 1984). Sitka spruce, which proved more difficult to raise in the nursery than the commonly planted European species, was emphasized in the afforestation programs. By 1958, old contacts with Scandinavia had been reestablished and Scots pine (Pinus sylvestris L.) and Norway spruce were the most commonly planted species in Iceland. Attempts were also made to procure seed of Siberian larch and of Russian larch (Larix sukaczewii Dylis). It was difficult, however, to obtain larch seed of the desired provenances of the larches and the origins obtained were haphazard (Blöndal et al., 1986). The IFS increased its estate considerably during this period, from some 7,000 hectares in 1935 to over 25,000 hectares in 1958 (Bjarnason, 1966). Unfortunately, statistics were not maintained on the area of natural birch woodland inside these fences at the time of enclosure or on the area of land considered suitable for planting.

Fig. 2. Forty-eight-year-old subalpine fir (Abies lasiocarpa (Hook.) Nutt.) at Hallormstaður, East Iceland. Provenance: White River National Forest, Colorado, U.S.A.

By 1958, public interest in forestry was assured and a new era began. Forestry societies had been established in most districts, and by 1955 the Icelandic Forestry Association included 30 district societies. Furthermore, sufficient experience had been gained so that it was possible to determine which species would most likely succeed. Scots pine, Norway spruce, Sitka spruce, and the larches were widely planted along with the native birch, which private landowners favoured over the exotics.

CULTIVATION OF THE COMMON EXOTICS

Exotic tree species have been planted in Iceland for over a century. The oldest surviving trees are a Wych elm (Ulmus glabra Huds.) and a sycamore (Acer pseudoplatanus L.) that were planted in Reykjavík in 1888 (Bjarnason, 1979). Since then, more than a hundred exotic tree species and probably over a thousand different provenances have been planted in Iceland (Blöndal, in press). It is pertinent to review the experiences Icelanders have had with cultivation of the common exotic species from the early history of tree planting.

Scots Pine

Initially, Scots pine showed great promise and was planted throughout Iceland. By the end of the 1950s, however, most plantations were infested by the woolly aphid (Pineus pini Macq.). Plantations were so heavily damaged that in 1962, the IFS and the Forestry Society decided to abandon all further planting of the species (Ottósson, 1988). Today, Scots pine has nearly disappeared from Icelandic plantations; ten years of work and several hundred hectares of Scots pine plantations were lost. Forestry had received a severe blow, one of several yet to come.

Sitka Spruce

The pre-World War II plantings of Sitka spruce initially showed considerable promise for commercial purposes. By 1951, following the return of normal relations with Canada and the United States after World War II, Sitka spruce was planted on a large scale. By 1960, several million seedlings had been planted. The early optimism for this species was dampened, however, following widespread damage from a severe spring frost in 1963. The most extensive damage occurred in the southern lowlands where a mild winter prior to the frost had encouraged early flushing and shoot elongation (Ragnarsson, 1964).

Despite this calamity, Sitka spruce has shown exceptional growth on productive sites in the high rainfall areas of south and west Iceland. Recent infestations by the green spruce aphid (Elatobium abietinum Walker), a species endemic to Europe, has further dampened the enthusiasm for planting Sitka spruce. The aphid, which was first observed in 1959, has spread over most of the country and has caused considerable defoliation and loss of wood yields since 1964 (Ottósson, 1985).

Norway Spruce

Because of close cultural connections with Scandinavia and the relative ease of procuring seed, Norway spruce was the most commonly planted species from 1958 to 1972 (Benedikz, 1973). Norway spruce is relatively easy to cultivate in the nursery and during the establishment stage. These characteristics probably explain the popularity for planting Norway spruce as a Christmas tree. Christmas trees yield a high return in a relatively short time.

Later experience showed that Norway spruce thrives only in a few places in Iceland. It cannot tolerate exposure to cold winds and atmospheric salts, and it requires sheltered sites and fertile soils for normal growth. It was planted, all too often, on unsuitable sites. Today, large areas of unthrifty, yellow-foliaged spruce can be found throughout the country. These areas are more reminiscent of northern timberline populations than of plantation forests.

Lodgepole Pine

Lodgepole pine (*Pinus contorta* Dougl. ex Loud.) is a comparatively late introduction. A small stand of the inland variety (*P. contorta* var. *latifolia* Engelm.) from Smithers in British Columbia was planted at Hallormsstaður in 1940. Further introductions were not made until the late 1950s (Bjarnason, 1978). The majority of the plantations in Iceland, however, have been established with seed of North Coastal shore pine variety (*P. contorta* Dougl. ex Loud. var. *contorta*) from the Skagway area in southeast Alaska. The shore pine variety has been successful in both the coastal and inland areas, although there are indications that it is not completely hardy in the interior valleys of northern Iceland, where the growing season is short and unseasonal frosts often occur. In those areas, Yukon Territory provenances are considered more suitable than the southern and coastal provenances (Sigurgeirsson, 1988). Provenance trials were established in 1984 but have not yet yielded any practicable results. Use of lodgepole pine as a plantation species is increasing because it is popular as a Christmas tree and because it grows on relatively poor soils.

As yet, lodgepole pine has not suffered serious injury from insects or fungal diseases, but it has been damaged by snow due to the heavy-branching habit of the provenance most commonly planted. Furthermore, instability has been observed in pole stands, where trees have been toppled during wind storms. Tree toppling can be traced to root deformation caused by improper cultural practices in the nursery or at planting.

Siberian and Russian Larches

Siberian and Russian larches are usually considered geographical varieties or subspecies of *L. sibirica*. Dylis (1945) considered morphological differences between the varieties sufficient to separate *L. sibirica* into two species. Their behaviour in Iceland certainly warrants keeping the two sources separate (Snorrason, 1987). Both species prefer the more arid climate of north and east Iceland, but they also grow in the inland valleys of other districts. Russian larch is much superior in form and seems to tolerate the unstable climate in Iceland better than Siberian larch. Siberian larch is more prone to damage from unseasonal frosts however, than Russian larch.

Generally, all seed sources that trace their ancestry to the renowned larch plantation at Raivola (northwest of St. Petersburg, Russia, at latitude 60°14′ N. and longitude 29°35′ E.) will succeed in Iceland. The Raivola stand was established in 1738 with seed that had been collected in the neighbourhood of the city of Archangel (Heikinheimo, 1927). Russian larch of northern origins has also shown great promise in Iceland, but seed from natural stands has been very difficult to obtain.

Larch is the most popular species in Iceland today, both for a large afforestation program in east Iceland and, due to its ability to grow on infertile and eroded soils, for reclaiming devastated land. It grows on infertile and eroded soils and is one of the species

Fig. 3. Twenty-year old Russian larch (Larix sukaczewii Dylis.), east Iceland. Provenance: Pinega, Arkhangelsk Oblast, northern Russia.

recommended by the Iceland Forestry Society for land reclamation projects (Blöndal, in press).

Broadleaved Species

In the past, Icelandic foresters have neglected broadleaved species. Broadleaved species were used mainly for their horticultural properties and not for plantation forestry. This has changed, and today the Alaskan black cottonwood (Populus trichocarpa Torr. et Gray) is the fastest growing species in Iceland. Alaskan black cottonwood thrives in most parts of the country and is popular as a plantation species. Provenance and clonal trials with both Alaskan black cottonwood and the closely related balsam poplar (Populus balsamifera L.) have recently been initiated (Anonymous, 1991). Clonal material from the Gulf of Alaska and the Kenai Peninsula have shown great promise. Kenai origins can be planted in most regions of Iceland (Sævarsdóttir and Óskarsson, 1990). On sheltered sites that are fertile, moist, and well-drained, Alaskan black cottonwood will yield industrial timber on 30-year rotations (Gunnarsson et al., 1987). In comparison, conifers are expected to grow industrial timber in 60- to 100-year rotations.

Other broadleaved species, such as silver birch (Betula pendula Roth), quaking aspen (Populus tremula L.), and alder, notably grey alder (Alnus incana (L.) Moench.), Sitka alder (A. sinuata (Reg.) Rydb.), and thin-leaved alder (A. tenuifolia Nutt.), have potential for special wood products and site rehabilitation. Alders, with their symbiotic nitrogen-fixing bacteria, have potential for soil amelioration on eroded or unvegetated areas in Iceland.

The alder species have wide geographical distribution and many provenances should grow well in Iceland. Species and provenance research on broadleaved trees has been neglected until now, but in view of their rapid growth in cold climates and potential for adapting to Icelandic conditions, the broadleaved species have been targeted for intensive investigation.

THE PRESENT SITUATION

The pioneers of Icelandic forestry have demonstrated to the public that forests can be grown successfully in Iceland. The earliest plantations were small and scattered but provided valuable information on which present-day practices are based. Emphasis on tree species and management practices will change, but it is certain that forestry has a bright future in Iceland.

The present objectives for planting forests in Iceland, in order of priority, are forest protection, recreation, and wood production. A large afforestation project has been initiated in east Iceland, where, during the next four decades, 80 landowners propose to establish some 15,000 hectares of production forests. For this project to succeed, the state must subsidize it completely. State support is justified because unborn generations will reap the harvest of the forests. This forest development project was possible only because the recent agricultural recession released land for planting that otherwise would be maintained for grazing.

FORESTRY RESEARCH

Although forestry has been practiced since the turn of the century, formal research was not initiated until after 1950. A forestry graduate was appointed by the IFS to initiate silvicultural research in 1955 and a Forestry Service Research Station was officially opened at Mógilsá in 1967. At first, it was both understaffed and grossly underfinanced. Gradually, it was expanded into the modern Forest Research Institute with a core of well-educated and competent scientists who should be able to improve afforestation practices in Iceland.

One of the first research projects of the Forest Research Institute was to build a forest site classification system, based on climatic factors (Ragnarsson, 1977). The site classification system divides the country into districts according to their potential for forest development. The classification system is now being improved by filling the climatic information gaps with results from a wide range of experiments and observations.

CONCLUSION

Icelandic forestry is at an exciting and fascinating turning point in its history. The public has accepted forestry as a realistic land use alternative to agriculture. Public awareness of the shortcomings of over-grazing and the realization that grazing must be controlled have helped to make forestry acceptable to most Icelanders. Probably the most important reason for the increasing interest in forest development, however, is that the trees have not only grown beyond our wildest hopes, but they are colonizing their surroundings naturally. That exotic species are capable of maintaining themselves unaided in marginal habitats tells us that we are on the right path. In time, the gaping holes in Iceland's green mantle will once again be forested with species from cold climates in Northwest America and Eurasia.

REFERENCES

Anonymous, 1991, Verkefnaskrá (Research projects), Iceland Forest Research Station, Mógilsá, 12 and 26-27.

Benedikz, Þ., 1973, Experiences with Norway spruce (Picea abies (L.) Karst.) in Iceland, Papers to IUFRO meeting of the Working Party for provenance research on Norway spruce held at Biri, Norway, 22 p.

Bjarnason, H., 1942, Ábúð og örtröð (Use and Abuse), Ársrit Skógræktarfélags Íslands (Yearbook of the Icelandic Forestry Association), Reykjavík, 8-40.

Bjarnason, H., 1946, Alaskaför haustið 1945 (A trip to Alaska in the fall of 1945), Ársrit Skógræktarfélags Íslands (Yearbook of the Icelandic Forestry Association), Reykjavík, 5-52.

Bjarnason, H., 1957, Lög um Skógrækt 50 ára (Icelandic Forestry Legislation—50 Years Anniversary), Prentsmiðjan Oddi, Reykjavík, 88 p.

Bjarnason, H., 1966, Skrár yfir skógræktargirðingar og gróðursetningu (List of forestry enclosures and plantation areas), Ársrit Skógræktarfélags Íslands (Yearbook of the Icelandic Forestry Association), Reykjavík, 30-32.

Bjarnason, H., 1978, Stafafura, Pinus contorta Dougl. (Lodgepole pine), Ársrit Skógræktarfélags Íslands (Yearbook of the Icelandic Forestry Association), Reykjavík, 16-18.

Bjarnason, H., 1979, Ræktaðu garðinn þinn—Leiðbeiningar um trjárækt (Grow your garden—Hints on cultivating trees), Iðunn, Reykjavík, 9-14.

Bjarnason, H., 1984, Gróðrarstöðin og trjágarðurinn í Múlakoti (The nursery and arboretum at Múlakot), Forestry Report 7, Iceland Forestry Service, Reykjavík, 4-18.

Blöndal, S., [In press], Socioeconomic Importance of Forests in Iceland. In: Alden, John N., Ødum, S., and Mastrantonio, J. L., eds., Forest Development in Cold Climates, Proceedings of an international symposium held at Laugarvatn, Iceland, 1991 June 18-24, NATO, Plenum, New York.

Blöndal, S., Benedikz, T., and Ottósson, J. G., 1986, Forestry in Iceland, A brief description of its history and present status, Forestry Report 8, Iceland Forestry Service, Reykjavík, 32 p.

Dylis, N. V., 1945, Novye dannye po sistematike i istorii sibirskoj listvennicy, Doklady Akademii Nauk SSSR, 1945, Vol. 50, No.4: 489-492, quoted in Simak, M., 1979, Larix sukaczewii: Naturlig utbredning, biologi, ekologi och frö-anskaffningsproblem, Sveriges Lantbruksuniversitet, Institutionen för skogsskötsel, Umeå, Sweden, Rapporter nr 1-1979, p. 7.

Gunnarsson, E., Guðmundsson, E., and Árnason, R., 1987, Hagkvæmni nytjaskógræktar (Feasibility of Production Forestry), Skógræktarrit 10, (Sérprent úr ritinu "Auðlindir um aldamót," Sérrit 3 úr Framtíðarkönnun ríkisstjórnar um Skógrækt), (Forestry Report 10, Iceland Forestry Service, Reprint from the Government Paper No. 3, "Resources for the Future"), Reykjavík, 50 p.

Heikinheimo, O., 1927, Beskrivning över Raivola lärkskogsomrade (On the larch plantation at Raivola), State Printing Office, Helsinki, 5-8.

Ottósson, J. G., 1985, Sitkalús (Elatobium abietinum Walker) (Green spruce aphid), Ársrit Skógræktarfélags Íslands (Yearbook of the Icelandic Forestry Association), Reykjavík, 8-16.

Ottósson, J. G., 1988, Furulús (Pine woolly aphid), Ársrit Skógræktarfélags Íslands (Yearbook of the Icelandic Forestry Association), Reykjavík, 41-46.

Ragnarsson, H., 1964, Trjáskemmdir vorið 1963 (Tree damage in the spring of 1963), Ársrit Skógræktarfélags Íslands (Yearbook of the Icelandic Forestry Association), Reykjavík, 25-27.

Ragnarsson, H., 1977, Um skógræktarskilyrði á Íslandi (Climatic conditions for forestry in Iceland), Skógarmál, Private publisher, Reykjavík, 224-247.

Sigurðsson, S., 1977, Birki á Íslandi (Útbreiðsla og ástand) (Birch woodlands in Iceland (Distribution and condition)), Skógarmál, Private publisher, Reykjavík, 146-172.

Sigurgeirsson, A., 1988, Stafafura á Íslandi—Vöxtur, ástand og möguleikar (Lodgepole pine in Iceland—Yield, status and potential), Ársrit Skógræktarfélags Íslands (Yearbook of the Icelandic Forestry Association), 3-36.

Snorrason, A., 1987, Lerki á Íslandi (Larch in Iceland), Ársrit Skógræktarfélags Íslands, Reykjavík (Yearbook of the Icelandic Forestry Association), Reykjavík, 3-22.

Sævarsdóttir, L. A., and Óskarsson, Ú., 1990, Ættbók alaskaaspar á Íslandi I: Safnið frá 1963 (Clone register for Alaskan black cottonwood I: Collection of 1963), Rannsóknastöð Skógræktar ríkisins, Mógilsá, Rit 10 (Iceland Forest Research Station Report No. 10), 2-16.

POTENTIAL TREE LINE IN THE FAROE ISLANDS

Tróndur G. Leivsson

Forestry Service of the Faroe Islands
Tórshavn, Faroe Islands

INTRODUCTION

The Faroe Islands consist of 18 islands and skerries in the Atlantic Ocean midway between Norway and Iceland at 62° N. latitude (Fig. 1). The land area totals 1,400 square kilometers. Elevation averages about 300 meters above sea level (masl) and the highest peak is 882 masl. Because of the steep topography, a relatively high proportion of the islands is unused. The soils are of basaltic origin and the lowland is covered with peat. The soil is continuously moist and strongly acidic (Rutherford and Taylor, 1981). Deposits from the Quaternary Period are often compact and podsolization can form an iron pan in the B horizon of the soil. The climate is extremely maritime (Table 1), and its low temperatures during the growing season, stormy winter weather, and variable winter temperatures are unfavourable for tree growth.

Apart from the dwarf juniper (Juniperus communis L.) and willow (Salix L.) species found in inaccessible gorges and cliffs, trees and shrubs are absent from the natural vegetation of the Faroe Islands (Hansen, 1966).

Jóhansen's (1985) studies of the vegetational history of the Faroes show that shrubs were abundant in the lowlands from the Boreal period (9,000 B.P.) until human activity and sheep grazing began about 650 A.D. The climate was unfavourable for seedling establishment, however, and by the 20th Century over-exploitation and grazing had eradicated almost all woodland vegetation.

From its phytogeographic position, Tuhkanen (1987) concluded that the Faroes represent a highly oceanic sector of the hemiboreal sub-zone. Although Tuhkanen gives examples of similar phytogeographic positions that have mainland connections and lack trees and shrubs, his paper indicates that conditions in the Faroes include niches for trees and shrubs.

Tree planting and afforestation attempts have taken place in the Faroes since the last decades of the 18th Century. The extent of tree planting is revealed in the work outlined by Hansen (1976) and Leivsson (1989a). Afforestation trials cover 80 hectares, 0.06 percent of

Forest Development in Cold Climates, Edited by J. Alden *et al.*
Plenum Press, New York, 1993

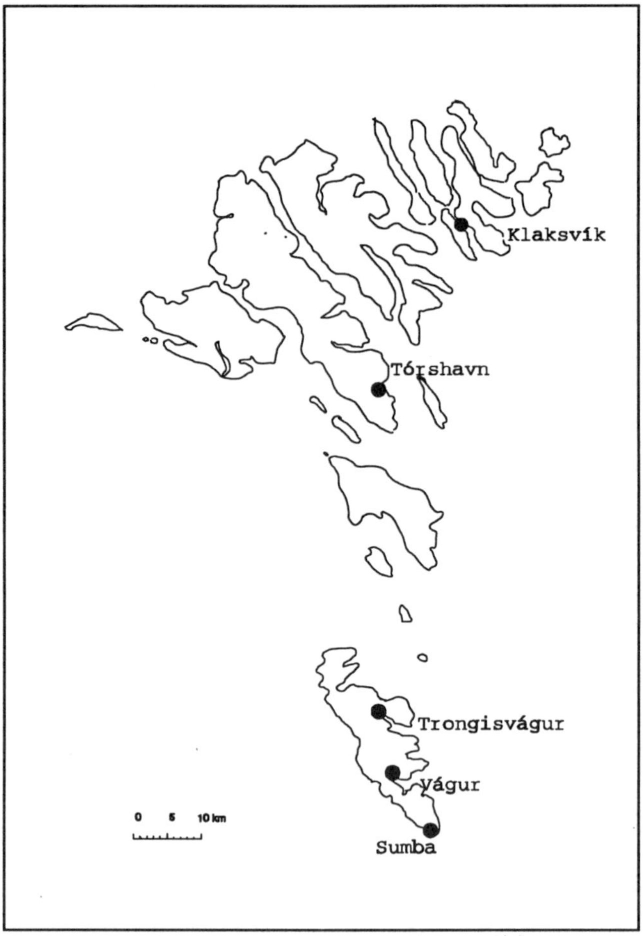

Fig. 1. The Faroe Islands and locations of tatter flag stations

the total land area. In a lowland plantation at 20 to 40 masl, the mean annual increment of 50-year-old shore pine (Pinus contorta Dougl. ex Loud. var. contorta) is at least seven cubic meters per hectare per annum (Leivsson, 1989b). In Vágur (Fig. 1), young shore pine are growing satisfactorily in topographic depressions at about 180 masl. In the Faroes, variable topography provides shelter from prevailing winds and salt spray deposition. A thorough inventory of trees and shrubs in gardens and plantations accounted for about 200 successful or promising species and cultivars (Ødum et al., 1989).

Geographical regions with promising species and origins for arboriculture and for further afforestation trials in the Faroes include coastal Alaska, coastal Scandinavia and Britain, regions of western European mountains with high precipitation, Tierra del Fuego and other regions of southernmost South America, the high altitudes of Tasmania and the South Island of New Zealand, and the high cloud forests of west Sichuan, China, Japan, and South Korea (Ødum, 1991).

Table 1. Climatic Data from Hoyvík (now inside Tórshavn city) for the Normal Period 1931-60 (Lysgaard, 1969)

	Jan.	Febr.	Mar.	Apr.	May	June	July	Aug.	Sept.	Oct.	Nov.	Dec.	Annual
Temp. °C													
avg. min.	0.9	1.6	1.2	3.2	5.0	7.5	10.1	9.3	8.1	6.1	3.9	2.0	6.3
mean temp.	3.9	3.7	4.6	5.4	7.3	9.2	11.0	11.1	10.0	7.9	6.1	5.0	7.1
avg. max.	6.2	6.4	7.4	6.9	8.8	11.1	12.1	12.4	11.3	9.7	7.8	7.0	7.9
Number of days with frost (T. min. <0) (mean)	9.4	8.9	7.7	3.9	1.1	-	-	-	-	1.1	2.6	6.1	40.8
Hours with sunshine (mean)	15	41	76	113	137	140	109	102	82	58	22	7	902
Precipitation mm (mean)	149	136	114	106	67	74	80	96	132	157	156	168	1,434
Rel. humidity % (mean)	82	82	82	82	83	84	86	87	86	83	84	83	84
Wind velocity m/s (mean)	6.3	5.8	5.5	5.1	4.5	4.4	3.9	3.9	4.9	5.6	5.8	5.9	5.1
Freq. of prevailing wind direction %	SW	N=SW	S=SW	N	N=E	SW	SW	SW	SW	SW	SW	SW	SW
(mean)	19	18	17	18	16	16	17	18	20	20	19	21	18

Because trees survive and grow normally in sheltered locations in the Faroe Islands, planting sites need to be assessed for exposure and severity of the local climate. An absence of natural woodlands and the indifference of ground vegetation to climatic change precludes the use of native flora as an index of site severity for trees and shrubs. Tatter flags have been developed and used in the British Isles for more than three decades as a practical means of assessing site severity in the absence of exposure-sensitive vegetation (Lines and Howell, 1963). This inexpensive, convenient, and fairly accurate tool is a standardized 38-centimeter x 30.5-centimeter cotton flag. The flags are exposed to the weather in situ for one or two months each for three years. The tatter of the cotton is measured regularly, and the tatter rate, expressed as square centimeters of tatter per day, is used to determine the exposure class on a particular site.

In the first extensive trials, conducted in Britain in 1957, tatter rates were related to tree survival and growth. The method for establishing and maintaining field stations and interpreting tatter rate is basically the same today as it was in 1957 (Reynard and Low, 1984). The empirical relation between daily tatter rate and tree growth is based on a vast number of records. The rate of tatter varies among regions within larger areas. Exposure classes were established for a range of tatter rates, as shown in Table 2. According to Reynard and Low (1984), a daily tatter rate of 12 to 13 square centimeters is the upper limit for commercial afforestation of Sitka spruce (Picea sitchensis (Bong.) Carr.) on upland sites in Britain. A tatter rate of 12 to 13 square centimeters corresponds to the General Yield Class of six to eight cubic meters of stem wood per hectare per annum.

Table 2. An Exposure Classification for Afforestation of Sites in Britain Based on "Tatter" Rates (Reynard and Low, 1984)

Tatter rate cm^2 per day	Exposure Class
>13.0	Severely exposed
10.1 - 13.0	Very exposed
6.6 - 10.0	Moderately exposed
4.0 - 6.5	Moderately sheltered
<4.0	Very sheltered

From surveys of the physical mechanisms affecting tatter rates on exposed sites in Britain, Rutter (1966, 1968) concluded that tatter flags are reliable instruments for afforestation purposes in windy climates.

In an attempt to assess the exposure levels for afforestation and tree growth in the Faroe Islands, a small tatter flag trial was established and maintained throughout 1984.

MATERIAL AND METHODS

Thirty flag stations were established at pre-planned intervals from 20 to 300 masl at five sites beginning on December 30, 1983 (Table 3). Tatter flag specifications and instructions for mounting and maintaining the flags and recording the data were provided by Reynard and Low (1984). The first flag stations were operational on December 30, 1983, and the last stations were closed on January 19, 1985.

Fig. 2. Flag stations 1 to 3 on the northeast slopes of the mountain "Háfjall" (646 masl) in Klaksvík.

Fig. 3. Flag station 7.

Fig. 4. Flag stations 11 and 12 on the southwest slopes of the mountain "Høgaklettur" (328 masl) in Trongisvágur.

Table 3. Location, Site, and Tatter Rates for Thirty Flag Stations in the Faroe Islands

Station No.	Location and altitude (masl)	No. of exposure days per period	Site	Tatter rate (cm² per day) Within each period of exposure	Average for the last 3 periods	Average for the year
1	Klaksvík undir Háfjalli 300 m	75	N-NE slope, steep. Solifluction. Upland vegetation.	11	10	10
		112		9		
		60		10		
		71		9		
		66		11		
2	Klaksvík undir Háfjalli 200 m	75	N-NE level area in cirque valley. Slight raw humus-like peat formation. Grass heath.	-	10	—
		112		9		
		60		10		
		71		11		
		66		10		
3	Klaksvík undir Háfjalli 100 m	75	NE slope, gentle. Peat. Dwarf shrub heath.	11	10	9
		112		6		
		60		10		
		71		9		
		66		10		
4	Klaksvík niðan í Gravardal 300 m	75	SW slope, steep. Slight raw humus-like peat formation. Grass heath	-	10	—
		112		10		
		60		8		
		71		11		
		66		10		
5	Klaksvík Gravardalur 200 m	75	SW slope. At bottom of cirque valley. Peat, >1 m horizon. Dwarf shrub heath.	-	7	—
		112		-		
		60		6		
		71		7		
		66		8		
6	Klaksvík Gravará 100 m	75	SW slope, gentle. Peat, >1 m horizon. Dwarf shrub heath.	10	7	8
		112		9		
		60		8		
		71		7		
		66		5		
7	Klaksvík Viðarlundin 20 m	75	SW level area. Well drained till. Former pasture land.	6	4	4
		112		4		
		60		3		
		71		4		
		66		5		
8	Trongisvágur Rangibotnur 300 m	75	N-aspect level area. At bottom of cirque valley. Waterlogged. Solifluction. Upland vegetation.	-	8	—
		118		9		
		59		6		
		67		11		
		59		7		

Table 3 (continued). Location, Site, and Tatter Rates for Thirty Flag Stations in the Faroe Islands

Station No.	Location and Altitude (masl)	No. of exposure days per period	Site	Tatter rate (cm² per day) Within each period of exposure	Average for the last 3 periods	Average for the year
9	Trongisvágur Ranghagi 200 m	75 118 59 67 59	NE slope, gentle. Peat, >1 m horizon, and rocks. Grass heath.	10 9 8 11 9	9	9
10	Trongisvágur niðri við Stórá 100 m	75 118 59 67 59	NE slope, gentle. Peat, <1 m Grass heath.	10 8 10 8 7	8	9
11	Trongisvágur Húsgarðshagi 200 m	75 118 59 67 59	SE slope, gentle. Peat. Grass heath.	9 9 7 8 6	7	8
12	Trongisvágur Húsgarðshagi 300 m	75 118 59 67 59	SW slope, steep. Solifluction. Upland vegetation.	14 7 5 11 11	9	10
13	Trongisvágur Viðarlundin 40 m	75 118 59 67 59	S-SE area in the bottom of a valley. A glade inside a shore pine plantation approximately 5 m tall	- 1 1 - -	—	—
14	Trongisvágur innanfyri viðar-lundina. 40 m	- - -	Due to wanton destruction this station was abolished.			
15	Sumba Skálin 300 m	75 118 59 67 59	SW slope, fairly steep. Mineral soil, rather deep. Grass heath.	15 9 15 14 17	15	14
16	Sumba Lambaklettur 200 m	75 118 59 67 59	S hillside. Peat, >1 m horizon. Grass heath.	- 9 9 12 15	12	—

Table 3 (continued). Location, Site, and Tatter Rates for Thirty Flag Stations in the Faroe Islands

Station No.	Location and Altitude (masl)	No. of exposure days per period	Site	Tatter rate (cm^2 per day) Within each period of exposure	Average for the last 3 periods	Average for the year
17	Sumba	75	S hillside. Peat, >1 m	14	10	10
	Vitin í Akrabyrgi	118	horizon. Pasture land.	9		
	100 m	59		4		
	Meteorological	67		11		
	station	59		14		
18	Vágur	75	SW slope. Grass heath. At	3	2	2
	Viðarlundin	118	the fringe of a shore pine	2		
	180 m	59	plantation approximately 3 m	2		
		67	tall.	2		
		59		3		
19	Vágur	75	SW slope. Grass heath. At	2	2	2
	Viðarlundin	118	the fringe of a shore pine	1		
	100 m	59	plantation approximately 3 m	2		
		67	tall.	2		
		59		2		
20	Tórshavn	83	Detached hill. Slight raw	-	10	—
	Veðurstøðin	93	humus-like peat formation.	-		
	55 m	79	Dwarf shrub heath.	6		
	Meteorological	60		13		
	station	61		12		
21	Tórshavn			11	11	11
	a duplicate of No.			11		
	20			7		
				13		
				12		
22	Tórshavn	83	E slope, gentle. Raw humus-	2	1	—
	Viðarlundin í Nið-	93	like peat. A glade in a shore	-		
	ara Hoydali	79	pine plantation approximately	1		
	35 m	60	10 m tall.	2		
		61		1		
23	Tórshavn			1	1	—
	a triplicate of No.			-		
	22			1		
				2		
				2		

Table 3 (continued). Location, Site and Tatter Rates for Thirty Flag Stations in the Faroe Islands

Station No.	Location and Altitude (masl)	No. of exposure days per period	Site	Tatter rate (cm^2 per day) Within each period of exposure	Average for the last 3 periods	Average for the year
24	Tórshavn a triplicate of No. 22			2	2	—
				-		
				1		
				2		
				2		
25	Tórshavn Gróðurstøðin 30 m	82	Detached plain. Riverbank. Cultivated land.	3	2	2
		93		2		
		79		1		
		62		2		
		59		3		
26	Tórshavn a triplicate of No. 25			3	2	2
				2		
				2		
				2		
				3		
27	Tórshavn a triplicate of No. 25			3	2	2
				2		
				1		
				2		
				3		
28	Tórshavn Gróðurstøðin 40 m	82	Detached plain. Slight raw humus-like peat formation. Dwarf shrub heath.	9	8	8
		93		8		
		79		4		
		62		12		
		59		8		
29	Tórshavn a triplicate of No. 28			9	8	8
				8		
				4		
				12		
				8		
30	Tórshavn a triplicate of No. 28			8	7	7
				7		
				2		
				11		
				7		

Each flag was exposed for three months during the first half of 1984, because Madapolam cloth (manufactured by Whitely Products, Ltd., Northamptonshire, England) was difficult to obtain. Exposure was reduced to two months per flag in the second half of 1984 as prescribed by Reynard and Low (1984).

Major goals of the trials were to:

1. Measure the exposure at sites from 100 to 300 meters above sea level,
2. Determine if sites with forest or trees have lower exposure levels than sites without forest or trees, and
3. Establish relations between tatter rates and anemometer records at official meteorological stations.

A "Topex score," a numerical measure of shelter afforded by the surrounding topography at a defined location (Wilson, 1984), was also calculated for each site from maps scaled at 1:20,000. Topex scores, however, failed to correlate with actual exposure (as measured by tatter flags) or the exposure perceived by local people, and were abandoned. The Topex score is assessed by measuring the angle of elevation in degrees from a fixed point to the horizon for a predetermined number of compass directions.

The field stations were distributed as shown in Table 3. Their locations are shown in Figure 1. Figures 2, 3, 4, and 5 are 1984 photographs of the landscape and locations of some of the flag stations.

Fig. 5. Flag station 10.

RESULTS

Results from the 30 flag stations are summarized in Table 3. Average daily tatter rates exceeded 10 square centimeters per day for only three of 30 stations during the last three periods of exposure and for only two stations during the entire year. Only one station had an average daily tatter rate greater than 13 square centimeters per day. All stations sheltered by trees had tatter rates less than or equal to three square centimeters per day throughout the year.

Tatter rates and anemometer records at stations 20 and 21 were not related, but the correlation between tatter rate and wind speed at station 17 was 0.79. Tatter rates and wind speed were not compared for the other stations.

DISCUSSION

Topography and exposure more than elevation determine the tatter rate of a given site.

Results from Stations 18, 19, and 22 to 24 show the extent of sheltering effects of forest plantations even at a relatively high elevation (Stations 18 and 19). Tatter rates during different periods of the year varied remarkably little and were inconsistent with anemometer records at the same site.

The benefits of shelter on crops growing in marginal sites, such as those at northern species limits, is discussed by Olesen (1979). The shelter provided by forests also improves the outdoor climate for human activity. The reduction in exposure afforded by trees at Stations 18 and 19 and 22 to 24 is consistent with the information of Olesen (1979).

Discrepancies exist in assessments of the severity of specific effects between traditional meteorological parameters and tatter rates. For example, the results of Lines and Howell (1963) and Jack and Savill (1973) for the British Isles differ from those of McAdam (1980) for the Falkland Islands. There is general agreement, however, that tatter flags are a useful method of assessing potential tree survival and growth.

Afforestation trials to determine altitudinal treelines in the Faroe Islands have not been attempted. The highest forest trial in the Faroes is at Vágur, 180 masl (tatter flag Station 18). Although this site is exposed to the prevailing winds, tatter rates averaged only two square centimeters per day. Growth was acceptable and indicates that treeline could be at even higher elevations. Ødum (1991) estimated that treeline in the Faroes might be around 150 masl in sheltered areas. His judgement was based on observations of the Faroese dendroflora in general, and he emphasized that recently introduced trees and shrubs from borderline forests of the world have potential for long survival and growth at tree limits.

Despite the failure to replicate this study (that is, the trials were conducted for one year instead of the usual three), the results indicate that potential treeline is close to 300 meters above sea level on topographically-sheltered sites in the Faroe Islands.

ACKNOWLEDGEMENTS

I conducted the field trial presented in this paper as a student at the Agricultural University of Norway, Ås. The idea for the study was encouraged by Professor Haveraaen, Department of Silviculture. Thanks are also extended to many people in Scotland and the Faroe Islands for valuable assistance in carrying out the study.

REFERENCES

Hansen, K., 1966, Vascular Plants in the Faeroes, Horizontal and Vertical Distribution, Dansk Botanisk Arkiv 24(3):1-141.

Hansen,,L., 1976, Trjárækt í Færeyjum, Sérprentun úr ársriti Skógræktarfélags Íslands 1976, 39-42. English summary.

Jack, W. H., and Savill, P. S., 1973, The causes of tattering of flags under natural conditions, Int. J. Biometeor., 17(2):185-192.

Jóhansen, J., 1985, Studies in the vegetational history of the Faroe and Shetland Islands, Annales Societatis Scientiarum Færoensis, Supplementum XI:1-117.

Leivsson, T. G., 1989a, Øki løgd til viðarlundir í Føroyum 1885-1985 (Areas laid out for afforestation 1885-1985 in the Faroe Islands). In: Højgaard, A., Jóhansen, J., and Ødum, S., eds., A century of tree-planting in the Faroe Islands, Annales Societatis Scientiarum Færoensis, Supplementum XIV, 35-49.

Leivsson, T. G., 1989b, Production in a 50 years old Pinus contorta stand. In: Højgaard, A., Jóhansen, J., and Ødum, S., eds., A century of tree-planting in the Faroe Islands, Annales Societatis Scientiarum Færoensis, Supplementum XIV.

Lines, R., and Howell, R. S., 1963, The use of flags to estimate the relative exposure of trial plantations, Forestry Commission, Forest Record No 51, H.M.S.O., London, 31 p.

Lysgaard, L., 1969, Foreløbig oversigt over klimaet på Færøerne, Det danske Meteorologiske Institut, Meddelelser 20:1-19.

McAdam, J. H., 1980, Tatter flags and climate in the Falkland Islands, Weather 35, (11):321-327.

Ødum, S., 1991, Choice of species and origins for arboriculture in Greenland and the Faroe Islands, Dansk Dendrologisk Årsskrift, IX:3-78.

Ødum, S., Hansen, L., and Rasmussen, S., 1989, Annotated inventory of trees and shrubs in gardens and plantations. In: Højgaard, A., Jóhansen, J., and Ødum, S., eds., A century of tree-planting in the Faroe Islands, Annales Societatis Scientiarum Færoensis, Supplementum XIV.

Olesen, F., 1979, Læplantning, Landhusholdningsselskabets Forlag, København, 138 p.

Reynard, B. R., and Low, A. J., 1984, The use of 'tatter' flags for exposure assessment in upland forestry, Research Information Note 96-84 SILN, Forestry Commission, Research and Development Division, Surrey, 4 p.

Rutherford, G. K., and Taylor, C. E. B., 1981, The soils of the Faeroe Islands, Geoderma 25:231-244.

Rutter, N., 1966, Tattering of flags under controlled conditions, Agr. Meteorol., 3:153-165.

Rutter, N., 1968, Tattering of flags at different sites in relation to wind and weather, Agr. Meteorol., 5:163-181.

Tuhkanen, S., 1987, The phytogeographical position of the Faeroe Islands and their eco-climatic correspondences on the other continents: Problems associated with highly oceanic areas, Ann. Bot. Fenn., 24:111-135.

Wilson, J. D., 1984, Determining a Topex Score, Scottish Forestry, 38(4):251-256.

FOREST TRIALS AT HIGH ELEVATIONS IN BRITAIN

C. J. Nixon and A. L. Tyler

Forestry Authority Commission
Northern Research Station
Roslin, Midlothian, Scotland

INTRODUCTION

The potential treeline in continental climates is controlled mainly by low temperature. In contrast, wind is the principal factor limiting tree growth in oceanic climates like that of Britain (Tranquillini, 1979). The natural treeline has recently been investigated in Britain and is higher than previously believed.

In a study of the upper elevational limit of Scots pine (Pinus sylvestris L.) in central Scotland, Pears (1967) concluded that although artificial treeline was around 490 meters above sea level, the potential for natural treeline was 610 to 685 meters above sea level, depending on exposure. This estimate was supported by an apparently natural Scots pine stand at an elevation of 640 meters.

Human activities are often cited to explain the artificially low treelines observed over much of upland Britain. During the 16th and 19th Centuries, these activities, especially tree felling, thinning, and subsequent wind damage, reduced the forest limits. More recently, the deer population has increased and severely restricted the success of natural regeneration.

Recent government incentives for restocking the early forests have increased interest in afforestation of native Scots pine (Pinus sylvestris L.) in Scotland. Many of the original Scots pine woodlands were at high elevations.

The accepted elevational limit for planted woodlands in Scotland and Ireland has steadily increased since the first large-scale afforestation schemes of the early 18th Century. The total upland area afforested in the 18th to early 20th Centuries was small, but the plantations were successful. An afforestation research programme was initiated by the Forestry Commission soon after it was founded in 1919, and controlled experiments and empirical results soon advanced techniques for reforesting upland sites.

The main developments were site amelioration and choice of species. Exotic species capable of survival on harsh sites were widely planted. On wet sites, Sitka spruce (Picea

Forest Development in Cold Climates, Edited by J. Alden *et al.*
Plenum Press, New York, 1993

sitchensis (Bong.) Carr.) demonstrated faster growth than pines (Pinus L. spp.) and larches (Larix Mill. spp.), especially on exposed upland sites.

In 1958, the Forestry Commission embarked on a program to establish a series of plantations for high-elevation trials. The plantations were laid out by Forest Districts and monitored by the Research Division of the Forestry Commission (Neustein, 1965). The objective was to determine the elevation of local tree limits and extended the land area economical for forestry. As altitudinal tree limits increased, some of the original plots were soon surrounded by commercial plantations. Provenances were also selected for their tolerance to exposure and for rapid growth.

In 1965, a series of trials was initiated on more exposed sites. The performance of several provenances of the two most promising species, Sitka spruce (Picea sitchensis (Bong.) Carr.) and lodgepole pine (Pinus contorta Dougl. ex Loud.), was compared to the performance of other species. Site amelioration treatments such as fertilization were also investigated to test the hypothesis that well-nourished plants can withstand exposure and poor soils better than nutrient-impoverished plots. To counteract edge effects, each plot was at least two hectares. Wind exposure at each site was also assessed with tatter flags.

In this paper, the effects of site and exposure on growth of Sitka spruce and Scots pine are discussed for a number of high elevation plantings. Sitka spruce and Scots pine are likely to be the major conifers of future commercial and native woodlands at high elevations in Britain. Results of site amelioration treatments such as fertilizer and nursing mixtures are also discussed.

METHODS

High-elevation plots were established at or above commercial treeline in Britain to cover a wide range of upland sites (Table 1). Three years of tatter-flag data were available for each site or comparable site sampled in the survey. To minimize the effects of improved establishment practice and provenance selection, only plots planted after 1950 were analyzed.

Mean annual height increment was used to assess tree growth. To minimize the effect of age on growth, height was recorded at varying ages between years 10 and 20 for each site.

Tatter flags are widely used to assess wind and other climatic effects on tree survival and growth in Britain. The rate of flag attrition correlates well with wind speeds in both field and controlled conditions (Jack and Savill, 1973; Rutter, 1966). Rain amplifies the effect of wind speed on tatter rates (Rutter, 1968), which are related to tree growth on exposed sites (Savill, 1974). Between 1954 and 1984, the Forestry Commission flew more than 1,100 flags at more than 100 sites in Britain. To average the annual variation in tatter rate, the flags were flown for at least two years (Miller et al., 1987). The combination of tatter rate and wind speed indicated the relative exposure at each location.

Mean annual cumulative temperatures greater than 5.6 °C were calculated for each site following the method of Worrel (1987).[1] Mean cumulative temperature at sea level was estimated from interpolated isotherms for Scotland, taken from Birse and Dry (1970) for England and Wales, and adjusted for elevation.

Table 1. Location, Exposure (Tatter Rate), Annual Temperature Sum above 5.6 °C, and Mean Annual Height Growth of Elevation Trials for Sitka Spruce and Scots Pine

Location	Plot Number[1]	Elevation (m)	Tatter Rate (cm^2/day)	Temp. Sum (°C)	Height Growth (cm/yr)
		Sitka Spruce			
Balblair	1	305	8.00	833	30.7
Balblair	2	305	8.00	833	18.6
Drumtochty	3	350	7.48	883	12.4
Drumtochty	5	320	7.50	923	15.4
Carrick	4	427	8.26	987	24.0
Clatteringshaws	3	560	13.80	791	6.7
Clatteringshaws	3	610	15.54	760	4.7
Clatteringshaws	3	680	16.65	689	5.7
Mynydd Ddu	6	732	16.90	825	9.3
S. Laggan	9	420	*	845	20.9
Garrarie	1	555	13.80	788	13.8
Garrarie	1	610	15.54	730	7.0
Garrarie	1	512	13.80	836	10.6
Hamsterley	15	648	14.45	778	12.4
Hamsterley	13	648	14.45	778	15.2
Hamsterley	14	610	13.74	819	18.3
Helmsdale	1	294	11.61	793	25.8
Helmsdale	2	270	10.71	1071	30.8
Helmsdale	2	315	11.60	1010	20.5
Kielder	68	440	9.74	921	17.5
Kielder	67	460	9.80	897	15.8
Kielder	62	412	8.26	955	18.1
Queens	17	633	5.76	636	16.4
Lewis	1	114	11.00	1118	20.2
Lewis	2	53	10.39	1214	24.0
Lewis	3	46	7.61	1225	20.3
Strathy	2	115	6.32	1027	21.0
Strathy	4	74	*	1091	15.0
Skiall	2	91	8.86	1045	13.2
Glencoe	1	320	7.83	1023	20.4
		Scots pine			
Carrick	1	366	11.20	–	6.2
Carrick	2	427	9.00	–	4.0
Carrick	7	427	15.50	–	3.9
Drumtochty	3	350	7.48	–	4.5
Hoy	1	76	9.61	–	7.0

Table 1 (*Cont'd*). Location, Exposure (Tatter Tate), Annual Temperature Sum above 5.6 °C, and Mean Annual Height Growth of Elevation Trials for Sitka Spruce and Scots Pine

Location	Plot Number[1]	Elevation (m)	Tatter Rate (cm^2/day)	Temp. Sum (°C)	Height Growth (cm/yr)
		Scots Pine			
Hoy	4	43	4.60	–	9.1
Kielder	55	320	8.20	–	9.1
Kielder	56	335	8.26	–	10.2
Kielder	68	440	9.74	–	5.7
Lael	14	282	4.50	–	15.7
Queens	17	633	5.76	–	8.5
Strathy	1	114	6.32	–	15.0
Strathy	2	115	6.32	–	17.5
Skiall	1	91	7.20	–	8.7
Skiall	2	91	8.86	–	8.4

[1] Provenance of all plots: Queen Charlotte Islands

* Exact location unknown

From a large number of climatic and edaphic variables, Worrell (1987) demonstrated that the productivity of Sitka spruce was significantly correlated with elevation, summer temperature, cumulative temperatures, tatter rate, and aspect, but not with wind. Tatter rates and cumulative temperature were strongly correlated with height growth of Sitka spruce. For this reason, both parameters were used as an index of exposure. Therefore, exposure is the combined effect of wind and low temperature sums.

Relationships between growth of Sitka spruce and Scots pine and climate, and between growth and edaphic factors, were investigated at altitude tree limits using regression analysis.

RESULTS

Growth and Survival of Sitka spruce

Mean annual height growth of Sitka spruce was described by a logarithmic function of

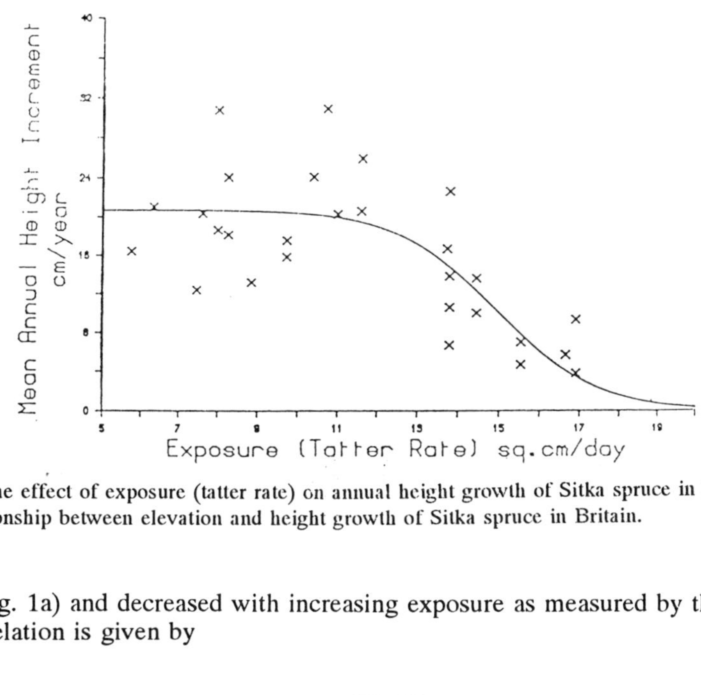

Fig. 1. (a) The effect of exposure (tatter rate) on annual height growth of Sitka spruce in Britain; (b) The relationship between elevation and height growth of Sitka spruce in Britain.

tatter rate (Fig. 1a) and decreased with increasing exposure as measured by the rate of flag tatter. The relation is given by

$$y = \frac{20.68}{1+e^{0.81(x-14.92)}},$$

where y is mean annual growth increment in centimeters and x is tatter rate in square centimeters per day. The squared correlation coefficient of the relationship, R^2, was 55.2 percent.

---

1 Worrell, R., 1987, Geographical variation in Sitka spruce productivity and its dependence on environmental factors, Edinburgh University, Scotland, UK, unpublished Ph.D. thesis.

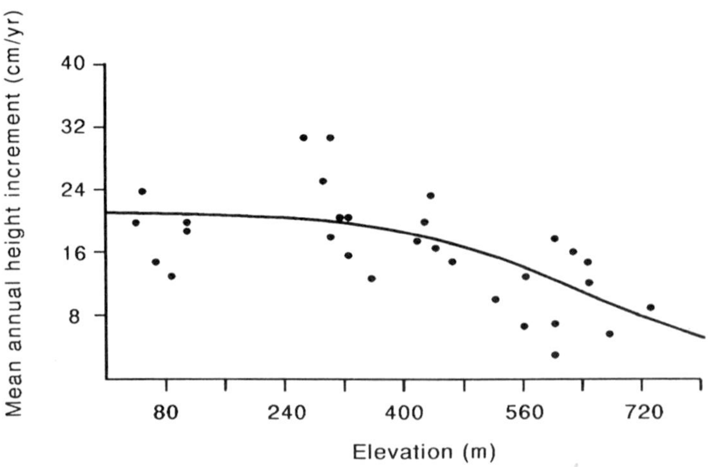

Fig. 2. The effect of accumulated temperature above 5.6 °C on height growth of Sitka spruce in Britain.

The fitted curve relates annual height growth to the tatter rate. At tatter rates greater than 12 square centimeters per day, mean annual height increment decreased rapidly and virtually ceased at rates above 16 square centimeters per day. Below 12 square centimeters, annual height increment did not increase significantly; factors other than wind appeared to limit growth. Considering the sites, growth-limiting factors could include low temperatures, short growing seasons, or poor soils.

Height growth at high altitudes is also restricted by adverse conditions associated with elevation (Fig. 1b). Growth was less well-correlated with elevation than with exposure. The relation between growth rate and elevation was best described by a logarithmic function:

$$y = \frac{21.51}{1+e^{0.008(x-636.3)}},$$

where y is mean annual increment in centimeters and x is elevation above mean sea level in meters. R^2 was 42 percent.

Temperature and forest productivity are related for both conifers and broadleaved trees on a variety of sites. Worrell (1987)[1] found a linear relationship between mean cumulative temperatures greater than 5.6 °C and height growth of Sitka spruce. Although growth varied

---

[1] Worrell, op. cit.

with cumulative temperatures at each site, a linear decline was apparent for the high elevation plots (Fig. 2) according to the function

$$y = -5.85 + 0.0249\ x,$$

where y = mean annual increment in centimetres and x = cumulative temperatures above 5.6 °C. R^2 was 28.6%.

In contrast to the response of tree growth to elevation and tatter rates, height growth did not decline sharply at any particular threshold temperature. Temperature declines at a constant rate with elevation (Smith, 1984). Thus, if temperature were the overriding reason for reduced growth at high elevations, forest productivity would decline more gradually than is shown in Figure 2.

Lack of uniformity in the timing of plot assessments precluded a detailed study of the effects of increasing elevation and exposure on tree survival. High overall survival rates, however, demonstrated that Sitka spruce is remarkably tolerant of adverse conditions at high elevations.

Growth and Survival of Scots Pine

The relationship between annual height growth of Scots pine and exposure could be linear. As with Sitka spruce, however, the non-random distribution of residual error indicates that a logarithmic function best describes the decrease in annual height growth with increasing daily flag tatter (Fig. 3):

$$y = 3.0 + \frac{11.53}{1 + e^{0.57(x-8.02)}},$$

where y is the mean annual height growth in centimeters and x is the tatter rate in square centimeters per day. R^2 was 43.5 percent.

The mean annual height growth of Scots pine is less than four centimeters on sites with mean daily tatter rates greater than 12 square centimeters per day. On sites with tatter rates below seven square centimeters per day, growth was more variable, less restricted by exposure, and more dependant on other environmental variables than on sites with tatter rates greater than 7 cm^2 per day. At age 10, a mean annual height increment of nine centimeters per year is equivalent to stem wood yields of approximately four cubic meters per hectare per year at full stocking (Edwards and Christie, 1981). Of 16 sites sampled, seven produced four or more cubic meters of stem wood per hectare, and of these only two had adequate stocking.

Differences in height growth between provenances from Scotland and those from southeast England were not significant, possibly because the number of plots was inadequate to sample site variability and estimate provenance means accurately.

After the first growing season, survival of Scots pine was recorded at all sites, but differences among sites and provenances were not apparent. Differences among sites in survival became apparent in later years, however (Table 2).

Inconsistencies in the data prevented in-depth analysis of the effect of exposure on survival. Nevertheless, trends in the survival rate of the various origins were apparent. Survival of all material from southeast England was less than 50 percent after 15 years, while the survival of Scottish provenances was more than 50 percent. Low survival of the Ballochbuie origin in the 1968 Kielder experiment was an exception. The poor performance of the Ballochbuie origin was noted in several early provenance trials. The seed weighed appreciably less than that of other origins, which caused poor height growth and survival (Lines, 1963).

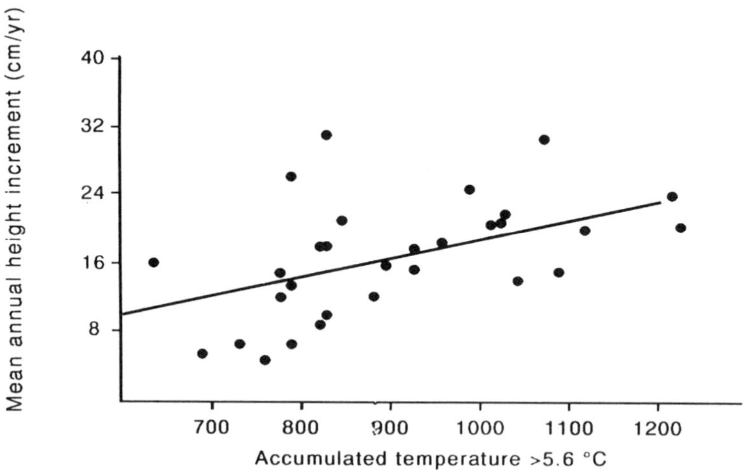

Fig. 3. The effect of exposure (tatter rate) on height growth of Scots pine in Britain.

DISCUSSION

A linear relationship between the growth of Sitka spruce and elevation was described by Malcolm and Studholm (1972), Mayhead (1973), and Worrell (1987).[1] These authors, however, studied growth rates at elevations below the environmental extremes for tree growth. Data from the Forestry Commission's high elevation plots in this study revealed a different pattern in the growth near altitude treeline, described best as a logarithmic function of elevation and exposure. The logarithmic function suggests that the severity of adverse factors accelerates near altitude tree limits, producing a curvilinear change in tree growth. This supposition agrees with the results of Oswald (1969), who recorded a similar pattern of tree growth in Norway spruce plantations in central France.

1 Worrell, op.cit.

Table 2. Exposure (Tatter Rate) and Survival of Scots Pine During Plantation Establishment

Location	Plot No.	Tatter Rate (cm^2/day)	Origin of Plant Material[1]	Percent Survival Year 1	~Year 6
Carrick	1	11.2	Thetford	98	27
Carrick	2	9.0	Thetford	100	16
Carrick	5	14.2	Elveden	86	30
Carrick	7	15.5	Elveden	94	46
Kielde	56	8.3	Thetford	91	Very Poor
Lael	14	4.5	Thetford	94	Very Poor
Hoy	1	9.6	Thetford	84	2
Hoy	4	4.6	Thetford	93	50
Strathy	1	6.3	Achnashellach	95	Good
Strathy	2	6.3	Achnashellach	93	Good
Skiall	1	7.2	Achnashellach	100	95 (Yr 3)
Skiall	2	8.9	Moray	64	60 (Yr 15)
Drumtochty	3	7.5	Moray	79	72 (Yr 5)
Queens	17	5.8	Speymouth	85	83 (Yr 4)
Kielder	68	9.7	Ballochbuie	89	Poor
Glencoe	1	7.3	Achnashellach	90	Good

[1]Thetford and Elveden are relatively sheltered, low-lying areas of eastern England. Achnashellach, Moray, Speymouth, and Ballochbuie are at varying elevations and exposures in Scotland.

The effect of site-specific factors on growth of Scots pine has been studied in Great Britain by Day (1946), Adu (1968),[1] Cook et al. (1977), Morgan (1972),[2] and White (1982) at sites at or below commercial planting limits. Site conditions appear to have different effects on the growth of Scots pine and Sitka spruce as exposure increases. At mean daily tatter rates of 12 square centimeters per day or more, the growth of Sitka spruce declines sharply. In contrast, growth of Scots pine is slower and declines less than Sitka spruce with increasing site exposure. Growth of Scots pine decreased rapidly between flag tatter rates of six and 10 square centimeters per day.

The sharp decline in growth of Sitka spruce at tatter rates greater than 12 square centimeters per day indicates that wind speed reduction on exposed sites would markedly improve forest productivity. The mutual shelter of trees has a very significant impact on wind speeds within a forest block. Papesch (1984) found that wind speed inside a Sitka spruce stand was only 35 to 70 percent of the wind speed at an equivalent height at the forest edge. Subsequent thinning of the trees increased the wind speed within the stand to 75 to 85 percent of that at the edge. Other studies also show that wind speeds quickly decline with distance from the forest edge. Wood (1990) showed that mutual shelter at a distance of 2.5 tree heights from the leading edge of a uniform Sitka spruce stand reduced wind speeds by 50 percent and enhanced survival and tree growth.

---

[1] Adu, S. V., 1968, Studies of land capability assessment for Scots pine in Strathdon, University of Aberdeen, Scotland, UK, unpublished Ph.D. thesis.

[2] Morgan, A. L., 1972, The effects of soil, site and climatic factors on the growth of Scots pine (Pinus sylvestris) and heather (Calluna vulgaris), University of Aberdeen, unpublished Ph.D. thesis.

Given these findings, Sitka spruce is unlikely to survive on exposed and elevated sites when planted in small and scattered blocks. Once a plantation is established, however, mutual shelter significantly improves the growth of trees some distance from the windward edge of the stand. The productivity of second rotation crops might also be improved by retaining part of the initial crop for shelter.

Nutrition

Because of its outstanding success in upland Britain, Sitka spruce has been established on sites of varying fertility. The value of adding phosphate fertilizer at planting was recognized soon after the first high-altitude plantations were established (Zehetmayr, 1954). As plantations aged, however, further application of phosphate was necessary at some sites, and deficiencies of nitrogen and potassium became apparent. More recently, interest has grown in increasing the productivity of established forests and in using less fertile sites for production forestry.

To quantify the growth response of Sitka spruce to fertilizer and increasing elevation, the results of three high-elevation nutrition experiments were analyzed. The location, elevation, and exposure of the three high-elevation experiments are shown in Table 3a. Although effects of the major elements [nitrogen (N), phosphorus (P), and potassium (K)] could not be separated, the growth effects of combined NPK applications at standard and luxury rates (Table 3b) were assessed and compared with growth of Sitka spruce on nutrient-deficient control plots.

Results of the high-elevation nutrient experiments showed that as site exposure increased, height growth and other benefits from fertilization decreased (Fig. 4). At high exposures, factors other than nutrition appeared to limit tree growth. The results indicated that at tatter rates between 8 and 11.6 square centimeters per day, tree growth no longer increased at luxury fertilizer rates.

At Balblair, the most sheltered site with a tatter rate of 8 square centimeters per day, fertilization promoted growth. In contrast, at Hamsterley Forest, the most exposed site (tatter rate = 14 square centimeters per day), application of NPK fertilizer at either standard or luxury rates did not affect height growth. Visual observations suggested that the standard

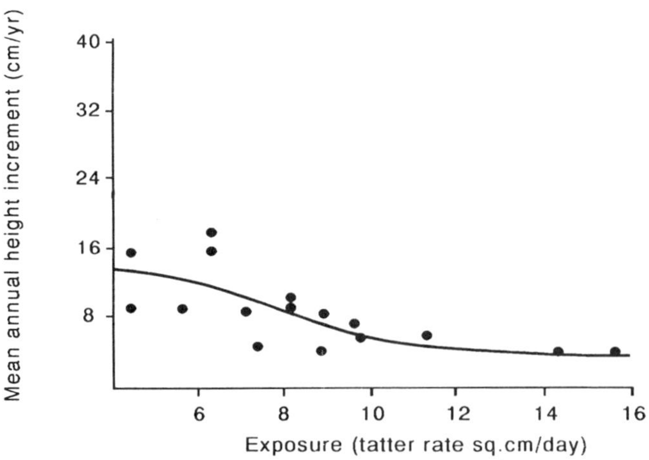

Fig. 4. The effect of exposure and nutrition on height growth of Sitka spruce in Britain.

Table 3a. Location and Exposure (Tatter Rate) of Sitka Spruce[1] Nutrition Trials at High Elevations in Britain

Location	Elevation (m)	Tatter Rate (cm^2/day)
Balblair	305	8.00
Hamsterley	648	14.45
Helmsdale	315	11.60

[1] Provenance: Queen Charlotte Island

Table 3b. Fertilizer Regimes for 20 Years After Plantation Establishment

Element	Application Rate Kg/ha		
	Deficient	Standard	Luxury
Nitrogen	0	700	1400
Phosphorus	150	150	300
Potassium	100	300	600

and luxury fertilizer treatments may have increased diameter growth and promoted crown development, however.

Trees on the most exposed sites have a higher foliar nitrogen content than do those on sheltered sites, presumably because trees are restricted in their ability to use nitrogen for growth (J. C. Dutch, Forestry Authority Commission, Northern Research Station, Roslin, Scotland, personal communication). High foliar nitrogen on exposed sites indicates that trees take up nutrients from the soil but are restricted in their nitrogen metabolism and growth response. This observation is supported by Nordmeyer (1980), who found that nutrient concentrations in conifer needles were independent of elevation.

In summary, the application of fertilizer, even at luxurious rates, is ineffective when climate limits growth. Climate and edaphic factors should be kept in mind when determining upper limits for planting, or when contemplating top dressing of severely exposed crops (Taylor, 1991). Both experimental and empirical data for high-elevation sites demonstrate that there is a threshold level of exposure above which fertilization is not justified financially.

Species Mixtures

As commercial afforestation expanded onto the heather (*Calluna vulgaris* (L.) Hull.)-dominated uplands of eastern Scotland and northern England in the 1930s, Sitka spruce

Table 4. Location[1] and Exposure of Mixed Sitka Spruce and Lodgepole Pine Species Trials

Location	Plot Number	Elevation (m)	Exposure (cm^2/day)
Balblair	1	305	8.00
Hamsterley	13	648	14.45
Helmsdale	1	294	11.61

[1] Sitka spruce is from the Queen Charlotte Islands and lodgepole pine is from south coastal origins.

often went into "check" following good early growth (Kerr et al., in press). Heather quickly reinvaded cultivated sites, and effective control was found to improve tree growth for only a short period before the spruce became nutrient deficient. Later observations showed that the nutritional status of spruce planted near pine or larch was much improved. The rapid early growth of pine and larch suppressed the heather and increased available nitrogen, allowing the spruce to become established.

Similar nursing practices also led to the widespread use of Sitka spruce and lodgepole pine mixtures during the late 1950s and 1960s on nutrient-deficient peats, peaty gley, and peaty ironpan soils. Unfortunately, many of the early mixtures failed because seed origins of lodgepole pine were unsuitable. Pine from south coastal regions in Washington and Oregon were often used because they grew fast and were relatively resistant to wind damage. In many cases, the early fast growth of pine suppressed the spruce. Pine provenances from coastal Washington and Oregon have coarse heavy branches, butt sweep, and generally poor form. At high elevations, the open crown structure of pine makes it vulnerable to severe snow damage, particularly on hill plateaus.

The nursing and nutritional benefits led to an increase in the use of mixed crop species at high elevations. A number of the high-elevation trial plots established by the Forestry Commission incorporated mixtures of spruce, commonly with lodgepole pine. Unfortunately, as with other early mixed plantings, unsuitable provenances of lodgepole pine were often chosen. A study of three high elevation trials of both mixed spruce and pine (Table 4) and of pure fertilized spruce crops revealed that, as shown in the fertilizer trials, the ability of spruce to respond to standard nitrogen applications declines with increasing exposure (Fig. 5).

At high exposure, the benefits of nursing practices decrease and mixed crops perform little better than pure ones. The uneven canopy of mixtures, particularly where heavily branched pine is a component, may increase air turbulence and the adverse effects of exposure. This effect has been noted by Savill (1983) in mixed stands of broadleaved species and spruce.

Other Species

Japanese larch (Larix kaempferi Sarg.), hybrid larch (Larix xeurolepis Henry), noble fir (Abies procera Rehd.), Serbian spruce (Picea omorika (Panic) Purkyne), Scots pine, and western hemlock (Tsuga heterophylla (Raf.) Sarg.) were also planted in a matrix of lodgepole pine at a number of the high-elevation sites. Most of the sites proved to be

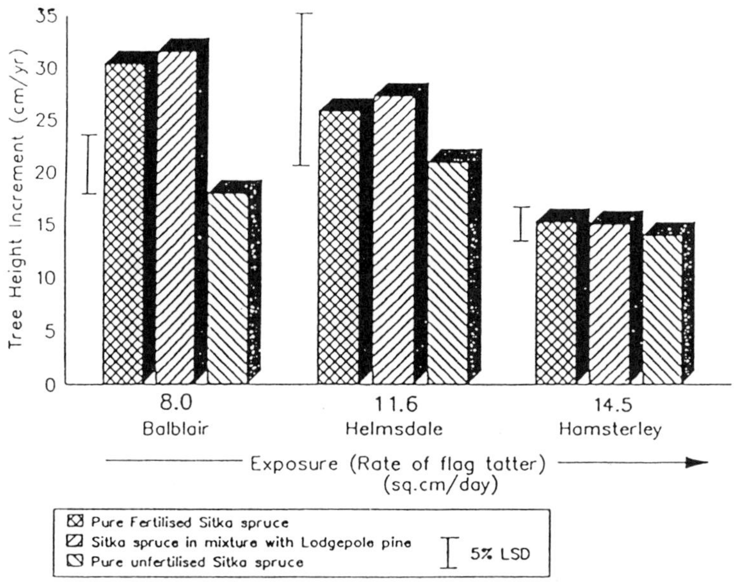

Fig. 5. The effect of exposure on height growth of Sitka spruce mixed with lodgepole pine.

severe tests for these species. Survival was generally poor and height growth was variable. Most species did not achieve the growth required for commercial afforestation in Britain. The larches showed the greatest potential. In particular, hybrid larch showed good height growth and often grew taller than Sitka spruce and lodgepole pine. But it usually had multiple leaders, damaged tops, leaning stems, and poor form. On some sites, Japanese larch reached heights comparable to Sitka spruce. European larch did not grow well at all and may not be adapted to the oceanic climate of Britain.

Some noble fir survived and grew, albeit slowly, on exposed sites. Western hemlock and Serbian spruce performed poorly and appear unsuited to exposed sites in Britain.

CONCLUSIONS

The Forestry Commission's trial plantations provided invaluable afforestation experience and knowledge of the effects of exposure on tree growth at the commercial treeline in Britain. Although many studies have shown that the relationship of tree growth to elevation and exposure at low elevations is linear, the plots at high elevations in this study indicate that the decline in growth near treeline is curvilinear.

The reduction in growth of Sitka spruce and Scots pine on severe sites is best described as a logarithmic function of exposure. The mutual shelter between both individual trees and forest blocks can significantly reduce wind speeds. Thus, large contiguous blocks of woodland in exposed areas are likely to be more successful than scattered blocks of trees, especially at high elevations. Large plantations will facilitate afforestation of upland areas from which woodland has been cleared in the past.

Severe exposure at high elevations negates the benefit of fertilizing commercial stands of Sitka spruce. As a result, fertilization at commercial tree limits cannot be justified in terms of increased tree growth. Further work, including an economic analysis, is required to predict threshold exposure levels.

Although species mixtures, especially Sitka spruce and lodgepole pine or larch, provide nursing and nutritional benefits, the advantages of both are less apparent at high elevations. Conversely, at extreme exposures, the mixed species stands, with their uneven canopy, can increase wind turbulence and provide less protection, and hence may exhibit less height growth than pure species stands.

Results of the species trials in the high-elevation experiments confirm that Sitka spruce and lodgepole pine are the most exposure-resistant exotic conifers currently used in upland afforestation in Britain. Future studies should investigate the application of continuous-cover silvicultural systems in high-altitude forests. Continuous-cover silviculture could aid the introduction of species not currently used as pioneer crops.

REFERENCES

Birse, E. L., and Dry, F. T., 1970, Assessment of climatic conditions in Scotland, Aberdeen, Scotland, UK, Macaulay Institute for Soil Science.

Cook, A., Court, M. N., and Macleod, D. A., 1977, The prediction of Scots pine growth in northeast Scotland using readily accessible site characteristics, Scottish Forestry, 31:251-264.

Day, W. R., 1946, On the effects of changes in elevation, aspect, slope and depth of free rooting material on the growth of European larch, Japanese larch, Sitka spruce and Scots pine in Mynydd Du forest, Forestry, 20:7-20.

Edwards, P. N., and Christie, J. M., 1981, Yield Models for Forest Management, Forestry Commission Booklet 48, HMSO, London, England, UK.

Jack, W. H., and Savill, P. S., 1973, The causes of tattering of flags under natural conditions, Int. J. Biometeor., 17:185-192.

Kerr, G., Nixon, C. J., and Matthews, R. W., [In press], The silviculture and yield of mixed species stands: The UK experience, Proceedings of the British Ecological Society/IUFRO Special Symposium, 1990 September 2–6, [Edinburgh, Scotland, UK], Blackwell Scientific Publications, Oxford, England.

Lines, R., 1963, East Scotland Scots pine provenance trial of 1952, Forestry Commission Report on Forest Research 1963, HMSO, London, England, 127-136.

Malcolm, D. C., and Studholm, W. P., 1972, Yield and form in high elevation stands of Sitka spruce and European larch in Scotland, Scottish Forestry, 26:296-308.

Mayhead, G. J., 1973, The effect of altitude above sea level on the yield class of Sitka spruce, Scottish Forestry, 27(3):231-237.

McIntosh, R., 1981, Fertilizer treatment of Sitka spruce in the establishment phase in upland Britain, Scottish Forestry, 35(1):313.

Miller, K. F., Quine, C. P., and Hunt, J., 1987, Exposure assessment for forestry in upland Britain, Forestry, 60(2).

Neustein, S. A., 1965, Trial plantations at high elevations, Forestry Commission Report on Forestry Research, HMSO, London, England, p. 169-172.

Nordmeyer, A. H., 1980, Tree nutrient concentrations near the treeline, Craigieburn Range, New Zealand. In: Benecke, U., and Davis, M. R., eds., New Zealand Forestry Service Technical Paper, Christchurch, New Zealand, 70:83-94.

Oswald, H., 1969, Conditions forestieres et potentialite de l'epicea en Haute Ardeche (English trans.), Ann. Sci. Forest., 26:183-224.

Papesch, A., 1984, Wind and its effects on Canterbury forests, University of Canterbury, Christchurch, New Zealand, p. 92–94. Ph.D. thesis.

Pears, N. V., 1967, Present tree-lines of the Cairngorm mountains, Scotland, Journal of Ecology, 55:815-829.

Rutter, N., 1966, Tattering of flags under controlled conditions, Agricultural Meteorology, 3:153-165.

Rutter, N., 1968, Tattering of flags in relation to wind and weather, Agricultural Meteorology, 5:163-181.

Savill, P. S., 1974, Assessment of the economic limit of plantability, Irish Forestry, 31:22-35.

Savill, P. S., 1983, Silviculture in windy climates, Commonwealth Agricultural Bureau, Forestry Abstracts, 44(8):473–488.

Smith, L. P., 1984, The agricultural climate of England and Wales, Ministry of Agriculture, Fisheries and Food, Technical Bulletin 35, HMSO, London, England, UK.

Taylor, C. M. A., 1991, Forest fertilization in Britain, Forestry Commission Bulletin 95, HMSO, London, England, UK.

Tranquillini, W., 1979, Physiological ecology of the alpine timberline - tree existence at high altitudes with special reference to the European alps, Springer-Verlag, Berlin, Germany.

Wood, C. J., 1990, Aeroblastic modelling of trees susceptible to windthrow damage, European Commission, unpublished contract report MA1B-0063-UK (BA).

White, E. J., 1982, The relationship between the height growth of Scots pine (Pinus sylvestris) and site factors in Great Britain, Journal of Ecology and Management, 4:225-245.

Zehetmayr, J. W. L., 1954, Experiments in tree planting on peat, Forestry Commission Bulletin 22, HMSO, London, England, UK.

WIND AND WAVE FORESTS: A CASE STUDY AND IMPLICATIONS FOR SILVICULTURE

Alexander Robertson

Forestry Canada
Newfoundland and Labrador Region
St. John's, Newfoundland, Canada

INTRODUCTION

The perception among scientists is that wave forests are rare. This is probably because few wave forests have been described in the scientific and technical literature, and are, therefore, largely outside the purview of forest managers. The number of reports of wave forests is growing, however. For example, until 1985, wave forests in Newfoundland were virtually unknown, although a few foresters mistook them for strange cutting patterns. Understandably, few have heard of a clonal wave krummholz, which I discovered by chance in southern Labrador in 1988.

Dieback in forests and plantations is a general problem in many parts of the world. Some patterns of forest decline are characterized by strips of dead trees and are usually non-cyclical. Non-cyclical wave dieback is often initiated by silvicultural and harvesting practices. Patch and strip cutting is the most common cause of wave dieback and timber losses in residual stands. Dieback strips often regenerate with weed species. In addition, wave-shaped forest profiles result from shoot mortality on windward sections of the canopy (Robertson, 1987a). Quite often, intervention with standard silvicultural routines, such as thinning, increases the risk of endemic windthrow. Finally, infestations of bark beetles and infections by pathogens in gaps created by spot fires and windthrow can also lead to wave dieback.

This paper is a brief synopsis of natural (cyclical) wave forests. A coastal wave forest at Spirity Cove, Newfoundland, and examples of non-cyclical wave dieback are also highlighted.

NATURAL WAVE FORESTS

Balsam fir (Abies balsamea (L.) Mill.) forests in the subalpine zone at 800 to 1,000 meters above sea level in the Appalachian Mountains of the northeast United States, and at all elevations in the coastal oceanic boreal zone of the Maritime Provinces of eastern

Forest Development in Cold Climates, Edited by J. Alden *et al.*
Plenum Press, New York, 1993

Fig. 1. Examples of wave forests on a coastal plain at Spirity Cove (a) and on a hilltop near Rocky Harbor in western Newfoundland (b); a recently discovered wave krummholz in southern Labrador (c); wave dieback induced by clearcutting at Northwest Arm, Hare Bay, northern Newfoundland (d); patch cutting at Ship Cove, Placentia Bay, southeastern Newfoundland (e) and in a larch plantation near Bethel, North Wales (f).

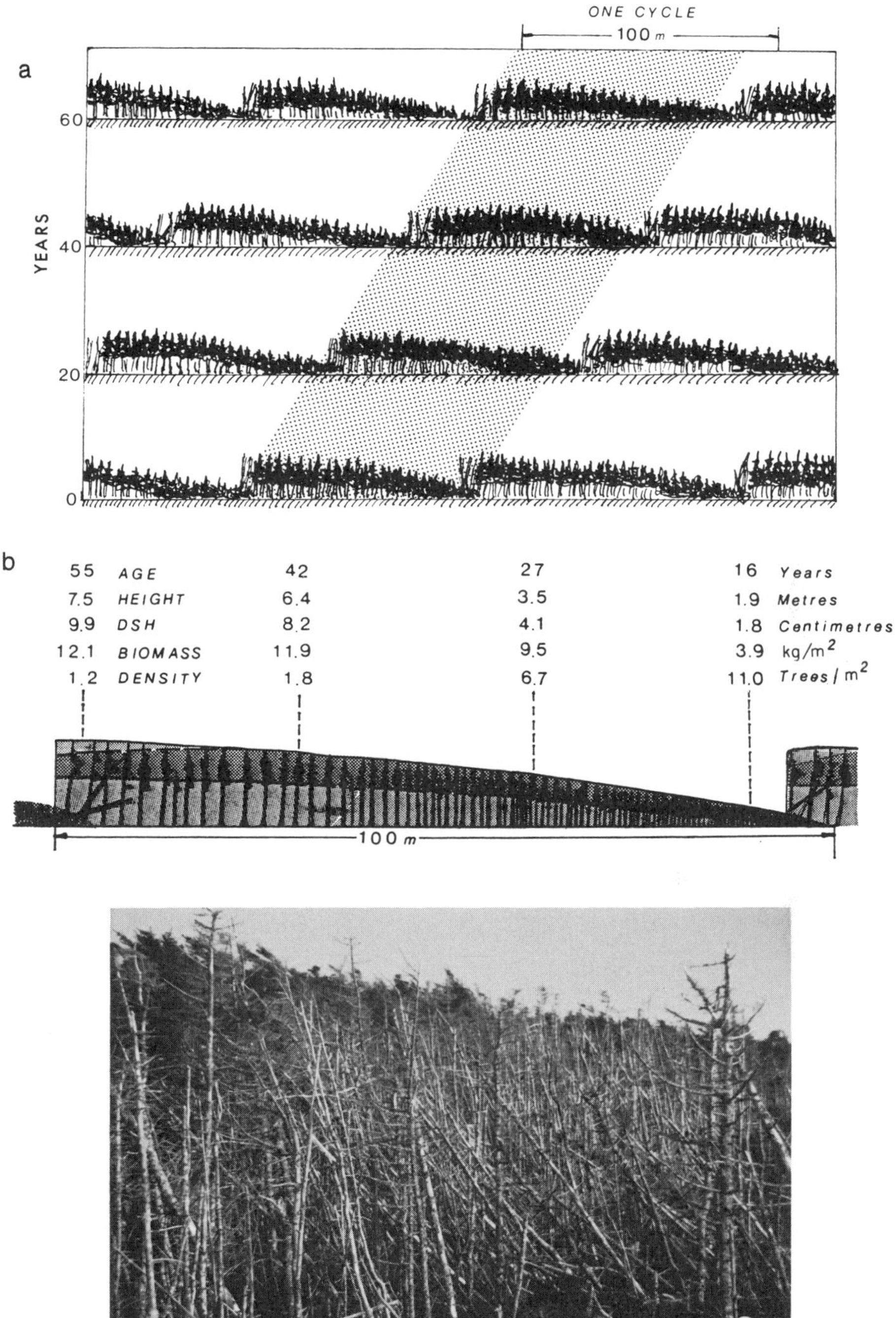

Fig. 2. Profile of a wave forest on a coastal plain at Spirity Cove, northwestern Newfoundland (a, b) with its characteristic dead-tree strip (c). Note dense regeneration.

Canada, are among the windiest in the world. Wind-induced crown and root damage in these forests (Rizzo and Harrington, 1988) produces patch disturbance known as "wave forests" (Robertson, 1987b). The term wave forest is synonymous with the ancient wave dieback in subalpine fir forests in central Japan (Iwaki and Totsuka, 1959), and wave regeneration in subalpine fir forests in the Appalachians (Sprugel, 1976).

Wave forests consist of dead tree strips that "move" across the landscape in wave-like formation in 55- to 80-year cycles in northeastern America and 60- to 100-year cycles in Japan. The profile of a wave forest resembles saw-teeth as the curvature of each stand progresses from dense regeneration to mature decadence and finally, to standing dead-tree strips. In all cases, sexual regeneration is the principal mode of rejuvenation. For the many aspects of Japanese and American wave forests, papers by Kimura (1984) and Sprugel (1984a) on stand dynamics, including population structure, biomass, and nutrient cycling, are of particular interest.

In Japan, Veitchi fir (Abies veitchii Lindl.) and Maria's fir (A. mariesii Mast.) subalpine wave forests on Mt. Shimagare (meaning "mountain with the dead-tree strips")

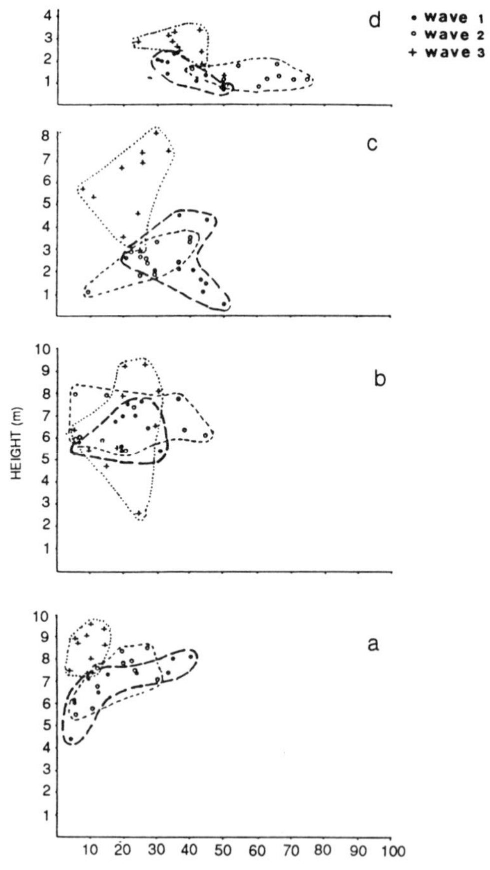

Fig.3. Chronosequence of height in relation to crown biomass as a percentage of total above-ground biomass based on a sample of 10 trees in four plots representing sapling a), immature (b), semi-mature (c), and mature (d) stages, respectively, within three contiguous wave cycles in the Spirity Cove wave forest. Wave cycle #1 is 46 meters long, #2 is 100 meters long, and #3 is 126 meters long.

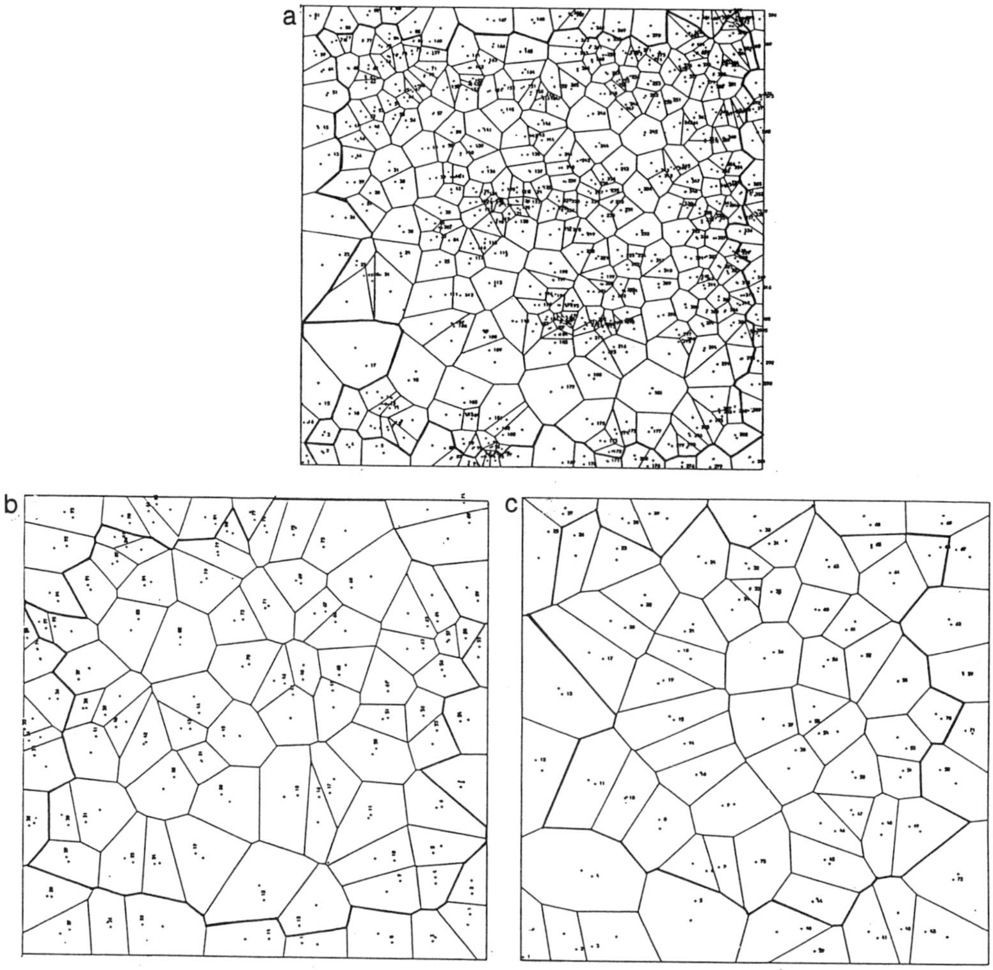

Fig. 4. Spatial point patterns of trees in the Spirity Cove wave forest from Dirichlet tessellations. The chronosequence progresses from clustered in a 31-year-old stand (a), to a Poisson forest (random) in a 39-year-old stand (b), and, finally, towards regularity in a senescent 56-year-old stand (c).

have been known for centuries (Oshima et al., 1958). The dynamics and principle features of Japanese wave forests have been studied by Iwaki and Totsuka (1959), Kimura (1984), and Todha and Kikuchi (1981).

In the northeastern United States, wave forests restricted to subalpine balsam fir forests, notably on the southwestern slopes of the Adirondack Mountains, are major scientific attractions (Lang, 1984; Marchand, 1984; Sprugel, 1984b).

Sprugel (1984b) notes that wave forests "...are apparently restricted to high-altitude areas of northeastern U.S.A. and central Japan, and have not been described elsewhere in the world." Although never formally described in the scientific literature, fir wave forests on coastal hills in the maritime provinces of eastern Canada and on the French Islands of St. Pierre et Miquelon have been known for decades.

The wave forest at Spirity Cove (longitude 57°20′ W., latitude 50°36′ N.) was first recognized as such in 1985. Until then, local woodsmen considered wave forests to be strange cutting patterns. The wave forest complex is discontinuous over an area of approximately 100 square kilometers and is probably the largest and certainly the most unique wave forest known (Fig. 1a). Since 1985, I have observed hundreds of wave forests on hills, plains, and valleys throughout Newfoundland (Fig. 1b). In addition, a clonal "wave krummholz," which regenerates entirely by layering, was found at L'Anse-aue-Clair in southern Labrador (Fig. 1c).

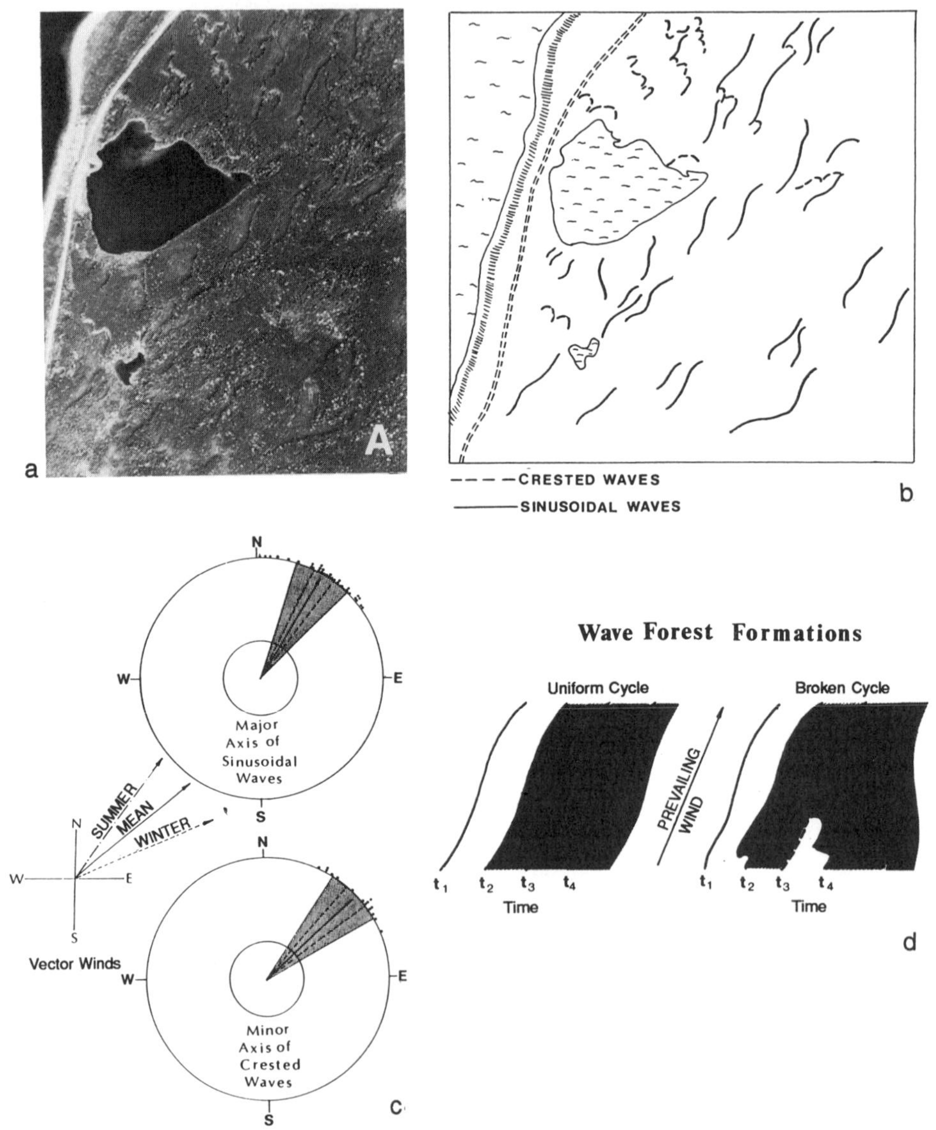

Fig. 5. Aerial photograph of Spirity Cove wave forest (a); map highlighting location of dead-tree strips (b); circular statistical analysis of the directionality of crested and sinusoidal dead-tree strips (c); and the time course of wave formations including uniform (steady-state) and broken (discontinuous) wave cycles (Robertson, 1987b).

Table 1. Age, density, height, diameter and biomass of sample stands in three contiguous wave forest cycles at Spirity Cove, Newfoundland

Stems per square meter	Age (years)		Height (meters)		Biomass (kilograms per tree)	
	Mean	SD	Mean	SD	Mean	SD
Wave Cycle #1 (length 48 meters)						
1.0	55	3.6	6.81	1.1	11.9	10.8
1.4	44	2.6	6.48	0.1	5.5	3.8
5.9	25	1.3	2.24	1.2	0.1	1.3
12.5	17	1.2	1.47	0.5	0.2	0.2
Wave Cycle #2 (Length 100 meters)						
1.4	54	3.0	7.31	1.0	8.2	6.6
1.9	44	2.3	6.09	1.6	6.7	4.6
11.8	25	1.0	2.47	0.7	0.7	0.8
14.1	12	0.7	1.34	0.3	0.2	0.2
Wave Cycle #3 (Length 126 meters)						
0.7	57	3.3	8.44	0.8	9.29	3.9
1.1	39	5.1	6.59	2.0	7.76	8.1
2.2	32	3.0	5.82	1.7	5.74	5.7
6.4	20	1.3	2.8	0.9	1.22	1.4

Wave Dieback

Wave dieback on forest edges is usually non-cyclical and is initiated by human disturbance such as clearing for roads, power transmission lines, farming, settlement, and logging. Cutting and wave dieback are common throughout the boreal and temperate forests of North America (Fig. 1d and 1e). Wave dieback also occurs in plantations (Fig. 1f).

SPIRITY COVE WAVE FOREST

General Features

The most striking feature of the Spirity Cove wave forest is the sharply defined mature canopy edges fringed with a narrow strip of standing dead trees (Fig. 2). Underneath the standing dead tree strips is very dense balsam fir regeneration, some of which may have been suppressed for a decade or more. Age varies little at any point in a wave forest profile. Regardless of length, all wave cycles reach a maximum average age at 55 to 60 years. The wave forest cycle has five distinct stages:

1. 0–15 years: mass regeneration (up to 140,000 seedlings per hectares), canopy contiguous;

2. 15–25 years: influx of new seedlings ceases, stands strongly self-thinning with contiguous crowns;

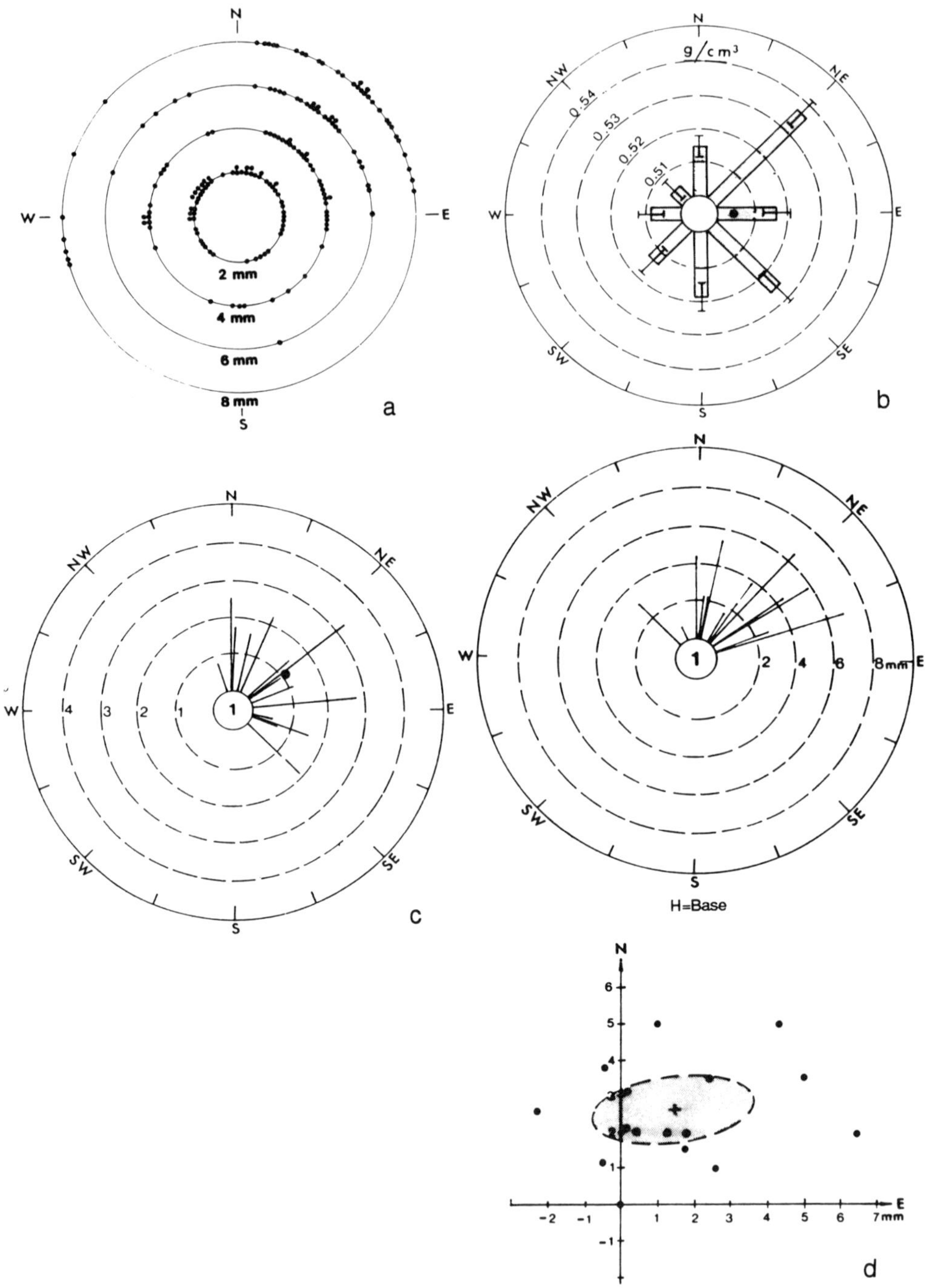

Fig. 6. Bole wood properties in Spirity Cove wave forest trees: (a) circular distribution of compression wood zones at two millimeter intervals from the pith; (b) centroid (center of mass) of wood density variation across the bole at 1/6th tree-height; (c) vector distribution of mean density weighted by tree ring asymmetry of 16 sample discs (the dot denotes the location of the mean vector); (d) vector eccentricity (tree ring asymmetry) distribution of 16 basal discs and Hotelling's confidence ellipse confirming the statistical significance of directionality at P = 0.05 level of significance.

3. 25–40 years: trees mature, stands lightly self-thinning with widely separated crowns;

4. 40–50 years: senescence begins, self-thinning ceases, crowns become deformed with little change in crown separation;

5. 50–60 years: main canopy dies standing, with highly directional sporadic windthrow in final stages.

Stand Structure

Table 1 gives basic statistics on the structure of three contiguous wave forest cycles at Spirity Cove, Newfoundland. Stand density includes all stems greater than one centimeter in diameter at the base, and ranges from a maximum of 14 stems per square meter when the influx of new cohorts ceases, to one stem per square meter in the final stages of decadence. Stand density of the Spirity Cove wave forest is described by the linear equation

$$\log D = -3.142 + 233.1\ (1/\text{age}),$$

where D = stems per square meter. From this equation, mortality is constant at roughly nine percent per year. A hand-fitted linear curve, however, suggests that mortality rate decreases from twenty-seven percent per year in the initial self-thinning stage, to three percent per year in the final stages of forest decline. Furthermore, the relationship between tree mass and stand density, represented by the equation

$$\log_{10} W = 2.7 - 1.5 \log_{10} D,$$

where W = above-ground dry weight (kilograms per tree) and D = density (trees per square meter), has the same slope as the well-known -3/2 self-thinning rule (Yoda et al., 1963).

Within the self-thinning zone, from 15 to 40 years, the initial size structure is strongly skewed, with very weak evidence of bimodality at age 55 years. The apparent bimodality means that larger trees would form a distinct sub-population. The best-fit equation relating height to diameter is

$$\text{Height} = 3.311\ (\text{DSH})^{0.3507},$$

where DSH = diameter at stump height.

The height-weight relationship is

$$Height = 8.447(1 - e^{-0.1786W})^{0.4301}$$

The above-ground biomass yield (W) is a reflection of wave speed; in other words, in the shorter wave cycle length (48 meters), which moves slower, the total above-ground biomass yield curve is exponential. Within the longer wave cycle lengths (100 and 120 meters, respectively), the maximum biomass peaks around 30 years, decreases rapidly until age 40 years, and then levels off before ending abruptly as the stand dies synchronously at 55 years.

Canopy Structure

Cannell (1985) contends that after canopy closure in plantations, more of the growth is partitioned into bole wood and relatively less into crown biomass as the tree declines with age. While this may be true of plantations, it may not necessarily hold for natural forests. The first stage in the cycle of a natural wave forest, that is, mass regeneration, ensures that crown closure is complete within three to five years after seed germination, and that five- to ten-meter tall saplings are released from suppression under a decadent overstory. Neighbouring crowns in the first stage of a wave cycle are contiguous. Because the overlapping crown structure allows very little light to penetrate to the forest floor, the forest floor is littered with necromass, a few clumps of moss, herbs, an occasional creeping or dwarf shrub, and suppressed fir seedlings one to two centimeters tall. Also, the very high density of the fir overstory restricts the invasion of shade-intolerant trees such as birch (Betula L.) and herbs; those established at the very earliest regeneration phase are soon shaded out by the fast-growing fir.

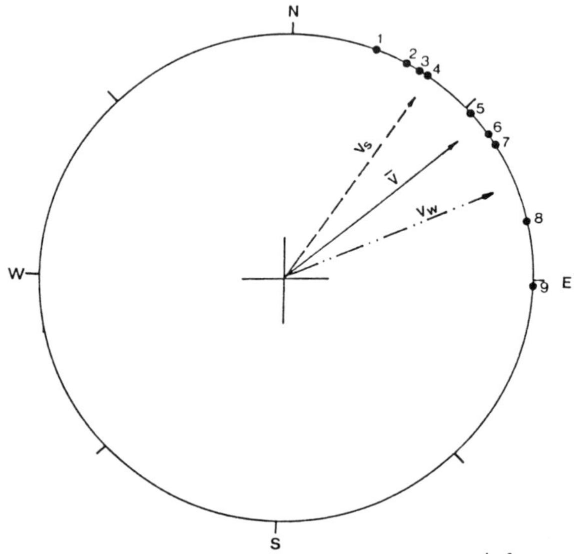

Figure . Directional relationship between vector wind directions and asymmetric tree variables.

1. Direction of compression wood zones at a distance of 6 mm and over from the pith.
2. Mean centroid of tree crowns in 55 year old stands (corrected by adding 90°).
3. Direction of bole (disc) eccentricity.
4. Longitudinal axis of sinusoidal wave fronts.
5. Perpendicular axis of crested wave fronts.
6. Mean vector weighted mean wood density.
7. Mean vector ring width.
8. Mean direction of wind-thrown trees in sinusoidal wave fronts.
9. Centroid of total mean wood density.

Fig. 7. Angular distribution of 9 variables in relation to prevailing mean vector (V), summer vector (V_s), and winter vector (V_w) winds, respectively: 1=compression wood, 2=centroid of tree crowns, 3=bole eccentricity, 4=sinusoidal waves, 5=crested waves, 6=vector weighted mean wood density, 7=vector ring width, 8=direction of wind thrown trees, and 9=centroid means of wood density (Robertson 1987b, 1990, 1991).

Fig. 8. The degenerating effects of cutting patterns on residual stands: clear cutting in a Sitka spruce (Picea sitchensis (Bong.) Carr.) plantation (A) exposes edges of residual stands to windthrow (B) at Kershope Forest, northern England. Typical patch cutting in a boreal forest (C) has a similar effect. While regeneration may be excellent, dieback of saplings exposed above the snow-line (D) may result in krummholz. Centuries of indiscriminate cutting on coastal sites (E) in the harsh coastal sites around Newfoundland (F) have converted forests to tundra. Reforestation of the barrens is usually not successful, as shown by an attempt to reforest this barren near Arnold's Cove, eastern Newfoundland.

After the second stage (no influx of new cohorts and onset of vigorous self-thinning), the canopy opens and remains at a fairly constant level of crown cover until senescence. Furthermore, the crown basal area to the drip-line of mature trees is virtually the same as the crown basal area of dominant young saplings; in fact, the proportion of crown biomass per tree actually decreases with age (Fig. 3). Apparently, the decrease in relative biomass with increasing height is mainly a function of abiotic stresses, such as damage to the lower crowns by constant tree sway and beating against neighboring crowns, and stripping of young twigs by glaze and wet snow (Robertson, 1987b). On approaching senescence, the crown canopy opens and more light penetrates to the forest floor on which carpets of moss, principally Pleurozium schreberi (BSG) Mitt., become barriers to fir seed germination and seedling survival. Disturbance of moss carpets by a fallen branch or an uprooted tree (usually a dead one), and even large animal imprints and fecal deposits, provide microniches for the occasional fir sapling.

Cyclical Spatial Point Patterns

Analysis of sequential spatial point patterns across wave forest cycles was based on the distribution of contact numbers (number of sides or neighbors) of Dirichlet tessellations. Although spatial pattern of trees may be described by more complex methods, the Dirichlet (also known as Voronoi), tessellations (tiles) were used because of their simplicity. The contact number of a tile represents the number of nearest neighbors for each tree. A tile is constructed by drawing perpendicular bisectors midway on the line connecting a subject tree and its nearest neighbor and linking the bisectors. Dirichlet tessellations for three sequential plots in one wave forest cycle (Fig. 4) show an obvious sequential change in point patterns from clustered to random, and finally towards slight uniformity at 31, 39, and 56 years, respectively.

With the exception of contact numbers for age class 55 years, the distribution of contact numbers representing spatial point patterns of 15- to 45-year-old stands is not statistically significant from a Poisson process generated by Crain (1972). The curves show a higher proportion of trees with fewer neighbors, indicating a tendency towards clustering, and a tendency toward random distribution during the self-thinning phase close to a Poisson forest. Finally, towards senescence (55 years), the contact number has a leptokurtic curve peaking at six, indicating a trend toward uniformity. The analysis indicates that throughout the living parts of a wave cycle, wind has little influence on spatial patterns beyond that considered to be normal self-thinning. This is surprising, considering the persistent abiotic damage inflicted on the crowns (primarily by swaying).

Regardless of environmental factors, spatial analysis demonstrates that seedlings die from competition when a stand reaches its maximum density and natural thinning processes set in.

Orientation of Wave Cycles

Direction of the wave fronts and their movement at Spirity Cove are strongly correlated with prevailing southwesterly winds.

Two basic shapes of dead tree strips, crested and sinusoidal, are easily identified on an aerial photograph (Fig. 5a) and highlighted to show that crested wave fronts are located close to the shore, while sinusoidal wave fronts predominate farther inland (Fig 5b). Also, crested wave fronts move directly upwind (to the northeast), while sinusoidal wave fronts move roughly parallel to the prevailing wind (eastward) (Fig. 5c).

Windthrow within dead-tree strips is oriented in an easterly direction by storm winds from the west. The majority of dead trees, however, usually remain standing for several years before they are broken or uprooted by storm winds.

The wave motion at Spirity Cove ranges from 0.6 to 3.2 meters per year for sinusoidal (steady-state) wave fronts, and 3.1 to 4.5 meters per year for crested wave fronts. The slowest moving and shortest wave cycles are found on the most exposed landscape, that is, close to shore.

Apparently, crested wave fronts are "broken" cycles, or sporadic events that originate from gaps within, or at the edge of, a sinusoidal wave cycle, whereas sinusoidal wave fronts are "uniform" (steady-state) cycles. Over time, a leading edge develops on the east side of a crested wave that evolves into a steady-state sinusoidal wave cycle, and synchronizes its motion with a series of neighboring leading edges (Fig. 5d).

The precise aeromechanics controlling the shape and movement of wave cycles are unknown, although small-scale helical roll vortices generated by the land-sea interface and perhaps turbulence similar to "honami" winds may be involved. Honami winds are downward gusts that cause the wave motion across fields of grass and wheat (Finnegan, 1979).

Cone Production

In Japanese subalpine wave forests, cone production is constant once stands reach maturity (Kohyami, 1982). Most adult trees produce cones every four years, and the net reproductive effort peaks mid-way through a wave cycle at about 50 years of age.

Cone production in the Spirity Cove wave forests suggests that the age of peak net reproductive effort is not constant and may be controlled more by wave speed or distance than by age. For example, although the majority of cones are produced on taller trees, the net reproductive effort peaks in immature stands of the longest wave cycle. And although the number of cones per tree peaks in older stands in the other two wave cycles, the pattern is not consistent with the actual number of cones produced. As to the distribution of cones on a tree, each age group has a whorl that produces the most cones. The distribution of cones is coincident for the 27- to 42-year-age group. Incidentally, the quantity of cones does not necessarily reflect the quality of seed, because the incidence of insect damage is far less in cones borne on very young stands than in cones on mature trees.

The soil seed bank was not studied and, as a consequence, actual germination processes in the Spirity Cove wave forest are unknown.

Forest Turnover and Soil Dynamics

Neither forest turnover nor soil dynamics have been studied at Spirity Cove. Nevertheless, basic features such as the mosaic of microtopography throughout a wave forest have distinctive cyclical changes in both forest turnover and soil taxonomy.

Only on occasion over the length of a cycle between, but not including the dead-tree strips, are self-thinned trees windthrown or knocked down by large animals. Thus, dead trees disintegrate while standing and bole and bark falls at an exponential rate. Litter fall (needles and twigs), on the other hand, is much more sporadic because of the variation in duration and severity of mechanical damage by wind, glaze, and heavy snow that strip twigs, shoots, and foliage from various parts of the crown, especially from windward sides.

Within the dead tree strips, bole and bark fall is more complex. Bark and small twigs are added to the forest floor necromass first. This is followed by tops of skeletons snapped off by wind and glaze. Occasionally strong westerly gales blow down some trees; a few will fall flat on the ground, but most hang in snags for several years before they disintegrate at roughly the same rate as upright, dead trees. Litter from needle and shoot fall is minimal because few trees have live crowns in the dead-tree strips. A considerable volume of dead branches and twigs complements the bolewood and bark on the forest floor.

The contrasting modes of bole and litter fall within the live canopy and dead tree strips produce very different forest floor microtopographies. Within the live canopy of stands between 20 and 40 years of age, the microtopography is remarkably smooth. Fallen trees, usually no more than three and five centimeters in diameter at the base, rot quickly and are covered by feather moss. Although the microtopography may appear homogeneous, the distribution of necromass is quite variable and produces a mosaic of soil types. As the stand density decreases with age, especially after 45 years, the forest floor microtopography is transformed by pedoturbation from a relatively smooth surface to undulations between large tree buttresses. The change is caused by either increasing penetration of soil frost as the canopy opens, or movement of the root plate by wind loads that exceed the stress tolerance of trees to aeromechanical forces, or both. Soil frost action may raise the level of depressions in the forest floor from five to 10 centimeters in winter, particularly if the soil is very moist. In addition to frost heaving, windthrow in dead-tree strips increases pedoturbation, particularly in redistributing the soil mechanical fraction through "hinge" and "rotational" uprooting. Hinge uprooting lifts soil vertically and leaves a pit and mound. Rotational uprooting, on the other hand, is a ball and socket action that causes root shearing and pit-scouring around the edges, and leaves the pit topped with a rill. Frost action tends to smooth out the mounds, litter fall tends to fill in the pits. The majority of root plates in the dead tree strips, however, remain in their original position as a contribution to the subterranean necromass.

There is a heterogeneous "ribbon" of rapid soil taxonomic changes under a dead-tree strip, in sharp contrast to the comparatively benign changes under the live canopy. As the soil taxonomic cycle begins again in the early stages of a dead-tree strip, regeneration, necromass accumulations from the litter of self-thinning, and the humus of rotted boles and bark fill in the pits and smooth the microtopography. The next period of forest floor perturbation begins when the canopy is predisposed to synchronous decline by recurring wind movement of the root mass. Such movement contributes to the formation of depressions between trees as shifting root plates rise to new heights, depending on the degree and frequency of the wind. Soil frost-lift also decreases root-plate stability, because soils and root masses are shallow. Under these circumstances, wind stress rather than nutrient deficiency is the primary triggering mechanism for decline of trees in a wave forest.

Wood Properties

Wind as the primary "trigger" for wave forest decline at Spirity Cove is apparent in the asymmetric formation of crown and bole wood of aging stands. In fact, the prevailing wind direction and vectors of crown and wood formations are closely correlated. For example, centroid (center of mass) plots of crown mass in the Spirity Cove wave forest show that as tree height increases, crowns are more susceptible to deformation by prevailing southwesterly winds (Robertson, 1987b).

Similarly, the effects of wind on wood properties and bole taper increase as stands open in early senescence. Surprisingly, compression wood was found in only two thirds of

90 boles sampled in mature stands at Spirity Cove and was restricted to the base and generally within 10 centimeters of the pith. When present, compression wood zones were distributed randomly within two millimeters of the pith (Fig 6a). As the stem grew radially beyond two millimeters, however, compression wood zones were distributed northeast of the pith, presumably because of the prevailing southwesterly wind. For this reason, only boles that lean develop extensive compression wood. The low frequency of compression wood in many boles suggests that, under wind loading, boles are less subject to flexural stresses than root plates are to tilting (Robertson, 1990). Few boles developed a permanent lean, even when exposed at the dead-tree strips. But both boles and rootplates move on windy days, which explains why most of the flexural stress, and most of the compression wood, occurs in the vicinity of the root collar.

Wood density of Spirity Cove wave forest trees ranged from 0.50 to 0.54 grams per cubic centimeter. This density is very high when compared to 0.35-0.40 grams per cubic centimeter for normal balsam fir wood. Also, the centroid of wood density across the bole shifted to the leeward (northeast) of center (Fig 6b).

The most obvious relationship between prevailing wind direction and bole asymmetry is seen in asymmetric tree rings, expressed in terms of mean ring width (Fig. 6c) and centroid of tree ring width in individual discs (Fig. 6d).

The directional relationship of crown asymmetry, and various wood characteristics and orientation of dead tree strips in relation to the prevailing wind, are summarized in Figure 7.

WAVE KRUMMHOLZ

In 1988, I found a series of lateral dead-tree strips in a balsam fir krummholz at L'Anse-au-Clair, on the south coast of Labrador. The wave krummholz rejuvenated entirely by layering, as does much of the fir krummholz in the region. Krummholz is a manifestation of degeneration from forest to epicormic shrubs of the same species, caused by indiscriminate cutting during the past 450 years of European settlement. The lateral dead-krummholz strips in the vicinity of L'Anse-au-Clair (Fig. 1c) is the only known wave krummholz (Robertson, 1988). Unfortunately, because of its location, it has not been formally surveyed or studied, and little hard data on the dynamics of the wave krummholz exist.

Because no seedlings were observed anywhere in the krummholz, it is assumed the wave krummholz is clonal. Unlike wave forests, however, it is uncertain whether the dieback is cyclical or not.

A cursory examination, based on measurement of tree rings of four discs taken from a layered section, suggests that the subsurface layering mass is at least 113 to 120 years old. Dieback of the canopy may occur at about 80 to 90 years of age. This indicates that layered branches grow exceedingly slowly for several decades before radial increments are appreciable, presumably because aerial shoots must grow through the previous layering mass. Considerable work, however, is needed to understand the dynamics of the wave krummholz.

It is clear that the krummholz in the general area of southern Labrador exists because of two factors: degeneration of the forest by centuries of dysgenic cutting dating back to the 16th Century Basque whaling activities at Red Bay, Labrador, and the inhibiting effects of cold on-shore winds on tree growth.

In summer, the live canopy in the wave krummholz is rejuvenated by uneven-aged layering to one to two meters tall. In the early growth stages, aerial shoots are long, slender, and supple. Limber shoots are compressed to within 0.5 meter to one meter off the ground in winter by heavy snow pack and are protected from desiccating winds. Some of the lowest branches, which are particularly long and slender, eventually touch the ground and become embedded in the subsurface humus and layer. In other cases, vertical shoots lose their elasticity as they mature and acquire the habit of small erect trees that protrude above the snowline and are exposed to winter winds. The principal aeromechanical causes of dieback in the wave krummholz are ice abrasion, "winter burn" mortality of needles, and damage to one-year-old shoots from wet snow and glaze. One-year-old shoots often litter the forest floor or the surface of snow packs after gales.

Patches of dieback in krummholz are common, but dieback strips characteristic of wave krummholz are found only on the middle and upper parts of west-facing slopes. The direction of crown flagging of tall trees scattered in the broad valleys is seaward (south), indicating a localized prevailing surface wind flow that is northeasterly. This suggests that northeasterly winds, flowing over a north-south ridge, generate lee vortices sufficiently strong and persistent to cause lateral bands of krummholz dieback.

The principal aerodynamic features of patch krummholz dieback are unknown. Apart from sporadic violent gusts, certain turbulent wind patterns can consistently affect some patches more than others, depending on their age and physiognomy. Among the possibilities are downward momentum gusts of honami winds. Honamis, which occur at all intensities, may be partially responsible for the mosaic of krummholz patterns and patch dieback.

Apart from scientific curiosity, the coastal krummholz displays the fragility of forest ecosystems in climates that are marginal for tree growth. Such forests, reduced to krummholz by human exploitation, fire, and abnormal climatic variations, can take centuries to regain their former stature. Furthermore, reforestation efforts are uneconomical in regions where layering is the principal mode of natural regeneration. This is an indication that environments are too harsh for silviculture.

HUMAN-INDUCED WAVE DIEBACK

Wave dieback is a common feature at the edge of forests and plantations even in climates that are not windy. Wind accounts for substantial timber loss and considerable modification of natural habitat (Fig. 8). Examples of human-induced wave dieback are progressive windthrow on the edges of a clearcut in monocultured Sitka spruce (Picea sitchensis (Bong.) Carr.) plantations (Fig. 8a and 8b), wave dieback following patch cutting of balsam fir (Fig. 8c and 8d), and conversion of coastal balsam fir forest to tundra by generations of indiscriminate cutting (Fig. 8e and 8f). Furthermore, regeneration problems are evident at each site. At Kershope, northern England, the site is colonized mainly by sedges and grasses (Fig. 8b) that require costly replanting and weed control. Following wave dieback in western Newfoundland, natural balsam fir regeneration is abundant, but the young saplings sustained severe winter injury above the snowline (Fig. 8d) and may form krummholz. In harsh environments of the Avalon Peninsula, eastern Newfoundland, efforts to restore the barrens to productive forest have failed and furrowing has resulted in serious soil erosion by wind, rain, frost, and snow melt (Fig 8f).

Among a number of windthrow hazard classification systems intended primarily for plantations, the one designed for the British Isles (Miller, 1985) is probably the best known. In addition, many models are available for predicting the stability of single trees (Couttes, 1986; Galinski, 1989; Ossenbruggen et al., 1986). Windthrow hazard classifications specific to forest edges of residual stands of cutover areas in the boreal region were developed by Elling and Verry (1978) and Fleming and Crossfield (1983). Models for predicting wave dieback, which often precedes windthrow and, especially in harsh climates, leads to forest degeneration, have not been written.

Silvicultural and harvesting systems, notably strip-harvest systems, reforestation with species mixtures, windward thinning, canopy profiling, and soil mounding, for example (Mathews, 1989), have been designed to eliminate or reduce dieback of coniferous forests. Few of these methods are successful in creating wind-firm edges (Savill and Evans, 1986) or eliminating wave dieback.

With the trend towards multiple land use, diversity of habitats is a major criterion for forest design. Urban forests that provide year-long shelter from wind, fuelwood, aesthetics, wildlife habitat, and watershed protection are prime examples of multiple use forests. These forests can also serve as indicators of appropriate forest composition and design for windy climates. For example, in a cool, windy climate with a short growing season, the average height of natural balsam fir forests near St. John's, Newfoundland, ranges from 10 to 15 meters. On the other hand, height of mature conifers such as tamarack (Larix laricina (Du Roi) K. Koch) and exotic blue spruce (Picea pungens Engelm.) in the reasonably sheltered inland location of Bowring Park, range from 15 to 20 meters (Fig. 9a), and the height of mature broadleaves, such as exotic maples (Acer L. spp.) and beech (Fagus L. spp.) on the more exposed Government House grounds located near the coast (seven kilometers east of

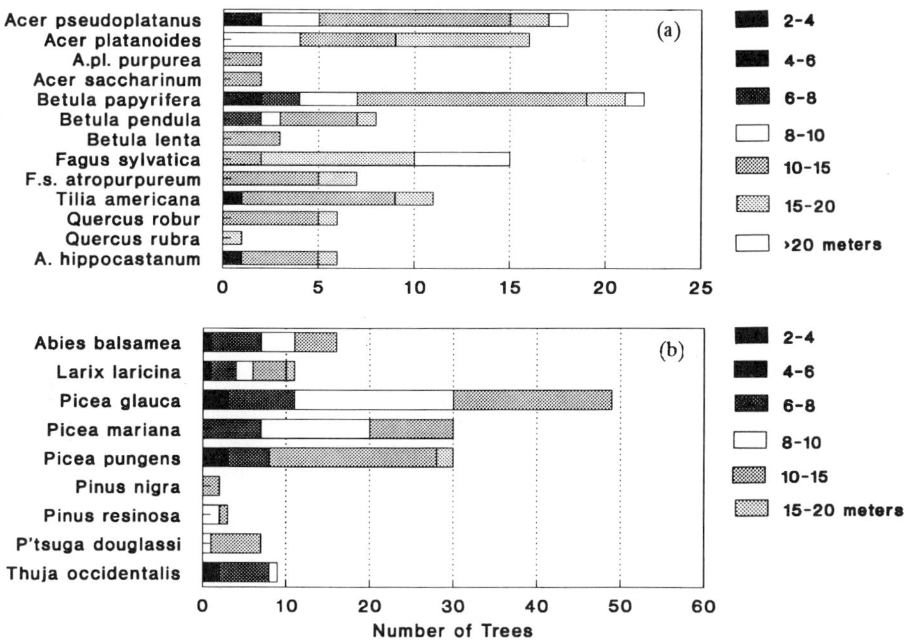

Fig. 9. Height growth of most of the mature broadleaves on the grounds of Government House (a) and a sample of conifers in Bowring Park (b). (Data provided by the Parks Department, City of St. John's, Newfoundland.)

Bowring Park) exceed 20 meters (Fig. 9b). Windthrow is virtually unknown at both sites and dieback involves only sporadic trees of short-lived species such as native birch (Betula papyrifera Marsh.) and balsam fir.

Unfortunately, potential wind-resistant plantations have not been formally studied in the boreal region. Nevertheless, urban and forest plantations provide valuable information about the adaptability of exotic species to harsh climates.

CONCLUSIONS

Wave forests provide valuable insight into the chronosequence of forest development in windy climates and are more common than the scientific literature suggests.

Wave krummholz forests regenerate vegetatively and reflect a natural restoration process in a harsh climate following degeneration of forest cover over many decades of indiscriminate cutting, and by fire and abnormal climate.

Wave dieback in forest edges of residual stands within patch and strip cutovers mimics wave forests to some extent. Wave dieback incurs substantial losses in timber resources and transformation of natural habitats including the degeneration of forests to krummholz and tundra. While many models have been developed for predicting windthrow on forests edges, few, if any, pertain to wave dieback. Although a number of silvicultural options are available for reducing windthrow and wave dieback on forest edges, most pertain to plantations in temperate regions and are unsuccessful in boreal regions.

Beyond this brief synopsis of natural wave forests and human-induced wave dieback, research applicable to wave forest management, guidelines for managing forest edges more conservatively, and criteria for designing plantations appropriate to windy climates has barely begun.

ACKNOWLEDGEMENTS

I wish to thank the organizing committee of the NATO-Advanced Research Workshop for the opportunity to present this paper at Laugarvatn and my colleagues in Forests Canada for their support in this study. I am indebted to Dr. John Alden and the editors for their valuable comments.

REFERENCES

Cannell, M., 1985, Dry matter partitioning in tree crops. In: Cannell, M. G. R., and Jackson, J. E., eds., Attributes of trees as crop plants, Institute of Terrestrial Ecology, Abbots Ripton, Huntingdon, p. 160-193.

Couttes, M., 1986, Components of tree stability in Sitka spruce on peaty gley soil, Forestry, 59:173-197.

Crain, I., 1972, Monte Carlo simulation of random Voronoi polygons: Preliminary results, Search, 3:220-221.

Elling, A., and Verry, E., 1978, Predicting wind-caused mortality in strip cut stands of black spruce, For. Chron., 54:249-252.

Finnegan, J., 1979, Turbulence in waving wheat. II. Structure and momentum transfer, Boundary-Layer Meteorol., 16:213-236.

Fleming, R., and Crossfield, R., 1983, Strip cutting in shallow-soil upland black spruce near Nipigon, Ontario. III. Windfall and mortality in the leave strips: Preliminary results, Information Report O-X-354, Great Lakes Forest Research Centre, Sault Ste. Marie, Michigan, 27 p.

Galinski, W., 1989, A windthrow-risk estimation of coniferous trees, Forestry, 62(2):139-146.

Iwaki, H., and Totsuka, T., 1959, Ecological and physiological studies of Mt. Shimagare. II. On the crescent-shaped "dead-tree" strips in the Yatsugatake and Chichubu Mountains, Botanical Magazine Tokyo, 72:255-260.

Kimura, M., 1984, Changes in population structure, productivity, and dry matter allocation with the progress of wave regeneration of Abies stands in Japanese subalpine regions. In: Wareing, R.H., ed., Carbon uptake and allocation in subalpine ecosystems as a key to management, Oregon State University, Oregon, p. 57-63.

Kohyami, T., 1982, Studies of the Abies population of Mt. Shimagare. II. Reproductive and life history traits, Botanical Magazine Tokyo, 95:167-182.

Lang, G., 1984, Forest turnover and the dynamics of bole wood litter in the subalpine balsam fir forest, Canadian Journal of Forest Research, 15:262-268.

Marchand, P. J., 1984, Dendrochronology of a fir wave, Canadian Journal of Forest Research, 14:51-56.

Mathews, J., 1989, Silvicultural systems, Oxford University Press, Oxford, 284 pp.

Miller, K., 1985, Windthrow hazard classification, Forestry Commission Leaflet No. 85, HMSO, London, 14 p.

Oshima, Y., Kimura, M., Iwaki, H., and Kuroiwa, S., 1958, Ecological and physiological studies of the vegetation of Mt. Shimagare. I. Preliminary survey of the vegetation of Mt. Shimagare, Botanical Magazine Tokyo, 71:289-300.

Ossenbruggen, P., Peters, M., and Shigo, A., 1986, Potential failure of a decayed tree under wind loading, Wood Fibre Sci., 18(1):169-186.

Rizzo, D., and Harrington, T., 1988, Root movement and root damage of red spruce and balsam fir on subalpine sites in the White Mountains, New Hampshire, Canadian Journal of Forest Research, 18:991-1001.

Robertson, A., 1987a, Use of trees to study wind, Arboricultural Journal, 11:127-143.

Robertson, A., 1987b, The centroid of tree crowns as an indicator of abiotic processes in a balsam fir wave forest, Canadian Journal of Forest Research, 17(7):746-755.

Robertson, A., 1988, New type of wave forest discovered in southern Labrador, Woody Points, 17(3):3-4.

Robertson, A., 1990, Directionality of compression wood in balsam fir wave forest trees, Canadian Journal of Forest Research, 20(8):1143-1148.

Robertson, A., 1991, Centroid of wood density, bole eccentricity and tree ring width in relation to vector winds in wave forests, Canadian Journal of Forest Research, 18(1):73-82.

Savill, P., and Evans, J., 1986, Plantation silviculture in temperate regions. Oxford University Press, Oxford, 246 p.

Sprugel, D., 1976, Dynamic structure of wave-regenerated Abies balsamea forests in northeastern United States, J. Ecology, 64:889-911.

Sprugel, D., 1984a, Density, biomass, productivity, and nutrient-cycling changes during stand development in wave-regenerated balsam fir forests, Ecol. Mon., 54(2):165-186.

Sprugel, D., 1984b, Characteristics, causes and consequences of wave regeneration in high-altitude forests. In: Turner, H., and Tranquillini, W., eds., Establishment and tending of subalpine forests: Research and management, Proceedings 3rd Workshop P 1.07-00, 1984, Anst. forstl. Versuchswes., Ber. 270:179-188.

Tohda, H., and Kikuchi, T., 1981, Regeneration of Abies mariesii forests in the Hakkoda Mountains. In: Ecological studies on the Abies mariesii forest, The Mtn. Hakkoda Botanical Laboratory, Sendal.

Yoda, K., Kira, T., Ogawa, H., and Hozumi, K., 1963, Self-thinning in overcrowded pure stands under cultivation and natural stand conditions, J. Inst. Polytechnics, Osaka City University, Series D14:107-129.

THE ALPINE TIMBERLINE OF TIBET

Li Bosheng

Institute of Botany
Academia Sinica
Beijing, China

TIMBERLINE IN TIBET

The Tibetan Plateau, surrounded by high mountains, lies mostly above altitudinal tree limit. Above timberline, the climate is arid and frigid, and is obviously not suited for forest development. East of the plateau and in many southern valleys, however, dense forest vegetation is quite well developed. The total forested area is about 6,130,000 hectares, approximately 5.1 percent of the total area of Tibet (Li et al., 1985). Evidently the favorable climate is caused by warm and moist air currents that travel inland along river valleys from the Indian Ocean; consequently, the forest vegetation extends inside the plateau and forms a distinct horizontal treeline with shrub-meadow vegetation. The eastern treeline is located near 92° E. longitude and extends from Gyayü through Nangxian to Gongbogyamda, Lhari, Biru, Dêngqêng, and finally to Riwoqê. To the south, treeline is near 28°30′ N. latitude, extending from Gongdang and Gyirong County through Nyalam, Karda, Yadong, and Lhozhag to Gyayü (Fig. 1).

With exception of major species differences, the forest-shrub ecotone at horizontal tree limits is very similar to the taiga-to-tundra transition zone in subarctic Eurasia and North America (Walter, 1979). The horizontal treeline is caused by loss of moisture and heat from the warm air currents that rise from the inland valleys to the high plateau (Chang, 1981; Zheng et al., 1979). As altitude increases, the forest ecosystem changes to "dry-cold" shrub-meadow vegetation. The taiga-to-tundra transition, on the other hand, is caused by low temperature as the angle between the sun and earth and seasonal variation in photoperiod increase with latitude.

Vertical Distribution

Forest vegetation in Tibet is mainly restricted to the extremely variable topography of the high mountains. Moisture and temperature decrease sharply with increasing elevation, so the forest vegetation gradually changes from tall trees to high mountain shrub-steppe vegetation forming a clear vertical (altitudinal) timberline. The elevation of timberline in

Forest Development in Cold Climates, Edited by J. Alden *et al.*
Plenum Press, New York, 1993

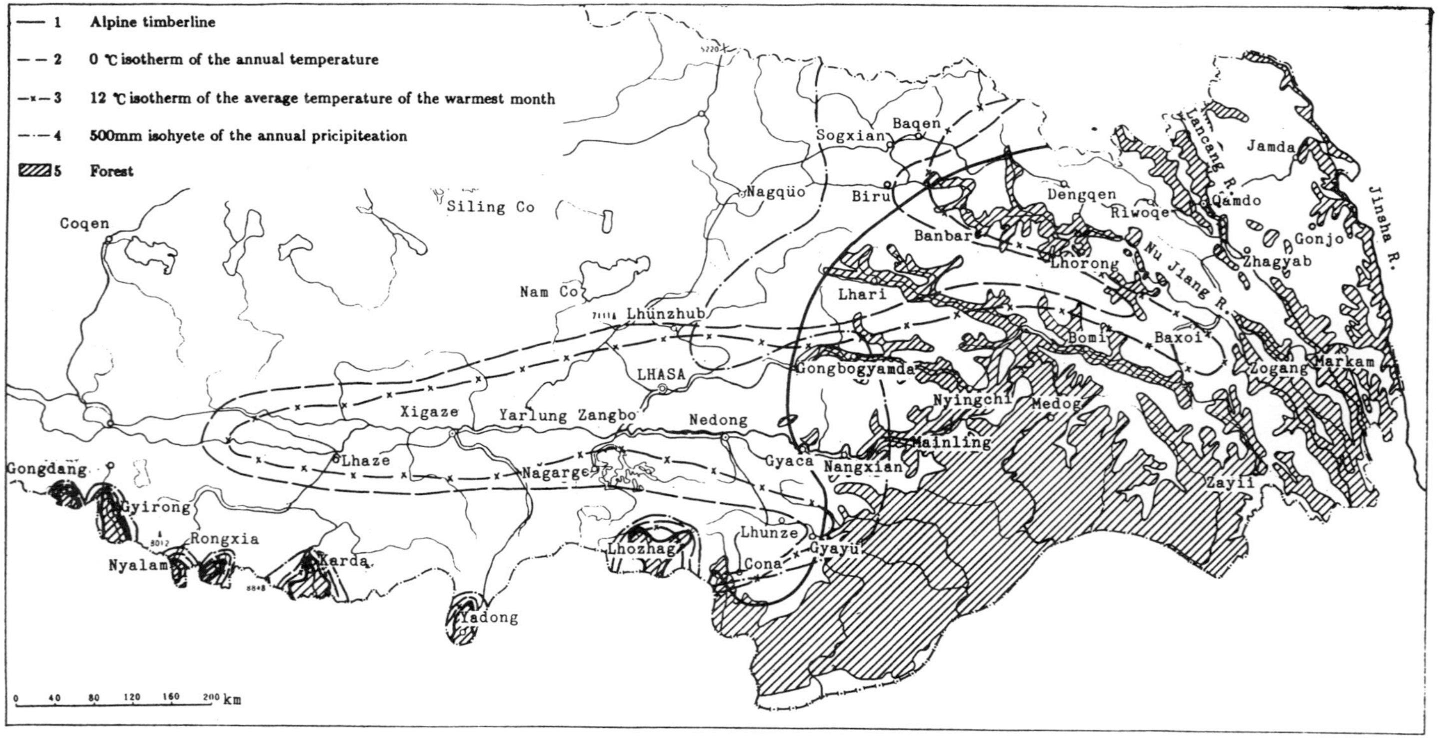

Fig. 1. The alpine timber line of Tibet.

Tibet varies geographically. Timberline is lowest in the East Himalaya montane region of southeast Tibet, where it is only 3,600 to 3,800 meters above sea level (masl). Along the Himalaya Mountain Range westward, timberline gradually increases from 3,900 to 4,100 masl and is accompanied by an increase in drought.

North of the East Himalaya montane region, timberline also increases with altitude (Zheng et al., 1975). Near Mainling, it is 4,000 to 4,100 masl, and at Nyingchi and Gongbujiangda it is between 4,400 and 4,600 masl. This is the highest timberline in the world. Further north at Dêngqêng and Jamda, however, the timberline decreases to between 4,000 and 4,200 masl (Fig. 2).

DIVERSITY OF TIMBERLINE COMMUNITIES IN TIBET

Because of variation in climate among individual valleys, the complex landforms in the periphery of the plateau region, and the unique history of the geological uplift of Tibet, the plant communities of the timberline forests are extremely diverse (Fig. 3). Timberline vegetation in Tibet includes not only the common evergreen and deciduous coniferous forests, but rare deciduous broadleaf and the unique sclerophyllous evergreen broadleaf forests as well (Zhang et al., 1988).

Evergreen Coniferous Forest

The evergreen coniferous forest is the major timberline vegetation of Tibet (Li et al., 1985). Major timberline conifers belong to the fir (Abies Mill.), spruce (Picea A. Dietr.), and juniper (Juniperus L.) genera. Fir forests are mostly flaky-barked fir (Abies squamata Mast.), Smith fir (A. georgei (Orr) Melville var. smithii (Vig. & Gauss.) Cheng & L. K. Fu), motuo fir (A. delavayi Franch. var. motuoensis Cheng & L. K. Fu), and Himalayan fir (A. spectabilis (D. Don) Mirb. or A. densa W. Griff. ex Parker) communities. Spruce forests are mainly Balfour spruce (P. likiangensis (Franch.) Pritz var. balfouriana (Rehd. & Wils.) Hillier ex Slavin) and Linzhi spruce (P. likiangensis var. linzhiensis Cheng & L. K. Fu) communities; big cypress (Cupressus gigantea Cheng & L. K. Fu), dense branchlet juniper (Juniperus convallium (Rehd. & Wils.) Cheng & Wang), black juniper (J. wallichiana Hook. f. ex Brandis), blackseed juniper (J. saltuaria (Rehd. & Wils.), Himalayan juniper (J. recurva (Buch.-Ham. ex D. Don) Antoine), and Tibet juniper (J. tibetica Kom.) communities constitute the juniper and cypress forests.

Fir forests are usually found near the moist mountain timberline of southeast Tibet and on the south slopes of the Himalayas, where annual rainfall averages 800 to 1,000 millimeters. Spruce timberline forests, on the other hand, are usually found on the northern slopes of the Hengduan Mountains (the Transverse Mountains), where annual precipitation averages 600 to 800 millimeters, while the juniper and cypress timberline forests are found on the dry sunny slopes of central and northern Tibet that average 400 to 600 millimeters of precipitation per year.

Deciduous Coniferous Forest

Only a few pure deciduous conifers exist within timberline vegetation types. They interweave to form a mosaic in the evergreen coniferous and deciduous broadleaf forests. Major deciduous conifers of treeline communities are the larches, potanin larch (Larix

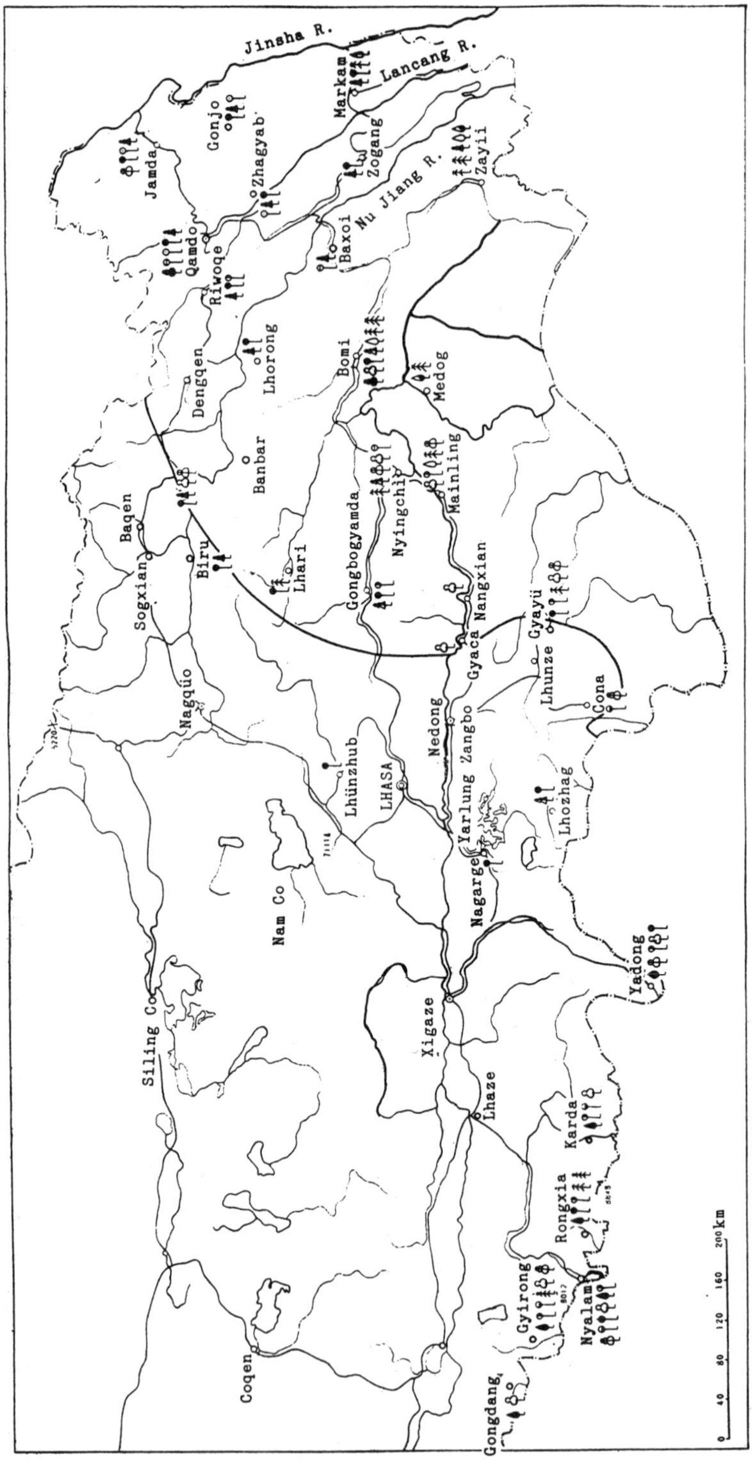

Jinsha R.
Lancang R.
Nu Jiang R.
Markam
Gonjo
Jamda
Zhagyab
Zogang
Zayü
Baxoi
Qamdo
Riwoqe
Lhorong
Dengqen
Bomi
Medog
Banbar
Bagen
Biru
Sogxian
Lhari
Gongbogyamda
Nyingchi
Mainling
Nangxian
Gyaca
Gyayü
Lhunze
Cona
Nagqu
Nedong
Lhünzhub
LHASA
Yarlung Zangbo
Lhozhag
Nagarge
Nam Co
Yadong
Siling Co
Xigaze
Lhaze
Karda
Rongxia
Gyirong
Nyalam
Gongdang
Coqen
0 40 80 120 160 200 km

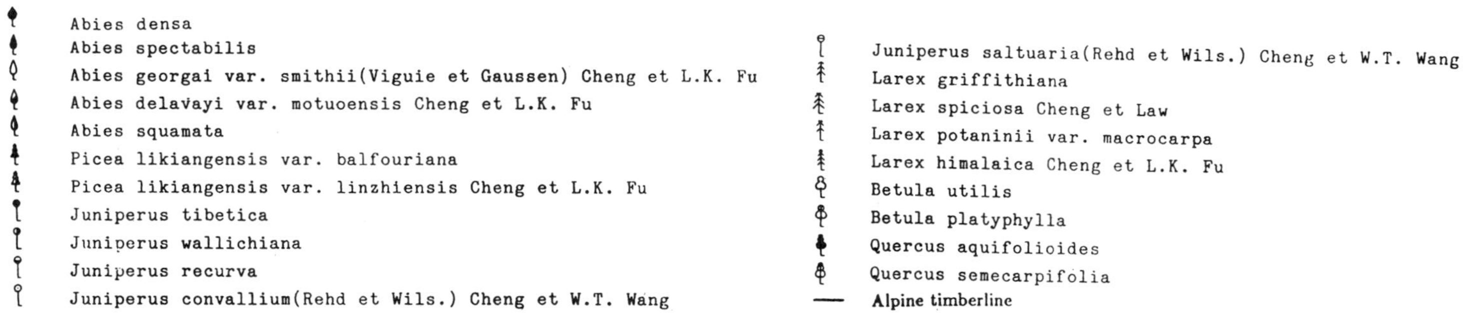

Fig. 2. Distribution of tree species in Tibet's timberline forests.

Fig. 3. The alpine timberline of southeast Tibet. Major species of timberline communities are Picea likiangensis var. linzhiensis, Juniperus tibetica, Abies georgai var. smithii, and Betula utilis.

potaninii Batalin var. macrocarpa Law), distributed at Gyayü and Markam among other treelines; Himalaya larch (L. griffithiana (Lindl. & Gord.) ex Carr.), distributed at Bomi, Nyingchi, Gongbogyamda, Mailing, Gyayü, Yadong, Karda, and Rongxia; Himalayan larch (L. himalaica Cheng & L. K. Fu), distributed at Gyirong; and Nujiang larch (L. speciosa Cheng & Law), found at Mêdog.

Deciduous Broadleaf Forest

The deciduous broadleaf forest communities are mainly composed of Himalaya birch (Betula utilis D. Don) and Asian white birch (B. platyphylla Sukacz.). Both species form timberline in the central Himalaya and the south slopes of the East Himalaya montane region. Betula constitutes the secondary timberline from natural succession after Picea likiangensis var. balfouriana and the A. spectabilis (or A. densa) and A. delavayi var. motuoensis forests, found in the Hengduan Mountains and at Bomi, Nyingchi and Gongbogyamda, have been damaged by cutting and fire.

Sclerophyllous Evergreen Broadleaf Forests

The hollyleaf-like oak (Quercus aquifolioides Rehd. & Wils.) and alpine oak (Q. semecarpifolia Smith) communities are the main components of the sclerophyllous evergreen broadleaf forests that form timberlines on dry and sunny mountains, and at timberline habitats too severe for evergreen coniferous and deciduous broadleaf forests. Q. aquifolioides is the principal sclerophyllous evergreen timberline species in eastern Tibet; Q. semecarpifolia is found only on the southern slopes of the central Himalaya (Fig. 2).

FACTORS THAT DETERMINE THE TIMBERLINE IN TIBET

Climate

The leading climatic factors that position the timberline in Tibet, and determine timberline plant communities, are temperature, moisture, and deep, persistent snow.

According to observation and calculation, the mean annual temperature of Tibet's timberline is approximately 0 °C (Fig. 4). The average temperature at timberline for the coldest month is -9 °C and the lowest is -27 °C. All temperature indices at timberlines in Tibet are far higher than the same indices of altitudinal timberline at equivalent latitudes elsewhere in the Northern Hemisphere. But the average temperature of the warmest month (10 °C) is universal for all altitude timberlines of temperate climates. Average temperature of the warmest month at horizontal tree limits in Tibet ranges from 10 to 20 °C (Gao et al., 1985) (Fig. 5) Obviously, then, the temperature during the growing season is not always the most important environmental element determining the timberline of the Tibet Plateau.

While considering the effects of heat on the position of timberlines, one anomaly is worth pondering: in general, calculations show that the average annual temperature (T) decreases by about 0.6 °C per 100 meter increase in altitude (H), according to the relationship

$$T = 26.04 - 0.61H, \quad r = 0.987.$$

Theoretically then, timberline in Tibet should be at 3,100 to 3,300 masl. But the actual timberline is found at more than 4,000 masl, where mean annual temperatures range from 2 to 8 °C and the mean temperature of the warmest month (T_W), calculated from

$$T_W = 32.28 - 0.57H, \quad r=0.987,$$

ranges from 10 to 15 °C. Mean temperatures calculated for treeline are even colder, ranging from -1 to +2 °C on an annual basis and 7 to 10 °C for the warmest month.[1]

The basic cause of such a unique phenomenon is that timberline communities of the Qinghai-Tibet Plateau react to temperatures of the central troposphere, which is close to the earth's surface at timberline altitudes. On the south slope of the East Himalaya, the relatively low tree line at 3600 to 3800 masl is caused by heat loss due to excessive evapotranspiration; whereas in Ningchi, Gongbogyamda, and Lhorong, where the timberline is 4600 masl, warm and moist air currents from the Indian Ocean cool and dry as they ascend the river valleys. In Tibet, moisture as well as temperature is a major element in determining timberline. Recent research in atmospheric physics indicates that the Qinghai-Tibet Plateau is a huge "hot island" that rises into the troposphere of East Asia (Ye and Gao, 1979). The powerful thermal and orographic effects of the plateau hinder atmospheric circulation in the Northern Hemisphere, especially in Asia (Gao et al., 1984). Therefore, the moist air masses from the Indian Ocean and the heat sink of the Tibet Plateau bring together the best combination of moisture and temperature for forest growth at high altitudes, and the highest timberline in the world.

[1] Li Wenhua, Commissioner for Integrated Survey of Natural Resources, Chinese Academy of Sciences, Beijing, China.

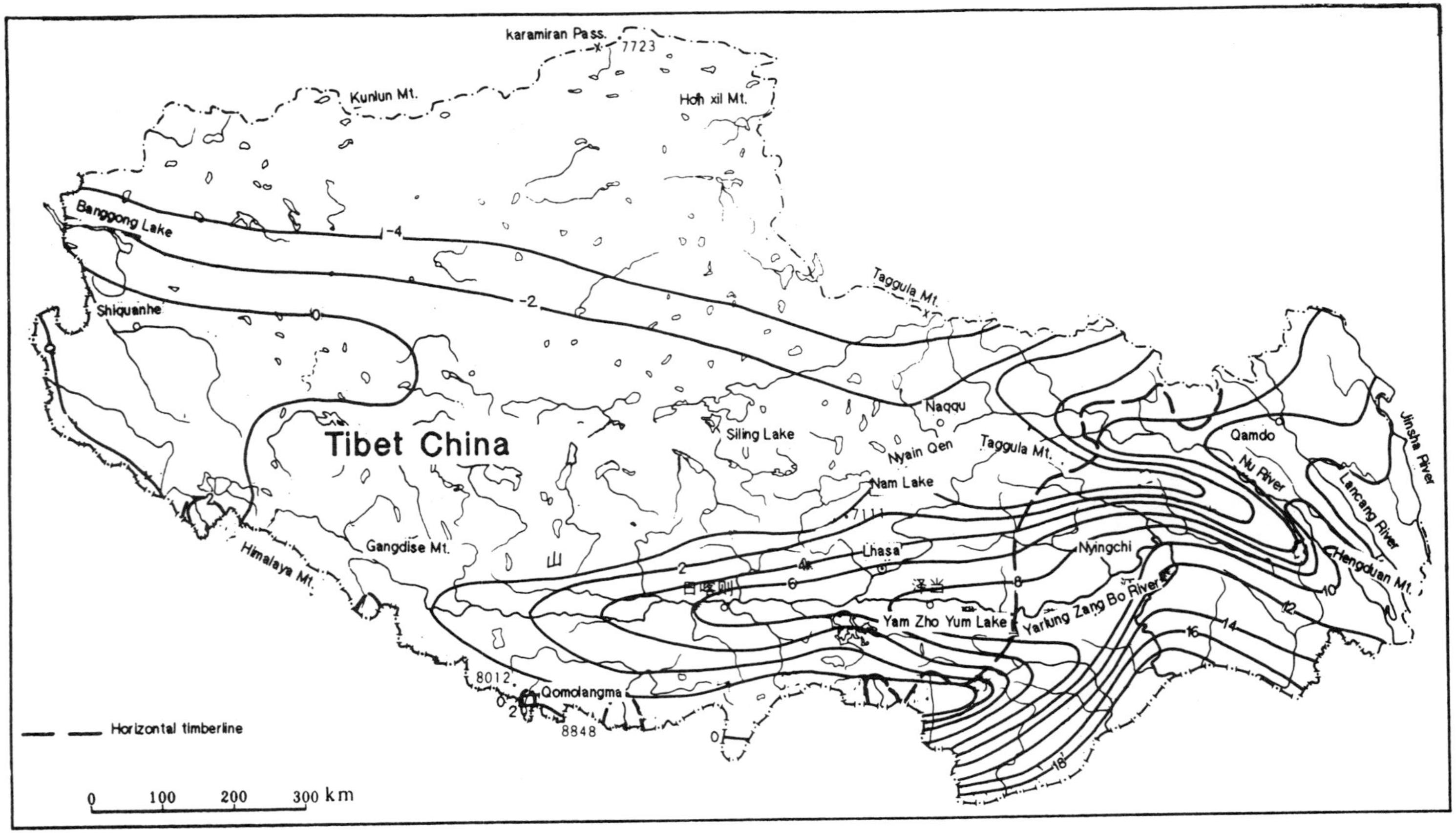

Fig. 4. Mean annual temperatures (°C) of Tibet.

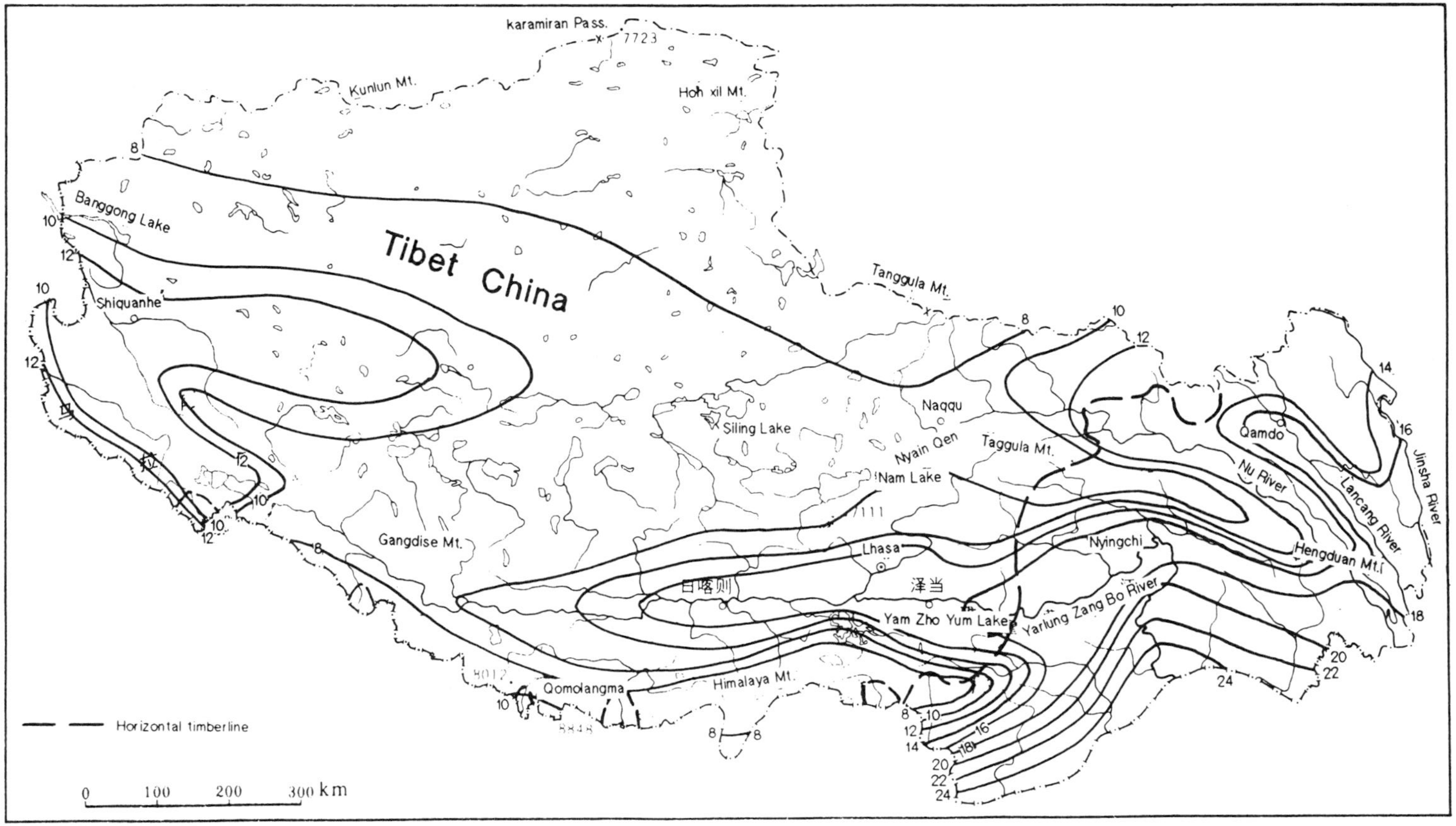

Fig. 5. Mean July temperature (°C) of Tibet.

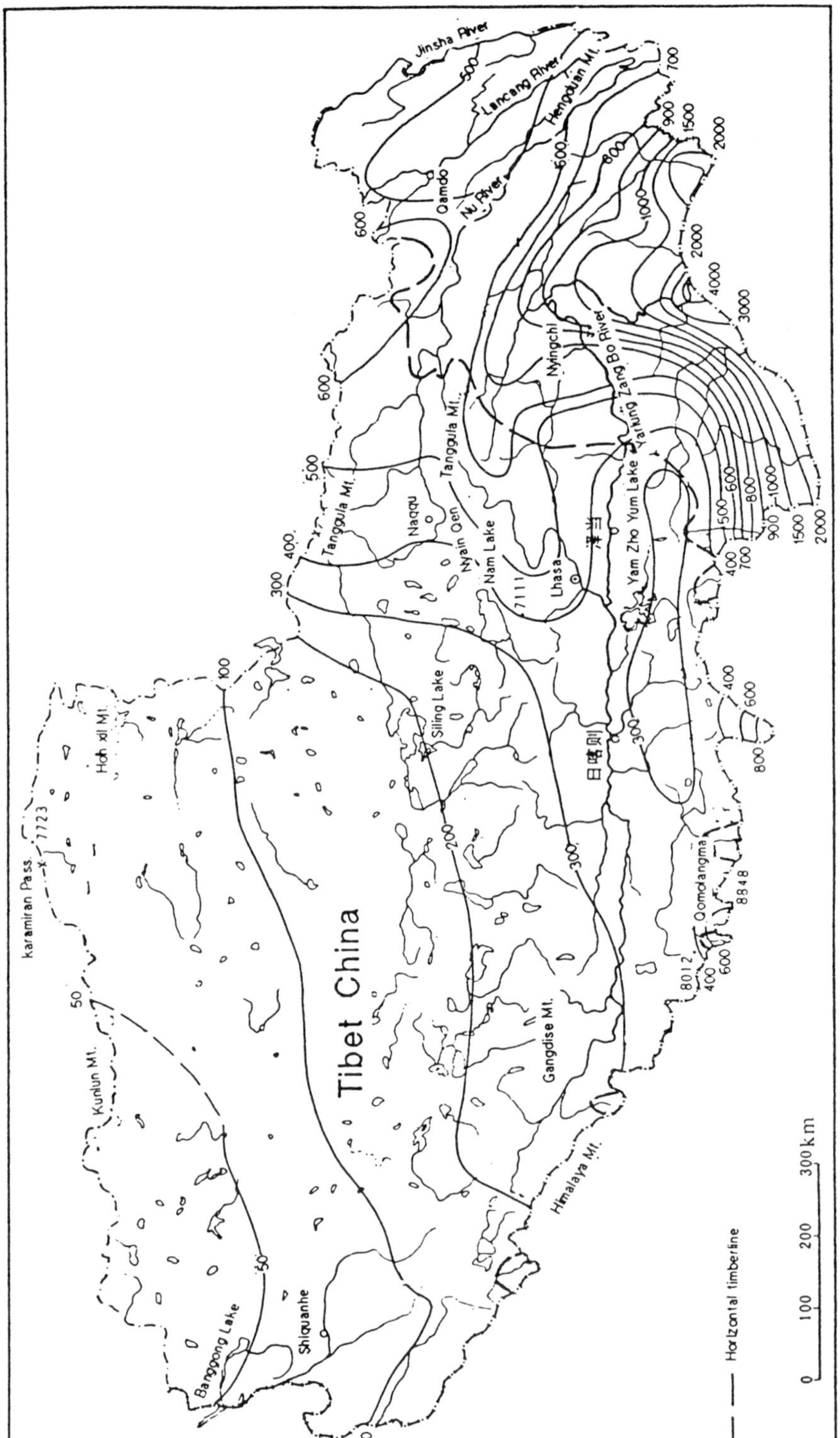

Fig. 6. Mean annual precipitation isohyetes (mm) in Tibet.

In contrast to the frigid montane and arctic timberline zones of Eurasia and North America, moisture is also a major factor in the position and species composition of the timberlines of Tibet. Forests of the eastern Qinghai-Tibet Plateau benefit from the ameliorating effect of moisture from the warm air currents that enter the plateau valleys from the Indian Ocean. Therefore, the distribution of forest vegetation from the southern slope of the Himalaya and the valleys of the Hengduan (Transverse) Mountains in eastern Tibet toward the Plateau depends on the strength of air masses from the Indian Ocean. For example, the forest vegetation distributed along the Yarlung Zangbo River Valleys only reaches nearby Nangxian. The valley from Nangxian to Lhazê is approximately 540 kilometers long, lies under 4,000 masl, and is more than warm enough for forest development, but it contains no natural forests. The reason is moisture: the main stream of the Indian Ocean air mass, after passing over the Himalayan Mountains, continues to flow to the northeast. Only a branch of the air mass travels west along the Yarlung Zangbo River valley, and it carries insufficient moisture for forest growth. This phenomenon can be seen in many other valleys such as the Xibaxiaqu, Pengqu, and Douga'erqu. If annual precipitation is used as an index, the horizontal distribution of the Tibetan forest lies within the 500 millimeter isohyete (Fig. 6).

In addition to heat and moisture, snow also affects the position of timberline on the southern slope of the East Himalaya. Timberline there often has a sawtooth pattern. On the main ridge of the mountains, timberline is relatively high; in the lower valleys, it is lower, because the snow cover is several meters deep in winter and melts late in summer. The extended snow pack reduces the growth of young seedlings and lowers timberline.

Geological History

Tibet's unique geological history has a major impact on the distribution and composition of the present timberline. Information from paleontological finds indicates that during the late Cretaceous Period, approximately seventy million years ago when Tibet rose from the ocean, Eucalyptus L'Herit species occupied the sclerophyllous evergreen broadleaf and montane evergreen coniferous forests. In the Quaternary Period, following many cycles of global cooling and warming, forests had receded and expanded repeatedly. After the last major Pleistocene glaciation and during the beginning of the Holocene epoch, temperatures increased significantly. Forests benefitted from the warm, moist climate and expanded extensively in most regions of the world. But, due to reduced precipitation, the forests of Tibet receded. At the present time, remnants of only two Tibet juniper (Juniperus tibetica) forests exist outside of the horizontal timberline, one each in the Rezhen Monastery of Lhünzhub County and northeast of Nagarzê (Fig. 7a,b). From Sage to Lang Counties, many treeline plants, such as Wilson juniper (Juniperus pingii (Cheng ex Feri) Cheng & W. T. Wang) var. wilsonii (Rehd.) Cheng et L.K. Fu), puberulent rhododendron (Rhododendron primulaeflorum Bur. & Franch. var. cepalanthoides (Balf. f. & W. W. Smith) Cowan Davidian), and glacier rhododendron (R. nivale Hook. f.) grow above timberline, indicating that not long ago, timberline existed above present limits. The historical factor provides an explanation of the diversity of the forest plant community and the unique phenomenon of oak timberlines in the eastern and southern slopes of the Himalayas.

REFORESTATION AT AND ABOVE TIMBERLINE

The History of Reforestation in Tibet

The history of reforestation at and afforestation above timberline in Tibet covers more than thirteen hundred years. This is the earliest record of afforestation in China and might even be the earliest in the world. It is known that when Princess Wencheng of the Tang Dynasty married the Tibetan king in the seventh century and learned that Tibet was barren, she decided to bring a variety of grain and vegetable seeds. But the horses and mules also carried many willow cuttings to the interior of Tibet. When she arrived in Lhasa, the Princess ordered the willow cuttings planted on the Lhasa Plain. When the Jokang Temple, the summer palace of the Dalai Lama, was built in 643 A.D., she also ordered one willow planted on each side. An old willow tree, said to have been planted by Princess Wencheng herself, still stands in front of the Jokang. The local people call it the Tang willow.

a

b

Fig. 7. (a) Remnants of an ancient Juniperus tibetica forest in the Rezhen Monastery of Lhunzhub County. This population is one of only two pure stands in existence. (b) An interior view of the Juniperus tibetica forest above.

If one wants to give the Tibetans credit for afforestation above timberline, the Tibetan concept of "linka" is very important. "Linka" has two parts: lin, meaning "woods," and ka—"to play." Thus, "linka" means "woods where people play." In Tibet, the temperatures in many valleys exceed the heat requirement for growth of tree-limit species, but drought has eliminated large areas of forests. Trees or shrubs such as poplar (Populus L.), willow (Salix L.), or seabuckthorn (Hippophae L.) exist only near a few river plains where moisture is sufficient for growth. The Tibetans treasure such wooded areas on the plateau, fencing them for protection from grazing animals. They propagate them over large areas with cuttings. As a result, linkas are found in villages and on flood plains, farm land, and roadsides above timberline. Linkas improve not only the ecological environment but provide areas for recreation and firewood cutting as well.

Among the numerous linkas in Tibet, some have very long histories. The famous Norbu Linka (Precious Linka) in Lhasa was established by the seventh Dalai Lama in 1735 (Fig. 8), and through continuous expansion, it became the site of the summer palace for all succeeding Dalai. The Norbu Linka is not only thick with poplar species such as Szechuan poplar (Populus szechuanica Schneid. var. tibetica Schneid.), white aspen (P. alba L.), and P. rotundifolia Griff. var. duclouxiana (Dode) Gomb., and willow such as false bayleaf willow (Salix paraplesia Schneid. var. subintegra Z. Wang & P. Y. Fu and S. paraplesia Schneid.), longistamen willow (S. longistaminea Z. Wang & P. Y. Fu), white willow (S. alba L.), and weeping willow (S. babylonica L.), but it also includes many exotic species such as Armand pine (Pinus armandii Franch.), alpine pine or Himalayan pine (Pinus densata Mast., or P. tabulaeformis Carr. var. densata (Mast.) Rehd.), blue pine (Pinus griffithii McClelland), Chinese or Oriental arborvitae (Platycladus orientalis (L.) Franch.), Linzhi spruce (Picea likiangensis Franch. Pritz. var. linzhiensis Cheng & L.K. Fu), Larix griffithiana, Juniperus tibetica, deodar or Himalayan cedar (Cedrus deodara (Roxb.) D. Don), Siberian elm (Ulmus pumila L.), common apple (Malus pumila Mill.), common apricot, (Prunus armeniaca L.), Mazzard cherry (Prunus avium (L.) L.), and Japanese plum (Prunus salicina Lindl.) that were transplanted from both within and outside Tibet (Fig. 9). The Norbu Linka is truly a high-plateau botanical garden.

Fig. 8. The Norbu Linka (Precious Linka) in Lhasa.

Plans for mass afforestation above timberline in the valleys of the Nianchu, Lhasa, and Yarlung Zangbo Rivers call for engineering of irrigation systems and terracing of large areas. These man-made forests will have major ecological effects, the most important of which is protection of large farmlands from wind erosion of fertile soils.

Selection of Suitable Forest Land

The major elements that restrict the growth of forests in Tibet are heat and moisture. Heat has a major impact on the vertical (altitudinal) distributions of forests, whereas moisture limits their horizontal distribution. The heat factor is very difficult to modify artificially. Therefore, reforestation near the vertical timberline in Tibet is extremely difficult. Survival rates for the seedlings are extremely low and growth is slow. On the contrary, in many of the valleys where precipitation is adequate, forests will not only survive and grow rapidly, but will also provide an enormous economical and ecological benefit. For example, beyond the growth zones of the natural forests in the valleys of the Yaluzangbu, Nianchu, and Lhasa Rivers and their tributaries, local Tibetans successfully established many linkas for recreation in villages and townships below 4,400 meters. After construction of irrigation systems to provide water for drought-sensitive species, successful plantations demonstrated that all of the valleys above natural timberline from Lhazê to Nangxian and in the Yarlung Zangbo's tributaries could be afforested. Below 4,200 masl in eastern Tibet, the climate is warm enough for survival and growth of natural forests.

In the regions between 4,200 and 4400 masl, growth rate is slow and many trees become shrubs. It is therefore very important for the local communities to plant trees near flood plains where a steady source of water is available to save the cost of constructing expensive irrigation systems. Afforestation of flood plains will reduce the damage to local farms and villages from wind and sand.

Fig. 9. An Ulmus pumila forest planted in the Norbu Linka.

Selection of Tree Species for Reforestation

Many tree species are suitable for afforestation of tree limit environments in Tibet. Choices among the indigenous species will enable forestation of the most suitable land both within and outside natural timberlines.

In the vicinity of the vertical timberline, the terrain is open to sufficient sunlight for shade-intolerant and drought-hardy species. The vertical timberline, however, is not suitable for survival and growth of shade- and moisture-demanding spruce and fir species. To be successful, sun-loving species such as *Juniperus tibetica*, *Larix potaninii* var. *macrocarpa*, and *L. griffithiana* should be selected for reforestation at altitudinal tree limits. On the moist and shady slopes, *Betula platyphylla* and *B. utilis* are preferred species.

In the valleys of the central Yarlung Zangbo River, flood-plain habitats should be reforested with the indigenous *Populus szechuanica* var. *tibetica*. This species is well-adapted to the timberline environment where a 29-year-old tree grew to 24.8 centimeters in diameter at breast height and 15.8 meters tall, and increased nine percent in stem volume annually. David poplar (*Populus davidiana* Dode) would be another good choice among the indigenous species.

At higher elevations, many indigenous willows such as *Salix paraplesia*, sand willow (*S. cheilophila* Schneid. var. *microstachyoides* (Z. Wang & P. Y. Fu) C. F. Wang), and *S. longistaminea* are well-adapted, and foreign species such as *Populus alba*, Lombardy poplar (*P. nigra* L. var. *italica* Muenchh.), and Canadian poplar (*P.* x*canadensis* Moench.)[2] have been introduced with good success. In some linkas, parks, and other small habitats, conifers such as Chinese pine (*Pinus tabulaeformis* Carr.), Mongolian Scotch pine (*P. sylvestris* L. var. *mongolica* Litvin.), *P. densata*, *P. armandi*, and *Juniperus tibetica* can be planted as ornamental species. On the flood plain where water is sufficient, Gyantsen seabuckthorn (*Hippophae rhamnoides* L. ssp. *gyantsensis* Rousi) is the best species for forestation up to 4,400 masl (Fig. 10). *H. rhamnoides* is the most valuable species of the flood-plain forests of the high plateau where its fruit is used to make a very suitable drink. Only below 4,200 masl will most trees grow taller than five meters; here the tallest trees reach ten meters.

Reforestation Problems

The most serious reforestation problems at timberline in Tibet are the limited choice of commercial species and the low economic value of production forests. Although many species are suitable for afforestation at timberline, most of the local people choose only the indigenous Beijing polar (*Populus* x*beijingensis* W. Y. Hsu),[3] *P. szechuanica*, *Salix alba*, *S. babylonica*, and *S. paraplesia* var. *subintegra*. Among these species, the wood of *Populus* is quite brittle and the *Salix* do not produce lumber. Neither genus provides wood building materials for construction purposes. Therefore, the linkas do not greatly affect the local economies.

Species that are well-adapted, fast-growing, and which produce choice quality wood should be selected for afforestation at the Tibet timberline. Because local residents have

---

[2] *P.* x*canadensis* is a hybrid between eastern cottonwood (*P. deltoides* Bartr. ex Marsh.) and black poplar (*P. nigra* L.).

[3] *P. beijingensis* is a hybrid of Lombardy poplar (*P. nigra* L. var. *italica* (Muenchh.) Koehne) and *P. cathayana* Rehd.

Fig. 10. A natural Hippophae rhamnoides forest at 4400 masl in Tingri County.

planted cuttings of the same clones of the indigenous species for many years, tree growth and wood quality have regressed. To reinforce the genetic resource of the indigenous species, outcrossed seedlings of the most productive clones must be planted.

In addition, limited availability of suitable land and water resources has restricted the establishment of linkas that improve the environment of many cities and villages. An increase in reforestation of land around farms and villages will provide wind breaks and favorable changes in local climate. Afforestation projects at timberline should bring the amenities and ecological and economic benefits of alpine forests to the people of Tibet.

ACKNOWLEDGEMENT

I wish to thank my colleagues Ms. Tian Xin-zhi, Mr. He Shi-qiang, and Ms. Li Xia for their help in preparing the manuscript. I also wish to thank Mrs. Chun-wei Su Chien of Woodlands Mountain Institute, West Virginia, United States, for translating the text. Finally, I thank Dr. John Alden for his thorough revision of the text.

REFERENCES

Chang, Hisn-shih, 1981, The Vegetation Zonation of the Tibetan Plateau, Mountain and Development, Mountain Research and Development, 1(1):29-48.

Gao You-xi, Jiang Shi-da, Zhang Yi-guang, Li Ji-you, Lin Zhen-yao, Wu Xian-ding, Sheng Zhi-bao, Yuan Fu-mao, Huang Chao-ying, and Li Cheng-fang, 1984, Climate of Xizang (Tibet), Science Press, Beijing, p. 102-122. In Chinese.

Li Wen-hua, Han Yu-feng, Shen Mo-mei, Xu You-ji, Li Yu-jiu, Chen Hong-zhao, and Luo Ju-chun, 1985, Forests of Xizang (Tibet), Science Press, Beijing, p. 3-14. In Chinese.

Walter, H., 1979, Vegetation of the Earth, Springer, New York, p. 216-222.

Ye Du-zheng, Gao You-xi, 1979, The Meteorology of the Quingha-Xizang (Tibet), Science Press, Beijing, p. 41-146. In Chinese.

Zhang, Jing-wel, Wang Jin-ting, Chen Wei-lei, Li Bo-sheng, and Zhao Kui-yi, 1988, Vegetation of Xizang (Tibet), Science Press, Beijing, p. 41-146. In Chinese.

Zheng Du, Hu Chao-bing, and Zhang Rong-zu, 1975, Natural Zonation of the Mount Qomolangma Region, Physical Geography Fascicle: 147-202, Science Press, Beijing, p. 147-202. In Chinese.

Zheng Du, Zhang Rong-zu, Yang Chin-ye, 1979, On the National Zonation of the Qinghai-Xizang Plateau, Acta Geogr. Sinica, 34(1):1-11. In Chinese.

FORESTRY IN NEW ZEALAND'S SOUTHERN HIGH COUNTRY

N. J. Ledgard

Forest Research Institute
Christchurch, New Zealand

INTRODUCTION

New Zealand is an insular country of 27 million hectares lying between latitudes 32° S. and 47° S. in the Southern Hemisphere. It has a mostly urban population of 3.3 million. There are two main islands, the North Island (12 million hectares) and the South Island (15 million hectares). Agricultural production dominates land use, and more than 14 million hectares are farmed, mainly for pasture.

Mountains and High Country

Mountain ranges and hill country dominate the landscape. Approximately half the country is mountainous with "characteristic vegetation and soil processes generally associated with altitudes above 700 metres" (Anonymous, 1982).

Most of this area is in the South Island, where the Southern Alps form a backbone that contains the country's highest mountains, the tallest of which is Mt. Cook at 3,664 metres above sea level. To the east of the Alps mountainous foothills and intermontane basins occupy 3.5 million hectares. This area is known as the South Island high country (Fig. 1) and is the subject of this review.

Forests

Forests covered 85 to 90 percent of New Zealand 3,000 years ago (McGlone, 1989). Only 15 percent of the South Island and 1.5 percent of the North Island was above timberline. Apart from limited areas of very poor soils, wetlands, and frost- and flood-prone valley floors, the forest cover was virtually unbroken. In nearly all lowland areas, the major forest type was tall conifer (Podocarpaceae) mixed with broadleaved hardwood forest. The axial ranges of both islands were predominantly covered with beech (Nothofagus species) or mixtures of beech, conifer, and broadleaved hardwood species. Beech formed the timberline

Forest Development in Cold Climates, Edited by J. Alden *et al.*
Plenum Press, New York, 1993

nearly everywhere, the major exception being the central portion of the Southern Alps where montane conifer-hardwood forests graded into subalpine shrublands at timberline.

Today, only 23 percent (6.2 million hectares) of the original native forest remains. Fortunately, the decline in native forest resources was recognized relatively early, and by the 1930s a large exotic planting program was underway (Roche, 1990). As a result, there are now 1.2 million hectares of introduced plantation forests of which Monterey pine (Pinus radiata D. Don) is 90 percent. Exotic timbers supply most of the country's domestic timber needs and 10 percent of its total export receipts.

Forest Clearance

Intermontane basins throughout Central Otago, South Canterbury, and Southland were marginal for forest establishment, and natural fires had reduced forest cover to grassland or shrubland in many of these areas by the time of Polynesian settlement, about 1,000 years ago (McGlone, 1983). Overall, about 75 percent of the New Zealand land area was forested. As the Polynesian population increased and the major food sources of large birds

Fig. 1. The high country of New Zealand's South Island.

(such as the moa) and sea mammals decreased 800 to 500 years ago, forest clearing accelerated and became widespread. By about 400 years ago, the pattern of forested versus non-forested open country had been established (McGlone, 1989). By the time of European settlement in the mid 1800s, the forested land was reduced to about 53 percent of the total land area. After European pastoralists settled the cleared areas, forest destruction again accelerated. Except in the northern kauri (Agathis australis D. Don) forests, only about 10 percent of the trees felled were used for timber production. Forests were an impediment to farming and most of the areas cleared were burned rather than logged. Since 1840, but mainly between 1870 and 1920, the indigenous forest cover was reduced to 23 percent of the total land area. Wood species were kept from reforesting land cleared for farming by repeated burning and grazing of sheep and other domesticated animals. Sheep and cattle numbers currently total nearly 70 million.

THE SOUTH ISLAND HIGH COUNTRY

High-country Vegetation

Present day high-country vegetation is dominated by native snow-tussock grasslands (Chionochloa species) that give way to short tussocks, introduced grasses, and herbs at lower altitudes. In the moister and less intensively grazed regions in the north and west, mountain beech (Nothofagus solandri (Hook.) Oerst. var. cliffortioides (Hook. f.) Poole) and scrub associations still exist.

A reduction in fire frequency over the last 50 years has allowed areas of scrub to increase (O'Connor, 1982). Scrub associations extend to 1450 metres above sea level and are commonly associated at higher elevations with extensive areas of tall snow tussock (Chionochloa flavescens Zotov), which often extends the vegetation line close to 2,000 metres. Despite the low percentage of woody cover, the environment favors forest growth, and forest species rapidly invade their former habitat where seed sources exist and land management practices permit.

Climate

The high country climate has both continental and oceanic influences. Accordingly, it is variable, with summer and winter temperatures moderated by the oceanic influence (Morris, 1965). The east-west rainfall gradient is steep, ranging from a mean of 500 millimetres to more than 8,000 millimetres annually over a distance of less than 40 kilometres. Despite these extremes, the high-country climate is generally favorable to tree growth. Day temperatures are mild, nights are cool, and moisture is adequate over most of the area.

The South Island high country stretches between latitude 42° and 46° S. and is closer to the equator than the "cold climate" forest-growing countries in the Northern Hemisphere. Comparisons with these areas are difficult, although the coastal warm temperature areas of British Columbia, Norway, and Sweden do have similarities with the South Island high country (Strahler, 1987). The mean temperature of the coldest month is above -3 °C, the warmest month is below 22 °C, and the mean temperature of the four warmest months is over 10 °C. All four areas receive adequate rainfall for plant growth. In New Zealand, rainfall is evenly distributed throughout the year except for dry summers in Central Otago (Fig. 1). In this area, the plateau tops of the main ranges, such as Old Man Range, are very exposed and have almost tundra-like vegetation and climate (Table 1).

Table 1. Comparison of Climate in the South Island High Country of New Zealand with the Austrian Alps (after McCracken, 1980)

Site	Latitude	Altitude (m)	Precipitation (mm/yr)	Mean annual temp. (°C)	Warmest month temp. (°C)	Coldest month temp. (°C)	Annual no. of screen frosts*	Wind speed (m/sec)
New Zealand†								
Old Man Range (Otago)	45° S.	1,590	1,600	0.4	6.5	-6.0	-	5.7
Ski Basin (Craigieburn Range)	43° S.	1,550	1,586	3.8	9.7	-1.4	179	2.7
Craigieburn Forest (Craigieburn Range)	43° S.	914	1,447	8.0	13.9	2.0	121	1.3
Austria†								
Rinn	47° N.	900	915	6.3	15.0	-3.2	136	-
Patscherkofel	47° N.	2,045	951	1.3	9.1	-6.3	199	4.1

* Days when temperatures at 1.5 m in Stevenson screen drop below 0 °C.
† Treelines at 1,370 m (N.Z.) and 1,950 m (Austria).

McCracken (1980) also made comparisons with continental mountain climates at similar latitudes north of the equator. The climate at Rinn in the Austrian Alps is much colder in winter and somewhat warmer in summer than the climate at similar altitudes in the Craigieburn Range of New Zealand (Table 1). However, at the Old Man Range site in the continental climate of the Central Otago plateau, winter temperatures are as cold and summer temperatures are cooler than at Patscherkofel in the Austrian Alps. Although the treeline at Patscherkofel is almost 600 metres lower than in the Craigieburns, mean temperatures of the warmest months are only 9.5 °C and 10.9 °C, respectively (McCracken, 1980), and close to the 10 °C threshold associated with treeline in different parts of the world (Wardle, 1965).

For forest growth, there are major differences between the climate of the New Zealand high country and the climate at similar altitudes in continental mountains:

1. Mild winters with short periods of snow cause a high incidence of frost heave on exposed surfaces. Unseasonal temperature fluctuations involving warm spells in winter encourage premature physiological activity and frost injury of maladapted species. The possibility of frosts during every month of the year contributes to frost injury during the growing season. Higher adiabatic lapse rate, which markedly influences species performance relative to altitude (Benecke and Havranek, 1980), is probably a major reason for New Zealand's low treeline (Wardle, 1965) relative to latitude.

2. Snow packs are dense, often weighing 140 to 160 kilograms per square meter, and damage to woody plants occurs more frequently.

3. Frequent strong, gusty foehn winds are very dry as they descend from the western mountains. In the Craigieburn Range at 1,400 metres, gusts exceeding 160 kilometres per hour are not uncommon in the spring.

Land Use and Tenure

The South Island high country has had the least development of any farmed region in New Zealand over the last 100 years. Extensive sheep grazing still dominates land use. Nearly 300 land owners manage approximately two million sheep on properties that average about 10,000 hectares (Kerr et al., 1979). Only tourism rivals pastoralism. Employment in tourism is now greater than in farming, but the impact on the landscape is generally small and localized. The erection of dams for hydroelectricity has had significant physical impact on two major catchments. In the 1960s and 1970s, many trees were planted along the hydro lake margins and in associated riparian areas. Few people are employed in power generation, but the lakes attract many seasonal visitors.

Less than seven percent of the high country is freehold or privately owned (Kerr et al., 1979). The rest is government owned and rented under a range of leasehold agreements. Most leases have automatic rights of renewal and are secure of tenure so that rented properties are bought and sold at freehold prices.

High-country Forestry

Remnant native forests are concentrated on the moist slopes in the west where fires are infrequent. They cover 34,278 hectares, or just over one percent of the high-country farms (Kerr et al., 1979).

Introduced trees have been planted in the high country since the middle of the 19th Century. In a largely treeless environment, lease-holders planted trees for shelter, timber, and firewood, and occasionally as a requirement of land tenure (O'Connor, 1986). With the exception of the 2,500 hectare Naseby Forest established in inland Otago in the early 1900s, large-scale plantation forestry has not been practiced.

Although the good growth of some introduced trees was recognized many decades ago (Morrison, 1919), plantation forests occupy less than one percent of the high country in 1978 (Kerr et al., 1979). In the Canterbury region, introduced trees occupied less than 1,500 hectares in the early 1980s (Ledgard and Belton, 1985).

Although many species have been tried in the 130 years of European settlement, the most successful have been conifers. The main conifers now present are Corsican pine (Pinus nigra Arnold subsp. laricio (Poir.) Maire), ponderosa pine (Pinus ponderosa Dougl. ex Laws), and radiata pine (Pinus radiata D. Don), Douglas-fir (Pseudotsuga menziesii (Mirb.) Franco), and European larch (Larix decidua Mill.). Other less frequent conifers are lodgepole pine (Pinus contorta Dougl. ex Loud. var. contorta), Scots pine (P. sylvestris L.), Bishops pine (P. muricata D. Don), and maritime pine (P. pinaster Ait.).

Few conifers have been planted above 1,300 metres. Most of the higher altitude planting has been done in the Craigieburn Range where species such as dwarf mountain pine (Pinus mugo Turra) have survived at altitudes up to 1,800 metres, and lodgepole pine has naturally regenerated above native tree line at 1,400 metres.

Natural regeneration from existing introduced plantings is increasing (Hunter and Douglas, 1984; Ledgard, 1988). Natural regeneration is so successful in places that O'Connor (1981) wrote: "The success of woody revegetation cultures forces land users to choose between forests and improved pastures." He considered the regenerating forests to be one of "the principal landscape planning issues of the tussock grasslands and mountain lands for the next twenty years."

Exotic broadleaved species are less common and are usually found in sheltered moist sites. Willows, mainly crack willow (Salix fragilis L.), and poplars, mainly Lombardy (Populus nigra L. var. italica Muenchh.), are the most common broadleaved species, and are followed in popularity by sycamore (Acer pseudoplatanus L.) and silver birch (Betula pendula Roth). Other broadleaved trees often encountered on better sites and adjacent to homesteads, are rowan (Sorbus aucuparia L.), oak, mainly pedunculata oak (Quercus robur L.), English elm (Ulmus procera Salisb.), common ash (Fraxinus excelsior L.), common lime (Tilia europae L.), copper beech (Fagus sylvatica L. 'purpurea'), golden chain (Laburnum anagyroides Med.), common walnut (Juglans regia L.), black alder (Alnus glutinosa (L.) Gaertn.), Gean cherry (Prunus avium L.), weeping willow (Salix babylonica L.), and silver and eastern poplars (Populus alba L. and P. deltoides Bartr. ex Marsh).

Very few broadleaved trees have been planted above 1,300 metres. In the Craigieburn Range, only green alder (Alnus virdis Chaix D.G.) and silver birch have survived near native treeline at 1,400 metres.

Plantation Performance

In the early 1980s, the Forest Research Institute assessed the potential of exotic forestry in the South Island high country. Plantations of introduced trees occupied less than 1,500 hectares between 500 to 800 metres in the Canterbury high country (Ledgard and

Belton, 1986). The main species, in decreasing order of frequency, are Corsican, ponderosa, and radiata pine, European larch, and Douglas-fir. Growth rates for these species compare well with the growth rates of other New Zealand and international forest-growing regions. Of all the factors affecting growth rates, moisture is by far the most influential (Fig. 2). Some of the best plots in high rainfall areas had basal areas of more than 130 square metres per hectare, standing whole-tree timber volumes of more than 1,500 cubic metres per hectare, and mean annual increments (MAI) of more than 25 cubic metres per hectare. In the driest areas (500 to 600 millimetres of rainfall), potential MAI's dropped to between five and ten cubic metres per hectare (Ledgard and Belton, 1986).

Few plantations exist above 800 metres and the rates of growth decline with increasing altitude cannot be determined accurately. In the Craigieburn Range, however (annual rainfall 1,400 to 1,600 millimetres), growth rates of 4.5 to 8.1 cubic metres per hectare per annum have been recorded in small experimental plots of 20- to 23-year-old lodgepole, Corsican, and Scots pine between 1,180 and 1,320 metres elevation.

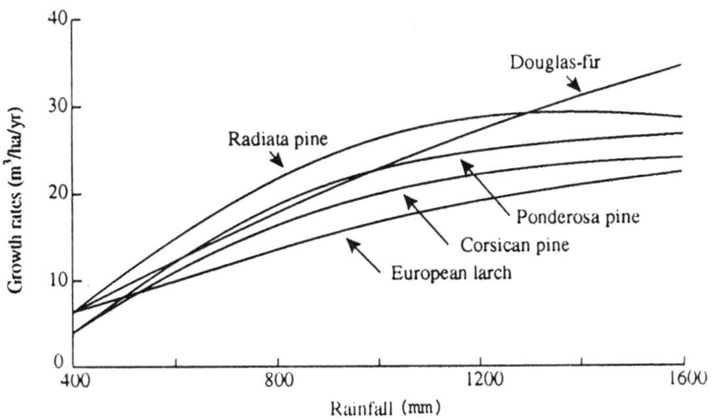

Fig. 2. The relationship of rainfall and the growth of five introduced conifers in New Zealand's South Island high country (Source: Ledgard and Belton, 1986).

In addition to ample moisture, the main reasons for high volume growth are longer retention of foliage (as a consequence of low incidence of pathogens), maximum site occupation, and low night temperature, which results in low loss by respiration of carbon fixed by photosynthesis during warm days.

Species Performance

Early research into the performance of forest species at higher altitudes (800 to 1,700 metres) was concentrated in the Craigieburn Range, 100 kilometres west of Christchurch (Fig. 1) (Ledgard and Baker, 1988). The major initial emphasis was to find species and techniques for reintroducing a vegetation cover to depleted and eroding slopes above 800 metres.

Several hundred native and introduced plant species were planted on sites varying from well-covered tussock slopes to steep running shingle screes from 800 to 1,700 metres (Ledgard and Baker, 1988; Wendelken, 1956). Most were woody species, but much research also went into utilizing grasses and leguminous herbs (Davis, 1978; Nordmeyer and Davis, 1976, 1978). The successful tree species were mostly conifers, particularly pines. The best growth on exposed eroded soils was produced by lodgepole pine, dwarf mountain pine, mountain pine (P. uncinata Mirb.), ponderosa pine, and Corsican pine. On the more favorable, sheltered slopes, Douglas-fir and European larch performed well. Few hardwood species were successful on eroded sites. Willows, particularly those with bitter, unpalatable foliage such as the Booth willow (Salix purpurea L.), and some poplars grew well in the rocky gullies. One of the snow gums (Eucalyptus pauciflora Sieb., ex Spreng.) grew well on more favorable sites. Silver birch grew slowly but steadily on bare sites, even near treeline, and the green alder was successful on moist screes (Table 2).

Generally, native species failed completely except where the vegetative cover and topsoil were intact. Survivors of the dominant native forest species, mountain beech grew less than the better introduced species, except when the beech seedlings were sheltered by a nurse crop. Compared with introduced conifers, particularly the pines, native mountain beech seedlings are less vigorous as pioneers, and less tolerant to cold temperatures and infertile soils. Once established, however, mountain beech grows vigorously over much of the high country, except in the driest areas.

Performance of the introduced species varied with the origin or provenance of the parent material (Benecke and Morris, 1978). For example, lodgepole pine from 1,450 metres in Oregon, United States, grew considerably faster than lodgepole pine from 3,000 metres in Utah (Ledgard, 1980). In another seedling trial, a New Zealand source of limber pine (Pinus flexilis James) of unknown overseas origin significantly out-performed two provenances imported from North America. Provenance trials for the successful species in the Craigieburn environment (Benecke and Morris, 1978; Ledgard, 1980) revealed three dominant trends.

As expected, oceanic provenances were superior to continental provenances. The best provenances originated from mountainous regions north of the equator and at latitudes similar to New Zealand. Finally and most significantly, seed collected from New Zealand sources of unselected and often unrecorded overseas origins frequently out-performed provenances from native habitats. Selection processes during propagation and establishment of the original seedlings may have adapted local sources (land races) to the New Zealand environment, but it is difficult to understand how land races can accrue advantages over one generation when deliberate selection pressures were rarely applied.

Many stands of introduced trees have been sampled to obtain information on phytomass and nutrient status of high country forests (Nordmeyer, 1980a, 1980b; Schonenberger, 1984). The above-ground production of two Pinus (introduced) and one Nothofagus species at treeline is shown in Table 3. High leaf-surface areas and nutrient levels may increase photosynthesis of trees at high altitudes and explain the relatively good growth of some cold climate species such as lodgepole pine (Benecke and Havranek, 1980).

The physiological response of trees growing on exposed mountain slopes has been studied intensively. A mobile tree laboratory (Anonymous, 1976) was used to investigate the gas exchange for a range of plants, including dwarf mountain pine. The long winter dormancy in dwarf mountain pine from continental climates was not evident in the Craigieburn environment (Benecke et al., 1978; Benecke and Havranek, 1980). The high leaf area

Table 2. Growth of the Most Successful Tree Species Introduced in the Craigieburn Range, New Zealand (modified from Ledgard and Baker, 1988)

Species	Altitude (m)	Age (yr)	MTH* (m)	DBH (cm)	TSV* (m^3/ha)	MAI* (m^3/ha)	Substrate
Alnus glutinosa	850	12	6.7	8.1	NA	NA	Topsoil
Alnus rubra	850	12	6.8	7.9	NA	NA	Topsoil
Alnus viridis	1,250	18	1.7	NA	NA	NA	Scree
Betula pendula	850	12	5.9	6.8	NA	NA	Topsoil
Betula papyrifera	850	12	5.7	7.1	NA	NA	Topsoil
Eucalyptus pauciflora	900	28	16.0	25.0	NA	NA	Topsoil
Larix decidua	900	20	10.7	13.9	146	7.3	Topsoil
Pinus contorta	900	19	14.3	15.4	259	13.6	Topsoil
Pinus contorta	1,180	23	7.9	15.6	158	6.7	Scree
Pinus nigra	1,200	23	8.1	16.6	143	6.2	Topsoil
Pinus ponderosa	900	20	10.9	20.0	275	13.7	Topsoil
Pinus ponderosa	960	25	12.3	18.3	209	8.4	Subsoil
Pinus sylvestris	886	20	10.2	16.7	156	7.8	Topsoil
Pinus uncinata	900	19	7.4	12.7	65	3.4	Topsoil
Pseudotsuga menziesii	750	18	14.1	21.5	188	10.5	Subsoil
Thuja plicata	750	36	14.0	38.0	NA	NA	Topsoil
Tsuga heterophylla	760	40	20.0	45.0	NA	NA	Topsoil

* MTH = mean top height
TSV = total standing volume
MAI = mean annual increment (TSV + age)
DBH = diameter at breast height (1.4 m)

Table 3. Component Dry Weights (metric tons/ha) of Pinus contorta, P. mugo, and Nothofagus solandri var. cliffortioides near Timberline in the Craigieburn Range, New Zealand

	Pinus contorta[1]	Pinus mugo[2]	N. solandri var. cliffortioides[3]
Needles or leaves	21.4	20.0	9.7
Branches	40.0	37.0	21.5
Stem and bark	138.2	52.0	116.5
Total oven-dry weight	200.0	109.0	148.0

Ages (in years) at harvest: [1] 17; [2] 16; [3] 55.

of some species may result from the high light and warm day/cool night climates of the South Island high country. The Craigieburn ecophysiological work has been integrated with detailed biomass and carbon balance studies (Benecke and Nordmeyer, 1982). It is now known how treeline stands maintain a positive carbon balance and control their growth patterns. Subalpine stands, for example, divert more biomass into leaves and roots than montane stands, and mountain beech uses more of its carbon for seasonal leaf and branch growth than does pine.

Planting Stock

Soon after founding a research field base in the Craigieburn Range, a nursery was established at 900 metres to produce seedlings conditioned for survival and early growth in the high country. It was soon found, however, that seedlings of appropriate dimensions and hardiness could be produced quicker and cheaper at lower altitudes. The high-altitude nursery was therefore abandoned, and all stock has since been produced at 50 metres above sea level in Rangiora.

Forest growers in New Zealand have always favored bare-root seedlings (Chavasse, 1981). Bare-root seedlings are easy to grow at reasonable cost, easy to handle in bulk, and easy to condition for planting on harsh sites. Above all, field trials in the South Island mountains have shown that properly prepared bare-root seedlings will survive and grow better than containerized stock (Baker and Ledgard, 1991).

Containerized seedling trials showed that large containers were cumbersome to handle, particularly in transport and planting. Tree and shrub seedlings raised in small (20 to 60 square centimetres) containers had much lower survival and growth rates than bare-root seedlings (Anonymous, 1982). Small containers restricted root growth before and after planting and increased frost lifting of seedlings during the winter.

Stock Condition and Quality

Objectives of nursery programs for the high country are to produce conditioned, fibrous-rooted seedlings 20 to 40 centimetres tall, 8 to 10 millimetres in diameter at the root-collar (Table 4), and with root-to-shoot ratios of about 0.5 but not less than 0.3. Of the above criteria, root-collar diameter is the best determinant of seedling quality (Chavasse, 1980). Conditioning hardens seedlings physiologically to withstand frost and drought (Rook, 1971). Conditioning is mainly achieved by undercutting and wrenching the roots of 1- to 2-year-old seedlings (van Dorsser and Rook, 1972). Undercutting involves passing a horizontal, flat,

Table 4. Recommended Shoot Length, Root-collar Diameter, and Maximum Cool-storage Periods for Bare-rooted Tree Stocks for Mountainland Planting in New Zealand

	Age (yr)	Minimum root-collar diameter (mm)	Desirable shoot length (cm)	Cool storage (weeks at 2 °C)
Alnus viridis	2	10	30	12
Larix decidua	2	10	50	8
Nothofagus solandri var. *cliffortioides*	3	6	60	1
Pinus nigra	2	8	20	10
Pinus mugo	2–3	8	20–25	12
Pinus ponderosa	2	10	25	10
Pinus radiata	1	5	30	4
Pseudotsuga menziesii	2	10	45	6

Table 5. Recommended Seedling Age for Undercutting, Wrenching, and Lifting of Tree Stock for Planting in New Zealand's Mountain Lands[1]

Species	Age 1st undercut[2] (months)	Age 2nd undercut (months)	Depth of undercut (cm)	Frequency of wrenching[3] (weeks)	Age at lifting[4] (months)
Alnus viridis	18	-	10	4–6	22
Larix decidua	10	18	10	4–6	22
Nothofagus solandri var. *cliffortioides*	22	30	12	6–8	34
Pinus nigra	18	-	10	4–6	22
Pinus mugo	18	-	10	4–6	22
Pinus ponderosa	10	18	10	4–6	22
Pinus radiata	6	-	9	4	9
Pseudotsuga menziesii	10	15	10	4	22

[1] Forest Research Institute, Rangiora nursery regime
[2] Sowing date is 1 October
[3] From last undercut to time of lifting
[4] Stock is lifted in July and August

thin, sharpened blade beneath the seedbed at a predetermined age and depth (Table 5). Wrenching is a similar process, but a thicker, tilted, generally blunt blade is used at a slightly deeper depth. Seedlings of mountain land conifers are conditioned by undercutting at ten centimetres. Usually this is done once, in late summer, although some species, for example, Douglas-fir, can be undercut earlier to control height growth. Wrenching is carried out at a slightly deeper depth of 12 centimetres, to avoid damaging callouses formed on the tips of the previously undercut roots. Undercutting and wrenching have to be timed accurately to produce good-quality stock for afforestation of mountain land (Table 5).

In the New Zealand high-country, proper handling of stock between the nursery and the planting site is prerequisite for top-quality forest establishment (Balneaves and Menzies, 1990) because snow and frozen ground often prevent planting until late spring, long after seedlings in lowland nurseries lose dormancy. Early failures were attributed to stock that flushed and began root extension before planting. Consequently, research was initiated into seedling physiology during storage and into packaging and storage procedures (McCracken 1978, 1979). As a result of the research, seedlings can now be planted in a healthy dormant state. Seedlings are lifted at the normal time in mid-winter, packed into multi-wall Kraft paper bags lined with a thin layer of polythene, and held at 1.0 to 3.0 °C until the planting sites are accessible (Table 5).

Plantation Establishment

Sites are prepared with ground cultivation and weed control to minimize competition from vegetation. Fencing to protect seedlings from domestic livestock is standard practice, at least over the initial years. Fencing must be durable and high quality (Baker and Ledgard, 1981). Electric fences save material and labor, and use of solar-powered energizers for fencing large remote areas is increasing.

Damage by smaller animals such as hares, possums, and rabbits is often more difficult to prevent than damage from livestock and large wild animals. Where populations are high, small animals can be controlled effectively by shooting, trapping, and poisoning. Repellents have been successful where animal numbers are low (Crozier, 1991).

Most high-country plantations have been established without ground cultivation. After five years on some sites, growth of trees on cultivated and ripped soils is often the same as on unprepared ground (Baker and Ledgard, 1991). On other high-altitude sites, however, cultivation improves growth (Balneaves and Hetherington, 1973), and it is commonly accepted as a beneficial establishment practice elsewhere in the country (Cullen and Mason, 1981). The usual method of cultivation is to rip the soil in a single line to a depth of at least 60 centimetres. Ripped soils are easier to plant than uncultivated soils and the quality of planting improves.

Machine planting has been successful on many high-country sites. In two trials, five-year survival and growth were the same for machine-planted and hand-planted seedlings on ripped and unripped ground (Baker and Ledgard, 1991). Costs were 7.6 cents (New Zealand) per seedling for machine planting and 7.0 cents per seedling for hand planting in unprepared ground. An advantage of machine planting is the ease with which planting fits into the busy late winter and spring farming schedule. In favorable weather, farmers can quickly attach the machine to a tractor, and with one assistant efficiently plant 750 to 1,000 trees per hour.

Although the herbaceous cover is often light in the high-country, vegetation control within a 75-centimeter radius of each planting spot is generally advocated to aid seedling establishment. Control of dense vegetation will significantly improve seedling survival and growth by increasing moisture (Balneaves, 1975), and minimize damage from radiation frosts. Exposed soils absorb more solar radiation and radiate more energy during the night than pale-colored grasses. On black mineral soils, surface temperatures can be 4 °C higher than surface temperatures on grass-covered soils (Menzies et al., 1981). Where vegetative cover has been improved for grazing, competition can lead to almost total failure of sensitive species such as Douglas-fir. For vegetation control, a mixture of simazine, dalapon, and atrazine applied before planting will usually suppress the competing vegetation for up to 18 months.

Normally, spring planting cannot be attempted in the high country before September and October. Autumn planting is not recommended because the planting stock is not established before winter and is subject to frost lift. The usual planting tool is a specially constructed straight-bladed spade; mattocks are used in the most stony country. If herbicide use is not justified, planting spots are often scalped (or screefed) to reduce competition during the critical period for early growth. Screefing involves lightly clearing the existing weeds from a 25 centimeter square around the planting spot. The middle of the screefed area is then cultivated to a depth of at least 20 centimetres, and the seedling is planted at a soil level of three to five centimetres above the root collar. Fertilizers increase weed competition (Baker and Ledgard, 1991) and are not generally recommended. An exception is boron, which is deficient in many high-country soils causing leader dieback, especially on dry sites (Will, 1985). Applications of 0.7 grams of elemental boron per tree two years after planting appear to correct the deficiency for at least five years after treatment.

Weed control is necessary for good seedling growth during the first two years, especially for species such as Douglas-fir (Baker and Ledgard, 1991) that are sensitive to competition. Selective herbicides are usually applied over the top of seedlings, either by spot, strip, or broadcast spraying (Zabkiewicz and Richardson, 1989). Herbicides are applied in early spring when soil moisture is high and soil temperatures are starting to rise, but not before the seedlings are actively growing and under moisture stress. Animals are not allowed to graze around the trees until seedlings grow above browse height. After that, grazing is possible if carefully monitored.

Direct seeding of conifers has been studied for both site protection (Ledgard, 1978) and wood production (Davis, 1990). On eroded sites, few seedlings survive frost heave during the first winter (Ledgard, 1979). Frost-heave can be significantly reduced with a light cover of herbaceous species during the first two or three years after planting. For this reason, a mixture of grass and legume seed and a balanced fertilizer are often applied with the tree seed (Ledgard, 1978). On sites with little weed competition, seeds of lodgepole, Corsican, and ponderosa pine, but not Douglas-fir, have been direct drilled successfully (Davis, 1990). Broadcast seeding of introduced conifers can also be successful, but it is influenced by climate, site variability, and weed competition.

A Comparison to the Northern Hemisphere

Because of the relatively mild climate of the high country in New Zealand, foresters have fewer problems afforesting marginal sites there than in the Northern Hemisphere. However, colder conditions in the high country than elsewhere in New Zealand mean that species and seedling conditioning, handling, and establishment techniques are similar to those used in cold climates overseas.

In New Zealand, the true firs and spruces do not perform as well as the pines and are not as important on marginal sites as they are in North America, Northern Europe, and Asia. Species such as lodgepole pine grow well on cold sites in all four areas, however.

Over the last 20 years, containerized stock has become more popular than bare-rooted stock in the Northern Hemisphere, particularly in the colder environments. For example, in British Columbia, container stock accounted for 91 percent of the 299 million trees sown in 1990 (Arnott and White, 1991). In the field, container stock had significantly higher survival than bare-root stock and grew at comparable rates for all species except Douglas-fir. In cold climates, growing conditions can be controlled in container nurseries and larger stock can be produced in less time than in bare-root nurseries. In recent years, however, research has shown that well-prepared bare-root stock of some species such as Douglas-fir is suitable for some cold sites (Nelson, 1989).

In New Zealand, most plantations established with bare-root stock have been so successful that container systems have not been tested fully. Bare-root stock requires skilled nursery techniques and more handling and storage than container stock. For this reason, more research has been conducted on seedling quality and handling in New Zealand. On the other hand, site preparation has received more attention in the Northern Hemisphere (Orlander et al., 1990), where growing seasons are short and seedlings must start growing as soon as temperatures are favorable. Soil temperatures significantly influence early root growth and the uptake of water and nutrients (Orlander, 1990). Site preparation treatments that increase soil temperature in cold climates will increase seedling survival and growth. Soils in the New Zealand high country are not as cold in winter and, therefore, reach threshold temperatures for early growth sooner than soils in the Northern Hemisphere.

Silvicultural Systems

Most high-country plantations were established with more than 2,000 stems per hectare and have not received any silvicultural treatments between planting and felling. In the future, species are not likely to change, but the initial spacing, thinning, and pruning of new forests will receive more attention as agroforestry increases (Hughes, 1991; O'Connor, 1986). Research and experience in forest grazing has increased in New Zealand (Percival and Knowles, 1983), but more agroforestry research is needed. Spacing trials for Douglas-fir and Corsican pine were established in the early 1980s to determine tree form and growth response to stocking (particularly for Corsican pine roundwood production), and the effect of tree species and densities on nutrient cycling. The influence of species and stocking on the growth and composition of understory grasses, forage shrubs, and legumes needs to be determined (Belton, 1991). High-country farmers traditionally graze livestock, but as farm incomes from pastoral systems decline, agroforestry options are increasingly attractive alternatives.

Forestry Potential

Improved seedling production and establishment practices have increased tree survival and growth rates in the New Zealand high country. A survey in the Canterbury high country showed good potential for forestry: large areas of gentle topography, easy access, and low management and harvesting costs; minimal weed competition, good access to export ports, and limited alternative land uses (Ledgard and Belton, 1985). It was also found that introduced conifers can improve the nutrient status of depleted high-country soils (Davis, in press). Major objections to larger-scale forestry include natural migration of trees onto undeveloped land and potential conflicts with scenic values and other land uses.

The major technical problems associated with species and seedling quality, handling, and establishment have been solved. Large forest estates can be established in the high country and, indeed, interest in forestry has increased in recent years. Potential sawlog and small roundwood species include Douglas-fir on moist, sheltered sites, and Corsican pine on dry, exposed sites (Ledgard and Belton, 1985). Other introduced species such as lodgepole and Scots pine have faster early growth, but they grow slower in later years, have fewer end-uses, and produce unwanted regeneration at an early age.

Predicted internal rates of return are 8 to 10 percent of investment, and annualized returns are ten to 50 times higher from forestry than from pastoral farming on the same land (Belton, 1992). Environmental and commercial interests in forestry have increased as large areas of grassland continue to degrade under existing management practices. Plagues of rabbits (Oryctolagus cuniculus L.) and the spread of hawkweeds (Hieracium spp.) are both the cause and symptom of a widespread malaise in pastoral farming (Hughes, 1991). Tree planting could reverse the decline in soil organic matter, fertility, and structure.

Ten percent of all New Zealand's farm land could be converted to forests without adversely affecting agricultural production and other land values (Sutton, 1990). The South Island high country has not been fully evaluated, but about 300,000 hectares, or 10 percent of the farm land, is probably suited for forestry. Of the 1.8 million hectares of Canterbury high country, an estimated 200,000 hectares is suited for forestry (Ledgard and Belton, 1986). This estimate does not include the best grazing land, areas receiving less than 800 millimetres rainfall annually, and steep remote regions, all of which can support forests. At maturity, a 300,000-hectare forest estate in the high country should produce at least three million cubic metres of stemwood annually. Before significant forest planting can begin, however, important cultural and environmental concerns must be considered.

Integrating Forestry

The biggest challenge to forest development in the high country is the wise integration of trees into traditional land use patterns. Attitudes and opinions about forestry vary from full support for an extensive forest estate to total aversion to trees. Most high-country farmers agree that tree planting will improve farming and living conditions (O'Connor, 1986), but they do not want large-scale forestry and feel that tree planting should complement pastoral farming and the open landscape. The attitudes of visitors are not well documented, but sentiments are with the local farmers to preserve the openness and grand scale of the landscape. Environmental and cultural influences of forestry on other land uses are major public concerns.

Environmental Influences

Concerns about forestry center around the visual impact and the natural regeneration of introduced trees (Hunter and Douglas, 1984). Although the high country was covered with woody species centuries ago (McGlone, 1989), the present landscape has extensive open areas of grazing land set against a mountainous backdrop. Forestry, therefore, represents a major change in land use and will have significant visual impacts. Local governments are deciding how to integrate forestry into the high-country landscapes. Although wild seedlings (wildings) occupy less than one percent of the high country, uncontrolled migration of trees is also of public concern. Some fear that the vigorous adventive nature of a few conifers could lead to the encroachment of trees over large areas of the high country, threatening particular parks and conservation reserves (Harding, 1990). Mechanisms and guidelines for controlling the spread of trees (Ledgard, 1988) are widely recognized, but their successful application depends on people's attitudes and the competence of various administrative institutions.

Cultural Influences

Land managers in the high country have little background in forestry. Although many farmers have good access to sources of advice and service, most are inexperienced and lack funds and sufficient local skilled labor for undertaking an on-going forestry program. These limitations must be overcome before forestry is a major land use.

Institutional Constraints

Most of the high country is State Leasehold Land and is under many types of tenure. In addition, the indigenous Maori people have many high-country land claims before the government, and there are many other vested interests in high-country management. Traditionally, the unique visual, cultural, and recreational values of the high country are special in the folklore of New Zealand. Administration involves numerous state acts and laws, official policies, government agencies, private organizations, and special-interest groups. Until the legal provisions are reduced and the actions of administrative bodies co-ordinated, forestry will be difficult to plan and implement.

The Agroforestry Option

Decisions about the place of forestry in the New Zealand high country have yet to be made, but farm forestry appears to be compatible with existing conditions of use and tenure (Ledgard and Belton, 1985, O'Connor, 1986). Because funds, knowledge, and experience are limited, however, land managers are unlikely to adopt agroforestry in the near future. Thus, awareness programs and technology transfer of forestry information are urgently needed. Greater awareness of the benefits and technology of agroforestry, in particular, should make it easier to reach a consensus on how to wisely integrate forestry into the high-country pastoral system.

Hughes (1991) may have envisaged the most appropriate scenario for high-country farms when she wrote of

> an aesthetic blend of reshaped and robust pastoral properties grazing cattle, sheep, goats, deer, as appropriate to land type and management need; having areas in agroforestry with larger blocks in production forestry, either as wide shelterbelts or stands on severely degraded lnd or land with aluminium toxicity; and with some landholders operating on-farm activities for tourists and conservationists. Elsewhere private blocks of privately managed plantation forests across formerly desertified lands could blend in with managed conservation areas and rural townships designed to service and process 'products' from or for the land.

REFERENCES

Anonymous, 1976, Mobile laboratory, What's New in Forest Research No. 43, Forest Research Institute, Rotorua, New Zealand, 4 p.

Anonymous, 1982, Container seedlings in erosion control, What's New in Forest Research No. 116, Forest Research Institute, Rotorua, New Zealand, 4 p.

Anonymous, 1982, Review of mountainland revegetation, New Zealand Forest Service, Private Bag, Wellington, 64 p.

Arnott, J. T., and White, W., 1991, The influence of nursery systems on plantation establishment in British Columbia. In: Menzies, M. I, Parrott G., and Whitehouse, L. J., eds., The efficiency of stand establishment operations, Proceedings of an IUFRO Conference, 11-15 September 1989, Rotorua, New Zealand, Forest Research Institute Bulletin No. 156, 56-66.

Baker, G. C., and Ledgard, N. J., 1981, Growing trees in the high country. II. Establishment, Tussock Grasslands and Mountainlands Institute, Lincoln University, Canterbury, New Zealand, Review 40, 3-7.

Baker, G. C., and Ledgard, N. J., 1991, Douglas fir seedling quality, handling and establishment practices for the South Island mountainlands, New Zealand. In: Menzies, M. I., Parrott, G., and Whitehouse L. J., eds., The efficiency of stand establishment operations, Proceedings of an IUFRO Conference, 11-15 September 1989, Rotorua, New Zealand, Forest Research Institute Bulletin No. 156, 134-140.

Balneaves, J. M., 1975, The importance of weed control for successful establishment of radiata pine in low rainfall areas. In: Chavasse, G. C. R., ed., The use of herbicides in forestry in New Zealand, Proceedings of Forest Research Institute Symposium No. 18, October 1975, Forest Research Institute, Rotorua, New Zealand, 139-149.

Balneaves, J. M., and Hetherington, M. W., 1973, Ripping in tussock country improves radiata growth, Forest Industries Review, 4(12):5-7.

Balneaves, M. J., and Menzies, M. I., 1990, Influence of lifting, packaging and storage on water potential and subsequent growth of radiata pine seedlings, New Zealand Journal of Forestry Science, 21(2):257-267.

Belton, M. C., 1991, Forestry: A sustainable landuse for degraded high country lands, New Zealand Forestry, 35(5):19-22.

Belton, M. C., 1992, Land use options with trees and forests in the Mackenzie Basin Rabbit and Land Management Area, Ministry of Forestry, Christchurch, New Zealand, 50 p.

Benecke, U., Baker, G., and McCracken, I. J., 1978, Tree growth in the Craigieburn Range. In: Orwin, J., ed., Revegetation in the rehabilitation of mountain lands, New Zealand Forest Service, Forest Research Institute Symposium No. 16, 77-98.

Benecke, U., and Havranek, W. M., 1980, Phenological growth characteristics of trees with increasing altitude, Craigieburn Range, New Zealand. In: Benecke, U., and Davis, M. R., eds., Mountain environments and subalpine tree growth, New Zealand Forest Service, Forest Research Institute Technical Paper No. 70, 155-174.

Benecke, U., and Morris, J. Y., 1978, Tree provenance trials. In: Orwin, J., ed., Revegetation in the rehabilitation of mountain lands, New Zealand Forest Service, Forest Research Institute Symposium No. 16, 99-138.

Benecke, U., and Nordmeyer, A. H., 1982, Carbon uptake and allocation by Nothofagus solandri var. cliffortioides (Hook f.) and Pinus contorta Douglas ex. Loudon ssp. contorta at montane and subalpine altitudes. In: Waring, R. H., ed., Carbon uptake and allocation in subalpine ecosystems as a key to management, Oregon State University, Corvallis, Oregon, United States, 9-21.

Chavasse, G. C. R., 1980, Planting stock quality: A review of factors affecting performance, New Zealand Journal of Forestry, 25(2):144-171.

Chavasse, C. G. R., 1981, Forest nursery and establishment practice in New Zealand, New Zealand Forest Service, Rotorua, Forest Research Institute Symposium No. 22, Part 2—Establishment Practice, 330 p.

Crozier, E. R., 1991, Practical animal repellents for tree seedlings: A success story. In: Menzies, M. I., Parrott, G., and Whitehouse, L. J., eds., The efficiency of stand establishment operations, Proceedings of an IUFRO Conference, Rotorua, New Zealand, September 1989, Forest Research Institute Bulletin, 156:172-177.

Cullen, A. W. J., and Mason, E. G., 1981, The use of cultivation as an aid to forest establishment. In: Chevasse, C. G. R., ed., Forest nursery and establishment practice in New Zealand, New Zealand Forest Service, Forest Research Institute Symposium No. 22, Part 2—Establishment practice, 115-130.

Davis, M. R., 1978, Phosphate responses of grasses and legumes growing in subsoil. In: Orwin, J., ed., Revegetation in the rehabilitation of mountain lands, New Zealand Forest Service, Forest Research Institute Symposium No. 16, 218-238.

Davis, M. R., 1990, Establishment of conifer plantations in the South Island by direct drilling, New Zealand Forestry, 34(3):21-24.

Davis, M. R., [In press], Nutrient enrichment of topsoils under conifers in the South Island high country, New Zealand Journal of Forestry Science.

Harding, M., 1990, Wilding pines--a growing problem, New Zealand Forest and Bird, 21:38-41.

Hughes, H. H. R., 1991, Sustainable use for the dry tussock grasslands in the South Island, Parliamentary Commissioner for the Environment, Wellington, New Zealand, 76 p.

Hunter, G. G., and Douglas, M. H., 1984, Spread of exotic conifers in South Island rangelands, Journal of Forestry, 29(1):78-96.

Kerr, I. G. C., Lefever, K. R., and Costello, E. J., 1979, High country production surveys; 1965/67-1971/73-1976/78, Special Publication No. 15, Tussock Grasslands and Mountainlands Institute, Lincoln College, Canterbury, New Zealand, 63 p.

Ledgard, N. J., 1978, Direct seeding of trees and shrubs. In: Orwin, J., ed., Revegetation in the rehabilitation of mountain lands, New Zealand Forest Service, Forest Research Institute Symposium No. 16, 153-160.

Ledgard, N. J., 1979, First year losses of Pinus mugo seed and seedlings on an exposed high country subsoil, New Zealand Journal of Forestry, 24(1):90-100.

Ledgard, N. J., 1980, Specific and intra-specific growth differences in tree trials up to timber-line in the Craigieburn Range, New Zealand. In: Benecke, U., and Davis, M. R., eds., Mountain environments and subalpine tree growth, New Zealand Forest Service, Forest Research Institute Technical Paper No. 70, 137-154.

Ledgard, N. J., 1988, The spread of introduced trees in New Zealand's rangelands -- South Island high country experience, Tussock Grasslands and Mountainlands Institute, Lincoln College, Canterbury, New Zealand, Review 44, 1-8.

Ledgard, N. J., and Baker, G. C., 1988, Mountainland forestry -- 30 year's research in the Craigieburn Range, New Zealand, Forest Research Institute Bulletin No. 145, Ministry of Forestry, Christchurch, New Zealand, 64 p.

Ledgard, N. J., and Belton, M. C., 1985, Diversification and opportunities in forestry in the South Island high country, New Zealand Journal of Forestry, 30(1):133-143.

Ledgard, N. J., and Belton, M. C., 1986, Exotic trees in the Canterbury high country, New Zealand Journal of Forestry Science, 15(3):298-323.

McCracken, I. J., 1978, Cool storage. In: Orwin, J., ed., Revegetation in the rehabilitation of mountain lands, New Zealand Forest Service, Forest Research Institute Symposium No. 16, 195-209.

McCracken, I. J., 1979, Changes in the carbohydrate concentration of pine seedlings after cool storage, New Zealand Journal of Forestry Science, 9(1):4-43.

McCracken, I. J., 1980, Mountain climate in the Craigieburn Range, New Zealand. In: Benecke, U., and Davis, M. R., eds., Mountain environments and subalpine tree growth, New Zealand Forest Service, Forest Research Institute Technical Paper No. 70, 41-59.

McGlone, M. S., 1983, Polynesian deforestation in New Zealand: a preliminary synthesis, Archaeology of Oceania, 18(1):11-25.

McGlone, M. S., 1989, The Polynesian settlement of New Zealand in relation to environmental and biotic changes, New Zealand Journal of Ecology, 12:115-129.

Menzies, M. I., Chavasse, G. C. R., Bowles, G. P., and Balneaves, J. M., 1981, Nursery stock quality and establishment techniques for frosty sites. In: Chevasse, C. G. R., ed., Forest nursery and establishment practice in New Zealand, New Zealand Forest Service, Forest Research Institute Symposium No. 22, Part 2—Establishment practice, 115-130.

Morris, J. Y., 1965, Climate investigations in the Craigieburn Range, New Zealand, New Zealand Journal of Science, 8(4):556-582.

Morrison, W. G., 1919, Some proposals with regard to natural afforestation in a New Zealand mountain area, New Zealand Journal of Science and Technology, 415:339-349.

Nelson, D. G., 1989, Developing an efficient restocking system for conifers in the uplands of Great Britain. In: Menzies, M. I., Parrott, G., and Whitehouse, L. J., eds., The efficiency of stand establishment operations, Proceedings of an IUFRO Conference, 11-15 September 1989, Rotorua, New Zealand, Forest Research Institute Bulletin No. 156, 115-124.

Nordmeyer, A. H., 1980a, Phytomass in different tree stands near timberline. In: Benecke, U., and Davis, M. R., eds., Mountain environments and subalpine tree growth, New Zealand Forest Service, Forest Research Institute Technical Paper No. 70, 111-124.

Nordmeyer, A. H., 1980b, Tree nutrient concentrations near timberline, Craigieburn Range, New Zealand. In: Benecke, U., and Davis, M. R., eds., Mountain environments and subalpine tree growth, New Zealand Forest Service, Forest Research Institute Technical Paper No. 70, 83-94.

Nordmeyer, A. H., and Davis, M. R., 1976, Legumes in high country development, Proceedings of New Zealand Grasslands Association, 38:119-125.

Nordmeyer, A. H., Lang, M. H., and Roberts, Q., 1978, Legume establishment. In: Orwin, J., ed., Revegetation in the rehabilitation of mountain lands, New Zealand Forest Service, Forest Research Institute Symposium No. 16, 9-20.

O'Connor, K. F., 1981, Changes in tussock grasslands and mountainlands, Tussock Grasslands and Mountainlands Institute, Lincoln College, Canterbury, New Zealand, Review 40, 47-63.

O'Connor, K. F., 1982, Implications of past exploitation and current developments to the conservation of South Island tussock grasslands, New Zealand Journal of Ecology, 5:97-107.

O'Connor, K. F., 1986, Roles for forestry in high country land use, Tussock Grasslands and Mountainlands Institute, Lincoln College, Canterbury, New Zealand, Review 43, 83-94.

Orlander, G., Gemmel, P., and Hunt, J., 1990, Site preparation: a Swedish overview, Canada-British Columbia Forest Resource Development Agreement Report 105, British Columbian Ministry of Forests, Research Branch, Victoria, Canada, 62 p.

Percival, N. S., and Knowles, R. L., 1983, Agroforestry research. In: Robertson, B. T., ed., Proceedings of the 1983 Hill and High Country Seminar, Centre of Resource Management, Lincoln College, Canterbury, New Zealand, Special Publication No. 26, 97-102.

Roche, M., 1990, History of forestry, New Zealand Forestry Corporation Ltd., Wellington, New Zealand, 466 p.

Rook, D. A., 1971, Effect of undercutting and wrenching on growth of Pinus radiata D. Don seedlings, Journal of Applied Ecology, 8:477-490.

Schonenberger, W., 1984, Above-ground biomass of mountain beech (Nothofagus solandri var. cliffortioides) in different stand types near timberline in New Zealand, Forestry, 57(1):60-78.

Strahler, A. N., 1987, Physical geography, John Wiley and Sons, New York, 544 p.

Sutton, W. R. J., 1990, Submission to Ministry of Research, Science, and Technology on Forestry, New Zealand Forestry, 35(3):18-20.

van Dorsser, J. C., and Rook, D. A., 1972, Conditioning of radiata pine seedlings by undercutting and wrenching: description of methods, equipment, and seedling response, New Zealand Journal of Forestry, 17(1):61-73.

Wardle, P., 1965, A comparison of alpine timberlines in New Zealand and North America, New Zealand Journal of Botany, 3:113-135.

Wendelken, W. J., 1956, Trials with exotic tree species in high country, New Zealand Journal of Forestry, 7(3):76-86.

Will, G. M., 1985, Nutrient deficiency and fertiliser use in New Zealand exotic forests, New Zealand Forest Service, Forest Research Institute Bulletin No. 97, 53 p.

Zabkiewicz, J. A., and Richardson, B., 1989, Weed control in New Zealand: costs, constraints and future options. In: Bassett, C., Whitehouse, L. J., and Zabkiewicz, J. A., eds., Alternatives to the chemical control of weeds, Proceedings of an International Conference, July 1989, Forest Research Institute, Rotorua, New Zealand, Ministry of Forestry, Forest Research Institute Bulletin No. 155, 111-115.

SUMMARIES OF WORKING GROUP SESSIONS

INTRODUCTION: John N. Alden

A major workshop objective was to discuss and prioritize problems with forest development in cold climates and to secure international coöperation for their solution, based on strategies for establishing and growing trees at and beyond natural tree limits.

The key problems, as identified in review and research papers presented in plenary sessions preceding group discussions, were

* socioeconomics of afforestation and agroforestry,
* gene conservation and breeding of cold-climate tree species,
* adaptation of land races to treeless regions, and
* knowledge gaps in the physical and chemical properties of soils and their interaction with plants in the forest-tundra ecotone.

Workshop participants addressed these issues in a series of four concurrent working groups. Conclusions and highlights of each group's discussions follow.

I. SOCIAL AND ENVIRONMENTAL IMPACTS OF FORESTRY IN A TREELESS COUNTRY: Sigurður Blöndal, Chairman

Key Discussion Issue: Species Introductions

To provide a basis for decisions concerning afforestation of treeless countries and to assist in establishing productive forests, more research is needed on the controversial aspects of exotic species introductions. The group felt that all species capable of thriving in a new environment were acceptable for afforestation. However, the group agreed that introductions must be controlled so that forest pests and destructive agents such as pathological fungi, harmful insects, and weed species are not allowed to proliferate unchecked as has happened in the past.

There is some danger that research on species introductions and afforestation might be used as an excuse to delay or avoid making decisions on critical land-use issues. Hence, it is essential that environmental reviews be convened for all controversial land-use projects involving exotics, especially for those projects that could radically alter the environment.

Forest Development in Cold Climates, Edited by J. Alden *et al.*
Plenum Press, New York, 1993

The group agreed that careful planning would allow afforestation of exotic species without affronting the public's sense that land ought to be reserved for agriculture or diminishing other attractions or values of the land.

The danger that any single government agency or private land-use organizàtion might have the final say on controversial and large-scale land-use decisions must be avoided. The final decision-maker on major land use questions should be accountable to the general public, as for example an elected or government-appointed official (such as a Minister of Agriculture or Forestry) would be. It was pointed out that such checks exist in many European countries.

Finally, education concerning the benefits of tree introductions and large scale afforestation programs is essential. Environmental information and results of research on tree introductions must be made readily available to the public and, if necessary, distributed to all concerned citizens. Discussion between concerned parties can avert senseless conflict and the threat of ecological terrorism from radical groups.

II. PROVENANCE COLLECTION, GENE CONSERVATION, AND ACQUISITION AND EXCHANGE OF REPRODUCTIVE MATERIALS: Maximilian Hagman, Chairman

Key Discussion Issue: Provenance Collection

The discussion group found that with the recent provenance collections of the Nordic Forest Research Council and others, the major boreal tree species of western North America are at present reasonably well sampled. Additional collections are needed only for a few species in certain areas. On the other hand, very little knowledge and material are available for tree species from the colder regions of the Eurasian continent, particularly the extreme northern forest–tundra region. The group therefore recommended that efforts should be made to send seed and plant collection expeditions to the taiga regions of Eurasia. The group felt that the Arkhangel'sk region of Russia and the northwest Ural Mountains deserved highest priority. Far eastern localities should be visited wherever possible, and the Kamchatka Peninsula was given particularly high priority. Expedition parties should collect material for use in agroforestry, horticulture, and landscaping, as well as for use in forestry in cold climates.

The group expressed concern about conserving gene resources in situ, but strongly supported action taken to make it possible. For sources of high economic value, stands stands should be established ex situ for gene conservation, and existing stands managed and used as seed production sources, if possible. For valuable populations, research is necessary to determine the best methods of conserving gene resources without changing the original genetic structure (gene frequencies) of the populations.

The group could not see any serious obstacles to the exchange of materials among willing agencies, although care should be taken to avoid introduction of undesirable organisms associated with the materials.

To improve the testing of material and knowledge of what material is available for testing, the group recommended that institutions with provenance studies in place should disseminate their test results and inventories of available materials to other interested parties. Propagules and pollen of the most suitable provenances then could be ordered as desired.

III. AGROFORESTRY: Peter M.A. Tigerstedt, Chairman

Key Discussion Issues: Stability of Agricultural Systems, Multiple Land Use, and Tree Breeding in Iceland

The working group arrived at several conclusions regarding the concepts of agroforestry in Iceland. The strong influence of animal husbandry in Icelandic farming systems must be considered in planning any forestry program. The detrimental effects of overgrazing by sheep must be reduced and sheep populations must be managed before launching an agroforestry program.

System Stability

Based on the concepts of sustainable systems in agriculture and forestry from the UN "Brundtland Commission" report,[1] the working group strongly recommends sustained-yield management for production systems in marginal areas. "Sustained yield" requires close integration of research and development in forestry, agriculture, and land reclamation. Forests must be viewed as an important component of all land-use systems in Iceland. Since Iceland does not have a forestry industry at present, the working group suggests that resource management programs include trees as a component of all land-use systems. In this context, the use of indigenous birch forests deserves special priority, because birch species constitute an endemic and well-adapted part of the ecosystem.

Multiple Use of Trees

In the absence of a forest industry, the industrial forest ideology is of secondary value. Multiple use of forests for wind control, soil amelioration, watershed management, improvement of local climate, landscaping, and control of soil erosion must be emphasized. The value of the indigenous rowan (Sorbus aucuparia L.) as food for wildlife also should be kept in mind.

Trees have aesthetic value for urban populations and provide shelter for community dwellings and vacation homes. Forests and forestry are important in planning parks, recreation areas, and recreation activities, and are important for tourism. Tourism and demands for forests will continue to grow as urban citizens and Europeans use the countryside for vacations and recreation.

Competition between Trees and Agriculture

Trees should be an integral part of all land use systems, but forests should not compete with agriculture or with other land uses. The importance of fences to exclude grazing animals during the critical period of forest establishment is well recognized.

Species and Breeding Methods

The survival and productivity of artificial forests in marginal habitats depend on finding adapted species and provenances. In Iceland, this work has continued for 30 to 40 years, but more species and provenances must be selected to insure that forests are well-adapted to the local environment and are of good quality. Although tree breeding (i.e.,

---

[1] Brundtland, Gro Harlem, 1987, Our Common Future, World Commission on Environmental Development, Oxford University Press, Oxford, Great Britain.

population and hybrid breeding) to improve the fitness of populations in cold climates has not yet been initiated, it should be focused on the best-adapted species and provenances. Tree breeding is expensive, so costs and benefits ought to be estimated before an active program is initiated.

The working group produced a prioritized list of species for active breeding in Iceland. The species are ranked (Table 1) according to how well they meet the following four objectives (listed in order of importance): (a) usefulness for reclamation, (b) industrial potential, (c) amenity values, and (d) genetic resource values.

Other important forest tree species that have less priority for breeding at this time include Sorbus aucuparia L., Alnus (Mill.) spp., Populus tremula L., Populus balsamifera L., Picea xlutzii Little, P. abies (L.) Karst., P. engelmannii Parry ex Engelm., P. glauca (Moench) Voss, Abies lasiocarpa (Hook.) Nutt., and Pinus cembra L. (Pinus sibirica Du Tour).

Table 1. Species for Active Breeding Programs in Iceland, Ranked according to Their Usefulness in Fulfilling Afforestation Objectives

Species	Uses[1]	Breeding Methods
1. Larix sukaczewii N. Dyl. "Raivola" strain	a, b, c	Seed orchards, intraspecific and interspecific hybrid breeding, and general and specific combining ability (G&S-CA) tests
2. Betula L. spp.	a, b, c, d	Phenotypic selection, cloning, breeding and seed orchards in plastic greenhouses (controlled-environment seed orchards), interspecific breeding, and G&SCA tests
3a. Populus trichocarpa Torr. & Gray	b, c	Provenance testing, family and within-family selection, selection within provenances, and cloning
3b. Salix L. spp.	a, c, d	Provenance testing, selection within provenances, and cloning
4. Picea sitchensis (Bong.) Carr.	b, c	Provenance selection, registration of Icelandic seed sources and plus tree selection in Iceland, and breeding and seed orchard establishment
5. Pinus contorta Dougl. ex Loud.	b, c	Provenance testing and plus tree selection in Iceland, and breeding and seed orchard establishment

[1] a = usefulness for reclamation; b = industrial potential; c = amenity value; and d = genetic resource value

In summary, economic constraints must be considered before entering into active breeding programs. The working group agrees on the importance of integrating new forests with agriculture. The forests will be composed of mixed species with the highest possible sustainable yields and tolerance of environmental extremes.

IV. THE FUNCTIONAL ECOLOGY OF THE FOREST–TUNDRA BORDER: Bjartmar Sveinbjörnsson and Peter Wardle, Co-Chairmen

Key Discussion Issue: Functional Ecology of the Forest-Tundra Boundary

The group agreed that information is spotty on the relationship between climate and the functioning of individuals, populations, and ecosystems across the forest–tundra boundary. It was felt that more ecological and biophysical research on the gradient is needed because of the role of forests in climate amelioration and as carbon sinks, and because of the biological diversity of the boundary. The group then identified major knowledge gaps.

Knowledge Gaps

The major knowledge gap relates to the soil and to plant–soil interactions. It was deemed important to study:

1. Physical parameters of soils across the forest-tundra gradient, including soil temperature and moisture.

2. Chemistry of soils and soil microbial activity, such as nutrient mineralization.

3. Plant partitioning and root function, including mycorrhizal associations.

4. Moss–soil interactions and their effects on other components of the system.

Approach

It was pointed out that manipulative research is needed to predict potential vegetative change in the forest-tundra transition zone. Because low biological diversity might reduce ecosystem stability at forest limits, the group suggested that species additions and removals should also be studied.

There was agreement that models of vegetation response to climate change at the forest-tundra transition zone should serve as tools for exchange and synthesis of scientific information. The need for economically-feasible and ecologically-relevant "big science" sites was discussed. Abisko, Sweden and Churchill, Manitoba, Canada were suggested as possible sites for studying climate and tundra-forest ecosystem relationships.

Cooperation

The participants expressed interest in future cooperation aimed at filling some of the gaps identified above. The research should be multi-disciplinary, including both biological and atmospheric sciences.

SPECIES INDEX[1]

Trees and Shrubs

Abies, 89, 255, 269, 302, 397, 513
A. amabilis, 92, 280, 411
A. balsamea, 206, 256, 491
A. delavayi var. motuoensis, 513
A. densa, 513
A. georgei var. smithii, 513
A. grandis,63, 271
A. lasiocarpa, 9, 92, 121, 245, 255, 268, 279, 308, 421, 450, 453
A. magnifica var. shastensis, 273
A. mariesii, 494
A. procera, 269, 419, 486
A. sibirica, 255
"A. sitchensis", 397
A. spectabilis, 513
A. squamata, 513
A. veitchii, 494
Acer, 507
A. negundo, 421
A. platanoides, 211
A. pseudoplatanus, 122, 293, 456, 534
A. rubrum, 180
Aesculus hippocastanus, 293
Agathis australis, 531
Alnus, 8, 85, 91, 257, 280, 296, 302
A. acuminatus, 91
A. crispa, 10, 285
A. glutinosa, 70, 534
A. incana, 10, 447, 458
A. rubra, 412, 394, 537
A. sinuata, 10, 271, 285, 394, 458
A. tenuifolia, 285, 458
A. viridis, 534
Amelanchier alnifolia, 285
A. florida, 285
Arabidopsis thaliana, 365
Araucaria araucana, 90
Arctostaphylos, 109

Betula, 90, 120, 245, 280, 302, 358, 397, 447, 500
B. glandulosa, 285
B. glandulosa x papyrifera, 285
B. nana, 65, 81, 359, 395, 447
B. papyrifera, 283, 359, 420, 508, 537
B. papyrifera var. cordifolia, 208
B. papyrifera var. neoalaskana, 80
B. pendula, 10, 69, 83, 248, 255, 359, 458, 534
B. pendula var. carelica, 420
B. platyphylla, 516
B. pubescens, 1, 44, 65, 79, 207, 256, 294, 359, 447, 453
B. pubescens ssp. tortuosa, 117, 358, 439
B. tortuosa, 305, 358
B. utilis, 516
Brassica, 364

Calluna vulgaris, 269, 485
Carex, 447
Carya glabra, 180
Castanea dentata, 180
Cedrus deodara, 523
Chamaecyparis lawsoniana, 123
C. nootkatensis, 92, 280, 421
Chionochloa, 531
Chionochloa flavescens, 531
Chosenia, 120
C. bracteosa, 296
Cornus, 420
C. stolonifera, 285
Cryptomeria japonica, 271, 411
xCupressocyparis leylandii, 421
Cupressus gigantea, 513
C. macrocarpa, 421
Cyathea, 91

[1] Listings indicate only the first reference to a species in each chapter.

Dendrosenecio, 91
D. keniodendron, 91
Dracophyllum longifolium, 123
D. scoparium, 123
D. traversii, 92
Drimys winteri, 119
Dryas, 448

Elaeagnus commutata, 285
Embothrium coccineum, 119
Empetrum, 445
Erica, 91
E. arborea, 91
Espeletia, 91
Eucalyptus, 90, 169, 249, 521
E. pauciflora, 90, 536

Fagus, 206, 507
F. crenata, 128
F. grandifolia, 207
F. sylvatica, 293, 534
Fraxinus excelsior, 534

Hagenia abyssinica, 91
Halocarpus biformis, 94
Hieracium, 5, 543
Hippophae, 523
H. rhamnoides ssp. gyantsensis, 525
Hoheria, 90

Ilex aquifolium, 122

Juglans regia, 534
Juniperus, 90, 120, 302, 513
J. communis, 437, 463
J. convallium, 513
J. horizontalis, 284
J. pingii var. wilsonii, 521
J. recurva, 513
J. saltuaria, 513
J. tibetica, 513
J. wallichiana, 513

Laburnum anagyroides, 122, 534
Larix, 204, 242, 255, 269, 302, 476
L. czekanowskii, 255
L. dahurica, 254, 420
L. decidua, 10, 453, 534
L. xeurolepis, 268, 421, 486
L. gmelini, 255
L. griffithiana, 516
L. himalaica, 516
L. kaempferi, 242, 270, 486
L. laricina, 10, 213, 255, 280, 507
L. leptolepis, 421
L. lyallii, 247
L. occidentalis, 420
L. potaninii var. macrocarpa, 513
L. sibirica, 6, 120, 219, 242, 254, 267, 294, 420, 451, 453
L. speciosa, 516

Larix (*cont'd*)
L. sukaczewii, 4, 171, 248, 254, 271, 421, 455
Leymus arenarius, 6
Lobelia, 91
Lonicera, 295
L. deflexicalyx, 295
L. hispida, 295
Lupinus nootkatensis, 269

Malus pumila, 523
Metasequoias, 394
Metrosideros lucida, 123

Nothofagus, 90, 117, 135, 529
N. antarctica, 90, 122, 137, 268
N. betuloides, 90, 122, 145, 271, 421
N. brassii, 143
N. cunninghamii, 156
N. dombeyi, 156
N. fusca, 143
N. aff. gunnii, 90, 143, 145
N. menziesii, 92, 98, 123
N. obliqua, 268
N. procera, 268
N. pumilio, 90, 122, 146
N. solandri, 92
N. solandri var. cliffortioides, 123, 156, 531

Olearia lacunosa, 92
Oplopanax horridus, 285
Oxyria digyna, 72

Phylica nitida, 123
Picea, 89, 302, 346, 370, 447, 51
P. abies, 9, 33, 44, 62, 65, 245, 253, 269, 293, 300, 354, 373, 383, 411, 428, 447, 453
P. abies ssp. obovata, 247, 321, 377
P. asperata, 373
P. asperata var. aurantiaca, 373
P. aurantiaca, 373
P. bicolor, 373
P. brachytyla, 373
P. breweriana, 371
P. chihuahuana, 373
P. engelmannii, 9, 92, 129, 247, 257, 294, 373, 429, 450, 453, 552
P. glauca 15, 45, 80, 121, 205, 245, 256, 271, 277, 293, 371, 383, 418, 427, 552
P. glauca forma parva, 212
P. glauca forma parviflora, 212
P. glehnii, 373
P. jezoensis, 373, 422
P. koraiensis, 373
P. koyamai, 373
P. likiangensis, 513
P. likiangensis var. balfouriana, 513
P. likiangensis var. linzhiensis, 513, 523
P. likiangensis var. purpurea, 373
P. xlutzii, 9, 45, 271, 277, 376, 383, 450, 428, 552
P. xlutzii x P. mariana, 376
P. mariana, 67, 228, 245, 255, 280, 373, 416

Picea (*cont'd*)
P. mariana forma semiprostrata, 212
P. mariana x P. sitchensis (putative), 420
P. maximowiczii, 373
P. mexicana, 373
P. meyeri, 373
P. morrisonicola, 373
P. neoveitchii, 373
P. obovata, 377
P. omorika, 271, 373, 486
P. orientalis, 373
P. polita, 373
P. pungens, 373, 427, 507
P. purpurea, 373
P. rubens, 96, 373
P. schrenkiana, 373
P. sitchensis, 9, 33, 92, 120, 171, 217, 247, 265, 277, 294, 373, 383, 393, 427, 450, 455, 465, 476, 501
P. smithiana, 373
P. spinulosa, 373
P. wilsonii, 373
Pinaceae, subgenus Pinus, 371
Pinaceae, subgenus Strobus, 371
Pinus, 89, 269, 302, 371, 447, 476
P. albicaulis, 92
P. aristata, 10, 92, 453
P. armandii, 523
P. attenuata, 421
P. banksiana, 259, 278, 346
P. cembra, 10, 94, 371
P. contorta, 6, 95, 121, 248, 252, 267, 278, 294, 313, 343, 450, 457, 476, 538
P. contorta var. contorta, 8, 280, 346, 457, 464, 534
P. contorta var. latifolia, 8, 33, 277, 343, 457
P. contorta x P. banksiana, 420
P. densata, 523
P. flexilis, 10, 92, 534
P. griffithii, 523
P. hartwegii, 91
P. longaeva, 128
P. monticola, 420
P. mugo, 4, 421, 453, 534
P. muricata, 534
P. nigra, 421, 537
P. nigra ssp. laricio, 534
P. nigra var. maritima, 421
P. peuce, 271
P. pinaster, 123, 534
P. ponderosa, 534
P. pumila, 10, 92, 120, 420
P. radiata, 249, 267, 421, 530
P. sibirica, 453
P. sylvestris, 10, 45, 65, 94, 105, 178, 204, 242, 253, 265, 300, 321, 333, 344, 371, 383, 447, 455, 475, 525, 534
P. sylvestris var. mongolica, 525
P. tabulaeformis, 525
P. tabulaeformis var. densata, 523
P. thunbergiana, 420
P. uncinata, 536

Pittosporum crassifolium, 123
Platycladus orientalis, 523
Polylepis sericea, 91
Populus, 120, 245, 257, 296, 302, 397, 523
P. alba, 523, 534
P. balsamifera, 10, 247, 285, 397, 458, 552
P. balsamifera x trichocarpa, 285
P. xbeijingensis, 525
P. xcanadensis, 525
P. cathayana, 525
P. davidiana, 525
P. deltoides, 525, 534
P. xeuramericana, 48
P. nigra var. italica, 525, 534
P. rotundifolia var. duclouxiana, 523
P. szechuanica var. tibetica, 523
P. tremula, 9, 256, 447, 458, 552
P. tremuloides, 207, 257, 285, 420
P. trichocarpa, 9, 171, 228, 271, 280, 294, 417, 458, 552
Potentilla fruticosa, 285
Prunus, 91
P. amygdalus, 15
P. armeniaca, 523
P. avium, 523, 534
P. padus, 6, 447
P. salicina, 523
Pseudotsuga menziezii, 10, 61, 248, 252, 519, 534

Quercus, 206
Q. aquifolioides, 516
Q. coccinea, 180
Q. prinus, 180
Q. robur, 534
Q. rubra, 180
Q. semecarpifolia, 90, 516
Q. virginiana, 393

Rhododendron, 90
R. nivale, 521
R. primulaeflorum var. cepalanthoides, 521
Ribes, 295
R. bracteosum, 285
R. lacustre, 285
R. laxiflorum, 285
Rosa, 295
R. acicularis, 285
R. nutkana, 285
R. pimpinellifolia, 295
Rubus parviflorus, 285
R. pedatus, 285
R. spectabilis, 285, 413

Salix, 4, 280, 302, 394, 463, 523
S. cv. "Brekkuvíðir", 296
S. alaxensis, 10, 285, 296
S. alba, 523
S. arbusculoides, 285
S. babylonica, 523, 534
S. barclayi, 412
S. berberfolia, 296

Salix (*cont'd*)
S. borealis, 9
S. caprea, 10, 294, 449
S. cheilophila var. microstachyoides, 525
S. fragilis, 534
S. lanata, 6, 295, 450
S. longistaminea, 523
S. myrsinifolia var. borealis, 296
S. paraplesia, 523
S. paraplesia var. subintegra, 523
S. pentandra, 44, 296
S. phylicifolia, 8, 450
S. polaris, 447
S. purpurea, 536
S. sitchensis, 285
S. sphenophylla, 285
S. viminalis, 231
Shepherdia canadensis, 285
Sophora chrysophylla, 91
Sorbus, 295, 302
S. aria var. intermedia, 421
S. aucuparia, 9, 293, 412, 447, 534
S. xhostii, 295
S. xhybrida, 293
S. intermedia, 122, 293
S. sambucifolia, 394
S. sitchensis, 286
Spirea, 295
S. beauverdiana, 286
Syringa, 295

Thuja koraiensis, 10
T. occidentalis, 411
T. plicata, 121, 280, 537
Tilia europae, 534
Tsuga, 89, 269, 302, 411
T. canadensis, 207, 427
T. heterophylla, 10, 271, 280, 418, 486, 537
T. mertensiana, 10, 92, 261, 271, 280, 422

Ulmus glabra, 10, 66, 293, 456
U. procera, 534
U. pumila, 523

Vaccinium, 91
V. alaskaense, 286
V. ovalifolium, 286
V. parvifolium, 286
Viburnum edule, 286

Other Species

Chrysomyxa pirolata (inland cone rust), 289

Dendroctonus spp. (bark beetle), 406

Elatobium abietinum (green spruce aphid), 427, 456, 179
Endocronartium harknessii (western gall rust), 268
Eppirritera autumnata, formerly known as Oporinia autumnata or O. autumalis (autumn moth), 85, 128

Gremmeniella abietina (Gremminiella), 313

Laspeyresia youngana (spruce seed moth), 289

Oryctolagus cuniculus (rabbit), 543

Panolis flammea (pine beauty moth), 268
Pineus pini (woolly aphid), 420, 456
Pleurozium schreberi (moss), 502
Potebniamyces coniferarum (larch canker), 179

Sphagnum spp. (mosses), 448
S. subsecundum (moss), 66

Rhyacionia buoliana (pine shoot moth), 268

Zeiraphera spp. (spruce bud moth), 406

SUBJECT INDEX

Acclimation, defined, 15-16
Adaptation to cold climates, 49, 251-252, 299-314, 358-360. *See also* Cold acclimation
  "aftereffects," importance of, 258
  "bottleneck years," 253
  via developmental processes, 43
  in exotics, 248-249, 258-261
  and fitness, 243
  gene conservation, 241, 244
  gene flow, effects on, 306-308, 310-311
  and genetic engineering, 312
  genetics of, 241-245, 248, 302, 310-313, 359-360, 449-450
  indicators of, 242
  metabolic categories, 31-37
  natural selection, 243-246
  and nitrogen availability, 69
  to photoperiod, 49, 193, 359
  physiological conditioning, 308-309
  of Pinus sylvestris, 321-330, 335-340
  Pinus contorta, case history, 313
  selection coefficient, 243
  tree breeding, 241-247
Adaptation to tree limits, 127-128, 304, 312. *See also* Pinus sylvestris
Adaptation to wind. *See* Wind, adaptation to
Afforestation at tree limits. *See also* Tree planting, and references to individual species
  clonal materials for, 257
  collective gap models for predicting success, 169-171, 179-183
  continental climates, species for, 269-271
  and exposure, 465 (*see also* Tatter rate)
  in Faeroe Islands, 463-464 (*see also* Faeroe Islands)
  growth models for predicting success, 168-169
  methods for reclaiming birch woodlands, 4
  with preconditioned genotypes, 37
  shelterwood ("Vorwald"), uses for, 257
  site preparation for, in New Zealand, 540-541
  Sitka and Lutz spruce crosses, 434-435
  social and economic impacts, 549
Afforestation at tree limits (*cont'd*)
  species for, 118-123
Afforestation, in Iceland, ix, 4-12, 85, 228, 293-294, 300, 453-460
  broadleaf exotics, 293-294, 458-459
  coniferous exotics, 294, 453, 456-459
  factors affecting, 85
  "Land Reclamation Forest," 4
  objectives of, 459
  production forests, species for, 9
  protection forests, species for, 7
  provenance trials, results of, 456-459
  seed procurement for, 294, 455
Afforestation in maritime climates
  controlled environment tests, 273
  exotics, 8-10, 269, 453-458
  land races, 273, 449-450
  nurse crops, 269, 476, 485-486
  on Calluna vulgaris uplands, 485-486
  seed origins for provenance trials, 271-273
  species choice, 267-272, 476, 486
  Scots pine and Sitka spruce at high elevations, 476
  wind abatement, silvicultural techniques for, 217-219
Afforestation, in New Zealand, 530, 534-537
Afforestation, in Nordic countries, 4, 8, 10-12, 427-434
  Kenai Peninsula Picea provenances, 427-428
  Picea sitchensis x P. xlutzii, 427-428
  Picea sitchensis, P. xlutzii, and P. abies, compared, 429-434
  seed collections for, 258-260
  species for, 254-257, 277
  Picea xlutzii, 427-429, 434
  Pinus sylvestris, in Finland, 333-342
  Pinus sylvestris, in northern Sweden, 322-330
  in Vesteraalens, Norway, 383, 427-436
Afforestation, on Plateau of Tibet. *See* Tibet, Plateau of
Agroforestry, 531

Agroforestry (*cont'd*)
effects on landscape patterns, New Zealand, 543-544
and shelterbelts, 10
and understory, 4
Albedo, 168, 182, 190-191, 199
Aleutian Islands
climate, 393
pre-Quaternary forests, 394
tree-planting, 19th Century, 395-410
Antarctica. *See also* Transantarctic Mountains
displacement of Tertiary treeline in, 154-155
Pleistocene glaciations, 136, 141, 148
Pliocene-Pleistocene flora and response to climate change, 145-148, 151
Pliocene climate optimum, 135-136, 143
refugia for Miocene-Pliocene flora, 147-151
topography, compared with Arctic regions, 139
"Arctic front," 204-205, 218
Ari the Learned (Icelandic historian), viii

Betula, morphological and physiological variations, 358-360
Betula pubescens (Icelandic or Mountain birch), 44, 448-449. *See also* Icelandic birch and Mountain birch
growth-limiting factors, 67-76
seed viability, 65
as shelterwood for afforestation, 257
at tree limit, 79-83, 117, 128, 305, 439, 442-443
Biometeorological monitoring, 228-237
air and soil temperature (Iceland and Newfoundland), measurements of, 234, 236-237
Bowen ratio, 228-229, 233-235
energy balance of forests, 227-228, 236-237
heat flux, 228-229
in Iceland, 230-232, 234-235
in Newfoundland, 230-235
precipitation, Iceland and Newfoundland, 230, 235
"reversing temperature difference measurement," 229
Bjarnason, Hákon (former Director, Icelandic Forest Service), 6-7, 454-455
"Book of Icelanders," viii
Boreal (conifer) zone, 247
and Icelandic lowlands, 3
Buckner, General Simon B. (Aleutian Islands), 413-415

Carbon balance
at altitudinal treeline, 94, 100-101, 110, 538
in mountain birch, 68-69, 81
Climate
defined, 189-190
as experienced by vegetation, 192-195
and forest productivity, 227-228
Climate change, 136, 167, 179, 182, 248, 465
and altitudinal (Betula pubescens) treelines, 70-72, 76

Climate change (*cont'd*)
and growing seasons, 167-168
and latitudinal treelines, 49-50, 86, 205-206
and productivity of boreal forests, 168, 218, 314-317
human influences on, 191-192
and hydrological cycle, 168
predictions, magnitude of, 167-168
and tree limits, 70, 76, 205-206, 260
Climate change models, 207, 218
and forest productivity, 169-183, 207, 218
GCMs (General Circulation Models), 168, 195-201, 204-205
improving, 201
Clonal forestry, 245, 257
Clonal propogation, 31, 37, 312
Clones, 59, 491, 496, 505. *See also* Krummholz
for afforestation at tree limit, 245, 257, 309, 526
for hardiness tests of Norway spruce (Picea abies), 62-64, 311, 313
of Picea sitchensis and P. lutzii, 435
of Salix phylicifolia, 451
and seed orchards, 322-323, 325, 327-329, 336-340, 342, 435
as silvicultural strategy, 217
Cold acclimation. *See also* Adaptation to cold climates
and adaptive behavior of plants, 17-21, 29-35
and amino acid metabolism, 23-25, 34-39
developmental and physiological homeostasis, 21-24
effects of climate warming on, 50
genetic variation (Pinus sylvestris) in development of, 324-326, 329-330
growth cessation and, 49, 303-304
metabolic markers of, 34
metabolic tracers of, 20-21
molecular basis of, 16-17, 20
and nitrogen metabolism, 23-24
in Picea glauca, 15
in Pinus sylvestris, 325, 329-330
response to photoperiod, 47-48
selecting for, 326-327
tests for, 309, 324-326
Cold hardiness, 100-101, 253-254, 304, 388-390. *See also* Freeze resistance
relation to growth cessation, 47-50
combining ability of Pinus sylvestris, 325
Combining ability, general and specific, 325, 348, 352-354, 552
Cuticular development, 105, 108-109, 112
variation with elevation, 81
and summer temperatures, 105
thickness, 81, 112
and transpiration, 59, 63, 108-109, 112
and wind abrasion, 209
and winter dessication, 60-63, 94, 96, 304

Daylength. *See also* Adaptation
algorithm for, 316-317

Daylength (*cont'd*)
effect on hardiness, 47, 429
Deforestation,
of Antarctica, 139, 151
anthropogenic, 119, 205-208
of Finnmark Country, Norway, 442-443
of Iceland, viii, 1-2, 453
of New Zealand, 530-531
Defoliation of birch forests by Eppirritera autumnata (autumn moth), 85
Dendrological studies, Finland, 254-256
and Raivola larch (Larix sukaczweii), 254
and treeline parks, 255-256
Drought (winter) resistance, 59, 65. *See also* Frost drought; Winter dessication
breeding for, 435-436
selection for, 59-64

Ecotypes, climatic, 302-303, 309, 449
of Betula, 70-71, 76
clinal variation, 242, 307
and photoperiod, 49-50, 53
of Picea, 370
of Pinus sylvestris, 46
Elevation. *See also* Treelines, alpine, and *under* Growth
and growth of Sitka spruce and Scots pine, 477
and leaf area (in mountain birch), 81
and nitrogen fertilization of Betula, 83
and Scots pine survival in Sweden, 322
and needle characteristics of Scots pine, 107, 108-109
and temperature, 80, 92-93, 109-110, 315, 477-478, 517-518
Elfinwood. *See* Krummholz
Environmental forestry, Finnmark County, Norway, 437
concept of, 444-445
land races, selection of provenances for, 449-450
plantation establishment, 450-451
species for, 448-449
tree breeding for, 449-450
Exotics, 248-249, 253-261
breeding minor species for afforestation, 311
at high elevations, 476-479
in Iceland, 5-10, 171, 453-454, 456-459
in New Zealand, 93, 530, 534-539
productivity of, 249
in silviculture, 277, 485-487
on Tibet, Plateau of, 521-523
in United Kingdom, 267-273
Exposure. *See also* Tatter flags; Tatter rate
effects on tree growth and survival, 476-484

Faeroe Islands. *See also* Afforestation of
altitudinal treeline (potential), 473
climate of, 465
exposure stations on, 467-471
geography of, 463-464

Fertilizer, application. *See also* Growth, as affected by nitrogen
in Aleutians, 417
to birch, 82-85
simulation of, 176
to spruce, 484-485, 487
Finland, northern, and Pinus sylvestris, 333-342
Finnmark County, Norway, 437-439
climate, 445-447
coastal mountain birch zone, 448
forests, development of, 442-443
Nature Conservation Center, Svartnes, 437-439, 449-451
settlement, impact on vegetation, 442-443
settlement, early patterns, 439-441
soils of, 445, 448
Svartnes Environmental Forest, 451
Forest Mensuration Project (Norway), 429-436
Forests
and birch woodlands, 1-4
environmental, 444-445
production, 9, 11, 214-217, 459, 484, 531, 544
productivity, models of, 227-228, 237-238
protection, 7-8
recreation, 4, 9-10
Forest-tundra ecotone, 79, 300, 553. *See also* Kampfgürtel; Treeline
and seed influx, 65
knowledge gaps concerning, 553
in Labrador, 205, 207, 209-212, 218
Free and fixed growth patterns, 44-48
Freezing resistance (= frost tolerance), 100-101. *See also* Cold acclimation
in Picea spp., 383-390
Frost damage, 95, 254, 270, 303-304, 429
in birch, 74
Frost drought, 111-112, 124, 177, 209

Gene conservation, 273, 550
General Circulation Models, 204-205
and warming projections, 168
Genetic assimilation, 26
Genetic diversity. *See* Genetic variation
Genetic expression, 15-17
induced and predetermined, 20
reaction norm, 17-21
Genetic variation, 15, 17, 258, 309, 449
analysis of, 246
heterozygosity (allozyme), 245-246
in Icelandic birch, 357-360, 363-365
and linkage disequilibrium, 246
in lodgepole pine (Pinus contorta), 343-344, 350, 352-354
in Scots pine, 324, 327-329, 333-342
in Picea, 370-371, 379, 383, 388-389
in treeline populations, 246, 248
Genotypes
ideotype, 17, 36
idiotype, 17
Green belt, 241-242. *See also* Forest-tundra ecotone

Greenhouse gases, 189, 191-192, 195, 196-197
effects on climate warming, 69, 86, 167-169, 195-196, 200
Growth, as affected by—
elevation, 44, 80-83, 85-86, 93, 321, 477-484, 487-488
nitrogen, 69, 75, 82-83, 85, 98, 176, 269, 484-485
photoperiod, 45-46, 47-49, 50, 384
temperature, 43-49, 51-52, 67-76, 79, 105, 107, 170, 253, 445, 477-478, 480, 535
wind, 80, 94-95, 177, 203, 205, 209, 218-219, 266, 269, 324, 473, 475-480, 505
(*see also* individual species)
Growth, at tree limits, 242, 302, 304. *See also* Regeneration; Treelines, alpine
Growth regulators, 27, 31, 50-53
cytokinins, 53
giberellins, 51-53
phytochrome, 53

Hardiness. *See* Freezing resistance; Frost damage
Holdridge biotemperature, 125

Iceland
afforestation of (*see* Afforestation, Iceland)
"Arctic Front," 204
Ari the Learned (historian), viii
biometeorological studies in, 228, 231-232, 235-236
climate change, response of vegetation to, 168, 206
deforestation of, vii-ix, 44, 119, 139
exotics in, 6-10, 120, 171, 267, 271-272, 293-296, 453-458, 551-552
forestry in, 7-12, 453
Icelandic birch in, 79, 81-85, (*see also* Icelandic birch)
Icelandic Forestry Association, 4, 456
Icelandic Forestry Service, 453-455, 439
settlement of, viii
soil erosion in, ix, 2-3
tree breeding in, 311, 552
tree planting in, 3, 6-10, 217, 269-271, 420-421, 453-459
Iceland, horticulture in
climate, effects on, 293
history of, 293-294
seed collections for, 294-296
shrubs, 295-297
Icelandic birch (Betula pubescens). *See also* Betula pubescens; Iceland
adaptability and genecology, 358-360
afforestation with, in Iceland, 4
cytogenetics, 360-362
DNA polymorphism, 363-366
evolution of, 358-362, 365-366
genetic diversity in, 357-360
in situ hybridization of chromosomes, 362-363, 366
Icelandic birch (*cont'd*)
introgressive hybridization with B. nana, 66, 360-362, 365
polycormic forms of, 66, 359
taxonomy, 358
uses in Iceland, 357

Kampfgürtel (= Kampfzone), 79, 94, 98-100, 117, 147, 300
Krummholz, 66, 94, 98-100, 107, 109, 111, 115, 117, 123, 125, 293, 300,
clonal, 156, 212, 491, 505-506
and genetic assimilation, 26
migration of, 98-99
Nothofagus, during Tertiary, 144-145, 147-148
in wave forests, 496, 501, 505-506
wind-formed, 209, 211, 212
and winter dessication, 94-95

Land races, 241, 273, 310-311, 449-450, 549
of lodgepole pine, 272
Leaf growth
in birch (Betula pubescens), 46-47, 64-68, 72
in Scots pine (Pinus sylvestris), 46-47
Lofoten and Vesteraalen Islands, Norway
geography of, 429

Marginal limits. *See* Treelines
Marginal sites, species and provenance collections for, 280-289
"Massenerhebung" effect, 118
Modelling afforestation success, 168-183
FORENA, 169, 172-177, 179-183
FORET, 169-171, 174-176, 182
FORSKA, 169, 174-179, 182-183
JABOWA, 169-171, 174-176, 178, 182-183
LINKAGES, 169-171, 176-179, 182-183
LOKI, 171, 176, 179, 183
extrinsic growth constraints, 176-177
intrinsic growth constraints, 175-176
Modelling
tree establishment, 169-171, 178
tree growth, 169-178
tree mortality, 178-179
Molecular markers
and Betula pubescens, 363-365
and chloroplast DNA, 371-373, 375-379
random amplified polymorphic DNA (RAPD), 363-366
restriction fragment length polymorphism (RFLP), 363-364, 370-371
use in selection and breeding, 363-366
Mork Growth Unit, 253
Mountain birch (Betula pubescens), 79-86. *See also* Betula pubescens
biogeography of, in northern Fennoscandia, 117, 128, 442-443, 445, 448-451
Coastal Mountain Birch Zone, 448
effects of temperature on, 65-75
growth studies of, 44-45
Nordic Birch Forest Zone, 437, 439

Mountain birch (*cont'd*)
as tree limit species, 305
Mycorrhizae, ectotrophic
and Nothofagus forests, 148
relation to Nothofagus treeline, 97

New Zealand, southern high country of, 529-544
climate, 531-533
forest growth, 533-538
forest nurseries, 538-540
geography of, 529
pastoral farming in, 531, 533, 543-544
plantation establishment in, 534-535, 540-542
pre-Polynesian forests, 529-531
treeline, altitudinal, 90-100, 535
vegetation, 531
Nitrogen
application (as fertilizer), 82-85, 484-485
availability in soil, 69, 75, 82-83, 85, 98, 176
demand for, in simulations, 176
effects on tree growth, 16, 69-70, 75, 82-85, 98
modelling, in stand simulations, 176
Nitrogen, foliar, effects on photosynthesis, 83
Nordic Birch Forest Zone, 437, 439
Nordic Forest Research Council, 277, 284
Nothofagus
afforestation with, in U.K., 151, 268, 271
in Antarctica, 135-137, 140-158
anthropogenic dispersal of, 139-140, 156-158
on the Faeroe Islands, 119, 156-157, 421
on the Falkland Islands, 421
on New Zealand, 93-99, 151-154, 529, 531, 536, 539
on Tierra del Fuego, 90, 122, 125, 157
Nurse crops, 4, 257, 486, 488. *See also* Shelterbelts

Paleoaustral vegetation, 139-141
Permafrost, 194-195
in forest simulation models, 167, 170, 176, 182
and treeline, 124
Pest resistance and cold-climate afforestation, 85
Phenotypic plasticity, 17, 28-36
adaptive, 31-36
and temperature extremes, effects on, 28-31
reaction norm, 17-18, 28
Photosynthesis in northern species, 65, 67-68
and foliar nitrogen, 83
in C-3 plants, 86
and climate change, 86
effects of elevation on, 80
and growth models, 183
and leaf area, 110, 536, 538
net, effects on growth, 68-69
as function of temperature, 93-94, 111
Picea
chloroplast DNA, 370-373, 375-376
clades, 373, 375
evolutionary history of, 369-377, 379
freeze resistance, tests of, 383-388
genetic diversity, preservation of, 378-379

Picea (*cont'd*)
growth, on Norwegian sites, 431-433
hybridization and introgression, northwest American species, 375-377, 383, 389-390
krummholz forms, 212
introduction of northwest American species to Norway, 383-390
on Kenai Peninsula, Alaska, 376-377
monophyletic groups (*see* clades)
morphological variation, 370-371
phylogeny, 372-375, 378-379
plastotypes (P. abies), 377-378
in Tibet, 513
Picea glauca
cold acclimation of, 15, 20-21, 31-34
at high latitudes, 256-257, 259-260, 427
pollen collections, 290-291
provenance research, 271
seed collections, for marginal sites, 280-281, 283, 286
Picea sitchensis. *See also* Afforestation
afforestation with, 120-121, 125, 171, 294, 393, 416
at maritime tree limits, 455, 476
for environmental forests (Finnmark County, Norway), 450
growth as affected by—
altitude and temperature, 480-481
exposure (*see also* tatter rate), 266, 465, 478-480, 483-486, 501, 506
nutrition, 484-486
provenance, 433
"maritime growth profile," example of, 247
for production forests (in Nordic countries), 9, 383, 427
seed collections (for marginal sites), 277-280, 283
Pinus contorta
afforestation with, 121, 294, 457
breeding of, 343-355
combining abilities for height, 348, 352-354
in the Faeroe Islands, 464
in Iceland, 6, 8, 294, 457
in New Zealand, 95, 98, 534, 537-538
in Norway, 450-451
P. contorta x banksiana, 420
provenance hybrids, 344-345, 350-355
for provenance research, 252, 277
seed collections (for marginal sites), 278, 280, 283-284, 286
in Sweden, 313-314, 343-344
"transitional growth pattern," example of, 248
Pinus sylvestris
breeding of, 321-330
carbon balance, 110
cold acclimatization of, 324-330, 383
cuticular development at treeline, 108-109
growth and exposure (in Great Britain), 481-482
growth and environment, 45-49, 178, 253
growth at treeline, 65, 94, 305

Pinus sylvestris (*cont'd*)
in Iceland, 10, 455
migration, 206
in New Zealand, 534, 537
in northern Finland, 333-342
provenance studies of, 242, 333-337, 340-342
seed production and maturation, 323, 333-334
in Scotland, 265
in Tibet, 525
wind, resistance to, 204, 210
Plant hormones. *See* Growth regulators
Plantation establishment
Finnmark Country, Norway, 450-451
New Zealand, 540-542
Planting stock, 268, 308-309
Falkland Islands, 421
Iceland, 4, 12, 454
New Zealand, 538-539, 541-542
Polar forests
Arctic and Antarctic, compared, 139-140
in Antarctic, during Cenozoic Period, 137-138
Precipitation, effects on tree limits, 50, 92 192, 194, 253
and climate change, 167-168, 170, 197-200
and growth of P. sylvestris, 106
in Faeroe Islands, 465
in modeling afforestation success, 170, 176, 178, 180
in Norway (Finnmark County), 446
in Tibet, Plateau of, 513, 520-521, 524
in United Kingdom, 265-266
in New Zealand, 532
Productivity. *See* Climate change models; Forest productivity
"Prose Edda" (Icelandic saga), vii
Provenance hybrids v. non-hybrids (Pinus contorta), 348-355
flowering, 352
growth, 350-355
moose damage, 344, 348, 350
prolepsis, 350-351, 354
weather damage, 348, 350-352
Provenance research, Nordic countries, 253-258, 260-261, 267, 277-284, 299, 333, 343-355, 433
broadleaf species, 458
drought resistance (Douglas fir), 61-62
freezing studies for Picea, 384-388
lodgepole pine, 255, 343-355
mountain birch, in Norway, 439, 451
for New Zealand, 536
North America collections, for Nordic countries, 258-260, 278-285
seed collection costs, 289
selection criteria for, 278-279

Qinghai-Tibet Plateau. *See* Tibet, Plateau of (China)

Radiation (solar), 190-194
absorption by soils, 541
Radiation (solar) (*cont'd*)
daily variation in, 231-236
modelling, 176, 182
partitioning of, 227-229, 232-234
at tree limits, 124
Reaction norms. *See* Phenotypic plasticity
Reforestation. *See* Afforestation
Regeneration, 211-212, 304-305, 311, 313, 497. *See also* Tree breeding; Tree planting
artificial, 248, 393, 416, 453-454, 521-526, 534, 543
and fertilization, 83
in Iceland, 3-4, 84
latitude and altitude effects on, 80, 83
modelling, 170-171
natural, 242-243, 411
and seed orchards, 248
and seed maturation, 242-243
and seed transfer, 244, 247-249
in Sweden, 83, 321-326
at tree limits, 99-100, 242-245, 321-322, 475-476
vegetative, 245
in wave forests, 494
wind, effects on, 205-208, 217-218
Respiration, 67-69
alternate path, 43
dark, effects on growth, 67-69
and elevation, 81
photosynthetic compensation point, 93, 183
and temperature, 80
Rezanov, Chamberlain Nikolai (in Aleutian Islands), 396-397

Sheep grazing
and birch forests in Iceland, 85
Shelikhov, Grigorii I (Aleutian Islands), 396
Shelterbelts, 9-11, 421, 450, 453
species for, in Iceland, 10
Silviculture, 169, 314, 491
birch shelterwood, 257
continuous cover (in Britain), 483-488
in New Zealand, 542
site classifications for (in Iceland), 459
and wave forests, 491, 506-508
in windy climates, 208, 214-219
Snorri Sturluson (historian and poet), vii
Snow damage
implications for evergreens, 80-81
and lodgepole pine, 272
Socioeconomics
importance of Icelandic forests for, 1
Species and provenance introductions. *See* Exotics
Sweden
afforestation of, 321-322 (*see also* Afforestation in Nordic countries)
climate warming, effects on forest productivity, 314-316
lodgepole pine in, 313, 343-346
exotics, introduction, 254, 258
forest land classification in, 314-316

Sweden (*cont'd*)
Institute of Forest Tree Improvement, 260
mountain birch performance in, 79, 81-84, 358
and Norway spruce, 269
and Scots pine, 321-330

Tatter flags, 476-481, 483, 485
in the Faeroe Islands, 465-471, 473
in the United Kingdom, 476-478
Tatter rate
and site exposure, 465, 466-473, 476-478
effects of planation forests on, 473
and growth of Picea sitchensis, 478-480
and growth of Pinus sylvestris, 481-483
Temperature. *See also* Adaptation to cold climates; Cold hardiness; Frost damage
birch ecotypes, effects on, 70-72
and cuticle maturation, 105, 109
and growth of Picea sylvestris, 106-107, 109
and photosynthesis, 67-68, 69-76
and polyembryony, 26, 28
physiological adaptation to, 31-35
tetraterm, 2, 445-447
Tibet, Plateau of
afforestation of, 521-523, 524-526
climate, at timberlines, 511, 517-521
deciduous conifers and broadleaf species, 513-516
evergreen coniferous species, 513
heat and precipitation, 517, 520-521, 524
linkas (parks), 523-526
sclerophyllous evergreen broadleaf species at treeline, 516
socioeconomic benefits of, 525-526
species for, 525-526
treeline
altitudinal, 511-513, 517, 524
latitudinal, 511, 517-521
species of, 521, 523, 525
Transantarctic land bridge, 142, 150-151
Transantarctic Mountains, 135-137, 141-143. *See also* Antarctica
as refugium for Tertiary vegetation, 141-142, 147-149
and Nothofagus during Cretaceous-Cenozoic, 140 (Fig. 5)-143
krummholz Nothofagus forests in, 144-145, 147-148
Pliocene flora of (in Dominion Range), 143-148, 150
Tree breeding
clone banks, 435
for cold climates, 299, 304-305, 314, 449-450
of P. sitchensis and P. xlutzii (in Norway), 434-436
Tree growth. *See also* Growth, as affected by
at tree limit, 304, 482-483, 487
cessation (bud dormancy), 47-50
initiation of, 44, 53, 242
patterns of, free and fixed, 44-46
Tree growth (*cont'd*)
phases of, 67
and phosphorous application, 82
and root-to-shoot ratio, 82
Tree limits in cold climates. *See also* Treelines
adaptations for, 304-306, 308-314
afforestation at, 305 (*see also* Afforestation)
environmental characteristics of, 124-126, 300-301
climate warming, effect on, 205-206
environment, described, 300-302
genera and species of, 242, 302
genetic variation at, 242, 246
Picea sitchensis, 482-483
on Plateau of Tibet, 511-521
reproductive strategies at, 244-245
seed maturation and viability at, 80, 97-98
survival ability, natural selection for, 243-244, 301-306
terminology of, 300
Tree planting, Aleutian Islands, Alaska
early history of, 393-411
Adak Island, 423-429, 430
on Amaknak Island, 395-399, 410, 416
on Expedition Island, 402-404
on Hog Island, 401-402, 404
and Russian-American Company, 396-400
Shemya Island, 419-420
tree growth and survival, Adak National Forest, 399, 415-419
Unalaska Bay, 395-410
in 20th Century, 411-420
from 1985-1990, 420
Tree planting on high-latitude islands
Falkland Islands, 421
Faeroe Islands, 421-422
Iceland (*see* Iceland, afforestation of)
Shetland Islands, 421
Treeline, alpine, 90-92, 97, 473. *See also* Krummholz; Tree limits; Treelines; Winter dessication
exposure, 476-481 (*see also* Tatter flags)
factors affecting, 100-101, 105-112, 124, 128-129
flora of, 89-92
and wind abrasion, 109, 111-112
and winter injury, 95-96
Treelines, 115-123
and air temperature, 79, 107-109, 124-125, 517
climatic characterizations of, 92-93, 124-129
climate warming, response to, 70, 76, 86, 155-157, 206-207
and cuticular development, 108
terminology, 115-118
fluctuations of, 127-129
as genetic test sites, 341
and global warming, 86, 168
and growing season, 92-93
growth models, 171
migration of, 205-206, 218
phytogeography of, 119-122, 125-126

Treelines (*cont'd*)
regulating factors, 123
seed viability at, 65
and summer temperature lows, 300-301
and 10 °C isotherm, 92, 105, 112, 124
tetraterm requirements of, 445, 447-449
and winter desiccation, 59-64, 94, 105

United Kingdom, 105-107, 111. *See also* Afforestation in maritime climates
afforestation of, 475-476, 485-488
climate, 265-266
tree limits in, 265-266
tree limits, species for, 267-273
wind exposure, 266, 475-486

Vegetative reproduction at tree limit, 66, 99, 156, 245, 312
Veniaminov, Father Ivan, 397-402, 405

Wave forest
crested, 496, 501-503
crown structure of, 500-502
cyclical and non-cyclical, 491, 496-500, 502-503
dead tree strips in, 494, 503-504
dieback, 216, 497-499, 502, 505-508
features and distribution of, 491-499
management of, 507-508
regeneration of, 497-503, 506-508
soil formation in, 503-504
wind effects on, 500, 504-505
windthrow in, 507-508

Weather, defined, 190 (*cont'd*)
Wind. *See also* Frost drought; Growth, as affected by; Windthrow; Winter dessication
adaptation to, 209, 211, 212-213, 218
as component of climate, 189, 192-195
forest productivity, 214-217, 478-481, 483-484
injuries from, 63, 209-210, 214-215
and krummholz, 209, 212, 505-506
modelling, 197-200
and shelterbelts, 9-10
silvicultural strategies for, 217, 483-484, 487-488
tatter rate, correlation with, 473
tree form, effects of wind on, 209-213
treelines, effects of wind on, 100, 204-205, 207, 218, 475
in U.K., 266, 269, 475-486
Windthrow, 214-217
Winter desiccation, 94-96, 105
cuticular water loss, 59, 63
testing injury from, 59-64
and treeline, 105
and water potential, 60-61